ATLAS OF WORLD HISTORY

OXFORD

ATLAS OF WORLD HISTORY

PATRICK K. O'BRIEN, GENERAL EDITOR
INSTITUTE OF HISTORICAL RESEARCH, UNIVERSITY OF LONDON

This edition first published in 1999 in the
United States of America by
Oxford University Press, Inc.
198 Madison Avenue,New York, N.Y. 10016

Oxford is a registered trademark of
Oxford University Press

Library of Congress Cataloging-in-Publication
Data available
ISBN 0-19-521567-2

First published in Great Britain in 1999
by George Philip Limited

Copyright © 1999 George Philip Limited

COMMISSIONING EDITOR Jane Edmonds

EDITORS   Jannet King
                       Petra Kopp
                       Martha Leyton
                       Richard Widdows

EDITORIAL ASSISTANT   Louise Jennett

PICTURE RESEARCH   Sarah Moule

PRODUCTION   Katherine Knowler
                       Sally Banner

CARTOGRAPHY BY Philip's Map Studio

ADDITIONAL CARTOGRAPHY BY Cosmographics, Watford

DESIGNED BY Design Revolution, Brighton

ADDITIONAL ARTWORK BY Full Circle Design

PRINTED AND BOUND BY Cayfosa, Spain

10, 9, 8, 7, 6, 5, 4, 3, 2, 1

CONTRIBUTORS

GENERAL CONSULTANT EDITOR
Patrick K. O'Brien FBA
Centennial Professor of Economic History
London School of Economics
Convenor of the Programme in Global History
Institute of Historical Research
University of London

CONSULTANT EDITOR: THE ANCIENT
WORLD
Jane McIntosh
University of Cambridge

CONSULTANT EDITOR: THE MEDIEVAL
WORLD
Peter Heather
Reader in Early Medieval History
University College London
University of London

CONSULTANT EDITOR: THE EARLY
MODERN WORLD
David Ormrod
Senior Lecturer in Economic and
Social History
University of Kent at Canterbury

CONSULTANT EDITOR: THE AGE
OF REVOLUTIONS
Roland Quinault
Reader in History
University of North London

CONSULTANT EDITOR: THE TWENTIETH
CENTURY
Pat Thane
Professor of Contemporary History
University of Sussex

Reuven Amitai
Senior Lecturer and Department Head
Department of Islamic and Middle
Eastern Studies
Hebrew University of Jerusalem

Lito Apostolakou
Visiting Research Fellow
Centre for Hellenic Studies
King's College
University of London

Dudley Baines
Reader in Economic History
London School of Economics
University of London

Ray Barrell
Senior Research Fellow
National Institute of Economic and
Social Research (NIESR), London

Antony Best
Lecturer in International History
London School of Economics
University of London

David Birmingham
Professor of Modern History
University of Kent at Canterbury

Ian Brown
Professor of the Economic History
of South East Asia
School of Oriental and African Studies
University of London

Larry Butler
Lecturer in Modern History
University of Luton

Peter Carey
Laithwaite Fellow and Tutor in
Modern History
Trinity College
University of Oxford

Evguenia Davidova
Research Associate
Institute of History
Bulgarian Academy of Sciences, Sofia

Kent G. Deng
Lecturer in Economic History
London School of Economics
University of London

Saul Dubow
Reader in History
University of Sussex

Ben Fowkes
Senior Lecturer in History
University of North London

Ulrike Freitag
Lecturer in History
School of Oriental and African Studies
University of London

Stephen Houston
University Professor of Anthropology
Brigham Young University

Janet E. Hunter
Saji Senior Lecturer in Japanese
Economic and Social History
London School of Economics
University of London

Robert Iliffe
Lecturer in the History of Science
Imperial College of Science, Technology
and Medicine
University of London

Timothy Insoll
Lecturer in Archaeology
University of Manchester

Liz James
Lecturer in Art History
University of Sussex

Simon Kaner
Senior Archaeologist
Cambridge County Council

Zdenek Kavan
Lecturer in International Relations
University of Sussex

Thomas Lorman
School of Slavonic and European Studies
University of London

Rachel MacLean
British Academy Post-Doctoral
Research Fellow in Archaeology
University of Cambridge

Patricia Mercer
Senior Lecturer in History
University of North London

Nicola Miller
Lecturer in Latin American History
University College London
University of London

David Morgan
Senior Lecturer in History
University College London
University of London

Jean Morrin
Lecturer in History
University of North London

R. C. Nash
Lecturer in Economic and Social History
University of Manchester

Colin Nicolson
Senior Lecturer in History
University of North London

Phillips O'Brien
Lecturer in Modern History
University of Glasgow

David Potter
Senior Lecturer in History
University of Kent at Canterbury

Max-Stephan Schulze
Lecturer in Economic History
London School of Economics
University of London

Ian Selby
Research Fellow
St Edmund's College
University of Cambridge

Caroline Steele
Lecturer in Iliad Program, Dartmouth College
Research Associate
State University of New York at Binghamton

Diura Thoden van Velzen
English Heritage

Jessica B. Thurlow
University of Sussex

Luke Treadwell
University Lecturer in Islamic Numismatics
Oriental Institute
University of Oxford

Nick von Tunzelmann
Professor of the Economics of Science
and Technology
Science and Technology Policy Research Unit
University of Sussex

Emily Umberger
Associate Professor of Art History
Arizona State University

Gabrielle Ward-Smith
University of Toronto

David Washbrook
Reader in Modern South Asian History
Professorial Fellow of St Antonys College
University of Oxford

Mark Whittow
Lecturer in Modern History
Fellow of St Peter's College
University of Oxford

Beryl J. Williams
Reader in History
University of Sussex

Richard Wiltshire
Senior Lecturer in Geography
School of Oriental and African Studies
University of London

Neville Wylie
Lecturer in Modern History
Acting Director of the Scottish Centre
for War Studies
University of Glasgow

CONTENTS

3 THE EARLY MODERN WORLD

CONTENTS CONTINUED

FOREWORD

There could be no more opportune time than the eve of the third millennium AD to produce an entirely new atlas of world history. Not only does this symbolic (if arbitrary) moment provoke a mood of public retrospection, but the pace of global change itself demands a greater awareness of "whole world" history. More than 20 years have passed since a major new atlas of this kind was published in the English language. In that period there has been an explosion of new research into the histories of regions outside Europe and North America, and a growing awareness of how parochial our traditional approach to history has been. In this changed environment, the demand for an unbiased overview of world history has steadily grown in schools, colleges and universities, and among the general reading public.

Several developments within the study of academic history promote the seriousness with which histories of the world are now taken. First the accumulation of knowledge about the past of different nations has engendered excessive specialization. The sheer volume of publications and data about details of the past stimulates demand from students, scholars and a wider public for guidelines, meaning and "big pictures" that world history, with its unconfined time frame and wider geographical focus, is positioned to meet.

Secondly the broadening of traditional history's central concerns (with states, warfare and diplomacy) in order to take account of modern concerns with, for example, ecology, evolutionary biology, botany, the health and wealth of populations, human rights, gender, family systems and private life, points the study of history towards comparisons between Western and non-Western cultures and histories.

Thirdly young people now arrive at universities with portfolios of knowledge and aroused curiosities about a variety of cultures. They are less likely than their predecessors to study national let alone regional and parochial histories. Schools and universities need to provide access to the kind of historical understanding that will satisfy their interests. To nourish the cosmopolitan sensibility required for the next millennium, history needs to be widened and repositioned to bring the subject into fruitful exchange with geography and the social sciences. Barriers between archaeology, ancient, classical, medieval, early modern, contemporary and other "packages" of traditional but now anachronistic histories are being dismantled.

Unsurprisingly, the implications of "globalization" for hitherto separated communities, disconnected economies and distinctive cultures have been analysed by social scientists. They serve governments who are uneasily aware that their powers to control economies and societies nominally under their jurisdiction are being eroded, both by radical improvements in the technologies for the transportation of goods and people around the world and by the vastly more efficient communications systems that diffuse commercial intelligence, political messages and cultural information between widely separated populations.

A NEW PERSPECTIVE ON WORLD HISTORY

As the world changes at an accelerated pace, for problem after problem and subject after subject, national frameworks for political action and academic enquiry are recognized as unsatisfactory. Historians are being asked for a deeper perspective on the technological, political and economic forces that are now transforming traditional frameworks for human behaviour, and reshaping personal identities around the world. *Philip's Atlas of World History* has been designed, constructed and written by a team of professional historians not only for the general reader but to help teachers of history in schools, colleges and universities to communicate that perspective to their pupils and students.

World histories cannot be taught or read without a clear comprehension of the chronologies and regional parameters within which different empires, states and peoples have evolved through time. A modern historical atlas is the ideal mode of presentation for ready reference and for the easy acquisition of basic facts upon which courses in world history can be built, delivered and studied. Such atlases unify history with geography. They "encapsulate" knowledge by illuminating the significance of locations for seminal events in world history. For example, a glance at maps on pages 78 and 116–17 will immediately reveal why explorers and ships from western Europe were more likely (before the advent of steam-powered ships) to reach the Americas than sailors from China or India. More than any other factor it was probably a matter of distance and the prevailing winds on the Atlantic that precluded Asian voyages to the Americas.

Historical atlases should be accurate, accessible and display the unfurling chronology of world history in memorable maps and captions. The team of historians, cartographers and editors who collaborated in the construction of *Philip's Atlas of World History* set out to produce a popular work of reference that could be adopted for university, college and school courses in world history. In the United States and Canada such courses are already commonplace and the subject is now spreading in Britain and the rest of Europe, Japan and China. New textbooks appear regularly. American journals dealing with world history publish debates on how histories designed to cover long chronologies and unconfined geographies might be as rigorous and as intellectually compelling as more orthodox histories dealing with individuals, parishes, towns, regions, countries and single continents.

The editors attempted to become familiar with as many course outlines as possible. Their plans for the atlas were informed by the ongoing, contemporary debate (largely North American) about the scale, scope and nature of world history. For example, they were aware that most "model" textbooks in world history are usually constructed around the grand themes of "connections" and "comparisons" across continents and civilizations, and that a scientifically informed appreciation of environmental, evolutionary and biological constraints on all human activity is regarded as basic to any understanding of world history.

Through its carefully designed system of cross-referencing, this atlas encourages the appreciation of "connections", "contacts" and "encounters" promoted through trade, transportation, conquest, colonization, disease and botanical exchanges and the diffusion of major religious beliefs. It also aims to facilitate "comparisons" across space and through time of the major forces at work in world history, including warfare, revolutions, state formation, religious conversion, industrial development, scientific and technological discoveries, demographic change, urbanization and migration. Histories or atlases of the world are potentially limitless in their geographical and chronological coverage. Publications in the field are inevitably selective and as William McNeill opined: "Knowing what to leave out is the hallmark of scholarship in world history".

HISTORY IN ITS BROADEST CONTEXT
As I write this foreword NATO troops are entering Kosovo. The Balkans crisis features in Part 5: "The Twentieth Century", but in the atlas it is also set in the context not just of our times, but of the whole span of history. The atlas opens with "The Human Revolution: 5 million years ago to 10,000 BC" placed within an innovative opening section dealing largely with archaeological evidence for the evolution of tools and other artefacts, as well as the famous transition from hunting to farming in Asia, Africa, Europe and the Americas from around 10,000 BC.

This first section also covers connections and comparisons across the first civilizations in Mesopotamia, the Indus Valley, Egypt, China and Mesoamerica and South America as well as those later and more familiar empires of Greece, India, China and Rome. Yet the editors have also ensured that small countries (such as Korea), important but often forgotten traders and explorers (such as the Vikings), and the nomadic peoples of Central Asia, the Americas and Africa have found their place in this comprehensive atlas of world history. Furthermore, coverage of the world wars of the 20th century, the Great Depression, the rise of communism and fascism, decolonization, the end of the Cold War and the events of the 1990s makes the atlas into a distinctive work of first reference for courses in current affairs and contemporary history.

Facts, brief analyses and illuminating maps of such seminal events in world history as the transition to settled agriculture, the inventions of writing and printing, the birth of religions, the Roman Empire, Song China, the discovery of the Americas, the Scientific, French and Industrial Revolutions, the foundation of the Soviet Union and of communist China are all carefully chronicled and represented on colourful maps drawn using the latest cartographic technology. Although any general atlas of world history will, and should, give prominence to such traditional historical themes as the rise and decline of empires, states and civilizations, a serious effort has been made wherever possible in the atlas to accord proper emphasis to the communal concerns of humankind, including religion, economic welfare, trade, technology, health, the status of women and human rights.

Through its individual spreads, the atlas can be used easily to find out about a significant event (*The American Revolution*), the history of defined places and populations (*India under the Mughals 1526–1765*), religious transitions (*The Reformation and Counter Reformation in Europe 1517–1648*), or social movements on a world scale (*World Population Growth and Urbanization 1800–1914*). Overall the atlas represents an exciting alternative to histories narrowly focused on the experience of national communities. World history offers chronologies, perspectives and geographical parameters which aim to attenuate the excesses of ethnicity, chauvinism and condescension. The length and breadth of an atlas of world history covering all continents, and a chronology going back twelve millennia, can work to separate the provincial from the universal, the episodic from the persistent. In so far as this atlas succeeds in these goals, and thus contributes to the widespread aspiration for an education in world history, it can also help nurture a cosmopolitan sensibility for the next millennium.

Patrick K. O'Brien FBA
Institute of Historical Research
University of London

THE ANCIENT WORLD

The first humans evolved in Africa around two million years ago. By 9000 BC their descendants had spread to most parts of the globe and in some areas were beginning to practise agriculture. From around 4000 BC the first civilizations developed, initially in the Near East and India and subsequently in China, Mesoamerica and South America. In the centuries that followed, to AD 500, many states and empires rose and fell.

S ome five to eight million years ago, a species of small African primates began walking upright. While there are many theories about the advantages conferred by moving on two legs rather than four, there is general agreement that the success of the hominid line (humans and their ancestors) is due in part to the adoption of this new method of locomotion. Between five and one million years ago, hominid species proliferated in East Africa and southern Africa, giving rise by 1.8 million years ago to the new genus, *Homo*, to which we ourselves belong (*map 1*).

The development by *Homo* of stone tools – and, we may presume, tools that have not survived, made of other materials such as bone and wood – was a major advance in human evolution, allowing our ancestors to engage in activities for which they lacked the physical capabilities. This ability to develop technology to overcome our physical limitations has enabled us to develop from a small and restricted population of African apes to a species that dominates every continent except Antarctica and has even reached the moon. Between 1.8 million and 300,000 years ago, members of our genus colonized much of temperate Europe and Asia as well as tropical areas, aided by their ability to use fire and create shelter. By 9000 BC the only parts of the globe which modern humans – *Homo sapiens* – had not reached were some remote islands and circumpolar regions.

▼ The world was not colonized in a single movement; there were at least two major episodes. In the first, between 1.8 million and 300,000 years ago, early *Homo* spread from Africa as far as China and western Europe. In the second, the descendants of early *Homo* were replaced by representatives of modern humans, *Homo sapiens*, who reached Australia by 60,000 and the Americas by 14,000 years ago. During the whole of this period the migration of humans was greatly affected by a number of ice ages, when sea levels fell to reveal land "bridges" that in later years became submerged.

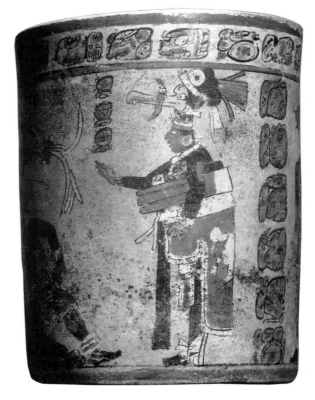

▲ With the development of agriculture and settled communities there was a growing need for storage. Pottery began to be made on a wide scale in order to meet this need, but it also served as a vehicle for human artistic activity. This Maya cylindrical pottery vessel depicts players in a ballgame that was an important ritual activity throughout the ancient civilizations of Mesoamerica. A standard but as yet undeciphered text in the complex Maya hieroglyphic writing runs round the top of the vessel.

1 COLONIZATION OF THE WORLD 1.8 MILLION YEARS AGO TO 10,000 BC

Area inhabited by early hominids 5 to 1 million years ago	Area colonized by early *Homo* 1.8 million to 300,000 years ago	Area colonized by modern humans 100,000 to 10,000 BC

FROM HUNTING TO FARMING

In 10,000 BC the world was inhabited solely by groups who lived by hunting and gathering wild foods. Within the succeeding 8,000 years, however, much of the world was transformed (*map 2*). People in many parts of the world began to produce their own food, domesticating and selectively breeding plants and animals. Farming supported larger and more settled communities, allowing the accumulation of stored food surpluses – albeit with the counterpoised risks involved in clearing areas of plants and animals that had formerly been a source of back-up food in lean years. Agricultural communities expanded in many regions, for example colonizing Europe and South Asia, and in doing so radically changed the landscape.

▲ Rock paintings, such as these "X-ray style" figures from Nourlangie in Australia's Northern Territory, provide a fascinating record of the everyday world of hunter-gatherers. They also give some insight into the rich spiritual and mythological life of the people who created them.

FIRST CIVILIZATIONS

As the millennia passed there was continuing innovation in agricultural techniques and tools, with the domestication of more plants and animals and the improvement by selective breeding of those already being exploited. These developments increased productivity and allowed the colonization of new areas. Specialist pastoral groups moved into previously uninhabited, inhospitable desert regions. Swamps were drained in Mesoamerica and South America and highly productive raised fields were constructed in their place. Irrigation techniques allowed the cultivation of river valleys in otherwise arid regions, such as Mesopotamia and Egypt.

High agricultural productivity supported high population densities, and towns and cities grew up, often with monumental public architecture. However, there were also limitations in these regions, such as an unreliable climate or river regime, or a scarcity of important raw materials (such as stone), and there was often conflict between neighbouring groups. Religious or secular leaders who could organize food storage and redistribution, craft production, trade, defence and social order became increasingly powerful. These factors led to the emergence of the first civilizations in many parts of the world between around 4000 and 200 BC (*maps 3 and 4 overleaf*). A surplus of agricultural produce was used in these civilizations to support a growing number of specialists who were not engaged in food production: craftsmen, traders, priests and rulers, as well as full-time warriors – although the majority of soldiers were normally farmers.

Specialists in some societies included scribes. The development of writing proved a major advance, enabling vast quantities of human knowledge and experience to be recorded, shared and passed on. Nevertheless, in most societies literacy was confined to an elite – priests, rulers and the scribes they employed – who used it as a means of religious, political or economic control. In most parts of the world, the belief that there should be universal access to knowledge recorded in writing is a recent phenomenon.

RITUAL AND RELIGION

Although without written records it is impossible to reconstruct details of the belief systems of past societies, evidence of religious beliefs and ritual activities abounds, particularly in works of art, monumental structures and grave offerings.

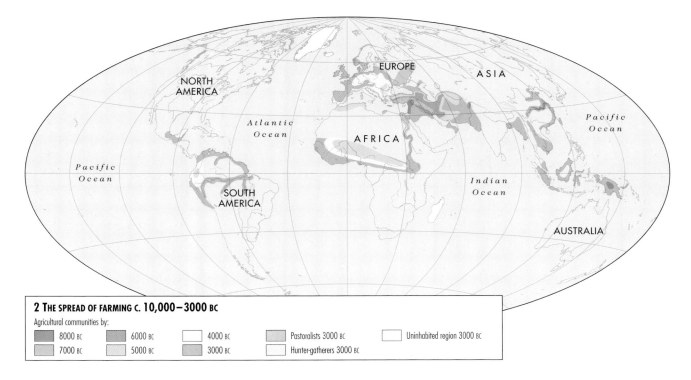

2 THE SPREAD OF FARMING C. 10,000–3000 BC

Agricultural communities by:

8000 BC	6000 BC	4000 BC	Pastoralists 3000 BC	Uninhabited region 3000 BC
7000 BC	5000 BC	3000 BC	Hunter-gatherers 3000 BC	

◄ Farming developed in many parts of the world from around 10,000 BC. Differences in the locally available plants and animals and in local conditions gave rise to much variation between regions. Domestic animals, for example, played an important part in Old World agriculture, whereas farmers in Mesoamerica and North America relied heavily on wild animals and crops such as beans for protein. A settled lifestyle usually depended on the practice of agriculture. However, in some areas, such as the Pacific coast of North America, an abundant supply of wild resources allowed settled communities to develop without agriculture.

Ritual and religion were a powerful spur to the creation of monumental architecture by literate urban societies such as the Egyptians, Greeks and Romans, but also in smaller societies dependent on agriculture, such as the prehistoric inhabitants of Europe who built the megalithic tombs, or the moundbuilders of North America. Monuments also reflected other factors, such as a desire for prestige or to affirm territorial rights. Although such building activity implied the ability to mobilize large numbers of people, this did not necessarily require hierarchical social control; it could be achieved within the framework of a community led by elders or priests.

▶ Intensive and highly productive agriculture gave rise to civilized societies in Mesopotamia, Egypt and northern India in the 4th and 3rd millennia BC and in China by 1700 BC.

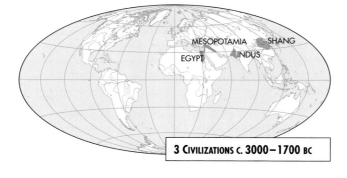

3 CIVILIZATIONS c. 3000–1700 BC

▲ Scenes from the life and "former lives" of Buddha (c. 563–483 BC) are among those decorating the *stupa* at Amaravati in southern India. The *stupa* dates mostly from the 2nd century AD, by which time several major religions – Hinduism, Zoroastrianism, Judaism, Buddhism and Christianity – had developed and begun to spread through Asia and Europe.

▶ Between 1200 and 500 BC civilized societies were established in the Americas. By this time the early states of Eurasia and Africa had declined and been replaced by others, such as the Persian Empire, Minoan and Mycenaean Greece and the Zhou state in China.

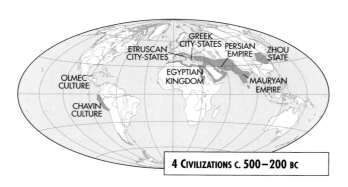

4 CIVILIZATIONS c. 500–200 BC

Concern with the proper disposal of the dead was displayed from Neanderthal times, more than 50,000 years ago. In the burial or other treatment of the body regarded as appropriate (such as cremation or exposure), the dead were often accompanied by grave offerings. These could range from food or small items of personal dress, to large numbers of sacrificed relatives or retainers as in tombs dating from the 3rd millennium BC in Egypt and the 2nd millennium BC in Shang China. The offerings might be related to life after death, for which the deceased needed to be equipped, but also frequently reflected aspects of the dead person's social position in life.

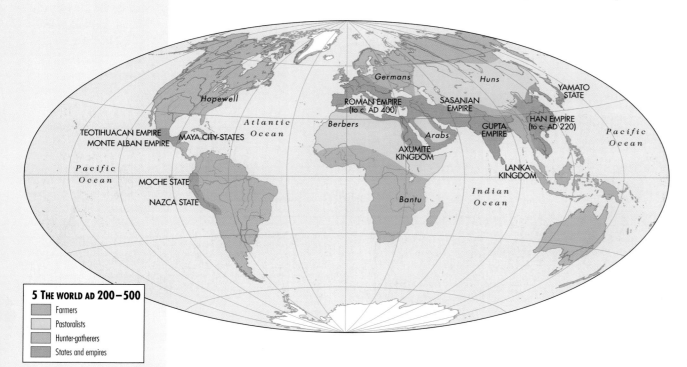

5 THE WORLD AD 200–500

- Farmers
- Pastoralists
- Hunter-gatherers
- States and empires

◀ New regions became caught up in the expansion of states: Korea and parts of Central Asia fell to the Chinese Han Empire, Europe was swept up by the Roman Empire, and the North American southwest came under the cultural influence of Mesoamerican states. Elsewhere, however, farmers, herders and hunter-gatherers continued their traditional lifestyle, affected to varying degrees by their civilized neighbours, who regarded them as "barbarians". Such "barbarians" could turn the tide of empires: Central Asian nomads were the periodic scourge of West, South and East Asia for thousands of years, and Germanic confederacies, with Central Asians, brought down the Western Roman Empire in the middle of the 1st millennium AD.

Grave offerings often provide valuable clues about past social organization. They also point to the important part played by artisans in the development of civilized communities, in particular producing prestige items for use by the elite and manufactured goods to be traded in exchange for vital raw materials. In developed agricultural societies, craft production was unlikely to be a full-time pursuit for more than a handful of individuals, but this did not prevent high standards being reached in many communities.

Unlike pottery, which was made by the majority of settled communities, and stone, used for tools worldwide from very early times, metalworking did not develop in all parts of the globe, due in part to the distribution of ores. Initially metal artefacts tended to be prestige objects, used to demonstrate individual or community status, but metal was soon used for producing tools as well. The development of techniques for working iron, in particular, was a major breakthrough, given the abundance and widespread distribution of iron ore.

STATES AND EMPIRES

By about 500 BC ironworking was well established in Europe, West and South Asia, and in parts of East Asia and Africa. States had developed in most of these regions at least a thousand years before, but for a variety of reasons the focal areas of these entities had changed over the course of time (*map 4*). The formerly fertile lower reaches of the Euphrates, cradle of the Mesopotamian civilization, had suffered salination, and so the focus had shifted north to the competing Assyrian and Babylonian empires. In India the primary civilization had emerged along the Indus river system; after its fall, the focus of power and prosperity shifted to the Ganges Valley, which by the 3rd century BC was the centre of the Mauryan Empire.

Europe was also developing native states, and by the 1st century AD much of Europe and adjacent

regions of Asia and Africa were united through military conquest by the Romans. The rise and expansion of the far-reaching Roman Empire was paralleled in the east by that of the equally vast Chinese Han Empire (*map 5*).

Military conquest was not, however, the only means by which large areas were united. The Andean region, for example, was dominated in the 1st millennium BC by the Chavin culture, seemingly related to a widely shared religious cult centred on a shrine at Chavin de Huantar. A complex interplay of political, economic, religious and social factors determined the pattern of the rise and fall of states.

On the fringes of the human world, pioneers continued to colonize new areas, developing ways of life to enable them to settle in the circumpolar regions and the deserts of Arabia and to venture huge distances across uncharted waters to settle on the most remote Pacific islands. By AD 500 the Antarctic was the only continent still unpeopled.

◀ The civilizations of the ancient world provided a milieu in which the sciences and technology thrived. The Babylonians, Indians and Greeks, for example, developed mathematics and astronomical knowledge to a high level, while the Chinese pioneered advances in a number of fields, among them metallurgy and mining technology. The Romans were also skilled innovators, particularly in engineering, where in the public domain they built magnificent roads and aqueducts, such as the Pont du Gard in France, pictured here.

▼ The burials of important people were often lavishly furnished with spectacular works of craftsmanship. The body of Princess Dou Wan of the Han kingdom of Zhongshan in China was buried in the 2nd century BC in this suit made of jade plaques bound together with gold thread. In Chinese belief, jade was linked to immortality, and suits such as this were intended to preserve the body of the deceased.

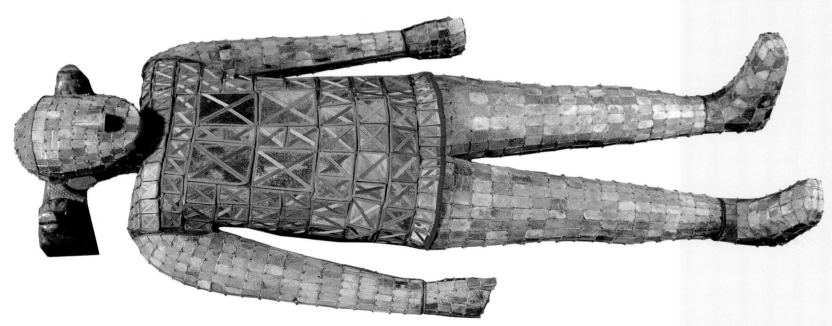

THE HUMAN REVOLUTION: 5 MILLION YEARS AGO TO 10,000 BC

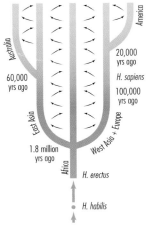

1 CONTINUOUS GENE FLOW MODEL

2 DISCRETE EVOLUTION MODEL

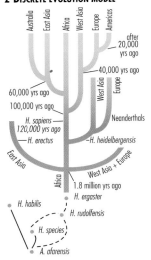

▲ Some experts believe that modern humans evolved from the early hominids in parallel in Africa, Asia and Europe (1). However, it is more generally accepted that they originated in Africa and then spread – at the expense of other hominid species (2).

► The last of the inhabited continents to be colonized by hominids was South America, probably between 14,000 and 11,000 years ago.

Traces of the earliest ancestors of humans, the Australopithecines, have been found in Africa, dating from between five and two million years ago when the forests had given way in places to more open savanna (*map 1*). A line of footprints discovered at Laetoli is vivid evidence that these now extinct early hominids (human ancestors belonging to the genera *Australopithecus* and *Homo*) walked upright. Hominid fossils from this remote period are rare, since the creatures themselves were not numerous. The remains that have been found probably belong to different species: some, such as *A. robustus* and *A. boisei*, lived on plant material; others, such as the smaller *A. africanus*, ate a more varied diet. By two million years ago the hominids included *Homo habilis*, small creatures whose diet probably included kills scavenged from carnivores. Unlike their Australopithecine cousins, *H. habilis* had begun to manufacture stone tools (called "Oldowan" after the key site of Olduvai), roughly chipped to form a serviceable edge for slicing through hide, digging and other activities which these small hominids could not perform with their inadequate teeth and nails. These developments, along with physical adaptation, were crucial in the amazing success of humans compared with other animal species.

THE MOVE INTO TEMPERATE REGIONS

By 1.8 million years ago this success was already becoming apparent in the rapid spread of hominids well outside their original tropical home, into temperate regions as far afield as East Asia (*map 2*). This move was made possible by a number of developments. Hominids began to make new and more efficient tools, including the multipurpose handaxe, which extended their physical capabilities. A substantial increase in body size allowed representatives of *Homo* to compete more successfully with other scavengers, and by 500,000 years ago our ancestors were hunting as well as scavenging, using wooden spears and probably fire. Fire was also important in providing warmth, light and protection against predators, and for cooking food, thus making it easier to chew and digest. To cope with the temperate climate, hominids used caves and rock shelters such as those found at the famous Chinese site of Zhoukoudian.

There had been a gradual cooling of the global climate, with ice sheets developing in the Arctic by 2.4 million years ago. Around 900,000 years ago this process had accelerated, giving rise to a pattern of short ice ages approximately every

100,000 years. These ice ages were interspersed with short phases of temperatures similar to or higher than those of today, and much longer periods of intermediate temperatures. The pattern of ice advance and retreat had a major effect not only on the distribution of hominids and other mammals but also on the preservation of their fossils, so the picture that we have today is at best partial. During warm periods, hominids penetrated as far north as southern England; in cooler periods, sea levels fell and many coastal areas that are now submerged became habitable.

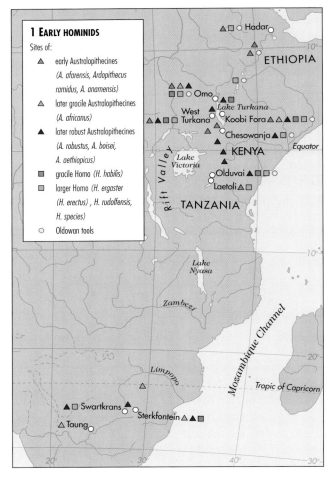

1 EARLY HOMINIDS
Sites of:
▲ early Australopithecines (A. afarensis, Ardopithecus ramidus, A. anamensis)
△ later gracile Australopithecines (A. africanus)
▲ later robust Australopithecines (A. robustus, A. boisei, A. aethiopicus)
▣ gracile Homo (H. habilis)
▢ larger Homo (H. ergaster (H. erectus) , H. rudolfensis, H. species)
○ Oldowan tools

▲ Many hominid species flourished in sub-Saharan Africa between five and one million years ago, but most died out. Modern humans are the only surviving descendants.

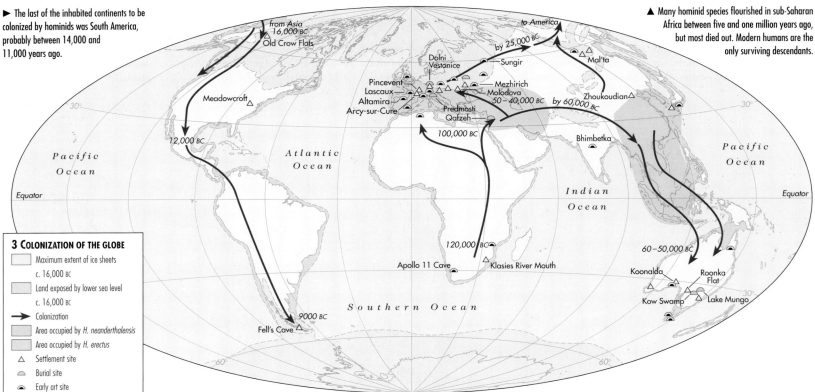

3 COLONIZATION OF THE GLOBE
- Maximum extent of ice sheets c. 16,000 BC
- Land exposed by lower sea level c. 16,000 BC
- → Colonization
- Area occupied by H. neanderthalensis
- Area occupied by H. erectus
- △ Settlement site
- ⌒ Burial site
- ◉ Early art site

16

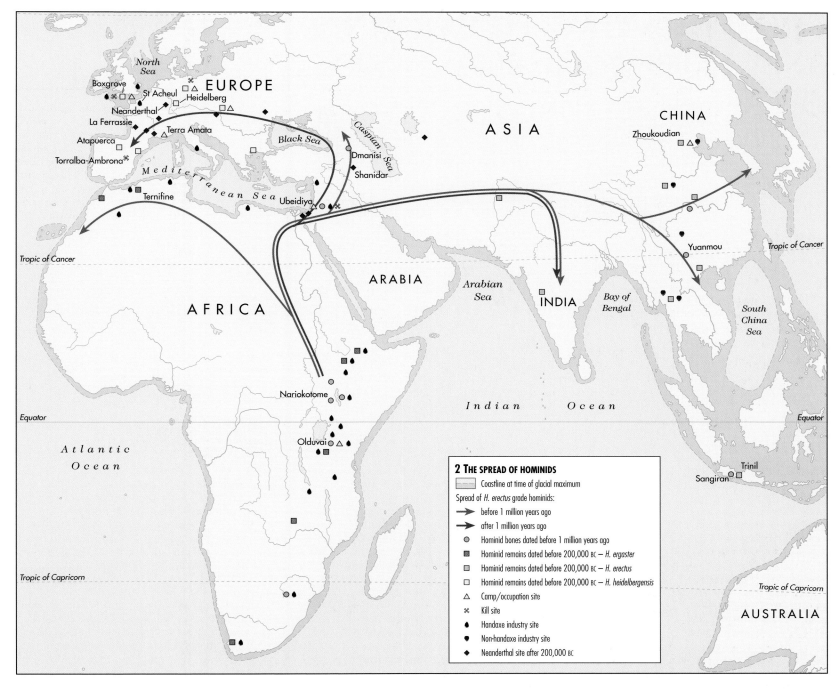

2 THE SPREAD OF HOMINIDS
☐ Coastline at time of glacial maximum
Spread of *H. erectus* grade hominids:
➤ before 1 million years ago
➤ after 1 million years ago
◯ Hominid bones dated before 1 million years ago
◼ Hominid remains dated before 200,000 BC – *H. ergaster*
◼ Hominid remains dated before 200,000 BC – *H. erectus*
☐ Hominid remains dated before 200,000 BC – *H. heidelbergensis*
△ Camp/occupation site
✳ Kill site
⬧ Handaxe industry site
⬧ Non-handaxe industry site
◆ Neanderthal site after 200,000 BC

THE EMERGENCE OF MODERN HUMANS

Around 100,000 years ago two hominid species were living in the eastern Mediterranean region. One was the Asian representative of the Neanderthals (*H. neanderthalensis*) – descended from *H. heidelbergensis* – who inhabited Europe and West Asia from some time after 200,000 BC; the other was an early form of *Homo sapiens* (modern humans) who had first appeared some 20,000 years earlier in southern Africa. By 40,000 BC modern humans were to be found throughout the previously inhabited world – Africa, Asia and Europe – and in Australia (*map 3*).

Opinions are divided as to how this came about. One school of thought holds that the descendants of the first hominids to colonize these various regions had evolved in parallel (*diagram 1*); there was continuous gene flow between adjacent regions, spreading adaptations and changes throughout the hominid world but with regional differences also present, as in the modern races. This view sees the emergence of modern humans as a global phenomenon.

The alternative and more generally accepted view is that the original colonists developed into different regional species (*diagram 2*). Modern humans emerged in Africa and were able to spread at the expense of other hominids, progressively colonizing West Asia by 100,000 BC, East Asia and Australia by 60,000 BC and Europe by 40,000 BC. Whether they interbred with the hominids they displaced or simply extinguished them is unclear, but almost certainly *Homo sapiens* was the only surviving hominid by about 30,000 BC.

From Asia modern humans moved into the Americas, crossing the Bering Strait during an ice age when the land bridge of Beringia was exposed, and migrating southwards later. The date of this colonization is still hotly debated, but the earliest incontrovertible evidence of humans in the Americas south of the glaciated area comes after the ice sheets began to retreat – about 14,000 years ago.

CULTURAL DEVELOPMENT

Early modern humans and their Neanderthal contemporaries used similar tools and seem to have been culturally related. However, although Neanderthals and even earlier hominids may have communicated with sounds to some extent, *H. sapiens* was the first hominid to be able to communicate in a fully developed spoken language. This was a critical development, making possible detailed planning and discussion of group activities and interactions and, more importantly, allowing the knowledge acquired through individual experience to be shared and transmitted from generation to generation.

From about 100,000 years ago many aspects of human consciousness and aesthetic sense began to evolve, as evidenced by the finely shaped and consciously planned stone tools of both Neanderthals and modern humans, and by the beginning of burial. The emergence of human consciousness becomes ever more apparent in the art that dates from about 35,000 BC, and very probably earlier in Australia. Archaeologists have found exquisite figurines depicting both humans and animals, as well as magnificent animal and abstract paintings and engravings on the walls of caves and rock shelters. The most famous of these finds are in southern France and adjacent Spain, but early art has been found all over the world, with fine concentrations in Australia, Africa and Russia.

▲ Until recently the immediate descendants of *Homo habilis* were all classified as *Homo erectus*, but it now seems more probable that there were a number of roughly contemporary hominid species: *H. ergaster* in Africa, *H. erectus* in East Asia and *H. heidelbergensis* in Europe. The paucity of hominid fossils makes their classification extremely difficult, and there are major and frequent changes in the interpretation of the limited evidence.

FROM HUNTING TO FARMING:
ASIA 12,000 BC–AD 500

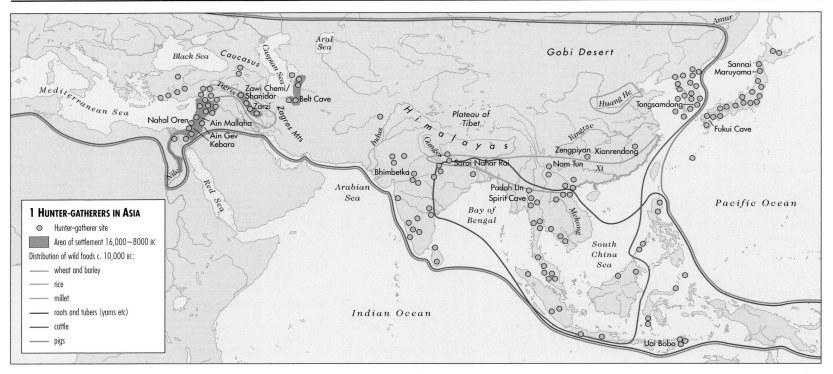

1 HUNTER-GATHERERS IN ASIA
- ○ Hunter-gatherer site
- ▨ Area of settlement 16,000–8000 BC

Distribution of wild foods c. 10,000 BC:
- —— wheat and barley
- —— rice
- —— millet
- —— roots and tubers (yams etc)
- —— cattle
- —— pigs

▲ Animal bones are much more likely to be preserved than plant remains, so the archaeologist's picture of past subsistence probably underestimates the importance of plant foods. This is particularly true of tubers, roots, leafy vegetables and fruits, which must have provided the bulk of the diet in areas such as Southeast Asia. We have a clearer picture of the development of early agriculture in areas such as China and West Asia, where cereals (rice, millet, wheat and barley) and pulses (beans, peas and the like) were the principal food plants.

▼ Living in sedentary settlements made it possible to store cereals and other plant foods, including nuts, to provide some insurance against lean seasons or years. It also enabled people to accumulate possessions that today provide valuable evidence of their way of life.

Evidence from many parts of the world indicates that during the final millennia of the last glacial age – between around 16,000 and 12,000 years ago – the range of foods eaten by humans broadened considerably. In the "Fertile Crescent" of West Asia (the arc of land comprising the Levant, Mesopotamia and the Zagros region) wild wheat and barley provided an abundant annual harvest that enabled hunter-gatherers to dwell year-round in permanent settlements such as Kebara (map 1). Nuts and other wild foods, particularly gazelle, were also important here.

Around 12,000 BC the global temperature began to rise, causing many changes. Sea levels rose, flooding many coastal regions; this deprived some areas of vital resources but in others, such as Japan and Southeast Asia, it created new opportunities for fishing and gathering shellfish. Changes occurred in regional vegetation, with associated changes in fauna. Throughout Asia, particularly in the southeast, plant foods became increasingly important.

In the Levant wild cereals at first spread to cover a much larger area, increasing the opportunities for sedentary communities to develop. A cold, dry interlude around 9000 to 8000 BC caused a decline in the availability of wild cereals

and the abandonment of many of these settlements, but communities in well-watered areas began to plant and cultivate the cereals they had formerly gathered from the wild (map 2). By 8000 BC, when conditions again became more favourable, these first farming communities had grown in size and number and they began to spread into other suitable areas. Initially these new economies combined cultivated cereals with wild animals, but around 7000 BC domesticated sheep and goats began to replace gazelle and other wild game as the main source of meat.

Subsequent millennia saw the rapid spread of farming communities into adjacent areas of West Asia (map 3). They appeared over much of Anatolia and northern Mesopotamia by about 7000 BC, largely confined to areas where rain-fed agriculture was possible. Agricultural communities also emerged around the southeastern shores of the Caspian Sea, and at Mehrgarh on the western edge of the Indus plains. Pottery, which began to be made in the Zagros region around this time, came into widespread use in the following centuries, and copper also began to be traded and worked. Cattle, domesticated from the aurochs (*Bos primigenius*) in the west and from native Indian cattle (*Bos namadicus*) in South Asia, were now also important. In Anatolia cattle seem to have played a part in religion as well as in the economy: for example, rooms in the massive settlement at Çatal Höyük in Anatolia were decorated with paintings of enormous cattle and had clay cattle-heads with real horns moulded onto the walls.

DIVERSIFICATION OF AGRICULTURE

By 5000 BC the development of more sophisticated agricultural techniques, such as irrigation and water control, had enabled farming communities to spread into southern Mesopotamia, much of the Iranian Plateau and the Indo-Iranian borderlands. It was not until the 4th millennium BC, however, that farmers growing wheat and keeping sheep, goats and cattle moved into the adjacent Indus Valley and thence southward into peninsular India. The development of rice and millet cultivation by the Indus civilization (pages 28–29) led to a further spread of agriculture into the Ganges Valley and the south of India.

Eastern India also saw the introduction of rice cultivation from Southeast Asia, while sites in the northeast may owe their development of agriculture to contact with northern China. In the latter region farming probably began around 7000 BC and was well established by 5000 BC (map 4). In two areas in the Huang He Basin, at sites such as Cishan and Banpo, communities emerged whose

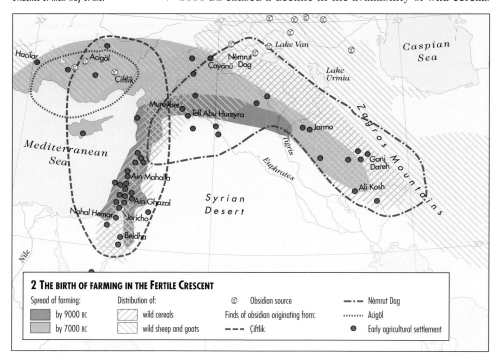

2 THE BIRTH OF FARMING IN THE FERTILE CRESCENT

Spread of farming:
- ▨ by 9000 BC
- ▨ by 7000 BC

Distribution of:
- ▨ wild cereals
- ▨ wild sheep and goats

- ⓣ Obsidian source

Finds of obsidian originating from:
- –·– Némrut Dag
- ········· Acigöl
- – – – Çiftlik

- –·– Némrut Dag
- ········· Acigöl
- ● Early agricultural settlement

economies depended on cultivated millet, along with fruits and vegetables, chickens and pigs, while further south, in the delta of the Yangtze River, wet rice cultivation began. Hemudu is the best known of these early rice-farming communities: here waterlogging has preserved finely constructed wooden houses and a range of bone tools used in cultivation, as well as carbonized rice husks and the remains of other water-loving plant foods such as lotus. Here also was found the first evidence of lacquerware: a red-lacquered wooden bowl. Although water buffalo and pigs were kept in this southern region, both hunted game and fish continued to play an important role in the economy.

By 3000 BC wet rice agriculture was becoming established in southern China, northern Thailand and Taiwan, and millet cultivation in northern China. Communities in the northwest also grew wheat and barley, introduced from the agricultural communities of West or Central Asia. In Southeast Asia tubers and fruits had probably been intensively exploited for millennia. By 3000 BC wet rice was also grown in this region and buffalo, pigs and chickens were raised, but wild resources remained important.

The inhabitants of Korea and Japan continued to rely on their abundant wild sources of food, including fish, shellfish, deer, nuts and tubers. Often they were able to live in permanent settlements. The world's earliest known pottery had been made in Japan in the late glacial period: a range of elaborately decorated pottery vessels and figurines was produced in the later hunter-gatherer settlements of the archipelago. Trade between communities circulated desirable materials such as jadeite and obsidian (volcanic glass). Around 1500 BC crops (in particular rice) and metallurgical techniques began spreading from China into these regions, reaching Korea via Manchuria and thence being taken to Japan. By AD 300 rice farming was established throughout the region with the exception of the northernmost island, Hokkaido, home of the Ainu people, where the traditional hunter-gatherer way of life continued into recent times.

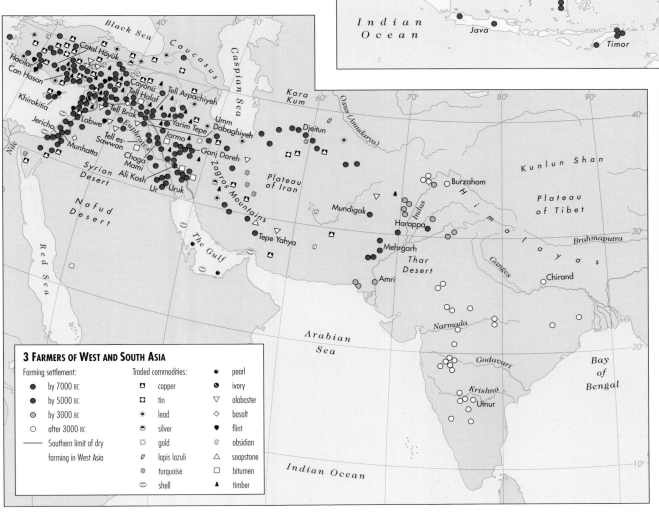

4 THE SPREAD OF FARMING IN EAST ASIA

Core area of cereal cultivation:
- rice
- millet
- → Spread of rice farming
- ● Initial farming settlement before 3000 BC
- ● Initial farming settlement after 3000 BC
- ● Developed farming settlement after 3000 BC

3 FARMERS OF WEST AND SOUTH ASIA

Farming settlement:
- ● by 7000 BC
- ● by 5000 BC
- ● by 3000 BC
- ○ after 3000 BC
- — Southern limit of dry farming in West Asia

Traded commodities:
- ◪ copper
- ◫ tin
- ◉ lead
- ◰ silver
- ◫ gold
- ∅ lapis lazuli
- ◉ turquoise
- ◡ shell
- ● pearl
- ◕ ivory
- ▽ alabaster
- ◇ basalt
- ◆ flint
- ◉ obsidian
- △ soapstone
- □ bitumen
- ▲ timber

▲ Banpo, a typical early Chinese farming settlement, contained dwellings, storage pits and animal pens, a communal hall, a cemetery and kilns in which finely decorated pottery was fired. The villagers were probably already keeping silkworms, although most textiles were made of hemp. By around 3000 BC settlements were often fortified with tamped earth walls, implying intercommunity warfare. Clear signs of developing social stratification appear at this time — for example, elite burials containing prestige goods of bronze and imported materials such as jade, made by an emerging class of specialist craftsmen. Following the introduction of metallurgy from China during the 1st millennium BC, Korea and Japan also developed a sophisticated bronze industry.

◄ By 4000 BC farming communities established in many areas of Asia were linked by trade. Areas of high agricultural productivity, such as southern Mesopotamia, were dependent on trade to obtain the basic raw materials lacking in the alluvial environment, such as wood and stone. They were, however, able to support full-time craft specialists producing goods for export, particularly textiles and fine pottery, as well as surplus agricultural produce.

● MESOPOTAMIA AND THE INDUS REGION 4000–1800 BC pages 28–29　● CHINA 1700–1050 BC pages 30–31

FROM HUNTING TO FARMING: EUROPE 8000–200 BC

▲ From the 6th century BC some Celtic chiefdoms began to benefit from trade with the Greeks and Etruscans, their increasing wealth being reflected in massive hillforts and splendidly furnished graves. Metal ores and other raw materials – goods previously circulated within Europe – were now syphoned off by the Mediterranean world in exchange for luxuries, especially wine and related artefacts, such as Greek pottery and Etruscan bronze flagons. These in turn provided inspiration for native Celtic craftsmen: this flagon came from a rich grave at Basse-Yutz, in northeastern France.

▶ By 7000 BC farming communities were spreading from Anatolia into southeast Europe, bringing wheat, barley, sheep and goats. Pigs and cattle, indigenous to Europe, were kept, and wild plants and animals were also exploited by these early farmers. Farming also spread into neighbouring areas and by 4000 BC was widespread across the continent, although the numbers of farmers were relatively small. The greater part of Europe was still sparsely inhabited forest, only gradually being cleared for farming settlement over succeeding millennia.

The postglacial conditions of the period 8000–4000 BC offered new opportunities to the hunter-gatherers of Europe. Activity concentrated on coasts, lake margins and rivers, where both aquatic and land plants and animals could be exploited; the ecologically less diverse forest interiors were generally avoided. Initially groups tended to move around on a seasonal basis, but later more permanent communities were established, with temporary special-purpose outstations. Dogs, domesticated from wolves, were kept to aid hunting. Some groups managed their woodlands by judicious use of fire to encourage hazel and other useful plants.

EUROPE'S FIRST FARMERS

From around 7000 BC farming communities began to appear in Europe (map 1). Early farmers in the southeast built villages of small square houses and made pottery, tools of polished stone and highly prized obsidian, as well as ornaments of spondylus shell obtained by trade. Once established, many of the sites in the southeast endured for thousands of years, gradually forming tells (mounds of settlement debris). By 5000 BC some communities were also using simple techniques to work copper.

Between 5500 and 4500 BC pioneering farming groups rapidly spread across central Europe, settling predominantly on the easily worked loess (wind-deposited) river valley soils. They kept cattle, raised crops and lived in large timber-framed long houses which often also sheltered their animals. At first these groups were culturally homogeneous, but after about 4500 BC regional groups developed and farming settlements increased in number, spreading out from the river valleys.

The hunter-gatherers in the central and western Mediterranean came into contact with early farmers colonizing southern parts of Italy. They acquired pottery-making skills and domestic sheep and goats from these colonists, and later they also began to raise some crops. By 3500 BC communities practising farming but still partly reliant on wild resources were established over most of western Europe. Huge megalithic ("large stone") tombs were erected, which acted as territorial

markers affirming community ties to ancestral lands. These tombs took many forms over the centuries and were associated with a variety of rites, generally housing the bones of many individuals, usually without grave goods.

THE USE OF METALS

By 3500 BC a new economic pattern had developed as innovations emanating from West Asia spread through Europe via farming communities in the southeast and the east, on the fringes of the steppe. These included the use of animals for traction, transport and milk, woolly sheep, wheeled vehicles and the plough. Plough agriculture allowed new areas and less easily worked soils to be cultivated, and there was a general increase in animal husbandry; specialist herders also appeared (map 2). Trade, already well established, now grew in importance, carrying fine flint and hard stone for axes over long distances in a series of short steps between communities. Major social changes were reflected by a significant shift in the treatment of the dead: in many regions communal burial in monumental tombs gave way to individual burials with personal grave goods, often under a barrow. New types of monuments erected in western areas suggest a change in religious practices, with a new emphasis on astronomical matters.

From around 2500 BC copper was alloyed with tin to form bronze. The need for tin, a rare and sparsely distributed metal, provided a stimulus to the further development of international trade in prestige materials (map 3). These were particularly used as grave goods and votive offerings, emphasizing the status achieved by their owners. Chiefs were now buried under massive barrows with splendid gold and bronze grave offerings, while lesser members of society were interred under barrows in substantial cemeteries. Command of metal ore sources gave certain communities pre-eminence, while others derived their importance from a key position at the nodes of trade routes. The Carpathian

1 THE SPREAD OF FARMING IN EUROPE 7000–3500 BC

Regions of dense hunter-gatherer settlement to 4500 BC

▲ Megalith/longbarrow

Spread of farming communities:

southeastern 7000–5500 BC

Mediterranean 7000–4500 BC

central 5500–4500 BC

Mediterranean 4500–3500 BC

western 4500–3500 BC

northern 4500–3500 BC

eastern 4500–3500 BC

● Early farmers

☐ Developed farmers working copper from 5500 BC

Sources of traded materials:

♦ stone axe factory/flint mine

spondylus shells

obsidian

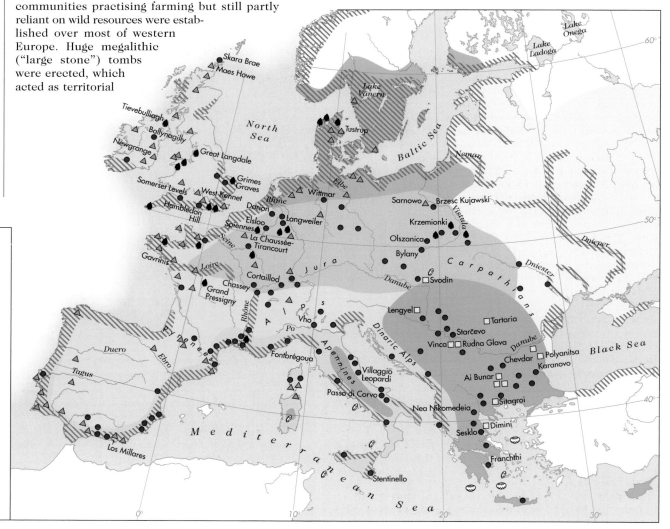

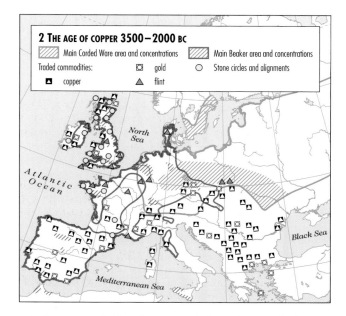

▲ By 3000 BC copper and gold metallurgy were practised across most of Europe. These metals were used to make prestige goods that enhanced the status of high-ranking individuals. Drinking vessels for alcohol were also status symbols – Corded Ware in eastern and northern Europe and, later, Beakers in central and western Europe.

region enjoyed particular prosperity around this time; Scandinavia, which lacked indigenous metal ores, nevertheless now became involved in international trade, and by the late 2nd millennium developed a major bronze industry based on metal imported in exchange for furs and amber. Agriculture and livestock also brought wealth to favoured areas, and there was a major expansion of farming onto light soils formerly under forest. Substantial field systems mark the organization of the agrarian landscape in at least some regions. By the start of the 1st millennium, however, many of the more marginal areas for agriculture had become scoured or exhausted and were abandoned.

WARFARE AND RELIGION

By the late 2nd millennium warfare was becoming a more serious business. Often settlements were located in defensible positions and fortified. (In previous centuries fortified centres had been far fewer and more scattered.) However, until the late centuries BC armed conflict between individual leaders or raids by small groups remained the established pattern, rather than large-scale fighting.

A greater range of weapons was now in use, especially spears and swords, their forms changing frequently in response to technical improvements and fashion. Bronze was in abundant supply and made into tools for everyday use by itinerant smiths. Iron came into use from around 1000 BC and by 600 BC it had largely replaced bronze for tools and everyday weapons, freeing it for use in elaborate jewellery and ceremonial armour and weaponry.

Major changes occurred in burial practices and religious rites. In most areas burial, often under large mounds, was replaced by cremation, the ashes being interred in urns within flat graves (urnfields). Funerary rites became more varied in the Iron Age and many graves – particularly in wealthy areas – contained lavish goods, as in the cemetery at Hallstatt in western Austria, which profited from the trade in salt from local mines. Substantial religious monuments were no longer built, religion now focusing on natural locations such as rivers and lakes.

CELTIC EUROPE

During the 1st millennium BC much of France, Germany and the Alpine region came to be dominated by the Celtic peoples (*map 4*), who also settled in parts of Britain, Spain northern Italy and Anatolia. By the 3rd century BC towns (known to the Romans as *oppida*) were emerging in many parts of Europe, reflecting both increased prosperity and more complex and larger-scale political organization. In the west this development was short-lived as Europe west of the Rhine progressively fell to Roman expansion. In the east and north, however, Germanic and other peoples continued the life of peasant agriculture, trade, localized industry and warfare that had characterized much of the continent for many centuries.

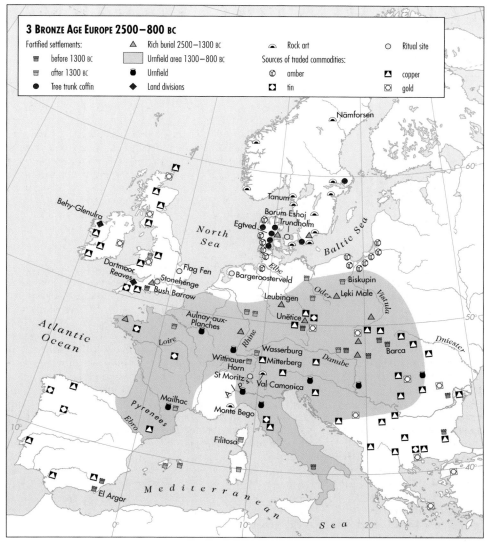

▲ Small-scale chiefdoms emerged in many parts of Europe during the 2nd millennium BC, but their leaders' power was limited. From around 1300 BC, however, this situation began to change, culminating in the larger groupings of the Iron Age. ▼ Metalwork and, occasionally, people were sacrificed by the Celts at their sacred European sites – rivers, lakes and woods.

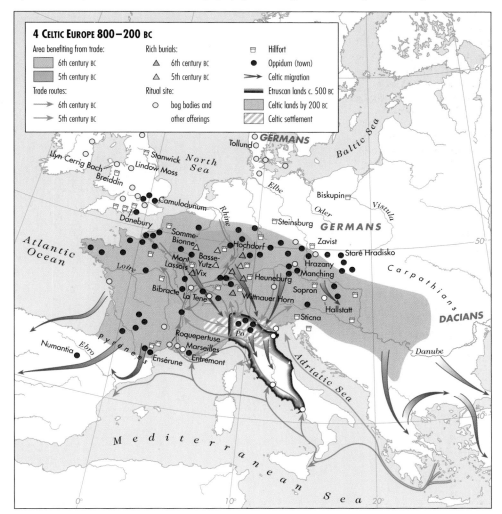

FROM HUNTING TO FARMING: AFRICA 10,000 BC–AD 500

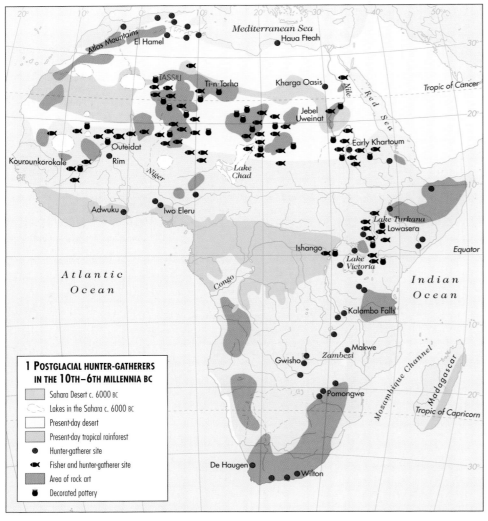

1 POSTGLACIAL HUNTER-GATHERERS IN THE 10TH–6TH MILLENNIA BC

- Sahara Desert c. 6000 BC
- Lakes in the Sahara c. 6000 BC
- Present-day desert
- Present-day tropical rainforest
- Hunter-gatherer site
- Fisher and hunter-gatherer site
- Area of rock art
- Decorated pottery

By 10,000 BC most of Africa was inhabited by hunter-gatherer groups (*map 1*). Although generally only their stone tools survive, the majority of their artefacts would have been made of perishable materials such as wood, leather and plant fibres. At Gwisho in Zambia a large find of organic objects, including wooden bows and arrows, bark containers, and leather bags and clothes, provides us with some insight into what is normally lost. Further information on the lives of African hunter-gatherers comes from their rich rock art, known in many areas of the continent but particularly in the Sahara and in southern Africa. This not only depicts aspects of everyday life, such as housing and clothing, but it also gives a picture of archaeologically intangible activities such as dancing and traditional beliefs.

With the retreat of the ice sheets around this time conditions became both warmer and wetter, creating new opportunities for hunter-gatherer communities. Rising sea levels encouraged the utilization of coastal resources, such as shellfish in southern Africa. Many groups moved between the coast and inland sites, exploiting seasonally available food resources, and people also began to hunt smaller game in the forests that were spreading into former savanna regions. In the Sahara belt, largely uninhabited during the arid glacial period, extensive areas of grassland now developed and the existing restricted bodies of water expanded into great lakes, swamps and rivers. These became favoured areas of occupation, often supporting large permanent settlements whose inhabitants derived much of their livelihood from fish, aquatic mammals (such as hippos), waterfowl and water plants, as well as locally hunted game. Similar lakeside or riverine communities developed in other parts of the continent, for example around Lake Turkana in East Africa.

EARLY FARMING IN AFRICA

Some communities began to manage their resources more closely: they weeded, watered and tended preferred plants, and perhaps planted them, and they herded local animals, particularly cattle but also species such as eland and giraffe

▲ During glacial periods tropical regions such as Africa experienced considerable aridity. With the retreat of the ice sheets in temperate regions by about 10,000 BC, parts of Africa became warmer and wetter, offering new ecological opportunities to the continent's population. Postglacial changes were particularly marked in northern Africa, where increased humidity provided conditions favouring permanent settlements. At many places pottery (too fragile to be used by mobile groups) was being made from around 7500 BC.

▶ A broad band eastwards from West Africa was the original home of many of the plant species that were taken into cultivation. Here farming had become well established by around 1000 BC.

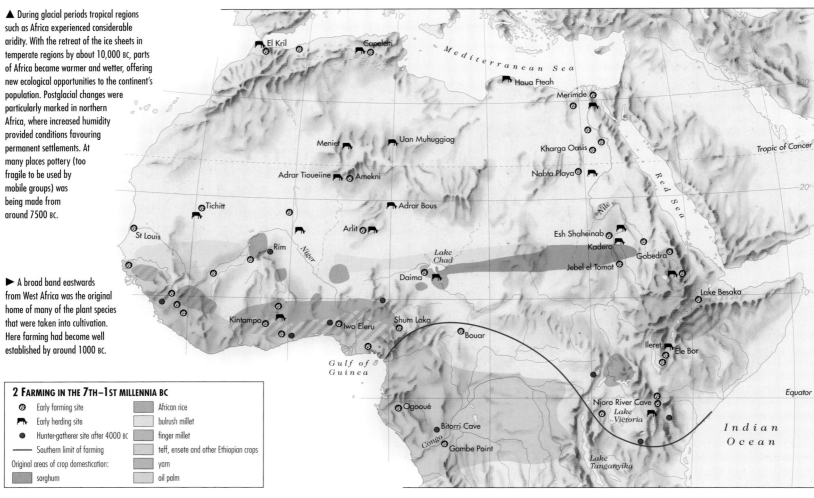

2 FARMING IN THE 7TH–1ST MILLENNIA BC

- Early farming site
- Early herding site
- Hunter-gatherer site after 4000 BC
- Southern limit of farming
- Original areas of crop domestication:
  - sorghum
- African rice
- bulrush millet
- finger millet
- teff, ensete and other Ethiopian crops
- yam
- oil palm

(*map 2*). In the Nile Valley, nut-grass tubers had been intensively exploited since glacial times, and by 11,000 BC cereals such as sorghum and probably barley were also managed. Sheep and goats, and some crop plants such as wheat, were introduced, probably from West Asia. By about 5000 BC many communities in northern Africa were raising indigenous crop plants such as sorghum and keeping domestic cattle, sheep and goats, though they also continued to hunt and fish and to gather wild plant foods. Dependence on agriculture intensified, domestic resources grew in importance, and the number of farming communities increased.

From around 4000 BC, however, the Sahara region became increasingly dry; lakes and rivers shrank and the desert expanded, reducing the areas attractive for settlement. Many farmers moved southwards into West Africa. Although harder to document than cereal agriculture, the cultivation of tubers such as yams and of tree crops such as oil palm nuts probably began around this time. Local bulrush millet was cultivated and African rice, also indigenous to this region, may well have been grown, although at present the earliest evidence for its cultivation is from Jenne-jeno around the 1st century BC. By around 3000 BC farming communities also began to appear in northern parts of East Africa.

THE SPREAD OF METALWORKING
Around 500 BC metalworking began in parts of West Africa (*map 3*). Carthaginians and Greeks had by this time established colonies on the North African coast (*pages 40–41*). They were familiar with the working of bronze, iron and gold and were involved in trade across the Sahara, and this may have been the means by which knowledge of metallurgy reached sub-Saharan Africa. Sites with early evidence of copperworking, notably Akjoujt, have also yielded objects imported from North Africa. Egypt, Nubia and Ethiopia were now working metals and may also have been a source of technological expertise. Alternatively, the working of gold and iron may have been indigenous developments: the impressive terracotta heads and figurines from Nok were produced by people well versed in smelting and using iron.

Although iron tools were very useful for forest clearance, agriculture, woodworking and other everyday activities, the spread of ironworking was at first extremely patchy. While some areas in both East and West Africa were working iron as early as the Nok culture around 500 BC, other adjacent regions did not begin to do so until the early or middle centuries of the first millennium AD (*pages 80–81*). In some cases, however, such as the equatorial forests of the Congo Basin, the absence of early evidence of metallurgy is likely to reflect the poor preservation of iron objects: ironworking was probably well established there by the late centuries BC.

EARLY FARMING IN SOUTHERN AFRICA
The early centuries AD saw the spread into much of the rest of Africa of ironworking, along with pottery, permanent settlements, domestic animals and agriculture (*map 4*). By the 2nd century the eastern settlers had reached northern Tanzania, from where they quickly spread through the coastal lowlands and inland regions of southeastern Africa, reaching Natal by the 3rd century. Depending on local conditions and their own antecedents, groups established different patterns of existence within the broad agricultural framework: those on the southeastern coast, for example, derived much of their protein from marine resources such as shellfish rather than from their few domestic animals; other groups included specialist pastoralists and broadly

▶ Archaeological data and linguistic evidence combine to indicate that a number of radical innovations – including agriculture, herding, metalworking and permanent settlement – were introduced to the southern half of the continent by the spread of people from the north who spoke Bantu languages. Originating in part of southern West Africa (now eastern Nigeria and Cameroon), Bantu languages progressively spread southwards along two main routes, in the east and west. The areas these farmers penetrated were inhabited by hunter-gatherer communities, speaking Khoisan languages in the south and probably in other areas.

based mixed farmers growing cereals that included sorghum and millet, plus other plants such as cowpeas, beans, squashes and probably yams.

The interrelations of these settlers with the native hunter-gatherer groups were varied. Some hunter-gatherers in areas suitable for agriculture were totally displaced by the newcomers; others established mutually beneficial relations, adopting aspects of the intrusive culture, such as pottery or domestic animals; some groups raided the new farming communities to lift cattle, sheep or goats. The southwest was unsuited to the cultivation of the introduced crops, but hunter-gatherers there began to herd domestic sheep.

By the late 1st millennium AD iron tools had largely replaced stone tools throughout most of Africa. In some areas – the Copperbelt in Zambia and Zaire, for example – copper was being made into ornaments such as bangles, though gold would not be worked in the southern half of the continent before the close of the millennium.

▼ The Greek historian Herodotus reported attempts by Persian and Phoenician sailors to circumnavigate Africa in the early 1st millennium BC. The Carthaginians also penetrated southwards by sea, establishing outposts as far south as Mogador and probably reaching Cerne (Herne Island). Paintings of chariots characteristic of the 1st millennium BC have been found in the Sahara. Although these do not mark the actual routes taken by traders across the desert, they do provide evidence of their presence. Trans-Saharan trade was facilitated in the late centuries BC by the introduction of camels for transport.

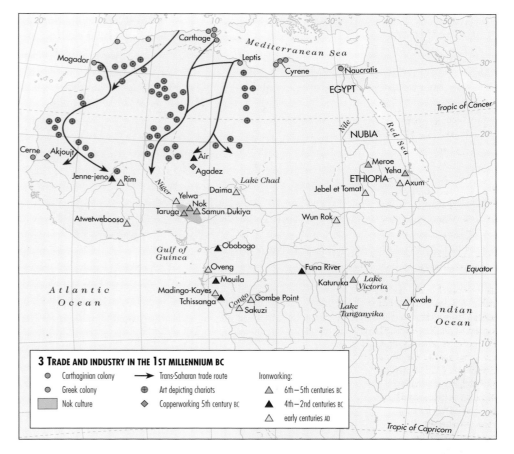

3 TRADE AND INDUSTRY IN THE 1ST MILLENNIUM BC

- ● Carthaginian colony
- ● Greek colony
- (Nok culture)
- → Trans-Saharan trade route
- ⊛ Art depicting chariots
- ◆ Copperworking 5th century BC

Ironworking:
- △ 6th–5th centuries BC
- ▲ 4th–2nd centuries BC
- △ early centuries AD

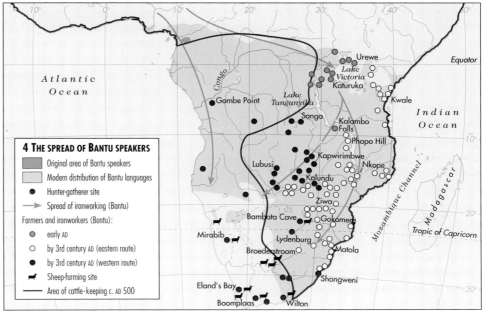

4 THE SPREAD OF BANTU SPEAKERS

- Original area of Bantu speakers
- Modern distribution of Bantu languages
- ● Hunter-gatherer site
- → Spread of ironworking (Bantu)

Farmers and ironworkers (Bantu):
- ○ early AD
- ○ by 3rd century AD (eastern route)
- ● by 3rd century AD (western route)
- 🐃 Sheep-farming site
- —— Area of cattle-keeping c. AD 500

◗ STATES AND TRADE IN WEST AFRICA 500–1500 *pages 80–81*   ◗ STATES AND TRADE IN EAST AFRICA 500–1500 *pages 82–83*

FROM HUNTING TO FARMING: THE AMERICAS 12,000–1000 BC

Controversy surrounds the date of human colonization of the Americas (*map 1*). During glacial periods when sea levels fell, the Bering Strait became dry land (Beringia), allowing humans living in Siberia to move across into the northernmost part of the Americas. However, substantial ice sheets would then have prevented further overland penetration of the continent. Only subsequently, when the ice sheets melted, could further advances occur – although it is conceivable that migration into the Americas took place by sea, down the Pacific coast.

Several glacial cycles occurred following the emergence of modern humans (*pages 16–17*), during which, at least hypothetically, such a migration could have taken place. Nevertheless, despite (as yet unsubstantiated) claims for early dates, humans probably reached the far north of the Americas about 16,000 BC, during the most recent glacial episode, and spread south when the ice sheets retreated around 12,000 BC. Not only do the earliest incontrovertibly dated sites belong to the period 12–10,000 BC, but biological and linguistic evidence also supports an arrival at this time. In addition, the adjacent regions of Asia from which colonists must have come seem not to have been inhabited until around 18,000 BC.

The colonization of the Americas after 10,000 BC was extremely rapid, taking place within a thousand years. The first Americans were mainly big-game hunters, although occasional finds of plant material show that they had a varied diet. Their prey were mostly large herbivores: bison and mammoths in the north, giant sloths and mastodons further south, as well as horses, camels and others. By about

7000 BC many of these animals had become extinct (except the bison, which became much smaller in size). Humans probably played some part in these extinctions, although changes in climate and environment are also likely factors.

HUNTER-GATHERERS AND EARLY FARMERS

After 8000 BC bison hunting became the main subsistence base of the inhabitants of the Great Plains of North America (*map 2*). Hunting was generally an individual activity, but occasionally groups of hunters and their families combined in a great drive to stampede bison over a cliff or into a natural corral, so that huge numbers could be slaughtered at once. Elsewhere in North America, a great range of regional variations developed on the theme of hunting and gathering, and in many areas these ways of life survived until the appearance of European settlers in recent centuries.

The people of the Arctic regions led a harsh existence. Their inventiveness enabled them to develop equipment such as the igloo and the kayak to withstand the intense cold of winter and of the Arctic seas, and to hunt large blubber-rich sea mammals such as whales and seals. Other northern groups relied more on land mammals, notably caribou. The inhabitants of the Pacific Coast region grew prosperous on their annual catch of salmon and other marine and riverine resources. They acquired slaves, constructed spectacular wooden structures and gave magnificent feasts. In the deserts of the southwest, seasonal migration enabled people to obtain a diversity of plant, animal and aquatic foodstuffs at different times of the year, while the wooded environment of the east also

▼ The antiquity of the first Americans is still a controversial issue. A few sites, such as Meadowcroft in North America and Monte Verde in South America, are sometimes claimed to have been occupied well before 12,000 BC. However, undisputed evidence of people at these and other sites dates from 12,000 BC onwards, with Fell's Cave in the extreme southern tip of the continent being occupied by 9000 BC.

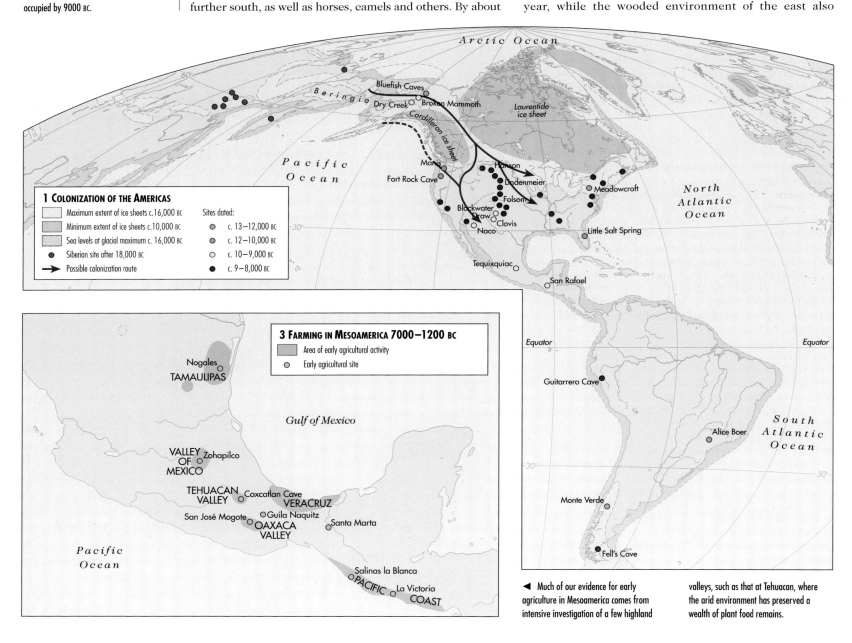

1 COLONIZATION OF THE AMERICAS
- Maximum extent of ice sheets c.16,000 BC
- Minimum extent of ice sheets c.10,000 BC
- Sea levels at glacial maximum c.16,000 BC
- ● Siberian site after 18,000 BC
- → Possible colonization route

Sites dated:
- ◉ c. 13–12,000 BC
- ◉ c. 12–10,000 BC
- ○ c. 10–9,000 BC
- ● c. 9–8,000 BC

3 FARMING IN MESOAMERICA 7000–1200 BC
- Area of early agricultural activity
- ○ Early agricultural site

◀ Much of our evidence for early agriculture in Mesoamerica comes from intensive investigation of a few highland valleys, such as that at Tehuacan, where the arid environment has preserved a wealth of plant food remains.

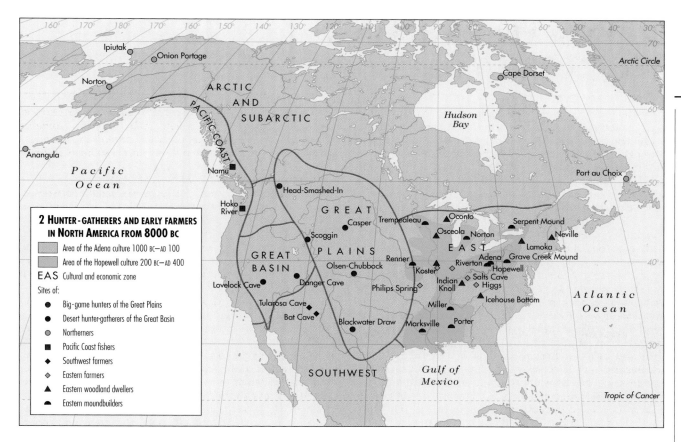

2 HUNTER-GATHERERS AND EARLY FARMERS IN NORTH AMERICA FROM 8000 BC

- Area of the Adena culture 1000 BC–AD 100
- Area of the Hopewell culture 200 BC–AD 400
- EAS Cultural and economic zone

Sites of:
- ● Big-game hunters of the Great Plains
- ● Desert hunter-gatherers of the Great Basin
- ● Northerners
- ■ Pacific Coast fishers
- ◆ Southwest farmers
- ◇ Eastern farmers
- ▲ Eastern woodland dwellers
- ⬟ Eastern moundbuilders

◀ The initial inhabitants of North America were big-game hunters, but after 8000 BC many regional groups began to develop their own individual ways of life based on locally available resources. Later, many groups also participated in regional trade networks, obtaining valued commodities such as turquoise and obsidian in the southwest. The rich diversity of North American life is reflected in the surviving art and artefacts: exquisite ivory figurines of animals from the Arctic; vivacious rock paintings from many areas showing hunting, dancing and musicians; beautifully made decoy ducks of reeds and feathers from the Great Basin; and carvings in mica, copper and soapstone from the Hopewell mounds of the east.

provided a diverse range of such foods. In areas of abundance, some eastern groups were able to settle in camps for much of the year, burying their dead in large cemeteries.

These woodland folk also developed long-distance trade networks, exchanging such prized commodities as copper, marine shells and fine-quality stone for tool-making. Later, groups in the Ohio Valley and adjacent areas (the Adena and Hopewell cultures) elaborated their exchange networks and raised substantial mounds over their dead. By about 2500–2000 BC some groups in the eastern region were cultivating local plants, such as sunflowers and squashes. In the southwest similar developments were encouraged by the introduction around 1000 BC from Mesoamerica of maize, a high-yielding crop which did not reach the eastern communities until around AD 800 (*pages 108–9*).

DEVELOPMENTS IN MESOAMERICA
After 7000 BC hunter-gatherer bands in highland valleys of Mesoamerica supplemented the foodstuffs they obtained through seasonal migration by sowing and tending a number of local plants such as squashes and chillies (*map 3*). By 5000 BC they were also cultivating plants acquired from other regions of Mesoamerica. Among these was maize, at first an insignificant plant with cobs barely 3 cm (1.2 in) long. However, genetic changes progressively increased the size of the cobs, and by 2000–1500 BC maize had become the staple of Mesoamerican agriculture, supplemented by beans and other vegetables. Villages in the highlands could now depend entirely on agriculture for their plant foods and were occupied all year round. As there were no suitable herd animals for domestication, hunting remained important into colonial times; the only domestic animals eaten were dogs, ducks and turkeys (introduced from North America). Lowland regions of Mesoamerica followed a somewhat different pattern: coastal and riverine locations provided abundant wild foods throughout the year, making year-round occupation possible at an early date. Agriculture, adopted in these regions later than in the highlands, provided high yields, particularly in the Veracruz region where the Olmec culture emerged around 1200 BC (*pages 32–33*).

EARLY FARMING IN SOUTH AMERICA
Preserved organic remains from arid caves in the Andes provide evidence that plants were cultivated in South America by around 6500 BC (*map 4*). Along with local varieties like potatoes, these included plants (such as beans and chillies) native to the jungle lowlands to the east. It is therefore likely that South American agriculture began in the Amazon Basin, although humid conditions in this area precluded the preservation of ancient plant remains. Pottery

and other equipment used to process manioc (cassava) offer indirect evidence that this important American staple food was grown in South America by 2000 BC.

By this time village communities were established throughout the Andean region and had developed strategies to exploit a variety of local resources. The coast provided exceptionally rich fisheries, while inland crops were cultivated using irrigation, with cotton particularly important. The lower slopes of the Andes were also cultivated, with crops such as potatoes at higher altitudes, while the llamas and alpacas of the high pastures provided meat and wool.

Apart from residential villages, often furnished with substantial cemeteries, early South Americans also built religious centres with monumental structures. By 1200 BC the Chavin cult, centred on the great religious monuments of Chavin de Huantar and marked by characteristic art, architecture and iconography, had united peoples along much of the Peruvian coast (*pages 34–35*).

▼ From about 6500 BC agriculture in South America included not only the cultivation of plants native to the local area but also crops from other regions. Maize was probably introduced from Mesoamerica: it appeared in Ecuadorian farming villages and in the Andean highlands around 5000 BC, then spread from 800 into the Amazon Basin, where it supported rapid population growth.

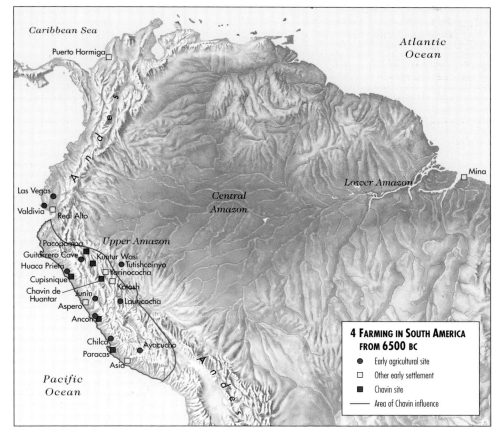

4 FARMING IN SOUTH AMERICA FROM 6500 BC
- ● Early agricultural site
- □ Other early settlement
- ■ Chavin site
- ── Area of Chavin influence

FROM HUNTING TO FARMING: AUSTRALIA AND THE PACIFIC 10,000 BC–AD 1000

The Pacific was one of the last regions on Earth to be colonized by people. Modern humans spread into Southeast Asia and from there crossed the sea to New Guinea and Australia (which formed a single landmass at that time) by about 60,000 BC. A few of the islands adjacent to New Guinea were also settled before 30,000 BC, but expansion into the rest of the Pacific only began around 1500 BC and was not completed until AD 1000 (*map 1*).

THE FIRST COLONIZATION OF AUSTRALIA

The early inhabitants of Australia were confined initially to the coast and inland river valleys, spreading to colonize the south by 40,000 BC (*map 2*). They gathered a variety of wild resources and hunted the local fauna, which at that time included a number of large species such as a giant kangaroo, *Procoptodon*. Between 25,000 and 15,000 these huge creatures became extinct: humans may have been partly to blame, although increasing aridity was probably also responsible. By 23,000 BC ground-stone tools were being made – the earliest known in the world – and by 13,000 BC people had learnt to process the toxic but highly nutritious cycad nuts to remove their poison. The harsh desert interior of Australia was colonized by groups who adapted their lifestyle to cope with this challenging environment.

By 3000 BC further major changes had taken place. New tools were now in use, including the boomerang (invented by 8000 BC) and small, fine stone tools suited to a variety of tasks, of which wood-working was of prime importance. The dingo, a semi-wild dog, had been introduced into Australia, perhaps brought in by a new wave of immigrants from Southeast Asia. Dingoes outcompeted the native predators such as the thylacine (Tasmanian tiger), a carnivorous marsupial which became extinct.

Although they never adopted farming Australia's aborigines exercised considerable control over the wild resources at their disposal, clearing the bush by firesetting in order to encourage new growth and attract or drive game, and replanting certain preferred plant species. New Guinea's first inhabitants were also hunters and gatherers, but by 7000 BC some communities here had begun cultivating local plants like sugar cane, yam, taro and banana, and keeping pigs (*map 1*). At Kuk, in the highlands, there is evidence at this early date for a network of drainage channels to allow crops to be grown in swampland.

MIGRATION AFTER 1500 BC

Farming communities were also developing in East and Southeast Asia; around 1500 BC a new wave of colonists began to spread out from this area, moving from the mainland into Taiwan and the Philippines, then into the islands of Southeast Asia and from here into the Pacific. By 1000 BC they had reached the Marianas in the north and, much further afield, Tonga and Samoa in Polynesia to the east. The movement of these people can be traced from the distribution of their distinctive pottery, known as Lapita ware, a red-slipped ware decorated with elaborate stamped designs. They also used obsidian (volcanic glass) and shell for making tools, and brought with them a range of Southeast Asian domestic animals, including dogs and chickens.

By this time the colonists had become skilled navigators, sailing in double canoes or outriggers large enough to accommodate livestock as well as people, and capable of tacking into the wind. The uniformity of their artefacts shows that contacts were maintained throughout the area, with return as well as outward journeys. The Polynesians used the stars, ocean currents, winds and other natural phenomena as navigational guides, and they made ocean charts of palm sticks with the islands marked by cowrie shells.

▲ Among the stones which the early Maori settlers of New Zealand became skilled in carving was jade, from which this pendant is made.

▼ The rapid spread of the Asian peoples who colonized the Pacific islands after about 1500 BC is something of an enigma. Their motivation cannot have been solely an expanding population's need to find new territories to settle, since only small founding populations remained – well below the numbers that the islands could have supported. They carried with them all the plants and animals they required in order to establish horticultural communities, but marine resources also played an important role in their economies.

▼ The inhabitants of the eastern Polynesian islands erected stone platforms and courts with stone monoliths. These were shrines (*marae*) which were used for prayer and for human and animal sacrifice to the gods, as were the unique stone monuments – huge stone platforms (*ahu*) and colossal stone heads (*moai*) – of Easter Island. No Easter Island statues were erected after AD 1600 and by 1863 all existing ones had been deliberately toppled (to be re-erected from the 1950s), a development that reflects social upheaval related to deforestation and consequent pressure on resources.

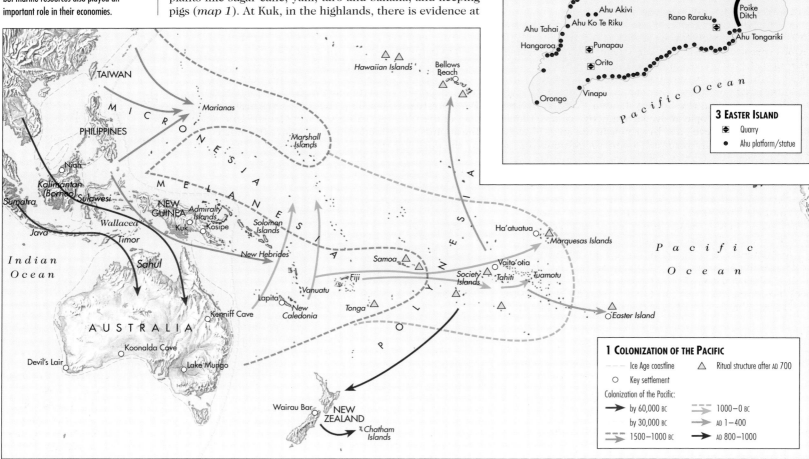

3 EASTER ISLAND
- ✠ Quarry
- ● Ahu platform/statue

Ahu Te Pito Kura
Ahu Akivi
Ahu Ko Te Riku
Ahu Tahai
Hangaroa
Rano Raraku
Poike Ditch
Punapau
Orito
Ahu Tongariki
Orongo
Vinapu

Pacific Ocean

1 COLONIZATION OF THE PACIFIC
- – – – Ice Age coastline
- ○ Key settlement
- △ Ritual structure after AD 700

Colonization of the Pacific:
- → by 60,000 BC
- → by 30,000 BC
- ⇒ 1500–1000 BC
- ⇒ 1000–0 BC
- → AD 1–400
- → AD 800–1000

TAIWAN
MICRONESIA
PHILIPPINES
Marianas
Hawaiian Islands
Bellows Beach
Marshall Islands
Niah
Kalimantan (Borneo)
Sulawesi
Sumatra
MELANESIA
NEW GUINEA
Admiralty Islands
Kuk
Kosipe
Solomon Islands
Wallacea
Java
Timor
New Hebrides
Ha'atuatua
Marquesas Islands
Indian Ocean
Sahul
Samoa
Vaito'otia
Society Islands
Tahiti
Tuamotu
Fiji
New Caledonia
Lapita
Vanuatu
Tonga
POLYNESIA
Easter Island
Pacific Ocean
Kenniff Cave
AUSTRALIA
Koonalda Cave
Devil's Lair
Lake Mungo
Wairau Bar
NEW ZEALAND
Chatham Islands

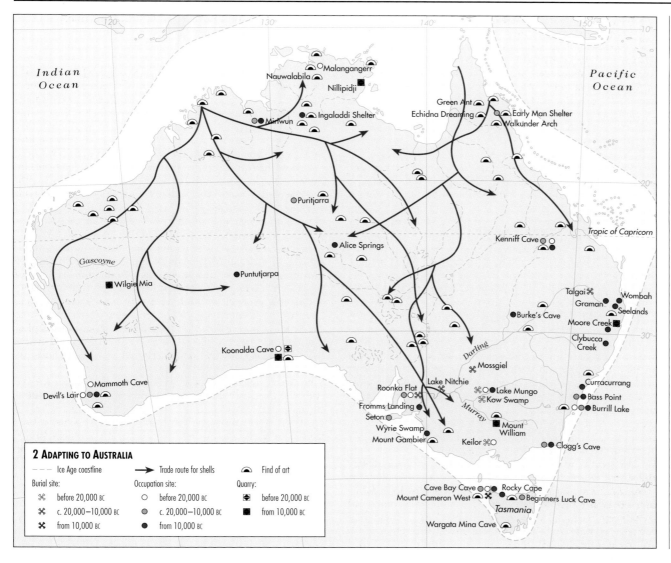

The complex social and cultural life of Australia's Aboriginal inhabitants is reflected in painted and engraved art (which appeared almost as early as the first settlement of Australia), in burials with an array of grave goods, in a variety of ritual sites, and in the Aborigines' rich oral traditions. Links between communities based on kinship were enhanced by long-distance trade: commodities such as coastal shells were taken into the interior while roughed-out stone axes from quarries in the interior moved in the opposite direction.

2 ADAPTING TO AUSTRALIA

- - - Ice Age coastline → Trade route for shells ◠ Find of art

Burial site:
- ✼ before 20,000 BC
- ✖ c. 20,000–10,000 BC
- ✖ from 10,000 BC

Occupation site:
- ○ before 20,000 BC
- ⬤ c. 20,000–10,000 BC
- ● from 10,000 BC

Quarry:
- ⊕ before 20,000 BC
- ■ from 10,000 BC

THE COLONIZATION OF EASTERN POLYNESIA

This wave of colonization came to a standstill around 1000 BC in western Polynesia. Groups from the colonized regions spread north and east to complete the settlement of Micronesia from that time, but it was not until about 200 BC that a new surge of eastward colonization took place, establishing populations on the more scattered islands of eastern Polynesia, including the Society Islands, Tahiti and the Marquesas. These people evolved a distinctive culture which differed from that developed by groups in the areas already settled – areas that were still open to influence from Southeast Asia. By now the Polynesians had almost entirely abandoned pottery: eastern Polynesians began making distinctive new types of stone adze, shell fish-hooks and jewellery. They also built stone religious monuments.

The best known and most striking of these were the Easter Island statues. Easter Island and Hawaii were settled in a further colonizing movement by around AD 400. Nearly 2,000 kilometres (1,250 miles) from Pitcairn, its nearest neighbour, Easter Island was probably never revisited after its initial settlement. The resulting isolation allowed its people to develop a unique form of general Polynesian culture, notable for its mysterious stone heads (*map 3*).

NEW ZEALAND'S FIRST SETTLERS

Between AD 800 and 1000 a final wave of Polynesian voyagers colonized New Zealand (*map 4*) and the Chatham Islands to the east. Here new challenges and opportunities awaited them.

New Zealand is unique in the Pacific in enjoying a temperate climate; most of the tropical plants cultivated by Polynesians elsewhere in the Pacific could not grow here, although sweet potatoes (introduced into Polynesia from South America) flourished. In compensation there were rich marine resources and a wide range of edible plants indigenous to the islands – of which one, the root of the bracket fern, became an important cultivated plant on North Island.

There was also a large population of huge flightless birds (moa), which had evolved in great diversity due to the absence of mammals and predators. Reverting to their distantly ancestral hunter-gatherer way of life, the new settlers (early Maori) hunted these birds to extinction within 500 years, aided by the dogs and rats they had introduced. The native flora also became depleted. As South Island was unsuited to agriculture its population declined, and on North Island increased reliance on horticulture went hand in hand with growing warfare between the communities, accompanied by the building of fortified settlements, trophy head-hunting and cannibalism.

▼ The culture of the early Maori settlers in New Zealand differed from that of other Polynesians in the emphasis it placed on long-distance trade. Among the items traded were various types of stone used for making tools and weapons, including greenstone for warclubs and amulets, and materials such as obsidian (volcanic glass), argillite (white clay rock) and shells.

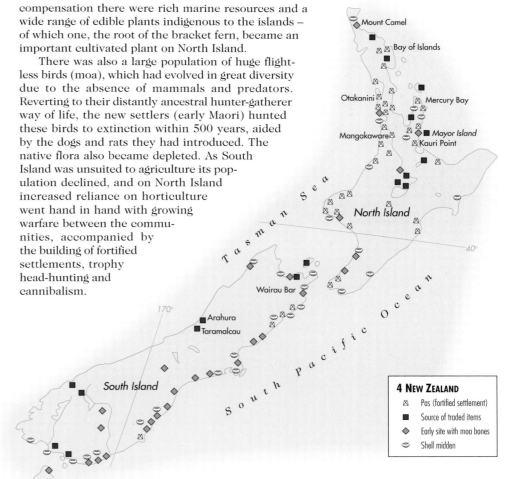

4 NEW ZEALAND
- ⊠ Pas (fortified settlement)
- ■ Source of traded items
- ◆ Early site with moa bones
- ◠ Shell midden

THE FIRST CIVILIZATIONS: MESOPOTAMIA AND THE INDUS REGION 4000–1800 BC

1 MESOPOTAMIA IN THE EARLY DYNASTIC PERIOD C. 2900 BC
- - - Ancient coastline
- - - - Course of river in 3rd millennium BC
▨ Alluvium

▲ The unstable physical environment of Mesopotamia caused many radical changes in the pattern of settlement. Sediments from the Tigris and Euphrates filled in the head of the Gulf, isolating ancient ports. Moreover, the courses of the rivers also changed, taking precious river water away from settlements. Since rainfall was inadequate to sustain crops, these settlements were usually soon abandoned.

▼ Early Mesopotamian cities varied in size and importance, from 10-hectare (25-acre) Abu Salabikh to Warka (Uruk), which covered over 400 hectares (1,000 acres) and had a population of 40–50,000 people. Warka's 9-kilometre (6-mile) city wall enclosed temples, palaces and houses, sometimes grouped into specialized craft quarters, as well as open spaces for gardens, burials and waste disposal. Indus cities, by contrast, generally comprised a large planned residential area and a raised citadel with public buildings and, probably, accommodation for the rulers. In the largest, Mohenjo-daro, the lower town contained both spacious private houses and industrial areas hosting the full range of Indus crafts.

Agricultural communities had emerged in many parts of the world by the 4th millennium BC. In some areas high productivity supported high population densities and the emergence of cities, necessitating more complex social organization and giving rise to more elaborate public architecture. These developments encouraged trade in essential and luxury goods as well as craft and other occupational specialization. Such "civilized" communities appeared first in Mesopotamia, around 4000 BC.

MESOPOTAMIA
By 4500 BC the advent of irrigation agriculture had enabled the settlement of the dry southern Mesopotamian alluvium (*map 1*). A social world comprising groups of agriculturalist kinsfolk living in hamlets, villages or towns evolved, to be transformed around 600 years later into one of specialists living in complex and hierarchical social arrangements in an urban milieu. Religion played an important part in this process: while religious structures are recognizable in the earlier archaeological record, palaces and other large secular buildings appear only later in the 4th millennium. Religious complexes became larger and increasingly elaborate throughout the period.

A number of urban centres emerged, of which one in particular stands out – ancient Warka (map 2A), also called Uruk. The city had at least two very large religious precincts – Eanna and Kullaba. In the Eanna Precinct the earliest written records, dating from around 3100 BC, have been found: tablets of clay or gypsum inscribed with ideographic characters. These first texts were economic in nature, comprising lists and amounts of goods and payments.

By 2900 BC there were also other important urban centres in southern Mesopotamia – city-states ruled by individual kings who negotiated shifting economic and political alliances among themselves and with polities outside Mesopotamia. The wealth and power of the Early Dynastic rulers can be seen in the elaborate burials in the Royal Cemetery of Ur, some including human sacrifices as well as objects of gold, silver and lapis lazuli.

SUMER AND AKKAD
From the fragmented historical record of this period it is apparent that the region was becoming divided between the lands of Akkad (from Abu Salabikh to the edge of the northern Mesopotamian plains) and of Sumer (from Nippur south to Eridu). Sumer and Akkad were not political entities but regions whose people spoke two different languages while sharing a common material culture. Around 2350 BC Sargon I, a charismatic and powerful Akkadian ruler, subjugated all Sumer and Akkad, also conquering lands to the northwest as far as Turkey and the Mediterranean, and to the east as far as Susa. His was perhaps the first empire to outlast the life of its founder, but by 2200 BC it had collapsed and was followed by a period of Sumerian revival.

At the close of the 3rd millennium BC Ur, long an important Sumerian city, came to dominate the region. The Third Dynasty of Ur ruled the cities of Sumer and Akkad and east beyond the Zagros Mountains, establishing a system of governors and tax collectors that formed the skeleton for the complex bureaucracy needed to control a large population. However, this last Sumerian flowering had lasted only 120 years when Ur was sacked in 2004 BC by the Elamites.

INTERNATIONAL TRADE
The literate Sumerians provide an invaluable source of information on contemporary cultures, from whom they obtained essential raw materials such as metals, wood and minerals, and luxuries including lapis lazuli. The most distant of their direct trading partners was the Indus region, known to them as Meluhha, the source of ivory, carnelian beads and gold; closer lay Magan, a major source of copper, and Dilmun (Bahrain), long known to the Sumerians as the source of "sweet water" and "fish-eyes" (pearls) (map 3). Dilmun acted as an entrepot in this trade, but there were also Meluhhan merchants resident in some Sumerian cities. Sumer exported textiles, oil and barley to its trading partners, but the Indus people were probably most interested in receiving silver obtained by Sumer from further west. It is likely that Magan was an intermediary for trade along the Arabian coast with Africa, the source of several types of millet introduced into India at this time. The Indus people also had writing, but the surviving texts – brief inscriptions on seals and copper tablets – have yet to be deciphered, and probably contain little beyond names and titles.

2 URBAN DEVELOPMENT

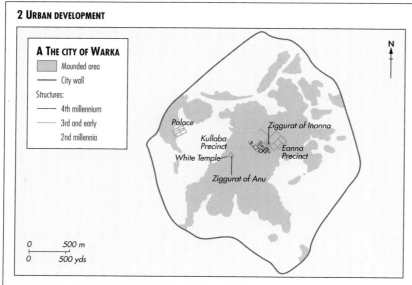

A THE CITY OF WARKA
▨ Mounded area
— City wall
Structures:
— 4th millennium
— 3rd and early 2nd millennia

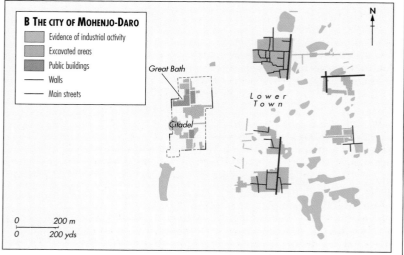

B THE CITY OF MOHENJO-DARO
▨ Evidence of industrial activity
▨ Excavated areas
▨ Public buildings
— Walls
— Main streets

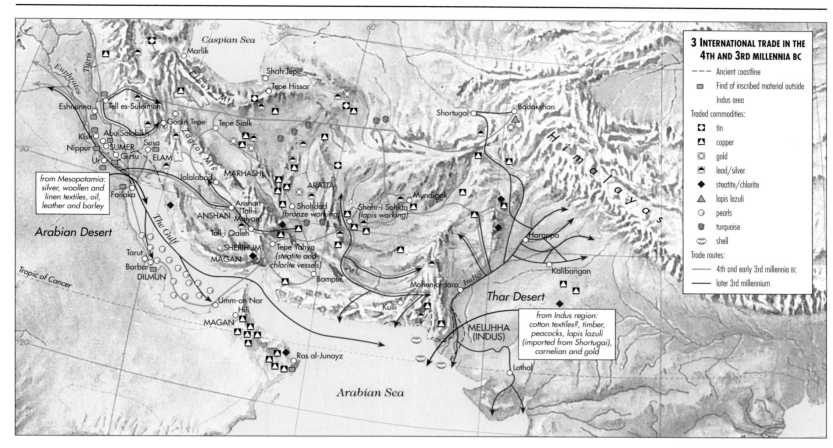

3 INTERNATIONAL TRADE IN THE 4TH AND 3RD MILLENNIA BC

- - - Ancient coastline
▢ Find of inscribed material outside Indus area

Traded commodities:
▯ tin
▲ copper
▢ gold
▭ lead/silver
◆ steatite/chlorite
△ lapis lazuli
○ pearls
❋ turquoise
◗ shell

Trade routes:
— 4th and early 3rd millennium BC
— later 3rd millennium

from Mesopotamia: silver, woollen and linen textiles, oil, leather and barley

from Indus region: cotton textiles?, timber, peacocks, lapis lazuli (imported from Shortugai), carnelian and gold

▲ In the 4th and early 3rd millennia BC Sumerians traded with towns across the Iranian Plateau. By the later 3rd millennium BC, however, they were trading directly with the Indus region by sea, and trade in lapis lazuli had become an Indus monopoly.

▶ Indus settlements in what is now desert point to a time when a network of rivers flowed parallel to the Indus, augmenting the area available for agriculture. The area at the mouth of these rivers was important in both local and international trade.

THE INDUS REGION

In the Indus region, colonized by farmers in the later 4th millennium BC, many settlements were replaced by planned towns and cities around 2600 BC (*map 4*). Within their overall similarity of plan there was considerable local variation, particularly in the layout of the citadel, probably reflecting heterogeneity in religious and cultural practices. For example, the citadel at Mohenjo-daro was dominated by a Great Bath, suggesting ritual bathing, important in later Indian religion (*map 2B*). In contrast, those of Kalibangan and Lothal had pits where sacrificial material was burnt.

Despite some regional variation, uniformity was a keynote of the Indus civilization. Throughout the Indus realms high-quality goods such as pottery, flint blades and copper objects, shell and stone beads and bangles, and steatite seals were manufactured from the best materials available, such as flint from the Rohri Hills. Although the Indus people owed much of their prosperity to the rich agricultural potential of their river valleys, a significant proportion of the population were mobile pastoralists, their flocks and herds grazing in the adjacent forests and grassy uplands; it is probable they acted as carriers in the internal trade networks that ensured the distribution of goods.

Outside the heartland of the civilization, mobile hunter-gatherers provided the means by which the Indus people obtained goods and materials (such as ivory, carnelian and gold) from other regions of the subcontinent, in exchange for cultivated grain, domestic animals and manufactured goods such as copper fish-hooks. The fishers of the Arawalli Hills also participated in this network, trading their locally mined copper.

Around 1800 BC the Indus civilization went into decline. A probable cause was the drying up of some of the rivers, but other factors may have included disease, changes in agricultural practices, and perhaps the depredations of Indo-Aryan nomads on the Indus periphery.

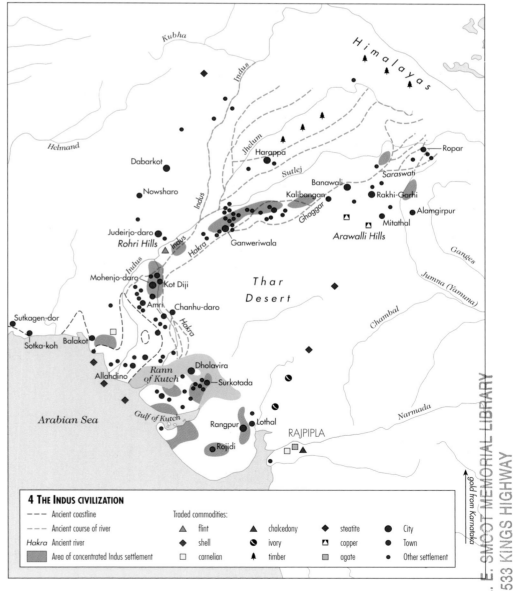

4 THE INDUS CIVILIZATION

- - - Ancient coastline
- - - Ancient course of river
Hakra Ancient river
▨ Area of concentrated Indus settlement

Traded commodities:
△ flint
◆ shell
▢ carnelian
▲ chalcedony
◔ ivory
◆ steatite
▱ copper
▦ agate

● City
● Town
· Other settlement

gold from Karnataka

◀ HUNTING TO FARMING: ASIA 12,000 BC–AD 500 *pages 18–19*　▶ THE MEDITERRANEAN 2000–1000 BC *pages 36–37*　▶ INDIA 600 BC–AD 500 *pages 46–47*

THE FIRST CIVILIZATIONS:
EGYPT 3500–2180 BC AND CHINA 1700–1050 BC

The first civilizations emerged in areas where high agricultural productivity was possible, supporting dense populations. In the Old World they appeared along the rivers in Mesopotamia, northern India, Egypt and northern China. Craft specialization developed, trade flourished, writing began and rulers were often given elaborate burials. However, each civilization also had unique features rooted in its own cultural background and environment.

Life in Ancient Egypt evolved around the Nile, which provided a regular water supply and fertile soils and thus, by contrast with the surrounding desert regions, made agricultural production possible. Navigation on the river was easy, as boats could travel northwards with the current or sail southwards on the northerly winds. From the 5th millennium BC farming communities along the Nile gradually began to merge into a cultural, political and economic unit. This process of unification was encouraged by trading contacts and the need to control the floodwaters of the Nile. To reap the benefits of the yearly inundation of the river, communities had to work together to build dams, flood basins and irrigation channels over large areas. In around 3000 BC this co-operation resulted in the establishment of a single kingdom and the First Dynasty: according to tradition, in 3100 BC King Menes united the delta region (Lower Egypt) and the river valley (Upper Egypt) and founded a capital at Memphis.

THE EARLY DYNASTIC AND OLD KINGDOM PERIODS
The period of the first Egyptian dynasties was one of great cultural and economic significance, when hieroglyphic script was developed and administrative centres established. During the succeeding period of the Old Kingdom (2686–2181 BC), Egyptian culture flourished and the great pyramids were built as spectacular royal tombs (*map 1*). The first was the step pyramid constructed for Pharaoh (or King) Djoser (2667–2648 BC) at Saqqara: over 60 metres (200 feet) high, it was the largest stone building of its time. The first true pyramids, with sloping sides, were constructed at Giza, and the largest, built for Pharaoh Khufu (2589–2566 BC), reached a height of nearly 150 metres (500 feet). Eventually the rule of the Old Kingdom dynasties collapsed, possibly because of the expanding power of the provincial governors, or perhaps because scarce rainfall led to famine and unrest. Central government would be restored with new dynasties during the Middle Kingdom (2055–1650 BC) and the New Kingdom (1550–1069 BC) periods (*pages 36–37*).

THE GROWTH OF EGYPTIAN TRADE
In search of building materials, gold and luxury items, the pharaohs established a wide trade network. During the Old Kingdom period links were forged with many areas of West Asia, including Byblos on the Lebanese coast, predominantly in a search for timber, and expeditions were sent to mine turquoise, copper and malachite in the Sinai Desert. The Eastern Desert yielded copper and stone and gave access to the harbours on the Red Sea, from where trade with East Africa and Arabia was conducted. While these trading missions were mainly peaceful, the area to the south of the First Cataract along the Nile became a prime target for expansion. This land, called Nubia or Kush, offered large quantities of gold as well as connections with the African hinterland, which was an important source of spices, ebony, ivory and other luxury goods. During the Old Kingdom period, a mining settlement was established at Buhen – the first step in a process of southward expansion which would peak in the 15th century BC.

Arts and crafts flourished in Ancient Egypt, particularly in the service of religion and in providing for the dead. Religion also played a major role in northern China, where ancestors were given the greatest respect and were consulted by divination using oracle bones prior to important events such as hunting trips, childbirth and military campaigns.

THE RISE OF THE SHANG CIVILIZATION
Around 1700 BC the Shang civilization emerged as a powerful new state in the northern plains of China. It is known from later historical sources, from magnificent archaeological remains of cities and great tombs, and from written inscriptions carved on oracle bones and cast on splendid ritual bronze vessels. Bronze-working was important to Shang culture and to many other peoples in China, and several different traditions can be recognized (*map 2*). However, it is the use of writing that sets the Shang civilization apart: although ideographic pictograms were used as potters' marks as early as the 3rd millennium BC, the Shang inscriptions provide the first evidence of the development of a literate civilization in China.

During the latter half of the 2nd millennium BC the Shang dynasty conquered and controlled large parts of northern China (*map 3*). The first Shang king, Tang, achieved dominance by defeating 11 other peoples and then winning over 36 more by his fair rule and moral leadership.

Shang rule reached its greatest extent under Wu Ding, one of Tang's successors, who was renowned for his wisdom and led a series of successful military campaigns. Wu Ding was supported in his campaigns by his consort Fu Hao, who herself led armies into battle against the hostile Fang people.

▲ Ancient Egypt became the world's first large, centrally ruled state. It was headed by a divine king (pharaoh) who was known as the son of Ra, the sun god. According to some experts, pyramids represented the staircase along which the pharaoh would return to the heavens after his death. The most famous pyramids are those at Giza, angled at a perfect 52°. Close by is Khafre's Sphinx, 73 metres (240 feet) in length and carved from a limestone outcrop. Originally it was plastered and brightly painted, the bearded face wearing a spectacular headdress sporting a cobra motif.

▶ "Gift of the Nile" was the name given by the Greek historian Herodotus (c.. 485–425 BC) to the country where Ancient Egyptian civilization flourished without rival for over 2,000 years. While the Nile Valley provided fertile soils, the surrounding deserts yielded the precious metals and building stone used in ambitious artistic and architectural endeavours such as the pyramids. These won such acclaim in Ancient Greece that they became known as one of the "Seven Wonders of the World".

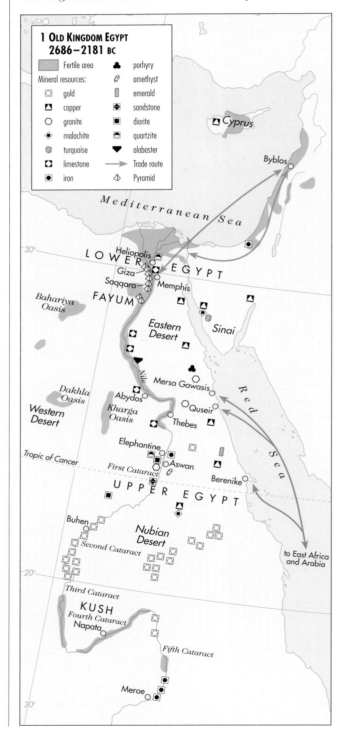

1 OLD KINGDOM EGYPT
2686–2181 BC

	Fertile area	♣	porphyry
Mineral resources:		⬭	amethyst
▢	gold	▯	emerald
⬛	copper	✥	sandstone
◯	granite	■	diorite
◉	malachite	⬓	quartzite
✺	turquoise	▼	alabaster
⬡	limestone	→	Trade route
⬛	iron	△	Pyramid

Cyprus

Byblos

Mediterranean Sea

LOWER EGYPT
Heliopolis
Giza
Saqqara — Memphis

Bahariya Oasis

FAYUM

Eastern Desert

Sinai

Nile

Mersa Gawasis

Dakhla Oasis

Abydos

Kharga Oasis

Quseir

Western Desert

Thebes

Red Sea

Elephantine

Tropic of Cancer

First Cataract

Aswan

Berenike

UPPER EGYPT

to East Africa and Arabia

Buhen

Nubian Desert

Second Cataract

Third Cataract

KUSH

Fourth Cataract

Napata

Fifth Cataract

Meroe

The secret of Shang military success was the use of war chariots, which were so prized that they were sometimes included in burials. Fu Hao's sumptuous tomb is the richest known Shang burial, containing over 400 bronze treasures, 2,000 cowrie shells and more than 500 jade artefacts. Most of the other great tombs, however, were looted in antiquity.

ROYAL CHINESE CITIES

Walled towns or cities ruled by royal lineages were central to early Chinese states, but they were often "moved": eight such transfers are recorded for the Shang capital before the reign of the first king (the beginning of the "dynastic period") and a further seven for the 30 kings of the dynastic period. We know most about the last capital, Yin (near modern Anyang), which was founded by Pan Geng in about 1400 BC.

Yin was located on the marshy plains of the Huang He River, at that time a warmer and moister environment than now exists. The coast was considerably closer and the region was fertile, supporting two crops a year of rice and millet. Water buffalo and wild boar roamed the luxuriant forests which have long since disappeared. Yin sprawled over a large area in which residential compounds for the ruling elite and clusters of commoners' dwellings were interspersed with bronze foundries and workshops producing jade and lacquer ware and pottery. At its centre lay the royal palaces and ancestor temples set atop platforms of pounded earth, and a royal cemetery where kings lay in magnificent shaft graves.

We know little about the later Shang rulers, except for the debaucheries of the last king, the tyrannical Chou. Such were Chou's excesses and tortures that the Shang people welcomed his defeat at the hands of the Zhou in the Battle of Chaoge, traditionally dated 1122 BC but probably closer to 1050 BC. The Zhou were to become China's longest-ruling dynasty, governing the region until 256 BC (*pages 48–49*).

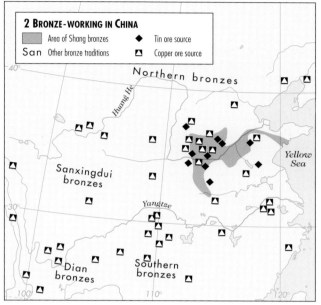

2 BRONZE-WORKING IN CHINA

Area of Shang bronzes ◆ Tin ore source
San Other bronze traditions ◻ Copper ore source

Northern bronzes

Sanxingdui bronzes

Dian bronzes

Southern bronzes

Yellow Sea

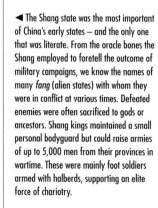

◄ The immediate predecessors of the Shang began working in bronze – a craft reaching great heights under both the Shang and their neighbours. Cast bronze vessels, used to serve food and drink in ceremonies honouring ancestors, followed the traditional shapes previously made in pottery, often intricately decorated and featuring the face of a monster known as *taotie*. The discovery of many fine bronzes at Sanxingdui in Sechuan proves the existence of excellent bronze-working traditions outside the Shang area. Working in bronze probably began earlier in Southeast Asia and south China.

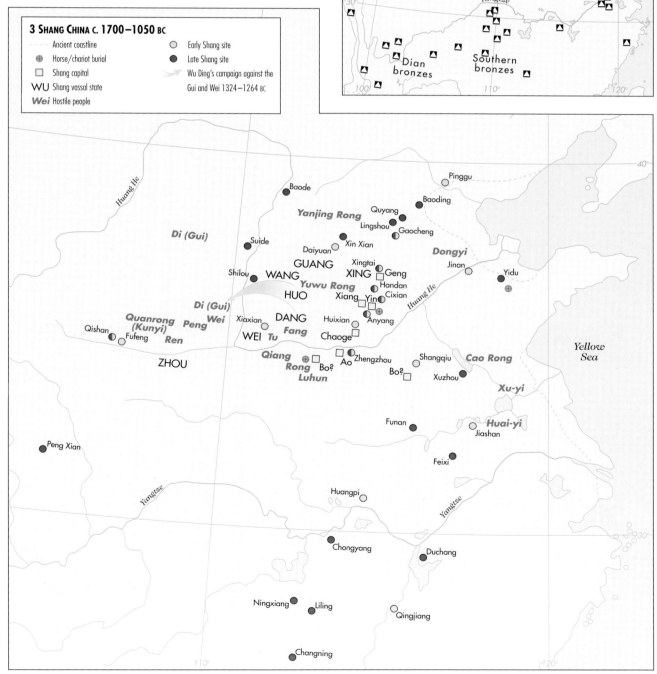

3 SHANG CHINA C. 1700–1050 BC

--- Ancient coastline ○ Early Shang site
⊛ Horse/chariot burial ● Late Shang site
◻ Shang capital ⇒ Wu Ding's campaign against the
WU Shang vassal state Gui and Wei 1324–1264 BC
Wei Hostile people

Yellow Sea

◄ The Shang state was the most important of China's early states – and the only one that was literate. From the oracle bones the Shang employed to foretell the outcome of military campaigns, we know the names of many *fang* (alien states) with whom they were in conflict at various times. Defeated enemies were often sacrificed to gods or ancestors. Shang kings maintained a small personal bodyguard but could raise armies of up to 5,000 men from their provinces in wartime. These were mainly foot soldiers armed with halberds, supporting an elite force of chariotry.

▼ Many bronze vessels produced in Shang China were decorated with animal motifs. The lid of this *gong* (lidded jar) is in the form of an imaginary animal combining features of birds and tigers. *Gongs* were used during the time of Fu Hao around 1200 BC, but were soon replaced by animal-shaped jars.

CIVILIZATIONS IN MESOAMERICA
1200 BC–AD 700

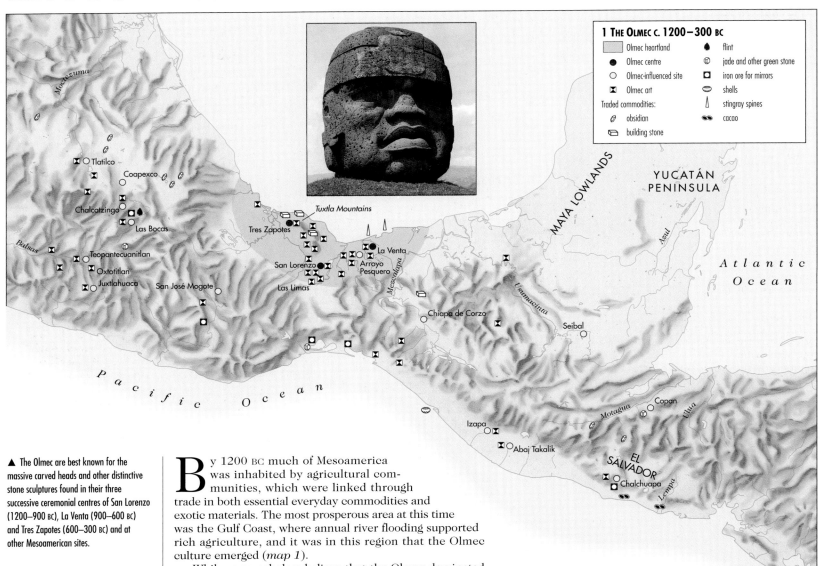

▲ The Olmec are best known for the massive carved heads and other distinctive stone sculptures found in their three successive ceremonial centres of San Lorenzo (1200–900 BC), La Venta (900–600 BC) and Tres Zapotes (600–300 BC) and at other Mesoamerican sites.

▼ Teotihuacan influenced and probably dominated much of the extensive area with which it traded, including the Maya city of Kaminaljuyu. It is unclear to what extent this dominance was achieved and maintained by military force: although Teotihuacan art rarely shows its people as warriors, this is how they appear in the art of their powerful neighbours, the Maya and Monte Alban.

By 1200 BC much of Mesoamerica was inhabited by agricultural communities, which were linked through trade in both essential everyday commodities and exotic materials. The most prosperous area at this time was the Gulf Coast, where annual river flooding supported rich agriculture, and it was in this region that the Olmec culture emerged (*map 1*).

While some scholars believe that the Olmec dominated Mesoamerica, controlling the settlements in which their distinctive artefacts have been found, others see the Olmec as the religious leaders of the time, with their successive ceremonial centres acting as places of pilgrimage. Another school of thought views the Olmec as the most visible and most easily identified of a number of contemporary regional cultures that were mutually influential.

Much that is characteristic of later Mesoamerican civilization is already evident in the Olmec culture. The dangerous animals (in particular the jaguar) and the natural phenomena (such as rain) which feature prominently in Olmec art reappear in various guises in later religious art. The concern with the movements of sun, moon and stars that underlies much Mesoamerican religion is apparent in the astronomically aligned layout of the Olmec ceremonial centres, where the first temple pyramids and plazas, as well as caches of precious offerings to the gods, have been found. The characteristic colossal carved heads, which may be portraits of Olmec rulers, wear helmets for the ritual ballgame, a dangerous sport with religious significance that was part of most Mesoamerican cultures and often involved the sacrifice of members of the losing team.

Personal blood sacrifice, practised in later Mesoamerican religions, also appears to have been a feature of Olmec life, as stingray spines and other objects used to draw blood have been found at Olmec sites. These items were widely traded – as were both jade, which had great ritual importance, and obsidian (volcanic rock glass), used to make exceptionally sharp tools but also fine ritual or status objects. The widespread distribution of these materials reflects not only their religious significance throughout Mesoamerica but also their role as indicators of status in communities where social hierarchies were beginning to emerge. Prestigious Olmec pottery and figurines (including the characteristic "were-jaguar" babies) served the same purpose.

THE TEOTIHUACAN AND MONTE ALBAN EMPIRES
By about 300 BC the Olmec had lost their pre-eminent position and other civilizations were developing in the highland zone, particularly the Teotihuacan Empire in the Basin of Mexico and the Monte Alban Empire of the Zapotec people in the Oaxaca Valley (*map 2*). This was the beginning of

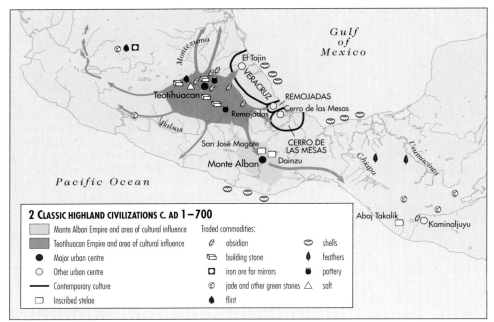

what is known as the Classic Period, which lasted until around AD 900. Agricultural productivity now greatly increased in this region as irrigation techniques using wells and canals were developed to supplement rain-fed farming. Raised fields may also have been cultivated.

Like the Olmec, all these civilizations were heavily involved in trade. The city of Teotihuacan (*map 3A*), founded before 300 BC, was well placed to control widespread trading networks. It contained over 600 workshops manufacturing goods for local use and for export – objects of obsidian (400 workshops), basalt (a building stone), shell and other materials, as well as distinctive pottery.

The city of Monte Alban was founded around 500 BC. Like Teotihuacan, it was the ceremonial and political centre of its state, but in contrast it was not the centre for regional craft production. Evidence shows that initially the Monte Alban state grew by military conquest, but by AD 300 its expansion had been checked by that of the Teotihuacan Empire, although the people of Monte Alban seem to have been on friendly terms with their neighbour.

Ballcourts and depictions of sacrificial victims at Monte Alban show the continuation in the highland zone of the religious practices of Olmec times. Also continued was the use of written symbols (glyphs) to record dates and related information. Concern with the movements of heavenly bodies and the related calendar had led to the development of glyphs by the Olmec; by 500 BC the people of the Oaxaca Valley were recording dates and names on their carved stone slabs (*stelae*). However, the only region where a complete writing system developed in the Classic Period was the Maya lowlands (*map 4*).

THE EARLY MAYA CIVILIZATION

The Maya writing system was extremely complex, with many variations in the form of individual glyphs and in the way in which a word could be expressed. It was also used to record an extremely elaborate calendric system, involving interlocking and independent cycles of time, including the 52-year repeating cycle used throughout Mesoamerica and the Maya Long Count, a cycle beginning in 3114 BC according to our present-day dating system. These depended both on a detailed knowledge of astronomical patterns and on sophisticated mathematics, including the concept of zero.

Although the Maya script is still not fully deciphered, scholars are now able to read many inscriptions on carved *stelae*, temple stairs and lintels and have pieced together the dynastic history of many of the Maya kingdoms. (Unlike the two highland empires, the Maya were not politically unified, although they were united culturally.) Maya inscriptions record the descent of each ruler from a founding ancestor, his performance of appropriate ritual activities on dates of significance in the astronomical religious calendar, and his victories over neighbouring rulers. Although wars of conquest did occur at this time – Uaxactun's takeover by Tikal (*map 3B*) in AD 378 is the prime example – the main motive for warfare was to capture high-ranking individuals to be used as sacrificial victims.

Blood sacrifice was of central importance in Maya and other Mesoamerican religions, based on the belief that human blood both nourished divine beings and opened a pathway through which humans could communicate with the spirit world. While personal sacrifices could be made by any member of Maya society, it was largely the responsibility of each king to ensure the well-being of his state through the provision of sacrificial victims and by letting his own blood. Members of the king's family were appointed as provincial governors of lesser centres within the kingdom, and they also acted in other official capacities including that of scribe.

The 7th century saw the demise of Teotihuacan and Monte Alban and the rise of other highland states, while in the Maya region important changes had already occurred (*pages 84–85*). The pattern of existence that had emerged in Olmec times continued, however, as the template for the Mesoamerican way of life up to the time of the European conquest in the 16th century.

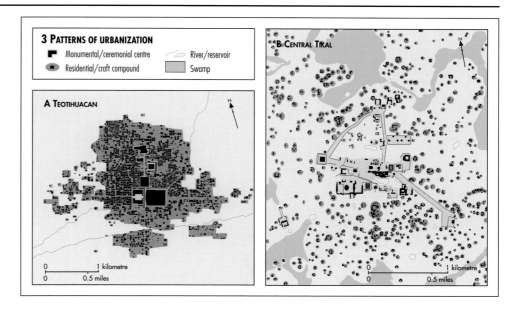

▲ The cities of Teotihuacan and Tikal highlight the contrasting patterns of life in the highland and lowland civilizations. Tikal, in the Maya lowlands, covered more than 120 square kilometres (47 square miles) with an estimated population of 50,000, while Teotihuacan in the highlands housed two to four times as many people in a sixth of the area. House compounds in Maya cities were interspersed with doorstep gardens and raised fields in swamp areas, and a great variety of crops were grown in both. By contrast the agricultural lands supporting Teotihuacan lay outside the city, in the Basin of Mexico. Highland and lowland cities alike, however, focused on a ceremonial centre containing temples and the residences and burial places of the elite.

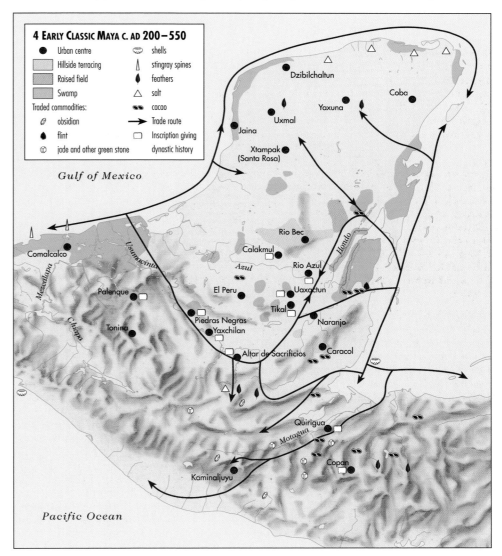

▲ Recent discoveries have shown that the Maya employed intensive farming techniques, including hillside terracing to counteract erosion, and canals dug along rivers and in *bajos* (seasonal swamps) for drainage, water storage and probably fish-farming and communications. Highly productive raised fields were constructed between grids of canals – although the known extent of these fields is likely to represent only a fraction of what once existed. As in other Mesoamerican civilizations, trade played an important role in Maya life, providing materials for daily living, religious rituals and status symbols.

CULTURES IN SOUTH AMERICA
1400 BC–AD 1000

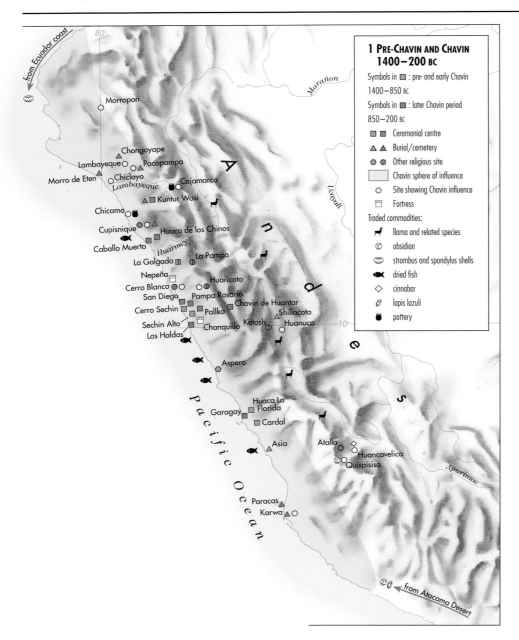

Tiwanaku. Other supernatural creatures included jaguars, caymans and composite beasts; shamans were also depicted and they were believed to be able to transform themselves into exotic birds and animals.

Traded objects, such as goldwork, were included as grave goods in the elaborate burials of the Chavin elite. These burials were often placed in shaft tombs within the platforms of the Chavin ceremonial centres, another practice that endured down the ages – for example in the magnificent burials found in the few unlooted Moche *huacas* (sacred pyramids) such as that at Sipan.

THE PARACAS AND NAZCA CULTURES
The distinctive Paracas culture emerged in Chavin times, around 600 BC. Their craftsmanship survived in an extensive cemetery (*map 1*) containing numerous mummies of elite individuals wrapped in beautifully embroidered cotton textiles and accompanied by fine pottery, goldwork and other offerings. By around 375 BC the Paracas culture had developed into the Nazca culture (*maps 2 and 4B*), also renowned for its textiles and fine polychrome pottery. Some vessels were designed in the form of trophy heads, and real heads – pierced for suspension on a rope – have been recovered from Nazca cemeteries, in particular that at the chief Nazca ceremonial centre of Cahuachi.

Unlike Chavin de Huantar and the ceremonial centres of other Andean civilizations, Cahuachi seems not to have functioned as a town, though it was probably a place occupied briefly by thousands of pilgrims during religious ceremonies and festivals. In its neighbourhood are the enigmatic Nazca Lines, designs on a gigantic scale which were created by removing stones to expose the light desert soil beneath and depict animals, birds and geometric shapes familiar from the Nazca pottery. Their form can only be appreciated from the air, so they are thought to have been intended for the gods to view and to have been used in the performance of religious activities.

▼ The Moche culture was centred on the site of Moche, in northwest Peru. Its adobe pyramids, among the largest in the New World, contained temples and rich tombs later desecrated by other Andean peoples and the Spanish. Through time, the Moche spread to most of the northern coast of Peru, from the Huarmey Valley in the south, and, in the latest phase, to the Lambayeque Valley in the north. Further south, the Nazca culture is well represented by large cemeteries and substantial religious structures of mudbrick. The culture is best known, however, for the Nazca Lines.

▲ Spondylus and strombus shells, widely regarded as food for the gods, featured prominently in Chavin and later Andean art. Imported from the coast of Ecuador, they were an important commodity in the exchange networks that ensured the distribution of foodstuffs and other raw materials (such as obsidian, or volcanic glass) and manufactured goods (notably pottery and textiles) between the different regions of the Andean zone during the Chavin period. Chavin de Huantar probably owed its pre-eminent position to its location at the centre of trade routes running both north–south and east–west. In some areas roads were built to facilitate trade and communications, and these networks (and the commodities they carried) changed little in later periods.

By the late 2nd millennium BC a patchwork of interrelated farming settlements existed throughout the Andean region, from coasts and lowland valleys to high pastures. In addition to residential villages, the Andean people were constructing religious centres which took various forms (*map 1*). Those in coastal regions were characteristically built in the shape of a U, with terraced mounds laid out along three sides of a rectangular plaza, and a pyramid often stood on the central mound. Some of these temple complexes – notably Cerro Sechin, where graphic carvings of victims survive – give evidence of human sacrifice as a part of the rites performed. Thus they foreshadow the practices of later Andean cultures, which included a widespread trophy head cult (for example among the Nazca) and warfare to obtain captives for sacrifice (particularly evident among the Moche).

CHAVIN DE HUANTAR
Around 850 BC a similar U-shaped ceremonial centre was constructed in the mountains at Chavin de Huantar. Housing the shrine of an oracular fanged deity set within labyrinthine passages, Chavin de Huantar became a place of pilgrimage, the centre of a cult that was widespread in its influence, as demonstrated by the distribution of artefacts in the characteristic Chavin style. Carvings decorating the temple mounds focused on religious themes, as did designs on pottery, jewellery and other objects. Chief among these was the Chavin deity, which continued to be worshipped down the ages in various forms, such as the Staff God of

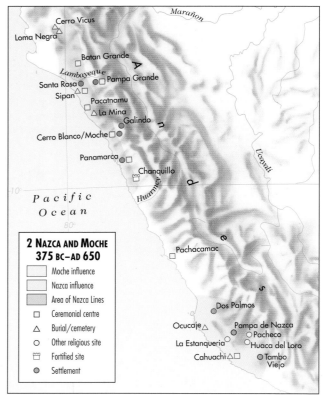

▼ Irrigation played an important role in South American agriculture, and water control was well developed during the Chavin period (1200–200 BC), when a series of canals was skilfully used to provide awe-inspiring sound effects in the great ceremonial centre of Chavin de Huantar. Later civilizations in the Andean region employed a variety of different techniques appropriate to local conditions. The Moche supplemented perennial and seasonal watercourses by creating a network of canals (B). To the south, in the Nazca region, underground aqueducts designed to prevent water loss by evaporation (A) were probably constructed after AD 600 when the region fell to the Huari, who also built sophisticated hillside irrigation terraces. The Tiwanaku state undertook a large-scale programme of swamp drainage and canal construction in the Pampa Koani region of Lake Titicaca to establish a complex network of fertile raised fields (C). Some of these irrigation systems (such as the Nazca underground aqueducts) have survived into modern times; others have recently been revived and are proving far more successful than modern methods.

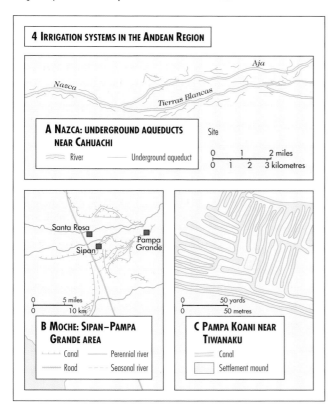

4 IRRIGATION SYSTEMS IN THE ANDEAN REGION

Aja

Nazca

Tierras Blancas

A NAZCA: UNDERGROUND AQUEDUCTS NEAR CAHUACHI

— River —— Underground aqueduct

Site

0   1   2 miles
0 1 2 3 kilometres

Santa Rosa

Sipan

Pampa Grande

0   5 miles
0   10 km

B MOCHE: SIPAN–PAMPA GRANDE AREA

—— Canal —— Perennial river
—— Road ---- Seasonal river

0   50 yards
0   50 metres

C PAMPA KOANI NEAR TIWANAKU

—— Canal
☐ Settlement mound

3 TIWANAKU AND HUARI AD 400–1000

▨ Tiwanaku state
▨ Huari Empire
☐ Ceremonial centre
○ Other religious site
⬤ Settlement
○ Site of uncertain type

▶ In the period AD 600–1000 Andean South America contained at least three expansive political entities embracing distinct ecological zones and ethnic groups. The city of Tiwanaku extended its control from the rich farmlands around Lake Titicaca to lower valleys in adjacent areas of southern Peru, northern Chile and northern Argentina. At about the same period, during the so-called "Middle Horizon", a related (but probably rival) polity flourished around the city of Huari in Peru, displacing the coastal culture of Moche around AD 650.

THE MOCHE CULTURE

Partially contemporary with the Nazca culture, which flourished until around AD 600, was the Moche culture of c. AD 1–650, *maps 2 and 4B*. Their ceramics, painted with exceptionally fine calligraphy, reveal a ceremonial life focused on mountain worship, royal mortuary cults, warfare and the dismemberment of captives. The recent discovery of an unlooted pyramid (*huaca*) at Sipan, containing the burials of two Moche lords, has given us a vivid picture of Moche burial practices. Accompanied by a number of sacrificed men, women and dogs, these lords were lavishly robed in garments decorated with gold and silver, copper and feathers; they were provided with rich grave goods in the same materials, along with spondylus and strombus shells.

Details of these burials are familiar from decoration on the painted or moulded pottery. Moche ceramics also included some of the first (and only) portrait effigies in the Americas, all cast from moulds and often into the stirrup-handled vessels common to Peru. Although heavy in religious imagery, these ceramics are unusually narrative for South American art, leading some scholars to postulate influence from other areas such as Mesoamerica.

THE CITIES OF HUARI AND TIWANAKU

Around AD 650 the Moche culture was eclipsed by new art styles emanating from Huari, near Ayacucho in the southern highlands of Peru (*map 3*). More distant still lay a city of comparable complexity, Tiwanaku, near Lake Titicaca. Although both cities had emerged c. 400, the connection between them remains enigmatic. Most archaeologists believe that they were not so much dual capitals of one empire (an older theory) as antagonistic polities, one – Huari – oriented to the north, the other – Tiwanaku – to the high timberless plains known as the *altiplano*.

While recent political instability in the region of Huari has made it difficult to study, Tiwanaku has been intensively investigated, unveiling elaborate raised fields (*map 4C*). Whether the fields around Lake Titicaca were systematically organized and harvested by the Tiwanaku state continues to be controversial. Field research in the Moquegua Valley indicates late Tiwanaku expansion into a number of enclaves, with maize in particular being cultivated. Also subject to Huari influence, this valley was important as the source of many prized materials which included lapis lazuli, turquoise, obsidian (volcanic glass) and copper.

◀ The Nazca pottery vessel (*left*) depicting a seated warrior holding a trophy head is representative of the cult of trophy heads which was widespread in South America. The container with a funerary effigy (*right*) is characteristic of the Chavin style.

THE MEDITERRANEAN AND THE GULF REGION 2000–1000 BC

▲ Nefertiti – the subject of this bust carved by the royal sculptor Thutmose – was the powerful wife of the heretical pharaoh Akhenaten (r. 1352–36 BC). Ascending the throne as Amenhotep IV, the king changed his name when he introduced the monotheistic worship of Aten, the sun god. He founded a new captial, Akhetaten (modern Amarna), but this, like his religion, was abandoned after his death.

▼ During the New Kingdom period a flow of goods such as gold, timber and ivory from Egypt reached Phoenicia, Cyprus, Crete and, further afield, the interiors of the Near East. In return Asiatic products such as copper and tin – and, before 1450, pottery from Crete – were imported into Egypt. While the Egyptian and Hittite empires played key roles in the extensive Mediterranean trade networks of the 2nd millennium BC, behind the coast there were other powerful states – those of the Assyrians, Babylonians (the Kassite kingdom), Hurrians (the kingdom of Mitanni) and Elamites. Much of their economic power derived from control of important overland routes – as well as those in the Gulf.

The eastern Mediterranean became extremely affluent during the Bronze Age. This prosperity was largely based on a booming international trade in which the Egyptians and later the Hittites played key roles (*map 1*). During the period of the Middle Kingdom (2055–1650 BC), Egypt experienced stability under a central government led by dynasties from Thebes. Dominion over Nubia, which had been lost during the political disintegration of the First Intermediate Period (2181–2055 BC), was restored, guaranteeing access to products from the African heartland. Royal missions were sent to re-establish diplomatic contacts with Syria and Palestine, a move that further encouraged trade in the eastern Mediterranean.

THE MINOAN AND MYCENAEAN CIVILIZATIONS
From approximately 2000 BC the Minoan civilization flourished on the island of Crete, centred around palaces such as Knossos, Phaistos and Mallia, and the island developed its own script. Initially pictographs resembling the Hittite signary and Egyptian hieroglyphs were used, but around 1700 BC a linear script was invented, the so-called "Linear A".

Around 1450 BC most Minoan palaces were destroyed by fire. This was once considered to be linked to the massive volcanic eruption on the nearby island of Thíra (Santorini), but the eruption is now thought to have taken place around 1628 BC. One possibility is that the destruction was due to occupation by mainland Greeks, the so-called Mycenaeans, who extended the already far-flung trading networks of the Minoans and adapted the Minoan script to suit their language, an early form of Greek. This "Linear B" script can be read, unlike the still undeciphered Linear A. Tablets written in this new script were found on the mainland and on Crete. While the Mycenaean culture showed great affinity with that of Minoan Crete, it also displayed a far more warlike character: Mycenaean palaces were reinforced with enormous fortifications and the theme of warfare dominated their wall paintings.

KINGDOMS AND CITY-STATES OF MESOPOTAMIA
The mighty states of the Assyrians, Babylonians, Hurrians and Elamites flourished by controlling hinterland connections (*map 1*). In southern Mesopotamia (Babylonia) foreign trade was increasingly in the hands of private individuals, in contrast to earlier periods when trade was controlled by temples or the government. Luxury items such as gold, lapis lazuli, ivory and pearls were exchanged for Mesopotamian textiles, sesame oil and resin.

At the beginning of the 2nd millennium there was a struggle for ascendancy and control among the southern cities, in which Isin and Larsa were early players. Later the city of Babylon under King Hammurabi (r. 1792–50 BC) conquered most of the cities of southern Mesopotamia and up the Euphrates to Mari. Although this empire was relatively short-lived, it transformed southern Mesopotamia into a single state. Hammurabi is most famous for his Law Code which, although not the earliest known in Mesopotamia, is the first for which we have the complete text.

While these changes were occurring in the south, in northern Mesopotamia the inhabitants of the core Assyrian city of Ashur were creating trading networks with cities in Anatolia up to 800 kilometres (500 miles) away, where they established trading outposts to exchange Assyrian textiles and "annakum" (probably tin) for silver and gold.

HITTITE EXPANSION AND CONTRACTION
To the north and east of Mesopotamia there were, by the mid-2nd millennium BC, numerous small Hurrian (sometimes called Mitannian) principalities, while the Hittites controlled much of Anatolia. Texts written in the wedge-shaped characters of the cuneiform script tell us there were other kingdoms in Anatolia such as Arzawa, Assuwa, Ahhiyawa and Lukka, but their exact location is uncertain.

In 1595 BC the Hittites under King Mursili defeated Babylon. Soon afterwards, however, the Hittites were beset by internal dissension and revolts, and lost much of their extended territory until they were left controlling only central Anatolia. For about a century very little is known about events in Mesopotamia and Anatolia. In 1480 BC the Hurrian kingdoms were united by King Parrattarna as the kingdom of Mitanni, and by 1415 BC the Kassites, a people who had been slowly moving into Babylonia, had established dominance in the area. The Hittites once again controlled much of the Anatolian plateau and were heavily involved in Mediterranean trade, receiving commodities such as copper, gold and grain as tribute from the cities under their influence or control. At the same time they were spreading southwards into the Levant, an area where the Egyptians under the New Kingdom dynasties were also expanding.

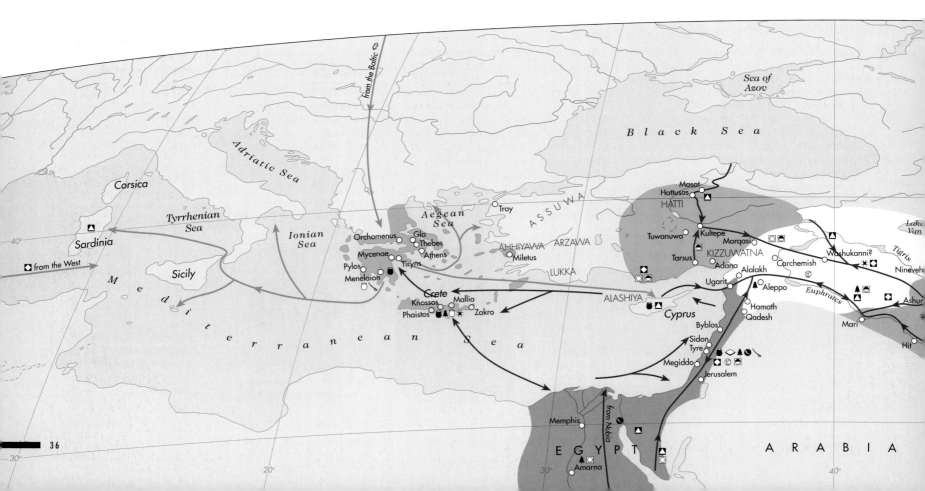

NEW KINGDOM EGYPT

Egyptian unity had once again been destroyed when the Hyksos, an Asiatic tribe, seized part of the country around 1650 BC. Their rule lasted for about 100 years until Ahmose (r. 1550–25 BC) drove them out and established the New Kingdom (1550–1069 BC), a period of great cultural flowering (*map 2*). This was also the time of the greatest Egyptian expansion, predominantly geared towards securing resources from Nubia and West Asia. Thutmose I (r. 1504–1492 BC) campaigned as far as the Euphrates River, and Thutmose III (r. 1479–25 BC) reclaimed Syria, thus extending the empire to Carchemish. He also established Egyptian control over Nubia up to the Fourth Cataract.

Egyptian domination over Palestine and Syria once again lapsed until Sety I (r. 1294–79 BC) recovered Palestine. He initiated a period of fierce competition with the Hittites for control of the Levant, which came to a head at the Battle of Qadesh in 1275 BC. Although the Egyptians claimed victory the Hittites probably gained the upper hand, as the area around and south of Damascus came under Hittite influence.

Soon after this battle the resurgent Assyrians under King Adad-nirari I (r. 1305–1274 BC) captured the Mitannian capital of Washukanni (whose location is still unknown) and, with the collapse of the Mitanni kingdom, established themselves as a power equal to Egypt. In response the Hittites formed a pact of non-aggression with the Egyptians that led to a period of stability in the region.

THE "SEA PEOPLES"

Early in the 12th century BC large movements of peoples around the eastern Mediterranean coincided with the social and economic collapse of many of the Late Bronze Age kingdoms (*map 3*). A wave of destruction was wrought by tribes known collectively as the "Sea Peoples": cities on the Syrian coast and Cyprus were sacked, along with Hittite settlements and Mycenaean palaces, and the Hittite Empire and Mycenaean civilization both came to an end.

The Assyrians were not directly affected by these upheavals and continued to expand. They invaded Babylon as well as the Levant, where they took advantage of the collapse of the Hittite Empire. However, by the close of the 2nd millennium Assyrian dominance was also fading and the kingdom of Elam to the east now became the most powerful player in the region.

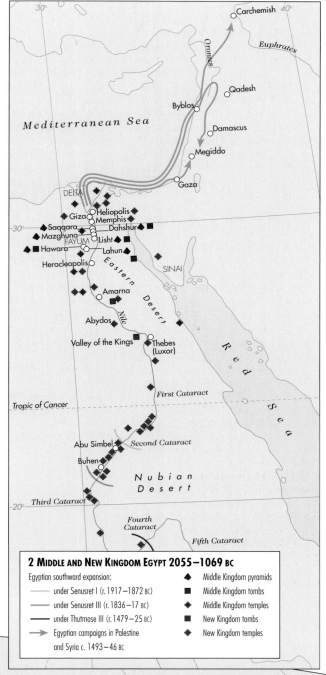

2 MIDDLE AND NEW KINGDOM EGYPT 2055–1069 BC

Egyptian southward expansion:
- under Senusret I (r. 1917–1872 BC)
- under Senusret III (r. 1836–17 BC)
- under Thutmose III (r. 1479–25 BC)
- ➤ Egyptian campaigns in Palestine and Syria c. 1493–46 BC
- ◆ Middle Kingdom pyramids
- ■ Middle Kingdom tombs
- ◆ Middle Kingdom temples
- ■ New Kingdom tombs
- ✦ New Kingdom temples

◀ While the Old Kingdom period is known as the "Age of the Pyramids", the New Kingdom was the era of the vast temples and lavishly painted tombs of pharaohs and nobles in the Valley of the Kings and the adjacent areas around Thebes. The Valley of the Kings alone hosted 62 rock-cut tombs, of which the most famous is that of Tutankhamun. His grave was the only one which archaeologists found largely intact and it contained, besides his mummy, an astounding wealth of grave goods including dismantled chariots, beds, masks, games and musical instruments.

▼ The movements of the "Sea Peoples" – bands who roamed the Mediterranean during the 13th century BC – have been reconstructed on the basis of few written sources and little archaeological evidence. In Egypt two attacks by these tribes have been documented. Merenptah (r. 1213–1203 BC) withstood an attack on the Nile delta by a united force of Libyans and the Sea Peoples. They returned during the reign of Rameses III (1184–53 BC), attacking by land and sea. They were defeated, but later some settled peacefully in Egypt, others in Palestine. Egyptian pharaohs triumphantly recorded their victories over the Sea Peoples, exaggerating the threat posed by groups whom at other times they often employed as mercenaries. It has been assumed that the razed cities elsewhere in the Mediterranean were caused by the same Sea Peoples, although internal unrest and earthquakes were probably among other factors involved.

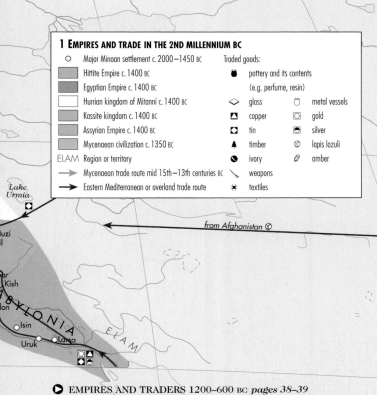

1 EMPIRES AND TRADE IN THE 2ND MILLENNIUM BC

- ○ Major Minoan settlement c. 2000–1450 BC
- Hittite Empire c. 1400 BC
- Egyptian Empire c. 1400 BC
- Hurrian kingdom of Mitanni c. 1400 BC
- Kassite kingdom c. 1400 BC
- Assyrian Empire c. 1400 BC
- Mycenaean civilization c. 1350 BC
- ELAM Region or territory
- → Mycenaean trade route mid 15th–13th centuries BC
- → Eastern Mediterranean or overland trade route

Traded goods:
- pottery and its contents (e.g. perfume, resin)
- ◇ glass
- ▲ copper
- ◻ tin
- ▲ timber
- ⬤ ivory
- ⬭ metal vessels
- ⬜ gold
- ⬛ silver
- ⊚ lapis lazuli
- ⬭ amber
- ＼ weapons
- ✴ textiles

3 INVASIONS AND MIGRATIONS IN THE MEDITERRANEAN C. 1200 BC

- Egyptian Empire
- Hittite Empire
- Mycenaean Greece
- ▨ Area of conflict between Hittites and Egyptians
- ✿ Destroyed site
- ➤ Movement of peoples

a Northern tribes, including the Lukka, Sherden and Teresh, attack Egypt but are defeated

b Peleset, Shekelesh, Denyen, Tjeker and Weshesh launch second unsuccessful attack on Egypt

c Ugarit and Cypriot towns possibly destroyed by Sea Peoples

d Greece and Crete subjected to widespread destruction

e Troy and Hittite cities destroyed, possibly by Armenians and Phrygians

f Possibly Teresh settle in Etruria, Sherden in Sardinia

⬤ EMPIRES AND TRADERS 1200–600 BC *pages 38–39*

EMPIRES AND TRADERS
1200–600 BC

From approximately 1200 to 900 BC West Asia was in an economic and political downswing. Both the archaeological and textual evidence indicates that there was no longer the vast wealth that had supported the lavish royal lifestyles and military campaigns of the Late Bronze Age. Although major cities remained occupied, the empires of the Egyptians, Hurrians, Hittites, Elamites and Assyrians no longer held sway over the region. However, beginning in 911 BC, Adad-nirari II (r. 911–891 BC) started to re-establish central authority in Assyria (*map 1*). After securing Assyria he sacked but did not conquer Babylon and subsequently conducted a successful series of campaigns in the Habur region. Expansion of the Assyrian Empire continued throughout much of the 9th century BC, and with their mighty armies the Assyrians were to dominate West Asia almost continuously for 200 years until their defeat by the Medes and Babylonians in 612 BC.

ASSYRIAN EXPANSION
The Assyrians did not have a policy of uniform military conquest and incorporation; instead they established a pattern of conquest that entailed first receiving gifts from independent rulers, who were considered as "clients". If the client state subsequently failed to provide "gifts" (tribute), the Assyrians treated this as an act of rebellion and conquered the state. A local ruler was then appointed, or the country was annexed and ruled by a provincial governor. This method of domination and control channelled all the tributes of clients and booty of conquered countries into the heartland of Assyria. Thus the Assyrians not only acquired an extensive empire but also great wealth, enabling their rulers to build fabulous palaces, establish several new capitals and commission works of art ranging from exquisite ivory carvings to monumental stone reliefs.

ISRAEL AND JUDAH
The Levant was one of the main areas to suffer the effects of Assyrian expansion. The Israelites had settled in Palestine, their traditional "promised land", around 1250 BC (*map 2*). A little later, around 1200 BC, the Philistines occupied the adjacent area of Philistia. Increasing pressure from this and other neighbouring tribes forced the Israelites to unite under one king during the 11th century BC. The first, Saul, was defeated by the Philistines, but his successor David

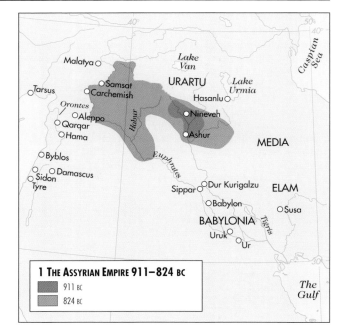

1 THE ASSYRIAN EMPIRE 911–824 BC
- 911 BC
- 824 BC

The Gulf

▲ The Assyrians controlled their empire by installing local rulers or provincial governors and a system of tribute. From the late 9th century onwards they sometimes enslaved and resettled thousands of conquered people in areas far from their homelands.

(r. 1006–966 BC) expanded the kingdom and chose Jerusalem as its religious and political centre. Under David and his son Solomon (r. 966–26 BC) the kingdom prospered, becoming an international power and a centre of culture and trade. Tensions between the northern and the southern tribes mounted, however, and after Solomon's death the kingdom was divided into two parts, Israel and Judah.

THE AGE OF THE PHOENICIANS
To the north Phoenicia had become a major trading empire after the collapse of the Mycenaean civilization around 1200 BC (*pages 36–37*). Phoenicia consisted of autonomous city-states such as Byblos, Sidon and Tyre, which established new trade routes and from the end of the 9th century BC founded colonies in North Africa, Spain and Sardinia (*map 3*). Carthage was a wealthy Phoenician

3 THE PHOENICIANS c. 800 BC
- Area of Greek settlement
- Area of Phoenician settlement
- ● Phoenician colony
- ← Phoenician trade route

Traded goods:
◇ glass	▲ copper	
◕ ivory	◉ lead	
● oil	△ salt	
◲ silver	♠ cedarwood	
◫ gold	✿ grain	

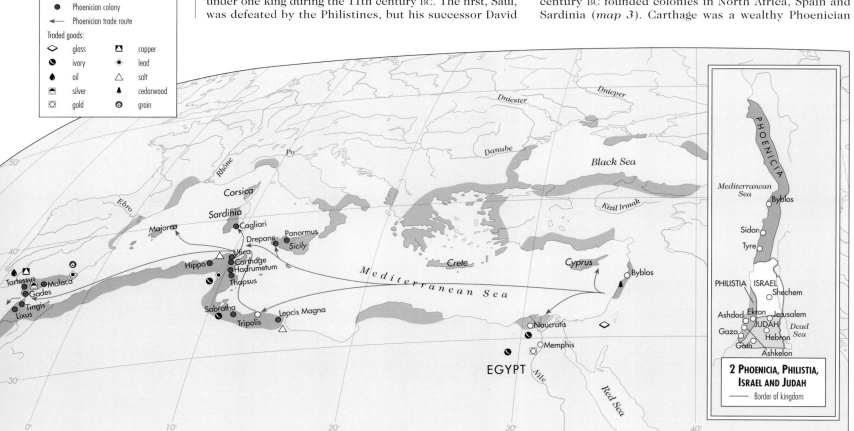

2 PHOENICIA, PHILISTIA, ISRAEL AND JUDAH
— Border of kingdom

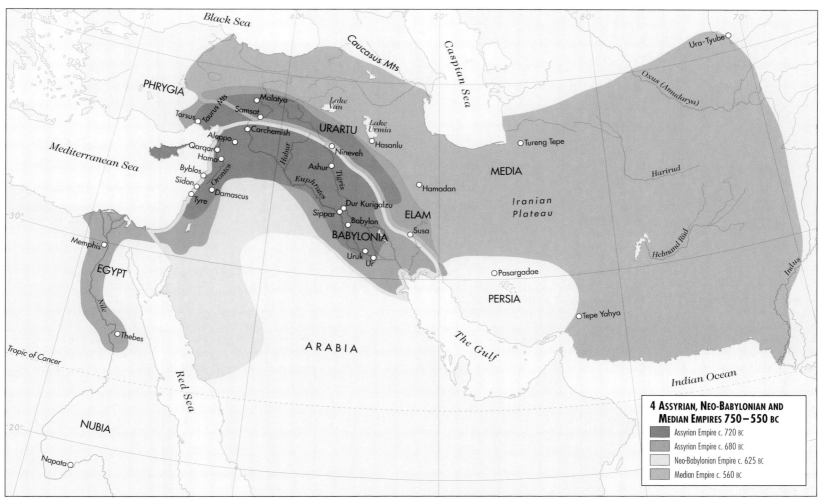

4 ASSYRIAN, NEO-BABYLONIAN AND MEDIAN EMPIRES 750–550 BC

- Assyrian Empire c. 720 BC
- Assyrian Empire c. 680 BC
- Neo-Babylonian Empire c. 625 BC
- Median Empire c. 560 BC

trading centre and gradually established its own empire. Phoenician interest in the western Mediterranean led to clashes with Greeks in southern France and Corsica, while the Carthaginians later engaged in a power struggle with the Romans that ended with their city's destruction in 146 BC.

EGYPT AND ASSYRIA

After the central government of the Egyptian New Kingdom collapsed around 1069 BC, the country was ruled by two competing dynasties based in the Nile delta and Thebes. Nubia, parts of which had been colonized by Egypt from Old Kingdom times (*pages 30–31*), now became independent (*map 4*). A family of local lords established itself as a powerful dynasty, governing from Napata. When the rulers based in the delta threatened Thebes, the priest of the state god Amun sought the protection of the Nubian king Piy (r. 746–716 BC), granting him the title Pharaoh of Egypt. Piy conquered Thebes and went northwards to put down opposition by the delta rulers. His successor completed the conquest of Egypt, reversing centuries of Egyptian domination of Nubia. The start of the Nubian dynasty marks the beginning of the so-called Later Period (747–332 BC).

In the early 8th century the powerful Assyrians suffered a period of weakness, which allowed the kingdoms of other peoples to thrive, among them the Urartians in eastern Anatolia and the Chaldeans in southern Mesopotamia (Babylonia). However, by the middle of the century the Assyrians were once again expanding, for the first time campaigning north of the Euphrates – where they conquered a number of city-states which had formed after the collapse of the Hittite Empire 600 years earlier.

The process continued under Sargon II (r. 721–705 BC), who expanded the boundaries of the empire beyond those of the 9th century BC (*map 4*). By 701 BC the Assyrians had annexed Phoenicia, Israel and Judah, and in the 7th century BC they turned their attention to Babylon, where they were confronted by a powerful culture that would successfully hold its own against the Assyrian might. Although

eventually defeating the Babylonians and their Elamite allies in 694 BC, Assyria always considered Babylon special because of its history, its culture and the power of its ancient gods. Thus Babylon was ruled by a member of the Assyrian royal family as co-king rather than as governor.

In 671 BC the Assyrians launched an attack against the Egyptians and, after initial setbacks, secured domination of the country. However, they never completely controlled it and, after a number of additional campaigns, they withdrew to leave friendly "client kings" in place. During this period Egyptian culture flourished, with Greek Classical and Hellenistic influences becoming increasingly prominent. The Nubians, meanwhile, retreated southwards.

THE NEO-BABYLONIAN EMPIRE

In 626 BC, after 60 years of stability and growth under Assyrian co-kings, a Chaldean who took the royal name of Nabopolassar seized power in Babylonia and established what is known as the Chaldean or Neo-Babylonian Empire. Ten years of civil war between the Babylonians and the Assyrians followed, but by 616 BC Nabopolassar was strong enough to take his armies north, where he defeated the Assyrians and their Egyptian allies. In 615 BC the Medes, who originated from the area around Hamadan, sacked the Assyrian capital Ashur. In 612 BC the combined forces of the Medes and Babylonians besieged and sacked Nineveh, effectively bringing the Assyrian Empire to an end.

Soon afterwards Nabopolassar was succeeded by his son, the biblical conqueror Nebuchadnezzar, and the Medes began their extensive conquest of the Iranian Plateau. They were eventually defeated around 550 BC by the Persian leader Cyrus, who went on to conquer Babylon in 539 BC. The fall of Nineveh in 612 BC can be seen as a turning point between the millennia that saw the old empires of Egypt, the Hittites, Babylon and Assyria rise, fall and rise again, and the arrival of new players on the world stage: these were the Persians and the Greeks, who also went on to establish extremely powerful entities that finally clashed.

▲ In the early 8th century BC waning Assyrian power allowed neighbouring kingdoms to prosper. The Urartians, centred in eastern Anatolia around Lake Van, greatly expanded their territory, notably to the south. They had adopted a number of ideas from the Assyrians – including the use of cuneiform writing – but they had their own distinctive culture and were skilled in working both bronze and iron.

In Babylonia the Chaldeans, an Amorite tribe, became prominent. The languishing Gulf trade revived under their auspices, and the resulting wealth and stability enabled Babylonian cultural life to continue, assuring the survival of Mesopotamian literary and scientific traditions.

Assyrian power grew once again in the late 8th century BC, and after gaining control of Babylonia and the Levant the empire was soon in conflict with Egypt. Assyria made a partially successful attack on Egypt in 671 BC, returning in 663 BC and attacking Memphis, prompting the Nubian ruler Taharqo to flee south to Thebes. Within just 40 years, however, Assyria itself was attacked and subdued by the Babylonians, who continued to rule in Mesopotamia until 539 BC, when Babylon fell to Cyrus of Persia.

◀ THE MEDITERRANEAN AND THE GULF REGION 2000–1000 BC *pages 36–37*   ▶ THE ACHAEMENID AND HELLENISTIC WORLD 600–30 BC *pages 42–43*

CLASSICAL GREECE
750–400 BC

▲ Greek art and architecture had a profound effect on the Romans. This Roman marble copy of Athena, goddess of war and wisdom, was based on a statue by the Greek sculptor Myron in the 5th century BC. The original would have been made of bronze using the "lost-wax" technique, a method that enabled the Greeks to portray the most lifelike of figures.

▼ During the 8th and 7th centuries BC the Greeks came to play a pivotal role in the growing Mediterranean trade. However, their ambitions also led to confrontations with rival merchant forces, notably the Phoenicians.

2 COLONIZATION AND TRADE 750–550 BC

- ○ Principal colony-founding city
- ● Colony established before 700 BC
- ● Colony established 700–600 BC
- ○ Colony established after 600 BC
- ◆ Phoenician colony
- — Principal trade route

Traded goods:
- △ copper
- ⊙ gold
- ◉ iron
- ▣ silver
- ◈ tin
- ⚙ metalwork
- ◉ grain
- ◊ oil
- ○ perfume
- ∾ slaves
- ⚚ timber
- ⬛ pottery
- ⚱ wine

M ore than 700 years after the fall of Mycenae (*pages 36–37*), a new civilization flourished in Greece. The cultural and political life of Greece, and particularly of Athens, in the 5th century BC was to have a profound impact on Western civilization. In Athens the principles of democracy were established and scientific and philosophical reasoning taken to unprecedented heights. The Athenian literary tradition – exemplified by the tragedies of Sophocles and the comedies of Aristophanes – formed a central part of its legacy. Also in Athens, architecture and forms of art such as sculpture and vase painting took on the Classical styles that still influence the Western sense of aesthetics.

The Greek landscape is dominated by the sea and by mountains, which cover 80 per cent of the mainland and reach heights of over 2,000 metres (6,000 feet) (*map 1*). Authors such as Plato glorified a past when the countryside was lush and densely wooded, but by the 1st millennium BC poor soil and the scarce rainfall during the summer months limited the possibilities for growing crops. Modern botanical and geological studies reveal a remarkable stability in the Greek countryside during the last 3–4,000 years, until the recent industrialization of agriculture. Today's farmers grow labour-intensive crops such as apricots and grapes in the valleys along the coast, cultivate cereals and olives on the less fertile mountain slopes, and use the mountain pastures as grazing land. It is likely that the ancient rural population of Greece practised a similar mixed agriculture, supplemented with marine resources.

THE GREEK CITY-STATES
Whereas the many islands in the Aegean Sea provide secure points for navigation and promote maritime traffic, cross-country communication is hindered by the mountains, which leave many areas isolated. In these mountain pockets independent, self-governing city-states, or *poleis*, developed during the 8th century BC. Their focal point was usually an urban centre positioned on a defensible rock: the *acropolis* (literally the "high town"). This functioned as the political, administrative and religious centre for the surrounding countryside. Some city-states expanded their influence and came to dominate; others remained on a more equal footing with neighbouring cities, with whom they acted as a federal unit in matters such as foreign policy. During the 8th century BC a sense of a Greek identity emerged, primarily based on language and religion – and expressed in the pan-Hellenic (all-Greek) festivals such as the Olympic Games and the shared oracles at Delphi and Dodona.

From around 750 BC food shortages, political unrest and trade interests prompted the Greeks to venture out and

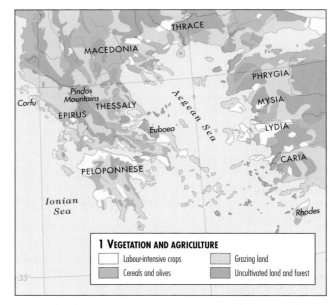

1 VEGETATION AND AGRICULTURE
- □ Labour-intensive crops
- ▨ Grazing land
- ▨ Cereals and olives
- ▨ Uncultivated land and forest

▲ Geography and natural resources set the parameters for the political and cultural development of Classical Greece. Often separated from each other by mountains, the city-states evolved independently, many of them relying on travel by sea. A lack of high-quality agricultural land further encouraged expansion overseas.

establish new city-states well away from home (*map 2*). These colonies retained the culture and religion of the mother cities, yet in a political sense functioned independently. The earliest colonies in Syria (Al Mina) and Italy (Ischia), founded by Eretria and Chalcis, were primarily trading posts, but the quest for arable land probably played a key role in the colonization of Sicily and the Black Sea area, mostly by Chalcis, Corinth and Miletus. While these trade connections and colonies were of great cultural significance, promoting an exchange between the eastern and western Mediterranean areas, they also led to major conflicts, for example with the Phoenicians (*pages 38–39*).

WAR WITH PERSIA
In the east the expansion of Persia's Achaemenid Empire (*pages 42–43*) led to confrontations with the Greek cities of Asia Minor (*map 3*). With the support of Athens and Eretria these cities rebelled against the Persian king Darius I in 499 BC, and the rebellions were not finally suppressed until 493 BC. Darius then demanded the submission of all the mainland Greek cities, but Athens and Sparta refused. In 492 BC Darius sent out a punitive mission, which backfired

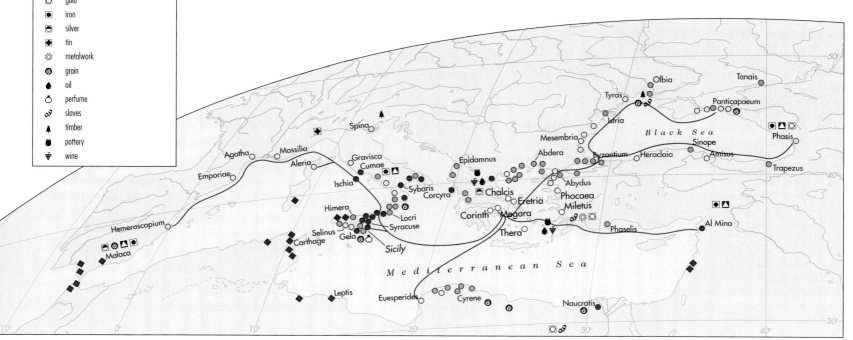

after most of the Persian fleet was lost in storms around Mount Athos. When Eretria was sacked in 490 BC Greece was divided on how to respond, but the Athenians and a small Plataean force took the initiative and defeated the Persians at Marathon that year. Infuriated, Darius's successor Xerxes prepared an even larger invasion, to which many of the Greek city-states responded by mounting their first united force, led by Sparta. The Athenian leader Themistocles interpreted the oracular pronouncement that they should rely on Athens's wooden walls to mean strengthening their navy, and he enlarged the fleet to 180 ships.

The first confrontation took place in 480 BC at Thermopylae, where the Spartan King Leonidas held out bravely but was defeated. After inflicting considerable losses on the Persian navy at Artemisium in 480 BC, the Athenians withdrew to the Bay of Salamis. They knew they could not defeat the Persians on land and so left their city to the enemy, who burned Athens to the ground. The huge Persian fleet followed the Athenian navy to Salamis but was unable to manoeuvre within the narrow straits there and was obliterated in 480 BC. The following year, at Plataea, the Persian land army suffered a similar fate at the hands of the Spartans, and the Greeks dealt the Persians the final blow in 479 BC at Mount Mycale, where the Persian troops had taken refuge. The small and independent Greek city-states had managed to defeat the greatest empire at that time.

ATHENS AND SPARTA

Athens gained tremendous prestige through its contributions to the victory over the Persians and, when Sparta declined, seemed the obvious leader of an anti-Persian pact. Although the main aims of this confederacy, the Delian League, were protection against the Persians and seeking compensation for the incurred losses, the Athenians soon used the alliance to build an empire. They imposed heavy tributes on their allies and punished revolts mercilessly. In 454 BC the Delian League's treasury was moved to Athens and funds were overtly channelled into the city's coffers. A grand building scheme was launched to restore the city, crowned by the construction of the Parthenon (477–438 BC) and the Erechtheum (421–406 BC). This was Athens's Golden Age, much of it masterminded by Pericles.

Sparta and other Greek cities watched the growth of Athens with suspicion. Not only did they fear Athens's military power, but they were also wary of democracy, Athens's radical contribution to political innovation. This rule of the people (women, slaves and foreigners excepted) was perceived as posing a direct threat to Sparta's ruling upper classes and, after mounting tension, war broke out in 431 BC (map 4). It was a costly conflict: Attica's countryside was sacked annually and the population, withdrawn within the city's walls, suffered famine and plague that killed a quarter of its number, including Pericles. The Peloponnesian War lasted 27 years, ending with Athens's downfall in 404 BC.

3 THE PERSIAN WARS 492–479 BC

- → Movement of Persian fleet 490 BC
- → Movement of Persian fleet and army 480 BC
- ✕ Major battle 492–479 BC
- Persian Empire 493 BC
- Persian vassal state 492 BC
- Neutral area
- Area at war with Persia

▲ The Persian kings Darius I and Xerxes planned three invasions in their attempts to subdue mainland Greece. While the first failed in 492 BC, the second and third (490 and 480 BC) posed such a serious threat that Greece responded as a united force.

▼ The unity displayed by Greece during the Persian Wars was short-lived. Athenian imperialist policy led to war with Sparta and its Peloponnesian allies – described by the historian Thucydides as the most appalling of all the Greek wars in losses and suffering.

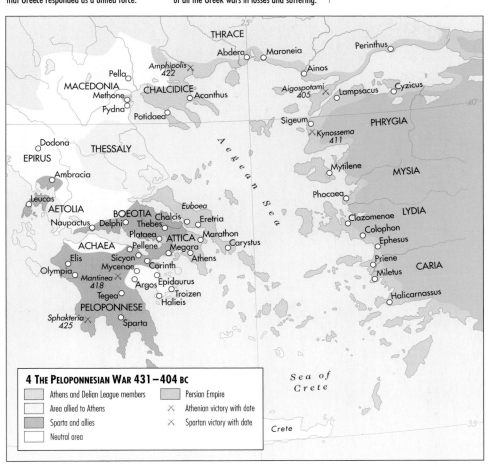

4 THE PELOPONNESIAN WAR 431–404 BC

- Athens and Delian League members
- Area allied to Athens
- Sparta and allies
- Neutral area
- Persian Empire
- ✕ Athenian victory with date
- ✕ Spartan victory with date

▲ The Greeks exported their political and social ideas alongside their art, and various colonies around the northern shores of the Mediterranean are still littered with temples, theatres, gymnasia and *agoras*, or marketplaces. The remains of this late 5th-century temple are at Segesta in Sicily – a focal point for Greek trade. Its columns are in the simple Doric style, first of the three major orders of Classical architecture; the progressively more complex and ornate Ionic and Corinthian styles followed later.

◀ THE MEDITERRANEAN AND THE GULF REGION 2000–1000 BC *pages 36–37* ▶ THE ACHAEMENID AND HELLENISTIC WORLD 600–30 BC *pages 42–43*

THE ACHAEMENID AND HELLENISTIC WORLD
600–30 BC

Following the fall of the Assyrian capital, Nineveh, in 612 BC, the former Assyrian Empire was divided between the Babylonians and the Medes, with a small corner of the extensive new Median territory occupied by a dependent related Indo-Iranian tribe, the Persians. In 550 BC the Persian King Cyrus, of the Achaemenid family, rose against his overlord and occupied the Median territory. Learning of this, King Croesus of Lydia (a country rich in goldmines) saw an opportunity to enlarge his empire to the east. He consulted the Delphic oracle, which prophesied that he would destroy a great kingdom and, confident of his success, Croesus faced Cyrus at Hattusas. The battle ended in stalemate, however, and Croesus retreated to Sardis, followed by Cyrus, who besieged the city until Croesus's surrender in 547 BC – when Croesus realized that the kingdom whose destruction the oracle had referred to was his own.

The Persian Achaemenid Empire (*map 1*) now encompassed the Lydian territory, including the Greek cities on the coast of Asia Minor which Croesus had annexed in 585 BC. In 539 BC Cyrus also conquered Babylon. He was said to have been a just ruler who allowed his subjects religious freedom and did not impose excessively harsh taxes.

THE PERSIAN SATRAPIES

In 530 BC Cyrus was killed on campaign and was succeeded by his son Cambyses, whose greatest military feat was the annexation of Egypt in 525 BC. After Cambyses and his brother mysteriously died, Darius I (a cousin of Achaemenid descent) came to the throne in 521 BC. Rather than accepting the existing administrative structures as his predecessors had done, Darius organized the empire into 20 provinces or "satrapies", each ruled by one of his relatives. To ensure efficient government he created a road network and installed a regular system of taxation based on the gold Daric coin.

Darius added the Indus province to the empire and brought Thrace under Persian rule in 512 BC, but his attack on the Scythians in the Danube area was unsuccessful. Darius suffered another setback in 499 BC, when Cyprus

▼ On his succession in 359 BC Philip II was master of a tiny kingdom, yet he transformed the Macedonian army into a formidable fighting machine – increasing the numbers of aristocratic cavalry, introducing the heavy infantry phalanx armed with *sarissas* (long pikes), and mounting sieges of unprecedented efficiency. By his death in 336 BC Macedonia was a major power, dominating Greece and threatening the Persian Achaemenid Empire. His son Alexander, charismatic leader and military genius, inherited Philip's ambitions as well as his army, and he conquered not only the Persian Empire but also lands well beyond. However, his attempts to weld his vast conquests into a unified empire under combined Macedonian and local rulers ended with his early death in Babylon at the age of 32.

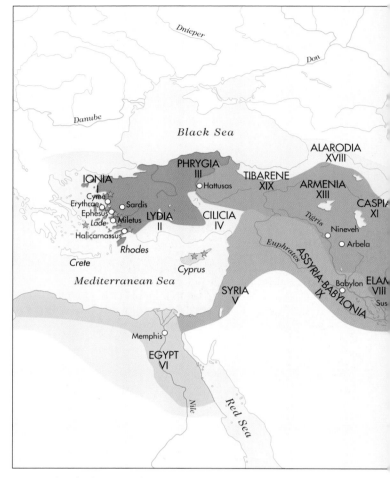

▲ Persian rule combined an empire-wide legal and administrative system with an acceptance of local customs, practices and religions. Trade prospered under the Achaemenids, facilitated by the efficient road network, a standardized system of weights and measures, and the innovative use of coinage. Sophisticated irrigation works using underground watercourses and canals increased agricultural productivity.

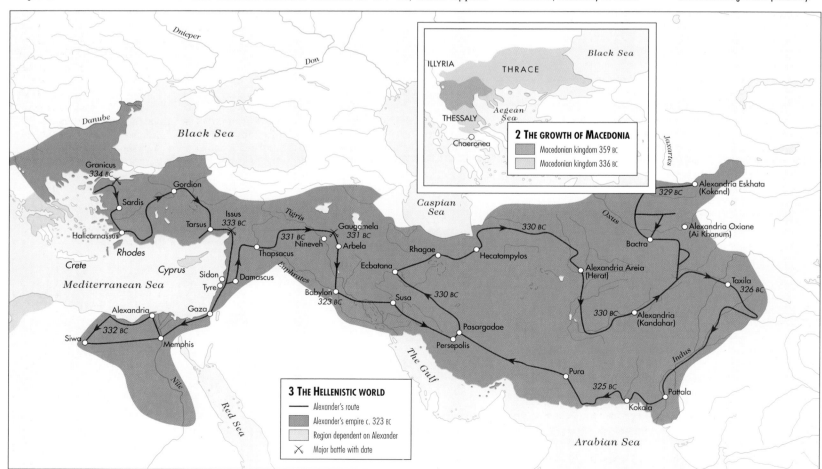

2 THE GROWTH OF MACEDONIA
- Macedonian kingdom 359 BC
- Macedonian kingdom 336 BC

3 THE HELLENISTIC WORLD
- Alexander's route
- Alexander's empire c. 323 BC
- Region dependent on Alexander
- ✕ Major battle with date

1 THE EXPANSION OF THE ACHAEMENID EMPIRE

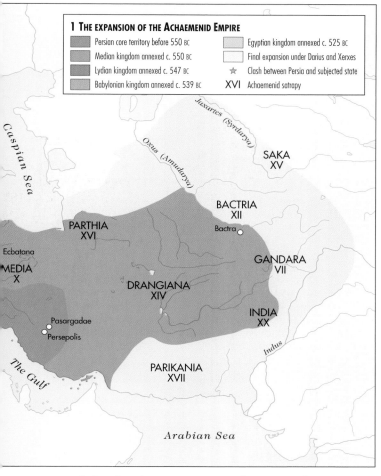

- Persian core territory before 550 BC
- Median kingdom annexed c. 550 BC
- Lydian kingdom annexed c. 547 BC
- Babylonian kingdom annexed c. 539 BC
- Egyptian kingdom annexed c. 525 BC
- Final expansion under Darius and Xerxes
- ★ Clash between Persia and subjected state
- XVI Achaemenid satrapy

◄ Alexander's army met the Persian forces of Darius III at Issus in 333 BC — and scored a victory that both heralded his conquest of southwest Asia and signalled the beginning of the end for the 220-year-old Achaemenid dynasty, rulers of the first Persian empire. This graphic detail, modelled on a 4th-century BC Hellenistic painting — commissioned by Alexander's own generals — is taken from the mosaic at the House of the Faun in Pompeii. It was created in the late 2nd or early 1st century BC — clear evidence of Alexander's enduring reputation among the Romans.

former Persian Empire, and he crossed the River Indus in 326 BC; he hoped to proceed to the River Ganges, regarded as the eastern limit of the inhabited world, but was stopped by mutiny in his tired army. Instead he subdued the tribes along the River Indus and returned to Babylon, where he died in 323 BC of fever, exhaustion or possibly poison.

Alexander the Great had forged an empire which stretched from Greece to the River Indus (*map 3*) and which merged Greek and Oriental cultures. Greek became the common language, and Greek gods were venerated side by side with local deities. Both Macedonians and Persians ruled as satraps, and Alexander encouraged his generals to marry Persian women, as he himself had done. He founded 70 new cities, many called Alexandria, which acted as military but also cultural centres of the new cosmopolitan society. Alexander's success was rooted in his prowess as a military leader, a role in which he displayed great personal courage, and in clever propaganda, such as the construction of a myth proclaiming his divinity – a belief which he himself seemed to share.

ALEXANDER'S SUCCESSORS

After Alexander's death a long power struggle ensued between his generals, the so-called "War of the Diadochi" (successors). The main contenders were Antigonus of Phrygia, Seleucus of Babylonia, Ptolemy of Egypt, and Antipatros, in charge of Macedonia and Greece. Macedonia, generally regarded as the seat of legitimate rule, became the centre of continuous conflict. After the murder of Alexander's son by Cassander, son of Antipatros, the various successors all proclaimed themselves kings between 306 and 303 BC (*map 4*).

While this marked the definite end of Alexander's empire, the war was not yet over: after renewed hostilities three kingdoms (later called the Hellenistic Kingdoms) were securely established by 275 BC. The Antigonids ruled in Macedonia, the Seleucids in Syria and the Ptolemies in Egypt, but their reigns ended when the Romans captured their territories (in 148, 64 and 30 BC respectively). Meanwhile the successors of Chandragupta – who, after Alexander's death, had founded the Mauryan Empire and taken control of the Punjab region – remained in power until approximately 186 BC (*pages 46–47*).

and the Greek city-states on the coast of Asia Minor revolted. Although Cyprus was swiftly brought back under Persian rule, the Greek rebellion persisted until 493 BC. The missions sent by Darius and his successor Xerxes to punish the mainland Greeks for their support ended in Persian defeats in 490, 480 and 479 BC (*pages 40–41*). The rest of the empire remained intact until it was conquered by Alexander the Great.

MACEDONIAN EXPANSION

When Darius invaded Thrace, Macedonia had little choice but to become a Persian vassal, and it remained a marginal state on the international political scene until Philip II ascended the Macedonian throne in 359 BC. Philip forged a professional army, unified Macedonia and, having gained control of Thessaly, expanded into Illyria and Thrace, bringing important harbours and goldmines into the empire.

His expansion (*map 2*) met with hostility from Athens and Thebes, whose military power had greatly diminished during the Peloponnesian War. After his victory over a combined Theban–Athenian army at Chaeronea in 338 BC, Philip was the undisputed master of Greece until his assassination in 336 BC – just as he was preparing to invade Persia. His 20-year-old son Alexander III succeeded him, and after crushing opposition to his reign in Macedonia he joined the remainder of his father's army in Persian territory. Having defeated the army of the Persian satraps at Granicus in 334 BC, Alexander faced Darius III (r. 335–330 BC) at Issus in 333 BC. On a narrow coastal plain he dealt the Persians a devastating defeat and captured Darius's family.

He then conquered Syria, Egypt and Mesopotamia before confronting Darius again in 331 BC on the plains of the Tigris near Arbela. After a long battle, Darius fled and Alexander moved on to sack Persepolis in retribution for the destruction of Athens in the Persian Wars some 150 years earlier.

In the east, Alexander's self-proclaimed status as King of Asia was threatened by rebel satraps. However, in 327 BC he crushed remaining opposition in eastern Iran and Afghanistan, before invading northern India. His ambition had now shifted to expanding beyond the boundaries of the

▼ Throughout the lands of Alexander's short-lived empire, Greek culture blossomed under Hellenistic rule, usually enriched by indigenous cultures; even in India, at the very limit of Alexander's conquests, it had a lasting effect. Developments in astronomy, medicine, mathematics and engineering took place alongside patronage of the arts, the building of libraries and the encouragement of education. With the Roman Empire acting as intermediary, these achievements laid the basis for a later European civilization.

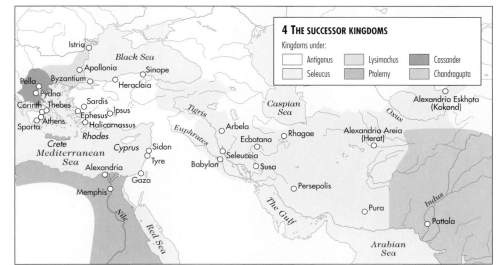

4 THE SUCCESSOR KINGDOMS

Kingdoms under:
- Antigonus
- Seleucus
- Lysimachus
- Ptolemy
- Cassander
- Chandragupta

◀ THE MEDITERRANEAN AND THE GULF REGION 2000–1000 BC *pages 36–37* ▶ THE ROMAN EMPIRE 500 BC–AD 400 *pages 54–55*

THE BIRTH OF WORLD RELIGIONS
1500 BC–AD 600

By 600 AD a series of major religions had spread throughout Eurasia (*map 1*). Distinguished from other, more local beliefs by a focus on holy writings, or scriptures, most of them continue to flourish today.

The oldest religion is Hinduism. Its sacred writings, the *Vedas*, were first compiled by seers and priests, or *rishis*, and were based on myths, legends and hymns passed down from antiquity. Many of the beliefs and rituals of Hinduism had their origins in the sacrificial cults introduced to India by the Indo-Aryans from around 1500 BC, while others were indigenous and can be traced back to the Indus civilization (*pages 28–29*); indeed it derives its name from the river.

Central to Hinduism are a belief in the transmigration of souls, the worship of many deities (who eventually came to be seen as aspects of one god), the religious sanction of strict social stratification, the caste system, and the ability to assimilate rather than exclude different religious beliefs. Unlike most of the later major religions, Hinduism never really spread beyond the bounds of its home country, although it was very influential in some of the early states of Southeast Asia (*pages 64–65*).

THE SPREAD OF BUDDHISM

Siddhartha Gautama (c. 563–483 BC), the founder of Buddhism, was born a wealthy prince in northeastern India (*map 2*). Renouncing worldly trappings and achieving enlightenment, or *nirvana*, he became known as the Buddha (the Enlightened). Gautama lived at a time of great religious ferment in India, and Buddhism was one of a number of sects that aimed to reform Hinduism. Another, more extreme, reform movement was Jainism, whose asceticism was a reaction to the rigid ritualism of Hinduism.

Buddhism shared with Hinduism the belief in the cycle of rebirth, but differed in the way in which escape from the cycle could be achieved. Indeed the appearance of Buddhism stimulated a resurgence in Hinduism, which may be why Buddhism failed to take a permanent hold in India.

▼ Several founders of world religions – notably Buddha, Confucius, Zoroaster and Christ – lived in the 1st millennium BC or immediately after it. Judaism and Hinduism had their roots in earlier times, when many peoples worshipped local gods.

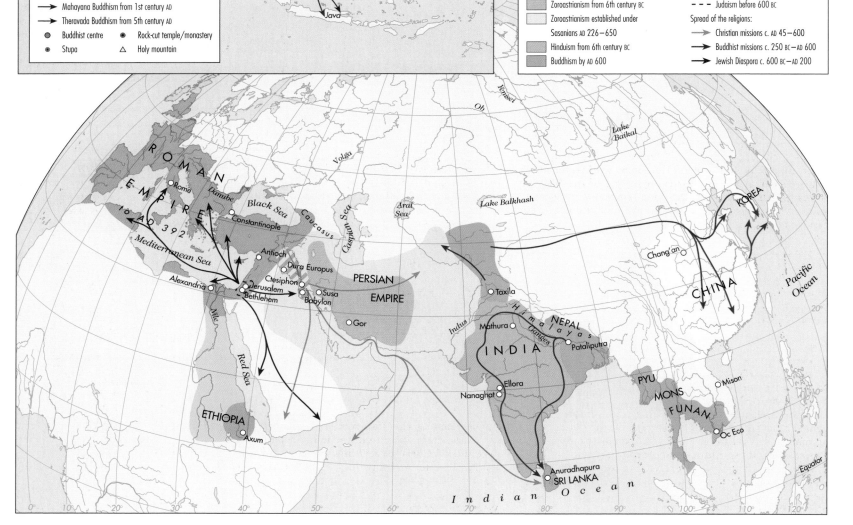

2 THE SPREAD OF BUDDHISM TO AD 600

- Original core area of Buddhism 6th century BC

Spread of:
- → Buddhism by 1st century AD
- → Mahayana Buddhism from 1st century AD
- → Theravada Buddhism from 5th century AD
- ● Buddhist centre
- ● Rock-cut temple/monastery
- ✴ Stupa
- △ Holy mountain

1 WORLD RELIGIONS TO AD 600

- Christianity by AD 392
- Christianity established AD 392–600
- Zoroastrianism from 6th century BC
- Zoroastrianism established under Sasanians AD 226–650
- Hinduism from 6th century BC
- Buddhism by AD 600
- Confucianism and Daoism 3rd century BC to 1st century AD
- - - - Judaism before 600 BC

Spread of the religions:
- → Christian missions c. AD 45–600
- → Buddhist missions c. 250 BC–AD 600
- → Jewish Diaspora c. 600 BC–AD 200

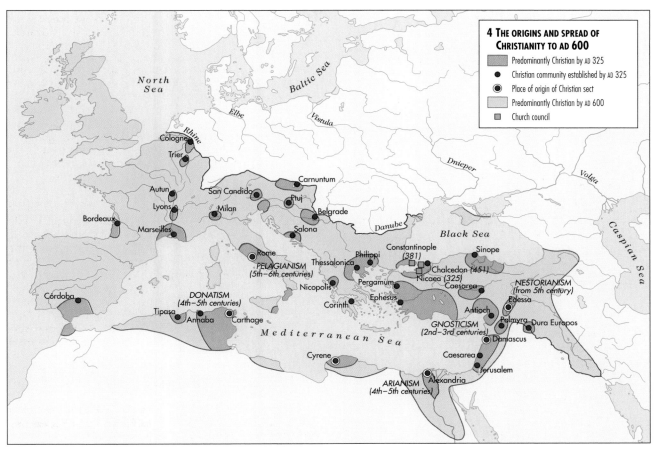

4 THE ORIGINS AND SPREAD OF CHRISTIANITY TO AD 600
- Predominantly Christian by AD 325
- ● Christian community established by AD 325
- ◉ Place of origin of Christian sect
- Predominantly Christian by AD 600
- ■ Church council

◄ Early Christians were often persecuted by the Romans, who saw them as a threat to the stability of the empire because they refused to acknowledge the divinity of the Roman emperor. By AD 64 Nero used Christians as victims in the imperial arenas, and in the early 4th century Diocletian organized campaigns against them. However, Diocletian's successor Constantine legalized Christianity, and at the first "Ecumenical Council" (held at Nicaea in 325) he brought church and state together. Constantine had converted to Christianity after a key victory over his rivals in 312, a victory he ascribed to the power invested in him as the servant of the Highest Divinity, which he equated with the Christian god.

Many sects emerged during this early spread of Christianity, and councils were periodically held to discuss the doctrinal disagreements raised – with some sects declared heretical as a result.

Buddhism was given official backing by the Mauryan Emperor Ashoka (r. 272–231 BC), and Buddhist monuments, such as the great *stupa* at Sanchi, were built. Over the following centuries Buddhism – with its emphasis on overcoming suffering and breaking out of the endless cycle of rebirth through discipline, meditation, good works and the banishing of desire – spread throughout much of Asia, reaching Japan in the 6th century AD. Great Buddhist centres, based around religious communities, developed.

CONFUCIANISM AND DAOISM
Two philosophical traditions were dominant in China when Buddhist monks arrived there in the 4th century AD. Confucianism, named after the author of the Classics, Kongzi, or Confucius (551–479 BC), propounded a set of morals encouraging a way of life ruled by the principles of order, hierarchy and respect. Confucius worked for much of his career as an administrator in one of the Warring States (*pages 48–49*), and his ideas subsequently greatly influenced political philosophy in China and many other parts of East Asia.

The other tradition, Daoism, or "the Way", called for people to find ways of being in harmony with the world. It was based on the teachings of the philosopher Lao-tze, written down in the *Dao De Jing* (probably in the 3rd century BC). In its combination of cosmology and the sanctification of nature, certain mountains were considered especially sacred and became the focus of worship.

ZOROASTRIANISM AND JUDAISM
In West Asia a new religion developed out of the ancient Indo-Iranian belief systems during the 1st millennium BC. Zarathrustra, known to the Greek world as Zoroaster, lived in Persia, probably during the 10th century BC, though some date him from 628 to 551 BC. Zoroastrianism, the religion named after him, had a major impact on the development of many other religious traditions, including Judaism and Christianity. Its scriptures, the *Avesta*, set out the Zoroastrian belief that life is a constant struggle between good and evil. Zoroaster rejected the pantheism of the Indo-Iranian religions and proclaimed one of the ancient deities, Ahura Mazda (the "Wise Lord") as the one supreme god.

Zoroaster believed that the end of the world was imminent, and that only the righteous would survive the great conflagration to share in the new creation.

Following the death of Zoroaster his teachings spread throughout the Persian Achaemenid Empire of 550–330 BC (*pages 42–43*) until the conquests of Alexander displaced Zoroastrianism with Hellenistic beliefs. Renewed interest in Zoroastrianism developed towards the end of the Parthian Empire (238 BC–AD 224), and it was taken up as the official religion of the Sasanian Empire, where it flourished until the arrival of Islam in the 7th century.

Zoroastrianism had considerable influence on the development of Judaism (*map 3*), which had originated with the people of Abraham – nomad groups living in the northern Arabian Desert in the 2nd millennium BC. Jewish tradition holds that these Hebrew people spent time in slavery in pharaonic Egypt before leaving under the leadership of Moses around 1250 BC. They settled in Canaan and fought with the local inhabitants, particularly the Philistines, until peace was achieved under King David around 1000 BC.

Jewish communities were established in Egypt in the 2nd century BC, in Italy from the 1st century AD, in Spain by AD 200 and in Germany by AD 300. The teachings of Judaism form the Old Testament of the Bible; in addition, Jewish law is recorded in the *Talmud*, the first codification being the *Mishnah*, written down about AD 200.

THE RISE OF CHRISTIANITY
Named after its founding figure, Jesus Christ (c. 4 BC–AD 29), Christianity (*map 4*) developed from Judaic roots. Christians believe in one God and that Jesus, born in Nazareth, is the Son of God – the Messiah whose arrival on Earth had long been promised in the Jewish tradition. Jesus's radical teachings and disregard for the establishment led to his death by crucifixion, an event Christians believe he overcame in the Resurrection. In the first few centuries AD Christianity was introduced in many parts of the Roman world, and Christ's teachings (written down in the New Testament) were spread by apostolic figures such as Paul of Tarsus. By 600 it had spread from its origins in the eastern Mediterranean as far as the western shores of the Caspian Sea in the east and the British Isles in the northwest.

▼ After the death of David's son Solomon in 926 BC, the Jewish lands were divided into the kingdoms of Israel and Judah, which then had a turbulent history of division and conquest by Assyria, Babylonia and, lastly, by Rome. Between AD 66 and 73 rebellion against Roman rule broke out, but the empire reconquered Jerusalem in 70, destroying the Jewish temple. Following a long siege at Masada the last of the rebels were crushed in 73, and after a second revolt was brutally put down (132–35) many Jews left Judah (called Judaea by the Romans).

3 THE HOLY LAND
- David's kingdom c. 1000 BC
- ■ Capital city after division of kingdom 9th century
- ○ Philistine city

FIRST EMPIRES IN INDIA
600 BC–AD 500

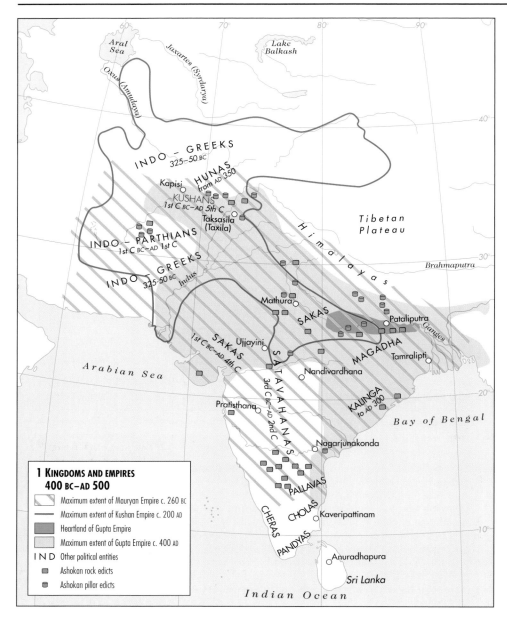

1 KINGDOMS AND EMPIRES
400 BC–AD 500

Maximum extent of Mauryan Empire c. 260 BC

Maximum extent of Kushan Empire c. 200 AD

Heartland of Gupta Empire

Maximum extent of Gupta Empire c. 400 AD

I N D   Other political entities

Ashokan rock edicts

Ashokan pillar edicts

▲ By the 6th century BC prosperous states in the Ganges Valley were competing for dominance, expanding not only by military conquest but also through dynastic marriages and political alliances – a trend that set the pattern for the rise and fall of states in subsequent centuries. Strong rulers such as the early Mauryas and the Guptas succeeded in uniting large areas to form empires, but weak successors were unable to hold them together.

► Despite their diverse origins and different political histories, the invaders of the subcontinent followed a common pattern. Each group introduced new cultural elements – seen, for example, in art styles influenced by the Hellenistic world – but far more marked was their "Indianization". Most of them readily adopted Indian culture, settling in towns such as Taksasila (Taxila) or Mathura, converting to Buddhism or other Indian religions, patronizing art and architecture, profiting from South Asia's flourishing international trade, and on the whole becoming socially assimilated.

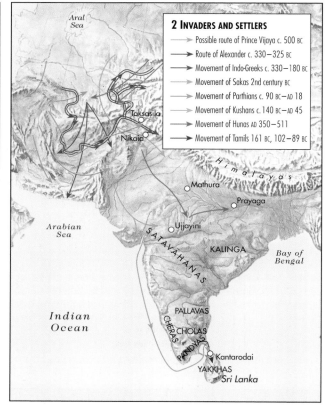

2 INVADERS AND SETTLERS

Possible route of Prince Vijaya c. 500 BC

Route of Alexander c. 330–325 BC

Movement of Indo-Greeks c. 330–180 BC

Movement of Sakas 2nd century BC

Movement of Parthians c. 90 BC–AD 18

Movement of Kushans c. 140 BC–AD 45

Movement of Hunas AD 350–511

Movement of Tamils 161 BC, 102–89 BC

During the 2nd millennium BC Indo-Aryan nomads were the first of many groups from Iran or Central Asia to invade the Indian subcontinent. Initially they spread only into the Ganges Valley, but according to legend (given support by recent archaeological work), around 500 BC a group led by Prince Vijaya also gained control of Sri Lanka. In 530 BC the Persians conquered the northwest, but the area subsequently fell to Alexander the Great (*pages 42–43*) and the Indo-Greek kingdoms that emerged after his death dominated the region for several centuries. However, neither Persians nor Greeks ever penetrated deeper into the subcontinent, due to the strength of native dynasties.

KINGDOMS AND EMPIRES

By 500 BC kingdoms existed throughout the Ganges region. Chief among these was Magadha, favourably located for control both of riverborne trade and of the sources of raw materials such as iron. Magadha gradually expanded at the expense of its neighbours and before 297 BC its king, Chandragupta Maurya, ruled most of north India (*map 1*). His grandson Ashoka (r. 272–231 BC) further extended the empire, conquering Kalinga in 261 BC, and only the extreme south retained its independence. Pillar and rock edicts mark the extent of Mauryan political authority: these proclaimed Ashoka's ethical code of social responsibility and toleration. It was an age of peace and prosperity.

The political unity of the Mauryan Empire did not long survive Ashoka's death in about 231 BC. Numerous independent kingdoms emerged, such as the Satavahana realms in western India, but none was strong enough to resist the waves of foreign invaders (*map 2*). The Sakas, arriving from Central Asia around 130 BC, gradually gained control of much of the north and west. They were succeeded by the Parthians from the Iranian Plateau and the Central Asian Kushans, who loosely united the Ganges Valley and the northwest until the mid-3rd century AD. From the 5th century AD onwards, the north was prey to attacks by the ferocious Hunas (White Huns) who swept in from the east.

By the time they reached the Ganges Valley or the Deccan, the force of foreign invasions was spent, and Sri Lanka and the south were generally spared. Instead they suffered periodic attacks by native groups such as the Mauryans, Tamils and Guptas. In the 4th century AD the Guptas, who ruled a small kingdom in the Ganges region, began to expand, gaining control of adjacent regions through military conquest, diplomacy and dynastic marriages. Unlike the earlier Mauryan Empire, however, they established only indirect political authority over much of this area, local rulers usually acting under their suzerainty.

RURAL AND URBAN DEVELOPMENT

Much of the subcontinent, such as the jungle regions, was unsuited to agriculture and was inhabited by hunter-gatherers. In addition to the wild produce they collected for their own needs they obtained materials for settled farmers, such as honey, venison and lac (used for lacquer), exchanging these for cultivated foodstuffs and manufactured goods.

Throughout this period the majority of South Asians dwelt in villages. Rice was the main staple in the east and Sri Lanka, millet in the south and wheat in the north; animals, particularly cattle, were kept. By around 500 BC irrigation works such as canals, dams and tanks were being constructed to increase agricultural productivity. Rulers – particularly the Mauryas, who exercised strong centralized control over their realms – also encouraged the cultivation of wasteland, often by the forced resettlement of groups of low-caste cultivators. In Sri Lanka sophisticated hydraulic engineering developed from around 300 BC, using sluice pits and long canals. Land taxes and levies on produce provided the main income for states throughout the period, although trade also yielded considerable revenues.

Many towns and cities developed as centres of trade and industry, and they flourished even during periods of weak political control (*map 3*). Many, especially in the west and south, were ports for seaborne trade. They contained

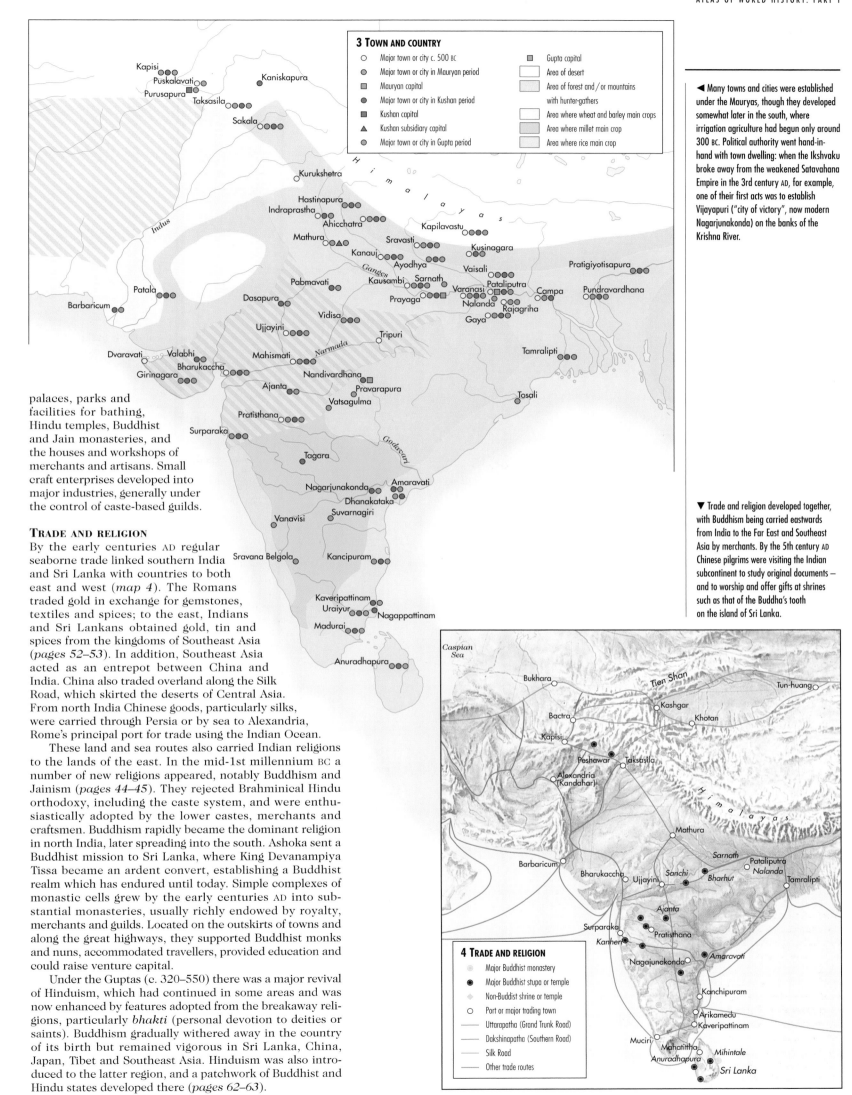

3 TOWN AND COUNTRY

○ Major town or city c. 500 BC
● Major town or city in Mauryan period
■ Mauryan capital
● Major town or city in Kushan period
■ Kushan capital
▲ Kushan subsidiary capital
● Major town or city in Gupta period
■ Gupta capital
☐ Area of desert
☐ Area of forest and / or mountains with hunter-gathers
☐ Area where wheat and barley main crops
☐ Area where millet main crop
☐ Area where rice main crop

◄ Many towns and cities were established under the Mauryas, though they developed somewhat later in the south, where irrigation agriculture had begun only around 300 BC. Political authority went hand-in-hand with town dwelling: when the Ikshvaku broke away from the weakened Satavahana Empire in the 3rd century AD, for example, one of their first acts was to establish Vijayapuri ("city of victory", now modern Nagarjunakonda) on the banks of the Krishna River.

▼ Trade and religion developed together, with Buddhism being carried eastwards from India to the Far East and Southeast Asia by merchants. By the 5th century AD Chinese pilgrims were visiting the Indian subcontinent to study original documents — and to worship and offer gifts at shrines such as that of the Buddha's tooth on the island of Sri Lanka.

palaces, parks and facilities for bathing, Hindu temples, Buddhist and Jain monasteries, and the houses and workshops of merchants and artisans. Small craft enterprises developed into major industries, generally under the control of caste-based guilds.

TRADE AND RELIGION

By the early centuries AD regular seaborne trade linked southern India and Sri Lanka with countries to both east and west (*map 4*). The Romans traded gold in exchange for gemstones, textiles and spices; to the east, Indians and Sri Lankans obtained gold, tin and spices from the kingdoms of Southeast Asia (*pages 52–53*). In addition, Southeast Asia acted as an entrepot between China and India. China also traded overland along the Silk Road, which skirted the deserts of Central Asia. From north India Chinese goods, particularly silks, were carried through Persia or by sea to Alexandria, Rome's principal port for trade using the Indian Ocean.

These land and sea routes also carried Indian religions to the lands of the east. In the mid-1st millennium BC a number of new religions appeared, notably Buddhism and Jainism (*pages 44–45*). They rejected Brahminical Hindu orthodoxy, including the caste system, and were enthusiastically adopted by the lower castes, merchants and craftsmen. Buddhism rapidly became the dominant religion in north India, later spreading into the south. Ashoka sent a Buddhist mission to Sri Lanka, where King Devanampiya Tissa became an ardent convert, establishing a Buddhist realm which has endured until today. Simple complexes of monastic cells grew by the early centuries AD into substantial monasteries, usually richly endowed by royalty, merchants and guilds. Located on the outskirts of towns and along the great highways, they supported Buddhist monks and nuns, accommodated travellers, provided education and could raise venture capital.

Under the Guptas (c. 320–550) there was a major revival of Hinduism, which had continued in some areas and was now enhanced by features adopted from the breakaway religions, particularly *bhakti* (personal devotion to deities or saints). Buddhism gradually withered away in the country of its birth but remained vigorous in Sri Lanka, China, Japan, Tibet and Southeast Asia. Hinduism was also introduced to the latter region, and a patchwork of Buddhist and Hindu states developed there (*pages 62–63*).

4 TRADE AND RELIGION

● Major Buddhist monastery
● Major Buddhist stupa or temple
◆ Non-Buddist shrine or temple
○ Port or major trading town
— Uttarapatha (Grand Trunk Road)
— Dakshinapatha (Southern Road)
— Silk Road
— Other trade routes

FIRST EMPIRES IN CHINA
1100 BC–AD 220

► In the 8th century BC regional entities began to assert their independence from the Zhou state, fighting among themselves for dominance as well as fending off attacks from barbarian neighbours. By the late 5th century power was concentrated in seven principal states – Han, Wei, Zhao, Qin, Chu, Yan and Qi. They all built enormous walls to protect their borders, fortified their cities and even their villages, and constructed roads and canals to expedite the movement of troops and supplies. As military technology and the science of warfare flourished, the organization, weaponry and ferocity of the Qin army combined to give them superiority over the other Warring States, and in 221 BC the Qin united the whole area to form the first Chinese empire.

▼ The conquests in Central Asia of the Han emperor Wu Di and his embassies to the west opened up a major trade route linking East and West. Merchant caravans took Chinese goods (especially silk) as far as the Roman Empire in exchange for Western luxury goods. Well-preserved documents from northwestern China and along this "Silk Road" record the everyday life in garrison towns.

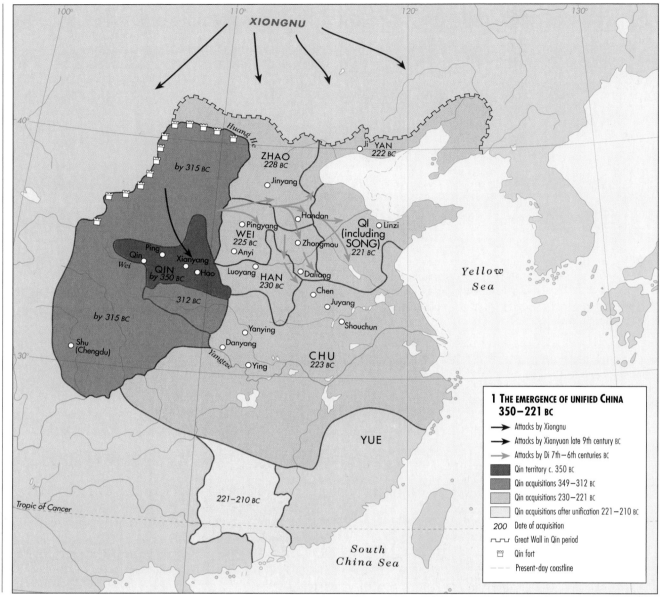

1 THE EMERGENCE OF UNIFIED CHINA 350–221 BC

→ Attacks by Xiongnu
→ Attacks by Xianyuan late 9th century BC
→ Attacks by Di 7th–6th centuries BC
▓ Qin territory c. 350 BC
▓ Qin acquisitions 349–312 BC
░ Qin acquisitions 230–221 BC
░ Qin acquisitions after unification 221–210 BC
200 Date of acquisition
⊓⊔ Great Wall in Qin period
⊞ Qin fort
--- Present-day coastline

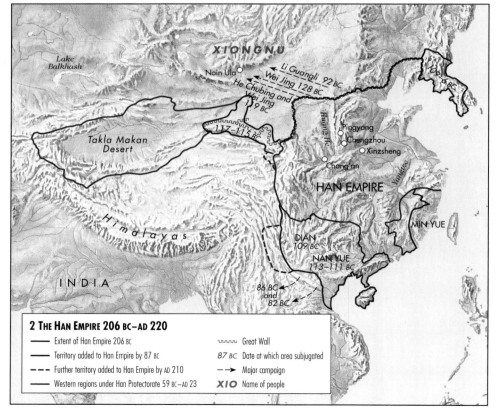

2 THE HAN EMPIRE 206 BC–AD 220

— Extent of Han Empire 206 BC
— Territory added to Han Empire by 87 BC
--- Further territory added to Han Empire by AD 210
— Western regions under Han Protectorate 59 BC–AD 23
〜〜 Great Wall
87 BC Date at which area subjugated
→ Major campaign
XIO Name of people

In the period between the victory of the Zhou king Wu over the Shang in the mid-11th century BC and the downfall of the last Han emperor, Xian Di, in AD 220, China underwent a series of political, economic and philosophical transformations that were to lay the foundations for Chinese government and society until the 20th century.

THE FIRST CHINESE DYNASTIES

The Zhou, possibly descended from nomads, established their royal capital at Hao in their ancestral heartland in the Wei River valley. For 250 years Zhou rulers held sway over a unified domain, their rule legitimated by the Mandate of Heaven – the divine right to rule China – which they claimed to have inherited from the Shang. Long inscriptions on fine bronze vessels record their achievements. By 770 BC, however, the empire had begun to fragment, and under pressure from barbarian tribes to the northwest the Zhou capital was moved east to Luoyang. Despite the continued claim of Zhou kings to the Mandate of Heaven, real power slipped away to a multitude of regional states.

By 403 BC seven major "Warring States" were competing for control of China (*map 1*). Through a series of tactical victories beginning in 280 BC, and under King Zheng from 246 BC, the state of Qin achieved supremacy by 221 BC. Zheng had reformed Qin, replacing the old kinship-based government with an efficient bureaucratic state. Proclaiming himself Shi Huang Di, "the First Emperor", he established his new capital at Xianyang. Despite an early death in 210 BC, he left a legacy that paved the way for Liu Bang, the founder of the Han dynasty four years later, to

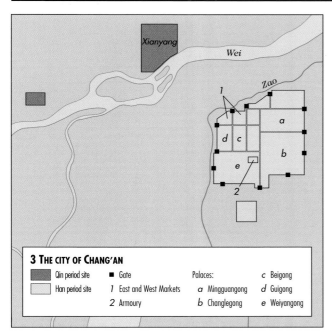

3 THE CITY OF CHANG'AN

		Palaces:	c Beigong
■ Qin period site	■ Gate		
▫ Han period site	1 East and West Markets	a Mingguangong	d Guigong
	2 Armoury	b Changlegong	e Weiyangong

▲ Chang'an, the capital of the Han from 206 BC to AD 23, had a population of about 250,000. Famed for its towers, it boasted wide boulevards, immense walls and gates, religious buildings, palaces and royal pleasure gardens. Its great markets were at the centre of a network of trading emporia that stretched across the empire.

build the Han Empire (*map 2*). Liu Bang and his descendants ruled China from 206 BC to AD 220, with a brief interruption during Wang Mang's Xin dynasty (AD 6–23).

MOVEMENTS OF POPULATION

By AD 2, the date of the first national census, China had a recorded population of 57 million. This huge number was often mobilized for warfare or vast public works, and in the reign of Wu Di (141–87 BC), the "Martial Emperor" who greatly expanded the territories of the empire, some two million people were resettled in colonies in the north and northwest. However, the later part of the Han dynasty saw a major movement of population southwards – a process that was precipitated by a major shift in the course of the Huang He River between AD 2 and 11 that left much of north China, traditionally the centre of power, depopulated.

THE ART OF WAR IN EARLY CHINA

These mass population movements occurred in a country unified through major developments in the art of war. Under the warlords of the Warring States, both individual gallantry and mass brutality were displayed, and armies became professional. From the 6th century BC new weapons, notably iron swords and armour, had replaced the traditional bronze halberds. Cavalry outmanoeuvred chariots on the battlefield and the new cities became targets for siege warfare. The Zhao stronghold of Jinyang was besieged for a year before the attackers turned on each other in a classic piece of Warring States treachery. From the 5th century BC the states built pounded-earth walls along their frontiers.

While earlier rulers either mounted expeditions against the nomadic "barbarian non-Chinese" or were harassed by them, the Qin and Han were aggressively expansionist. To keep the nomads out of his new empire, Shi Huang Di joined the sections of walled defences earlier states had built, thus creating the Great Wall. The Xiongnu, among the most aggressive of the Central Asian peoples (*pages 50–51, 52–53*), were particularly troublesome for the early Chinese empires, and the Han emperor Wu Di's constant search for allies against them created new links with the middle of the continent. The nomads often had to be bought off as much as driven away by force, as shown by the Chinese treasures from the tomb of the Xiongnu chief at Noin Ula. Under the Han, military expansion was backed up by a programme of colonization, and commanderies were set up in areas as far-flung as modern Korea and Vietnam.

TOWN AND COUNTRY LIVING

A truly urban civilization developed in this period, with walled cities becoming the focus of trade, as in the case of Chang'an (*map 3*). Many modern Chinese cities are built on foundations laid in the Zhou period, and the earliest Chinese coins, miniature bronze knives and spades come from Zhou cities. Coinage was standardized by the First Emperor and the multitude of local mints was finally brought under central control in 119 BC.

The empire depended on the production of a wide range of goods and services, and in particular stable agriculture (*map 4*). Agricultural productivity was increased by government reforms and the use of more efficient tools, especially new ploughs made of iron. The importance of iron was recognized through the introduction, again in 119 BC, of state monopolies over its production, along with control of the production of salt and alcohol.

POLITICS AND THE END OF THE HAN EMPIRE

In the period of the Warring States, a political philosophy developed that recognized the uplifting nature of public life, but also viewed politics as ultimately corrupting. Clashes resonate throughout the history of the early Chinese empires between, on the one hand, the authoritarian politics of many of the rulers and, on the other, the high ideals of Confucius (551–479 BC) – perhaps the most influential of all Chinese philosophers – and his Reformist successors, which placed emphasis on virtue and fair government. Unlike their Shang predecessors, rulers were bound more by codes of human conduct than the demands of the spirits. Laws were first codified in the state of Wei under the rule of Duke Wen (r. 424–387 BC). Although much criticized, these formed the model for the Han law code. It was, however, peasant revolts inspired by messianic beliefs, often drawing on Daoism, that disrupted and weakened the Han Empire towards the end of its life. Movements such as the revolt of the Yellow Turbans in 184 AD, punished by the slaughter of over 500,000 people, left the empire open to the ambitions of powerful independent generals who divided up its territories between them.

▲ The massive mausoleum of Shi Huang Di, "the First Emperor", located at the Qin capital of Xianyang (later Chang'an under the Han dynasty), took 700,000 conscripted labourers 35 years to build. The life-size terracotta soldiers pictured here were among the 7,500 that guarded the vast burial pits surrounding the elaborate tomb.

▼ While rice, millet and wheat were the staples of Han agriculture, supplemented by vegetables, many areas also produced other commodities such as timber or fruit. Hemp was grown to make clothing for the majority, while silk supplied the elite. Iron was produced from the 6th century BC and was used for the majority of tools and weapons. Salt production was another major industry, obtained from the sea in coastal regions but elsewhere mined from brine deposits often found deep underground.

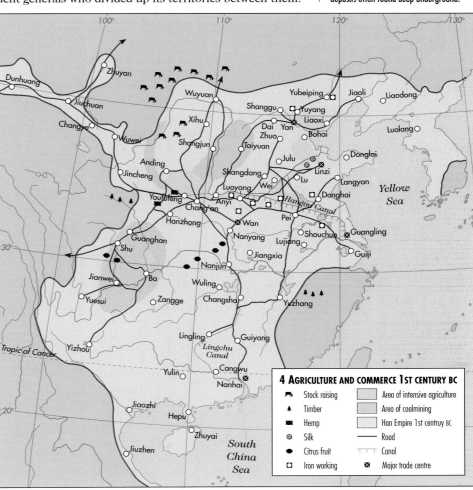

4 AGRICULTURE AND COMMERCE 1ST CENTURY BC

↣ Stock raising	▨ Area of intensive agriculture
▲ Timber	▨ Area of coalmining
⌐ Hemp	▫ Han Empire 1st century BC
◎ Silk	— Road
● Citrus fruit	⊤⊤⊤ Canal
▫ Iron working	⊗ Major trade centre

◀ CHINA 1700–1050 BC *pages 30–31* ▶ EAST ASIA IN THE TANG PERIOD 618–907 *pages 72–73*

PEOPLES OF CENTRAL ASIA
6000 BC–AD 500

Central Asia is a vast arid zone of steppe grasslands, looming mountains and inhospitable deserts. On its southwestern mountain fringes an agricultural way of life developed as early as the 6th millennium BC at sites like Djeitun, and some of these communities later developed into towns and cities (*map 1*). For example, Altyn Depe was first occupied in the 6th millennium, was enclosed by a wall in the 4th millennium, and by the 3rd millennium covered an area of nearly 30 hectares (74 acres) with craft production areas, elite compounds, fine burials and large platforms reminiscent of the great Mesopotamian *ziggurats* (pages 28–29). Agriculture in this region depended on a precarious irrigation system that collapsed around 2000 BC. However, later inhabitants such as the Persians (later 1st millennium BC) and Sasanians (from the 3rd century AD) devised more complex underground irrigation canals (*qanats*) which again brought prosperity to the region.

Up to the 5th millennium BC settlements were scattered along the rivers of Central Asia. These often consisted of partially subterranean houses and were home to small groups of hunter-gatherers who caught fish and a variety of game and collected plant foods. Later these hunter-gatherer communities began to adopt pottery and aspects of food production from the agricultural or pastoral groups with whom they came into contact (*map 2*).

SETTLEMENT AND PASTORALISM

By 4500 BC small permanent communities had appeared in favoured regions of Central Asia on the margins of Europe and West Asia, growing crops and, more particularly, herding livestock. Some of these were among the first to domesticate the horse, initially for meat. Their successors used wheeled vehicles: indeed four-wheeled wagons appeared in burials in

▼ Between 1500 and 800 BC copper- and bronze-working were taken up and refined across the Central Asian steppe — at the same time as a new way of life appeared, linking European Russia with the western borders of China (*map 2*). Horses and wheeled transport allowed people to exploit areas where pasture was too sparse to support herds in one place all the year round. Encouraged partly by changes in climate and vegetation, people took up a nomadic existence, moving with their herds. These animals, formerly kept for meat, were now mainly reared for milk which was made into a variety of foods, including cheese, yoghurt and fermented drinks.

Among the nomads were groups speaking Indo-European languages (*map 3*). They probably included Tocharian speakers in the Tarim Basin, where there have been finds of desiccated mummies of individuals with a strongly European appearance which date from this period. In West Asia, texts that include Indo-European terms identify other Indo-European-speaking groups, including the leaders of the non-Indo-European-speaking Mitanni.

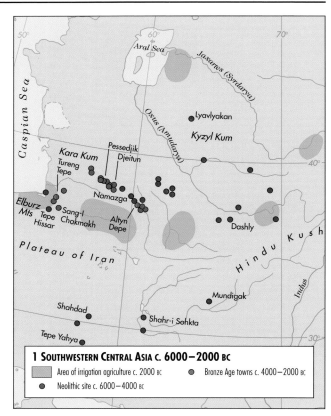

1 SOUTHWESTERN CENTRAL ASIA C. 6000–2000 BC
- Area of irrigation agriculture c. 2000 BC
- Neolithic site c. 6000–4000 BC
- Bronze Age towns c. 4000–2000 BC

▲ Southern Turkmenia was one of the regions in which agricultural communities had developed by 6000 BC. Part of the urban revolution, the later towns and cities of Turkmenia were centres of technological excellence and trading entrepots.

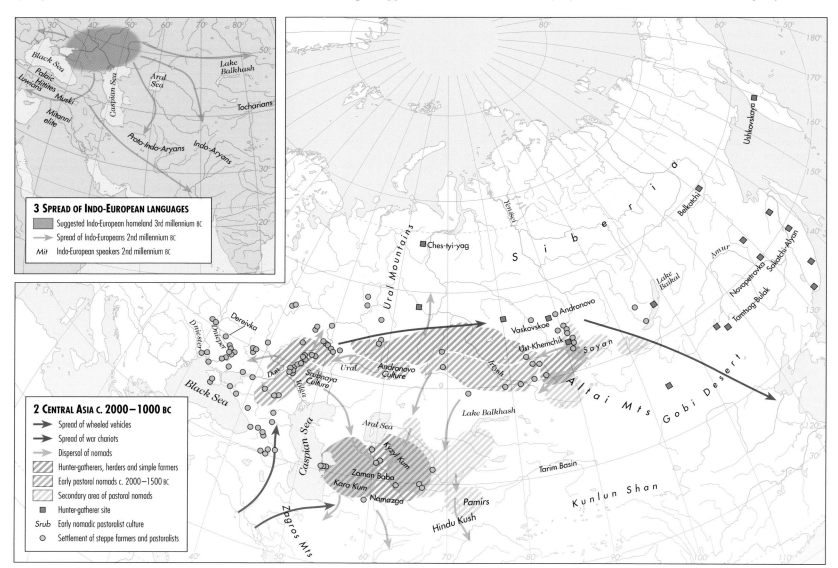

3 SPREAD OF INDO-EUROPEAN LANGUAGES
- Suggested Indo-European homeland 3rd millennium BC
- Spread of Indo-Europeans 2nd millennium BC
- *Mit* Indo-European speakers 2nd millennium BC

2 CENTRAL ASIA C. 2000–1000 BC
- Spread of wheeled vehicles
- Spread of war chariots
- Dispersal of nomads
- Hunter-gatherers, herders and simple farmers
- Early pastoral nomads c. 2000–1500 BC
- Secondary area of pastoral nomads
- Hunter-gatherer site
- *Srub* Early nomadic pastoralist culture
- Settlement of steppe farmers and pastoralists

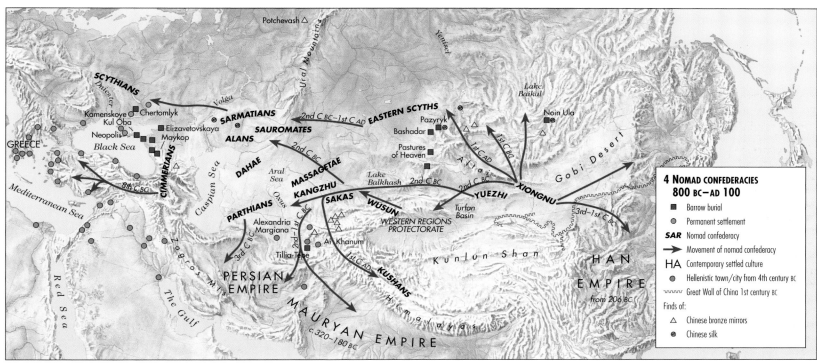

4 NOMAD CONFEDERACIES 800 BC–AD 100

- ■ Barrow burial
- ● Permanent settlement
- **SAR** Nomad confederacy
- → Movement of nomad confederacy
- **HA** Contemporary settled culture
- ● Hellenistic town/city from 4th century BC
- ∿∿∿ Great Wall of China 1st century BC

Finds of:
- △ Chinese bronze mirrors
- ◉ Chinese silk

southern Russia in the 4th millennium BC, and by 2000 BC the chariot dominated battlefields from Mesopotamia to China. The introduction of the spoked wheel (replacing the heavy solid wheel) made these vehicles much more manoeuvrable. Horse-riding was first adopted around 2000 BC by peoples dwelling north of the Caspian Sea. By 1000 BC full nomadic pastoralism had developed, from which emerged the horse-riding warriors who were to become the scourge of the Classical world.

While the origins of Indo-European speakers are still a matter of heated debate, many scholars would now place them among the groups dwelling between the Black Sea and Caspian Sea in the 4th and 3rd millennia BC. These are archaeologically identifed as the Srubnaya and Andronovo cultures and their predecessors. During the 2nd millennium BC groups speaking Indo-European languages can be identified in adjacent areas (*map 3*). By the beginning of the 1st millennium AD Indo-European languages were spoken in Europe as well as much of West Asia, Iran, South Asia and parts of Central Asia.

By the 1st millennium BC a fusion of nomadic and sedentary cultures gave rise to several kingdoms in southwestern Central Asia, which by the mid-6th century BC were largely under Persian control. The Achaemenid kings of the Persian Empire built roads, fortified cities and developed irrigation systems, and the influence of Persian culture was felt deep into Central Asia. Persian rule came to an end with the campaigns of Alexander the Great, and Hellenistic systems of administration and culture spread throughout the region (*pages 42–43*). The Graeco-Bactrian kings were the first to establish links across Central Asia with China.

THE NOMAD CONFEDERACIES

In the later centuries BC a series of powerful confederacies emerged among the nomad peoples. Historical accounts of these nomad societies and the threat they posed to the Classical civilizations have been left behind by Greek, Roman, Chinese and other authors, who named great tribal confederacies, including the Xiongnu and Yuezhi in the east, and the Scythians, Sakas, Cimmerians and Sarmatians further west (*map 4*). These nomad groups buried their elite in great mounds such as those at Noin Ula, Pazyryk and Kul Oba. Horses, central to the nomadic way of life, often played a major role in burial rituals, sacrificed to accompany their owners, along with much gold and silver and lavishly decorated textiles, some of which have been marvellously preserved in the frozen conditions of the tundra. Such rich burials are described by the Greek historian Herodotus,

whose accounts closely match the archaeological finds. These nomads wore highly decorated clothes and ornamented their bodies with tattoos. Hemp was not only used for textiles but was also smoked, as evidenced by remains of smoking paraphernalia. Stringed instruments also found in the tombs attest a love of music and song.

The Xiongnu formed one of the greatest of the nomad confederacies. Originating on the Mongolian plateau, they conquered and ruled the oasis cities of the Turfan Basin in the 2nd century BC. While they sometimes harried the borders of the Chinese Empire, on other occasions they enjoyed good trading relationships with China (*pages 52–53*), as can be seen in the presence of exquisite Chinese silks and other manufactured treasures, such as bronzes and lacquer, in the burial of a Xiongnu chief at Noin Ula. Xiongnu expansion drove other nomad groups further west, including the Yuezhi, who settled on the Oxus (Amudarya) River. One branch of the Yuezhi, the Kushans, later established an empire in northern India (*pages 46–47*).

The Xiongnu and other nomad peoples developed a distinctive culture, marked particularly by a splendid tradition of zoomorphic art. Other shared practices included binding children's heads in infancy to produce an elongated shape. They also developed major innovations in equestrian and military equipment, such as the composite bow or the scale-armour which made Sarmatian cavalry such formidable opponents of the Romans. Similarly the Huns, mounted steppe warriors armed with powerful reflex bows, wrought havoc in 5th-century Europe and northern India (*map 5*).

▲ From the 1st millennium BC substantial population movements took place in the steppe region. Groups often spilled over into adjacent settled lands, in some cases laying waste settled communities before being driven off, as with the 8th-century incursions of the Cimmerians into West Asia. Sometimes the invaders settled and became incorporated into the civilization of the lands they overran – the Sakas and Kushans in South Asia, for example. China successfully resisted many nomad incursions – partly by erecting massive defences that culminated in the Great Wall – though its western provinces fell for a period to the might of the Xiongnu nomads.

▼ The Huns moved through Central Asia during the 4th century AD, as evidenced by finds of their typical large bronze cauldrons, bows and artificially deformed skulls. One branch entered Europe in the 5th century, briefly wreaking havoc under the charismatic leadership of Attila, while the Hephtalites (Hunas or White Huns) overran the Sasanian Empire and laid waste the cities of northern India, where they established a short-lived empire.

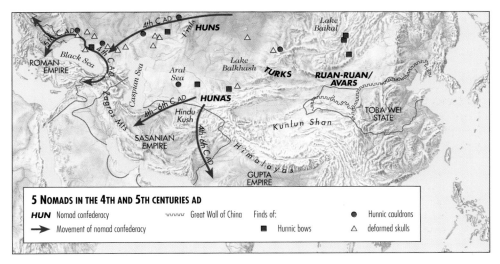

5 NOMADS IN THE 4TH AND 5TH CENTURIES AD

HUN Nomad confederacy | ∿∿∿ Great Wall of China | Finds of: | ● Hunnic cauldrons
→ Movement of nomad confederacy | | ■ Hunnic bows | △ deformed skulls

◄ FROM HUNTING TO FARMING: ASIA 12,000 BC–AD 500 *pages 18–19* ▶ EAST ASIA IN THE TANG PERIOD 618–907 *pages 72–73*

EURASIAN TRADE
150 BC–AD 500

In the early 2nd century BC the Xiongnu nomads drove their Yuezhi neighbours westwards, in the process making the Yuezhi king's skull into a drinking cup. In 138 BC the Han Chinese emperor Wudi sent Zhang Qian to the Yuezhi, hoping to make common cause with them against their mutual Xiongnu enemies. After enormous difficulties and numerous adventures, Zhang Qian reached the Yuezhi in the Oxus Valley – and although he failed to persuade them to renew their conflict with the Xiongnu, he took back to China detailed accounts of the lands he visited and the new opportunities for trade that they offered.

Over the following century Han China established trade routes through Central Asia which, despite passing through some of the most inhospitable terrain in Eurasia, soon provided access to West and South Asia and indirectly to the Roman world (*map 1*). For a time the Chinese controlled this "Silk Road" through Central Asia, establishing the Western Regions Protectorate with garrisons in the caravan towns, but the area was always menaced and often controlled by barbarian groups such as the Wusun and, especially, the Xiongnu. During the first three centuries AD the western portion was ruled by the Kushans, who had established an empire in northern India (*pages 46–47*).

Dependent largely on the hardy Bactrian camel, the Silk Road trade took Chinese silks (a prized commodity in the Roman Empire) and other luxuries to India and thence to the markets of the West. In exchange, many Roman manufactured goods found their way to China, along with the highly valued "heavenly horses" of Ferghana, gems from India, and grapes, saffron, beans and pomegranates from Central Asia. Ideas travelled, too: by the 1st century AD Buddhism was spreading from its Indian home to the oasis towns of the Silk Road, later becoming established in China, Korea and Japan (*pages 44–45*).

A number of possible routes linked China and the West, their course channelled by lofty mountains and freezing deserts, but political and military factors were also important in determining which routes were in use at any time. The oasis towns along the Silk Road rose and fell in prosperity with the fluctuating importance of the various routes. The collapse of the Han Empire in the 3rd century AD, the decline of the Kushans and the break-up of the Roman

▼ Bronze-working cultures had developed in mainland Southeast Asia during the 3rd millennium BC, and by 500 BC the bronze objects that were produced included the famous Dong Son drums. The drums were placed in elite burials and probably had a ritual significance. Made using a "lost wax" casting technique, they were widely distributed and reached the islands of Southeast Asia, where metallurgy was also being practised. By the 2nd century BC the area was linked to both India and China by sea routes which were used by Hindu Brahmin priests and Buddhist missionaries as well as merchants. As a result, new ideas of astronomy, art, science, medicine, government and religion were spread, and Buddhist and Hindu states were established in the region. One of the greatest was Funan, reputedly founded in the 2nd century BC by the Brahmin Kaundinya and reaching its peak in the 3rd century AD. The remains of a major Funan trading city have been excavated at Oc Eo.

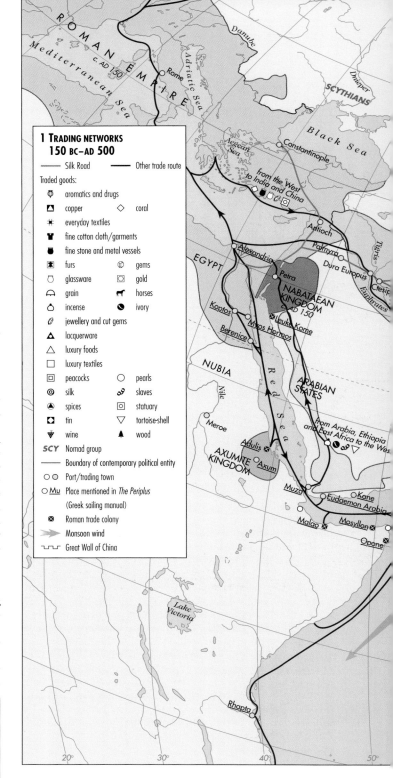

1 TRADING NETWORKS 150 BC–AD 500

— Silk Road   ▬ Other trade route

Traded goods:
- ⚱ aromatics and drugs
- ▲ copper   ◇ coral
- ✳ everyday textiles
- ♈ fine cotton cloth/garments
- ▮ fine stone and metal vessels
- ▦ furs   ⊚ gems
- ⊓ glassware   ▣ gold
- ⌂ grain   🐎 horses
- ⬡ incense   🐘 ivory
- ∅ jewellery and cut gems
- △ lacquerware
- △ luxury foods
- □ luxury textiles
- ▣ peacocks   ○ pearls
- ◎ silk   ❧ slaves
- ⬢ spices   ▢ statuary
- ⬓ tin   ▽ tortoise-shell
- ▽ wine   ▲ wood

SCY Nomad group

— Boundary of contemporary political entity

○ ○ Port/trading town

○ Mu Place mentioned in *The Periplus* (Greek sailing manual)

⊠ Roman trade colony

➜ Monsoon wind

〰 Great Wall of China

Empire all had their impact on the Silk Road, though links between East and West continued – for example, taking Chinese pilgrims to visit the Buddhist holy places in India.

SOUTHEAST ASIA

By the 2nd century BC sea routes linking India with China via Southeast Asia were also in common use. While Indian literature makes only vague references to trade with Southeast Asia, finds of Indian beads and Western objects in the region – such as Roman coins and cut gems – and of Southeast Asian tin in south Indian sites, attest to the region's contacts with India. The seaborne trade grew in the early centuries AD, a period when urban centres and states were appearing in much of Southeast Asia (*map 2*).

Riverborne trade linked China and mainland Southeast Asia during the 1st millennium BC, and sea traffic developed during the period of the Han Empire. In 111 BC Han armies conquered the formerly independent state of Nan Yue, establishing colonies and, from AD 40, directly administering the province. At this time the area to its south was probably home to a number of small independent chiefdoms united in opposition to Chinese territorial aggression. Chinese interest in Southeast Asian trade burgeoned after the fall of the Han in AD 220, when the Chinese elite fled south, and trade with the West along the Silk Road was largely replaced by maritime trade via Southeast Asia to India.

2 SOUTHEAST ASIA 150 BC–AD 500

- ⚙ Early metal-using settlement
- △ Buddhist remains from 5th–6th centuries AD

Finds of:
- ⊟ Dong Son drums
- ◆ Brahminical remains from 5th–6th centuries AD
- ■ Indian inscriptions

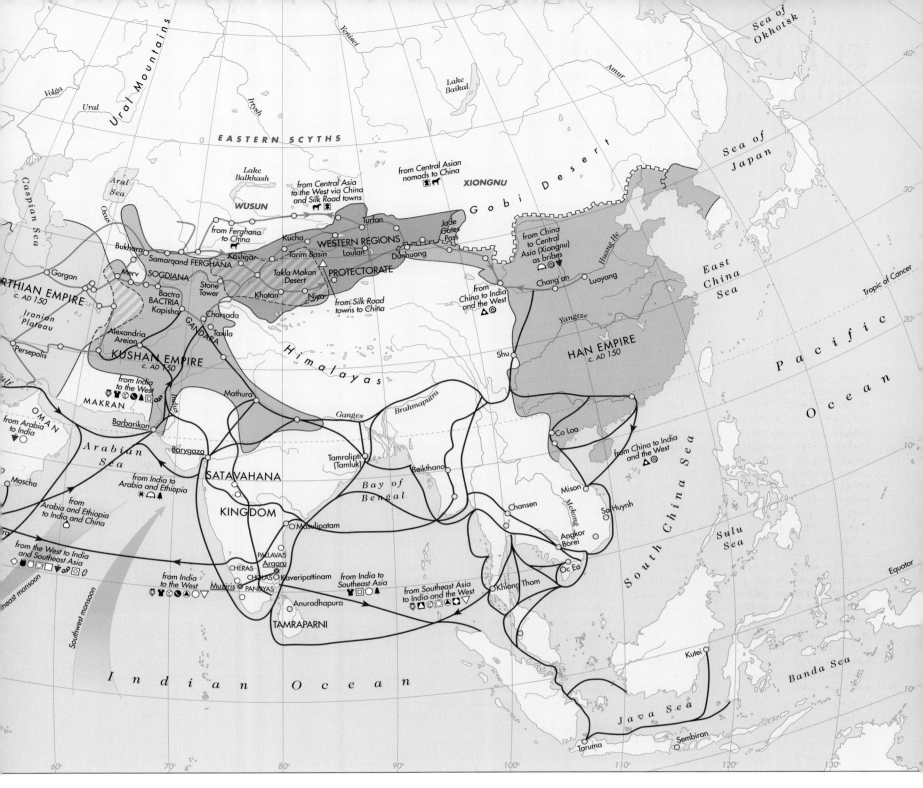

TRADE ACROSS THE INDIAN OCEAN

Trade links had been operating around the coasts of the Indian Ocean from the later 3rd millennium BC. Regular seaborne trade took place in the Gulf, Sumerians trading directly with the Indus civilization, along with the coastal inhabitants of Oman and Makran and the seafaring traders of Bahrain. Land or coast-hugging sea routes also brought African plants and Arabian incense to India and the lands of the Gulf (*pages 28–29*). Egypt was economically and politically involved with Nubia to its south along the River Nile (*pages 30–31*), and seaborne expeditions through the Red Sea were mounted by Egypt to bring back exotic materials from the Land of Punt, probably situated in Ethiopia.

In subsequent centuries the rise and fall of Mediterranean, western Asiatic and Indian Ocean states and cultures brought a variety of participants into this network, including Persians, Phoenicians and Greeks. By the 1st millennium BC both Arabians and Indians were familiar with, and exploiting, the monsoon winds to cross the Indian Ocean instead of laboriously following the coast. These winds carried them east in the summer, down the Red Sea and across to India, while the northeast monsoon in the autumn carried vessels westward from India and down the African coast. It was not until the final centuries BC, however, that the Greeks and Romans also became acquainted with the monsoon winds. The volume of Roman

traffic in the Indian Ocean greatly increased during the reign of Emperor Augustus (27 BC–AD 14), with perhaps over 100 ships setting out from the Red Sea in a single year.

A Greek sailing manual of around 60 AD, *The Periplus of the Erythraean Sea* (Indian Ocean), has provided a wealth of information on trade in this area. Alexandria was the starting point for most east- and southbound trade: here the bulk of cargoes were assembled and shipped down the Nile as far as Koptos, where they were taken by camel to either Myos Hormos or Berenice on the Red Sea. Some expeditions travelled south as far as Rhapta on the coast of East Africa, obtaining ivory, tortoise-shell and incense – a round trip of two years because of the timing of the winds.

Others made the more dangerous ocean crossing to India, where they exchanged gold, wine, manufactured goods and raw materials for gems, fine Indian cotton textiles and garments, Chinese silks, spices, aromatics and drugs. On the return journey they would stop at Kane and Muza to obtain frankincense and myrrh, reaching Alexandria within a year of departure. Arab and Indian merchants also still plied these routes. Unlike the Romans (whose trade was in low-bulk, high-value commodities, carried directly between their source and the Roman world), other Indian Ocean traders dealt in everyday commodities such as grain, foodstuffs and ordinary textiles and might trade in any port.

▲ A variety of routes linked the countries of Asia, East Africa and the Mediterranean. Long-established routes through the Gulf and across the Iranian Plateau flourished during the 1st millennium BC under the Achaemenids and their Hellenistic successors. From the 2nd century BC the newly established Chinese trade route across Central Asia linked with these existing routes, while Arabs and Indians operated sea trade across the Indian Ocean, and desert caravans carried incense from southern Arabia via the Nabataean state to Rome. By the 1st century AD hostility between the Parthian and Roman empires had closed the overland route through Persia, and the Romans became directly involved in Indian Ocean trade. Chinese goods reached India via the Silk Road and indirectly by sea via Southeast Asia; from here they were taken by Roman shipping across the Indian Ocean, along with Indian goods. The Axumite kingdom benefited from this shift, becoming a major producer of incense, while the Arab states that had operated the overland caravans declined.

THE ROMAN EMPIRE
500 BC–AD 400

▲ Skilful political manoeuvring helped Octavian (Augustus) to secure victory over his rivals in the struggle to succeed his uncle Julius Caesar. Augustus used his position of supreme power well, enacting a raft of important legal, economic, social and administrative reforms, reviving traditional religious beliefs, encouraging the arts, and constructing and restoring many public buildings in Rome.

▼ The Roman Empire was the first state to bring unity to much of Europe. From the cold hills of southern Scotland to the deserts of North Africa, Rome introduced a common culture, language and script, a political system that gave equal rights to all citizens, a prosperous urban way of life backed by flourishing trade and agriculture, and technical expertise that created roads, bridges, underfloor heating, public baths and impressive public buildings, some of which survive today. Roman culture also spread to lands beyond the imperial frontier, influencing among others the Germanic barbarians who later overran the empire – but who would eventually perpetuate many of its traditions and institutions, notably through the medium of the Christian Church.

The classical world was the cradle of European civilization: if Greece shaped Europe's culture, Rome laid its practical foundations. Throughout Rome's mighty empire, science was applied for utilitarian ends, from under-floor heating to watermills, aqueducts and an impressive road network. Rome bequeathed to posterity its efficient administration, codified laws, widespread literacy and a universally understood language. It also adopted and spread Christianity, for which it provided the institutional base.

The city of Rome developed in the 7th and 6th centuries BC from a number of settlements spread over seven low, flat-topped hills. Ruled by kings until about 500 BC, it then became a republic governed by two annually-elected consuls and an advisory body, the Senate. Around the same time Rome defeated the tribes in the surrounding area and gradually expanded through Italy: in the Latin War (498–493 BC) it crushed a rebellion of the Latin tribes, incorporating them in a pro-Roman League, and by the 3rd century BC it had overrun the Greek-influenced civilization of the Etruscans, famous for their fine pottery.

Victory over the Samnites in 290 BC led to a confrontation with the Greek colonies in southern Italy, whose defeat in 275 BC gave Rome control of the entire Italian peninsula. To strengthen its grip on the conquered territory, colonies were founded and settled by both Roman citizens and Latin allies. Swift access to these colonies was provided by an extensive road network, created from the late 4th century BC and greatly extended during the 2nd century BC.

EXPANSION BEYOND ITALY
The first confrontation outside Italy was against the Carthaginians, who saw their commercial interests in Sicily threatened by Rome's expansion. During the three Punic Wars (264–241, 218–201, 149–146 BC) Rome seized territory formerly held by the Carthaginians (Sardinia, Corsica, Spain and the tip of northern Africa), but also suffered its worst defeats. In 218 BC the Carthaginian general Hannibal crossed the Alps and obliterated the Roman army at Lake Trasimene (217 BC) and at Cannae (216 BC). To withstand the Carthaginians, Rome had constructed its first fleet around 260 BC and became a maritime power with control over a Mediterranean empire that incorporated the former Hellenistic kingdom of Macedonia (*pages 42–43*) from 148 BC and Pergamum from 133 BC. As a result, Greek culture began to exert a powerful influence on Roman life and art.

The newly acquired provinces (*map 1*) created the opportunity for individuals to make a fortune and forge a loyal army. One of these new powerful commanders, Pompey (106–48 BC), conquered Syria, Cilicia, Bithynia and Pontus, while Julius Caesar (100–44 BC) annexed Gaul and expanded the African province.

Caesar's influence had grown to such an extent that the Senate saw its position threatened and ordered him to disband his army in 49 BC.

Caesar disobeyed and crossed the Rubicon River – in defiance of the law that forbade a general to lead his army out of the province to which he was posted – and ruled Rome as a dictator until he was assassinated in 44 BC. Caesar's adoptive son Octavian (63 BC–AD 14) officially restored the Senate's powers, nominally taking up the position of *princeps* (first citizen) while gradually increasing his authority. In 27 BC he was awarded the title "Augustus" ("revered one"), and this date is usually taken as the start of the imperial period.

Augustus's reign brought a period of peace and stability, the so-called *Pax Romana,* which would last until AD 180. His main military efforts were aimed at creating a fixed and easily defensible border for his empire (*map 2*). Augustus conquered the entire area up to the River Danube, which, together with the River Rhine, formed his northern border. In the east the frontier was less well defined and was controlled more by political means, such as alliances with neighbouring kingdoms.

Augustus also annexed Egypt, Judaea and Galatia and reorganized the legions left by his predecessors, keeping a firm grip on those provinces that required a military presence by awarding them the status of imperial province. The emperor himself appointed the governors for these provinces, while the Senate selected the governors for the others. Augustus also reorganized the navy: he based his two main fleets at Misenum and Ravenna to patrol the Mediterranean against pirates, while smaller fleets were stationed within the maritime provinces to guard the borders.

ROMAN TRADE
Trade flourished under Augustus's rule. The military infrastructure such as sheltered harbours, lighthouses and roads greatly benefited commercial activity, and the presence of Roman soldiers in faraway provinces further encouraged long-distance trade (*map 3*). Gradually, however, the provinces became economically independent: they started to export their own products and eventually, during the 3rd century, began to deprive Rome of its export markets.

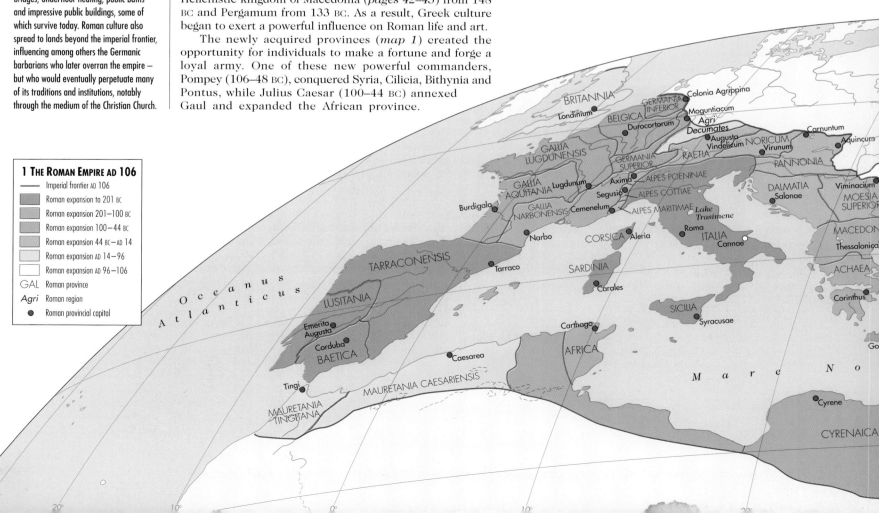

1 THE ROMAN EMPIRE AD 106
- Imperial frontier AD 106
- Roman expansion to 201 BC
- Roman expansion 201–100 BC
- Roman expansion 100–44 BC
- Roman expansion 44 BC–AD 14
- Roman expansion AD 14–96
- Roman expansion AD 96–106
- GAL Roman province
- *Agri* Roman region
- ● Roman provincial capital

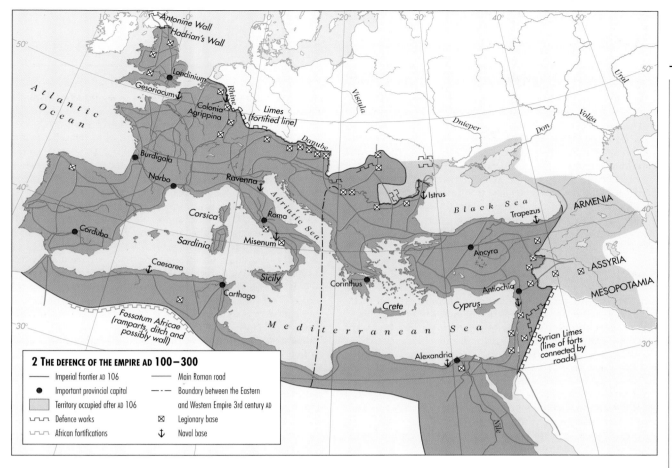

2 THE DEFENCE OF THE EMPIRE AD 100–300

—— Imperial frontier AD 106	—— Main Roman road
● Important provincial capital	—·—· Boundary between the Eastern
Territory occupied after AD 106	and Western Empire 3rd century AD
⌐⌐⌐ Defence works	⊠ Legionary base
⌐⌐⌐ African fortifications	↓ Naval base

◄ Unlike his acquisitive predecessor Trajan, Emperor Hadrian concentrated on reinforcing the previous Roman *limes*, or frontiers. He strengthened the *Agri Decumates limes* between the Rhine and the Danube with a wooden palisade and numerous forts and is thought to have started work on a mudbrick wall and ditch which was to become the African frontier, the *fossatum Africae*. He built the first stone wall to secure the British frontier – a second was later constructed by Antoninus (r. 138–161) – and also reinforced Trajan's work on the Syrian *limes*, a policy later continued by Diocletian.

THE EMPIRE AFTER AUGUSTUS

Some of Augustus's successors attempted to enlarge the empire, others to consolidate existing territory. Whereas Tiberius (r. AD 14–37) refrained from any expansion, Claudius (r. 41–54) annexed Mauretania, Thrace, Lycia and parts of Britain, while Vespasian (r. 69–79) conquered the "Agri Decumates" region. Under Trajan (r. 98–117) the empire reached its maximum extent, including Arabia and Dacia by 106. Trajan subsequently subjugated Armenia, Assyria and Mesopotamia, but these conquests were soon abandoned by Hadrian (r. 117–138).

Under Diocletian (r. 284–305) the empire was divided into Eastern and Western parts, each ruled by an "Augustus", while the provinces were replaced by a massive new bureaucracy and the army was greatly extended. However, the resignation of Diocletian in 305 was followed by chaos – out of which, in 312, Constantine (r. 306–337) emerged victorious in the West. In 324 he reunited the empire and made Christianity the official religion, and in 330 he established a new capital at Constantinople. Following his death in 337 the empire was divided and reunited several times before it was permanently split in 395. The sacking of Rome by the Visigoths in 410 (*pages 56–57*) signalled the end of the Western Empire; to the east, the empire was to continue in the guise of the Byzantine Empire until 1453.

▼ During the reign of Augustus trade became Rome's lifeline. To feed its rapidly expanding urban population, it depended on the import of corn – first from Sicily, later from Africa and Egypt – and to suit the tastes of Rome's "nouveaux riches" luxury goods were imported from even further afield – silk from China, hair for wigs from Germany, ivory from Africa. However, the traffic was two-way: during the 1st century AD, for example, Rome developed a lucrative business supplying the provinces with products such as wine and olive oil.

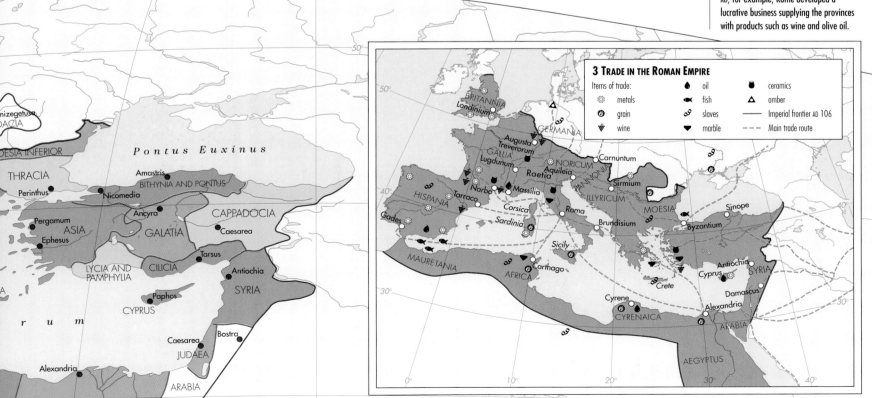

3 TRADE IN THE ROMAN EMPIRE

Items of trade:
💧 oil		▪ ceramics	
⚙ metals	⋈ fish	△ amber	
◉ grain	∽ slaves	—— Imperial frontier AD 106	
▽ wine	▼ marble	– – – Main trade route	

◀ EUROPE 8000–200 BC *pages 20–21*   ◀ BARBARIAN INVASIONS 100–500 *pages 56–57*   ◀ BYZANTINE EMPIRE 527–1025 *pages 66–67*

BARBARIAN INVASIONS OF THE ROMAN EMPIRE 100–500

▲ Roman legionaries were first called upon ro defend the empire against a serious threat from the Germanic tribes in the 2nd century AD – the date of this Roman stone relief.

▼ From the pages of *Germania* by the Roman historian Cornelius Tacitus (55–120) there emerges a clear picture of the Germanic world of the first century AD, comprising a multiplicity of small political units, with any larger structures being little more than temporary tribal confederations. By the 350s, however, long-term processes of social and economic change (largely the product of extensive contacts with the Roman Empire) had created a smaller number of much more powerful groupings. Of these the Gotones (Goths), then based in Poland, would have the biggest impact on Rome and its European dominions.

Throughout its history the Roman Empire suffered frequent small-scale raids along its European frontier, but major invasions were rare. In the early 1st century AD a defensive alliance to resist Roman aggression had been formed under the leadership of Arminius, a chieftain of the Cherusci – one of a host of minor political units that comprised the Germanic world at this time (*map 1*).

However, the first large-scale invasion of the Roman Empire did not occur until the 160s, when the movement of Gothic and other Germanic groups from northern Poland towards the Black Sea led to the Marcomannic War. Recent archaeological investigations have revealed the spread of the so-called Wielbark Culture south and east from northern Poland at precisely this period (*map 2*). Another time of turmoil followed in the mid-3rd century, associated with Goths, Herules and others in the east and Franks and Alemanni in the west. Archaeologically, the eastward moves are mirrored in the creation and spread of the Goth-dominated Cernjachov Culture in the later 3rd century. None of this, however, amounts to a picture of constant pressure on the Roman Empire.

Relations between the empire and the peoples beyond its borders, whom the Romans regarded as uncivilized "barbarians", were not all confined to skirmishing and warfare. Numerous individual Germans served in Roman armies, while Roman diplomatic subsidies supported favoured Germanic rulers. Some important trading routes also operated, such as the famous amber route to the Baltic (*pages 38–39*), and there was a steady flow of materials (timber, grain, livestock) and labour across the border.

These new sources of wealth – and in particular the struggle to control them – resulted in the social, economic and political transformation of the Germanic world. By the 4th century the many small-scale political units, which had relatively egalitarian social structures, had evolved into fewer, larger and more powerful associations that were dominated by a social elite increasingly based on inherited wealth. The main groups were the Saxons, Franks and Alemanni on the Rhine, the Burgundians and Quadi on the

middle Danube, and the Goths on the lower Danube (*map 2*). None had the power to stand up to the empire on their own, but neither was Roman domination of them total, the Alemanni even seeking to annex Roman territory in the 350s and dictate diplomatic terms.

THE ARRIVAL OF THE HUNS

The prevailing balance of power was transformed some time around 350 by the arrival on the fringes of Europe of the Huns, a nomadic group from the steppe to the east (*map 3*). By 376 the Hunnic invasions had made life intolerable for many Goths and they had started to move westwards. Three groups came to Rome's Danube frontier to seek asylum: one group was admitted by treaty, a second forced its way in, and the third, led by Athanaric, sought a new home in Transylvania. Goodwill was lacking on both sides, however, and the two admitted groups became embroiled in six years of warfare with the Roman Empire.

A huge Gothic victory won at Hadrianople in 378 convinced the Roman state of the need to recognize the Goths' right to an autonomous existence – a compromise confirmed by peace in 382. In the meantime the Goths under the leadership of Athanaric had in turn forced Sarmatians onto Roman soil, Taifali barbarians had crossed the Danube to be defeated in 377, and numerous groups of Alans had begun to move west, some being recruited into the Roman army in the early 380s. In 395 the Huns made their first direct attack on the empire, advancing from the area northeast of the Black Sea (where the majority were still based) through the Caucasus into Asia Minor.

The division of the Roman Empire into the Western and Eastern Empires in 395 (*pages 54–55*) was soon followed by further invasions (*map 3*). In 405–6 Goths under the leadership of Radagaisus invaded Italy, and while he was defeated and killed in the summer of 406, many of his followers survived to be sold into slavery or incorporated into the Roman army. At the end of 406 another large group of invaders – mainly Vandals, Alans and Sueves – crossed the Rhine. It is likely that, as with the invaders of the 370s, they were fleeing from the Huns, who by around 420 were established in modern Hungary, the subsequent centre of Hunnic power (*pages 76–77*).

THE COLLAPSE OF THE WESTERN EMPIRE

By around 410 numerous outsiders were established within the Roman Empire in western Europe. The Vandals, Alans and Sueves had pillaged their way to Spain (*map 3*), and

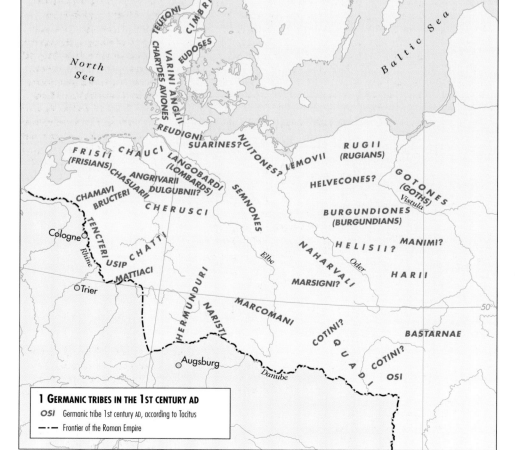

1 GERMANIC TRIBES IN THE 1ST CENTURY AD

OSI Germanic tribe 1st century AD, according to Tacitus

–·–· Frontier of the Roman Empire

2 BARBARIANS BEYOND THE FRONTIER 100–350

▨ Wielbark Culture 1st century AD		*GOT*	Germanic group 4th century
▒ Expansion of Wielbark Culture c. 150–230		*(AL)*	Non-Germanic group 4th century
▨ Cernjachov Culture 250–350		–·–·	Frontier of the Roman Empire c. 300

▲ The Romans regarded all peoples outside their empire as inferior, referring to them as "barbarians". There were two main groups: first, the largely Germanic-speaking settled agriculturalists of central

and eastern Europe; second, the nomadic steppe peoples belonging to various linguistic and ethnic groupings who periodically disturbed the eastern fringes of continental Europe.

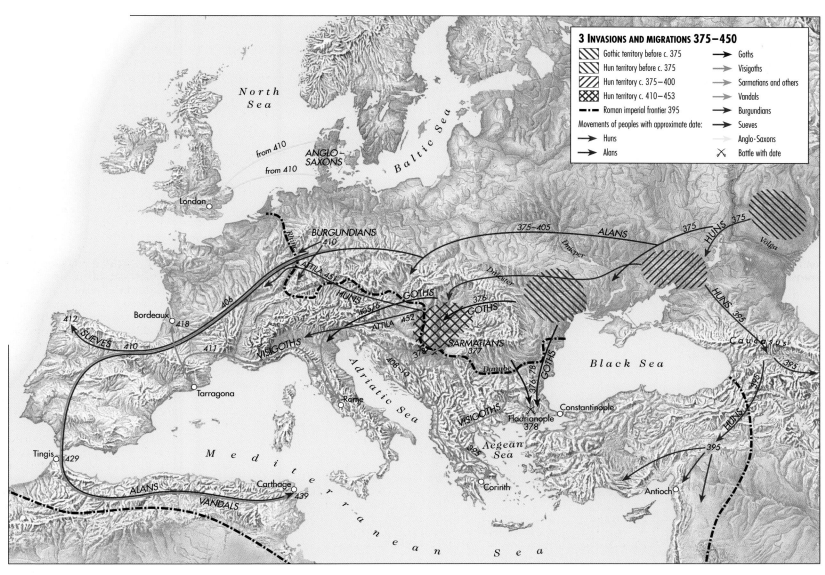

3 INVASIONS AND MIGRATIONS 375–450

Gothic territory before c. 375
Hun territory before c. 375
Hun territory c. 375–400
Hun territory c. 410–453
Roman imperial frontier 395
Movements of peoples with approximate date:
Huns
Alans

Goths
Visigoths
Sarmatians and others
Vandals
Burgundians
Sueves
Anglo-Saxons
Battle with date

the Goths, who had crossed the empire's frontier in 376, had moved to Italy under the leadership of Alaric. Here they were reinforced by the former followers of Radagaisus to create the Visigoths. They sacked Rome in August 410, but by 420 the Romans had forced them to accept settlement in Aquitaine on compromise terms. Rome had also counterattacked in Spain, where one of the two Vandal groups and many Alans were destroyed, before the death of Emperor Honorius in 423 led to ten years of internal political strife which crippled the empire's capacity for action. During this period the Vandals and Alans, now united under Geiseric, seized the rich lands of North Africa, while eastern Britain fell decisively under the sway of Anglo-Saxon invaders.

The losses in Britain, Aquitaine, Spain and North Africa fundamentally eroded the power of the Western Empire. Essentially, it maintained itself by taxing agricultural production, so that losses of land meant losses of revenue. Tax-raising in northern Gaul was periodically disrupted by Franks and others. By 440 the Western Empire had lost too much of its tax base to survive. It was propped up for a generation, however, through a combination of prestige (after 400 years it took time for the empire's contemporaries to realize that it was indeed at an end), support from the Eastern Empire, and temporary cohesion fuelled by fear of the Huns, whose empire reached its peak under Attila in the 440s.

The collapse of Hunnic power in the 450s, however, heralded Roman imperial collapse. New kingdoms quickly emerged around the Visigoths in southwestern Gaul and Spain, and the Burgundians in the Rhône Valley, where they had been resettled by the Romans in the 430s after being mauled by the Huns. At the same time the Franks, no longer controlled by the Romans, united to create a

kingdom either side of the Rhine (*pages 74–75*). The end of the Huns also freed more groups to take part in the share-out of land (*map 4*). Lombards and Gepids took territories in the middle Danube, and Theoderic the Amal united Gothic renegades from the Hunnic Empire with other Goths serving in the Eastern Roman army. This new force, the Ostrogoths, had conquered the whole of Italy by 493.

4 SUCCESSOR KINGDOMS C. 500
Approximate frontiers of "barbarian" kingdoms c. 500

▲ In the 5th century a combination of fear of the Huns (especially for the Visigoths, Vandals, Alans, Sueves and Burgundians) and opportunism (notably for the Anglo-Saxons, Franks and Ostrogoths), prompted a series of militarily powerful outsiders to carve out kingdoms from the territory of the waning Western Roman Empire. To protect their estates, the basis of their wealth, many local Roman landowners decided to come to terms with the invaders, with the result that the successor kingdoms all acquired some important vestiges of Roman institutions and culture.

◄ The frontiers that replaced the divisions of the Western Roman Empire by 500 were far from fixed. For example, in the 6th century as the Frankish kingdom grew apace, the Ostrogoths were destroyed by the Byzantine emperor Justinian, and the rise of the Avars prompted the Lombards to invade northern Italy in 568.

◀ THE ROMAN EMPIRE 500 BC–AD 400 *pages 54–55* ▶ FRANKISH KINGDOMS 200–900 *pages 74–75*

THE MEDIEVAL WORLD

Humans already occupied much of the globe by the year 500. Over the next thousand years the spread of intensive food production enabled their numbers to continue rising and a growing area to become more densely occupied. As a result, states and empires and other complex forms of socio-economic organization developed in almost every continent. Foremost in terms of wealth, population and technological achievement was China.

► Between 500 and 1500 intensive forms of agriculture developed in many parts of the world, but the vast grasslands of the Eurasian steppe continued to be populated by horse-breeding pastoralist nomads and semi-nomads. Riding eastwards and westwards from Central Asia, they frequently raided the lands of permanently settled peoples who increasingly used the plough to cultivate their fields.

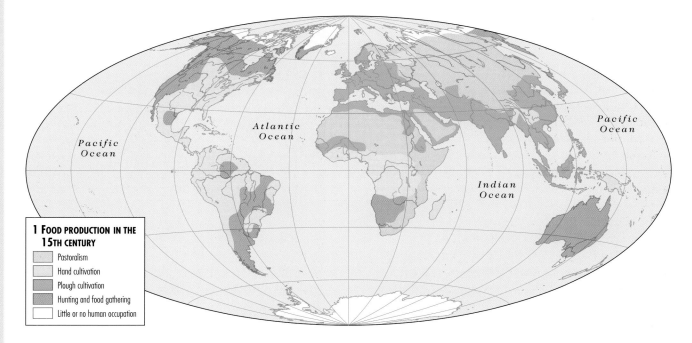

1 FOOD PRODUCTION IN THE 15TH CENTURY
- Pastoralism
- Hand cultivation
- Plough cultivation
- Hunting and food gathering
- Little or no human occupation

► The West African city-kingdom of Benin – renowned partly for the brass heads of which this is an example – developed from the 13th century as an important centre of trade. It was at the southern end of a network of trade routes across the Sahara, some of which had existed for many centuries but did not become important until the 9th century when Muslim merchants in North Africa began to travel southwards.

A number of intensive methods of cultivation had been developed before 500. However, the medieval period witnessed the spread of such methods over an ever-expanding area, dramatically increasing outputs in parts of Africa by the 8th century, in eastern Europe by the turn of the millennium, and in some regions of North America throughout the centuries up to 1500 (*map 1*). Depending on the environment, different crops were involved: sorghum and millet in Africa, wheat in Europe, and maize, beans and squash amongst others in North America.

At the same time new intensive farming regimes were developed which tackled the problem of sustaining soil fertility in the face of continuous use. In medieval Europe an unprecedented level of central planning evolved, based on the manor. This made possible economies of scale in the use of expensive items (such as draught animals and iron tools) and the implementation of a new strategy for raising production while maintaining fertility – the three-year rotation system. Wheat was grown in one year, beans and other legumes to restore nitrogen to the soil were grown in the next, and the land was allowed to lie fallow in the third.

On the basis of such advances, populations often grew dramatically. In England, for example, the figure of just over one million in about 500 nearly quadrupled to over four million before the Black Death (bubonic plague) took its dreadful toll across Europe in 1347–52, while China's population under the dynasties of the Tang (618–907) and Song (960–1279) increased from just over 50 million in the mid-8th century to over 100 million in the late 13th century.

Food production and populations did not always increase, however. Where a figure seems to have reached its optimum under a precise set of environmental conditions, a period of depletion often followed. In Mesoamerica, for example, the

"Maya Collapse" of the 9th century, when the population dropped dramatically from almost five million in the Yucatán Peninsula alone, can at least partly be attributed to degradation of the land caused by intensive agriculture coupled with a reduction in rainfall. In western Europe it is possible that the impact of the Black Death – which reduced the population by between a quarter and a half – may have been intensified because numbers had in places already passed the point of sustainability for the agriculture of the time.

THE SPREAD OF WORLD RELIGIONS

The Black Death was seen by the Christian population of Europe as God's punishment for their sins. Christianity won an increasing number of adherents in Europe during the medieval period, while Buddhism spread to East and Southeast Asia. In India, the land of Buddhism's birth, Hinduism revived, particularly in the south.

In the 630s the new religion of Islam emerged in the Arabian Peninsula and through military conquest rapidly took hold of the Middle East, North Africa and parts of Europe. It reached the limits of its westward expansion in 732, when a Muslim army was defeated at Poitiers in central France. However, over the following centuries the states and empires of Islam frequently inflicted defeats on Christendom. At the end of the 13th century the Mamluks of Egypt and Syria completed their recapture of the Holy Land (Palestine) from the Latin Church and in 1453 the Ottoman Turks finally succeeded in capturing Constantinople – capital of the Orthodox Church. Islam also eclipsed Zoroastrianism in southwest Asia, pushed Hinduism back in India from the 1190s, and spread into Central Asia through the conversion of the Mongols from the late 13th century.

TOWNS AND TRADE

In the ancient world much effort was devoted to building and adorning cultural and ceremonial capitals such as Babylon, Athens, Rome and Constantinople. The medieval period too saw the construction and expansion of such cities. In China, Chang'an was adopted by the Tang dynasty as their capital and was developed to cover an area of 77 square kilometres (30 square miles), with a population of about one million in the 7th century. With Baghdad, the Muslim Abbasids founded what was to become probably the world's largest city in the early 9th century, with an area of 90 square kilometres (35 square miles). The Muslims also oversaw the development of some of Europe's largest cities at this time – notably Córdoba and Seville in Spain and Palermo in Sicily. It was not until the 12th century that the towns of Latin Christendom really began to grow, the larger among them – such as Paris and Cologne – building magnificent churches, town halls and palaces.

By 1500 only a tiny proportion of the world's population lived in large cities. In Europe, for example, just three million out of an estimated total of 80 million lived in cities with over 10,000 inhabitants. The characteristic form of medieval urbanism everywhere was the modest market town, evolved as a service centre for the local agricultural economy. It was a place where surplus crops could be exchanged for other foodstuffs and goods, making it possible to grow a wider range of crops suited to local soils. It was also home to a variety of specialist craftsmen, whose various wares (tools, leather goods, ceramics, and so on) were made for sale to the rural population.

◄ Throughout the medieval period agriculture was the occupation of the vast majority of people. From the 10th century it was made more productive in Europe partly by the introduction of the three-year rotation system and improvements in the design of the plough. However, the pattern of life continued much as it always had, dictated by the seasons. This 15th-century illustration of ploughing the fields and sowing the winter grain in October is taken from a Book of Hours (*Les Très Riches Heures du Duc de Berry*), which was produced by the Franco-Flemish Limbourg brothers. Like many medieval calendars, the book illustrates the changing occupations of the months, from sowing to harvesting.

▼ China's cities were among the most impressive of the medieval world. A busy street scene is depicted in this 12th-century illustration of Kaifeng, capital of the Song dynasty between 960 and 1126. Attacks from the north by the Jurchen then led to the adoption of the more southern Hangzhou as the Song capital. With its estimated population of one and a half million, Hangzhou became a symbol of a golden age in China's history.

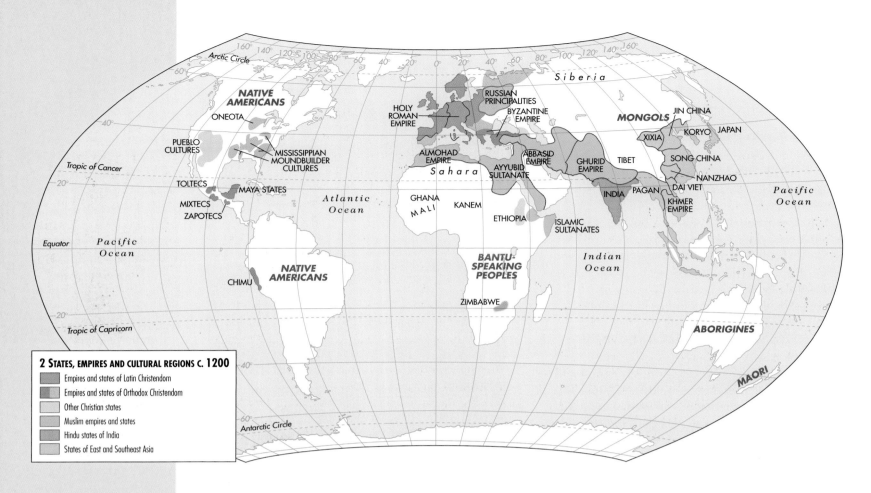

2 STATES, EMPIRES AND CULTURAL REGIONS C. 1200

Empires and states of Latin Christendom

Empires and states of Orthodox Christendom

Other Christian states

Muslim empires and states

Hindu states of India

States of East and Southeast Asia

▲ States and empires continued to rise and fall in the medieval period. Many of those in Eurasia in 1200 were to be overwhelmed by the destructive conquests of the Mongols in the 13th century.

▼ The Byzantine Emperor Justinian I (r. 527–65) attempted to recreate the Roman Empire of the 4th century, before it was divided into Eastern and Western parts. Among his conquests were Italy, where he adopted the city of Ravenna as the imperial capital and did much to adorn it. This 6th-century mosaic in the Church of San Vitale shows the Empress Theodora with her attendants.

The development of market towns was a clear sign of growing sophistication in rural economies, where specialization and exchange (developed in many parts of Asia, Europe, Mesoamerica and South America well before 500) replaced self-sufficiency as the basis of agricultural production. During the medieval period they spread across Europe and came to play an important role in the economies of both West and East Africa.

Some towns also serviced regional and long-distance trade based largely on linking contrasting ecological zones and dealing in items that were perceived as luxuries – notably metals, clothing materials and spices. From the later 8th century the Viking merchants of Scandinavia linked the fur-producing forests of subarctic regions with the wealthy cities of the Middle East, while from the 9th century a growing trans-Saharan trade moved gold, ivory and slaves between West Africa and the Muslim north African coast. Trade in a variety of

items, including metalwork, stones and cacao, continued to flourish in Mesoamerica, as did the movement of silks and spices along the highways of Central Asia until the nomadic Mongol hordes created havoc there in the 13th century.

STATES AND EMPIRES

Much of the new food surplus was now used to support people performing a range of specialist functions, many of which were not directly concerned with traditional forms of economic activity. The number of religious specialists grew as Christianity joined Buddhism in generating numerous monastic communities. Most specialists, however, were associated with the spread of states and empires (*map 2*). A class of literate bureaucrats – devising and administering laws and gathering taxes – became a feature of the majority of medieval states. Long established in parts of Asia, such people became central to the functioning of many European states from the 12th century.

Another specialist, even more widespread, was the warrior. The Chinese Song Empire was sustained by huge armies, supported by taxes raised from a dependent rural populace, while in Japan the *samurai* became a socially dominant military aristocracy in the first half of the 2nd millennium. The great empires of Mesoamerica and South America were similarly built around large bodies of specialist warriors. In Europe an elite knightly class developed from the late 11th century, eclipsing the more widely spread military obligations of earlier centuries. For 200 years these knights provided the backbone of the crusader armies that set out to recover and protect the Holy Land from the Muslims.

Medieval state structures took many forms. Some were extremely loose associations, such as the merchant communities of Viking Russia. While these did support a king, his rights were very limited and he and his fellow merchant oligarchs did little more than exact relatively small amounts of tribute from largely autonomous Slav subjects.

The feudal states of western Europe, by contrast, supported an oligarchic landowning elite who exercised tight controls over their peasantry. The kings, however, again had restricted powers; it was only the development of royal bureaucracies after about 1200 that allowed them to exploit their kingdoms' taxable resources more effectively.

The vast Chinese empires were organized on yet another basis, with an oligarchy of bureaucratic families competing for power and influence through a governmental system which they entered via civil service examinations. Some Mesoamerican states, such as those of the Maya, also had literate bureaucracies, while in the 15th century even the non-literate Incas in South America used their *quipus* (knotted strings) for the record-keeping vital to any dominant imperial power.

The history of medieval empires and states was never confined to armies, bureaucracies and dominant elites. Nearly all displayed progress in art, music, architecture, literature and education. Elites everywhere patronized the arts and sponsored entertainments, as surviving examples from imperial China, Moorish Spain, early Renaissance Italy and many other places testify.

Sometimes these cultural spin-offs marked advances in themselves. In the 8th century, for example, the monasteries of Carolingian Europe produced a cursive form of writing that accelerated manuscript production for the remainder of the medieval period, and in early 15th-century Korea the world's first system of moveable metal type for book printing was introduced.

BROADENING HORIZONS

During the prehistoric period humans had become widely dispersed as they had colonized the globe. Nevertheless, many groups had maintained contacts with their neighbours, exchanging ideas and materials. The development of civilizations from the 4th millennium BC saw the establishment of direct political and trade links between geographically distant regions. Such links increased very noticeably during the medieval period, in line with advances in nautical technology.

At the turn of the millennium Viking adventurers combined the sail power and hull strength of their ships to forge the first tenuous links across the Atlantic to America. More substantial connections were developed by Muslim traders who in their dhows exploited cyclical winds and currents to expand the triangular trade that had existed since the 1st century AD between the Red Sea, East Africa and India. Beyond India the trade network extended as far east as China, from where in the early 15th century expeditions sailed to Southeast Asia and Africa. Their ships were five times the size of the Portuguese caravels in which the northwest coast of Africa was explored from 1415.

While ocean travel would produce maritime empires outside the Mediterranean only after 1500, land empires continued to ebb and flow in the medieval period, with some covering vast areas. Successive Chinese dynasties controlled states often larger than modern China. In the 7th century the power of the Western Turks ran from the borders of China to the fringes of eastern Europe, and in the 13th century the nomadic Mongols conquered a vast area of Eurasia to create the largest land empire the world has ever seen.

Political, economic and cultural ties between states all burgeoned in the medieval period, accelerating the process of making the world a "smaller" place. However, as well as generating new wealth and cultural stimulation, interaction across Eurasia brought the plague to Europe – to particularly devastating effect in the 14th century. The medieval world was a place in which empires were established and sustained by bloodshed, great art often flourished because of unequal distributions of wealth, and the triumph of Christianity and Islam came at the cost of widespread persecution.

▲ In common with the other world religions, Islam generated its own style of art and craftsmanship – of which this 14th-century mosque lamp is an example. Geometric and floral patterns adorned the walls of mosques and secular buildings, as well as pottery, glass and metalwork.

▼ Ankgor Wat, built in the 12th century, is perhaps the most impressive of the Hindu and Buddhist temple complexes that survive among the ruins of Angkor in Cambodia. Angkor was the capital of the Khmer Empire, which emerged in the 9th century and dominated mainland Southeast Asia for over 400 years.

RELIGIONS OF THE MEDIEVAL WORLD
600–1500

▲ The magnificent temple complex of Borobudur in central Java was built between 750 and 850 as an expression of devotion to Mahayana Buddhism. This carving adorns one of the temple walls.

▼ The rise of Islam from the 630s cut a swathe across the Christian Mediterranean world. By way of compensation, missionary Christianity spread ever further into northern and eastern Europe, while minority Christian regions survived in Central Asia, the Middle East and northeast Africa. Meanwhile Buddhism, marginalized in the subcontinent of its birth, extended ever further north and east, into Tibet, China, Southeast Asia, Korea and, finally, Japan. In Southeast Asia it faced in turn a challenge from Hinduism and then from Islam.

In the period 600–1500 AD all the great world religions extended their sway. Buddhism, Christianity and Islam were ultimately the most successful (*map 1*), but the older tenets of Judaism and Brahmanical Hinduism still found converts. Other ancient systems were threatened: Hellenism, the sophisticated neo-Classical philosophy of the Mediterranean world, survived only in a subordinate role, while localized "pagan" traditions and preliterate belief systems often disappeared when challenged persistently by a missionary religion such as Buddhism or Christianity – particularly if it enjoyed the backing of a government.

THE IMPACT OF ISLAM
Islam emerged in the 7th century as a mass movement of devout converts to the Koranic revelation (*pages 68–69*), men who employed warfare to help win adherents from Christianity, Judaism, Hinduism, Buddhism and the older localized faiths. It fractured the cultural unity of the Christianized Roman Mediterranean and totally eclipsed Zoroastrianism in Persia. Islamic secular culture absorbed Classical, Zoroastrian and Hindu traditions as well as those of the Arabian Desert. However, the global expansion of the Islamic world (*Dar al-Islam*) brought subdivision and even schism. The Islamic *sunna* (code of law) was variously interpreted, often regionally, by four separate law schools. Shiite partisans of dynastic leadership split right away from the consensual Sunni tradition and developed their own conventions. By the time Islam reached the Danube in Europe, the Niger in West Africa and the Moluccas in Southeast Asia in the 15th century, it was far from cohesive.

THE CHANGING FACE OF CHRISTIANITY
Although Christian minorities held on in Egypt, the Middle East and Central Asia (*map 2*), "Christendom" became increasingly identified with Europe, where both the Western (Latin) and the Eastern (Greek or Orthodox) traditions compensated for their losses to Islam by vigorous and some-

times competitive missionary activity. Latin Christianity won over Germanic-speaking peoples and their central European neighbours, while large areas of the Balkans and eastern Europe were converted to Orthodoxy. After centuries of intermittent disagreement between the Latin and Greek Churches, the Great Schism of 1054 finally brought about the divide between Catholicism and Orthodoxy.

The crusades of 1095–1291 to the Holy Land were essentially counter-productive (*pages 94–95*). They put Muslims forever on their guard against Latin Christendom and may have added to the pressure on communities of oriental Christians to convert to Islam. Militant Latin Christendom was more successful in the Baltic region and the Iberian Peninsula, where the later medieval period saw the political reconquest of all Moorish territory. By 1500 Spain had become a launchpad for transatlantic ventures and the transmission of Christianity to the New World.

THE SPREAD OF BUDDHISM OUTSIDE INDIA
Buddhism lost its western lands to Islam and it never regained any large-scale presence in India, the subcontinent of its birth, where the mainstream Hindu tradition predominated alongside what remained of the Jain faith. Buddhist numbers were increasingly concentrated in lands to the east and north and, paradoxically, Buddhist strength was at its greatest where there was ideological power-sharing with other faiths – the case in both China and Japan (*map 3*).

In China the secular philosophy of Confucianism was revitalized during the Tang dynasty of the 7th to 9th centuries, retaining its classical status and control of the education system. It offered moral and intellectual guidelines for a life of public service, virtuous prosperity and happiness to members of the scholar gentry, including the "mandarins" of the Chinese civil service. Buddhism remained – like the indigenous Chinese philosophy or "way" of the Dao (Tao) – as an alternative, culturally sanctioned code, appealing to those who could never hope to

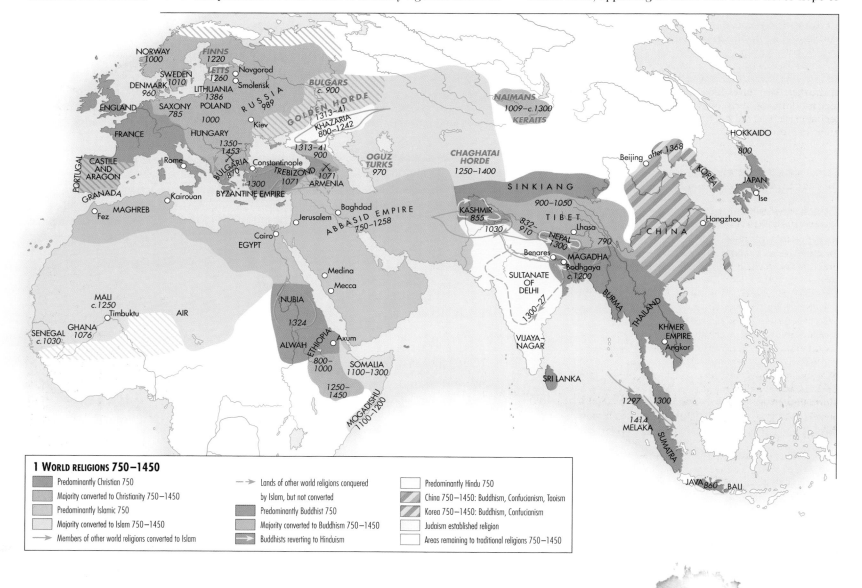

1 WORLD RELIGIONS 750–1450

- Predominantly Christian 750
- Majority converted to Christianity 750–1450
- Predominantly Islamic 750
- Majority converted to Islam 750–1450
- → Members of other world religions converted to Islam
- ⇢ Lands of other world religions conquered by Islam, but not converted
- Predominantly Buddhist 750
- Majority converted to Buddhism 750–1450
- Buddhists reverting to Hinduism
- Predominantly Hindu 750
- China 750–1450: Buddhism, Confucianism, Taoism
- Korea 750–1450: Buddhism, Confucianism
- Judaism established religion
- Areas remaining to traditional religions 750–1450

achieve the Confucian scholarly ideal or who found its secular priorities unsatisfying.

In Japan Buddhism had been adopted from China by the 6th century. It became remarkably pervasive and was intellectually and spiritually creative, bringing literacy to the whole country – but it never ousted Kami (Shinto), a traditionalist compendium of reverence for nature, land and state which remained intrinsic to Japanese cultural identity.

ORGANIZATIONAL AND CULTURAL PARALLELS

Despite profound divergences in creed and world outlook, the major medieval faiths had organizational and cultural parallels. All had "professional" adherents who adopted a consciously devout, disciplined or even ascetic way of life. While the reclusive tradition of withdrawal to the wilderness pervaded a range of religious cultures, hermits and wandering "holy men" were never as influential as members of disciplined religious orders and brotherhoods. The *Sangha* (monastic order) was central to the life of the Buddhist world and included nuns; the Persian *Sufi* movement was vital to the spread of Islam among the ordinary people; the great Benedictine houses of western Europe preserved a cultural and political inheritance through centuries of feudal disorder – as did, in a similar political context, the great Buddhist houses of medieval Japan. However, when mendicancy appeared in the West, with the establishment in the 13th century of wealthy orders of friars, it was very different from the contemplative and ascetic mendicancy of the East.

Medieval religions offered practical services to state and society. In many countries the educated clergy were the only people able to write and therefore worked as official scribes. Churches, mosques and temples operated a broadcasting system and communications network, and pilgrims and travellers could expect hospitality from religious foundations. Members of many religious communities were adept at acquiring communal or institutional (as distinct from personal) wealth. They could operate as financiers and at the same time expand their sphere of influence; thus Hindu temples were the banks of South India and 15th-century Portuguese overseas enterprise was funded by the crusading Order of Christ.

Much of the ritual year was defined by medieval religion and, where communal prayer was an obligation, the hours of the day. The spires, domes and towers of religious architecture dominated the skylines of major cities. Yet remote regions retained old beliefs and customs: there were fringe areas in Mesopotamia where sects clung to the traditions of the temples as late as the 11th century, and the 14th-century traveller Ibn Battutah found West African Muslims, even some of those who had made the pilgrimage to Mecca (the *hajj*), amazingly relaxed in their religious observance.

CHALLENGES TO THE ESTABLISHED RELIGIONS

Challenges to the established religions came from within rather than from residual "old beliefs". The Buddhist world, for example, saw the development of eccentric and magical practices on the margins of the Tantric tradition, while early Islam experienced a succession of breakaway movements from the mainstream Sunni community – Kharijite, Ibadhi and a range of Shiite alternatives. In the Christian world many "heresies" countered established orthodoxy. Medieval religious culture was not necessarily intolerant: pilgrimage, a universal form of devotion, could be a mind-broadening experience, and different religions were sometimes capable of coexistence and even co-operation. For example, in the 13th century, at the height of the Christian reconquest of Moorish territory in Spain, Santa Maria La Blanca in Toledo functioned peacefully as the mosque on Fridays, the synagogue on Saturdays and the church on Sundays.

▶ The Buddhist canonical divide between the Mahayana and Theravada traditions continued to follow Asia's cultural and ethnic faultlines. Wherever it took root in Southeast Asia, such as Annan (Vietnam), the Mahayana tradition was widely regarded as "Chinese" Buddhism, while recognition of the Theravada tradition was associated with independence from the influence of Chinese culture.

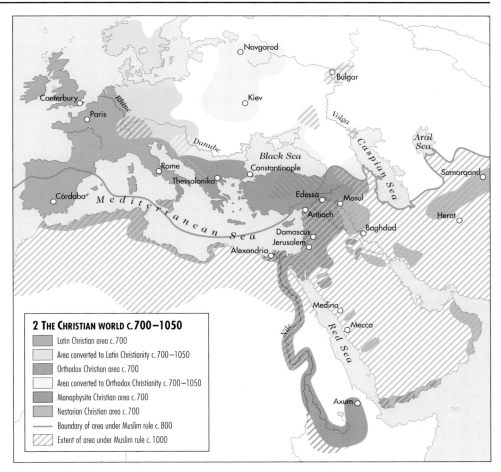

2 THE CHRISTIAN WORLD c. 700–1050

- Latin Christian area c. 700
- Area converted to Latin Christianity c. 700–1050
- Orthodox Christian area c. 700
- Area converted to Orthodox Christianity c. 700–1050
- Monophysite Christian area c. 700
- Nestorian Christian area c. 700
- Boundary of area under Muslim rule c. 800
- Extent of area under Muslim rule c. 1000

▲ The last three centuries of the first millennium AD saw the steady development of a deep and lasting cultural divide – between an Eastern, Greek-rooted Orthodox tradition and a Western, Latin-based Catholic culture. Both lost both lands and devotees to Islam in the Near East and North Africa, but resilient Christian communities continued to survive in these areas under Muslim rule.

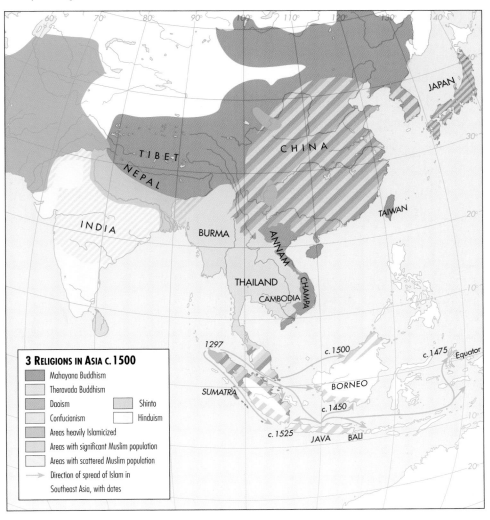

3 RELIGIONS IN ASIA c. 1500

- Mahayana Buddhism
- Theravada Buddhism
- Daoism
- Confucianism
- Areas heavily Islamicized
- Areas with significant Muslim population
- Areas with scattered Muslim population
- Shinto
- Hinduism
- Direction of spread of Islam in Southeast Asia, with dates

⬤ THE BIRTH OF WORLD RELIGIONS 1500 BC–AD 600 *pages 44–45* ⬤ THE REFORMATION AND COUNTER-REFORMATION IN EUROPE 1517–1648 *pages 154–55*

KINGDOMS OF SOUTHEAST ASIA
500–1500

▼ Angkorean power reached its greatest height during the reign of Jayavarman VII (r. 1181–c. 1218). His capital was Angkor, at the centre of which was Bayon, a huge pyramidical temple and one of more than 900 Buddhist temples built by Khmer rulers from the 9th century onwards. While the Angkhorean *mandala* dominated the mainland of Southeast Asia for four centuries, the empire of Srivijaya gradually gained control of many of the ports and polities scattered along the coasts of the archipelago. Although not the closest of these polities to the sources of major trade commodities – such as camphor, sandalwood, pepper, cloves and nutmeg – Srivijaya did have the advantage of possessing a rich agricultural hinterland.

In the 6th century Southeast Asia was a region in which warfare was endemic and the borders of political entities, known as *mandalas*, expanded and contracted with the power of their overlords. The influence of India was evident in the widespread practice of Hinduism and Buddhism (*pages 44–45*). Also evident was the influence of China, which under the Han dynasty had first begun to administer the area of Nam Viet (in what is now northern Vietnam) in 40 AD (*map 1*). In 679 the Chinese Tang government set up a protectorate-general in the area and the Chinese commanderies – in particular, that in Chiao-Chih – became important trade centres. There were, however, many rebellions, and in 938 independence from China was secured and the Dai Viet kingdom established. To the south of Nam Viet was Champa, where fishing, trade and piracy were more important economic activities than agriculture.

THE KHMER KINGDOMS
In about 550 the capital of the great Hindu kingdom of Funan, Vyadhapura, was conquered by King Bhavavarman of Chen-la. Regarded as the first state of the Khmers – one

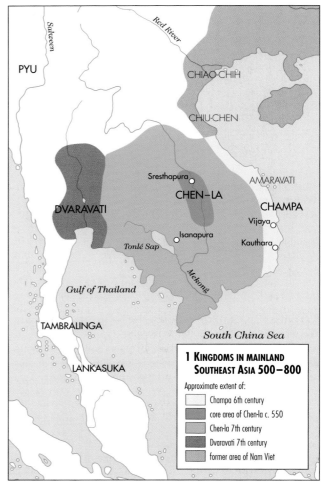

▲ By the 6th century Champa included areas that had previously been part of Nam Viet to the north and the great Hindu kingdom of Funan to the south. Funan was finally conquered in 550 by Chen-la, a kingdom that had once been its vassal.

of the many ethnic groups in the region – Chen-la had by the 7th century expanded its power throughout much of mainland Southeast Asia. In 802 the Khmer king Jayavarman II established the Angkorean *mandala*, the forerunner of modern Cambodia, which was to dominate central mainland Southeast Asia until the 13th century (*map 2*). His new capital at Hariharalaya was on the great inland sea of Tonlé Sap – the key to the floodwaters of the Mekong that were essential for the intensive rice irrigation schemes on which Angkor depended.

THAI AND BURMESE KINGDOMS
The hold of the Khmers over central mainland Southeast Asia was to be broken by the Thais. In the middle of the 7th century the Thais had formed the kingdom of Nanzhao in southwestern China. Perhaps partly due to pressure from the Chinese, they had moved south along the river valleys into Southeast Asia, conquering the Buddhist kingdom of Pyu in the middle of the 8th century. Around 860 a Thai polity in the area of modern Thailand was founded with its capital at Sukhothai (*map 2*). It was the first of three Thai kingdoms to emerge on the Chao Phraya River, displacing earlier Hindu kingdoms such as Dvaravati. The invasion of southwest China by Mongol forces under Qubilai Khan in 1253–54 pushed more Thais south – probably from the region of Nanzhao – and the Thai kingdom centred at Chiengmai was founded around 1275, followed further south by Ayuthia in 1350 (*map 3*).

The Burmese kingdom of Pagan was established shortly after Angkor emerged in Cambodia in the 9th century (*map 2*). In 1044 Anawratha ascended the throne and did much to extend the realm of the Pagan kings, the greatest of whom was Kyanzittha (r. 1082–1112). These kings built one of the most elaborate and extensive Buddhist monuments in the world in their capital at Pagan, where vast temple

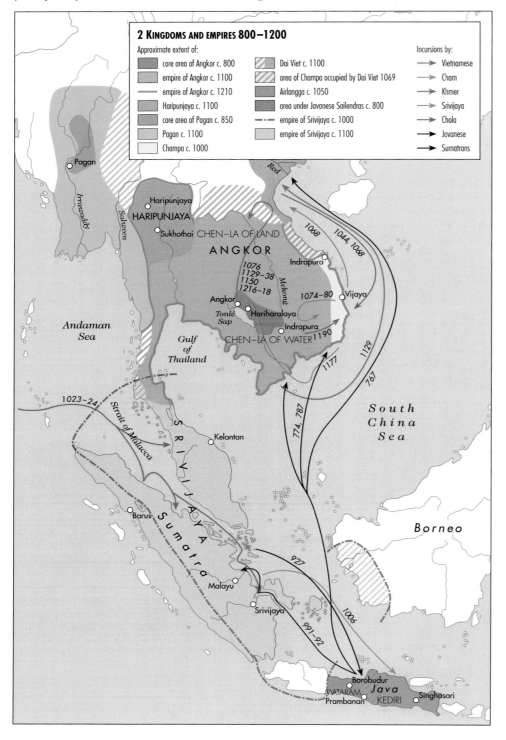

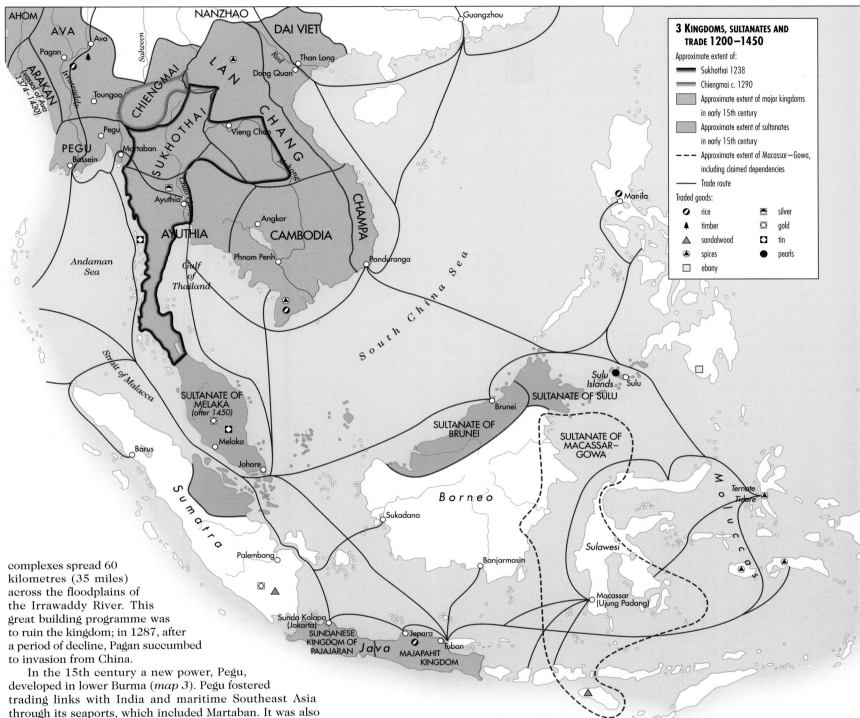

complexes spread 60 kilometres (35 miles) across the floodplains of the Irrawaddy River. This great building programme was to ruin the kingdom; in 1287, after a period of decline, Pagan succumbed to invasion from China.

In the 15th century a new power, Pegu, developed in lower Burma (*map 3*). Pegu fostered trading links with India and maritime Southeast Asia through its seaports, which included Martaban. It was also often in conflict with the inland agricultural state of Ava, which craved access to the ports controlled by Pegu. Despite occasional support from Ming China, the rulers of Ava were constantly harassed by the Shan hill peoples, culminating in the assassination of the king in 1426, and as a result Ava eventually gave up its ambitions regarding Pegu.

THE EMPIRE OF SRIVIJAYA

Throughout the Malaysian Peninsula and much of island Southeast Asia, maritime empires flourished. The empire of Srivijaya (c. 670–1025) (*map 2*), with its centre near the modern port of Palembang in Sumatra, was based on control of the resources of the forests and seas of the Indonesian archipelago. The city blossomed, its wealth reflected in ceremonial centres such as those described by the 7th-century Chinese traveller I Ching, where 1,000 priests served gold and silver Buddhas with lotus-shaped bowls.

In central Java, kingdoms had developed by the 6th century in which some of the greatest monuments of the ancient world were to be constructed (*map 2*). The Sailendras, one of the central Javanese royal lineages, supported Mahayana Buddhism, a patronage that found its greatest expression in the magnificent temple complex of Borobudur, built between 750 and 850. As well as being devout the Sailendras were aggressive warriors, and they mounted a series of seaborne expeditions against kingdoms on the mainland: Chiao-Chih in 767, Champa in 774 and Chen-la of Water in around 800. They kept control of Chen-la of Water until it was taken over by the Khmer Empire. They also held sway over large areas of Sumatra. However, after 860 control over Java moved from the Sailendras to Hindu lineages, including the builders of the great Hindu complex at Prambanan.

In the 11th century a new power emerged in east Java, and control of the international trade routes began to slip away from Srivijaya. In 1025 this process was hastened when the Srivijayan capital was sacked by Chola invaders from south India. Airlangga (c. 991–1049) was one of the most important of the rulers of this east Javanese realm, which came to dominate and grow wealthy on the burgeoning international trade in spices. Following Airlangga's death in 1049 the realm was divided in two, with Singharasi to the east and Kediri to the west. In the mid-13th century the rulers of Singhasari took over Kediri to lay the foundations of the great maritime empire of Majapahit, which controlled the region until the 15th century.

▲ The trade routes that had facilitated the spread of Hinduism and Buddhism to Southeast Asia also encouraged the spread of Islam. It reached the northern tip of Sumatra in the 13th century; by the 15th century it had reached Malaya and Java. A number of Muslim states were created at the expense of the faltering Majapahit kingdom, including one based on Melaka, a thriving commercial port which by the end of the 15th century controlled the Strait of Malacca. In 1511 Melaka fell to the Portuguese, thus ushering in an era during which Europeans wreaked great change on the Muslim, Buddhist and Hindu kingdoms and empires of Southeast Asia.

◐ EURASIAN TRADE 150 BC–AD 500 *pages 52–53* ▶ EUROPEANS IN ASIA 1500–1790 *pages 118–19*

THE BYZANTINE EMPIRE
527–1025

Throughout their history the Byzantines described themselves as Romans, and saw their empire as the continuation, without break, of the Roman Empire. Consequently, to give a starting date for the Byzantine Empire is a matter of debate among historians. The date of 527, when Justinian became emperor and launched a far-reaching campaign of conquest, is one of several options. Others include 330, when the Roman emperor Constantine the Great moved his capital to the city of Byzantium, naming it Constantinople, and 410, when Rome was sacked. Yet another is 476, when the Western Empire virtually ceased to exist, leaving Constantinople and the Eastern Empire as the last bastion of Christian civilization.

FLUCTUATING BORDERS
The history of the empire is one of constantly fluctuating borders as successive emperors campaigned, with varying degrees of success, against Persians and Arabs to the east, and Avars, Slavs, Bulgars and Russians to the north and west (*map 1*). Two of the most successful conquering emperors were Justinian (in power from 527 to 565) and Basil II (co-emperor from 960 and in sole authority from 985 to 1025). Justinian looked to the west to regain the old empire of Rome, and he and his general Belisarius conquered North Africa and Italy, while struggling to hold the eastern frontier. However, the resources of the empire were not sufficient to retain this ground, and during the 7th century most of these territorial gains were lost. The rise of Islam offered a new enemy with whom the empire was to be in

conflict until finally succumbing in 1453 (*pages 96–97*). In the four centuries between the reigns of Justinian and Basil, emperors never ceased both to fight and to negotiate for territory. However, it was in the 11th century that Byzantium made its greatest gains to the west, with Basil "the Bulgar-Slayer" bringing the entire Balkan peninsula under Byzantine control after defeating the Bulgarians. Basil also forged links with the Rus and Vikings to the north, employing them as troops in his wars of conquest.

ADMINISTRATIVE STRUCTURE
Totalitarian in ambition and ideology, absolute in his power to intervene directly in every aspect of both government and life itself, the emperor was the beginning and end of the political and administrative structure. Initially this was based on the Roman system of provincial government. In the 7th century, however, the traditional Roman provinces were reorganized into large units called "themes" (*map 2*), where the military commander also functioned as civil administrator and judge. The population of each theme provided the basis of recruitment for the army, which took the form of a peasant militia. Ordinary soldiers were given land in frontier regions and exempted from taxation in exchange for military service. By the 8th century the themes were the centres of revolts, with theme generals becoming pretenders to the imperial throne. Consequently, throughout the 8th and 9th centuries the central government worked to diminish the power of large themes, and by the 11th century the military commanders had been replaced by civil governors.

CHURCH AND STATE
Byzantium saw itself as the Christian empire under God, its mission to reduce the world to one empire. Church and state were inextricably linked. Ecclesiastical organization was as hierarchical as that of the state. Five patriarchates, based at Constantinople, Rome, Jerusalem, Alexandria and Antioch, marked out the centres of Christian worship in the Late Roman period and fought for supremacy in the Church. By the 11th century, however, the three oriental sees were no longer part of the empire, and in the ensuing centuries it was the struggle between Rome and Constantinople that affected the course of Byzantine history. Beneath the patriarchs was a system of bishoprics, within which the bishops derived considerable influence from their control of all ecclesiastical properties and charitable institutions. The empire also extended its influence through missionary expeditions, above all in the strategically important Balkan area (*map 3*).

▼ In the 7th century the traditional Roman provinces were reorganized into large *themes* that were ruled initially by military commanders. This was the first step to ending a system in which the expansion and defence of the empire depended on the deployment of mercenary armies and the imposition of high levels of taxation on the peasantry.

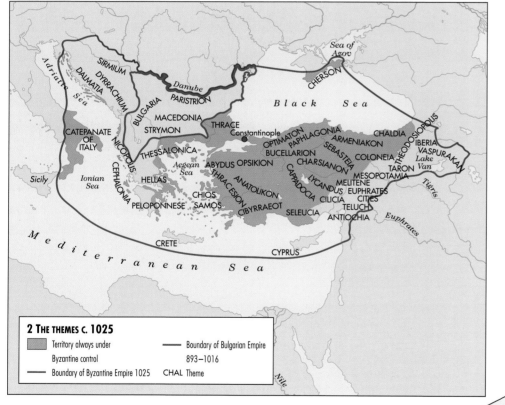

2 THE THEMES C. 1025

- [shaded] Territory always under Byzantine control
- [line] Boundary of Byzantine Empire 1025
- [line] Boundary of Bulgarian Empire 893–1016
- CHAL Theme

▶ Under Justinian the Roman provinces of Africa (533–34) and Italy (535–40) were reconquered. From the mid-6th century, however, defensive warfare became endemic, and in the early 7th century attacks by the Avars and Arabs led to the virtual extinction of the empire. A prolonged period of determined defence followed before Basil II succeeded in expanding the boundaries once more in the 11th century.

1 BOUNDARIES AND CAMPAIGNS OF CONQUEST 527–1025

- [shaded] Byzantine Empire 527
- [shaded] Conquests of Justinian 527–65
- [line] Boundary of empire 527
- [dashed line] Boundary of empire 565
- [line] Boundary of empire 1025
- [arrow] Byzantine campaigns of reconquest
- ✕ Arab raid
- ✕ Bulgarian raid
- ✕ Russian raid
- AV Empire's neighbours 565
- MA Empire's neighbours 1025
- ■ Theme capital c. 1025

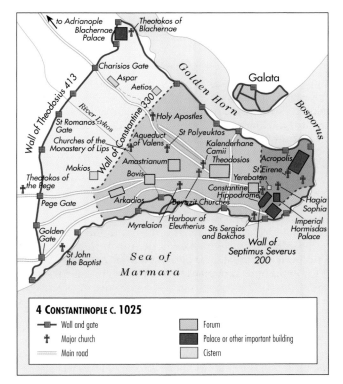

4 CONSTANTINOPLE C. 1025
- Wall and gate
- † Major church
- Main road
- Forum
- Palace or other important building
- Cistern

Simply called "the City" (*map 4*), Constantinople was the most important city in the empire. It was the emperor's base, and thus the centre of all civil, military and ecclesiastical administration. Its position was almost unassailable, as the Muslim armies who attempted to capture it in the 7th and 8th centuries discovered (*pages 66–67*). For almost 900 years it withstood all attacks by enemy forces until, in 1204, it was overrun and ransacked by the army of the Fourth Crusade.

▼ The main trade routes were sea or river-based and the chief centres of trade were on the coast. Dominant among them was Constantinople, which not only served as the emperor's capital but also as the heart of Christendom for many centuries.

▲ The transformation of the small town of Byzantium into the city of Constantinople was accomplished remarkably quickly. There is evidence that by the middle of the 4th century there were 14 palaces, 14 churches, 8 aqueducts, 2 theatres and a circus, as well as homes for the inhabitants who were forced to move to the city from nearby settlements. Comparatively little now survives of Byzantine Constantinople in present-day Istanbul, but Hagia Sophia, the great church built by Justinian as a centre of worship for all Christendom, can still be seen, along with a host of lesser churches. A handful of imperial monuments exist, the most obvious of which are the 5th-century city walls in the shape of an arc almost 6 kilometres (4 miles) long.

The importance of religion in the empire is reflected in its surviving artistic achievements. Churches and monasteries, often beautifully decorated with mosaics and wall paintings, are to be found throughout the empire's territories, along with portable works of art, such as enamels, books, metalwork and, above all, icons. The few secular buildings and objects that remain are often in Late Roman cities such as Ephesus – gradually abandoned in the 7th century – but most notably in Constantinople.

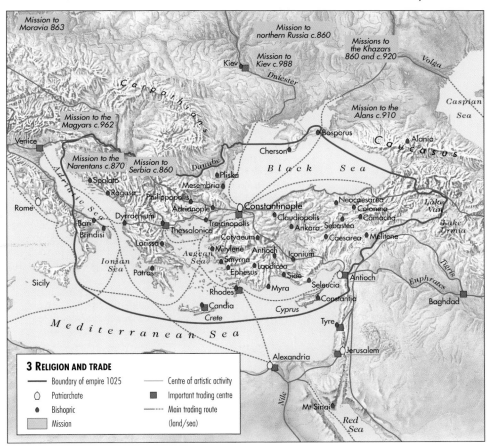

3 RELIGION AND TRADE
- Boundary of empire 1025
- ◯ Patriarchate
- Bishopric
- Mission
- Centre of artistic activity
- Important trading centre
- Main trading route (land/sea)

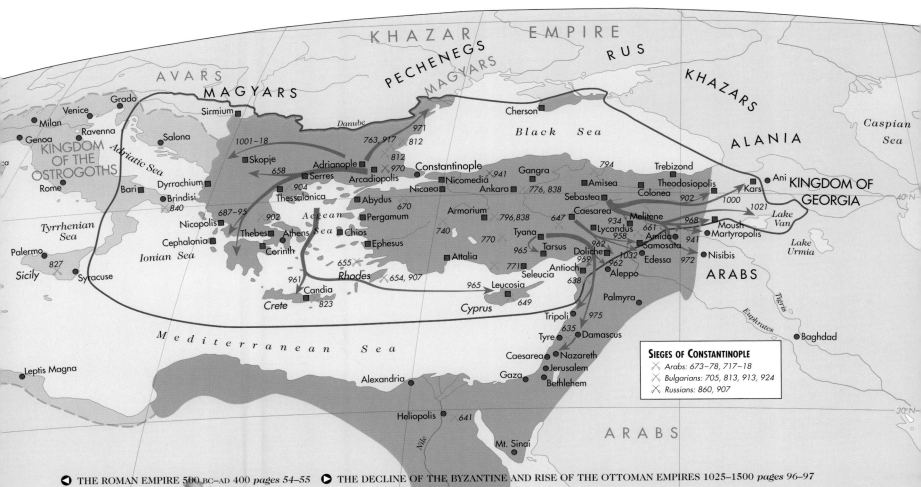

SIEGES OF CONSTANTINOPLE
- ✕ Arabs: 673–78, 717–18
- ✕ Bulgarians: 705, 813, 913, 924
- ✕ Russians: 860, 907

◀ THE ROMAN EMPIRE 500 BC–AD 400 *pages 54–55* ▶ THE DECLINE OF THE BYZANTINE AND RISE OF THE OTTOMAN EMPIRES 1025–1500 *pages 96–97*

THE SPREAD OF ISLAM
630–1000

In the second quarter of the 7th century AD the map of the world was abruptly and irreversibly changed by a series of events that astonished contemporary observers. From the 630s the tribes of the Arabian Peninsula, previously accorded little attention by the "civilized" world, burst out of their homelands and attacked the fertile regions to the north in a series of campaigns that resulted in the complete destruction of the Sasanian Empire and the end of Byzantine control of the Near East. They then set about forging a new social and cultural order in the conquered territories, based on the principles of the religion they brought with them – a force which has continued to exert a profound influence over the region to the present day.

MUHAMMAD: THE "PROPHET"

In the early years of the 7th century tribal Arabian society underwent a transformation: a new communal structure emerged to replace the traditional tribal divisions that had hitherto dominated the Arabian Peninsula. This community was largely the creation of a single man, Muhammad, a trader from Mecca, the main commercial town of western Arabia. Following divine revelations in which he identified himself as the "Seal of the Prophets" (after whom no others would come), Muhammad preached a new moral system that demanded the replacement of idol worship with submission to a common code of law and the unity of Muslims ("those who submit [to God]") against unbelievers.

Although he was persecuted by the Meccans in the early years of his mission, Muhammad later enjoyed rapid success in nearby Medina, where he made many converts and laid down the rules governing the conduct of the community. Thereafter he sent missionaries to spread his message throughout Arabia, and shortly before his death (probably in 632) he led his triumphant army back to reclaim Mecca.

THE VICTORY OF ISLAM

Within a decade of Muhammad's death the Muslim armies – inspired by zeal for their new faith and a desire for plunder – had inflicted defeat on both regional superpowers, the Byzantines and the Sasanians, already weakened by decades of conflict with each other. The Muslim victories at Yarmuk and Qadisiyya (in 636) opened the way to further expansion (*map 1*). In 642 the Muslim armies conquered Egypt, by the mid-640s Persia was theirs, and by the late 640s they had occupied Syria as far north as the border with Anatolia.

The wars of conquest continued, albeit at a lesser pace, for roughly a century after the humiliation of the Byzantines and Sasanians. After overrunning the whole of the North African coastal region and taking root in much of the Iberian Peninsula, the Muslim state reached the limits of its westward expansion into Europe at the Battle of Poitiers in central France in 732. The one realistic prize which always eluded these conquerors was Constantinople: in spite of several Muslim attempts to capture it by siege, it remained the Byzantine capital until 1453.

INTERNAL CONFLICT

The euphoria generated by these successes was tempered from the start by disagreements between Muslims concerning several matters – including, most crucially, the question of who was to lead the community. The Prophet had combined both religious and political authority in his own person and this model was followed for the first three centuries by the caliphs who led the community after him. However, Muhammad had made no arrangement for the succession, and more than once in the century after his

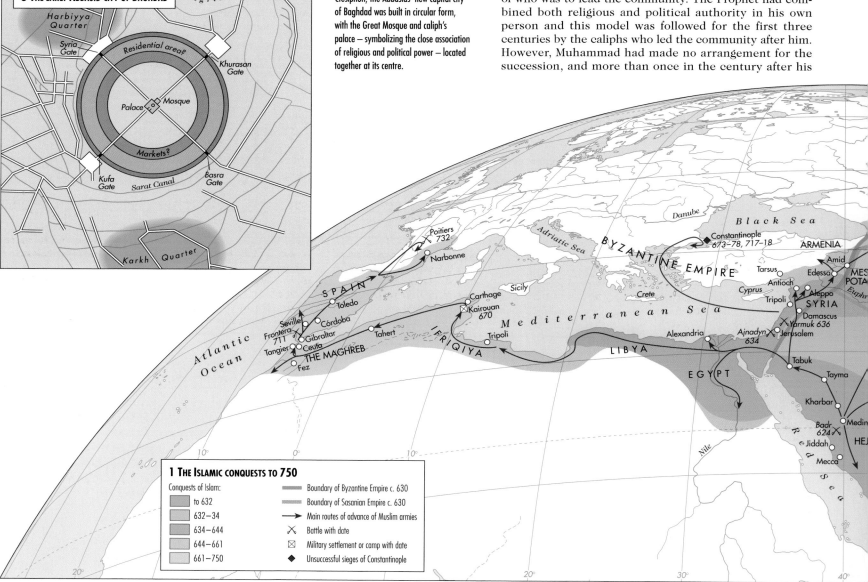

3 THE EARLY ABBASID CITY OF BAGHDAD

◄ Near the ancient Sasanian capital of Ctesiphon, the Abbasids' new capital city of Baghdad was built in circular form, with the Great Mosque and caliph's palace – symbolizing the close association of religious and political power – located together at its centre.

1 THE ISLAMIC CONQUESTS TO 750

Conquests of Islam:	
to 632	Boundary of Byzantine Empire c. 630
632–34	Boundary of Sasanian Empire c. 630
634–644	→ Main routes of advance of Muslim armies
644–661	✕ Battle with date
661–750	⊠ Military settlement or camp with date
	◆ Unsuccessful sieges of Constantinople

death the Islamic world was thrown into turmoil by fiercely contested civil wars fought over this issue.

In spite of such upheavals, political power was consolidated at an early stage in the hands of the first Islamic dynasty, the Umayyads, who ruled from their capital in Damascus for nearly 100 years (661–750). Although much maligned by later Muslim writers, this caliphal dynasty succeeded in giving an Arab Muslim identity to the state. The caliph Abd al-malik b. Marwan (d. 705) decreed that Arabic (instead of Greek or Pahlavi) should be the language of administration, began a programme of religious building, and instituted a uniform Islamic coinage. Trade flourished in the region, with Syria in particular benefiting from the revenues flowing into the caliph's coffers.

THE ABBASID DYNASTY

In the middle of the 8th century a new dynasty, the Abbasids, toppled the Umayyads, whom they accused of ruling like kings rather than caliphs – without the sanction of the community (*map 2*). Abbasid rule witnessed a real change in the Muslim state, with the caliphs constructing a grand new capital of Baghdad (also known as the City of Peace) in Mesopotamia (*map 3*). It is no coincidence that Abbasid courtly culture borrowed heavily from that of the Persian royalty, for the focus of Muslim culture now swung eastwards from Syria.

At the same time as Islam was expanding internally, Muslim eyes and minds began to be opened to a wider world, both through growing trade – in particular with the Far East – and through a burgeoning interest in ancient knowledge, primarily Greek, which was furthered by the translation into Arabic of foreign books.

Like their predecessors, however, the Abbasids failed to gain universal acceptance for their claim to be the legitimate leaders of the Muslim world. Although the caliphs continued to rule in Baghdad until they were deposed by the Mongols in the mid-13th century, they gradually lost their

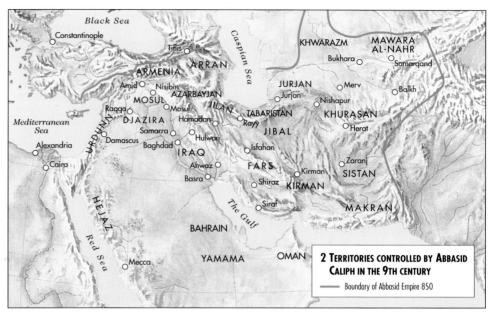

2 TERRITORIES CONTROLLED BY ABBASID CALIPH IN THE 9TH CENTURY
— Boundary of Abbasid Empire 850

territories to local warlords, rulers who governed independently while still proclaiming formal subservience to the caliph. Parts of North Africa, far from the seat of caliphal power, began to fall outside caliphal control practically from the first years of Abbasid rule. By the beginning of the 10th century a rival caliphate was set up in Egypt, and Iraq and Iran were divided into petty kingdoms, many ruled by Iranian kings (*map 4*). In the 11th century these kingdoms were swept away by the steppe Turks who invaded the Muslim world and changed the ethnic and cultural map as decisively as the Arabs had done four centuries earlier.

▲ Rapid urbanization followed the rise of the Abbasids, particularly in Iraq and Persia, as would-be converts flocked to the cities from the countryside. It has been estimated that while only 10 per cent of the population of these regions was Muslim when the Abbasids came to power, within a century this figure had grown to 50 per cent – and had reached 90 per cent by the beginning of the 10th century.

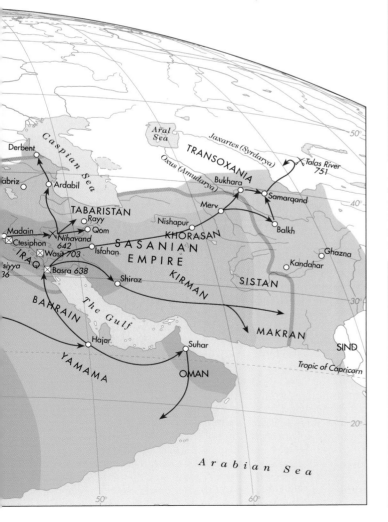

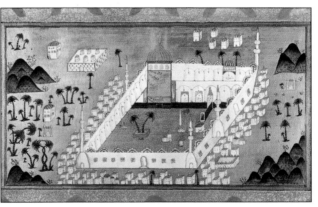

▲ By 750 Islam was the major civilization west of China and one in which there was a particularly close association between religion and culture. Mosques served not only as religious and social centres but also as centres of scholarship, which was overwhelmingly Arab in orientation, although influenced by Greek, Roman, Persian and Indian traditions. This painting of Medina, with the mosque of Muhammad at its centre, comes from an illustrated Persian text written in Arabic.

▼ As the political unity of the Muslim state began to disintegrate, local cultures reasserted themselves. The Samanid kings (819–1005) who ruled from their capital in Bukhara encouraged the composition of Persian poetry at their court, while their western rivals, the Buyid rulers of Iraq and Persia (932–1062), held the caliph captive in his palace and styled themselves Shahanshahs like the Persian kings of old.

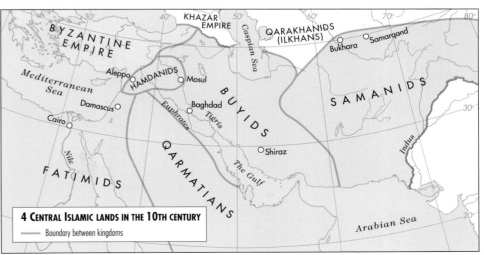

4 CENTRAL ISLAMIC LANDS IN THE 10TH CENTURY
— Boundary between kingdoms

◀ THE BYZANTINE EMPIRE 527–1025 *pages 66–67* ▶ THE MUSLIM WORLD 1000–1400 *pages 88–89*

THE FIRST SLAVIC STATES
400–1000

▲ An early Premyslid ruler of Bohemia, Prince Wenceslas in 925 overthrew his mother who, as regent, was persecuting the Christians. He continued the Christianization of Bohemia but this, together with his submission to the Germans, aroused opposition, and in 929 he was killed and succeeded by his brother, Boleslav I. This portrait of the prince, the patron saint of the Czechs, was painted by a member of the Czech School in the 16th century.

▼ In the 9th and early 10th centuries Slavic states formed in Moravia, Poland and Bohemia. Polish and Bohemian rulers used fortified administrative centres to dominate previously independent tribes. While Great Moravia was based on large urban centres on the River Morava, state formation among the Elbe Slavs was held in check by the power of the German duchies, notably Saxony under Otto I.

It is evident from first archaeological traces of the Slavs that in the 3rd and 4th centuries they lived in the fertile basins of the Vistula, Dniester, Bug and Dnieper rivers (*map 1*). In the early 5th century, however, the nomadic Huns conquered and drove out Germanic peoples to the west of this area (*pages 56–57*), allowing the Slavs to move as far as the Danube frontier of the Byzantine Empire by around 500. The subsequent victories over the Byzantines by a second nomadic people, the Avars (*pages 76–79*), meant that Slavic groups were able to penetrate southeastwards into the Balkans and even the Peloponnese. At the same time Slavs also moved north and west as Avars encroached on their territory.

As a result, most of central Europe as far west as the Elbe was settled by Slavs – Moravia and Bohemia had been settled by 550, and much of the Elbe region by 600. The process can be traced archaeologically in the emergence and distribution of various Slavic cultures, which are mainly distinguished by the pottery they produced.

In the 6th century the Slavs operated in numerous small and independent social units of a few thousand. Some had kings, but there were no established social hierarchies and no hereditary nobility – merely freemen and slaves. Slavs were particularly ready to adopt captured outsiders as full members of their groups, and this partly explains why they were able to Slavicize central and eastern Europe in such a relatively short period of time. They lived in small, unfortified villages, grew crops and raised animals.

However, from the 7th century, hillforts – each serving as a local centre of refuge for a small social unit – became the characteristic form of Slavic settlement, and several thousand have been found in central and eastern Europe. They subsequently merged into larger, more organized political entities, the first of which evolved in Moravia in the 9th century (*map 2*) but was swallowed up by Magyars moving westwards from around 900 (*pages 76–77*).

ECONOMIC TRANSFORMATION

After about 500 Slavic agriculture became more productive thanks to the adoption of Roman ploughs and crop rotation. This agricultural revolution was only one element in a wider process of economic development which, archaeologically, is reflected in the wide range of specialist manufactures, not least of silver jewellery, found on Slavic sites. Much of the Slavs' new wealth derived from contacts with economically more developed neighbours. Its greatest

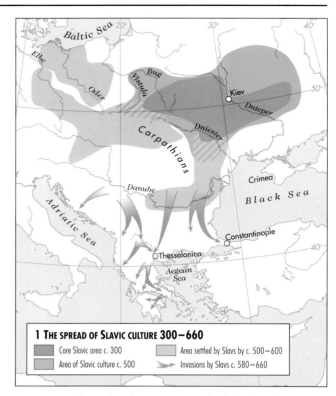

1 THE SPREAD OF SLAVIC CULTURE 300–660
- Core Slavic area c. 300
- Area of Slavic culture c. 500
- Area settled by Slavs by c. 500–600
- Invasions by Slavs c. 580–660

▲ Between around 400 and 650 Slavic-speaking groups came to dominate much of central and eastern Europe. Their spread in and around the Carpathians (to c. 550) is mirrored in the distribution of the so-called Prague Culture. Over the next century, large areas of the North European Plain were similarly colonized by Slavic peoples.

single source was the trade in slaves with the Muslim caliphates, conducted from the 8th century onwards and evidenced by hoards of Arab silver coins found in central Europe (*map 3*). Western Slavic groups and the Rus captured slaves from eastern Slavs living in the area between their respective territories. Some slaves were sold directly to Muslim (and some western) merchants in central Europe, notably in Prague, while many were shipped to the Muslim world by Scandinavian and other "middlemen". These intermediaries bought slaves at the trading centres of the south Baltic coast (such as Elbing, Wiskiauten and Grobin) and subsequently transported them down the river routes of eastern Europe, particularly the Volga, which gave direct access to the Caspian Sea and Muslim Mesopotamia.

THE FORMATION OF STATES

The slave trade played an important role in generating new political structures. Traders had to organize to procure slaves, and this, together with the new silver wealth, made possible new ambitions. In the first half of the 10th century, for example, Miesco I established the first Polish state with the help of his own armoured cavalry, which his wealth enabled him to maintain. Perhaps this force was first employed to capture slaves, but it soon took on the role of establishing and maintaining territorial control with the aid of a series of hillforts. The Premyslid dynasty of Bohemia, which originated around Prague, adopted a similar strategy, and by around 900 it controlled central Bohemia through a network of three central and five frontier hillforts. Over the following century the dynasty extended its influence much further afield and in its newly acquired territories it replaced existing hillforts, which had served for local self-defence, with fortified administrative centres in order to maintain its control.

To the east, the Rus of Kiev had by about 1000 created the first Russian state, extending their control over other, originally independent trading stations such as Smolensk, Novgorod, Izborsk and Staraia Ladoga (*map 4*). Each of these trading groups consisted of a relatively small number of originally Scandinavian traders and a much larger number of Slavs who produced the goods, shared in the profits – and quickly absorbed the Scandinavians.

2 STATE FORMATION C. 800–1000
- Polish hillfort 10th century
- Expansion of the Premyslid dynasty 9th–11th centuries
- Premyslid hillfort 9th–10th centuries
- Premyslid administrative centre 10th–11th centuries
- **HEV** Slav people

4 SLAVIC STATES C. 1000
- ☦ Episcopal see
- Dynastic core area of Poland
- **PEC** People

◄ By the year 1000 three dominant dynasties had emerged in the Slav lands of central and eastern Europe – in Bohemia, Poland and Russia – each centred on their respective capitals of Prague, Grezno and Kiev. While closely controlling their core areas, these new states also fought each other for control of the lands in between (Moravia, Volhynia, Silesia, Byelorussia), which repeatedly changed hands over several centuries. Dynastic unity in Poland and Russia was to collapse gradually in the 12th and 13th centuries, leading to partitions and the creation of less expansionist kingdoms. At the same time German expansion – at first demographic, then political – was to undermine the independence of the western Slavic states.

While Slavic state formation generally involved asserting aggressive dominion, this was not always the case. During the 10th century the Elbe Slavs – comprising the previously independent Abodrites, Hevellians and Sorbs – increasingly acted together to throw off the domination being exerted on them by Ottonian Saxony, which in the middle of the century had carved up their territories into a series of lordships or marches. However, the Elbe Slavs reasserted their independence in a great uprising of 983.

THE ADOPTION OF CHRISTIANITY
State formation also had a religious dimension. Franks and then Ottonians, the Papacy and Byzantium were all interested in sending missionaries to the Slavic lands, most famously in the mid-9th century when Cyril and Methodius went, with papal blessing, from Constantinople to Moravia.

There the brothers generated a written Slavic language to translate the Bible and Christian service materials. In the 10th century Rus, Polish and Bohemian leaders all adopted Christianity. Kiev, Gniezno and Prague, capitals of their respective states, all became archbishoprics, Kiev and Gniezno with their own episcopal networks.

Christianization allowed ambitious Slavic dynasts to sweep away not only the old Slavic gods but also the cults that were unique to each independent group and so reflected the old political order. The establishment of strong Christian churches thus contributed significantly to the process whereby the small, independent Slavic communities of the 6th century evolved into the new Slavic states of central Europe in the 9th and 10th centuries.

► From the 8th century hoards of Arab silver coins were deposited in Slavic central and eastern Europe – evidence of Slavic participation in the fur and slave trades conducted in the rich lands of the Abbasid Caliphate. Slavs also traded with the Frankish Carolingian world to the west.

3 TRADE C. 700–1000
- —— Major trade route
- ⊟ Major coin hoard
- Abbasid Empire c. 800
- Carolingian Empire c. 800
- Byzantine Empire c. 800

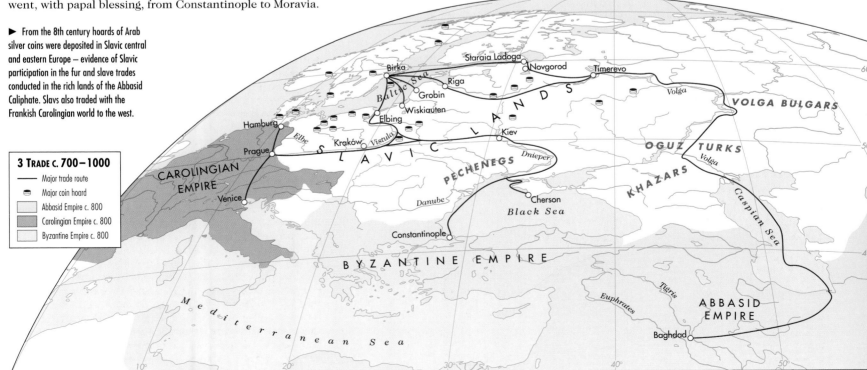

BARBARIAN INVASIONS OF THE ROMAN EMPIRE 100–500 *pages 56–57*   THE MONGOL EMPIRE 1206–1405 *pages 98–99*

EAST ASIA IN THE TANG PERIOD 618–907

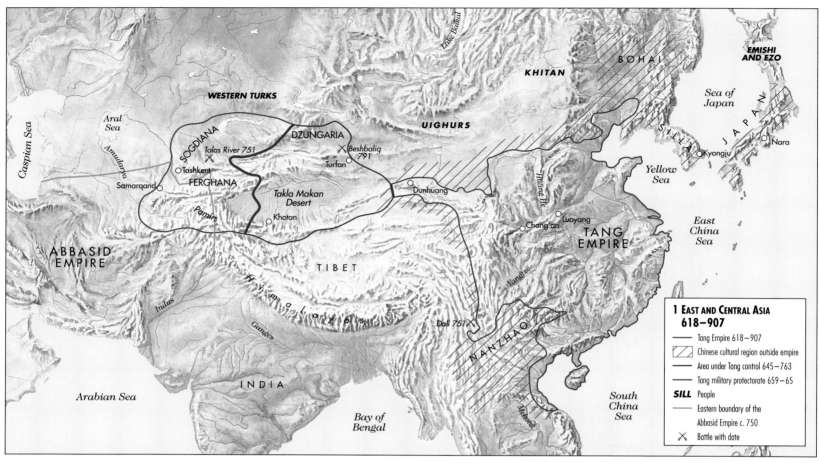

1 EAST AND CENTRAL ASIA
618–907
- Tang Empire 618–907
- Chinese cultural region outside empire
- Area under Tang control 645–763
- Tang military protectorate 659–65
- **SILL** People
- Eastern boundary of the Abbasid Empire c. 750
- × Battle with date

▲ The Tang dynasty established a vast empire – larger than any other Chinese empire before the conquests of the Manchus 1,000 years later. Throughout the empire Buddhism flourished, and Chinese pilgrims travelled along the trade routes of the Silk Road – firmly under Tang control between the mid-7th and mid-8th centuries – to visit *stupas* and shrines in India. The expansion

of the Tang was finally halted in 751 when two major defeats were inflicted on their armies – by the kingdom of Nanzhao at the Battle of Dali and by the Muslim Arabs at the Battle of Talas River. This last battle resulted in the Abbasid Empire gaining control of the area west of the Pamirs and established the boundary between the civilizations of Islam and China.

▼ The central administration controlled every province, using regular censuses to gather information about the available resources and population. (In 754 there were nearly 53 million people living in over 300 prefectures.) A network of canals linked the Yangtze Valley with areas to the north, supplying the huge army that defended the long imperial borders.

Following the collapse of the Han Empire in AD 220 China was divided into the three competing kingdoms of Shu, Wei and Wu. A brief period of unity was provided by the rule of the Western Jin between 265 and 316 before northern China fell under the control of non-Chinese chiefs, leaving the south in the hands of an elitist aristocracy. The country was reunited under the Sui dynasty – established in 581 – but the dynasty was short-lived. In 618, after four centuries of division and turmoil, the Tang dynasty took control (*map 1*).

The influence of Tang China was to be felt throughout Asia in the three centuries that followed. Its political stability and economic expansion led to the unprecedented development of links with many peoples throughout East and Central Asia, and these fostered a cultural renaissance and cosmopolitanism in China itself. Tang armies brought the trade routes of the Silk Road under Chinese control, with protectorates established as far west as Ferghana and Samarqand. In the middle of the 7th century, the Chinese Empire reached its maximum extent prior to the Manchu conquests a thousand years later. For a hundred years Tang armies were not seriously challenged, and Tang models of government were taken up by many neighbouring peoples – who in turn expanded their own spheres of influence. These included the kingdom of Nanzhao in the southwest, Bohai in the northeast, Silla in Korea and the early Japanese state centred on Heijo.

The Tang system of centralized government (*map 2*) was introduced by the second Tang emperor, Tai Zong (r. 626–649), and was supported by a professional bureaucracy of civil servants. The cities were linked to the countryside through a well-developed infrastructure of canals and roads. New agricultural land was opened up, especially in the south, and in the first part of the Tang period peasants owned their own land, paying for it in taxes and labour. Later on, however, as central power waned, wealthy and powerful landowners extended their area of control. Rural prosperity supported the growth of new industries, notably the production of fine pottery and luxury goods that were often inspired by fashionable foreign items.

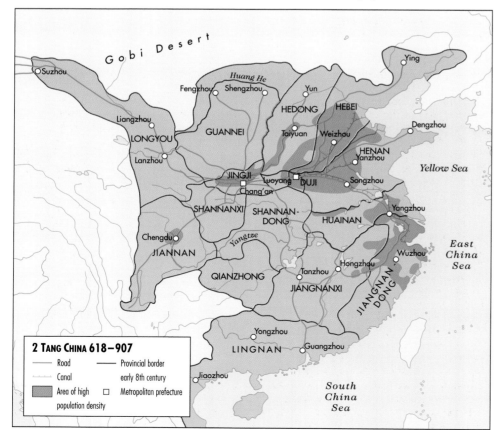

2 TANG CHINA 618–907
- Road
- Canal
- Area of high population density
- Provincial border early 8th century
- ☐ Metropolitan prefecture

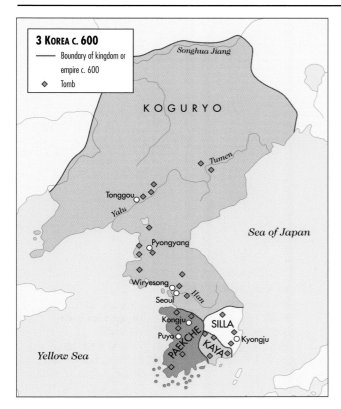

3 KOREA C. 600
— Boundary of kingdom or empire c. 600
◆ Tomb

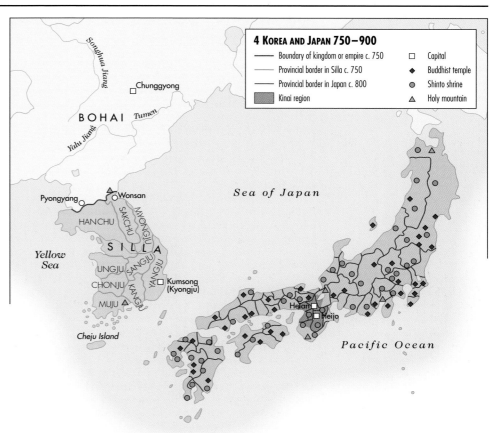

4 KOREA AND JAPAN 750–900
— Boundary of kingdom or empire c. 750
— Provincial border in Silla c. 750
— Provincial border in Japan c. 800
▨ Kinai region
□ Capital
◆ Buddhist temple
◉ Shinto shrine
▲ Holy mountain

▲ Lavishly furnished tombs, often adorned with fine paintings, housed the remains of the elite in Korean society, while the majority had simpler burials. Among the grave offerings were exquisite gold crowns and other jewellery made of gold foil and wire. Fine stoneware pottery made in the kingdom of Kaya was exported to Japan.

THE KOREAN PENINSULA

In the Korean Peninsula, Tang armies assisted the kingdom of Silla (*map 3*), which in its campaign of expansion had crushed Paekche in 660. The defeat of Koguryo in 668 marked the beginning of the unification of Korea. To the northeast the state of Bohai was established by Tae Cho-yong, a general from Koguryo who refused to surrender to Silla, and in 721 a wall was built to separate the two states (*map 4*). Silla finally compelled the Chinese to abandon their territorial claims in Korea in 735, but all through this period maintained good relations with the Chinese: Korean scholars, courtiers and Buddhist monks made frequent journeys to China, and Korean trading communities were established in eastern China. Many individual Koreans played important roles in the Tang Empire. In 747 a Chinese army was led to the upper ranges of the Indus by Ko Son-ji, a Korean military official.

THE ROLE OF BUDDHISM

Not only the Chinese and Koreans, but also the Japanese, were brought together by the spread of Buddhism from India throughout East Asia. Buddhism often received official support and many of the most spectacular Buddhist monuments in Asia were built at this time, from the cave temples at Dunhuang in China to the Horyuji and Todaiji temples in Nara in Japan. The Silla capital at Kumsong (modern Kyongju), which already boasted fine monuments such as the Ch'omsongdae observatory, was further embellished with great Buddhist structures including the Pulguk-sa temple (c. 682). However, the relationship between this new religion and the government was not always easy: in 845 Emperor Wu Zong ordered the closure of nearly 45,000 monasteries and temples throughout China in an attempt to restrict the influence of Buddhism.

DEVELOPMENTS IN JAPAN

On the Japanese archipelago a centralized bureaucratic government developed from a series of successive capitals in the Kinai region. In 710 the new capital at Heijo, near the present city of Nara, was designed by Emperor Gemmyo following Chinese principles of city planning. The subsequent Nara period saw major political, economic and land reforms as well as campaigns against the Emishi and Ezo peoples who lived north of the boundaries of the expanding Japanese state. In 794 the capital was moved to Heian (now Kyoto), ushering in the golden age of Heian civilization during which a sophisticated courtly lifestyle developed among the elite classes. In the later part of the Heian period (794–1185) the *samurai* culture, which placed great value on military prowess, also evolved.

THE DECLINE OF TANG POWER

The 9th century saw the waning of Tang influence and an ever-increasing independence in surrounding countries (*map 1*). In 751 Tang armies suffered two major defeats: at the Battle of Dali in the south, over 60,000 Tang soldiers perished at the hands of the troops of the kingdom of Nanzhao; in the west, Arabs took control of much of Central Asia in the Battle of the Talas River, which set the border between the Chinese and Abbasid empires.

The faltering of the Tang dynasty was symbolized by the rebellion of An Lushan, the commander of the northeastern armies, who gained great influence over Emperor Xuan Zong (r. 712–56) through the imperial concubine Yang Yuhuan. In 755 An Lushan rebelled against the emperor and led a force of over 100,000 men on the capital. Although the rebellion was eventually put down, the empire was greatly weakened and became vulnerable to external attacks. In 787 the Tibetans sacked the capital Chang'an, and in 791 defeated Chinese and Uighur forces near Beshbaliq, ending Chinese domination of Central Asia. As central control weakened and provinces became more powerful, China once again moved towards disintegration. Following more revolts, the last Tang emperor was deposed in 907.

China's relations with surrounding countries changed as these countries themselves changed. The last Japanese embassies were sent to China in 838, and in 894 the Japanese government, now dominated by the Fujiwara clan, officially banned travel to China. In the Korean Peninsula serious rebellions broke out in Silla in 889, and out of these rebellions was born the kingdom of Koryo, centred in the north, which was to control all of Korea from 936.

▲ Buddhism rapidly gained popularity in Japan following its introduction from Korea in the 6th century, but traditional Japanese Shinto religion was actively encouraged by 7th- and 8th-century rulers. The two creeds were brought together in the Tendai teachings of Saicho after the capital was moved from Heijo to Heian in 794, and the strong links between religion and government were subsequently severed.

▲ The long-established East Asian tradition of erecting lifesize stone terracotta guardian figures on and around tombs reached its apogee in the three-coloured glazed statues of the Tang period.

◀ FIRST EMPIRES IN CHINA 1100 BC–AD 220 *pages 48–49*   ▶ EAST ASIA 907–1600 *pages 86–87*

FRANKISH KINGDOMS
200–900

The Franks were created by the reorganization of a number of Germanic groups on the northern Rhine frontier of the Roman Empire in the 3rd century AD. They comprised several subgroups, most prominently the Salians and Ripuarians, which were further divided into warbands, each with their own king. The collapse of the Roman Empire after about 450 prompted further changes, with Childeric (d. 482) and his son Clovis (r. 482–511), uniting increasing numbers of Franks under their rule.

The two men belonged to a prominent Salian family – called the Merovingians after a legendary founder Merovech – but their careers turned the family into a royal dynasty for all Frankish peoples. At the same time, the newly united Franks were able to conquer more and more territory: Childeric started by taking over the Roman province of Belgica II, to which Clovis added the region around Paris (the kingdom of Aegidius and Syagrius), Alemannia and Aquitaine. Clovis's sons and grandsons further conquered Provence, Burgundy and Thuringian territory (map 1).

The Franks did not, however, evolve governmental structures of sufficient strength to hold this large new state together. The conquests had generated renewable wealth for kings to reward local landowners and hence attract their support, but when the conquests petered out kings had to buy support using their own landed resources, so that great men became wealthier at the expense of kings. By around 700 the real power had passed to a relatively small number of families in each of the regions of the kingdom: Austrasia, Neustria, Burgundy, Aquitaine and Provence (map 2).

In the 8th century the rulers of Austrasia in the northeast – called the Carolingian dynasty – reunited the whole Frankish world. Between about 695 and 805 their armies

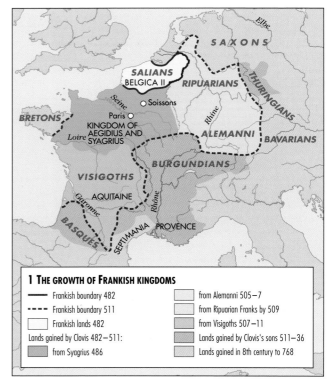

1 THE GROWTH OF FRANKISH KINGDOMS

- —— Frankish boundary 482
- ---- Frankish boundary 511
- Frankish lands 482
- Lands gained by Clovis 482–511:
  - from Syagrius 486
- from Alemanni 505–7
- from Ripuarian Franks by 509
- from Visigoths 507–11
- Lands gained by Clovis's sons 511–36
- Lands gained in 8th century to 768

▲ The collapse of Roman power in northern Gaul after about 450 facilitated the unification of the Franks and the extension of their dominion. The Romans had kept the tribes divided and weak, but Merovingian leaders Childeric and Clovis eliminated rival Frankish warlords to create a new dominant force in post-Roman western Europe.

2 THE EMPIRE OF CHARLEMAGNE AND HIS SUCCESSORS

- Kingdom of the Franks 768
- Charlemagne's empire at greatest extent c. 800
- —— Partition of Verdun 843
- —— Missatica of the Capitulary of Servais 853
- ✠ Royal palace
- ▽ Charlemagne, 1–4 stays
- ▽ Charlemagne, 5–7 stays
- ▽ Charlemagne, more than 7 stays

were on campaign for all but five years, taking advantage of an open frontier to the east. As a result, Austrasia's rulers could offer ongoing rewards to would-be supporters and thus outbid noble rivals from the other regions. In three generations – Charles Martel (d. 741), Pippin the Short (r. 741–68) and Charlemagne (r. 768–814) – the dynasty reunited Francia and conquered Lombard Italy, Saxony, Alemannia, Thuringia, Bavaria and the Avars (*map 3*). On Christmas Day 800 Charlemagne was crowned emperor in Rome.

THE STRUCTURE OF GOVERNMENT

The Merovingians based their rule on the existing Roman structures: the cities, or *civitates*, and their dependent territories. However, by about 800 the *civitates* had ceased to exist, and in their place was a patchwork of smaller counties. It was thus much easier to create continuous territories when the kingdom was divided, as between Charlemagne's grandsons in the Treaty of Verdun in 843.

The main governmental problem remained constant: how to exercise centralized control over a very large kingdom in an era of primitive communications. Powerful landowners were essential to a king's rule, but they had to be prevented from becoming too independent; continual royal travel was a central part of the strategy.

Royal finance still relied on conquest. Once expansion petered out after the conquest of Saxony (805), and especially when Louis the Pious (d. 840) was succeeded by a great number of quarrelling sons, Merovingian patterns reasserted themselves. Financial resources, above all land, were transferred by rival members of the dynasty in a bid to buy supporters. By 900 Carolingian power in West Francia was confined to the Paris region, while East Francia was run by non-Carolingians from 911 (*pages 92–93*).

THE CAROLINGIAN RENAISSANCE

Under Charlemagne determined efforts were made to revive Classical learning. Texts were gathered and copied, and the teaching of good Latin was made a priority in royally sponsored monasteries and cathedrals with *scriptoria* or writing offices (*map 3*). This Carolingian Renaissance was generated by the work of a relatively small number of institutions, and its central thrust was religious. Carolingian monks copied Classical texts because their language and contents were considered necessary for a full understanding of the Bible. Editing variant texts of the Bible to produce one orthodox version, codifying divergent sources of church law, providing service books in good Latin: all of these were basic tasks Charlemagne wanted his scholars to undertake. Charlemagne also wished – as he proclaimed in the *Admonitio Generalis* of 789 and the *Programmatic Capitulary* of 802 – to ensure higher standards of Christian religious observance and biblically guided morality in his realm. His bishops attempted to enforce this programme through a sequence of reforming councils designed to harmonize standards throughout the empire. Louis the Pious did the same with monastic practice through further councils between 817 and 819. The Papacy likewise received strong royal support, and was endowed with the lands which would form the basis of the papal state through to the 19th century.

THE FRANKISH ECONOMY

By around 600 the Merovingians had presided over the collapse of most of the more sophisticated elements of the Roman economy: taxation, substantial long-distance trade, towns, specialized manufacture and coins (apart from a very high-value gold coinage that was useless for everyday transactions). There were also associated declines in population and agricultural production. The 7th and 8th centuries, however, witnessed substantial recovery. New trading routes spread across the Channel and North Sea, their progress marked by the appearance of a series of trading stations or emporia (*map 4*). Monetary-based exchange also increased – using, from the later 7th century, a lower value silver currency. The quantity and quality of silver coins grew dramatically with the new coinage introduced by

Charlemagne in the 790s – a coinage that Charles the Bald later managed to his own profit; a dense network of mints allowed him periodically to change coin types, demand that people use new coins, and charge them fees for reminting.

CAROLINGIAN ACHIEVEMENTS

Politically the Carolingian period ended in failure. The united western European empire could not be held together, even if Charlemagne's resumption of an imperial title would directly inspire his Holy Roman successors (*pages 90–91*). In economic and cultural terms, however, the Carolingian period was deeply formative. Trade, a monetarized economy and more specialized production all began to flourish, providing the essential backdrop to the "take-off" of the western European economy which followed in the 11th century and after (*pages 100–1*). Carolingian scholars also set new standards in Christian belief, practice and intellectual development, with Latin Christendom growing from the seeds planted by Charlemagne.

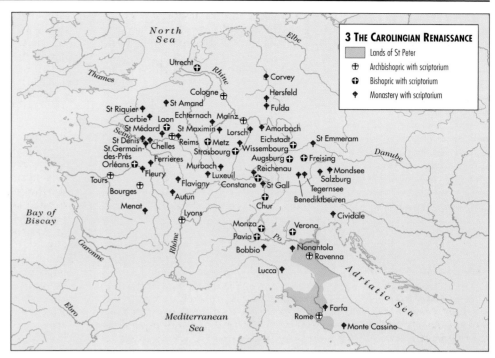

3 THE CAROLINGIAN RENAISSANCE
- Lands of St Peter
- ✠ Archbishopric with scriptorium
- ✟ Bishopric with scriptorium
- ✦ Monastery with scriptorium

▲ Carolingian scholars developed a new, easily written script – the Carolingian minuscule – which greatly speeded up the tedious process of book copying. They also revived Classical Latin from Classical texts, making it the language of medieval learning. Their strict choices helped define the limits of modern knowledge: they ignored texts whose contents they considered unnecessary or inappropriate for Latin Christendom, and consequently these works have failed to come down to us in the modern world.

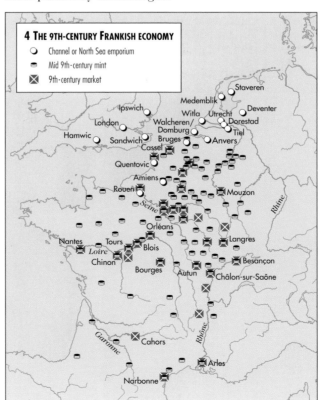

4 THE 9TH-CENTURY FRANKISH ECONOMY
- ○ Channel or North Sea emporium
- ▭ Mid 9th-century mint
- ✖ 9th-century market

◄ In the 7th and 8th centuries the Frankish economy recovered well from its Merovingian decline. Sea trading links flourished to the north and new centres of trade were established. Louis the Pious (r. 814–40), Charlemagne's only surviving son, ordered that there should be a market in every county, and they feature widely in the charters of Charles the Bald. The Carolingian period thus witnessed substantial moves away from locally focused subsistence agricultural economies towards greater specialization and exchange.

PEOPLES OF THE EUROPEAN STEPPE 350–1000

▶ By the mid-440s the Hunnic Empire dominated large numbers of Germanic groups in the middle Danube region and exercised a loose hold over large tracts of eastern and north-central Europe. The military success of the empire is evident from the large number of rich burials that have been found, particularly in the middle Danube region, which date from the Hunnic period. Some of these burials may have been of Huns, but many clearly belonged to the Germanic dynasts who first profited from the empire and subsequently led the independence movements which destroyed it after the death of Attila in 453.

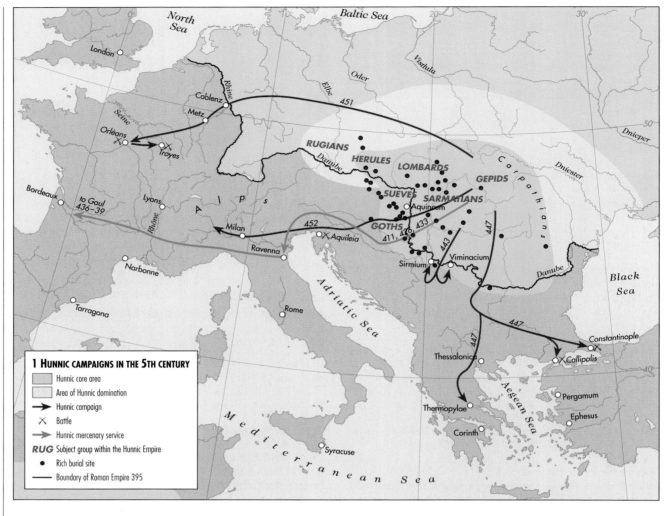

1 HUNNIC CAMPAIGNS IN THE 5TH CENTURY
- Hunnic core area
- Area of Hunnic domination
- → Hunnic campaign
- ✕ Battle
- ⭢ Hunnic mercenary service
- **RUG** Subject group within the Hunnic Empire
- ● Rich burial site
- —— Boundary of Roman Empire 395

▼ In the 560s the Avars established themselves in the area of modern Hungary and for the next 70 years raided territories from the Rhine to Constantinople. They nearly conquered Constantinople in 626 but in doing so suffered a defeat which greatly reduced their offensive military potential. While this allowed the defection of many of their subjects, they remained a dominant power in central Europe until being defeated by Charlemagne in 796.

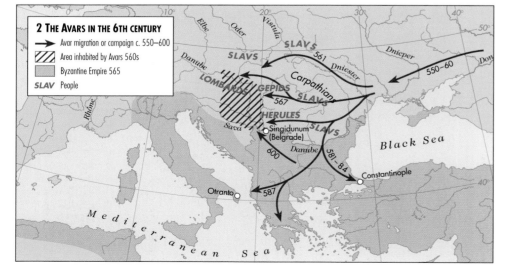

2 THE AVARS IN THE 6TH CENTURY
- → Avar migration or campaign c. 550–600
- ▨ Area inhabited by Avars 560s
- Byzantine Empire 565
- **SLAV** People

At the western end of the immense grasslands that run between China and Europe is the Volga and Ukrainian steppe, while further west are two regions of Europe that in soil and climate can be regarded as continuations of the steppe, the Dobrudja in modern Romania and the Great Hungarian Plain. In the 1st millennium AD the rich grazing lands of this area attracted successive waves of Asian nomads and semi-nomads who were from a variety of ethnic backgrounds and supported themselves by raising animals that were moved annually between upland summer and lowland winter pastures.

Among the most important of these westward-moving peoples were the Huns (from c. 350), whose ethnic affiliation is unknown, and the Turkic-speaking Avars (from around 560). In the latter half of the 6th century they were followed by further groups from the confederation of the so-called Western Turks (the Bulgars, Khazars and the Finno-Ugrian-speaking Magyars), and in the 9th century by independent Turkic-speaking groups, the Pechenegs and the Oguz. As more nomads moved onto the steppe, they drove the earlier arrivals further west and towards the lands around the Mediterranean – lands whose relative wealth could be tapped through raids and more sustained military campaigns, or through the extraction of annual tributes. In 395, for example, the Huns, who at this point were settled in the Ukrainian steppe, raided both the Roman and Persian empires (*pages 56–57*), and by the 410s they were established on the Great Hungarian Plain, supplying mercenaries to the Roman state. In the 440s, after a sequence of highly destructive campaigns, their feared leader Attila was receiving 900 kilograms (2,000 pounds) of gold a year in tributes. The Avars later mounted a series of campaigns against the Byzantines, particularly in the 580s, and extracted a steadily increasing tribute. In the 10th century the Magyars terrorized Europe with raids from the Baltic Sea to the Mediterranean coast of France.

THE BUILDING OF EMPIRES

The steppe peoples not only raided the empires of other peoples but also built empires of their own, either on the steppe or within Europe. On the Great Hungarian Plain the Huns established a powerful and aggressive empire between about 410 and 469 (*map 1*). They were succeeded by the Avars, who moved west from the Ukrainian steppe in around 560 to escape the Western Turks and established an empire that was to last until 796 (*map 2*).

Centred around the ruling clan of the Asina, the Western Turks built a huge empire stretching from the borders of China to the Ukrainian steppe, but it had collapsed by the 630s. During the following 40 years three of its constituent parts – the Bulgars, Khazars and Magyars – established longer-lived entities in the Dobrudja, Volga and Ukrainian steppe respectively. These empires remained relatively stable for over 200 years, until in the late 9th and

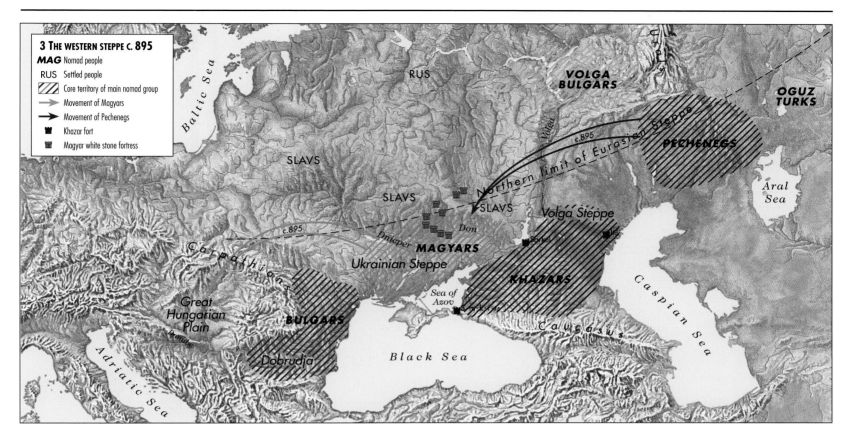

3 THE WESTERN STEPPE C. 895

early 10th centuries the Pechenegs moved west, expelling the Magyars and undermining Khazar power (*map 3*). The Pechenegs themselves would later fall victim to the Seljuk Turks, a dynasty who were to emerge from the Oguz in the 11th century (*pages 88–89*).

All these shifting empires were based on the conquest and exploitation of subject tribes, who were usually a mixture of nomadic peoples and more settled agricultural groups. Attila's Hunnic Empire of the 440s consisted of a dominant Hunnic core but with numerous, particularly Germanic, groups such as Goths, Gepids, Herules, Rugians, Sueves and Lombards. The Avar Empire of the later 6th century incorporated Gepids, Bulgars and numerous Slavic groups, and the Bulgar state in the Dobrudja and surrounding territories also incorporated many Slavic tribes. The Khazars on the Volga steppe exercised dominion over the nomadic Magyars before they established their own empire in the Ukraine, as well as over large Slavic and later Scandinavian Rus groups to the north.

Once they had achieved some degree of dominance, peoples of the steppe tended to cease being simple nomads and profound social evolution sometimes followed. For example, when the Huns first reached the Ukrainian steppe around 375, they were led in their continual search for new pastures by a multiplicity of chiefs. By the 430s, however, one dominant dynasty, that of Attila, had emerged, suppressing all rivals. With warfare dominating their lives, the Huns were able to use the wealth of the Roman Empire to create a new, more stratified social hierarchy under a single ruler.

THE IMPACT OF THE NOMADS ON EUROPE

The nature of these nomad empires explains much of their impact on Europe. Built on military dominance, they required continued military success to survive. In their campaigns they used soldiers and leaders recruited from the peoples they dominated, and their successes were to some extent shared with these peoples. A successful campaign both maintained a leader's prestige and provided booty to be distributed – not only among the nomad core but also to selected leaders among subject groups, whose loyalty was thus maintained. The campaigns led to a substantial degree of instability in Europe, as groups escaping from the intruders sought new homes. The collapse of the

Western Roman Empire in the 5th century was brought about by Germanic groups escaping the Huns, and Avar pressure later led to a great migration of Slavs into central and eastern Europe and Lombards into Italy.

Warfare, however, could not be successful forever. The Europeans eventually learned how to contain the steppe peoples, for whom the logistic problems of continuous warfare increased as closer targets were conquered. Once expansion stopped, decline quickly followed. Within 16 years of Attila's death in 453, the Huns had ceased to exist as an independent force in Europe. Without booty to distribute or prestige to inspire fear, Attila's sons lost control of the subject peoples. Similarly, when defeat by Constantinople had curbed the power of the Avars in the 7th century, numerous Slavs and Bulgars escaped from the Avar Empire. Long-term survival was only possible for steppe peoples by adopting the lives of sedentary landowners and embracing mainstream European culture, as the Magyars did after being defeated by the Saxons at the Battle of Lechfeld in 955 (*map 4*).

▲ In the 9th century the Khazars played a dominant role in trade throughout the Ukrainian steppe with both the Bulgars and Magyars. Directly or indirectly, their hegemony also extended to the Slavic and Rus groups of the neighbouring forested zone to the north.

▼ Driven into the heart of the continent by the arrival of the Pechenegs on the Ukrainian steppe around 895, the Magyars in turn terrorized central, southern and even parts of western Europe with widespread raids. Their expansion was first curbed in 936 and then halted in 955 by the newly powerful Saxon kings Henry I and his son Otto I.

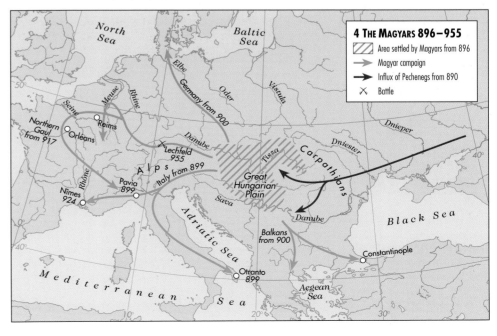

4 THE MAGYARS 896–955

◖ FROM HUNTING TO FARMING: EUROPE 8000–200 BC *pages 20–21* ◗ THE MONGOL EMPIRE 1206–1405 *pages 98–99*

THE VIKINGS
800–1100

► Viking raiders ranged widely, reaching the coast of Italy. So, too, did Viking traders, exchanging goods at towns in western Europe and following the river routes of western Russia to sell furs and slaves as far away as Baghdad. Both traders and raiders used the new ship technology to create new ways of making money out of the wealth of the great Carolingian and Abbasid empires.

► New ship technology, combining the use of sail power with a strong but flexible hull which could survive the impact of ocean waves, made extraordinary voyages of exploration possible for the adventurous Vikings. In 986 Bjarni Herjolfsson reached North America after being blown off course during a voyage from Iceland to Greenland. His discoveries along the coasts of Newfoundland and Labrador were followed up by Leif Eiriksson, who in about 1003 sailed from Greenland in order to follow Herjolfsson's route in reverse.

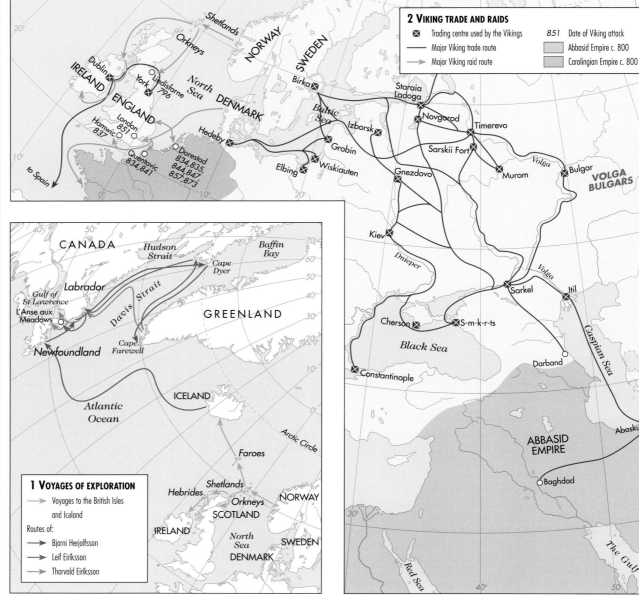

▲ This Viking silver dragon-headed amulet comes from Iceland, which was colonized by the Vikings in the late 9th century. Its cross shape may well have a Christian connotation: the inhabitants of Iceland – together with those of Denmark, Norway and Sweden – were converted to Christianity in the late 10th and early 11th centuries.

The Vikings first came to the attention of other Europeans when, at the end of the 8th century, they sailed from their Scandinavian homeland to launch a series of ferocious raids on the coasts of Britain, Ireland and France. However, in the 300 years that followed they not only plundered in western Europe but also embarked on voyages of exploration, established a far-reaching network of trading routes and created new states. During these years the term "Viking" was applied only to those who undertook expeditions of plunder, but it has since come to be used more widely to refer to all the inhabitants of Norway, Denmark and Sweden at that time.

VOYAGES OF EXPLORATION
In the late 8th century Norwegians sailed to the Shetlands and Orkneys, drawn across the North Sea by the prevailing winds and currents. This was a shorter journey than coasting round Scandinavia and led naturally on to the northern coast of Scotland, the Hebrides, Ireland and western Britain (*map 1*). The Norwegians then ranged further afield and reached the Faroes in the early 9th century and Iceland, another 1,600 kilometres (1,000 miles) northwest, in the 860s and 870s. Greenland was first visited in about 900, when the Norwegian Gunnbjörn was blown off course. Settlement there began in the late 10th century, bringing further explorers, such as Eirík the Red, who surveyed much of the new land. According to a 12th-century saga, it was during a voyage to Greenland in 986 that Bjarni Herjolfsson was storm-driven south to reach the shores of North America. He made three landfalls, one of which is thought to

coincide with the site of a permanent Norwegian settlement dating from around 1000 near L'Anse aux Meadows, on the northern tip of Newfoundland. Herjolfsson was followed by other voyagers, notably Leif Eíriksson (in 1003) and his brother Thorvald (between 1005 and 1012).

TRADING AND RAIDING
Most Vikings sailed in search of profit, whether as traders or raiders. They exchanged goods at trading centres (emporia) in northern Europe and followed the river routes of western Russia – chiefly the Volga route to the Caspian – to gain access to the rich Muslim world (*map 2*). Between the later 8th and 10th centuries the natural resources of the north – particularly furs but also honey, wax, falcons, walrus ivory and large numbers of slaves – were exchanged for Arab silver, mostly at a great emporium in the land of the Volga Bulgars (*pages 76–77*). During the 9th century Norwegians and Danes also moved west, taking slaves from Ireland and Scotland via new trading settlements at Dublin and York.

Commerce and plundering were linked: slaves were usually captured in raids and the trading centres became a natural target for raiders. Exploiting many of the established trading routes, Norwegians raided northern Britain from 796, and Danes quickly followed suit, moving along the Channel to attack southern England and northern France. Merchants were forced to pay protection money and many of the old emporia (especially Quentovic, Dorestad and Hamwic) were repeatedly sacked. In the 840s and 860s settlements along the western coasts of France and Spain, and along the Mediterranean coast as far as Italy, were also raided.

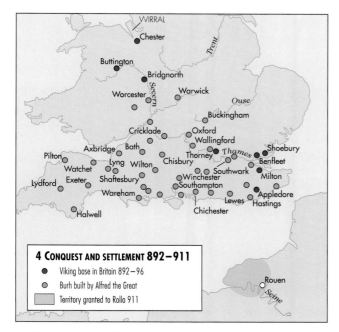

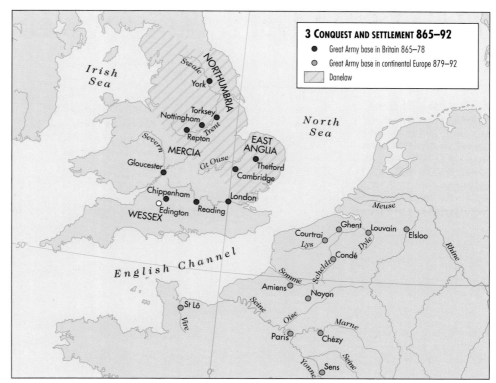

3 CONQUEST AND SETTLEMENT 865–92
- Great Army base in Britain 865–78
- Great Army base in continental Europe 879–92
- Danelaw

4 CONQUEST AND SETTLEMENT 892–911
- Viking base in Britain 892–96
- Burh built by Alfred the Great
- Territory granted to Rollo 911

▲ Alfred's newly constructed fortresses (the burhs) protected his kingdom from the second Great Army of 892–95. Many of its frustrated contingents then returned to the continent, creating chaos in Brittany and, under King Rollo, eventually being granted land to found the Duchy of Normandy at the mouth of the Seine in 911.

CONQUEST AND SETTLEMENT

A totally new level of activity unfolded in western Europe from the 860s with the arrival of the "Great Armies", independent (mostly Danish) groups led by their own kings but often totalling several thousand men and now enabling Vikings to settle in previously inpenetrable areas south of Scotland. The first Great Army landed in England in 865 and within five years had subdued Northumbria, Mercia and East Anglia. The next seven years saw a series of assaults on the one surviving kingdom, Wessex, which under Alfred the Great successfully resisted and defeated the Viking Guthrum at Edington in 878. The Vikings were given territory north of the River Thames, and this was formally established as Danelaw (*map 3*). Dissatisfied with this arrangement, some Vikings turned to continental Europe, and for 13 years (879–92) battles raged along the rivers of northern France, even reaching Paris. Following a serious defeat on the River Dyle the remaining Vikings returned to England in 892, but this time Alfred fended them off with ease (*map 4*).

THE FORMATION OF STATES

Danelaw never constituted a unified state, and when the Vikings no longer arrived in large numbers after 900 the Wessex monarchy swallowed up their territories to create the first united kingdom of England. By contrast, King Rollo's settlement in France eventually emerged as the independent Duchy of Normandy, and Viking trading stations in western Russia coalesced into a state in the 10th century (*pages 70–71*). However, the main forum of Viking state formation was Scandinavia itself. In about 800 no unified kingdoms existed there, but by around 1000 a dynasty with its capital at Jelling, led by Svein Forkbeard and his son Cnut, had established control over all of Denmark. Having suppressed their rivals they built fortresses, set up regional administrative centres, created the first native Scandinavian coinage and – because Svein and Cnut were also Christians – established a number of bishoprics (*map 5*).

Similar processes began in Norway in the 990s, when Olaf Tryggvasson, returning from extensive raiding in England as a rich man and a convert to Christianity, founded the Norwegian monarchy. The entity he created was far from stable, however, and Sweden also remained politically fragmented. Thus when Svein and Cnut gathered forces for the conquest of England (1003–17) they were joined not only by Danes but also by numerous independent groups from across Scandinavia. Cnut became a strong ruler of England, but his hold on Denmark and Norway was weak, and on his death in 1035 his empire disintegrated. Within 50 years the Vikings had been driven out of England by the Normans, and by the 12th century they were no longer a force to be feared outside the shores of Scandinavia.

▲ Numbering several thousand men, the "Great Armies" which started to collect in western Europe from about 865 marked a new era in Viking expansion. Mainly Danish, they were large enough to conquer and settle whole Anglo-Saxon kingdoms and – when checked by Alfred the Great of Wessex in 878 – to cause similar disruption on the Continent by exploiting the major river systems of France and the Low Countries.

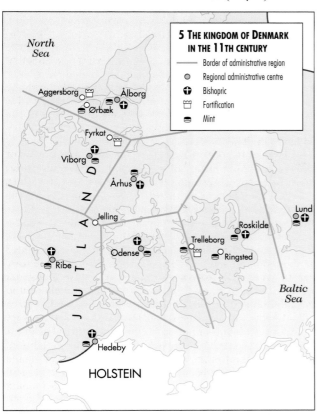

5 THE KINGDOM OF DENMARK IN THE 11TH CENTURY
- Border of administrative region
- Regional administrative centre
- Bishopric
- Fortification
- Mint

▲ By around 1000 the Jelling dynasty had created the first Danish kingdom. It reduced local autonomy and created new political structures, allowing it to exploit both human and other resources of Jutland and its neighbouring islands.

► Even to the modern eye the Viking longships are impressive. The 9th-century, 16-seater Gokstad ship, recovered by a Norwegian excavation, is 23.5 metres (just over 76 feet) long, clinker-planked with thin oak attached by a combination of lashings and small iron plates to 19 frames built up from a huge keel. An Atlantic crossing of 1893 in a replica of this ship – made in just 28 days from Bergen to Newfoundland – demonstrated the timeless efficiency of the design. It was, however, normally only used for coastal sailing; the broader and deeper halfship was considered more suitable for long-distance ocean crossings.

◗ FROM HUNTING TO FARMING: EUROPE 8000–200 BC *pages 20–21*   ◗ EUROPE 1350–1500 *pages 106–7*

STATES AND TRADE IN WEST AFRICA
500–1500

► By 1500 a number of rival states had emerged in West Africa, each governed by an elite whose wealth and power can be judged from their substantial towns, their rich burials and the fine works of craftsmanship created for them.

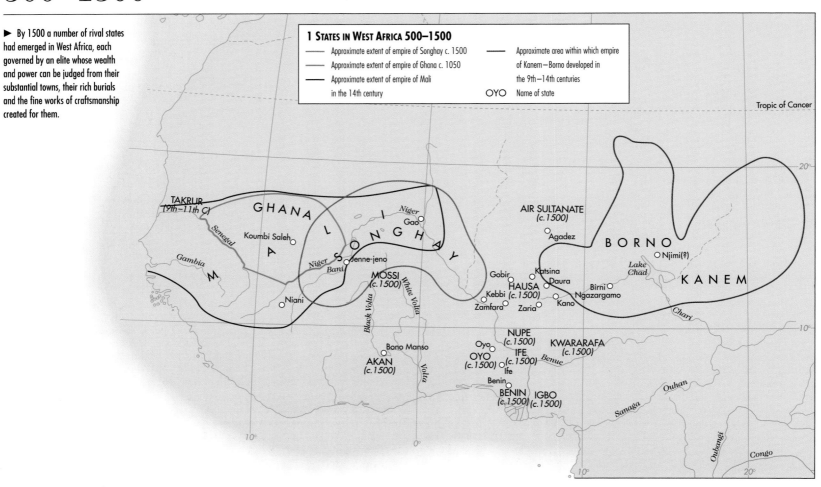

1 States in West Africa 500–1500

— Approximate extent of empire of Songhay c. 1500
— Approximate extent of empire of Ghana c. 1050
— Approximate extent of empire of Mali in the 14th century
— Approximate area within which empire of Kanem–Borno developed in the 9th–14th centuries
OYO Name of state

Tropic of Cancer

▼ The various vegetation zones of West Africa supported different agricultural regimes and produced different raw resources – such as gold from the savanna and forest, and salt from the desert. This diversity in turn helped stimulate the development of interregional trade.

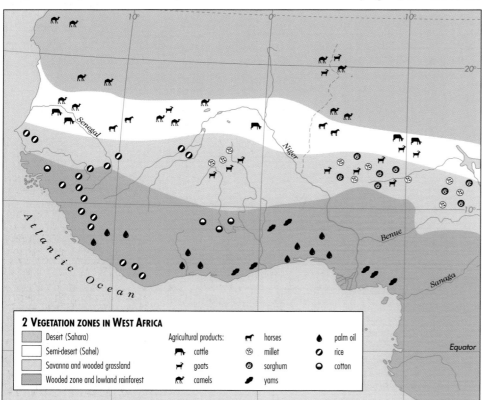

2 Vegetation Zones in West Africa

▨ Desert (Sahara)	
☐ Semi-desert (Sahel)	
▨ Savanna and wooded grassland	
▨ Wooded zone and lowland rainforest	

Agricultural products:
🐎 horses ◆ palm oil
🐂 cattle ⊛ millet ⊘ rice
🐐 goats ⊛ sorghum ⊖ cotton
🐪 camels ✍ yams

Equator

Early West African states took a number of forms, varying in size from the vast Songhay Empire, which held in its sway many different ethnic groups, to smaller, more ethnically homogeneous Hausa city-states such as Kano (*map 1*). Methods of government, too, were equally varied: the great medieval empires of the savanna and semi-desert Sahel regions employed often complex bureaucracies utilizing Muslim officials and the Arabic script, while in the forested region of the south, different systems existed which attached varying importance to the role of king. Among the Igbo in the Niger delta, for example, there was no king and loyalty to the state was maintained through religious ties, ceremonies and clans.

TRADE AND THE FORMATION OF STATES

Trade was intimately linked with the growth of states in West Africa, initially local and interregional in focus but later developing into long-distance trade across the Sahara. Trade flourished partly because of the existence of different environmental zones that stretched east-west across the continent and comprised the Sahara Desert, the Sahelian semi-desert, the Sudannic savanna and wooded grasslands, and finally the more heavily wooded region merging into the rainforest (*map 2*). The forms of agriculture practised varied between zones: for example, the yams cultivated in the southern wooded region could not be grown in the Sahelian or Saharan zones, whereas pastoralism or animal herding was viable in the Sahel. This variation resulted in a need to exchange commodities, often carried out by merchants from the Sahel or savanna regions (*map 3*).

Prosperity generated through trade, coupled with the growth of settlements at important trade centres, gradually led to urbanization and the foundation of states. Recent excavations have shown that the settlement of Jenne-jeno in Mali, the earliest town yet found in West Africa, was founded in about 300 BC and had developed into a thriving town by AD 500. Although Jenne-jeno never grew into a state, it served as a centre of trade where savanna commodities such as gold, iron and various foodstuffs were traded for Saharan salt and possibly – though this is less certain – for copper.

Another town founded in Mali by the 7th century was Gao, later to become the capital of the Songhay Empire. To the west, in Mauritania, the capital of the empire of Ghana also appears to have been in existence by this time, though only part of the settlement – the merchants' town of Koumbi Saleh – has so far been found. While Ghana was in all probability the first of the states founded in West Africa, events were also proceeding rapidly to the east of this area on the margins of Lake Chad. The kingdom of Kanem, east of the lake, was mentioned in an Arab document in the mid-9th

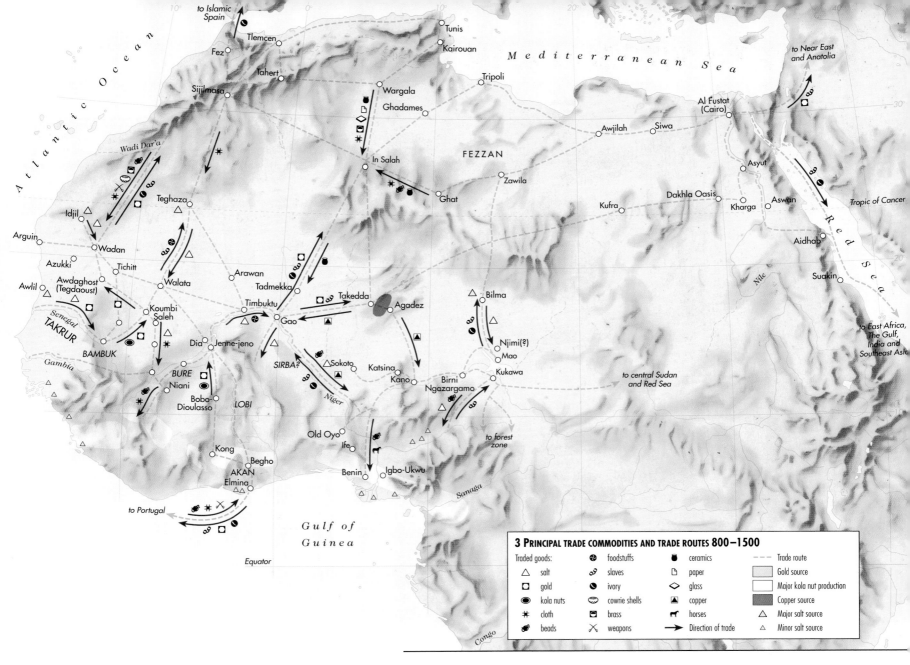

3 Principal trade commodities and trade routes 800–1500

Traded goods:

△	salt	🐚 foodstuffs	▯ ceramics
▢	gold	🐌 slaves	▭ paper
◉	kola nuts	● ivory	◇ glass
✳	cloth	🐚 cowrie shells	▲ copper
🐚	beads	▭ brass	🐎 horses
		✕ weapons	→ Direction of trade

- – – – Trade route
- ▢ Gold source
- ▢ Major kola nut production
- ▨ Copper source
- △ Major salt source
- △ Minor salt source

century and had certainly been in existence for some time before that. Later, apparently in the 14th century, this state shifted west of Lake Chad to Borno and became known as Kanem–Borno.

The forest regions, with their higher density of population than the savanna or Sahel regions, were a source of slaves, and states began to emerge in this area around the 12th century. Trade appears to have been linked with the growth of the Akan states in modern Ghana, an area rich in gold where trade centres such as Begho were founded perhaps as early as the 12th century. To the northeast the seven Hausa city-states, the Hausa Bakwai, were established during the 13th century. Together with a further seven related but non-Hausa states to the south, these formed a link in the 15th century between Kanem–Borno to the east and the Songhay Empire and the Akan states to the west.

The forest kingdoms also emerged comparatively early, with Benin (now famous on account of its bronze sculptures), occupied by the 13th century. Similar castings, predating those of Benin, were produced in Ife, birthplace of the Yoruba nation – a state with a well-developed tradition of forest farming, town living, crafts and government.

CONTACTS WITH THE MUSLIM WORLD

Indirect trans-Saharan trade is known to have occurred during the 1st millennium BC, but it is unlikely that caravans travelled right across the desert until the introduction of camels towards the end of that period. Archaeological evidence indicates that trans-Saharan trade became far more important with the consolidation of Islam in North Africa from the early 9th century AD, and from this time it had a major economic and social impact on the developing states of sub-Saharan Africa.

There was a great demand in the Muslim world for West African products, particularly gold, slaves and ivory. Among the items sent south in return were manufactured goods such as cloth, glazed pottery, glass vessels, beads, paper, brass and cowrie shells (later used as currency). Transport was by camel caravans, which travelled from well to well to the Sahelian trade centres of Koumbi Saleh, Tegdaoust and Gao. From there some of the goods were traded on further into West Africa – indicated, for example, by the discovery of many thousands of 9th-century coloured glass beads at the site of Igbo-Ukwu in the southern forest zone.

Through contacts with Muslim merchants, the Sahelian trade centres were exposed to Islam from the very beginnings of trans-Saharan trade with Muslim North Africa. Various local rulers of the empires of Ghana, Kanem–Borno, Mali and Songhay converted to Islam, which spread right across the region through the activities of local merchant groups such as the Mande or Wangara, who were responsible for much of the trade in gold and kola nuts from the Akan states. Hausa was also gradually Islamized but further south, in the forest states such as Ife or Benin, the traditional beliefs of animism were maintained, with religious and secular authority often intermixed.

THE ARRIVAL OF THE PORTUGUESE

Major events in the second half of the 15th century were to have far-reaching effects on the states, societies and trade systems of West Africa. Paramount among these was the arrival of the Portuguese on the west coast in the 1440s, followed by the establishment in 1482 of a Portuguese trading post at Elmina on the coast of modern Ghana. This meant that imported manufactured goods such as cloth could now be obtained directly from the coast and that another outlet for West African commodities was established. The slave trade across the Atlantic also began, starting with the first cargo of slaves from West Africa to the West Indies in 1518 – a momentous event with tragic consequences.

▲ Located on the inland Niger delta, the town of Jenne-jeno owed its prosperity to its great agricultural wealth, exporting rice, cereals, dried fish and fish oil to neighbouring regions by using the Niger as a transport highway. It was the first of many such towns that emerged in West Africa, all of them trading local raw materials and produce for everyday commodities and luxuries from other regions as far away as Muslim North Africa.

▼ Like the people of Benin, the Yoruba produced fine bronze heads and figurines. However, they are particularly renowned for their terracotta heads, such as this one of a 12th–13th century queen from Ife.

STATES AND TRADE IN EAST AFRICA
500–1500

▲ Soapstone brought from a source 24 kilometres (15 miles) away was used at Great Zimbabwe to carve ritual objects in the form of people and birds.

▶ The agricultural communities that had colonized East and southern Africa in the 1st millennium AD developed into kingdoms and states in the early centuries of the 2nd millennium. Both cattle-herding and command of raw materials – including gold, copper and ivory – were by now of major importance. In the north, following a mission of 543, Christianity had become established in the Axumite kingdom, while Muslim traders who settled on the coast from the 9th century were responsible not only for the introduction of Islam but also the development of Islamic states. Further inland elites emerged, marked by rich burials such as those at Sanga and by substantial centres such as Great Zimbabwe.

In the 6th century East Africa was a mosaic of very different cultural groups employing a variety of subsistence strategies. Though in many areas foraging was still the primary means of providing food, agriculture and stock-keeping had already spread throughout the length of the continent. In areas such as the arid far southwest and the forests of central Africa, nomadic hunter-gatherers, being so well adapted to these environments, were still thriving in 1500 AD. However, by the 8th century more settled communities had also begun to be established, which frequently controlled resources such as copper and ivory or acted as trading settlements. Some of these settled communities later developed into kingdoms and became integrated into extensive trading networks.

In Ethiopia the Christian Axumite kingdom had begun to decline in the 7th century after losing control of its ports to the Arabs, and was finally destroyed in the 10th century. Christianity nevertheless remained strong in Ethiopia, and the focus of Christian Ethiopia (*map 1*) shifted south from Axum to Lalibela (then called Adefa). While the Axumite kingdom had been urban in character, the empire which

replaced it was largely feudal, its rulers shifting their court when local resources had been exhausted. Rock-cut churches, created between the 10th and the 15th centuries, are the main legacy of the Christian Ethiopian Empire.

THE ISLAMIZATION OF EAST AFRICA
To the east and southeast of the Christian empire, Islamic trading settlements were established along the coast and along the trade routes leading into the interior from the major ports, of which Zeila was perhaps the most important. As the Muslim population increased, the creation of a number of Islamic sultanates led to conflict with the Christian Ethiopian Empire. During this period the Somali slowly expanded from around the Gulf of Aden – along the coast north to Zeila and south to Mogadishu, and into the interior – to occupy much of the Horn of Africa. By the 12th century Islamization of this area had become well advanced.

During the 9th century a series of trading settlements, united by a common religion, language and style of living, emerged along the East African coast. These Swahili-speaking Islamic communities, though African, lay on a branch of the great trade routes connecting the Red Sea, southern Arabia and India, and they adopted various aspects of the cultures with which they came in contact. By the 14th century Swahili towns and settlements had greatly expanded from the early sites of Manda and Shanga and stretched from Mogadishu south to Chibuene, with communities on the Comores and Madagascar. Towns such as Kilwa contained fine, multi-storied houses built of coral, and their inhabitants ate a diet containing rice, spices and coconut – cosmopolitan Indian Ocean tastes.

STATE FORMATION IN THE INTERIOR
Political developments also occurred in interior East Africa. In the region of the Great Lakes a series of huge earthwork enclosures was built: at Bigo over 10 kilometres (six miles) of ditches and ramparts enclosed almost 300 hectares (750 acres). It is thought that these enclosures were used for corralling cattle and that this kingdom, which later came to be known as Bunyoro, based its wealth and power on its control of cattle. Further south, control of the copper and goldfields (*map 2*) may have been a factor in the rise of other powerful elites. An excavated sequence of burials at Sanga illustrates the emergence of a hierarchical society by the 10th century and the development of a currency system of uniform small copper crosses. Although the main copper belt was 200 kilometres (125 miles) to the south, the society represented in the Sanga cemetery used copper to indicate wealth and status.

On the Zimbabwe Plateau, with its highland and lowland grazing areas and its gold, iron, copper and tin resources, a powerful elite emerged at the beginning of the present millennium. Its capital was located at Great Zimbabwe (*map 3*), a substantial complex of stone towers and enclosures surrounded by *daga* (mud structures), which may have had a population of some 18,000 people. Similar stone structures are found across the plateau, indicating the extent of the authority exercised by the Zimbabwe elite. Religion may have played a role in legitimizing this authority: many ritual objects have been found at Great Zimbabwe, in particular soapstone carvings and monoliths, some surmounted by birds.

EAST AFRICAN TRADE
The control and exploitation of particular resources or of trade routes played a role in the development of virtually every state and kingdom in East Africa. The area was rich in resources – in metals such as gold, copper and iron, and in exotic materials such as ivory. Whereas West Africa, with its treacherous winds and coasts, had to rely on the trans-Saharan trade routes until the end of the 15th century, East Africa was connected from an early date to the trade networks of the Red Sea and Indian Ocean (*pages 52–53*), and beyond as far east as Java and China (*map 2*). At the northern end of the coast, traders may have been active from as

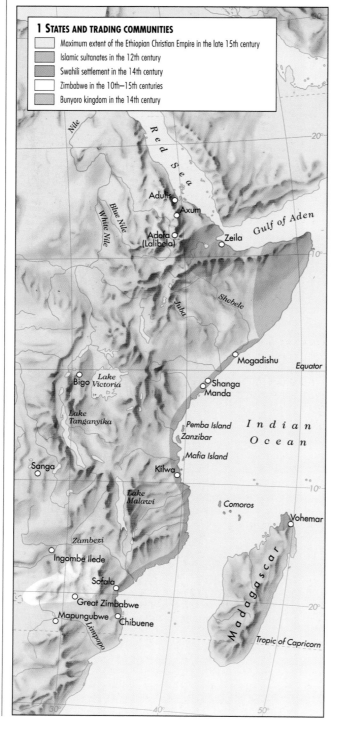

1 STATES AND TRADING COMMUNITIES
- Maximum extent of the Ethiopian Christian Empire in the late 15th century
- Islamic sultanates in the 12th century
- Swahili settlement in the 14th century
- Zimbabwe in the 10th–15th centuries
- Bunyoro kingdom in the 14th century

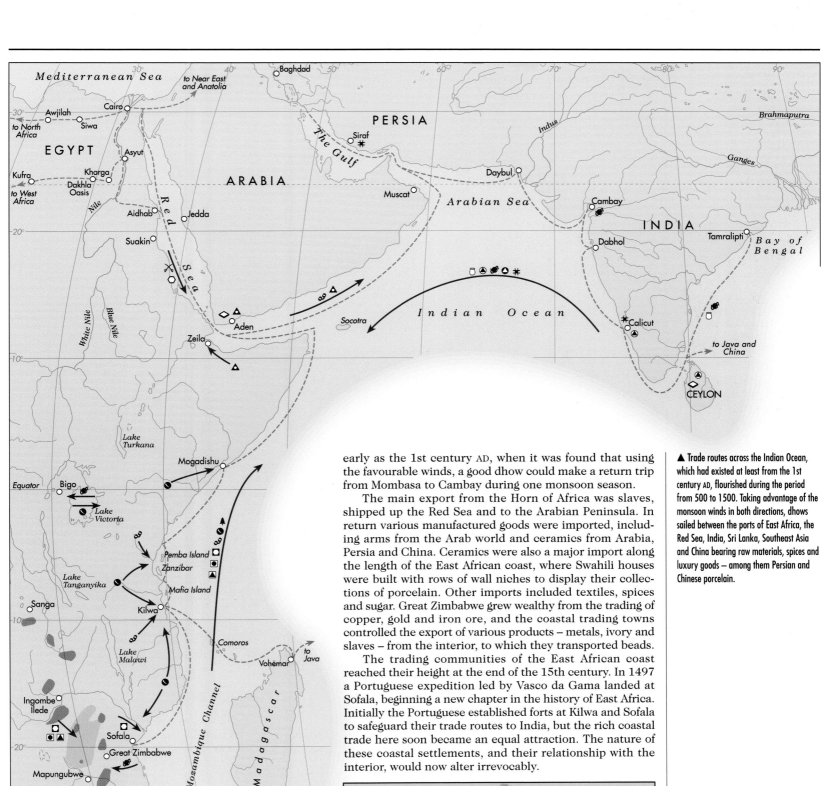

early as the 1st century AD, when it was found that using the favourable winds, a good dhow could make a return trip from Mombasa to Cambay during one monsoon season.

The main export from the Horn of Africa was slaves, shipped up the Red Sea and to the Arabian Peninsula. In return various manufactured goods were imported, including arms from the Arab world and ceramics from Arabia, Persia and China. Ceramics were also a major import along the length of the East African coast, where Swahili houses were built with rows of wall niches to display their collections of porcelain. Other imports included textiles, spices and sugar. Great Zimbabwe grew wealthy from the trading of copper, gold and iron ore, and the coastal trading towns controlled the export of various products – metals, ivory and slaves – from the interior, to which they transported beads.

The trading communities of the East African coast reached their height at the end of the 15th century. In 1497 a Portuguese expedition led by Vasco da Gama landed at Sofala, beginning a new chapter in the history of East Africa. Initially the Portuguese established forts at Kilwa and Sofala to safeguard their trade routes to India, but the rich coastal trade here soon became an equal attraction. The nature of these coastal settlements, and their relationship with the interior, would now alter irrevocably.

▲ Trade routes across the Indian Ocean, which had existed at least from the 1st century AD, flourished during the period from 500 to 1500. Taking advantage of the monsoon winds in both directions, dhows sailed between the ports of East Africa, the Red Sea, India, Sri Lanka, Southeast Asia and China bearing raw materials, spices and luxury goods – among them Persian and Chinese porcelain.

◄ In about 1250, stone structures began to be constructed at Great Zimbabwe, comprising drystone walls forming enclosures, platforms to support huts and a massive enclosure containing a conical tower. Great Zimbabwe was the capital of the rulers of a society that drew its wealth from both cattle-keeping and trading with the coastal states of East Africa. In the mid-15th century the settlement – like Kilwa on the coast – began to decline.

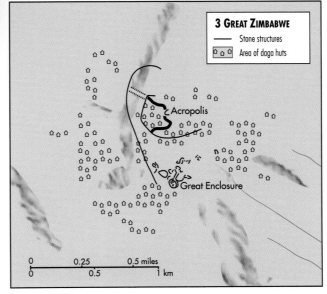

3 GREAT ZIMBABWE

— Stone structures

Area of daga huts

Acropolis

Great Enclosure

| 0 | 0.25 | 0.5 miles |
| 0 | 0.5 | 1 km |

2 TRADE ROUTES AND COMMODITIES

- – – – Major trade route
- → Direction of trade

Traded goods:

○ manufactured goods	✳ textiles
✕ arms	◉ spices
◗ ivory	◔ sugar
ᔏ slaves	● beads
◨ copper	▢ porcelain
◻ gold	◇ glass
▣ iron ore	△ incense
	▲ timber
	Gold field
	Copper deposit

CIVILIZATIONS IN MESOAMERICA AND SOUTH AMERICA 500–1500

▲ Gold – of which this Chimu *tumi*, or ceremonial knife, is made – was prized by many South American cultures for its symbolic connection with the sun.

▼ The Yucatán Peninsula and adjacent regions were home to the Maya. In the period 500 to 800 large cities, some containing as many as 100,000 people, dominated the smaller cities and kingdoms under divine rulers. Calakmul, in southeastern Campeche, was by far the most active in forging alliances and orchestrating battles. A persistent antagonism existed between Calakmul and the similarly large and prestigious kingdom of Tikal, with both apparently organized into state-like entities.

Mesoamerica and the Andes region of South America were home to some of the most sophisticated civilizations in ancient America – including, in the period from around 500 to 1500, the Later Maya, Toltec, Teuchitlan, Tarascan, Zapotec, Mixtec, Sican and Chimu. While some consisted of only one ethnic group, others occupied an ecologically distinct region, such as areas in the hot lowlands (*tierra caliente*) or cooler highlands (*tierra fría*). Most began in a heartland under tight dynastic control but then spread to more distant areas which were governed only indirectly, often through local rulers.

THE CHIMU CULTURE

To the west of the Andes the Chimu, a dynasty from the Moche Valley, gradually came to dominate a thin coastal strip in Peru between the 10th and 15th centuries (*map 1*). Iconographic clues suggest substantial continuity with the religion of the earlier Moche state (*pages 34–35*), although with a new twist: the capital city of Chan Chan contains ten immense enclosures thought to have served as mortuary temples for deceased Chimu emperors.

In three phases of expansion the Chimu lords extended control over and beyond the valleys once controlled by the Moche, with the same tendency of avoiding highland zones. Evidence of Chimu control in the south is patchy as local polities were incorporated by the Chimu without any substantial change to local government. By contrast, areas to the north may have been subjected to territorial conquest. Around 1350 the Chimu conquered the Lambayeque Valley, where the Sican culture with its rich burials and prosperous, irrigated settlements had succeeded the Moche. Chan Chan wielded heavy control until 1475, when the Chimu emperor was seized by the Incas (*pages 110–11*) and taken back to their highland capital of Cuzco.

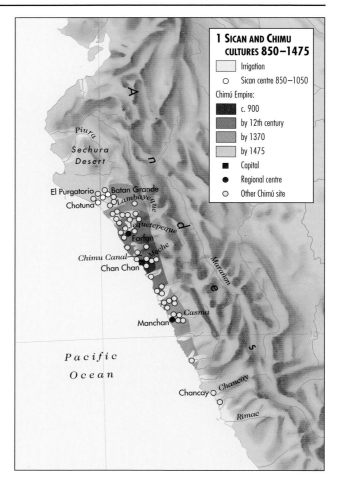

▲ At its height in 1475 the Chimu culture occupied a thin coastal strip from near present-day Lima to the Gulf of Guayaquil, in Ecuador. Sketchy historical evidence helps identify the the lords of Chimu and of its capital Chan Chan, who presided over an expansion that emanated from the Moche Valley. By 1200 this dynasty held sway over five valleys and by 1475, led by Emperor Minchancaman, it had vaulted over the Sechura Desert into a region formerly linked to the Amazonian cultural area. Great canals connecting river valleys facilitated irrigation agriculture and the growth of urban civilization in the heartland of the Chimu.

THE LATER MAYA

In Mesoamerica the Maya went through great changes in the period between 500 and the Spanish conquest in the 16th century. Until about 800, kingdoms ruled by "holy lords" and administered by courtiers waged war and created alliances against a backdrop of a rising population – one that approached five million in the central Yucatán Peninsula alone (*map 2*). However, between 800 and 900 the population plummeted dramatically for a variety of reasons, some of them agricultural and meteorological (such as environmental degradation) and others political, including intensified conflict between elites.

The so-called "Maya Collapse" was more pronounced in the centre of the peninsula than elsewhere, partly due to a lower birth rate and a higher mortality rate here than elsewhere, but also because of large-scale movements of people into more peaceful zones. Thus while the reduced population of the central area settled on defended islands in lakes, some Maya groups undoubtedly moved to cities in the northwest which had only just overcome a severe water shortage by developing a new means of collecting and storing rainwater in underground cisterns.

At the time of the collapse, the large city of Chichen Itza lorded over a confederacy that shaped the northern peninsula (*map 3*). In the late 13th century the smaller city of Mayapan took over, its rule lasting until around 1450. The final years before the Spanish conquest saw power disperse into small kingdoms – a development that made the Yucatán Peninsula far more resistant to Spanish incursions than Tenochtitlan, imperial city of the Aztecs in the Valley of Mexico (*pages 110–11*).

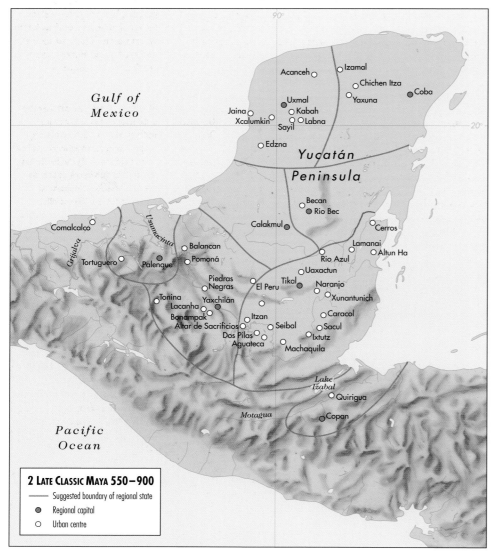

2 LATE CLASSIC MAYA 550–900
— Suggested boundary of regional state
● Regional capital
○ Urban centre

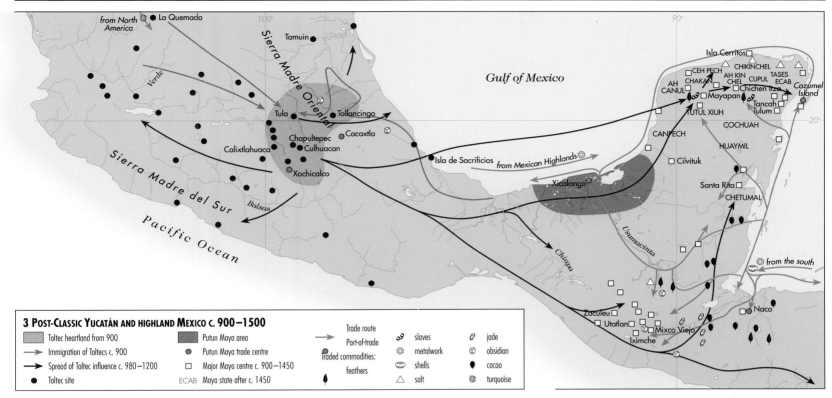

3 POST-CLASSIC YUCATÁN AND HIGHLAND MEXICO c. 900–1500

- Toltec heartland from 900
- → Immigration of Toltecs c. 900
- → Spread of Toltec influence c. 980–1200
- ● Toltec site
- Putun Maya area
- ● Putun Maya trade centre
- □ Major Maya centre c. 900–1450
- ECAB Maya state after c. 1450

Trade route
Port-of-trade
Traded commodities:
- feathers
- 🐚 slaves
- ⚙ metalwork
- 🐚 shells
- ▲ salt
- 🪶 jade
- 🔘 obsidian
- ● cacao
- ⚙ turquoise

THE TOLTECS

The emperor of the Aztecs was one of the 15th-century rulers in Mesoamerica who claimed descent from the Toltecs, a legendary people who had inhabited the semi-mythical paradise city of Tula. There is some historical evidence to support these legends, Tula having been identified with a major ruined city which was at its peak around the 10th century and was abandoned and destroyed around 1160 (*map 3*). Its inhabitants, the Toltecs, included groups from the Gulf coast as well as Nahuatl speakers originally from the "barbarian" lands to the north. Monumental sculptures and other artwork at Tula show the Toltecs as warriors – and practising the Mesoamerican rituals of captive sacrifice and the ballgame.

Major conflict around 980 may have led one group of Toltecs to flee to the Yucatán, where religious and perhaps dynastic elements typical of Tula appeared in Chichen Itza at this time. The Toltecs remaining at Tula then came to dominate a large area of central Mexico, playing a major role in trading networks which stretched as far north as the Pueblo area of southwestern North America (*pages 108–9*), the source of highly-prized turquoise. After the collapse of Tula there was probably a major dispersal of its inhabitants, introducing Toltec elements into the Valley of Mexico, Cholula and the Maya area.

THE TEUCHITLAN, TARASCAN, ZAPOTEC AND MIXTEC CIVILIZATIONS

Western Mexico (*map 4*) has often been described as the land of "enduring villages", each with deep-shaft tombs containing sculptures of everyday life. However, recent research has shown that from 500 to 900 this hilly, dry and remote part of Mesoamerica contained not only shaft tombs but also a distinctive temple type known as the *guachimonton*: a circular configuration of mounds around a central pyramid, often with a ballcourt extending out as an alley from the central group of buildings. The concentration of such features in the Teuchitlan Valley, together with raised field agriculture (*chinampas*) and fortified control points along valleys leading into this area, suggest a unitary state.

By the late pre-Conquest period a local people, speaking an isolated language known as Tarascan, controlled a large area of western Mexico around Lake Patzcuaro, from where they successfully harried the Aztecs. The Tarascans were exceptional craftsmen, particularly in their working of gold and silver. Their emperor, the *kasonsi*, commissioned

stepped platforms known as *yacatas*, probably the funerary monuments of his ancestors. In a dualistic pattern also common in central Mexico the *kasonsi* shared power with a powerful priest.

To the southeast of the Tarascan kingdom, in the Oaxaca Valley, were the Mixtecs. They had eclipsed the power of the Zapotecs, who around 700 had abandoned their great Classic centre of Monte Alban in the valley and later moved to a new base at Mitla. Here the Zapotecs constructed a fortified stronghold with fine palaces and continued to practise sacrificial rites until the arrival of the Spanish.

The Mixtecs, who were originally based in a series of small warring kingdoms in the north and west of the Oaxaca Valley, expanded their territory by warfare and dynastic marriages during the Post-Classic period (between 900 and the Spanish conquest). By 1350 they controlled the Oaxaca Valley and influenced neighbouring regions as far as Cholula. Both the Mixtecs and Zapotecs suffered at the hands of the Aztecs, but neither people was ever completely conquered; like the Tarascan Empire, both these cultures would soon be destroyed by powerful European invaders.

▲ After the "Maya Collapse" in the 9th century, Chichen Itza flourished before being replaced in the late 13th century by a political hegemony centred on the densely settled and walled city of Mayapan. Trading communities prospered both along the coast, particularly behind the protection of the barrier reef on the east coast of the Yucatán Peninsula, and in the southwest, home of the Putun Maya, who operated a major Post-Classic maritime network.

▼ From an original homeland somewhere in the Sonora Desert in the extreme northwest of Mexico, Nahuatl-speaking peoples – among them the ancestors of the Toltecs and Aztecs – migrated into central Mexico via western Mexico, an area that was subject to substantial population movements between 500 and 900.

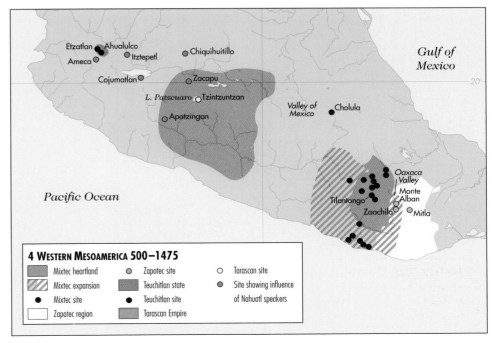

4 WESTERN MESOAMERICA 500–1475

- Mixtec heartland
- ▨ Mixtec expansion
- ● Mixtec site
- □ Zapotec region
- ● Zapotec site
- Teuchitlan state
- ● Teuchitlan site
- Tarascan Empire
- ○ Tarascan site
- ● Site showing influence of Nahuatl speakers

◀ MESOAMERICA 1200 BC–AD 700 *pages 32–33* ◀ SOUTH AMERICA 1400 BC–AD 1000 *pages 34–35* ▶ SPAIN AND THE AMERICAS 1492–1550 *pages 120–21*

EAST ASIA
907–1600

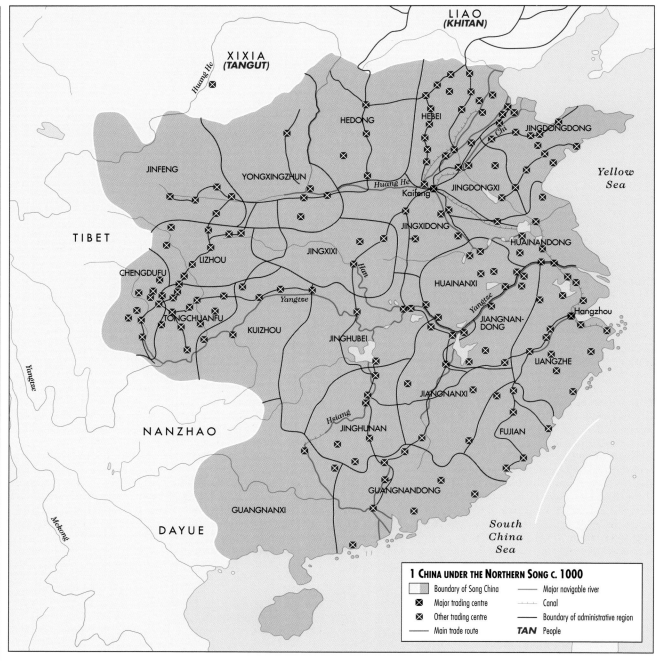

▲ A painted wooden carving of Buddha from Japan's Muromachi period (1335–1573) conveys a vastly different image to the traditional Buddhist figures of the Indian subcontinent. Arriving in Japan from China by the 6th century, Buddhism was hugely influential, notably in education, but it failed to replace the indigenous religion of Shinto.

1 CHINA UNDER THE NORTHERN SONG c. 1000
- Boundary of Song China
- ⊠ Major trading centre
- ⊠ Other trading centre
- —— Main trade route
- —— Major navigable river
- ···· Canal
- —— Boundary of administrative region
- **TAN** People

Following the fall of the Tang dynasty in 907 (*pages 72–73*), southern China was broken up into small "kingdoms" ruled over by warlords, while northern China was controlled by a rapid succession of "dynasties", the Later periods of the Liang, Tang, Jin, Han and Zhou. This period of disunity, known as the Ten Kingdoms and Five Dynasties, was ended in 960 by the general Zhao Kuangyin, who brought China under the control of the Song dynasty and reigned as Emperor Taizu until 976.

The reunified Chinese Empire (*map 1*) was rather different in character from its Tang predecessor. It was much smaller: Central Asia had been lost, and the Liao state in the northeast was controlled by the Khitan people, the Xixia state in the northwest by the Tangut people. The Khitan and the Tangut were non-Chinese, and the north presented a constant military threat to the Song. Initially the Song emperors established the northern city of Kaifeng as their capital. However, after the loss of much of northern China to Jurchen invaders, who created the Jin state, the Song established a second capital further south in Hangzhou.

CULTURE AND ECONOMY OF THE SONG PERIOD
The Song period saw a great revival in Confucianism, regarded as the native Chinese philosophy, at the expense of Buddhism, which had been imported from India during the Tang period. The class of scholar-officials burgeoned as great emphasis was placed on civil service examination, which began during the Han period and continued under the Tang rulers, as the method of recruiting the governing elite. By the end of the era some 400,000 candidates sat exams each year, sometimes with hundreds of aspirants chasing a single post. Scholarly families fuelled a demand for the many new books of all sorts that the improvements in printing, such as wood-block printing and the use of moveable type, allowed to be produced. The Song era also witnessed new artistic forms, notably the rise of landscape painting – and indeed the Emperor Huizong (r. 1100–1126) was blamed for the loss of the north because he allowed his interests in art to distract him from government.

The population of China rose to over 100 million by 1100, with a much higher increase in the south than in the north. This demographic growth was accompanied by great economic growth and an expansion in mercantile activity, notably in waterborne trade, facilitated by the world's first paper money. Vast new tracts of land were opened up for agriculture, and the development of an unregulated property market led to the appearance of huge estates. All across China new cities flourished, often starting out as bustling markets but with tea houses and shops soon added to attract traders and customers. In the 13th century the Italian traveller Marco Polo was to describe the later Song capital of Hangzhou as the finest and most splendid city in the world.

EVENTS IN THE NORTHEAST

The Liao state in the northeast was a union of a number of Khitan tribes – originally from the margins of the Manchurian steppe – brought together by the ruler Abaoji in the early 10th century. Their state comprised a solidly Khitan northern part and a southern part divided into 16 provinces and occupied mostly by the three million Chinese ruled over by the Khitan. From the late 10th century the Khitan repeatedly attacked the Koryo kingdom in Korea, capturing the capital Kaegyong in 1011. There were also frequent forays against the Tangut to the west.

By the 12th century a new power had emerged in the northeast – a confederation of Jurchen tribes from the mountains of eastern Manchuria. Following victory over the Liao state in 1125, the Jurchen seized north China two years later and established the Jin dynasty (*map 2*). The Song dynasty survived in the south until 1279, when the whole country fell to the Mongols (*pages 98–99*); they were, in their turn, to be replaced in 1368 by the Ming dynasty.

KOREA AND JAPAN

On the Korean Peninsula (*map 3*) the Koryo kingdom lasted until 1392. The later years of the dynasty were marked by repeated debilitating incursions by northern nomads and, from 1231, a series of invasions by Mongol armies. In 1232 the court was forced to flee the capital to Kanghwa Island and by 1259 the government had accepted Mongol domination. Rebellions and coups took their toll, and in 1388 General Yi Song-gye mounted a coup d'état, ushering in the Yi dynasty that was to last from 1392 until 1910 (*map 4*).

Hanyong, modern Seoul, replaced Kaegyong as the capital and in October 1446 Hangul, the new Korean script, was promulgated. Employing a phonetic alphabet, which can be learnt much more quickly than Chinese ideographs, this script brought literacy to the peasants and enabled the gradual appearance of a vernacular literature.

In Japan the seat of government shifted from Kyoto to Kamakura in 1185 as military overlords, or *shoguns*, took power from the emperor in Kyoto. The Kamakura period (1185–1335) saw the development of the militaristic *samurai* culture. In 1274 and 1281 two unsuccessful

2 EAST ASIA IN 1150
— Boundary of empire or kingdom c. 1150
MON People

expeditions were launched against Japan from Korea by the Mongols. Power returned to the imperial capital of Kyoto in the Ashikaga or Muromachi period (1335–1573), but during the Onin Wars, which began in 1467 and continued for over a century, the country was wracked by bloody civil conflict. Christianity arrived in 1543, accompanied by new tools of war, including castle architecture and flintlock guns.

The internal fighting was ended by two successive unifiers of the country, Oda Nobunaga and Toyotomi Hideyoshi, whose respective castles give their names to the Azuchi-Momoyama period (1573–1613). After winning control of most of Japan in 1590, Hideyoshi failed in his first invasion of Korea in 1592 when his force of 160,000 men – aiming to conquer China after subduing Korea – were thwarted after the Korean admiral Yi Sun-Sin famously cut his enemy's nautical supply lines.

Japanese incursions into Korea were met with counterattacks by combined Ming Chinese and Korean forces, and indeed Hideyoshi died in his second attempt at conquering Korea in 1597. Power passed to Tokugawa Ieyasu, who established the Tokugawa Shogunate (*pages 140–41*) and closed the doors of Japan to the outside world.

▲ In 1161 the Jin dynasty adopted Kaifeng, the old Song seat of government on the Huang He, as their capital, while the retreating Song set up a new capital further south at Hangzhou.

▼ The 16th century in Japan is known as the era of the Warring States, or Sengoku period, during which regional warlords fought each other to win control of the country. When it ended, the Japanese rulers set their sights on conquering Korea.

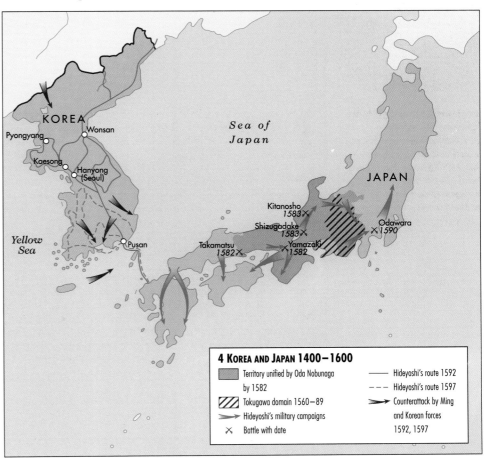

3 KOREA UNDER THE KORYO DYNASTY 936–1392
— Boundary of administrative district
■ Capital
▣ Garrison
▪ District administrative headquarters
᠆ᴸ᠊ Wall between Korea and Jin
● Koryo kiln site

4 KOREA AND JAPAN 1400–1600
▨ Territory unified by Oda Nobunaga by 1582
▨ Tokugawa domain 1560–89
➤ Hideyoshi's military campaigns
✕ Battle with date
— Hideyoshi's route 1592
--- Hideyoshi's route 1597
➤ Counterattack by Ming and Korean forces 1592, 1597

▲ Under the Koryo, pottery manufacture flourished. Cultural achievements included the publication of the first Korean histories, while among technical innovations was the use of moveable type, leading to the world's first casting of metal type in 1403.

◀ EAST ASIA IN THE TANG PERIOD 618–907 *pages 72–73* ▶ CHINA 1368–1800 *pages 138–39* ▶ TOKUGAWA JAPAN 1603–1867 *pages 140–41*

THE MUSLIM WORLD
1000–1400

▼ During the 10th century the political unity of the Muslim world collapsed. The Abbasid caliphs, previously dominant from the Atlantic to India, were replaced by a series of regional dynasties, and the caliph in Baghdad was reduced to little more than a religious figurehead.

At the beginning of the 11th century the Muslim world stretched from Spain in the west to the borders of Central Asia and India (*map 1*). Yet the political and religious unity provided for most of the Muslim world by the Abbasid Caliphate – with the notable exception of Umayyad Spain – had been lost by the 10th century. The Abbasid Empire had fragmented and the central lands of Egypt and Iraq were occupied by the Fatimids and the Buyids, both Shiite states that rejected the Sunni caliph's religious authority. The caliph himself now survived as no more than a powerless figurehead in Baghdad under the ignominious tutelage of a Buyid sultan. In the far west the Umayyad Caliphate was close to collapse and partition between a number of successor states – the *taifa* kingdoms – and the Maghreb (North Africa) was divided between several Berber dynasties. The major power in the east was the Ghaznavids, a Turkish dynasty of former slave soldiers whose only rivals were the recently converted Turkish Qarakhanids and the still largely non-Muslim Turkish nomads, especially the Oguz, on the steppe to the north. Muslim political weakness had already allowed the Byzantines to expand into Syria and Armenia, and it would soon open the way for Christian conquests in Spain and Sicily.

THE GREAT SELJUK EMPIRE
In the west the Muslim retreat was only temporarily halted by the occupation of Muslim Spain by Berber dynasties from the Maghreb – first the Almoravids (1086–1143) and later the Almohads (1150–1228). In the central and eastern lands the situation was transformed first by the conversion of the Oguz Turks to Sunni (rather than Shiite) Islam, and then in 1038 by the Oguz invasion of Iran, led by the Seljuk dynasty. Victory over the Ghaznavids at Dandankan in 1040, the conquest of Baghdad from the Buyids in 1055 and the defeat of the Byzantines at Manzikert in 1071 enabled the Seljuks to create a loose Sunni empire that stretched from the edge of the steppe to Anatolia and Palestine. The religious, if not the political, authority of the Abbasid caliph was restored, and the next target was Shiite Egypt.

The so-called Great Seljuk Empire (to distinguish it from the later Anatolian state of the Seljuks of Rum) reached its zenith under Malik Shah (*map 2*). His death in 1092 opened a new phase of political instability and fragmentation which provided the opportunity in 1098–99 for Latin Christians from western Europe to establish the Crusader States in Syria and Palestine (*pages 94–95*). The Seljuks continued to rule in parts of western Iran as late as 1194, but the Seljuk era was over in Syria by 1117, and in most of eastern Iran by 1156. Only in Anatolia did an independent branch of the Seljuk dynasty flourish into the 13th century.

One beneficiary of Seljuk decline were the Abbasid caliphs, who enjoyed a new-found political independence in southern Iraq, but otherwise the central and eastern lands

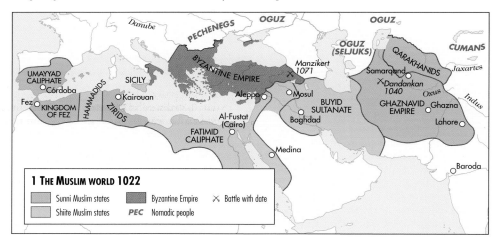

1 THE MUSLIM WORLD 1022

- ▢ Sunni Muslim states
- ▢ Shiite Muslim states
- ▢ Byzantine Empire
- **PEC** Nomadic people
- ✕ Battle with date

► Under Malik Shah, the Seljuk-led warbands of the Oguz Turks reunited much of the old Abbasid Empire. His authority was based loosely on a combination of personal prestige and the ability, furnished by his military successes, to distribute material reward to more or less autonomous subordinate rulers, each with his own warrior following.

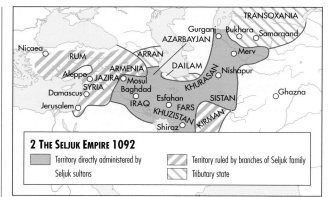

2 THE SELJUK EMPIRE 1092

- ▢ Territory directly administered by Seljuk sultans
- ▨ Territory ruled by branches of Seljuk family
- ▨ Tributary state

► The unity fostered by the Seljuks in the 11th century was illusory. Reliant on continued military expansion to provide the rewards coveted by local leaders, it was not sustainable in the long term. Instead, in the 12th century the Muslim world fragmented into a series of regional authorities — a localization of power which made possible gains by the Byzantines, crusaders, nomads and others at the expense of particular Muslim communities.

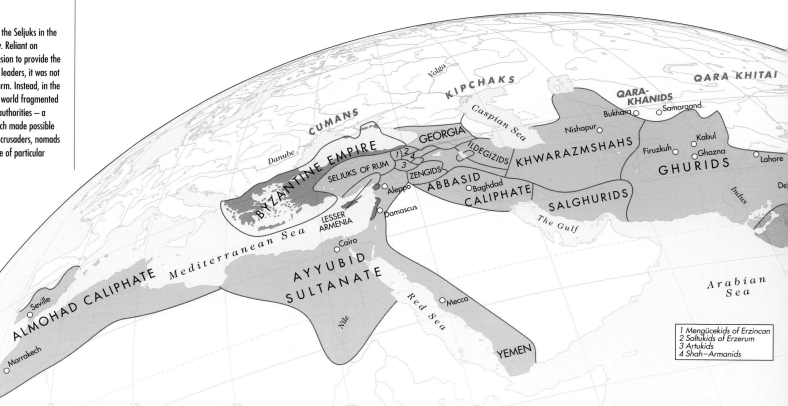

1 Mengücekids of Erzincan
2 Saltukids of Erzerum
3 Artukids
4 Shah–Armanids

of the Muslim world fell to Turkish dynasties. Several of these lineages, including the Zengids, the Ildegizids and the Salghurids, had their origins as *atabegs*, holders of delegated Seljuk authority (*map 3*), but there were two important exceptions – the Ayyubids and the Ghurids.

The Ayyubids were a Kurdish dynasty who began as soldiers serving the Zengids. The most famous Ayyubid, Saladin, overthrew the Fatimid Caliphate in 1171, so restoring Sunni authority in Egypt. Having expelled the Zengids from Damascus and Aleppo and retaken Jerusalem from the crusaders, he established himself as the dominant Muslim leader in the western Near East (*pages 94–95*).

The Ghurids were an Iranian dynasty from a tribal background in eastern Iran. They came to prominence serving the Ghaznavids and Seljuks – before, like the Ayyubids, taking over from their former masters as rulers in their own right. From the 1150s until their disastrous defeat by the nomad Qara Khitai in 1204, the Ghurids were the leading power in eastern Iran. Their conquests in India between 1192 and 1206, going beyond the earlier Ghaznavid territories based on Lahore, laid the foundation for the Turkish Sultanate of Delhi in 1211 and long-lasting Muslim rule in the subcontinent (*map 4*).

THE MONGOL INVASIONS

The late 12th century, the age of Saladin and the Ghurids, was a period of calm before a storm which threatened the complete destruction of Islam. From 1219 the pagan Mongols invaded and gradually conquered the area of modern-day Iran, Iraq and eastern Anatolia (*pages 98–99*). Baghdad was sacked in 1258, and the last generally recognized Abbasid caliph put to death. In the West, Christian armies were conquering most of what remained of Muslim Spain – and in 1217–21, and again in 1249–50, they threatened to seize Cairo and end Muslim rule in Egypt.

The Muslim world was saved partly by disunity among the Mongols. After 1242 the Mongols in the west were divided between the Golden Horde, the Ilkhanate and the Chaghatai Khanate, and they frequently fought one another as fiercely as they did their non-Mongol enemies (*map 5*). Islam as a religion and a culture also proved capable of converting some of its conquerors. Although the Spanish Christians proved resistant, both the Golden Horde and the Ilkhan Empire had converted to Islam by the early 14th century. Muslim survival was also due to fierce resistance – in India from the sultans of Delhi, in Syria and Palestine from the Mamluk rulers of Egypt.

THE MAMLUKS OF EGYPT

Slave soldiers or *mamluks* (usually Turks imported from the steppe) had been a feature of Muslim armies since the 8th century. The Egyptian mamluks serving the Ayyubids were mostly Kipchak Turks, brought as slaves from the Black Sea and

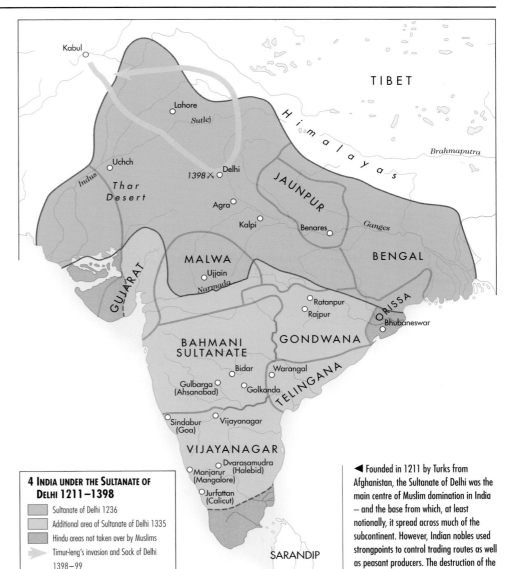

4 INDIA UNDER THE SULTANATE OF DELHI 1211–1398

- Sultanate of Delhi 1236
- Additional area of Sultanate of Delhi 1335
- Hindu areas not taken over by Muslims
- ➤ Timur-leng's invasion and Sack of Delhi 1398–99
- Boundaries of Hindu states re-established 1399 onwards
- ✕ Battle with date

◀ Founded in 1211 by Turks from Afghanistan, the Sultanate of Delhi was the main centre of Muslim domination in India – and the base from which, at least notionally, it spread across much of the subcontinent. However, Indian nobles used strongpoints to control trading routes as well as peasant producers. The destruction of the sultanate by the Mongol conqueror Timur-leng in 1398 paved the way for the decentralization of power into the hands of local Hindu and Muslim rulers.

▼ Mongol military power conquered much of the Muslim world in the 13th century. However, because the Mongols converted to Islam their fragmented empire failed to threaten Muslim religious and cultural domination of most of the lands of the former Abbasid Caliphate.

taken to Egypt, where they were converted to Islam and trained to become a formidable military force. In 1250, after the French crusader invasion landed, the leaders of one of the main mamluk regiments murdered the last Ayyubid sultan in Egypt and seized power. By the beginning of the 14th century the Mamluk regime had permanently halted the Mongol advance – and expelled the crusaders from their last coveted territories on the Levantine mainland.

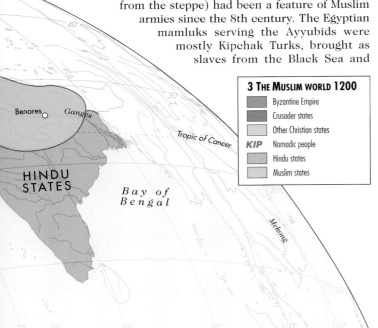

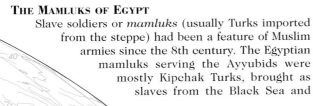

3 THE MUSLIM WORLD 1200

- Byzantine Empire
- Crusader states
- Other Christian states
- **KIP** Nomadic people
- Hindu states
- Muslim states

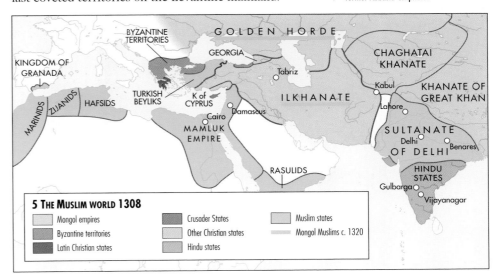

5 THE MUSLIM WORLD 1308

- Mongol empires
- Byzantine territories
- Latin Christian states
- Crusader States
- Other Christian states
- Hindu states
- Muslim states
- Mongol Muslims c. 1320

◀ THE SPREAD OF ISLAM 600–1000 *pages 66–67* ▶ THE BYZANTINE AND OTTOMAN EMPIRES 1025–1500 *pages 96–97*

THE HOLY ROMAN EMPIRE
962–1356

When the East Frankish king, Otto I, was crowned emperor by the Pope in Rome in 962, his empire comprised those lands north of the Alps which had formed the East Francia of the 843 Carolingian partition (*pages 74–75*) together with Lotharingia (the 843 "middle kingdom" to which Burgundy – the territories from Basel to Provence – was to be added in 1032–34), and Lombardy (*map 1*). This empire was passed on with relatively minor geographical alteration thereafter to his son and grandson (Otto II and Otto III) and then to his Salian, Staufen, Welf, Luxembourg and Habsburg successors.

By taking the imperial title, Otto was deliberately presenting himself as the successor of Charlemagne – restorer of the Christian empire in the west – in order to enhance his prestige. Two centuries later, when Frederick Barbarossa succeeded to the same kingship and imperial status, he reaffirmed the continuing tradition by instigating Charlemagne's canonization and by adding the word "holy" to the name of the empire. A further two centuries later, in 1355, Charles IV of Luxembourg secured his imperial

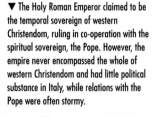

▼ The Holy Roman Emperor claimed to be the temporal sovereign of western Christendom, ruling in co-operation with the spiritual sovereign, the Pope. However, the empire never encompassed the whole of western Christendom and had little political substance in Italy, while relations with the Pope were often stormy.

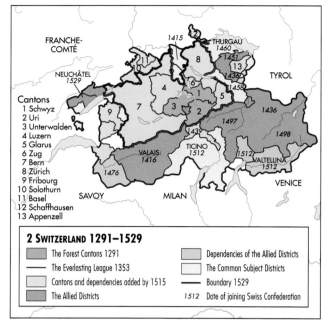

2 SWITZERLAND 1291–1529

▓ The Forest Cantons 1291	▓ Dependencies of the Allied Districts
— The Everlasting League 1353	░ The Common Subject Districts
░ Cantons and dependencies added by 1515	— Boundary 1529
▓ The Allied Districts	*1512* Date of joining Swiss Confederation

Cantons
1 Schwyz
2 Uri
3 Unterwalden
4 Luzern
5 Glarus
6 Zug
7 Bern
8 Zürich
9 Fribourg
10 Solothurn
11 Basel
12 Schaffhausen
13 Appenzell

▲ The Swiss Confederation grew from an initial "peace association" formed by the three Forest Cantons in 1291. It expanded in the mid-14th century to include the towns of Luzern, Bern and Zürich in a league which controlled the trade route from the Rhine Valley across the Alps via the St Gotthard Pass.

1 THE HOLY ROMAN EMPIRE C. 950–1360

— Boundary of empire 1356	▓ Kingdom of Italy, 12th and 13th centuries
— Boundary of East Francia and Italy c. 900	⚏ Alpine pass
▓ Kingdom of Otto I, 936	✕ Battle with date
▓ Kingdom of Burgundy c. 1032	TRIER Electorate

Places of residence of imperial dynasties:
● Liudolfing ◉ Habsburg
● Salian ◉ Wittelsbach
● Staufen ○ Luxembourg

coronation in Rome, and then, in 1356, issued the Golden Bull. This came to be viewed as the basic constitutional law of the empire, defining as it did the right of seven Electors meeting at Frankfurt – the archbishops of Mainz, Cologne and Trier, the Count Palatine of the Rhine, the Duke of Saxony, the Margrave of Brandenburg and the King of Bohemia – to designate the emperor-elect, also called "King of the Romans". In this form, the Empire continued until its dissolution in 1806.

THE ITALIAN KINGSHIP

Within the Empire the sense of two component kingships was maintained: the primary northern kingship comprising Franks, Saxons, Swabians, Bavarians and Lotharingians, and the southern secondary kingship of the Lombards. The emperor-elect, chosen by German princes, travelled south across the Alps to secure recognition in northern Italy and coronation by the Pope in Rome, but there was little governmental substance to his position in Italy. Intermittently, attempts were made to change this situation. Between the mid-10th and mid-11th centuries the Liudolfing and Salian emperors spent lengthy periods south of the Alps. In the years 1158–77 the Staufen emperor Frederick Barbarossa sought to benefit from the gathering pace of economic growth and north Italian trade (*pages 100–1*), but failed to win a decisive victory over the Lombard League of northern town communes. His son successfully took over Sicily and southern Italy in 1194, but his grandson's renewed attempt in 1236–50 to master Lombardy was thwarted by the alliance of communes and Papacy.

The pattern of northern intervention in Italy survived the Staufens' loss of the Sicilian as well as the German kingship in 1254–68. However, after the expeditions of Henry of Luxembourg in 1310–13 and Ludwig of Wittelsbach in 1328, imperial jurisdiction south of the Alps was merely theoretical. In practice, government and politics evolved as an autonomous system of local regimes – and the flowering of both Italian economic enterprise and Renaissance culture developed independently of the Empire (*pages 102–3*).

THE NORTHERN EMPIRE

In Germany the king's position was stronger than in Italy, yet here too the force of localism was of primary importance. Traditions of local lordship and identity were very powerfully entrenched, pre-dating the Carolingian "unification" of the region under a single kingship, and

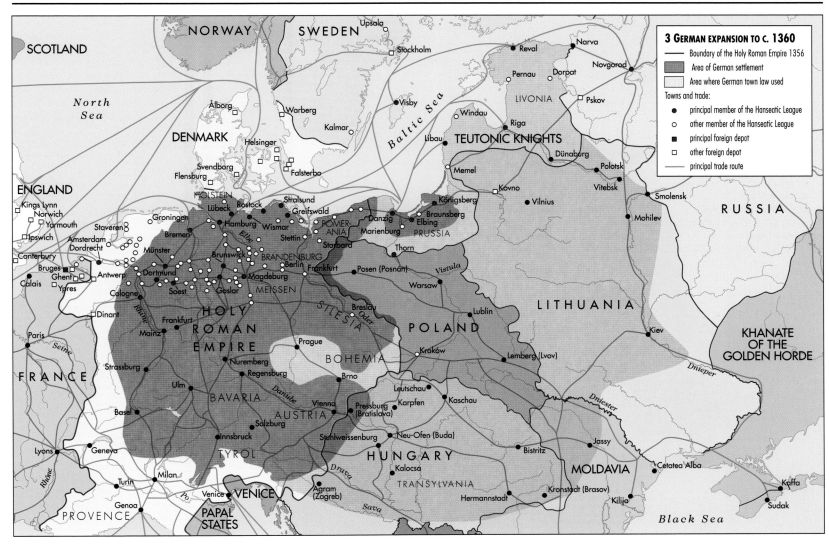

3 GERMAN EXPANSION TO C. 1360

—— Boundary of the Holy Roman Empire 1356

Area of German settlement

Area where German town law used

Towns and trade:

● principal member of the Hanseatic League

○ other member of the Hanseatic League

■ principal foreign depot

□ other foreign depot

—— principal trade route

remained the necessary framework of government. It was impossible for any single authority to exert control over so large and diverse an area and even when – in Germany as elsewhere in the 12th and 13th centuries – more bureaucratic governmental techniques were developed, they benefited local rulers rather than the emperor. These local rulers might be noble dynasts, communal associations in individual "free towns", or more varied groupings. Among the latter the Swiss cantons, which included both Alpine rural communities and towns, were the most successful in consolidating a separate existence (*map 2*).

EASTWARD EXPANSION

Both the diversity and the extent of German society were enhanced between the 10th and 14th centuries by large-scale expansion eastwards. In the 10th century the Saxon Liudolfings gained acceptance as kings through their successful military leadership in warfare against the Slavs east of the Elbe – and above all against the Magyars who, from 900, were raiding along the Danube Valley. The victories of Henry I in the north in 933 and Otto I in the south in 955 opened the way to German movement eastwards, in a number of permutations of tribute-taking and land-settling ventures (*map 3*).

After the 11th century, kings and emperors had little to do with such expansion. Instead, local dynasties – such as the Babenbergs in Austria or the Wettins in Meissen – recruited the necessary human resources of peasant farmers and urban traders and provided the local structure of military and juridical organization. This movement of eastward expansion far exceeded even the expanded limits of the Empire (Reich), whose princes attended the Reichstag and engaged in the politics of elective kingship. Throughout east-central Europe, with the active encouragement of local rulers, German communities, equipped with German customary law, were induced to settle alongside Slav and Magyar populations.

From the mid-12th century some of these local rulers were connected with crusading impulses (*pages 94–95*). The Wendish Crusade from 1147 to 1185, waged by German princes and Danish kings, brought forcible Christianization to Holstein, Mecklenburg and Pomerania. A further series of crusades developed after 1200 in the east Baltic area of Livonia, extending into Finland by the 1240s under the impetus of Swedish conquest. Most notably, from the 1220s the Teutonic Order (an organization of soldier-monks, founded in Palestine in the 1190s, whose members were recruited from the Rhineland and other parts of the Empire) acquired independent rule in Prussia and from there waged the "Perpetual Crusade" against the pagan Lithuanians.

THE HANSEATIC LEAGUE

The 12th and 13th centuries also saw the creation of a network of German maritime enterprise in the Baltic, from Novgorod to Flanders and England through the North Sea. The timber, furs and grain of Scandinavia, northern Russia, and the southern hinterland of the Baltic were shipped westwards, with return cargoes of cloth and other manufactured commodities. Merchants formed associations (*hanses*) to protect and enhance their trade and in the 13th century this trading network developed into the Hanseatic League (*map 3*). The League linked the newly founded German towns (dominated by the Hanseatic merchants) on the southern Baltic coast between Lübeck and Riga, both southwards to the German hinterland and the newly exploited lands to the east, and northwards to Scandinavia. Throughout this area local rulers awarded grants of privilege in return for profit-sharing arrangements, thus contributing to German economic and cultural expansion within Europe.

▲ By the 13th century the movement of Germans eastwards had advanced the limit of the Empire over a wide band of territory from Austria north to Meissen, Brandenburg, Holstein, Mecklenburg and Pomerania. In the 1220s the Teutonic Order contributed to the defence of Hungary and Poland against their pagan neighbours in Transylvania and Prussia, and in the following decades it established control over Prussia and Livonia. From here it waged the "Perpetual Crusade" against the pagan Lithuanians until 1410, when it was defeated at Tannenberg by the Poles and Lithuanians (whose conversion to Christianity was achieved in 1386–87 by the less violent method of dynastic marriage diplomacy).

FRANCE, SPAIN AND ENGLAND
900–1300

1 THE KINGDOMS OF FRANCE AND BURGUNDY C. 1050

--- Boundary of Kingdom of France
--- Boundary of Kingdom of Burgundy
▨ Royal domain of Capetian kings
▨ Episcopal lordships
NOR Important lay lordships
Co County
Visco Viscounty

▲ The more important regional powers in France and Burgundy around 1050 included Normandy, Flanders, Anjou and Toulouse as well as the Capetian kings. Their authority was no more stable than had been that of the Carolingians.

▼ The kings of Aragon were united in 1137 with the already powerful counts of Barcelona, and they used the growing commercial wealth of the port of Barcelona to extend their control to southern France through the imposition of feudal ties.

▼ The Christian kings in Spain strengthened their position by organizing opposition to the Muslim rulers in the south. Having held out against the Almohads and Almoravids, they overran much of the Muslim territory in the 13th century.

Between the 10th and 13th centuries much political control in France, Spain, England and other areas of western Europe was devolved to local landowning aristocracies who built castles and employed armoured knights to assert their power over the peasants. Depending on circumstances, these local magnates came more or less under the control of kings or regional lords. There was no simple pattern, but underlying changes in the economy meant that the power and influence of kings and regional lords, after declining during the 11th century, had generally grown by around 1300.

THE KINGDOM OF FRANCE
During the 8th and early 9th centuries the French Carolingian kings (*pages 74–75*) had been immensely successful in harnessing the aristocracy in a common enterprise. However, by the end of the 10th century royal power and the political structure of West Francia were undergoing a fundamental transformation. One reason for this was that in about 950 the economy of western Europe had entered a phase of steady growth, marked by rising population, new settlements and an increasing volume of exchange (*pages 100–1*). At the same time the Carolingian lands in West Francia had been given away or sold off in an attempt to buy support – and lacking any obvious foreign enemy either to plunder or unite against, the French kings had soon been reduced to comparative impotence. By 987, when Hugh Capet replaced the last Carolingian king, royal authority extended little beyond the small royal domain in the Île de France (*map 1*).

The extent to which power had devolved varied from area to area, and authority by no means remained stable. In the county of Mâcon, for example, the counts had largely thrown off the authority of the dukes of Burgundy by 980, only to then find their own authority steadily undermined. As a result, by about 1030 the local castle-holders (*castellans*) and great churches were in effect independent, with their own courts exercising private justice – "banal lordship" – over a large subject population.

THE CONSOLIDATION OF POWER
By the 12th century three factors tended to favour larger and more coherent political units. First, the growing profits arising from customs, tolls and urban expansion were more easily exploited by regional powers than by independent castellans. As trade across Europe increased, the taxation of its profits at regional level made kings and other greater lords a dominating social force. Second, the increasing use of written records and accounts gave rise to a new bureaucracy of clerks, accountants and lawyers whom only the wealthiest could afford to employ, but who in turn allowed a

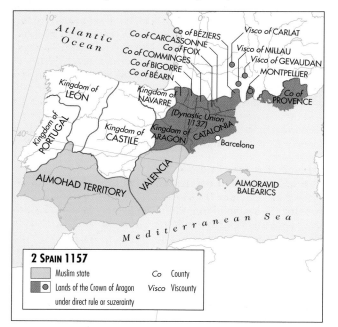

2 SPAIN 1157

▨ Muslim state
◉ Lands of the Crown of Aragon under direct rule or suzerainty
Co County
Visco Viscounty

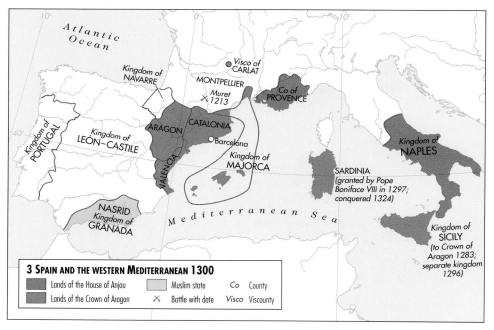

3 SPAIN AND THE WESTERN MEDITERRANEAN 1300

▨ Lands of the House of Anjou
▨ Lands of the Crown of Aragon
▨ Muslim state
✕ Battle with date
Co County
Visco Viscounty

much more effective exploitation of resources. Third, the spread of feudal relations enabled kings, on the basis of their growing wealth, to impose greater obligations on their castle-holding subjects.

SPAIN: THE RISE OF ARAGON

An example of these factors being turned to good effect is the rise of the House of Aragon. In the late 11th and 12th centuries the counts of Barcelona (from 1137 also kings of Aragon) imposed feudal ties on the aristocracy of Catalonia, and went on to do the same in the kingdom of Burgundy for the turbulent aristocracy of the county of Provence (*map 2*). Although Count Pere II's defeat and death at the Battle of Muret in 1213 brought an end to Aragonese power north of the Pyrenees, his successors had carved out a substantial Mediterranean empire by the end of the century (*map 3*).

Controlling and directing the reconquest of Muslim Spain was a further lever of power in the hands of Christian Spanish monarchs. During this period, the Christian kingdoms first terrorized the successor states (*taifas*) to the once-powerful Muslim Umayyads (*pages 88–89*), and then held out against the counterattack of the Berber Almoravids and Almohads before overrunning most of what was left of Muslim territory in the 13th century.

ENGLAND: A PROCESS OF CENTRALIZATION

During the 10th and early 11th centuries the Anglo-Saxon kings faced the threat of Viking conquest, and in the process forged a sophisticated and centrally controlled administrative machine. A network of shires was created, and royal mints enabled the Crown to enforce a standardized coinage and gain a considerable income through regular remintings.

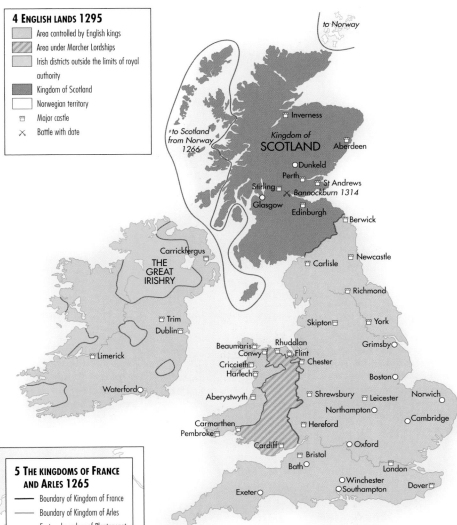

4 ENGLISH LANDS 1295
- Area controlled by English kings
- Area under Marcher Lordships
- Irish districts outside the limits of royal authority
- Kingdom of Scotland
- Norwegian territory
- Major castle
- Battle with date

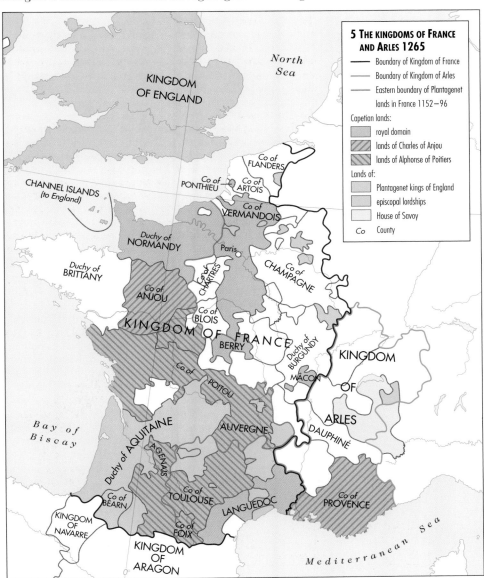

5 THE KINGDOMS OF FRANCE AND ARLES 1265
- Boundary of Kingdom of France
- Boundary of Kingdom of Arles
- Eastern boundary of Plantagenet lands in France 1152–96

Capetian lands:
- royal domain
- lands of Charles of Anjou
- lands of Alphonse of Poitiers

Lands of:
- Plantagenet kings of England
- episcopal lordships
- House of Savoy

Co   County

◄ By 1265 the Capetian kings directly or indirectly ruled large areas of France, and the extent of English-controlled territory had been greatly reduced.

▲ The English crown effectively controlled most of the British Isles by 1300. Its advance into Scotland came to a halt in 1314 at the Battle of Bannockburn.

The Norman Conquest in 1066 paradoxically reinforced the English state, sweeping away aristocratic rivals to the crown and leaving William I and his successors with the most centralized and best administered state in western Europe.

As in Spain, royal power in England benefited from controlled expansion and the distribution of any profits arising from it. Between the 11th and 13th centuries the English kings conquered Wales (complete by 1295) and Ireland (from 1169), and threatened to do the same to Scotland until their defeat at Bannockburn in 1314 (*map 4*). The English kings also extended their territory in France. By the time Henry II ascended the throne in 1154 he ruled, in addition to England and Normandy (which he had inherited from his mother), territory in western France (inherited from his Plantagenet father); further territory had come with his marriage to Eleanor of Aquitaine (*map 5*).

FRANCE: CAPETIAN DOMINANCE

In France, luck and political skill favoured the Capetians. The death of Henry II's son Richard I in 1199 opened the way for the French king, Philip Augustus (1180–1223), to deprive Richard's brother John of French lands, including Normandy and Anjou, in a series of campaigns between 1203 and 1206. Philip's achievements, confirmed by a decisive victory in 1214, transformed the political geography of western Europe, with the Capetian kings now dominant (*map 5*). Paris became the uncontested political and administrative hub of the kingdom, and an intellectual centre for the whole of Latin Christendom.

◀ FRANKISH KINGDOMS 200–900 *pages 74–75* ◀ EUROPE 1350–1500 *pages 106–7*

THE WORLD OF THE CRUSADERS
1095–1291

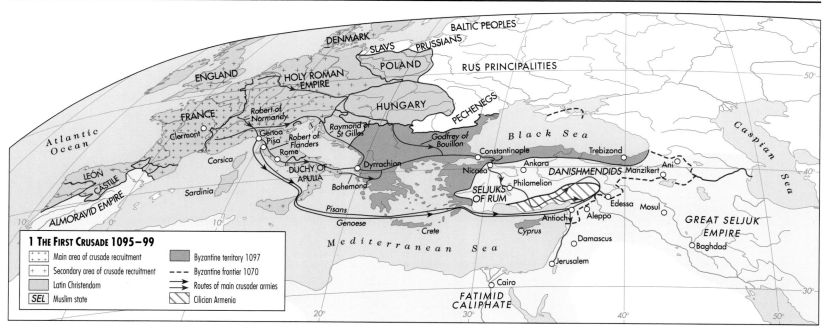

1 THE FIRST CRUSADE 1095–99

Main area of crusade recruitment	Byzantine territory 1097
Secondary area of crusade recruitment	Byzantine frontier 1070
Latin Christendom	Routes of main crusader armies
SEL Muslim state	Cilician Armenia

▲ The backbone of the armies of the First Crusade was provided by knights travelling as part of their lords' households. The capture of Jerusalem in July 1099 after two years' journeying – and a series of unlikely military victories – convinced survivors and contemporaries that the enterprise had been blessed by God.

Over the course of 200 years a total of five major and several minor crusades set out from Christian Europe with the declared aim of either recapturing or protecting the Holy Land (Palestine) from the Muslims. The first was launched at Clermont in central France on 27 November 1095 by Pope Urban II. A vast number of people – perhaps about 100,000 – were inspired to take part in a penitential military pilgrimage to recover the Holy Sepulchre in Jerusalem (*map 1*). For the Pope the expedition was a response to Byzantine appeals for help in the wake of the Turkish conquest of Anatolia, offering the opportunity to raise papal prestige through the leadership of Latin Christendom in such a spiritually beneficial enterprise. For the participants it was, perhaps above all else, an opportunity to earn salvation, their enthusiasm testifying to the degree to which Christian teaching had implanted in Western society a fear of the dreadful fate after death that awaited people who had not atoned for their sins. However, hopes of land, booty and fame were also important.

THE CRUSADER STATES

By the time the expedition reached Jerusalem there were barely 14,000 crusaders. They nevertheless managed to capture the city and, over the next 40 years, establish and expand the boundaries of four states in the surrounding region: the kingdom of Jerusalem, the county of Tripoli, the principality of Antioch and the county of Edessa (*map 2*).

Their initial success owed a great deal to the temporary political divisions in the Muslim world. The death of the powerful Seljuk sultan Malik Shah in 1092 had plunged the sultanate into a complex civil war. Ultimately Malik Shah's son Berkyaruk prevailed, keeping control of the area of present-day Iraq and Iran, but Ridwan and Dukak, the sons of his uncle and chief opponent, Tutush (d. 1095), still ruled in Aleppo and Damascus respectively. The brothers were loath to co-operate with each other, with Kerbogha (the Seljuk governor of Mosul whom Berkyaruk sent to bring help against the crusaders), or with the Shiite Fatimid Caliphate in Egypt. The Fatimids had ruled most of Syria and Palestine through the 11th century up to the 1070s, and had themselves recaptured Jerusalem from the Seljuks only a year before the crusaders entered the city in 1099.

The Second Crusade (1146–48) failed to take Damascus, and after 1154 the situation changed significantly. In that year Mosul, Aleppo and Damascus were united under the aggressive leadership of Nur al-Din, who deliberately underpinned his authority with an ideology of holy war against the crusaders. The decline of the Shiite Fatimid Caliphate also altered the balance of power. The agricultural and commercial riches of Egypt were potentially the key to domination of the Levant. However, attempts led by King Amalric of Jerusalem between 1163 and 1169 to conquer or control Egypt merely encouraged Nur al-Din to send one of his generals, a Kurd called Saladin, to keep the crusaders out. Saladin successfully fought off the crusaders, before putting an end to the Fatimid Caliphate in 1171 (*map 3*).

After Nur al-Din's death in 1174, Saladin gradually dispossessed his former master's heirs, and by 1186 they had been forced to recognize his overlordship. Saladin was now able to wage war with the combined resources of Egypt and Syria, and in July 1187 he inflicted a crushing defeat on the crusaders at the Battle of Hattin, near the Sea of Galilee.

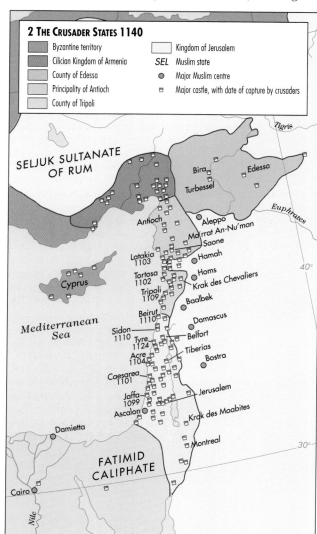

2 THE CRUSADER STATES 1140

Byzantine territory	Kingdom of Jerusalem
Cilician Kingdom of Armenia	SEL Muslim state
County of Edessa	Major Muslim centre
Principality of Antioch	Major castle, with date of capture by crusaders
County of Tripoli	

► Despite many appeals, the Christian rulers of the Crusader States were unable to attract sufficient military manpower to ensure the survival of their territories. Many western Europeans did settle in the East, but most regarded crusading activity as an extended penitential pilgrimage rather than the start of a new life as a colonial elite. Those who did settle gradually acclimatized to an extent that pilgrims and crusaders fresh from the West found disconcerting.

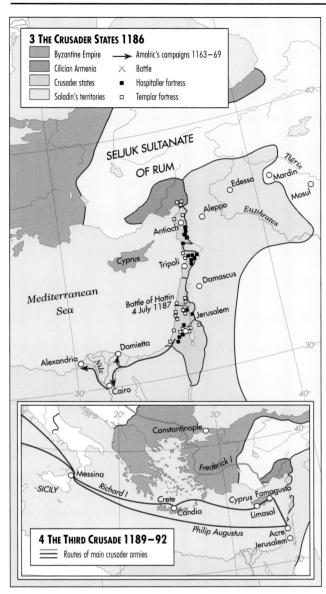

3 THE CRUSADER STATES 1186

Byzantine Empire	→ Amalric's campaigns 1163–69
Cilician Armenia	× Battle
Crusader states	■ Hospitaller fortress
Saladin's territories	□ Templar fortress

4 THE THIRD CRUSADE 1189–92

Routes of main crusader armies

▲ The crusaders' hold on the Holy Land was threatened by the rise of Saladin and the unification of Egypt and Syria. However, during the Third Crusade, Richard I of England came close to reversing Saladin's 1187 conquest of Jerusalem.

THE THIRD, FOURTH AND FIFTH CRUSADES

The Crusader States were saved from complete extinction by the arrival of the Third Crusade (1188–92) (*map 4*); political divisions among Saladin's Ayyubid heirs and then the growing Mongol threat to the world of Islam (*pages 98–99*) prolonged their existence. At the same time Western enthusiasm for crusading only continued to grow, and in fact Latin territories in the eastern Mediterranean reached their greatest extent in the early 13th century.

The Fourth Crusade (1198–1204) was diverted to conquer Constantinople, and its aftermath saw the creation of a series of Latin states on former Byzantine territory (*map 5*). The Fifth Crusade (1217–21), with contingents from Germany, Italy, Hungary, England and France, appeared close to success in Egypt before its final defeat in 1221. The French king Louis IX invested enormous resources on crusading in the east, but his Egyptian expedition of 1249–50 ended in disaster. The powerful Mamluk state which replaced the Ayyubids after 1250 (*pages 88–89*) was initially more concerned with the imminent threat from the Mongols, but as that receded the Mamluk advance proved relentless, culminating in 1291 in the fall of Acre, last of the major crusader strongholds in the Near East.

THE ESTABLISHMENT OF MILITARY ORDERS

The crusading movement between 1095 and 1291 is striking evidence of the militaristic nature of Western aristocratic culture. It also reflects the importance of European sea

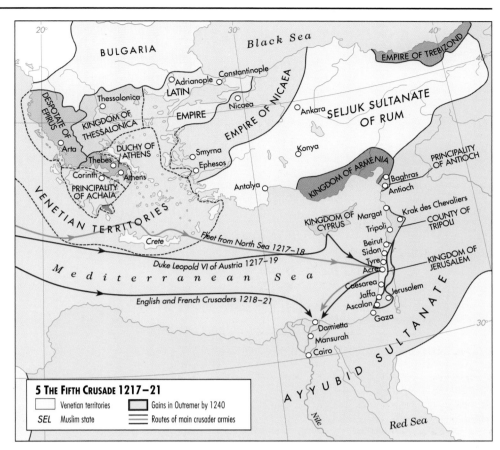

5 THE FIFTH CRUSADE 1217–21

Venetian territories	Gains in Outremer by 1240
SEL Muslim state	Routes of main crusader armies

power, especially that of Venice and Genoa, whose ships carried many of the crusaders to Palestine. During this period European maritime power grew to dominate the Mediterranean, creating a base of experience for later expansion to the Americas and the East. The failure to maintain crusader settlement in the Levant reflects the strength of Muslim opposition, but also the inadequacy of crusader manpower and resources. Even at their greatest extent in the 1140s the Crusader States amounted to little more than an embattled coastal strip.

One solution was the establishment by 1139 of the military orders of the Hospital of St John and the Knights Templar. Effectively knights living by monastic rule, both the Hospitallers and the Templars soon acquired extensive properties in the West which gave them the financial strength the settlers lacked. From the 1140s onwards many crusader lords found it necessary to hand over their more exposed strongholds to the military orders, who alone had the means to maintain and defend them.

Soon after its inception the crusading idea was transferred to other contexts. The war against the Muslims in Spain was now treated as a crusade, as was that against the pagan Slavs, Lithuanians and Balts in the north, where the Teutonic Knights – founded in the Levant in the 1190s – played a major role (*pages 90–91*). Also treated as crusades were expeditions to crush heresy, such as the Albigensian Crusade in southern France (1209–29) and those against the Hussites in Bohemia (1420–21, 1427, 1431), as well as those against political opponents of the Papacy. One such opponent was the Emperor Frederick II, who had actually taken part in a crusade in 1228–29, but himself became the target of a papal crusade in 1240–50.

Even after 1291 crusading remained deeply rooted in Western chivalric and popular culture through to the Reformation of the 16th century, and resistance to the Muslim Ottomans could still be seen in crusading terms in the 17th century. The Templars were suppressed in 1312 in the wake of heresy charges brought by Philip IV of France, but the Hospitallers survived (on Rhodes until 1522, on Malta until 1798), and do so still with their headquarters in Rome. In the modern Islamic world the crusading movement has come to be seen as evidence of the long and bloody past of Western Christian imperialism.

▲ The Fifth Crusade was an attempt to destroy Muslim power through the conquest of Egypt, whose commercial and agricultural wealth was the key to long-term control of the Near East. Ironically, more was achieved by the excommunicate crusader, Emperor Frederick II, who in 1229 recovered Jerusalem by negotiation.

▲ Captured from the Byzantines by the Seljuk Turks in 1084, Antioch was taken by the forces of the First Crusade in 1098. The principality it served – one of the four Crusader States – remained a Christian outpost for nearly two centuries.

◗ THE MUSLIM WORLD 1000–1400 *pages 88–89* ◗ THE DECLINE OF THE BYZANTINE AND RISE OF THE OTTOMAN EMPIRES 1025–1500 *pages 96–97*

THE DECLINE OF THE BYZANTINE AND RISE OF THE OTTOMAN EMPIRES 1025–1500

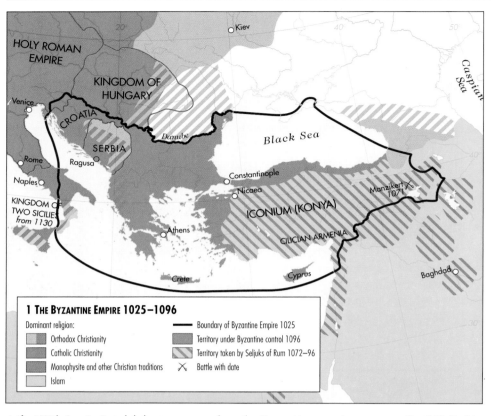

1 THE BYZANTINE EMPIRE 1025–1096

Dominant religion:

- Orthodox Christianity
- Catholic Christianity
- Monophysite and other Christian traditions
- Islam

— Boundary of Byzantine Empire 1025

Territory under Byzantine control 1096

Territory taken by Seljuks of Rum 1072–96

✕ Battle with date

▲ After 1025 the Byzantine Empire lacked the infrastructure and resources to maintain the boundaries that had been established under Basil II. In the east their defeat in the Battle of Manzikert in 1071 enabled the Seljuk Turks to establish themselves in Anatolia, while the Normans took over Byzantine territory in southern Italy.

▼ Following the sack of Constantinople by the Fourth Crusade in 1204, Byzantine lands were divided up. Territory in Europe came under the control of a Frankish emperor, who tried unsuccessfully to convert the populace to Catholicism, while the centre of Orthodox power shifted to Nicaea in northern Anatolia.

When the Byzantine warrior emperor Basil II died in 1025 he left an empire that had doubled in size during his reign and presented a serious challenge to its Muslim neighbours. Unfortunately for the Byzantines, subsequent emperors could not maintain the impetus achieved under Basil. They became embroiled in the ecclesiastical politics that provoked the "Great Schism" of 1054 – a theological split between the Orthodox and Western churches that has effectively lasted ever since. The schism invited hostility from the West at a time when Muslim power was regrouping. Norman adventurers took control of what was left of Byzantine southern Italy, just as a renewed Muslim offensive by Seljuk Turks culminated in the Battle of Manzikert (1071) – a Byzantine defeat that wiped out the eastern gains of Basil II and established the Muslim state of Iconium (Konya) in the heart of what had once been Christian Anatolia (*map 1*).

2 THE BALKANS AND ANATOLIA AFTER THE FALL OF CONSTANTINOPLE 1204

- Latin states

THE DECLINE OF THE BYZANTINE EMPIRE

The Byzantine Comnenian dynasty (1081–1185) attempted to cope with the aftermath of the Battle of Manzikert by rebuilding diplomatic bridges with the Latin West. A request by Alexius I Comnenus for modest Western military assistance was one of the factors that promoted the crusading movement (*pages 94–95*). The crusades temporarily transformed the politics of the Near East by taking Muslim pressure away from Constantinople – only to bring the city under increasing Western or Frankish influence.

In the 12th century Constantinople enjoyed a brief economic boom as a major staging post for western Europeans on the road to Jerusalem. However, the empire's finances were fundamentally weak and the Byzantines could meet their commitments only by granting commercial concessions to their erstwhile dependency, Venice. As a result the Byzantine economy became increasingly dominated by Venetian merchants in Constantinople – to the extent that from 1171 onwards Byzantine rulers attempted to cut back Venetian interests. This promoted tension and led ultimately to anti-Venetian riots in Constantinople at a time when the empire was increasingly threatened in the Balkans and Anatolia. Venice was now an enemy and took its revenge. In 1204 the old blind Venetian doge, Enrico Dandolo, successfully engineered the diversion of the Fourth Crusade away from Jerusalem and towards Constantinople. The sea walls were breached for the first time and the city was captured and systematically looted over a period of three days. This event was to mark the beginning of the Byzantine Empire's fragmentation.

Between 1204 and 1261 Constantinople was the seat of a Frankish emperor and Latin patriarch, ruling over subordinate Frankish fiefdoms: the kingdom of Thessalonica, duchy of Athens and despotate of Achaia (*map 2*). Venice dominated the Greek islands and made a particularly lasting mark in and around Naxos (where there was a Venetian duchy until 1566), although it proved impossible to graft Catholicism and an alien feudalism onto rural Greek society. Greek rule survived in Western Anatolia, based at Nicaea, and also in Epirus and in Trebizond on the Black Sea.

It was the Greek Emperor of Nicaea, Michael VIII Palaeologus, who recaptured Constantinople for Orthodoxy in 1261. The restored Byzantine Empire was, however, beset by the same problems as before: it was economically hamstrung, with Venetian and Genoese trading houses in control of its international commerce. Furthermore, it was hedged in by quarrelling rivals – threatened to the north by Balkan Slavic peoples and in Anatolia by the Turks. By the mid-14th century Greece had fallen to the Serbs (*map 3*), who were countered not by Byzantine forces but by advancing Muslim power. By 1354 the Ottoman Turks were in Europe. Thereafter the Byzantine polity dwindled into a diplomatic entity based on what was effectively the city-state of Constantinople.

THE RISE OF THE OTTOMAN EMPIRE

The Ottoman victors were the major Turkish force to emerge from the crisis of the Mongol invasions that devastated the Muslim world in the 13th century and eliminated Seljuk power (*pages 98–99*). Ottoman rulers claimed descent from Osman (Uthman), the most prominent of the Muslim "ghazis" who, in the 13th century, established independent fiefdoms amid the political ruins of what had formerly been Byzantine and Seljuk Anatolia. Ottoman society and culture were profoundly Islamic, but with a distinctive ethos derived from Central Asian nomadic antecedents. Politically, the Ottoman world was opportunist and expansionist. Osman's son, Orhan Ghazi, was able to move his capital as far west as Bursa and marry a daughter of the Byzantine Emperor John VI Cantacuzene. This marriage epitomized the steady increase of Turkish influence in medieval Anatolia – a process which led to Byzantine culture gradually losing, or abandoning, its long struggle with Islam in the interior of Asia Minor.

The Ottoman capture of Gallipoli in 1354 presaged a serious Ottoman invasion of Europe (*map 4*). By 1365 Adrianople had become the Ottoman capital Edirne. Advances into Serbia, culminating in the Battle of Kosovo Polje in 1389, put an end to Serbian expansion. At the same time the Ottomans consolidated their control of Asia Minor, and an Ottoman navy came into being, plying the waters of the Mediterranean, Aegean and Adriatic. Many of its captains were renegade Europeans. The first Ottoman siege of Constantinople itself was mounted in 1391. It was to be diverted only because of a renewed threat from the Mongols under the leadership of Timur-leng (*pages 98–99*).

THE DEFEAT OF CONSTANTINOPLE

It was now obvious that Byzantine Constantinople was living on borrowed time. It continued to function as a centre of scholarship and of an artistic style visible today in the remains of medieval Mistra in the Peloponnese. The Classical and Post-Classical heritage of Constantinople was still impressive, despite the ravages of 1204. However, its latter-day scholars were slipping away towards Renaissance Italy, taking their manuscripts with them. Meanwhile, the Ottoman Turks were developing their war machine. Since the 14th century Ottoman victories had been won with the aid of Balkan and other mercenaries. This recruitment of foreigners was formalized by the use of *devshirme* troops (recruited from Christian slaves taken into Islamic military training and educated as an elite corps).

Constantinople, as a Christian bastion, continued to receive the political sympathy of western Europe, although this was bedevilled by a mutual suspicion which the token reunion of the Greek and Latin churches in 1439 could not dispel. The Greeks feared papal aggrandisement and they had long seen unruly Western mercenaries and ambitious Italian merchants as more threatening than the Ottoman Turks. It was from the East, however, that the final blow was to fall when, in 1451, the Ottomans, under Mehmet II, laid

3 THE BYZANTINE EMPIRE: restoration and decline 1340–60

— Boundary of Byzantine Empire 1340

Territory controlled by:

- Byzantine Empire 1360
- Serbia 1360
- Ottoman Empire 1360
- Knights of St John
- Venice
- Genoa

siege to Constantinople. Powerfully armed with artillery, some of which was of Western manufacture, the Ottomans broke through the walls of the city on 29 May 1453 – the last day of the Roman Empire and the first day of a mature Ottoman Empire that would continue to expand until well into the 17th century.

◄ In 1361 an Orthodox ruler was restored in Constantinople in the form of the Emperor of Nicaea, but by the mid-14th century the Ottomans had taken control of northwest Anatolia and were making inroads into Europe. From the northwest the Serbs were also expanding, and the restored Byzantine Empire was powerless to resist.

▲ In their siege of Constantinople in 1453 the Ottomans successfully used cannon to break down the city's outer walls. They also gained access to the harbour (the Golden Horn), despite a Byzantine blockade, by the feat of dragging their ships out of the Bosporus and across a stretch of land. The Ottoman pillage of Constantinople – depicted here in a Romanian wall painting – lasted for three days and nights before Sultan Mehmet II restored order.

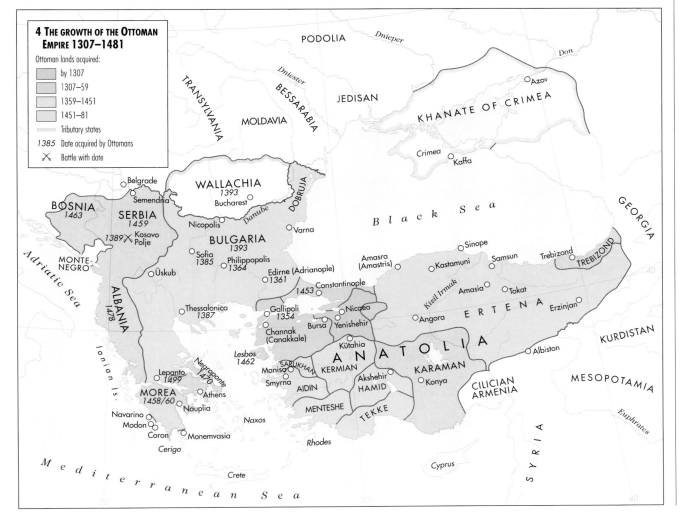

4 THE GROWTH OF THE OTTOMAN EMPIRE 1307–1481

Ottoman lands acquired:

- by 1307
- 1307–59
- 1359–1451
- 1451–81
- Tributary states

1385 Date acquired by Ottomans

✕ Battle with date

◄ As the Byzantine state declined, the Ottomans moved in to fill the resulting power vacuum, not only overcoming other Muslim states in Anatolia, but also establishing a stronghold in mainland Europe and defeating the Serbs in Kosovo in 1389. In 1453 they captured Constantinople and, strengthened by this success, they expanded westwards to control the Balkans as far north as Belgrade.

THE MONGOL EMPIRE
1206–1405

The largest land empire ever created, the Mongol Empire was founded by Temujin, who united the Mongolian and Turkish-speaking tribes roughly in the area known today as Mongolia. In 1206 he was acclaimed ruler by a council of tribal leaders and given the title of Chinggis (Genghis) Khan, usually translated loosely as "universal ruler". The following year he embarked on a series of raids into northern China, which were soon to turn into a full-scale campaign of conquest that was only completed by his successors over 70 years later (*map 1*).

Meanwhile, Mongol forces were expanding westwards along the steppe as far as the kingdom of the Muslim Khwarazm-shah (*pages 88–89*). Chinggis Khan decided to redirect the bulk of his army against the Islamic world, and in a campaign lasting from 1219 to 1223 he conquered most

▼ The empire created by Chinggis Khan between 1206 and his death in 1227 stretched from China to Persia (Iran). However, it did not survive as a united empire beyond 1260 when it split into a number of khanates whose rulers went on to conquer further territories — most notably China in 1279.

▶ The Mongols did not follow up the total victories they secured in 1241 at Liegnitz (in Poland) and Pest (in Hungary), and soon withdrew to the south Russian steppe.

This may have been because of the news of the death of the Great Khan Ogodei, but also perhaps due to a lack of sufficient pasture lands in this area.

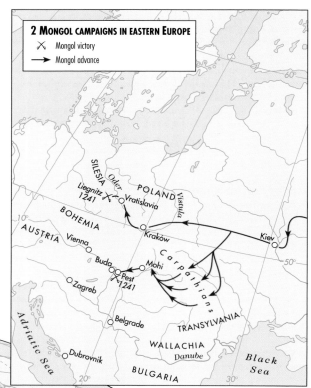

2 MONGOL CAMPAIGNS IN EASTERN EUROPE
✕ Mongol victory
→ Mongol advance

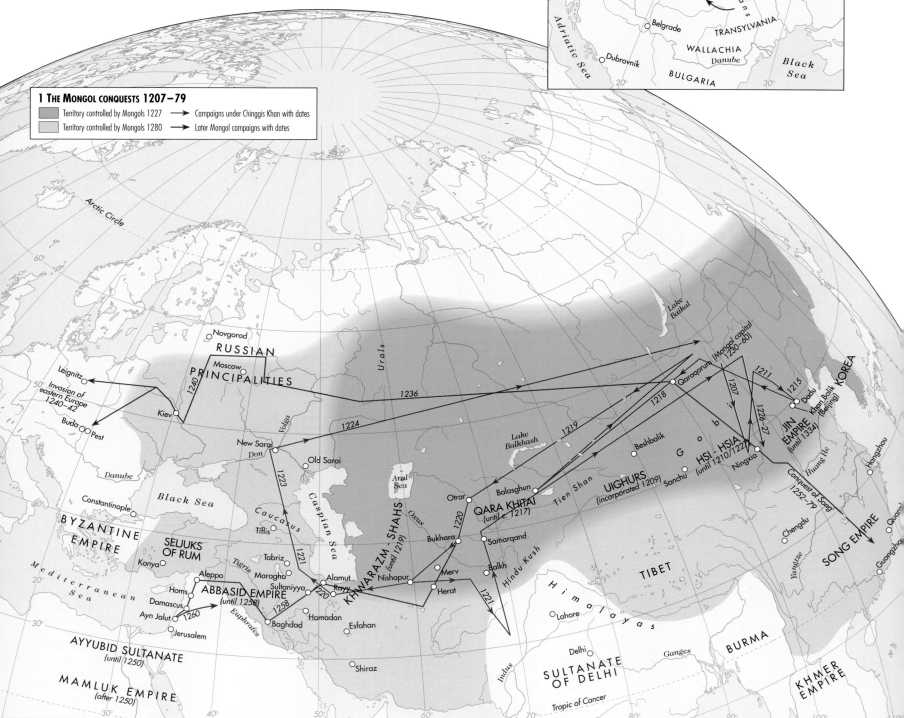

1 THE MONGOL CONQUESTS 1207–79
- Territory controlled by Mongols 1227
- → Campaigns under Chinggis Khan with dates
- Territory controlled by Mongols 1280
- → Later Mongol campaigns with dates

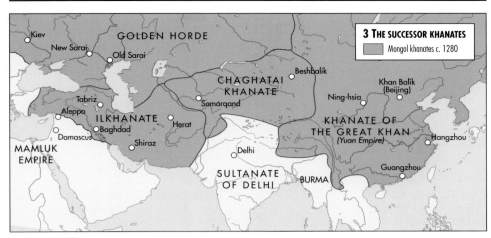

▲ After bringing the Turkic nomadic populations inhabiting the steppe north of the Caspian and Black seas under control, Mongol forces launched a devastating campaign in the winter of 1237–38 against the cities of the Russian principalities. In 1240 the Battle of Kozelsk – depicted in this illustration from a 16th-century Russian chronicle – resulted in the city of Kiev being razed to the ground.

of the kingdom of the Khwarazm-shah. Great destruction was wrought on the cities of Bukhara and Samarqand and in the area south of the Oxus. A rudimentary Mongol administrative apparatus was set up in Iran, which grew into the bureaucracy that ruled the country into the 14th century.

There were several reasons for Chinggis Khan's success in establishing a widespread tribal empire which long outlived him. He built a large army of top-quality soldiers – the traditional horse-archers of the Eurasian steppe, experts in the tactics of concerted mass assault, whom he infused with iron discipline. An effective military leader himself, he had the foresight and talent to cultivate a cadre of extremely capable and loyal generals. He introduced several changes that laid the groundwork for a long-term Mongol administration – the adoption of an alphabet for the Mongolian language, the basic tenets of a financial system, and a system of law known as the *Yasa*. Finally, he propagated an imperialist ideology, premised on the assumption that the Mongols had a heaven-given "mandate" to conquer the world. All those who resisted this mandate were rebels against the heavenly order and could be dealt with accordingly.

Chinggis Khan died in 1227, on campaign in China. He was followed as Great Khan by his second son, Ogodei (r. 1229–41), under whose rule the empire continued to expand. In China the Jin Empire was eliminated in 1234, and war began with the southern Song. In the Middle East all of Iran and the Caucasus were subjugated in the 1230s, and most of Anatolia followed in 1243. The most impressive campaigns, however, were those in Russia and then eastern Europe, where total victories were secured in April 1241 at Liegnitz (Legnica) and Pest (Budapest) (*map 2*).

THE SUCCESSOR KHANATES

In the aftermath of the death of the fourth Great Khan – Mongke, a grandson of Chinggis Khan – the Mongol Empire effectively split up into a number of successor states. In China and the Mongolian heartland, Qubilai (Kublai) – a brother of Mongke (d. 1294) – established the Yuan dynasty, and had conquered all of China by 1279. This conquest was accompanied by much destruction, particularly in the north, but not all aspects of Mongol rule were negative. Trade appears to have flourished and the country was united for the first time in centuries. From West Asia there was an influx of cultural influences in such areas as medicine, mathematics and

astronomy. Mongol rule lasted in China until a series of popular uprisings in the 1360s, from which emerged the first Ming emperor – at which point large numbers of Mongols left China for the steppe.

In Central Asia the Khanate of Chaghatai – Chinggis Khan's third son – gradually coalesced under his descendants, while further to the west the so-called Golden Horde, ruled by the descendants of Jochi, Chinggis's fourth son, evolved. Around 1260 there arose in Iran an additional Mongol state known as the Ilkhanate, from the title Ilkhan ("subject ruler") by which the rulers were known. This state was founded by Hulegu, the brother of Mongke and Qubilai, who conquered Baghdad in 1258 and brought to an end the Abbasid Caliphate which had existed for over 500 years. Hulegu's troops were stopped at Ayn Jalut in northern Palestine in 1260 by the Mamluks of Egypt (*pages 88–89*), and the border between the two states was stabilized along the Euphrates – though the war between them, at times intense, lasted until 1320. The Ilkhans, along with their subjects, converted to Islam around the beginning of the 14th century, leading to large-scale patronage of Islamic institutions. In Iran, as on the steppe to the north, the Mongols appear to have been absorbed by a larger nomadic Turkish population, whose size greatly increased during the period of Mongol domination.

In the late 14th century the Turkified and Muslim descendants of the Mongol tribesmen in Transoxania gathered around Timur-leng (Tamerlane), who created an empire stretching from Central Asia to western Iran (*map 4*). The empire did not survive his death in 1405 as he had failed to set up an efficient administration and made no serious provision for his succession.

THE LEGACY OF THE MONGOL EMPIRE

Looking at the history of the Mongol Empire as a whole – and without belittling the destructive effects of their conquests – one clear beneficial outcome can be seen: for the first time in history, most of Asia was under one rule, enabling the transfer of merchandise, ideas and other cultural elements. This legacy was to continue long after the demise of the united Mongol state in 1260.

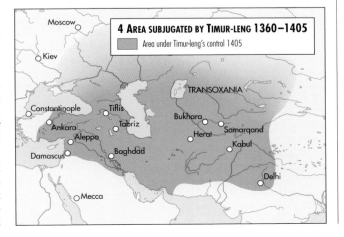

▲ Among the successor states of the Mongol Empire, the Khanate of Chaghatai and the Golden Horde had much in common: in both there were large permanently settled areas controlled by nomads living on the steppe. The relatively small number of Mongols, both elite and commoners, were gradually absorbed by the much larger Turkish tribal population, adopting Turkic languages while maintaining aspects of Mongol identity and culture. Around the same time they converted to Islam, although there were those who resisted the abandonment of traditional Mongol shamanism.

◄ Timur-leng's campaigns contributed to the collapse of the Golden Horde in around 1400. In its place a number of smaller hordes arose, which were gradually absorbed by the growing Russian state of Muscovy. The Tatar, Uzbek and Kazakh peoples were to emerge from the nomadic populations controlled by the Horde, the last two moving eastwards around 1500 to their current locations.

◄ SLAVIC STATES 400–1000 *pages 70–71* ◄ EAST ASIA 907–1600 *pages 86–87* ◄ CHINA 1368–1800 *pages 138–39* ► RUSSIA 1462–1795 *pages 148–49*

THE ECONOMY OF EUROPE
950–1300

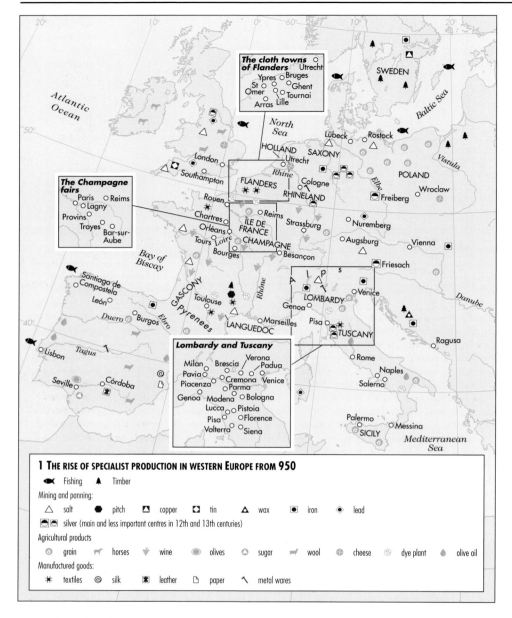

1 THE RISE OF SPECIALIST PRODUCTION IN WESTERN EUROPE FROM 950

- 🐟 Fishing 🌲 Timber

Mining and panning:

- △ salt ● pitch ◨ copper ◻ tin ▲ wax ◼ iron ◉ lead
- ◨◨ silver (main and less important centres in 12th and 13th centuries)

Agricultural products:

- ◎ grain 🐎 horses ⚘ wine ◉ olives ○ sugar ⌐ wool ◉ cheese ◉ dye plant ◉ olive oil

Manufactured goods:

- ✳ textiles ◉ silk ▣ leather ▯ paper ◣ metal wares

---

the market, confident that they could obtain food and clothing from the same source. Similarly, farmers aimed less at self-sufficiency and more at the production of cash crops such as grain, grapes or wool.

Regions and sub-regions also started to specialize. By the beginning of the 12th century Flanders had become a cloth economy, its towns dependent on wool from England, grain and wine from the Île de France and the Rhineland, and on access to customers. Indeed the cloth industry had made Flanders the richest, most densely populated and urbanized region of northern Europe. By the 13th century areas of specialist production included the wine trade in Gascony; grain in Sicily, southern Italy and eastern Europe; salt in the Bay of Biscay, the Alps, the west of England, Saxony and Languedoc; timber and fish in Scandinavia and the Baltic; fur in Russia; iron in Sweden, Westphalia and the Basque country; metalworking in the Rhineland; and cheese in eastern England, Holland and southern Poland (*map 1*).

MEDITERRANEAN COMMERCE

Italian merchants reached Flanders as early as the beginning of the 12th century, but at this date links between northern Europe and the Mediterranean were still fairly limited and it is more realistic to think in terms of European economies rather than an integrated whole. While the wealth and developing urban culture that characterized southern France, Catalonia and above all northern Italy was based partly on the same pattern of population growth and rural development occurring in Europe north of the Alps, the southern economies also benefited from access to the flourishing commercial world of the Mediterranean (*map 4*).

The documents of the *Cairo Geniza*, an extraordinary Jewish archive amassed from the 11th century onwards, vividly illustrate the growing involvement of Latin merchants, especially Italians, in Mediterranean commerce. From the mid-11th century their activities were increasingly backed by force, and during the 12th century Muslim, Jewish and Greek shipping and much of their trade were all

▼ More intensive agricultural regimes formed the backbone of economic expansion in Europe, providing sufficient surpluses in basic foodstuffs to feed the growing number of specialist producers offering their goods in exchange for the food produced by the peasantry. The development of the Chartres region, with its pattern of forest clearance and the subjugation of the landscape, is typical.

▲ During the central part of the Middle Ages, Europe moved decisively away from locally self-sufficient, "closed" economies. Trade was no longer limited to transporting relatively small quantities of high-value luxury items destined for consumption by a rich and privileged elite, but came instead to encompass a wide range of agricultural and manufactured goods.

Between about 950 and 1300 the European economy was transformed (*map 1*). The motors of economic growth were a growing population, a developing market structure, increasing regional and subregional specialization and growing monetarization, based partly on the discovery of major new silver mines and partly on the development of commercial instruments (such as bills of exchange and letters of credit) that allowed monetary transactions to extend beyond the immediate availability of coin.

RURAL AND URBAN GROWTH

The clearest evidence that the European population increased comes from the growing number of settlements of all types throughout the continent. Many mark the opening up of previously uncultivated land for agriculture: place-names and archaeology tell a story of forests cut back, marshes drained and former pasture lands brought under the plough (*map 2*). New markets also appeared and old towns expanded, with urban growth evidenced by new parishes, larger circuits of walls and new suburbs (*map 3*).

In France, Germany, Italy and England local secular and ecclesiastical lords played decisive roles in the creation of a hierarchy of new market towns. Founding a market town not only opened the prospect of a new source of revenue; it also made it possible for the lord either to take payments in kind and sell them on the market for cash, or to demand the payment of rents and dues in coin, which peasant producers could now obtain by entering the market themselves.

Markets encouraged specialization at all levels, and urban craftsmen produced a growing volume of goods for

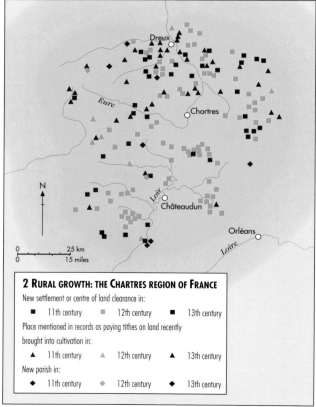

2 RURAL GROWTH: THE CHARTRES REGION OF FRANCE

New settlement or centre of land clearance in:

- ◼ 11th century ◼ 12th century ◼ 13th century

Place mentioned in records as paying tithes on land recently brought into cultivation in:

- ▲ 11th century ▲ 12th century ▲ 13th century

New parish in:

- ◆ 11th century ◆ 12th century ◆ 13th century

but driven from the Mediterranean Sea. When the Spanish Muslim scholar Ibn Jubayr went on a pilgrimage to Mecca in 1183–85 he travelled entirely on Genoese ships, apart from the small coaster which took him across the Strait of Gibraltar and the boat in which he crossed the Red Sea.

Between the 11th and 13th centuries a number of important developments took place in the Mediterranean region: Pisa and Genoa took over Corsica and Sardinia in 1015; the Normans conquered southern Italy and Sicily (secure by 1070), and Malta in 1091; the Crusader States were established in Syria and Palestine after 1099 (*pages 94–95*); Cyprus was conquered in 1191 by Richard I of England (who then gave the island to Guy of Lusignan, titular King of Jerusalem); a Venetian empire was created in the Aegean after 1204; and the Balearics, Valencia and Murcia were recaptured from the Muslims by 1243 (*map 4*). As a result the Latin states had complete control of the Mediterranean trunk routes by the mid-13th century. Trading networks were established that would continue to flourish for centuries to come.

Part of what passed along these routes was a trade in foodstuffs, bulk raw materials and textiles. Italian, French and Spanish merchants not only took European goods to North Africa, Egypt and the Byzantine world, but also played an increasingly dominant role in the internal trade of these societies. Profits from this involvement brought enough Islamic gold to Italy to enable Genoa and Florence in 1252, and then Venice in 1284, to strike a regular gold coinage for the first time in Latin Europe since the 8th century. However, the big profits of Mediterranean trade were to be made in the luxuries for which the West was offering a rapidly expanding market – the spices, silks dyestuffs and perfumes of the East – and here the balance was heavily in favour of Muslim sellers. To buy on the Egyptian markets, Latin merchants needed large supplies of coin and bullion.

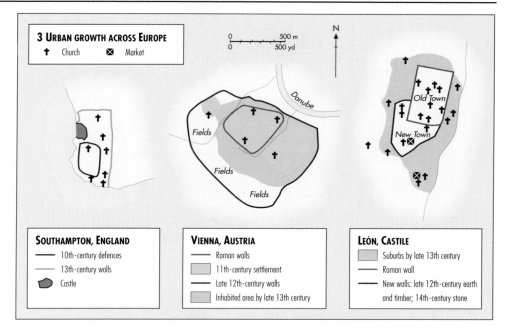

3 URBAN GROWTH ACROSS EUROPE

† Church   ⌧ Market

SOUTHAMPTON, ENGLAND	
——	10th-century defences
——	13th-century walls
◼	Castle

VIENNA, AUSTRIA	
——	Roman walls
▨	11th-century settlement
——	Late 12th-century walls
▨	Inhabited area by late 13th century

LEÓN, CASTILE	
▨	Suburbs by late 13th century
——	Roman wall
——	New walls: late 12th-century earth and timber; 14th-century stone

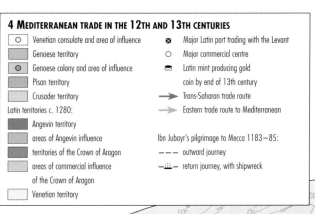

4 MEDITERRANEAN TRADE IN THE 12TH AND 13TH CENTURIES

○	Venetian consulate and area of influence
▨	Genoese territory
◉	Genoese colony and area of influence
▨	Pisan territory
▨	Crusader territory

Latin territories c. 1280:
- Angevin territory
- areas of Angevin influence
- territories of the Crown of Aragon
- areas of commercial influence of the Crown of Aragon
- Venetian territory

⌧	Major Latin port trading with the Levant
○	Major commercial centre
⬮	Latin mint producing gold coin by end of 13th century
⟶	Trans-Saharan trade route
⟶	Eastern trade route to Mediterranean

Ibn Jubayr's pilgrimage to Mecca 1183–85:
- – – – outward journey
- —⫿— return journey, with shipwreck

THE ROLE OF SILVER

A crucial development was the opening up from the 1160s of new European silver mines, of which the most important were in Germany. Interregional trade in northern Europe brought large quantities of German silver into the hands of Flemish, French, Rhenish and English merchants who then paid silver to southern merchants, mostly Italians, in exchange for goods from the East.

The linchpin of the new trans-Alpine economy was the Champagne fairs, held at Troyes, Bar-sur-Aube, Lagny and Provins, where the powerful counts of Champagne could guarantee security. These new ties brought a large amount of silver to the south – so large in fact that during the second half of the 12th century the Provins denier (the coinage of Champagne) became the standard coin for commercial payments in northern and central Italy. They also brought Mediterranean commercial techniques and firms of Italian bankers to the north. With the introduction of transferable bills of exchange, the European economy was no longer limited by the availability of precious metal. Bankers were willing to offer enormous credit facilities to reliable clients, so that the rulers of the major European states were now given the means to operate on an entirely new scale.

▲ Expansion in sectors of the European economy not geared to food production is strikingly demonstrated in the phenomenon of urban growth. Towns and cities provided manufacturing centres and markets for long-distance trade, whether interregional or international. They also serviced their local agricultural economies, providing the markets and goods that made possible local specialization and exchange.

▼ The era of the crusades was also one of growing Mediterranean commerce. European traders took some textiles and foodstuffs east, but above all they carried silver coins with which to purchase the valuable dyes and spices that came from India and the Far East.

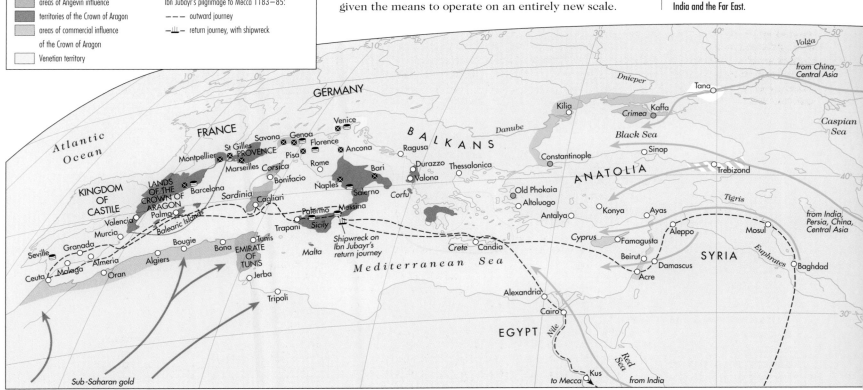

◗ FRANKISH KINGDOMS 200–900 *pages 74–75*   ◗ EUROPE 1350–1500 *pages 106–7*

URBAN COMMUNITIES IN WESTERN EUROPE 1000–1500

After the collapse of the Roman Empire at the end of the 4th century, towns in Europe had tended to decrease in size, complexity and autonomy, particularly within Latin Christendom. In 1000 Europe's five largest towns – Constantinople, Córdoba, Seville, Palermo and Kiev – were outside this area. However, by 1500 the pattern of urban development in Europe had undergone great changes: Constantinople was still one of the five largest towns, but the other four were now Paris, Milan, Venice and Naples. At this time around 70 per cent of the estimated 80 million inhabitants of Europe lived in the countryside, with a further 20 per cent in small market towns. Just three million people lived in the hundred or so towns of at least 10,000 inhabitants, but they represented a social, economic, cultural and political force of far greater importance than their number might suggest.

During the Middle Ages urban enterprise came to set the pace of social and cultural development in western Europe. By 1300, under the impulses of the new international economy of trade, finance and industry (*pages 100–1*), two main clusters of towns had developed: one in northern Italy, the other in northern France and Flanders, with London and Cologne in close proximity (*map 1*).

THE ITALIAN COMMUNES

Between 1050 and 1150 Italian towns from the Alps as far south as Rome were controlled by communal regimes made up of local men of property and high status. The communes achieved power partly by violent assertion but also by the formation of "peace associations", which had the declared aim of bringing peace and order to a locality. Once in charge, the communes directed their energies towards mastering the immediately surrounding territory (*contado*) – vital for maintaining food supplies and communications. In the later 12th and 13th centuries their local control was repeatedly challenged by the Staufen emperors, rulers of the Holy Roman Empire (*pages 90–91*).

The communes ultimately emerged victorious, but the strain of warfare, together with increasing social tensions generated by large-scale immigration from the countryside, frequently fuelled recurrent factional conflicts. This resulted in the subversion of communal government and the seizure of power by partisan cliques under so-called *signori*, such as the Visconti in Milan (dukes from 1395) or the Este family in Modena and Ferrara (dukes from 1452) (*map 2*).

TOWNS IN NORTHWEST EUROPE

In northwest Europe the forms of town government varied. Here too, from around 1100, communes were set up by local revolt, or by local lords granting jurisdictional privilege. Paris and London, however, developed as royal residences and capitals of kingdoms, while the towns of the Low Countries, although prone to turbulence, remained within the framework of territorial principalities. The county of Flanders was divided into four territorial-jurisdictional sectors known as the "Four Members", three of which were dominated by the towns of Ghent, Bruges and Ypres. Much of the business of government was transacted not by the count's officials, but in the regular meetings of representatives of the Four Members.

By the 1460s, 36 per cent of the population of Flanders were town dwellers, half of them resident in the three big

▼ In the 14th century all the towns in the two urban clusters that had developed in northern Italy and northern France and Flanders were to some degree self-governing, although only Venice asserted absolute freedom from outside authority.

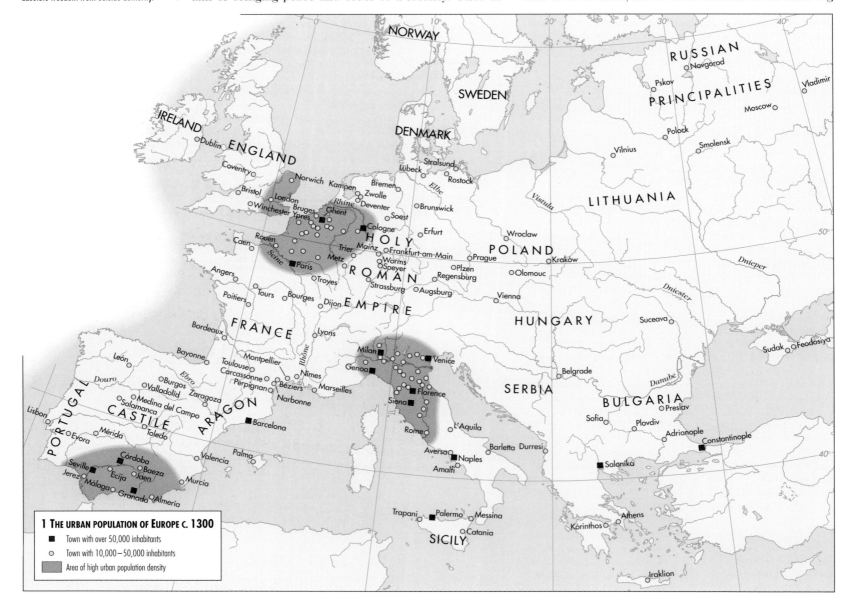

1 THE URBAN POPULATION OF EUROPE C. 1300
- ■ Town with over 50,000 inhabitants
- ○ Town with 10,000–50,000 inhabitants
- ▨ Area of high urban population density

► From the early 14th century only a few communes in Italy escaped princely control – notably Venice, intermittently Genoa and Lucca, and Florence before the Medici coup of 1434. Much of their internal organization was grounded in occupational guilds which exercised protectionist control of local vested interests.

towns, half in the 49 smaller towns (*map 3*). This demographic pattern was even more pronounced in Holland, where 45 per cent lived in towns but no single town exceeded 16,000 inhabitants.

THE GROWTH OF URBAN AUTONOMY IN GERMANY

By the 15th century urban development in Germany – although gathering force later than in some other regions – had produced some 35 communities with over 2,000 inhabitants and around 3,000 with some sort of recognized town status. About 50 of these were free cities under no princely jurisdiction. Unlike the Italian communes, some of which controlled whole regions, the German communities were more tightly focused on their urban centres; even Metz, one of the largest, held jurisdiction over only 250 surrounding villages. Also unlike their Italian counterparts, they rarely engaged in warfare. Even after trade guilds had occasionally asserted themselves forcefully in the 14th and 15th centuries, the towns remained under the control of a small number of noble families – 42 in Nuremberg, for example, and 76 in Frankfurt in around 1500.

By this date the German towns were enjoying a golden age of economic growth and cultural vitality – a vitality that had been a feature of European urban society since the 12th century. Among its achievements had been the Gothic architectural style of church building; secular buildings of equivalent scale, such as the town halls of Florence and Bruges; the spread of printing presses from the Rhineland to over 200 towns throughout Latin Christendom between 1450 and 1500; the "civic humanism" of post-communal Italy; and the "scholastic humanism" fostered by the foundation of some 80 universities – five by 1200, a further 14 by 1300, 26 in the 14th century, and 35 in the 15th century (*pages 134–35*).

THE EARLY RENAISSANCE

The great town halls of communal Italy were built mainly between 1260 and 1330 – around the lifetime of the civic-minded vernacular poet Dante (1265–1321), and of his fellow Florentine, Giotto (1266–1337), whose painting came to be seen as marking the beginning of a new sense of space and form. Over the following century Florence continued to loom especially large in the visual arts, with architecture and sculpture as well as painting coming to express a "classical" ideal inspired by the Graeco-Roman past. Florence also produced writers such as Boccaccio (1313–75), whose vernacular poems and prose rapidly influenced French and English writing, and Petrarch (1304–74), whose humanist Latin writings became formative in the education of the elite throughout Latin Christendom in the course of the 15th century.

The transmission of style, however, was not all one way. The "new art" of the painters and musicians of the towns of the Low Countries was much in demand in 15th-century Italy, and in 1500 artists and writers were, literally, citizens of a world of Renaissance culture. The career of the artist Dürer (1471–1528) moved between his native Nuremberg, Venice and Antwerp, while the humanist writer Erasmus (1469–1536) travelled constantly between Gouda, Deventer, Paris, London, Bologna, Rome, Leuven, Freiburg and Basel. Their achievement, in their own lifetimes, of Europe-wide fame beyond the span of their personal travels was itself an early product of the general spread of three urban inventions: the woodcut, the engraving and the printed book.

► By 1500 some 34 per cent of the population of the Low Countries lived in towns – an urban density equalled only in parts of northern Italy. Despite the protection of local interests by the occupational guilds, there was considerable economic and cultural exchange between towns – so much so that Antwerp had become the leading commercial and cultural centre of western Europe.

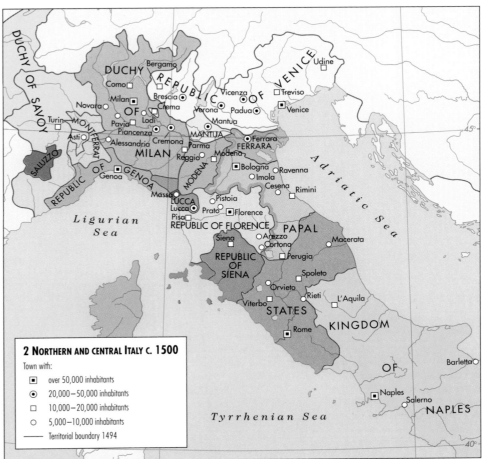

2 NORTHERN AND CENTRAL ITALY c. 1500

Town with:
- ▣ over 50,000 inhabitants
- ◉ 20,000–50,000 inhabitants
- ◻ 10,000–20,000 inhabitants
- ○ 5,000–10,000 inhabitants
- —— Territorial boundary 1494

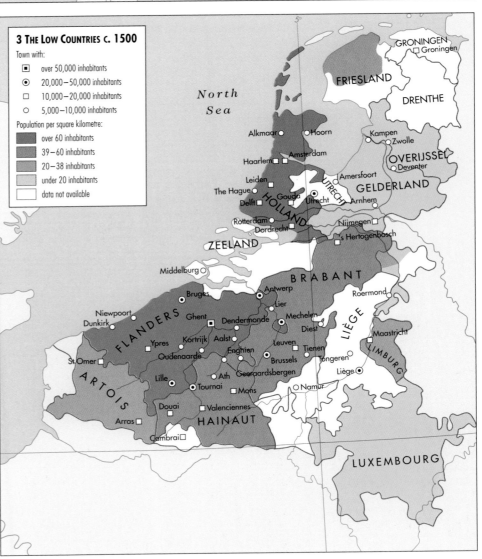

3 THE LOW COUNTRIES c. 1500

Town with:
- ▣ over 50,000 inhabitants
- ◉ 20,000–50,000 inhabitants
- ◻ 10,000–20,000 inhabitants
- ○ 5,000–10,000 inhabitants

Population per square kilometre:
- over 60 inhabitants
- 39–60 inhabitants
- 20–38 inhabitants
- under 20 inhabitants
- data not available

◑ THE ECONOMY OF EUROPE 950–1300 *pages 100–1* ◐ EUROPEAN URBANIZATION 1500–1800 *pages 132–33*

CRISIS IN EUROPE AND ASIA
1330–52

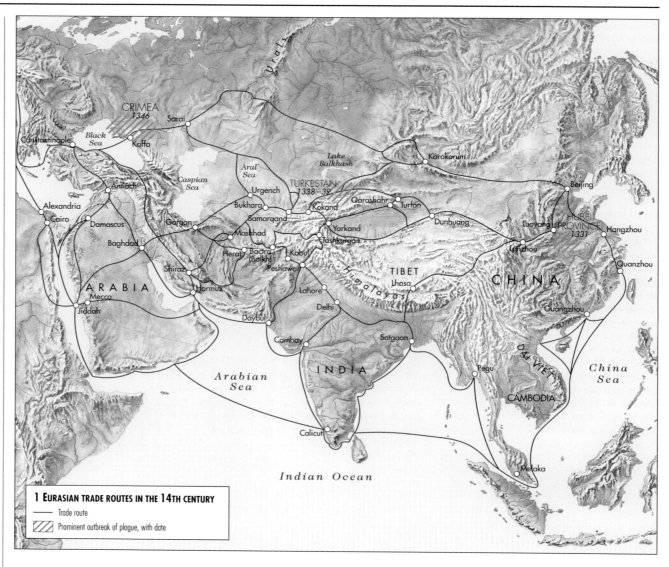

▶ The merchants' "Silk Roads", which doubled as military routes for invaders and mercenaries, and linked up with the seaways of the Indian Ocean and the Black and Mediterranean seas, were also highways for infection with the plague. Medieval international travel was slow and companionable: wayfarers carried huge quantities of supplies; they utilized ports, campsites, caravanserais and storehouses that were infested with black rats whose fleas carried the plague. They also dealt extensively in the bales of cloth which so often harboured fleaborne infection.

1 EURASIAN TRADE ROUTES IN THE 14TH CENTURY
— Trade route
/// Prominent outbreak of plague, with date

▶ Part of the response of western European culture to the plague was to personify death via various visual media. The *danse macabre* entered court entertainment, and artists and sculptors experimented with the grisly themes of the cadaver and the skull. This 15th-century fresco from the Italian School, entitled *The Triumph of Death*, is a direct descendant of the genre spawned by the terrifying disease a century earlier.

In the 14th century the "Old World" may have lost between a quarter and half of its population as a result of pandemic plague. The infective agent or plague bacillus was, and is, endemic to the ecology of certain remote areas of Asia. At times environmental factors or simple mutation can promote a dramatic rise in the numbers of the rodent fleas which are the plague's usual carriers. Facilities for transport and travel can then promote widespread person-to-person infection and turn an isolated outbreak of bubonic plague into an epidemic and ultimately a pandemic – without the intervention of rat or flea.

The "Black Death" of the 14th-century was not the first visitation of plague to the Middle East or to Europe. The Byzantine historian Procopius gave a chillingly precise account of the symptoms and progress of the disease as it struck the Persian and Byzantine empires in the 540s. This plague reached Britain in 546 and Ireland in 552, and its aftershocks extended late into the 7th century.

THE BLACK DEATH INVADES EUROPE
The medieval pandemics of the 6th and 14th centuries were the unpredicted side-effects of expanding horizons and increasing contact between East and West (*map 1*). The second scourge of the plague reached East Asia in the early 1330s and West Asia less than a decade later.

This time it may well have hit an already debilitated population. A run of rainy years and poor harvests in much of mid-1340s Europe had lowered resistance and led to the widespread consumption of suspect food supplies. Typically the plague was at its most virulent in congested urban areas, and dedicated professionals such as doctors and priests suffered disproportionately. Yet there were always survivors – as many as a quarter of sufferers may have lived through an attack of plague to become invested with an awe-inspiring immunity – and there were regions, even towns, that went largely unscathed (*map 2*).

While much plague history is anecdotal and local, such details can be just as telling as the massive mortality estimates. Pestilence halted work on the cathedral of Siena in Italy, and the building is still truncated today. The population of the Oxfordshire village of Tusmore in England was wiped out in 1348 and never restored. There were dramatic local responses to stress, such as episodes of penitential flagellation and vicious outbursts of scapegoating as vulnerable groups in society, notably the Jews, were targeted as the bringers of death. Such incidents were not, of course, unknown outside the plague years.

EFFECTS OF THE BLACK DEATH

The questions whether or to what extent the 14th century pandemic changed the course of world history can only be the subject of conjecture. In China, which suffered the first and perhaps the most serious wave of devastation, demographic collapse may have fostered the consensus that the ruling Mongol or Yuan dynasty had lost the "mandate of heaven". The Yuan were ousted in 1368 in favour of an indigenous Chinese dynasty, the Ming. In the West, the loss of manpower to pestilence may have left a declining Constantinople too weak to prevent Ottoman incursions into Europe: from 1354 there were Ottoman victories in the Balkans which reached a peak at Kosovo (1389) and established a lasting Muslim government in the midst of Orthodox Christendom. West Asia certainly saw a dramatic reduction in the population of its big Islamic cities and a reversion to nomadism outside them. Perhaps the effects of the plague facilitated a last Mongol invasion by the armies of Timur-leng (1369–1405), who briefly redrew the political map from Afghanistan to the Mediterranean (*pages 98–99*).

However, no western European states or societies collapsed in the wake of the plague. Great cities like Venice experienced short-lived administrative dislocation and then recovered. Social tensions were exacerbated as surviving craftsmen, labourers and servants now had the advantage of scarcity and might resist the demands of lords, masters or officialdom. There was an increase in the Mediterranean slave trade as one solution to the labour shortage.

There was also a demographic shift. Thousands of settlements in agricultural western Europe were abandoned in the two centuries that followed the population peak of the early 14th century. Very few of these "lost villages" were specifically eliminated by the plague or its accompanying panic, but in the aftermath of the plague, survivors from the fens and moorlands of the agricultural margins could move (with the encouragement of landowners who needed their labour) into the best of the farming land.

The "time of pestilence" was also a time of resilience. Survivors dutifully buried their dead and coped with the paperwork of mortality, probate and the ricocheting finances of societies which had lost, on average, a third of their taxpayers. The 14th century had none of the universal expectation of population growth and longevity which characterizes the modern era. Life expectancy was less than half that of today and even those who survived the plague years had a very limited chance of reaching 70. Eyewitness accounts of the plague years describe a society whose preachers used *memento mori* ("remember you must die") as a watchword and regularly portrayed earthly existence as a vale of tears. The plague, which served to underline this concept, was easily incorporated into Christian theological debate; it is also likely to have reinforced Islamic fatalism and possibly the cyclical view of history and society set out in the writings of the philosopher Ibn Khaldun (1332–1406).

Meanwhile, mainstream Western culture took refuge in the incorporation of mortality into art and personified death as a figure in popular stories and morality plays. Modern communicators still draw on this plague-time imagery of mortality to convey an apocalyptic warning.

▼ The plague reached East Asia in the mid-1330s and West Asia a decade later. The Crimean port of Kaffa was an important flashpoint for the transmission of the plague to Anatolia, the Levant and Europe. Kaffa was a Genoese trading base which in 1347 was under attack from the Kipchak Turks, in whose ranks the plague was raging. Kaffa's policy of "business as usual" in a corpse-strewn environment resulted in the flight of its business partners and they took the infection with them: a fleet of Genoese galleys from Kaffa carried the plague to Messina in Sicily and then, by January 1348, to Genoa itself. Genoa's commercial rivals Pisa and Venice succumbed shortly afterwards, and the pestilence went on to devastate most of Europe until it had reached Scandinavia via the Hanseatic seaways by 1350.

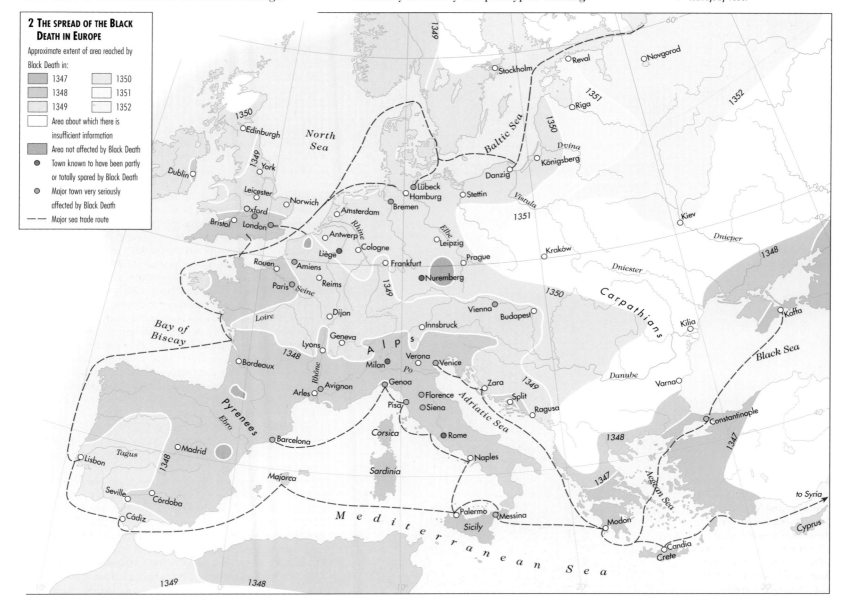

2 THE SPREAD OF THE BLACK DEATH IN EUROPE

Approximate extent of area reached by Black Death in:

1347	1350
1348	1351
1349	1352

Area about which there is insufficient information

Area not affected by Black Death

● Town known to have been partly or totally spared by Black Death

● Major town very seriously affected by Black Death

- - - Major sea trade route

EUROPE
1350–1500

The period 1350–1500 was one of major transition in the history of Europe. Constant warfare reshaped the boundaries of kingdoms and other political entities (*map 1*), while the loss of over a third of the population as a result of the Black Death of 1347–52 (*pages 104–5*) generated economic, social and political change. It was also a period of crisis in the Church, as papal schism let loose challenges to the old order of Latin Christendom.

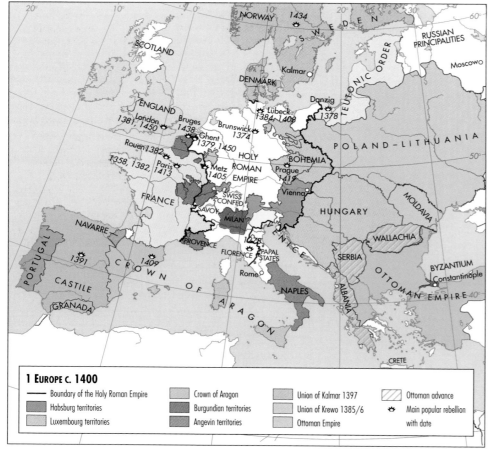

1 EUROPE C. 1400

- Boundary of the Holy Roman Empire
- Habsburg territories
- Luxemburg territories
- Crown of Aragon
- Burgundian territories
- Angevin territories
- Union of Kalmar 1397
- Union of Krewo 1385/6
- Ottoman Empire
- Ottoman advance
- Main popular rebellion with date

▲ In the wake of the Black Death there was an outbreak of popular revolts across Europe. The sudden, dramatic fall in the population resulted in the contraction of the labour force and a rise in wages. However, while living standards improved, there was an increase in the incidence of warfare – leading to higher taxation and social unrest.

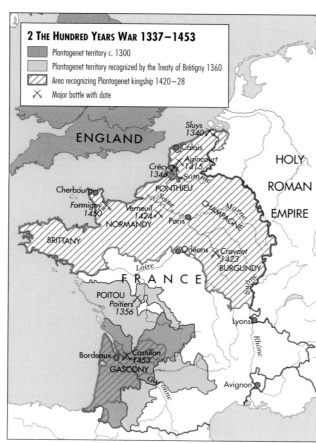

2 THE HUNDRED YEARS WAR 1337–1453

- Plantagenet territory c. 1300
- Plantagenet territory recognized by the Treaty of Brétigny 1360
- Area recognizing Plantagenet kingship 1420–28
- ✗ Major battle with date

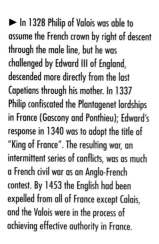

▶ In 1328 Philip of Valois was able to assume the French crown by right of descent through the male line, but he was challenged by Edward III of England, descended more directly from the last Capetians through his mother. In 1337 Philip confiscated the Plantagenet lordships in France (Gascony and Ponthieu); Edward's response in 1340 was to adopt the title of "King of France". The resulting war, an intermittent series of conflicts, was as much a French civil war as an Anglo-French contest. By 1453 the English had been expelled from all of France except Calais, and the Valois were in the process of achieving effective authority in France.

WESTERN AND CENTRAL EUROPE

From 1337 much of western Europe became the arena for a struggle between the the Valois princes and the Plantagenet kings of England for the succession to the Capetian kingship of France. The resulting Hundred Years War (*map 2*) gave rise to a network of alliances linking the Valois to Scotland and Castile, the Plantagenets to Portugal, and both at different times to the Wittelsbach and Luxembourg dynasties of the Holy Roman Empire. Such links helped to sustain Scotland's independence from England. They also stimulated the emergence of a more powerful Burgundy which brought together the territorial principalities of the Low Countries – first, in the 1360s, as a Valois satellite, then as a Plantagenet ally (1419–35 and 1468–77), and finally as a Habsburg inheritance.

The Hundred Years War network of alliances figured significantly in the warfare in the Iberian Peninsula which resulted in the establishment of the Trastámara dynasty in Castile in 1369 and the Aviz dynasty in Portugal in 1385. A century later, between 1474 and 1479, two autonomous monarchies emerged whose expansionist ambitions found expression, in the case of Portugal, in maritime expeditions along the coast of Africa, and, in the case of Castile and Aragon, in the conquest of Muslim Granada (1480–92).

Italy developed as an essentially self-contained political complex, with Milan, Venice and Florence expanding into regional territorial states by the mid-15th century. In the south, the Trastámaran Alfonso V of Aragon added the kingdom of Naples to his existing possession of Sicily in 1442, after conflict with a Valois claimant. This was followed half a century later by a renewed Valois-Trastámara struggle in the post-1494 wars which turned Italy into the battleground of Europe (*pages 146–47*). In the meantime, Naples along with Milan, Venice, Florence and the Papacy sought intermittently after 1455 to function as a league to secure "the concert of Italy" from outside intervention.

Germany and the Holy Roman Empire (*pages 90–91*), which were far less affected by large-scale warfare than other areas, came to function as a network of princely and urban local regimes, with relatively few moments of widespread disruption after the 1340s. The institution of elective kingship proved largely cohesive and peaceful, and the imperial title passed in virtually hereditary succession from the House of Luxembourg to the Habsburgs in 1438.

EASTERN AND NORTHERN EUROPE

In east central Europe the position of the Luxembourgs and Habsburgs as rulers of Bohemia (from 1310) and Hungary (from 1387) was intermittently challenged by the rise of the Lithuanian Jagiellon dynasty. To their rule of the Polish-Lithuanian commonwealth the Jagiellon dynasty added the kingship of Bohemia (1471–1526) and Hungary (1440–44 and 1490–1526). In the Baltic, attempts to unite the three kingships of Denmark, Norway and Sweden were briefly successful with the creation in 1397 of the Union of Kalmar. Nonetheless, from 1448 the Oldenburg dynasty maintained its control in Denmark and most of the western Norse world from Norway to Iceland. Flanking Latin Christendom, the Muslim Ottoman Empire (*pages 96–97*) and the Orthodox Christian Russian Empire (*pages 148–49*) emerged.

RELIGIOUS DEVELOPMENTS

In 1309 the French Pope Clement V had taken up residence in Avignon. The monarchical style of the Papacy had reached its peak when in 1378, shortly after its return to Rome, a disputed papal election caused the Church to split and two rival popes – based in Avignon and Rome – to operate simultaneously (*map 3*). This remained the situation until 1417, when the General Council at Constance (1414–18) secured the election of Pope Martin V.

At the same time parts of Europe were marked by dissent from established theological doctrine and by anti-clerical criticism. In England the Lollards, influenced by John Wycliffe, made no effective headway. However, in Bohemia the Hussite movement, launched by John Hus,

developed into a revolutionary challenge to the established order. In 1415 Hus was burned at the stake for heresy, an event that provoked the Hussite Wars against the Holy Roman Emperor. The Hussites achieved dramatic military victories in the 1420s, but their theological and political impact was contained after peace was agreed in 1434–36.

A great challenge to the Papacy came from the Conciliar movement. This developed into a constitutional struggle between reformist clergy seeking to use the church councils (such as that at Constance) to reduce the authority of the Pope, and the bid by the Papacy to reassert the pre-1378 order of church government. The Conciliarists eventually had to acknowledge defeat in 1449, the preference of lay rulers for a monarchical papal ideology proving decisive.

THE EFFECTS OF THE BLACK DEATH

The dramatic fall in population during the Black Death led to severe disruption of agricultural and industrial production and trade (*map 4*). It also led to smaller and more professional armies, although there was an increase in the incidence of warfare, which in turn induced social tension and revolts (among them the Jacquerie Revolt in northern France in 1358, the Peasants' Revolt in England in 1381, and a wave of urban revolts in northwest Europe, the Baltic region and Italy around 1375–85). The levy of war taxation, often the trigger of such unrest, was of fundamental importance in the development of representative institutions, which in the form of parliaments or "Estates" became the vehicle for a heightened sense of the political community throughout Europe.

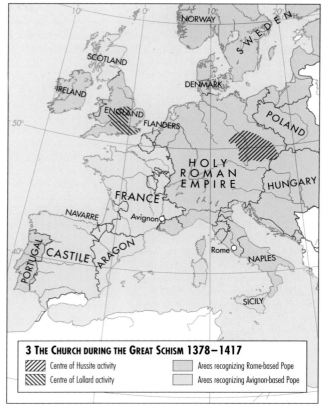

◀ The initial cause of the Great Schism was a disputed papal election in 1378. It lasted for almost 40 years (1378–1417) because lay political groups exploited the situation, rapidly aligning themselves behind the rival claimants to papal office. Thus Valois France and its allies in Scotland and Castile recognized the Pope resident (from 1379) in Avignon, while England and Portugal as well as most parts of the Holy Roman Empire and northern and eastern Europe recognized the Pope resident in Rome.

3 THE CHURCH DURING THE GREAT SCHISM 1378–1417

- Centre of Hussite activity
- Centre of Lollard activity
- Areas recognizing Rome-based Pope
- Areas recognizing Avignon-based Pope

▼ Between about 1370 and 1500 the rural world was marked by depressed grain prices, partly offset by increasing diversification from arable into pasture farming and horticulture. With the contraction of the labour force, wages rose and sustained the demand for a wide range of manufactured and other commodities, both staples and luxuries. The result was a more buoyant economy in the towns and the fostering of technological innovation in, for example, silk weaving, printing and metallurgical processes.

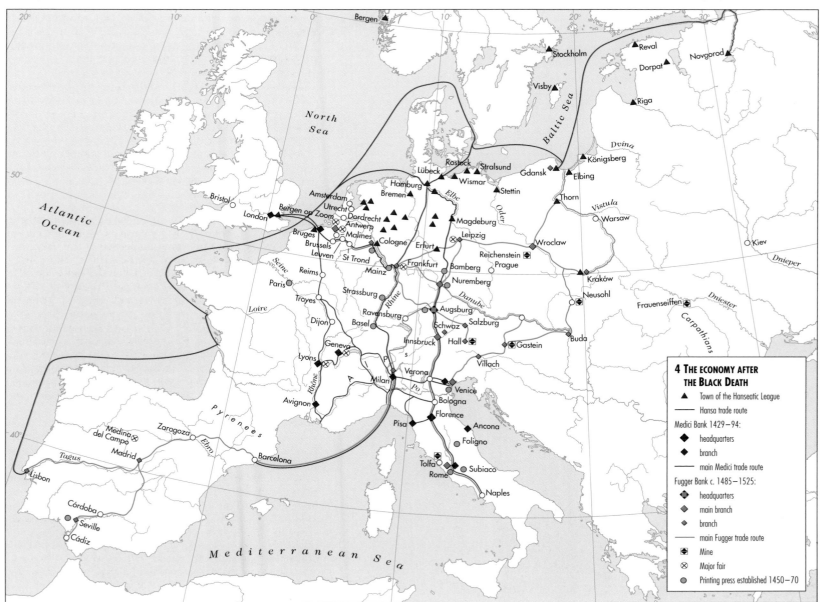

4 THE ECONOMY AFTER THE BLACK DEATH

- ▲ Town of the Hanseatic League
- —— Hansa trade route

Medici Bank 1429–94:
- ◆ headquarters
- ◆ branch
- —— main Medici trade route

Fugger Bank c. 1485–1525:
- ◆ headquarters
- ◆ main branch
- ◆ branch
- —— main Fugger trade route
- ⊞ Mine
- ⊗ Major fair
- ● Printing press established 1450–70

CULTURES IN NORTH AMERICA
500–1500

North America in the 6th century was home to many different cultural traditions. Farming communities, growing native or introduced crops, were established in some parts of the south. Elsewhere, richly diverse ways of life were based on natural resources.

THE SOUTHWEST

Between 200 and 900 settled communities developed in the American southwest (*map 1*), growing crops (especially maize, squash and beans) introduced from Mesoamerica. These communities also began to make pottery to supplement their traditional basket containers. Semi-subterranean houses were constructed. Plazas, mounds and ballcourts reminiscent of those of Mesoamerica appeared in the Hohokam area by 600, at settlements such as Snaketown; these public spaces were probably the focus of ceremonial and ritual activities. Smaller villages clustered around the main centres, which are thought to have been the homes of chiefs controlling the networks of irrigation canals that made two annual crops possible in this arid region.

Irrigation was also vitally important to the Anasazi and Mogollon peoples in the similarly arid areas to the north and east of Hohokam. Around 700 in the Anasazi area and 1000 among the Mogollon, villages of semi-subterranean houses gave way to villages built above ground but containing a subterranean ceremonial structure (*kiva*). These developed into larger and more elaborate complexes of adjoining rooms, called pueblos by the Spanish in the 16th century. Among the best known is Pueblo Bonito (*map 2*). Here a massive plaza containing two large *kivas* was surrounded by a semi-circular, five-storey, tiered complex of some 200 rooms and smaller *kivas*, housing up to 1,200 people.

Further north the pueblos of the Mesa Verde region had developed along different architectural lines. At first situated on plateaus, by 1150 most were constructed on natural or artificial platforms on the face of canyon cliffs, such as Cliff Palace. These cliff-side villages, many dominated by watchtowers, were probably designed for defence and reflect deteriorating environmental conditions at the time.

A major shift in trade patterns took place around the 14th century, when it appears that the Mogollon village of Casas Grandes was taken over by Mexican *pochtecas* (merchants). It grew into a town and became a trade and craft production centre, surrounded by a network of roads and forts, directly controlling the turquoise sources. Mexican architecture now appeared and sophisticated irrigation systems were constructed.

In other areas favourable climatic and environmental conditions had promoted the spread of farming into marginal regions in preceding centuries, but by the later 13th century conditions were deteriorating. There was widespread drought and many sites were abandoned, their inhabitants moving into more fertile areas, particularly along the banks of rivers. In the 1450s Apache and Navajo hunters began to make raids on the fringes of the area, and in 1528 a Spanish expedition signalled future domination by Europeans.

THE SOUTHEAST

By about 400 the extensive exchange networks of the Hopewell people (*pages 24–25*) were in decline and funerary moundbuilding was going out of fashion in all but the southern regions of the southeast. However, by 800 the introduction of maize, later supplemented by beans, allowed an increased reliance on agriculture, but concentrated settlement on the easily cultivated river floodplains (*map 3*). As before, communities were linked by a long-distance trade network. Many were autonomous small chiefdoms but in some areas a hierarchy developed, with subordinate chiefdoms answerable to a centralized authority operating from a major centre. The largest town in this emerging mosaic of Mississippian chiefdoms was Cahokia, a powerful and prosperous centre c. 1050–1250, which housed perhaps 30,000 people in dwellings clustered around the palisaded centre with its plaza and huge mounds.

OTHER NATIVE AMERICANS

From 800, horticulture based on beans, squash and maize spread through the mid- and northeast (*map 4*). Although hunting continued to be important, the increased reliance on agriculture encouraged settlement in semi-permanent villages. By the time the Europeans arrived in North America in the 16th century, the northeast was a patchwork of nations settled in small territories, constantly at war but also trading with one another. Later some settled their differences, uniting into the Iroquois Confederacy which became involved in the wars between rival European powers in the region.

The Great Plains had been home for thousands of years to small groups of buffalo (bison) hunters and small-scale horticulturalists. The introduction of the bow and arrow may have increased hunting efficiency and, possibly for this reason, several peoples moved onto the Great Plains from the surrounding areas. After about 900, colonists from the Mississippian cultures brought maize cultivation to the Missouri region of the Great Plains. The stockades and moats surrounding their settlements, along with evidence of massacres and scalpings, indicate that these groups were constantly at war.

Further west, in the Great Basin, hunter-gatherer groups continued their long-standing nomadic way of life (*map 5*) until it was destroyed by white settlers. Under influence from

▼ Among the pueblos built in the southwest were a group in Chaco Canyon. These may have housed members of the elite, or been craft and redistribution centres, or communal religious centres occupied only on ceremonial occasions. Chaco Canyon was connected to towns and villages several hundred kilometres away by a network of wide, straight roads (used only by travellers on foot, as there were neither wheeled vehicles nor pack animals). Trade was well developed, linking the early pueblo peoples with the north, the Pacific coast and Mesoamerica, from where they obtained copper bells and live scarlet macaws prized for their feathers. In exchange they provided the Mexicans with turquoise mined in the region immediately to the south of the Sangre de Cristo Mountains.

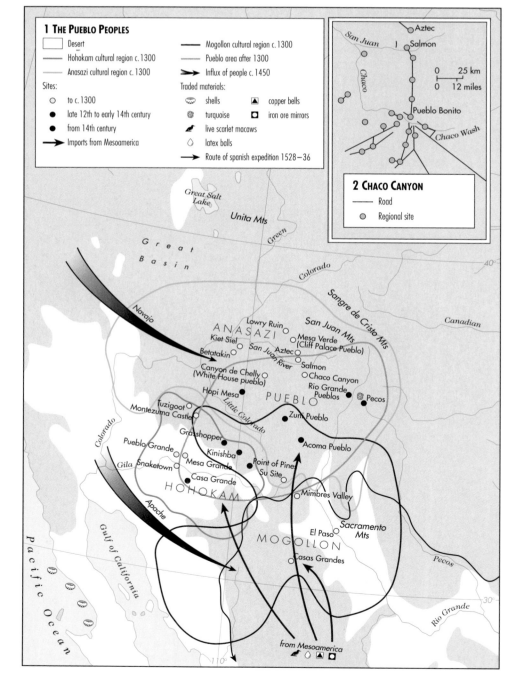

1 THE PUEBLO PEOPLES

- ☐ Desert
- ── Hohokam cultural region c. 1300
- ── Anasazi cultural region c. 1300

Sites:
- ○ to c. 1300
- ◑ late 12th to early 14th century
- ● from 14th century
- ➤ Imports from Mesoamerica

- ── Mogollon cultural region c. 1300
- ── Pueblo area after 1300
- ➤ Influx of people c. 1450

Traded materials:
- 🐚 shells
- ⬟ turquoise
- ◢ live scarlet macaws
- ◇ latex balls
- ➤ Route of spanish expedition 1528–36
- ▲ copper bells
- ☐ iron ore mirrors

2 CHACO CANYON
- ── Road
- ● Regional site

▼ Mississippian towns were the ceremonial centres for their surrounding communities, participating at this time in the religious tradition known as the "Southern Cult". Symbolic artefacts characteristic of this cult – such as copper pendants, seashells and figurines bearing distinctive designs (including snakes, hands and weeping faces) – were found at centres throughout the Mississippian cult area. Mounds in the heart of these centres were crowned by temples and sometimes the houses of the elite.

the Anasazi of the southwest, the Fremont – a number of culturally-related groups who practised horticulture and made distinctive figurines and other artefacts – flourished from around 500 until the late 13th century, when they were wiped out by droughts. Around 1450 Apache and Navajo from the far northwest reached the area and, after contact with the Spanish, took up horse-breeding and hunting on the western Great Plains.

The Pacific coast, with its wealth of game, wild plants and fish, enabled communities to live in villages all year round. The general abundance, coupled with periodic shortages, led to a stratified society: chiefs gained prestige by providing

lavish feasts and gift-giving displays, which might involve the deliberate destruction of valued objects (the "potlatch system"). Shells were used by some groups as a medium of exchange, and slave-raiding was also widespread. Expert woodcarvers, these coastal groups fashioned totem poles and extravagantly decorated houses and artefacts. A detailed insight into their life comes from Ozette, a village partly covered by a mudslide around 1550 (and thus preserved for posterity): here wooden houses and beautifully made wooden tools, nets and other objects were found, including a decorated wooden replica of a whale's fin.

In the far north, Inuit communities spread northwards and eastwards through the Arctic. This was made possible by a number of innovations that improved adaptation to life in extreme cold: igloos, snowshoes, snow goggles, dog sledges, kayaks and the larger umiaks, as well as harpoons capable of killing sea mammals as large as whales. During the warmer temperatures of the period from around 900 to 1300, the Inuit colonized Greenland, where they came into contact and sometimes conflict with the Vikings, who established a toehold there and on Newfoundland between 982 and 1400 (*pages 78–79*).

▼ Outside the southwest and southeast many different cultures flourished, depending to a varied extent on hunting, fishing, gathering and agriculture. The arrival of the Spanish in the 16th century brought horses to North America; rapidly

adopted by the Plains peoples, these animals revolutionized hunting techniques, enabling efficient slaughter of buffalo and easy long-distance movement. Many peoples soon abandoned agriculture in favour of a way of life based on horseback hunting.

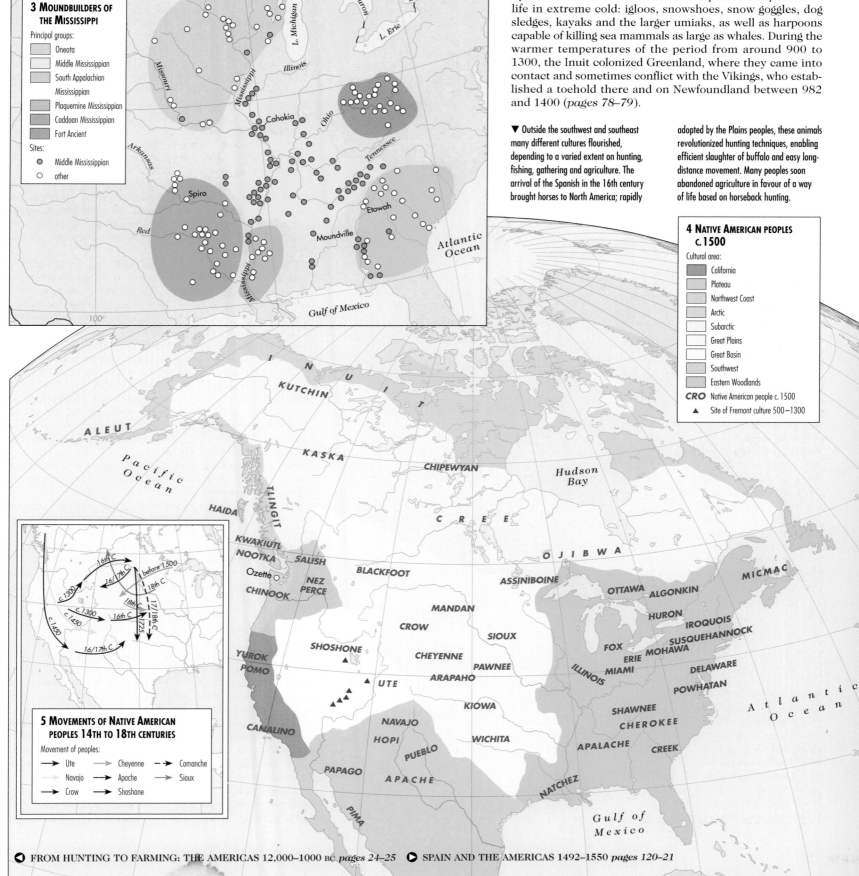

3 MOUNDBUILDERS OF THE MISSISSIPPI

Principal groups:
- Oneota
- Middle Mississippian
- South Appalachian Mississippian
- Plaquemine Mississippian
- Caddoan Mississippian
- Fort Ancient

Sites:
- ● Middle Mississippian
- ○ other

4 NATIVE AMERICAN PEOPLES c.1500

Cultural area:
- California
- Plateau
- Northwest Coast
- Arctic
- Subarctic
- Great Plains
- Great Basin
- Southwest
- Eastern Woodlands

CRO Native American people c. 1500
▲ Site of Fremont culture 500–1300

5 MOVEMENTS OF NATIVE AMERICAN PEOPLES 14TH TO 18TH CENTURIES

Movement of peoples:
- Ute
- Navajo
- Crow
- Cheyenne
- Apache
- Shoshone
- Comanche
- Sioux

◖ FROM HUNTING TO FARMING: THE AMERICAS 12,000–1000 BC *pages 24–25* ◗ SPAIN AND THE AMERICAS 1492–1550 *pages 120–21*

THE INCA AND AZTEC EMPIRES
1400–1540

► Also known as Tahuantinsuyu ("the land of the four quarters"), the Inca Empire extended from modern Ecuador to southern Chile. The rulers established their authority over the peoples they conquered by relocating large numbers, either sending them to work temporarily at nearby way-stations, or moving them permanently to more distant provinces. They also ensured that provincial heirs to power were educated in Cuzco and brought provincial cult objects to the capital. In the provinces sacred mountains such as Cerro El Plomo in Chile became the sites of state-dedicated child sacrifices, and oracular centres and ancient ruined cities were appropriated for Inca ceremonies.

▲ The Inca ruler was believed to be descended from the Sun God, one of a number of deities to whom offerings were made – as visualized in the painting on this wooden cup. Decorated with inlaid pigments, it represents the trophy head of an Anti, an uncivilized enemy from the Antisuyu tropical forest "quarter" of the empire. Made by Inca descendants in the colonial period and influenced by European art, it juxtaposes pre-Hispanic characters and activities with the abstract motifs (tokapu) of traditional Inca art.

► The Inca capital of Cuzco was literally the focal point of the empire. Four avenues emanating from the centre of the city were linked to the empire's road system and led to the symbolic four "quarters" of the empire. Two of these avenues also divided the city into ritually complementary northwest and southeast halves, Hanan and Hurin. The stone walls of Cuzco later served as the bases for Spanish colonial buildings.

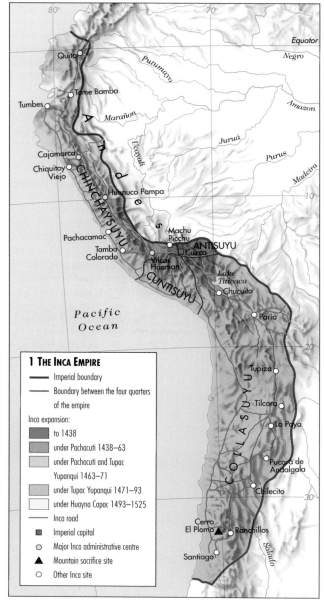

1 THE INCA EMPIRE
— Imperial boundary

— Boundary between the four quarters of the empire

Inca expansion:

▮ to 1438

▮ under Pachacuti 1438–63

▮ under Pachacuti and Tupac Yupanqui 1463–71

▮ under Tupac Yupanqui 1471–93

▯ under Huayna Capac 1493–1525

— Inca road

▪ Imperial capital

○ Major Inca administrative centre

▲ Mountain sacrifice site

○ Other Inca site

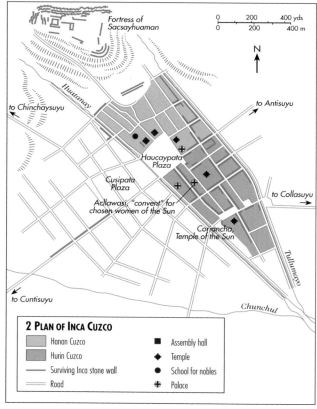

2 PLAN OF INCA CUZCO

▮ Hanan Cuzco

▮ Hurin Cuzco

— Surviving Inca stone wall

▒ Road

▪ Assembly hall

✚ Temple

● School for nobles

✚ Palace

The short-lived Inca Empire in the Andes and Aztec Empire in Mesoamerica were the last to dominate the two principal areas of urbanized culture which had developed over a period of 3,000 years before the arrival of the Spanish. Both mobilized labour for state projects and extracted valued materials and objects from their subjects, but while the Aztecs undertook most of their building and manufacturing projects in the imperial core – particularly in their capital city, Tenochtitlan, under present-day Mexico City – the Incas had broader control over their subjects and directed projects in distant territories. In Tenochtitlan the Aztecs created a remarkable assembly of large, finely carved stone sculptures in a mere 70-year period before the fall of their empire to the Spanish in 1521, but little can now be seen of these. In comparison, distinctive Inca architecture, ceramics and other remains have been found throughout their empire, the largest in pre-Spanish America.

THE INCA EMPIRE

Unlike the inhabitants of Mesoamerica, who recorded history in manuscripts with hieroglyphic dates and pictographic representations of rulers and their activities, the ancient Andeans used knotted strings (quipus) for record-keeping. The reconstruction of the history of the Inca Empire is therefore problematic. Inca conquests of local neighbours around the capital of Cuzco probably date from the 14th century (pages 84–85), and the period of greatest expansion began around 1440 under Pachacuti, who rebuilt the imperial capital, and his successor Tupac Yupanqui. At its height the empire covered a 4,200-kilometre (2,600-mile) strip along western South America, encompassing coastal and highland valleys from Quito in modern Ecuador to southern Chile (map 1).

The Incas were great builders, and the extent of their empire is still visible in an advanced road system of highland and lowland routes along which armies and caravans of llamas moved. At intervals there were settlements or way-stations built of distinctive Inca stonework, such as the well-studied site of Huanuco Pampa. These architectural complexes included accommodation for local artisans and labourers working for the state, feasting halls and ceremonial plazas for the wooing of the local elite, facilities for storage, and lodgings for imperial representatives. All aspects of production, from the acquisition of materials to the manufacture and distribution of finished items, were controlled by the state.

THE INCA CAPITAL OF CUZCO

Cuzco was the political, cultural and ritual focal point of the empire. It was surrounded by settlements of Inca commoners and members of the elite and their retainers, relocated from sometimes distant areas of the empire. Cuzco proper (map 2) was relatively small, containing only the residences of the living ruler and royal clans reputedly descended from previous kings (some fictitious), plus the temples, plazas, platforms and halls for imperial ritual. Palaces and temples consisted of rows of simple adobe or stone rooms with gabled straw roofs; where they differed from homes of commoners was in the quality of workmanship and materials, such as finely worked ashlar masonry, gold and silver sheets attached to walls, and elaborately dyed and plaited thatch.

THE AZTECS

Because the Aztecs kept written records, we have a better idea of their imperial history. The empire was founded in 1431, after the Aztec war of independence from the Tepanecs who had previously dominated the Valley of Mexico. It was formed by an alliance of three cities – Texcoco, Tlacopan and Tenochtitlan – the last of which quickly became the dominant city.

All Tenochca Aztec rulers were warriors, but the two responsible for the greatest expansions were Motecuhzoma, or Montezuma I (r. 1440–69), who also reorganized Aztec society and rebuilt the imperial capital, and Ahuitzotl (r. 1486–1502), who extended the empire to the border of

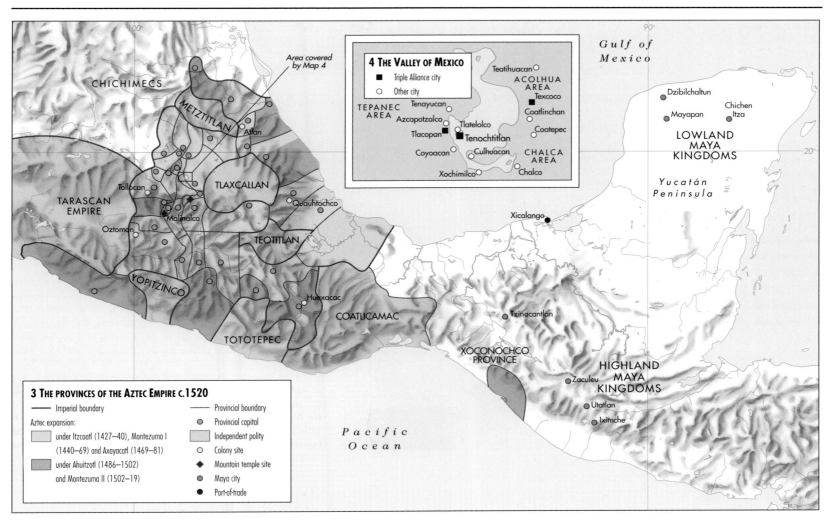

4 THE VALLEY OF MEXICO
- ■ Triple Alliance city
- ○ Other city

3 THE PROVINCES OF THE AZTEC EMPIRE c.1520

—— Imperial boundary —— Provincial boundary

Aztec expansion:

under Itzcoatl (1427–40), Montezuma I (1440–69) and Axayacatl (1469–81)	● Provincial capital
	Independent polity
under Ahuitzotl (1486–1502) and Montezuma II (1502–19)	○ Colony site
	◆ Mountain temple site
	● Maya city
	● Port-of-trade

modern Guatemala. Early expansion by Montezuma I and two other kings consolidated the highlands on all sides of the capital, while later thrusts by Ahuitzotl and Montezuma II (r. 1502–19) went into tropical coastal areas and temperate highlands to the south and east. The west and north were blocked by the enemy Tarascan Empire and by culturally less complex groups to whom the Aztecs applied the derogatory term "Chichimecs". At the time of the Spanish arrival in 1519, Aztec armies were reportedly poised to invade the northern Maya kingdoms on the Yucatán Peninsula from the port of Xicalango.

THE STRUCTURE OF THE AZTEC EMPIRE

The Aztec Empire extended from the Pacific to the Gulf coast, but imperial provinces were bordered by blocks of unconquered territories, keeping the people of Mesoamerica in a constant state of warfare. The region had well-developed market and long-distance trading systems centuries before the rise of the Aztecs, who tried to control these where they could; however, many networks continued to operate independently. The Aztecs did not put their energies into administrative structures, and their empire lacked the monumental road system of the Incas' polity. However, Aztec artisans were accomplished stone carvers, as evidenced by surviving temples at mountain sites like Malinalco to the southwest of the capital.

After conquest of a province, numerous captives of war were brought to the capital for sacrifice. As in Peru, captured deity images were put in Aztec temples, sacred mountain sites were appropriated for ceremonies and temples, and tribute was demanded. However, conquered groups were not relocated; instead, loyal subjects from Tenochtitlan and nearby areas were sent to strategically located colonies, while members of the foreign elite and traders spent time in the cities of the imperial centre.

At its height Tenochtitlan, which occupied an island in the shallow lake that dominated the Valley of Mexico, had a population of perhaps 200,000, four times that of its nearest rival. According to contemporary descriptions, it had a huge central precinct in which four great causeways met. The precinct contained many temples and was immediately surrounded by the palaces of rulers and the elite. Beyond were the neighbourhoods of commoners, where enclosed compounds and house gardens were organized in a grid of streets and canals.

Texcoco and Tlacopan on the east and west shores, along with numerous other towns as old as or older than Tenochtitlan, remained uneasy allies and potential enemies of the capital. Thus when the Spanish arrived in 1519 they found thousands of Indian allies both in the valley and throughout the empire ready to revolt against the Aztecs.

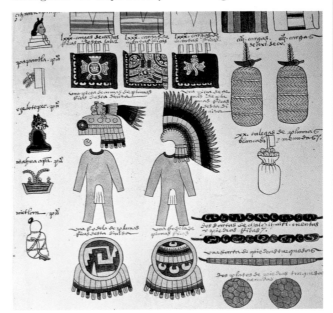

▲ The Aztec Empire covered much of what is now central Mexico, with one separate province adjacent to distant Maya territory. There were substantial unconquered areas next to and surrounded by imperial provinces. The empire's capital, Tenochtitlan, and its two uneasy allies – Tlacopan and Texcoco – were just three of some 50 cities with surrounding territories and satellite towns in the lake zone of the Valley of Mexico.

◀ Manuscripts of the Spanish colonial period have made it possible to reconstruct the Aztec Empire's structure. Among them is the *Codex Mendoza*, which includes pictures of the pre-Conquest tributes that were demanded from individual provinces – among them warriors' clothing, bags of feathers and dried chillies.

THE EARLY MODERN WORLD

Before 1500 there was a gradual overall increase in the world's population and economy, although epidemics and widespread famine sometimes caused a temporary decline. Then in the space of 300 years the population more than doubled, from 425 to 900 million, and the world economy expanded rapidly as Europe embarked on a process of exploration, colonization and domination of intercontinental commerce.

▶ Porcelain was amongst the Chinese products for which there was a great demand in Europe. Another was silk. The export of both products from China ensured that trade with the West continued to flourish throughout the 16th, 17th and 18th centuries, although Chinese merchants did not themselves venture outside Asia.

The Europeans' exploration and discovery of the world began in earnest in the second half of the 15th century when the desire to find a sea route to the East led to a series of Portuguese voyages down the west coast of Africa. The Cape of Good Hope was finally reached in 1488, just four years before Christopher Columbus set sail across the Atlantic, on behalf of Spain, in search of a westward route to China. His discovery of the West Indies was quickly followed by Spanish expeditions to the American mainland and the creation of Spanish and Portuguese colonies in the Caribbean and South America. New trade routes across the Atlantic and Indian oceans were pioneered by the Spanish and Portuguese, to be taken over in the 17th century by the Dutch, English and French.

Africa was both a survivor and a victim of this transoceanic transport revolution. The economies of its states – and the extensive trade network linking the north, east and west of the continent – were little affected by contact with the Europeans. However, from 1450 over 12 million Africans were forced to embark on a journey across the Atlantic as slaves destined to work in the plantations and gold and silver mines of Europe's colonies in the Americas and the Caribbean.

EUROPEAN TRADE WITH ASIA

The Europeans were to have a greater effect on the economies of Asia. In South and Southeast Asia the Portuguese combined plunder with trade, and by the 1560s they were importing about half the spices reaching Europe from the East. With overland Eurasian trade becoming increasingly hazardous – and also costly as local rulers extorted high protection costs – merchants from other European nations sought to establish themselves in the

▶ Despite periods of vigorous territorial and economic expansion, the great land empires failed to participate in the commercial revolution led by the countries of northern Europe in the 17th and 18th centuries. In 1700 they still covered vast areas, but in the following century the three Muslim empires – the Mughal, Safavid and Ottoman – declined as the commercial and military power of the Europeans expanded.

1 EURASIAN LAND EMPIRES C. 1700
Boundary of empire at greatest extent in 16th–18th centuries:
- Ottoman 1683
- Russian 1795
- Manchu Qing 1760
- Safavid 1514

RUSSIAN EMPIRE

AUSTRIAN HABSBURG EMPIRE

Black Sea

Caspian Sea

OTTOMAN EMPIRE

Mediterranean Sea

SAFAVID EMPIRE

MANCHU QING EMPIRE

Pacific Ocean

MUGHAL EMPIRE

Equator

Indian Ocean

oceanic Asian trade. In 1600 and 1602 the English and Dutch East India Companies were created, and within a few years the Dutch company had weakened Portuguese power in the Indian Ocean. However, local politics and rivalries between Hindu and Muslim entrepreneurs and courtier-traders continued to influence the patterns of European commerce and imperialism.

In the first half of the 17th century a struggle between Crown and Parliament in England, and a war of liberation in the Netherlands (from which the independent Dutch Republic emerged), placed merchant capitalists in both countries in more powerful positions. By the 1650s they were the leading economies of Europe. A century later trade outside Europe accounted for 20 to 25 per cent of the Dutch Republic's total trade, while the figure for England was as high as 50 per cent.

THE EMPIRES OF ASIA

The rapid growth of northern European trade was not closely related to technological achievement: in the 17th century Europe imported Asian manufactured goods rather than vice versa, and per capita productivity in India and China was probably greater than in Europe. However, the technological superiority of India and China was not matched by an urge towards overseas expansion and conquest. Under the Ming dynasty (1368–1644) Chinese voyages of exploration in the early 15th century had reached as far as the east coast of Africa. Yet while these voyages helped to consolidate China's sphere of influence in Asia, they did not lead to the creation of a far-reaching overseas trading network. Instead, trade with the rest of Asia and with Europe continued to flourish with the aid of overland routes, short-distance sea routes and foreign merchants, resulting in an outflow of ceramics and silk, and an inflow of silver.

China relied on intensive agriculture to support its ever-growing population, but in the 16th century it was stricken by harvest failures, droughts and famine, which in turn led to frequent rebellions. Insufficient resources were devoted to defence, and in 1644 the Ming dynasty gave way to Manchu conquerors from the north. Under the Manchus, China became preoccupied with defending its own borders, which by 1760 had expanded to encompass a greater area than ever before (*map 1*).

In India the Mughal Empire – established in 1526 by Muslim warrior descendants of the Mongols – was centred on cities in the country's heartland. Its rulers financed their administration, and the architectural achievements for which they are renowned, by taxing local agriculture and commerce. However, they had little interest in overseas trade beyond the existing involvement of the artisanal industries in the Muslim trading networks that stretched from Arabia to Indonesia. The Portuguese, who were intent on seizing control

of these networks, used their ships' guns to overcome opposition and established trading posts around the coast. They were followed by Dutch, English and French merchants.

The Mughal Empire was just one of three powerful Muslim empires in the 16th century. Another was that of the Ottoman Turks, who after their capture of Constantinople in 1453 had embarked on a process of territorial expansion in Africa, Asia and Europe. This was to continue until 1683 when their last major expedition was driven back from Vienna, the Austrian Habsburg capital.

Among the other great powers with which the Ottomans came into conflict in the 16th century was the third representative of the political and cultural achievements of Islam at this time – the Safavid Empire (1501–1736) in Iran. Despite a resounding Ottoman victory in 1514, it was not until 1639 that the border between the two empires – the present-day frontier between Iran and Iraq – was firmly established.

◄ The Mughal emperor Akbar is shown in this painting after riding an elephant over a bridge of boats across the River Jumna. Ruling between 1556 and 1605, Akbar was responsible for the considerable expansion of the Mughal Empire's territory and for creating a centralized and efficient administration.

During the Mughal period the Europeans established trading posts around the coast. They brought gold and silver from the Americas, and so in the short term they stimulated the Indian economy. However, in the 18th century their activities were to contribute to the decline of the Mughals and the beginning of British rule in India.

▼ The shahs of the Safavid Empire were great patrons of architecture and art – of which this picture made up of tiles is a fine example. Greatest of all artistic patrons was Abbas I (1587–1629). After his death the empire went into decline and finally collapsed in 1736.

THE MAJOR LAND EMPIRES OF EUROPE

The conflict with the Safavids temporarily diverted Ottoman attention away from Europe, where the power with which it most frequently came into direct confrontation in the 16th and 17th centuries was the Habsburg Empire. In the 1520s this empire was little more than the largest conglomeration of territories and rights in Europe – among them Spain, Austria, Hungary and the former lands of the Duchy of Burgundy – since the 9th century. It was not welded into a more coherent empire until the Thirty Years War of 1618–48, from which time the Habsburgs began the reconquest of Hungarian territory lost to the Ottomans and thus became the major dynastic power of central Europe.

To the northeast of the Habsburg Empire lay Poland – a kingdom which through much of the 17th and 18th centuries was in conflict with Russia. Under Muscovy's Grand Duke Ivan III (r. 1462–1505), Russia began a process of exploration and expansion on land comparable with that undertaken overseas by the western European maritime powers. By the end of the 18th century its empire stretched from the Baltic to the Pacific Ocean, and formed a world economy in miniature.

◀ The Europeans' "discovery" of the world gave an enormous stimulus to cartography and the improvement of optical instruments. It also heralded a new capacity for observation of the natural world which eventually surpassed even that of the Chinese. The sophisticated depiction of spatial relationships which evolved in art is exemplified in *The Artist's Studio* (c. 1660) by the Dutch painter Jan Vermeer.

▲ In 1607 an English colony was established in Virginia, where John White had painted this view of a Native American village in the 1580s. Further north the colony of Plymouth was established in 1620 by the Pilgrim Fathers, a Puritan group who had broken away from the Church of England. Many such separatist groups were to settle in North America.

COLONIZATION OF THE AMERICAS

Following the European discovery of the Americas – and the highly valued commodities to be found there – world demand for gold and silver ensured the gradual integration of the New World into the emerging European world economy. The Spanish conquest of Central and South America from the end of the 15th century was accompanied by the decimation of the native Indian population – not as a deliberate act of genocide but mainly as a result of diseases imported from Europe and a regime of forced labour. The estimated pre-conquest population of about 57 million was reduced to less than six million by the late 16th century. A similar fate awaited the smaller North American population when European colonists began to arrive in the 17th century. In order to replace native forced labour, slavery was introduced by the Spanish *conquistadores* and their successors. Between 1500 and 1650 about 500,000 African slaves were imported by the Spanish and Portuguese. Far greater numbers were subsequently imported when the slave system was extended to the Dutch, English and French colonies.

In the short term the Europeans' discovery of the New World drained resources away from Spain and Portugal, who pursued their expansionist strategies through conquest. Expansion in the Americas did not become profitable for the European powers until the later 17th century, when a thriving colonial economy began to develop, based on the plantation crops of sugar in the West Indies; tobacco, rice and indigo in the central and southern mainland colonies; and family farms,

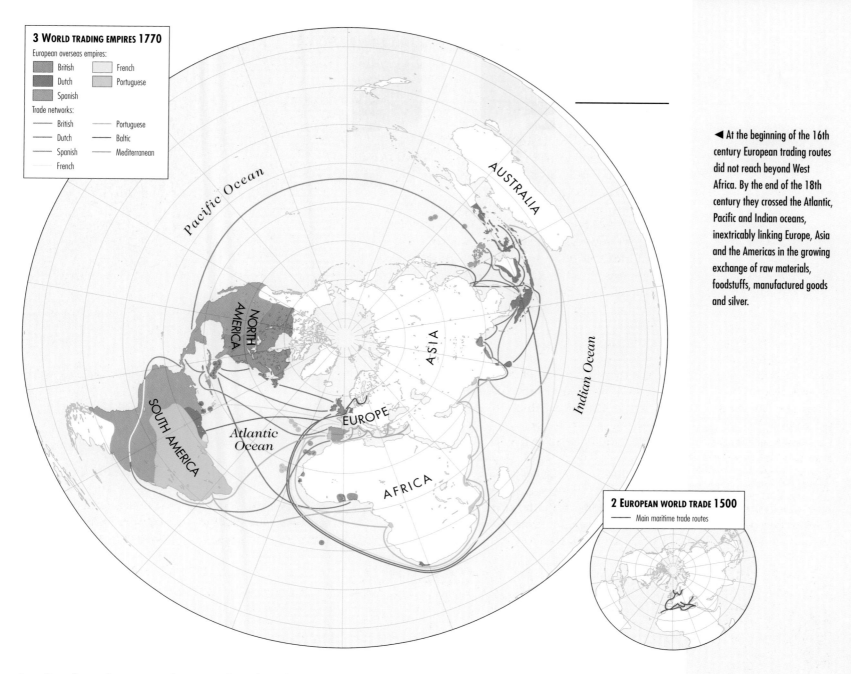

Pacific Ocean

AUSTRALIA

NORTH AMERICA

ASIA

SOUTH AMERICA

EUROPE

Atlantic Ocean

Indian Ocean

AFRICA

◄ At the beginning of the 16th century European trading routes did not reach beyond West Africa. By the end of the 18th century they crossed the Atlantic, Pacific and Indian oceans, inextricably linking Europe, Asia and the Americas in the growing exchange of raw materials, foodstuffs, manufactured goods and silver.

2 EUROPEAN WORLD TRADE 1500
— Main maritime trade routes

handicraft production and intra-colonial trade in New England and the other northern colonies. Profits from trade with the colonies at first went principally to the Dutch Republic, followed closely by England and then France.

EUROPEAN DOMINATION OF TRADE

The domination of the evolving global economy by Europe, rather than by China or the Islamic powers, was due to a number of convergent forces, including the development of maritime enterprise and, later, of scientific and technological innovations. The division of the Church during the 16th-century Reformation, between Catholic and Protestant believers, encouraged international rivalry and emigration to the New World. However, above all else, it was the existence of a competitive state system in Europe, and the willingness and capacity of European governments to mobilize military and naval power in support of trade, which secured European hegemony. By the mid-18th century the octopus-like grip of the European trade routes formed an interlocking whole, in which American bullion paid for Asian luxuries and for the supplies of timber and other naval stores from the Baltic countries that were essential for further commercial expansion (*maps 2 and 3*).

The growing European appetite for colonial and Asian goods – including tea, sugar, tobacco, spices, and silks – as well as luxury items produced within Europe, was to play a significant role in the industrialization of western Europe, and of Britain in particular. The spread of consumerism and the desire for market-bought products encouraged rural households to specialize in both food production and various types of cottage industry in order to enhance their purchasing power – with the result that an early "industrious revolution" operating at the level of the household economy took place.

At the same time the commercial revolution provided new overseas markets for manufactured goods, especially in North America after around 1750, as well as essential raw materials such as dyestuffs, raw cotton and silk, and iron ore. The struggle to protect overseas markets and colonial sources of supply stimulated war industries such as shipbuilding, armaments and metal-smelting, all of which saw major technological improvements in the 18th century. The expansion of the Europe-centred world economy thus paved the way for the Industrial Revolution which was to take place first in Britain, and then in Europe and the United States, with enormous repercussions for the world in the 19th century.

THE EUROPEAN DISCOVERY OF THE WORLD 1450–1600

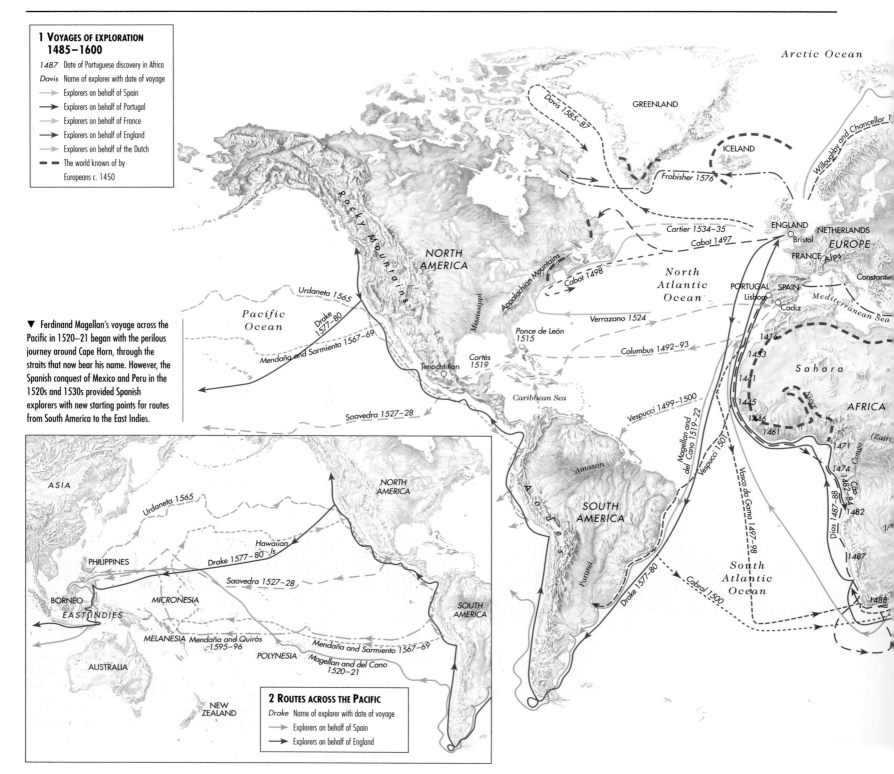

1 VOYAGES OF EXPLORATION 1485–1600

1487 Date of Portuguese discovery in Africa
Davis Name of explorer with date of voyage
→ Explorers on behalf of Spain
→ Explorers on behalf of Portugal
→ Explorers on behalf of France
→ Explorers on behalf of England
→ Explorers on behalf of the Dutch
– – – The world known of by Europeans c. 1450

▼ Ferdinand Magellan's voyage across the Pacific in 1520–21 began with the perilous journey around Cape Horn, through the straits that now bear his name. However, the Spanish conquest of Mexico and Peru in the 1520s and 1530s provided Spanish explorers with new starting points for routes from South America to the East Indies.

2 ROUTES ACROSS THE PACIFIC

Drake Name of explorer with date of voyage
→ Explorers on behalf of Spain
→ Explorers on behalf of England

Most civilizations knew something of the world outside their own territorial boundaries before Europeans discovered the existence of the Americas in the 1490s. The Greeks had circumnavigated Britain as early as 310 BC, by the 1st century AD Rome had established links with China, while the Chinese themselves had explored Central Asia, reaching the Euphrates by AD 360. However, the insularity of the Chinese court in the late 15th century (*pages 138–39*) – leading to the destruction of most of the official records of Zheng He's pioneering voyages of 1405–33 in the Pacific and Indian oceans – undermined any sustained contact with the wider world. The discoveries by European explorers were new and momentous in the sense that expanding geographical horizons were matched by new mental horizons.

The geographical discoveries of the late 15th century were neither isolated nor accidental historical events. Rather, they were part of a European expansionist phase, and were to some degree a response to the disruption of Eurasian commerce brought about by plague, the closure of the Silk Road and the caravan routes during the 1360s, and the fall of Constantinople to the Ottoman Turks in 1453. The need to find a direct route to the Far East, principally for trading silks and spices, provided a powerful impetus to exploration.

The Portuguese led the way with a series of expeditions from 1415 to explore the west coast of Africa (*pages 80–81*). In 1445 the westernmost tip of the continent was rounded, and by 1460 they had travelled 3,200 kilometres (2,000 miles) south as far as Sierra Leone, bringing back spices, gold and slaves. By 1474 the equator had been crossed, and in 1488 Bartholomew Dias reached the Cape of Good Hope (*map 1*) – an important step towards the establishment of a sea route to India, which was achieved by Vasco da Gama in 1497–98. After Dias's voyage, mapmakers were able to show the sea encompassing southern Africa, but the globe was still envisaged as a much smaller – and younger – planet than is actually the case, and was thought to be dominated by the Eurasian landmass.

Spanish to find a sea route to Asia encouraged further colonization and plunder. Mainland settlement began in 1509–10 on the isthmus of Panama. Hernán Cortés, the first of the *conquistadores*, established Spanish control over the Aztec Empire in Mexico in 1521, and in South America Francisco Pizarro subdued the empire of the Incas in Peru and Bolivia during the 1520s and early 1530s (*pages 120–21*). The conquest of Mexico and Peru provided new opportunities for transpacific exploration (*map 2*), and in 1527 Saavedra travelled across the Pacific from the coast of Mexico to the Moluccas. A viable return route, from the Philippines to Acapulco, was first navigated by Urdaneta in 1565 and was followed thereafter by Spanish galleons. In 1567 Mendaña and Sarmiento led an expedition in search of a great southern continent and found the Solomon Islands. Mendaña attempted to return there to establish a Christian colony in 1595, accompanied by the Portuguese navigator Quirós. They were unable to find the Solomons but instead stumbled on the Marquesas and Santa Cruz islands. However, it was not until the more scientific voyages of the 18th century that the full extent of the Pacific, from Alaska to New Zealand and the east coast of Australia, was to be explored.

THE ENGLISH, FRENCH AND DUTCH IN NORTH AMERICA

For much of the 16th century the Spanish and Portuguese attempted to exclude northern Europeans from their expanding colonial empires and the new sea routes across the southern hemisphere. As a result, the opening up of the north Atlantic world was mainly an English, French and Dutch enterprise, although it was more than a by-product of the quest for a northwestern route to the East. The first initiatives were probably undertaken as early as the 1420s by Bristol merchants involved in trade with Iceland. These traders were certainly exploring the coast of Newfoundland in 1481, some time before John Cabot made his historic voyage of 1497. Cabot, under commission from the English crown, discovered 640 kilometres (400 miles) of coastline from Newfoundland to Cape Breton, and by 1509 his son Sebastian had travelled as far south as Cape Cod.

In 1510 the English knew more about North America than any other European country did, but during the next half century the French moved into the lead. In 1524 Verrazano, in the service of France, sailed along the coast from Cape Fear to Newfoundland, thereby proving that the earlier discoveries of Columbus and Cabot were part of a single landmass. The first steps in exploring North America's interior were taken ten years later by Jacques Cartier, who travelled along the St Lawrence River as far as Montréal. It was not until the 1570s and 1580s that the English returned to the area, with the voyages of Frobisher and Davis, in search of a northwest passage via Newfoundland (*map 1*). The years 1577–80 also saw an important breakthrough in English efforts when Francis Drake circumnavigated the world in the search for a new transpacific route.

The northern maritime countries were fortunate to inherit the more sophisticated seamanship and navigational skills of the Portuguese and Spanish. The art of celestial navigation, using the quadrant and astrolabe, was improved by the Portuguese during the 1480s, when manuscript copies of the first navigational manual, the *Regimento*, became available prior to its publication in 1509. Sebastian Cabot, an expert cartographer, helped to spread knowledge of Spanish navigational techniques in England. Although ships gradually increased in size during the 16th century, improvements in ship design were not, of themselves, sufficient to stimulate the long-distance exploration which took place during this period. The Dutch introduced top masts and sails, as well as the *fluytschip* (a flat-bottomed cargo carrier), and these advances certainly facilitated commercial exploitation and colonization of a type that was markedly different from the plundering of the *conquistadores* and the privateering expeditions of Drake. However, the idea of European settlement in the Americas in order to exploit fully the land's natural resources was surprisingly slow to win acceptance and, when it did, was invariably difficult to sustain.

▲ When Christopher Columbus set sail across the Atlantic in 1492 he was guided by the assertion of the Greek geographer Ptolemy (c. AD 85–150) that the circumference of the Earth is about 11,000 kilometres (7,000 miles) shorter than it actually is and that, going west, there is no land between Europe and Asia. His belief that the West Indies were islands off the coast of China was quickly discredited when further Spanish expeditions began to explore the Americas and, beyond them, the Pacific Ocean.

THE SPANISH AND THE NEW WORLD

While Portuguese explorers searched for a passage to the East by a southeasterly route, the Spanish searched in a westerly and southwesterly direction. Although they were unsuccessful in reaching their immediate goal, the result was the discovery of the West Indies and the Venezuelan coast by Christopher Columbus between 1492 and 1502. Columbus, as his Spanish patrons realized, had greatly underestimated the distances involved in reaching Asia by a southwesterly route, but he nevertheless pressed on. The New World was Spain's unexpected prize, confirmed in the Treaty of Tordesillas of 1494, and first described by the explorer and writer Amerigo Vespucci in travel accounts published from 1507. By the 1520s the Old World recognized the Americas as an enormous "new" continent between Europe and Asia.

Spanish exploitation of the Caribbean islands began with the settlement of Hispaniola in 1493, followed by that of Cuba and Puerto Rico. These islands provided a base for the exploration of Central America, and the failure of the

EUROPEANS IN ASIA
1500–1790

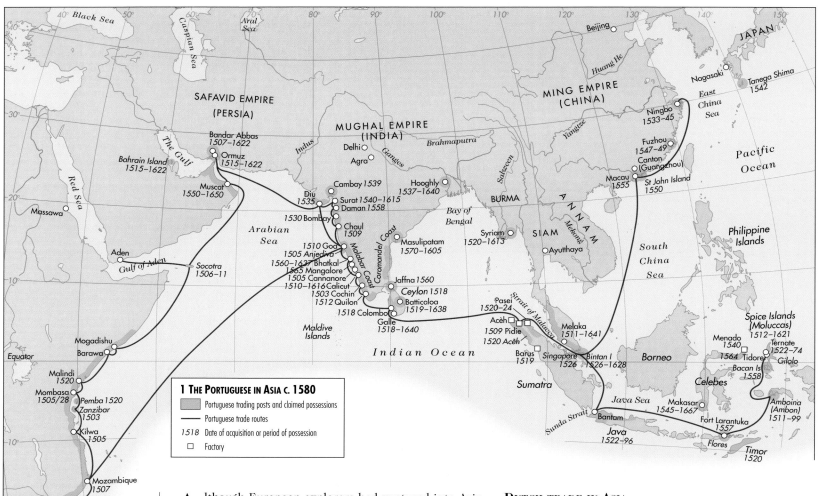

1 THE PORTUGUESE IN ASIA C. 1580

- ▨ Portuguese trading posts and claimed possessions
- ── Portuguese trade routes
- *1518* Date of acquisition or period of possession
- ☐ Factory

▲ The Portuguese seaborne empire was based on a series of forts linking together trading entrepots from the coast of Africa to South and Southeast Asia, and on to China and Japan. This system secured Portuguese trade with the East for nearly a century. The empire was governed from Goa, on the west coast of India, which had been captured for Portugal by Afonso de Albuquerque in 1510. Although the Portuguese were to lose most of their eastern possessions to the Dutch in the 17th century, they managed to hold on to Goa, surviving Dutch blockades of the city in 1603 and 1639.

Although European explorers had ventured into Asia in the 1st century AD, significant European contact with the continent only began on 27 May 1498 when the Portuguese fleet of Vasco da Gama landed at Calicut on the west coast of India. Da Gama had rounded the Cape of Good Hope in search of the valuable spices and silks which had long reached Europe only via expensive overland routes. Over the next hundred years a Portuguese "seaborne empire" spread around the coasts of the Indian Ocean, moving ever further east and developing a chain of forts linking Ormuz, Goa, Cochin, Ceylon (Sri Lanka), Melaka and Ternate (*map 1*). Japan was reached in 1542 and a settlement established in China, at Macau, in 1555.

PORTUGUESE TRADING EMPIRE
The motives of the Portuguese were both economic and religious. In the pursuit of wealth, they attempted to establish a monopoly over the spice trade to Europe and to force entry into an already extensive trading network within Asia. Previously, this commerce had been conducted by indigenous merchants along free-trade principles, but the Portuguese coerced local merchants into paying them licence fees and seized the most lucrative trade routes for themselves. In the service of God, they promoted Christianity. In some cases, the two objectives dovetailed neatly: in Japan, between 1542 and 1639, they made more than 100,000 converts while running a valuable silk trade from Macau and advising the rising power of the Tokugawa shogunate on military tactics.

Yet Portuguese influence in the East was to prove short-lived. In part, it suffered from problems at home. Rivalry with Spain was intense and after the crowns of the two Iberian countries were united in 1580 internecine strife became bitter. A further problem was caused by the revival of Asian empires, whose temporary weaknesses had been exploited by the Portuguese. In Japan, for example, once the Tokugawa (*pages 140–41*) had achieved victory in the civil wars, they expelled the Iberians and in 1639 outlawed Christianity as a danger to the stability of their new state.

DUTCH TRADE IN ASIA
For the most part, however, Portuguese influence was eclipsed by the rise of another European power. The Dutch had long been involved in war against Spain (*pages 152–53*) and took the unification of its throne with that of Portugal as a signal to penetrate Asian waters and attack the Portuguese Empire. Following the establishment of their East India Company in 1602, the Dutch progressively displaced the Portuguese in Asian trade and developed their own trading empire further east (*map 2*). They also expanded Asian trade with Europe, Africa and the Americas, bringing Chinese porcelain into Western markets and Indian cotton textiles to the slave coasts of Africa and plantations of the New World.

The success of the Dutch was based on superior mercantile and maritime skills, which enabled them to enforce trade monopolies with greater ruthlessness. It also owed something to religion since, as Protestants, they were less interested in making converts than their Catholic rivals and were thus perceived as less of a threat by the indigenous societies. Following the expulsion of the Iberians from Japan, for example, the Tokugawa invited the Dutch to conduct Japan's external trade at Nagasaki.

Dutch maritime influence grew during the 17th century and remained strong east of Ceylon throughout the 18th century. However, it too faced eventual eclipse. One reason for this was that the Dutch were drawn into the politics of the hinterlands behind their port settlements and spent scarce resources on local wars at great cost to their trade. However, the principal reason for their demise was the belated entry into Asian trade of the much stronger European states of England and France.

THE ENGLISH AND THE FRENCH IN INDIA
English merchants had initially tried to break into the spice trade of the Indonesian archipelago but after the Massacre of Amboina in 1623, when Dutch forces had destroyed their principal trading settlement, they were effectively excluded. Instead they concentrated on India, where the

authority of the Mughal Empire (*pages 144–45*) constrained the Dutch from gaining too tight a control and offered opportunities for competitive trade (*map 3*).

India was originally regarded as of limited mercantile importance because its spices were thought to be of lower quality than those found elsewhere. Yet this judgement was subsequently proved to be mistaken; India also possessed an enormous cotton textile industry, the significance of which became increasingly apparent as the 17th century advanced (*pages 194–95*). Cotton textiles were already established in the vast network of Asian trade, so the

English gained secondary access to markets from the Gulf to the China seas. There was also a growing demand in Europe for Indian textiles, and from the 1650s onwards the cloth trade became the main source of European profits in Asia. This, in turn, caught the attention of the French, whose first Asian settlement was established in India in 1664, and the two newcomers steadily reduced the Dutch presence around the shores of India. The English also used India as a staging post for ventures further east, forging a broad triangular trade with China, from which tea, raw silk and porcelain were exported to the West in return for Indian silver and opium.

From the second quarter of the 18th century trade relations between England, France and India began to change. Many European states put up tariff barriers against Indian textile imports in order to protect their own domestic industries. This increased the importance to the English of trade with China and, in turn, placed greater emphasis on their ability to gain access to Indian silver and opium. In addition the Mughal Empire, which had previously confined European activities to the coasts, began to break up. Its successor states were soon at war with one another, making demands for finance and armaments which the Europeans found too lucrative to ignore. From the 1740s England and France also began a series of wars against each other which were to last – with brief interruptions – for the rest of the century, and end in the domination by "British India" of a vast area of the world from Arabia to the China seas.

▼ The Europeans were drawn towards Asia by the lure of exotic consumer goods – tea, spices and silk – and by high-quality manufactures such as porcelain and printed cotton textiles (chintzes).

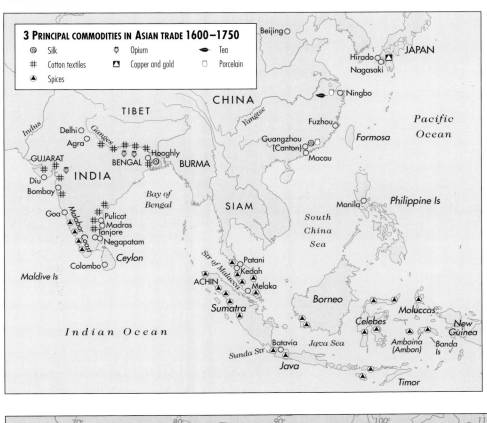

3 PRINCIPAL COMMODITIES IN ASIAN TRADE 1600–1750

◎ Silk ▯ Opium ◗ Tea
⌗ Cotton textiles ▲ Copper and gold ▢ Porcelain
▲ Spices

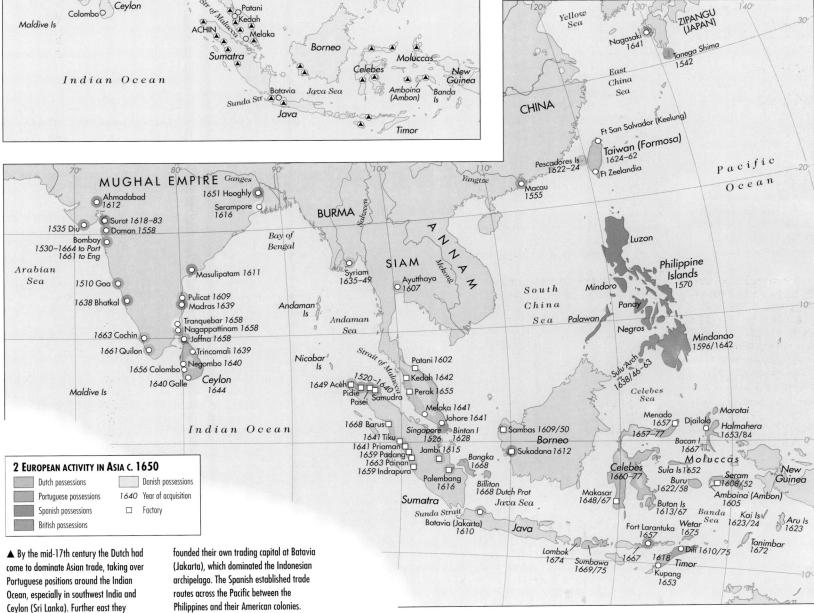

2 EUROPEAN ACTIVITY IN ASIA C. 1650

Dutch possessions		Danish possessions
Portuguese possessions	1640	Year of acquisition
Spanish possessions	□	Factory
British possessions		

▲ By the mid-17th century the Dutch had come to dominate Asian trade, taking over Portuguese positions around the Indian Ocean, especially in southwest India and Ceylon (Sri Lanka). Further east they founded their own trading capital at Batavia (Jakarta), which dominated the Indonesian archipelago. The Spanish established trade routes across the Pacific between the Philippines and their American colonies.

◀ KINGDOMS IN SOUTHEAST ASIA 500–1500 *pages 64–65* ▶ SOUTHEAST ASIA IN THE ERA OF IMPERIALISM 1790–1914 *pages 196–97*

SPAIN AND THE AMERICAS 1492–1550

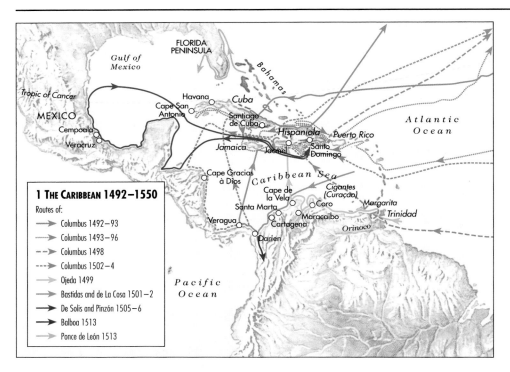

(pages 110–11)

Puerto Rico, using forced Indian labour in agriculture and goldmining. From 1510, however, the economy was undermined by the collapse of the indigenous workforce, caused by Spanish mistreatment and by the spread of European diseases to which the islanders had little natural resistance.

THE AZTEC AND INCA EMPIRES

Spanish interest therefore turned to the great civilizations of the mainland (*pages 110–11*) which, in the second and most important phase of Spanish colonization, were reconnoitred and eventually conquered in a two-pronged exploration from the islands (*map 2*).

In 1518 Hernán Cortés was sent by the governor of Cuba on a commercial and exploring expedition to the Yucatán Peninsula. Once ashore, Cortés repudiated the governor's mandate and henceforth acted on his own initiative, acknowledging only the authority of the King of Spain. His small army of military adventurers or *conquistadores*, having founded the town of Veracruz and symbolically scuttled its own boats, marched to Tlaxcala (*map 3*). Here they overcame initial resistance to form an alliance with the Tlaxcalans, themselves resentful of Aztec overlordship.

Cortés and his Tlaxcalan allies entered the Aztec capital, Tenochtitlan, in 1519, but early in 1520 Cortés was forced to return to the coast to meet and win over to his side a hostile Spanish army dispatched from Cuba under Narvaez. Unfortunately the greed of the Spanish left behind in Tenochtitlan had alienated the Aztecs and, on Cortés' return, the Spanish were driven from the city in a series of events which led to the death of the Atzec emperor Montezuma. Cortés' army retreated to Tlaxcala, and in 1521 they and their Tlaxcalan allies launched a successful campaign against Tenochtitlan. This victory brought under Spanish control the millions of central Mexicans who had formerly been Aztec tributaries.

Meanwhile, from Hispaniola, the Spanish had organized colonies in Darien and on the Pacific coast of the Panama isthmus, first crossed by Balboa in 1513. Panama was used as a base for expeditions into Nicaragua and beyond and,

1 THE CARIBBEAN 1492–1550

Routes of:

→ Columbus 1492–93
⋯ Columbus 1493–96
–– Columbus 1498
⋯⋯ Columbus 1502–4
→ Ojeda 1499
→ Bastidas and de La Cosa 1501–2
→ De Solis and Pinzón 1505–6
→ Balboa 1513
→ Ponce de León 1513

▲ Crucial to the first phase of Spanish colonization were the four voyages in which Columbus discovered the principal Caribbean islands and explored major sections of the mainland coast. These were followed by further naval expeditions mounted from Spain – involving many of Columbus's former companions.

C olumbus discovered America in the name of Spain in 1492, but this famous voyage was merely the initial step in the Spanish colonization of a large part of the continent, a process that took place in three stages.

THE CARIBBEAN AND THE GULF OF MEXICO

Until 1518 the Spanish undertook the exploration and settlement of the Caribbean and the Gulf of Mexico (*map 1*). However, Spanish attempts to exploit their new territories by establishing trading posts in the Caribbean were unsuccessful, because the simple agrarian societies of the islands could not sustain a trading economy. Instead, the Spanish established colonies of exploitation in Hispaniola, Cuba and

▶ The travels of Narváez, de Vaca, de Soto and Coronado were not considered successful since they brought neither wealth nor property to the Spanish crown. Information they provided, however, resulted in a new understanding of the main contours of the southern part of North America, which was reflected in contemporary maps of the area.

▼ Acting on information gleaned from earlier voyages around the Yucatán Peninsula, Hernán Cortés led a small army into Mexico in search of Aztec gold in 1519. On the way he formed an alliance with the Tlaxcalans, enemies of the Aztecs, and with their help he completed his conquest of the Aztec Empire in 1521.

3 CORTÉS' EXPEDITION TO TENOCHTITLAN

→ Cortés' route to Tenochtitlan 1519
→ Forced march 1520
⋯ Retreat and return 1520–21

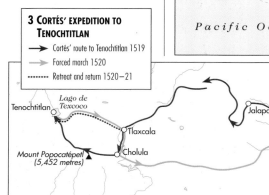

2 CENTRAL AND SOUTHERN NORTH AMERICA 1519–1550

Routes of:

→ Cortés 1519–21   → De Soto 1539–43
–– Cortés 1524–25   → Alarcón 1540
→ Narváez and de Vaca 1528–36   → Coronado 1540–42

► Pizarro's conquest of the empire of the Incas was the first stage of the Spanish colonization of South America. Rumours of gold inspired three separate expeditions in six years into the mountains of what is now Colombia.

4 SOUTH AMERICA 1526–50

Routes of:
Cabot 1526
Francisco Pizarro 1531–33
Amalgro 1535–37
Federmann 1537–39
Benalcazar 1538–39
Orrellana 1540
Gonzalo Pizarro 1540–42
Valdivia 1540–47
Quesada 1542

more importantly, for a series of exploratory voyages in the late 1520s along the Peruvian coast, organized by Francisco Pizarro and Diego de Amalgro (*map 4*). Between 1531 and 1533 Pizarro's small army conquered the Inca imperial cities of Cajamarca and Cuzco, put to death the Emperor Atahualpa and replaced him with a puppet ruler, the Emperor Manco. Victory in Peru, however, was not as clearcut as that in Mexico: the Incas rebelled under Manco and brutal civil wars broke out, both between the *conquistadores* themselves and later between the colonists and royal officials sent to govern them. Amalgro and all five Pizarro brothers were killed in these wars, and Peru was not brought under Spain's control until around 1560.

FURTHER INTO THE MAINLAND

Mexico and Peru provided the resources for the third and final stage of Spanish territorial gains between the mid-1520s and mid-1540s. Alvarado's and Cortés' expeditions from Mexico began the process by which Guatemala and the Yucatán were brought under Spanish control, while a number of other campaigns extended Spanish authority into northern Mexico. However, the protracted wanderings of the Narvaez, de Vaca, de Soto and Coronado bands in the southern United States were epic failures, establishing the northern limits of Spanish colonization. The expeditions of Amalgro, Valdivia and Benalcazar from Peru extended Spanish rule into Chile in the south and Ecuador and Colombia in the north, where the *conquistadores* encountered independent expeditions, such as Quesada's, pushing down from the Caribbean coast. South America also had its share of heroic failures, such as Orellana's descent of the Amazon (*map 4*).

The Spanish also tentatively explored the Plata region in naval expeditions mounted from Spain, the most notable of which was Sebastian Cabot's exploration of the Paraná and Paraguay rivers in 1526–30. From the mid-1540s the surge of conquests waned. By this time Spain had conquered the Americas nearly as far it was ever going to, although many areas were not intensively colonized until the 18th century.

The relentless courage, determination and energy which had been displayed by the Spanish *conquistadores* in acquiring land, wealth and subject populations in the Americas are probably without parallel in the history of European imperialism. However, the ferocious cruelty with which they treated the native populations is hard to square with their lofty claims that they were driven not just by the desire to get rich but also by the ideals of bringing Christianity and civilization to the American Indians. In practice they recognized no authority but their own, and their reckless disregard for their own lives was exceeded only by their callous indifference to the welfare of the peoples they conquered.

▼ Atahualpa, the Inca ruler, was captured by Francisco Pizarro after being enticed to a meeting in the main square of Cajamarca. His unarmed retinue was quickly overcome and slaughtered by the Spanish artillery.

THE COLONIZATION OF CENTRAL AND SOUTH AMERICA 1500–1780

▶ Silver mining, which was concentrated in Mexico and based on the forced labour of American Indian workers, accounted for over 90 per cent of Spanish-American exports between 1550 and 1640. In the Spanish Caribbean colonies of Cuba, Santo Domingo and Puerto Rico, however, African slave labour was used to work the sugar and coffee plantations.

1 MEXICO, CENTRAL AMERICA AND THE EASTERN CARIBBEAN 1520–1750

ARA Native people c. 1520
— Aztec Empire 1519
Territory colonized by the Spanish:
▓ by 1640
▒ by 1750
☐ frontier lands in 1750

1520 Date of foundation of town
--- Sea trade route
Economic activities:
▒ livestock
✴ leather

⬤ sugar
⬤ coffee
✳ cochineal
● indigo
⬛ silver
✷ textiles

2 SPANISH AND PORTUGUESE SOUTH AMERICA 1525–1750

▨ Inca Empire 1525
Spanish settlement:
▓ to 1640
▒ to 1750
☐ frontier lands 1750
Portuguese settlement:
▓ to 1640
▒ to 1750
☐ frontier lands 1750

▓ Dutch colony
▒ French colony
☐ Jesuit mission state
Economic activities:
⬤ coffee
△ sugar
△ mixed agriculture
▣ silver
◎ gold

⬥ mining
● cocoa
☐ mercury
▣ hides
♈ wine
--- Sea trade route
— Land trade route

The peoples conquered by the Spanish and Portuguese in the Americas embraced a very wide range of cultures. Within the Inca and Aztec empires there were urban and agricultural communities in which small-scale farmers produced ample surpluses for the noble and religious classes (*pages 110–11*). In other regions there were less stratified, semi-sedentary and nomadic societies in which people produced little beyond their own consumption needs. At the time of the Conquest it is probable that the indigenous population of Spanish America amounted to some 40–50 million, 60 per cent of which was found in Mexico and Peru, while Portuguese Brazil had a population of 2.5 million (*pie chart 1*). What is certain is that until around 1650 all American Indian societies suffered massive population losses – reducing the original totals by 90 per cent. These losses, once thought to be caused by Spanish brutality, are now largely attributed to the Indians' lack of resistance to European and African diseases. While the Indian population declined, the European, African and mixed populations rose sharply as a result of migration from Spain and the slave trade (*pie chart 2*). In the 18th century there was very rapid population growth among all racial groups, particularly the mixed and African populations.

THE SPANISH EMPIRE

The economic development of the Spanish Empire was concentrated in areas that had once been part of the Inca and Aztec empires in central Peru and central Mexico (*maps 1 and 2*). Here the Spanish introduced a system known as the *encomienda*, under which groups of American Indians were allotted to a Spanish overlord, or *encomendero*, to whom they supplied labour and tribute and from whom, supposedly, they received protection.

In practice, the *encomienda* system was highly exploitative and this, combined with the decline in the Indian population, led to its replacement by the *repartimiento* in Mexico and the *mita* in Peru. These were state-regulated labour systems under which the Indian communities were required to supply labour to private employers (and also to the state in Mexico) in three main activities: mining, agriculture and textiles. The mining of silver and mercury, which grew rapidly between 1550 and 1640, was of key importance: silver alone provided Spanish America with 90 per cent of its exports. The agricultural

◀ The Spanish crown claimed sovereignty over all American territory to the west of the line laid down at the Treaty of Tordesillas in 1494, while Portugal was given the territory to the east. This formed the basis of the two empires. In practice, however, Spanish wealth in South America was concentrated in Peru, while the Portuguese empire extended across the line along the Amazon and into the Mato Grosso region to the south.

1 DISTRIBUTION OF THE AMERICAN INDIAN POPULATION OF SPANISH AND PORTUGUESE AMERICA c. 1500

TOTAL POPULATION: 52,900,000

Mexico	Brazil
Central America	Peru
Caribbean	Other South America

3 ADMINISTRATIVE DIVISIONS OF SPANISH AND PORTUGUESE AMERICA 1780

Area added to New Spain (Audiencia of Cuba) 1763
LIMA Audiencia
—— Border between Portuguese and Spanish territory 1750
- - - Amended border 1778

◀ In the 18th century the structure of colonial government in Spanish America was reformed. The viceroyalty of New Granada was created in 1739 in the north of Peru, and in 1776 a fourth viceroyalty was established in the Rio de la Plata region.

sector also expanded as the Spanish set about producing commodities previously unknown to the Indians, principally wheat, cattle, sheep, wine and sugar. The production of wool and cotton textiles was concentrated in Mexico. Economic development outside Mexico and Peru was slow or even non-existent, and here the Spanish continued to use the *encomienda* system to appropriate the small surpluses of foodstuffs and cash crops, such as cochineal, which the depleted Indian populations could produce.

In the middle decades of the 17th century the decline in the number of Indians and in the international price of silver caused an economic recession in Spanish America. However, recovery began around 1670 and in the 18th century there was rapid economic growth. In Mexico and Peru this was based on the revival of the silver export industry and the expansion of agriculture and textile manufacturing. These activities used mainly wage labour.

However, the reluctance of Indians to work outside their communities led to the practice whereby Spanish employers advanced wages and credit to Indians and used the resulting debts, which the labourers could not repay, to bring them into the workforce. In the peripheral areas, expansion was driven by goldmining in Ecuador and Colombia and by the plantation production of sugar, coffee and indigo in Mexico, the Central American isthmus, Cuba, Venezuela, Colombia and Ecuador – all activities which depended on imported slave labour and external markets. These areas were integrated into the mainstream economy in the 18th century.

THE PORTUGUESE IN BRAZIL

In Brazil, which was developed much more slowly than Spanish America, the Portuguese began by bartering tools and trinkets for Indian-supplied dyewoods. However, the indigenous market for manufactures was soon saturated, and from c. 1550 the colonists turned to sugar production, the basis of the New World's first great plantation system.

The sugar industry depended entirely upon foreign markets and dominated Brazil's economic and social development until 1700. The early sugar plantations were worked by Indian labourers, most of them enslaved. However, their productivity was low because they came from cultures with little experience of settled agriculture, and their numbers were drastically reduced by exposure to European diseases, particularly during the 1550s and 1560s. Consequently, by the early 17th century the colonists had substituted imported African slaves. From around 1670 the sugar industry was checked by competition from English and French Caribbean producers, and thenceforth the main impetus to Brazilian economic growth came from the opening up of gold and diamond mines in the interior regions of Minas Gerais and Goias, which were also worked by imported slaves (*map 2*).

SPANISH AUTHORITY IN THE COLONIES

The economic and social development of the Spanish colonies did not take place in a political vacuum. In the early colonial period the Spanish crown had little authority in America. The colonists observed the legal forms, as when they founded new townships, but in effect they ruled themselves. They largely ignored their chief critics, the friars, who came to the Americas to christianize the Indians in the "spiritual conquest", and most of whom deplored the Spanish mistreatment of the indigenous population.

The Spanish crown, fearful that the *conquistadores* – the adventurers who had conquered Mexico, Central and South America – would form an autonomous and hereditary aristocracy, began from around 1550 to impose its authority on its American acquisitions. The government's main concern was to curb the colonists' virtually unlimited powers over the Indians, so it whittled away the quantities of tribute and labour extracted by the *encomenderos* and transferred numerous *encomiendas* from private to Crown jurisdiction. Furthermore, a royal bureaucracy was created to absorb the powers formerly held by the *conquistadores*. Spanish America was divided into viceroyalties (*map 3*), each subdivided into a small number of *audiencias* – substantial areas administered by a legal council – and a larger number of *corregimientos* – rural districts with urban centres governed by *corregidores*.

From around 1640 Spain's authority in the Americas weakened as important royal powers over the colonists were commuted in exchange for fiscal payments, and as the practice of selling official posts to American-born Spaniards became widespread. These posts were used to benefit their holders, and their extended family networks, rather than to enhance royal authority. Weak government led to a stagnation in Spain's revenues from the New World and a decline in the empire's capacity to defend itself. The consequences of these developments became all too apparent in the Seven Years War (1756–63), when Britain inflicted crushing defeats on the Spanish in North America (*pages 124–25*). This experience stimulated the "Bourbon Reforms", a programme of economic and political reorganization through which the Spanish crown attempted the bureaucratic reconquest of its American empire.

2 POPULATION OF SPANISH AMERICA c. 1800 (all ethnic groups)

TOTAL POPULATION: 12,600,000

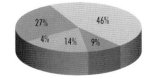

Mexico	Peru
Central America	Other
Caribbean	S. America

● SPAIN AND THE AMERICAS 1492–1550 *pages 120–21* ● INDEPENDENCE IN LATIN AMERICA AND THE CARIBBEAN 1780–1830 *pages 190–91*

THE COLONIZATION OF NORTH AMERICA AND THE CARIBBEAN 1600–1763

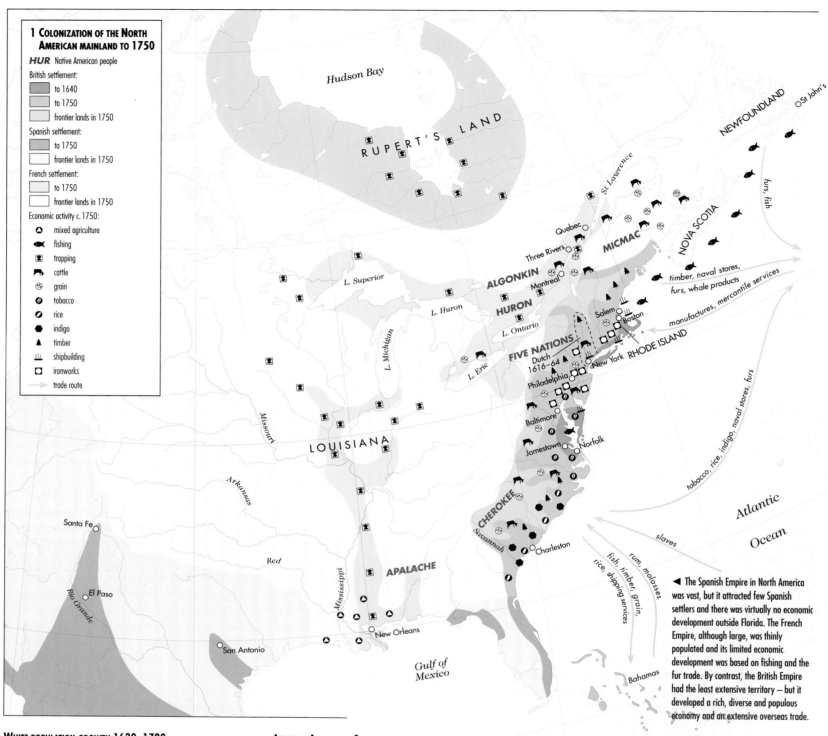

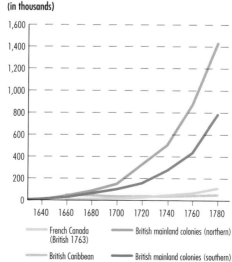

1 COLONIZATION OF THE NORTH AMERICAN MAINLAND TO 1750

HUR Native American people

British settlement:
- to 1640
- to 1750
- frontier lands in 1750

Spanish settlement:
- to 1750
- frontier lands in 1750

French settlement:
- to 1750
- frontier lands in 1750

Economic activity c. 1750:
- ◐ mixed agriculture
- ⤜ fishing
- ☒ trapping
- ⌐ cattle
- ✿ grain
- ◉ tobacco
- ⊘ rice
- ⬢ indigo
- ▲ timber
- ⊥⊥ shipbuilding
- ▢ ironworks
- → trade route

◀ The Spanish Empire in North America was vast, but it attracted few Spanish settlers and there was virtually no economic development outside Florida. The French Empire, although large, was thinly populated and its limited economic development was based on fishing and the fur trade. By contrast, the British Empire had the least extensive territory – but it developed a rich, diverse and populous economy and an extensive overseas trade.

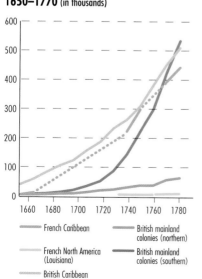

WHITE POPULATION GROWTH 1630–1780
(in thousands)

Legend:
- French Canada (British 1763)
- British Caribbean
- British mainland colonies (northern)
- British mainland colonies (southern)

AFRICAN–AMERICAN/CARIBBEAN POPULATION GROWTH 1650–1770 (in thousands)

Legend:
- French Caribbean
- French North America (Louisiana)
- British Caribbean
- British mainland colonies (northern)
- British mainland colonies (southern)

Following the discovery of the New World by European explorers at the end of the 15th century and beginning of the 16th, Spain and Portugal had laid claim to all of the Americas. However, this Iberian monopoly was not accepted by the other European powers and in the second half of the 16th century it was pierced by hundreds of voyages dispatched from northern Europe. Ships were sent to trade or pillage and even, in a few instances, to found colonies, although none of the latter survived. From these beginnings Britain, France and Holland founded empires in America and the Caribbean in the 17th century. British colonies were set up in two main waves: from 1607 to 1634, when settlements were established in Virginia, Maryland, New England and the eastern Caribbean; and from 1655 to 1680, when Jamaica was seized from the Spanish, the Carolinas and Pennsylvania were founded and New York was taken from the Dutch (*map 1*).

◀ Unlike the white population of the British mainland colonies, the population of French Canada grew slowly because its economy was based on furs and fish, which required much less labour than agriculture.

In the British mainland colonies the slave population increased rapidly, but in the Caribbean harsh treatment and tropical diseases prevented its natural growth and encouraged the slave trade with Africa.

In the early 17th century the French established fishing and fur-trading colonies in Canada at New France and Acadia (Nova Scotia) and settler colonies in the Caribbean and the western portion of Hispaniola (*map 2*). The Dutch established trading factories – as on Curaçao – rather than colonies, but they founded one major colony, Dutch Guiana, taken from the British in 1665 (*pages 122–23*).

THE NORTHERN COLONIES

Outside the southeast and southwest regions, the indigenous people of North America (*pages 108–9*) lived mainly in semi-sedentary or nomadic societies, and the North American colonists never seriously attempted to live from their labour as the Spanish colonists did in parts of South America. Some Native Americans were enslaved – as in South Carolina – but the main contacts between Europeans and Native Americans were through the fur trade, where furs were supplied by native trappers, and through warfare. In general the Native Americans responded to the arrival and settlement of the Europeans on the east coast by moving west, leaving depopulated regions to be settled by migrants from Europe. These migrants were mostly people seeking economic betterment or freedom from religious persecution. Taking advantage of the region's rich natural resources, they created prosperous farming communities specializing in the production of grain, livestock and timber, and benefiting from the relatively disease-free environment of the region.

THE PLANTATION COLONIES

Conditions in the plantation colonies of the southern mainland and the Caribbean were very different. Here disease was rife, discouraging free migration and killing many of those who did take the risks of settlement – mainly white indentured servants who had little choice over their destinations and provided several years of unpaid labour in exchange for their passage and a plot of land at the end of their service. Some 200,000 of these servants migrated to British plantation colonies, fewer to the French Caribbean, and they were employed in the production of tobacco and other plantation staples for export to Europe. From around 1650, however, there was a fundamental change in the labour system of the plantation colonies. The shift from tobacco to sugar caused an explosive increase in the demand for labour which could not be met by Britain and France. This led to the use of imported African slaves, first in the Caribbean and then, from 1680, in Virginia and Maryland (*pages 126–27*).

CONTINUED EXPANSION

In the 18th century the populations of all the British mainland colonies had fast natural rates of growth (*graphs*). In the northern colonies this pushed agricultural settlement into the interior. In the southern colonies the coastal regions intensified the slave-plantation production of tobacco, to which was added rice and indigo in South Carolina and Georgia. Settlement also spread into the southern "backcountries" – temperate mixed farming zones – whose economic and social development was akin to that of the northern colonies. The French mainland colonies in Canada and Louisiana achieved a massive territorial expansion to 1763, but their demographic and economic development was very slow. In the Caribbean, both the British and French slave-plantation economies grew rapidly.

COLONIAL GOVERNMENT

Neither Britain nor France exercised much political influence over their colonies until the 1660s, when France established an authoritarian system with military governors and powerful colonial officials accountable to the king. Britain also created royal bureaucracies but their power was shared with elected legislative assemblies. Both governments subjected imperial trade to strict mercantilist controls, requiring the colonies to trade exclusively with their mother countries. The benefits reaped by Britain and France were enormous because colonial trade was the fastest growing sector of international commerce in the period.

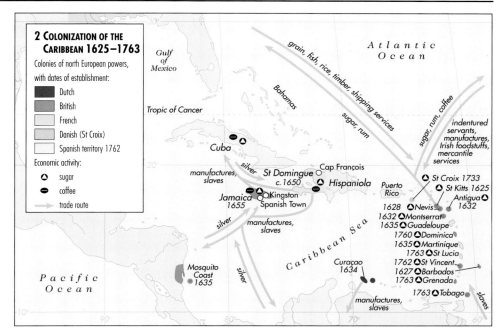

Colonial trade had two dimensions: the export of slave-produced staples such as tobacco and sugar from the plantation colonies to the metropolis, and a reverse stream of manufactured goods, services, and labour from Europe and Africa to the colonies. The British northern colonies exported relatively little to Britain, but they imported vast quantities of manufactured goods from Britain, covering their trade deficits by exporting foodstuffs, raw materials and shipping services to the Caribbean and southern Europe.

The strengthening of government in North America also had diplomatic consequences. Between 1689 and 1763 Britain and France fought four major wars – conflicts that became increasingly focused on colonial disputes. Britain got the better of these wars, especially the last, the Seven Years War of 1756 to 1763 (*map 3*). However, post-war British attempts to make their colonists share the burden of the huge military costs of these endeavours also precipitated the American Revolution (*pages 164–65*) and, with that, the collapse of British imperial power on the mainland.

▲ During the 17th century the British and the French made significant inroads into Spanish territory in the Caribbean, establishing colonies in Jamaica and St Domingue as well as on the islands of the Lesser Antilles. The economies of these colonies were based heavily on sugar plantations worked by African slaves.

▼ The Seven Years War, in which Britain inflicted a number of crushing military and naval defeats on France and Spain, brought an end to the French Empire in mainland America. Under the Treaty of Paris in 1763, Britain took Canada and all territory east of the Mississippi, while Spain acquired the vast territory of French Louisiana.

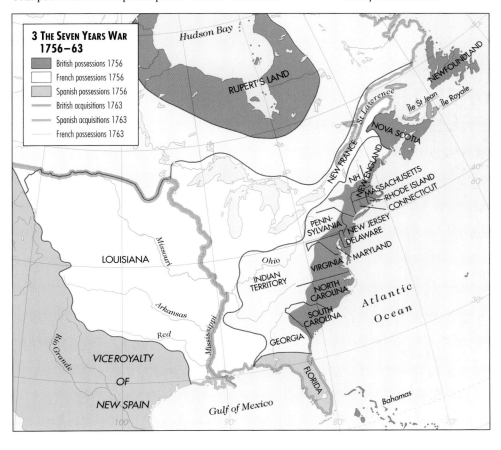

◗ EUROPEAN DISCOVERY OF THE WORLD 1450–1600 *pages 116–17* ◗ AMERICAN REVOLUTION 1775–83 *pages 164–65* ◗ CANADA 1763–1914 *pages 188–89*

SLAVE ECONOMIES OF THE WESTERN HEMISPHERE 1500–1880

▼ Between around 1500 and 1870 at least 9.5 million African slaves were forcibly transported to the European empires in the Americas. It has been estimated that over two million more died, mainly from disease, while crossing the Atlantic on grossly overcrowded and insanitary ships. Most were shipped to the Caribbean and Brazil, where high mortality rates among the slave populations meant that new slaves were constantly being imported to replenish the labour force. Fewer slaves were imported to British North America because better conditions there allowed slave populations to increase naturally.

Five major European empires were established in the Americas between the 16th and 19th centuries (*map 1*). In the economies of four of these empires – the Portuguese, Dutch, British and French – African slavery was the most important form of labour. In the fifth – the Spanish – African slaves played a significant and, in the 18th century, an increasing role. This occurred alongside the exploitation of the indigenous population.

Slavery was an important element of European imperialism in the Americas because of the scarcity of labour in relation to the region's abundant natural resources. Exploitation of the indigenous population was a strategy used in Spanish Mexico and Peru, where the sedentary and economically advanced American Indian societies provided labour and tribute payments to the Spanish as they had to their former Aztec and Inca overlords. However, the semi-sedentary and nomadic Native American peoples who occupied much of Spanish North America and overwhelmingly predominated in the other empires, could not satisfy the white colonists' demands for labour and commodities.

Attempts to enslave these peoples proved unsuccessful in the long run, partly because they exhibited fearful mortality rates in captivity and partly because colonial governments generally opposed such enslavement.

A second source of labour was the large number of European migrants to the more temperate zones, such as the mainland colonies of British America, but white migrants preferred to become independent farmers rather than wage labourers. The shortage of such labour was even more acute in the tropical colonies, where the hot and humid climate and the constant threat of disease discouraged free migrants from settling.

The colonists therefore turned to a third source of labour: slaves from Africa. Since the late 15th century African slaves had been used on plantations on European-colonized Atlantic islands such as Madeira and São Tomé. They proved to have two great advantages for the European colonists. First, they and their offspring, who were treated as chattels, could be coerced into almost any form of work; second, their supply was infinitely more elastic than the availability of labour from indigenous or European sources.

THE GROWTH OF THE SLAVE ECONOMIES

The first major slave economies were created in the Spanish and Portuguese empires, which imported about 500,000 slaves between around 1500 and 1650. The Portuguese concentrated their slaves in the sugar plantations of coastal Brazil, while the Spanish used theirs in a number of regional economies, the most important of which were the sugar and wine estates of the semi-tropical coastal lowlands of Peru and Mexico and the silver mines of northern Mexico.

The period between 1650 and 1810 saw a massive expansion of slavery in all the major European empires in the Americas (*map 2*). The Portuguese expanded their sugar plantation system in Brazil and, after 1700, imported hundreds of thousands of slaves to work the diamond and gold mines in the interior of the country in the Minas Gerais and Goias regions. The vast majority of the Spanish-owned slaves were employed not in Mexico and Peru but on the sugar and cocoa plantations of Cuba and Venezuela and in the gold-mines of Colombia. These formerly peripheral regions of the Spanish Empire became increasingly important, entering the mainstream of the Spanish-American economy in the 18th century. The British, Dutch and French poured slaves into their Caribbean and Guyanese colonies, where they produced sugar, coffee and other plantation staples. On the northern mainland the British and French colonists imported smaller numbers of slaves into the tobacco-producing colonies of Virginia and Maryland, the rice and indigo economies of South Carolina and Georgia and the sugar colony of Louisiana.

THE DEMOGRAPHICS OF SLAVERY

The conditions of life for slaves in the Americas, and in particular their relative ability to produce new generations of slaves, were determined by the labour requirements of the plantation crops that they cultivated and the disease environments in which they lived. Most were employed on large-scale sugar and coffee plantations in the tropical and semi-tropical zones, where their masters underfed and overworked them, and where they were ravaged by diseases such as dysentery and yellow fever. These slave populations experienced high mortality and low fertility rates, which meant that the expansion of labour forces depended on a swelling stream of human imports from Africa, from where over six million slaves were imported between c. 1650 and c. 1800 (*map 1*). The extent of the natural decline of slave populations can be gauged from the example of the British Caribbean colonies, which imported some 1.5 million slaves during this period, but by 1800 had an African-Caribbean population of just over 500,000. Natural increase was experienced by only a small number

Map and charts

1 THE TRANSATLANTIC SLAVE TRADE

European territories in the Americas c. 1770:

- British
- French
- Dutch
- Portuguese
- Spanish

560 Number of slaves imported, in thousands

- 1500–1600
- 1601–1700
- 1701–1810
- 1811–1870

British North America and United States: 51, 384

Spanish America including Cuba: 75, 293, 579, 606

British West Indies: 264, 1,401

Other West Indies: 44, 484

French West Indies: 156, 1,348, 96

Brazil: 50, 560, 1,891, 1,145, 3,596

Atlantic Ocean

Gulf of Mexico

Tropic of Cancer

Caribbean Sea

Equator

Tropic of Capricorn

TOTAL SLAVE IMPORTS, BY REGION 1500–1870 (in thousands)

- USA and British North America: 453
- British West Indies: 1,665
- French West Indies: 1,600
- Other West Indies: 528
- Spanish Empire: 1,553
- Brazil: 3,596

Southern United States
c. 1860
3,724

Southern United States
c. 1800
848

Mexico and
Spanish Central America

2 SLAVE ECONOMIES OF THE WESTERN HEMISPHERE
International boundary c.1830
Economic activity in which slaves employed:
cotton
rice and indigo
tobacco
coffee
sugar
mixed agriculture
mining
○ Slave population (in thousands) c.1800
○ Slave population (in thousands) c.1860
1731 Site of slave revolt with date

1712
Baltimore
1800
1831 Norfolk
1739
1822
Charleston
Savannah
Pensacola
1811
New Orleans
11 French
Louisiana
St Augustine

*Atlantic
Ocean*

MEXICO

*Gulf of
Mexico*

Bahamas

Spanish
West Indies

170.5

Havana
1503
1726 Cuba
1820
1805
1812
Cap
Francois
Hispaniola
17th C, 1734, 1760, St Domingue
1795, 1831 1791
Jamaica Kingston
1522
St Domingo
Puerto
Rico
1527
Danish
Antilles
20
Leeward Islands
Guadeloupe

French
West Indies
643.4

British
West Indies
523.9

Windward
Islands
1833 Martinique

Barbados
1816

Mexico City
Various
Veracruz
1725–35

Belize City

1548

Guatemala
City
UNITED PROVINCES
OF CENTRAL AMERICA

Caribbean Sea

Cartagena
1530
1550
1619
1532
1795
1552
1730s
Caracas

90

VENEZUELA

100

DUTCH
GUIANA
BRITISH 1823 FRENCH
GUIANA GUIANA
c. 1770 1731
12

PANAMA

1598
1820s
Bogotá
1540s–50s
COLOMBIA
1820s

Pacific Ocean

Equator

Northern
Brazil
347

Recife

NORTHERN
BRAZIL

ECUADOR
50
Ecuador
and
Colombia

1848 PERU

1578

EMPIRE OF
BRAZIL

1807–35
Salvador

GOIAS
MINAS
GERAIS
INTERIOR
BRAZIL
Interior
Brazil
238

MATO
GROSSO

1820
1850s

BOLIVIA

Rio de Janeiro

Southern
Brazil
142
SOUTHERN
BRAZIL

PARAGUAY

30

ARGENTINA

URUGUAY

Buenos Aires

10
CHILE

PATAGONIA

of the slave populations – for example, those in the tobacco colonies of Virginia and Maryland – who benefited from adequate food supplies, an environment less conducive to disease than was to be found in the tropical colonies, and a less demanding labour regime.

ABOLITION AND THE SLAVE TRADE
The period from 1810 to 1880 represented the final era of slavery in the Americas. Although a number of countries abolished their transatlantic slave trades (Britain in 1807 and the United States in 1810, for example), American slavery continued to expand. The plantations of Brazil and of the Spanish and French colonies in the Caribbean imported nearly two million slaves between 1810 and 1860. In Cuba the slave population more than doubled in these years, while in the same period the slave population of the southern United States, mainly engaged in cotton production, increased by natural means from 0.9 to 3.7 million.

The abolition of the institution of slavery, as opposed to that of the slave trade, was a long process which extended from the 1820s up to the 1880s. The number of slave revolts increased in the late 18th and early 19th centuries (*map 2*), but with the exception of the revolt in 1791 in French St Domingue (which was to become the independent state of Haiti in 1804), none succeeded in achieving local abolition. Instead, the end of slavery was brought about partly by the economic decline of the slave economies but largely by political events – in particular, war and revolution. Several of the newly independent Spanish-American republics outlawed slavery between 1824 and 1829; slavery in the British West Indies was abolished by a reforming British government in 1834; and in the United States slavery was ended in 1865 by the victory of the Union states over the Confederate states in the American Civil War.

◄ In the 17th and 18th centuries the largest slave populations were in Brazil, the Caribbean and the southern British mainland colonies (part of the United States from 1783). Slave populations in the vast area of Spanish mainland America were quite modest by comparison. The brutal conditions of slavery throughout the Americas caused frequent slave revolts which were suppressed with great ferocity.

● THE AMERICAN CIVIL WAR 1861–65 *pages 184–85* ● LATIN AMERICA AND THE CARIBBEAN POST-INDEPENDENCE 1830–1914 *pages 192–93*

THE GROWTH OF THE ATLANTIC ECONOMIES 1620–1775

After more than a century of economic growth, 1620 saw the beginning of a period of economic crisis and stagnation in many parts of Europe. The economic decline of Spain and Italy was accompanied by the migration of skilled labour and capital to the north. English and Dutch merchants broke into Mediterranean trade during the Eleven Years Truce with Spain, from 1609 to 1621 (*pages 156–57*). The Dutch retained and expanded their share of Baltic commerce to achieve a near-monopoly of the region's trade by 1650, while English trade with the Baltic grew significantly from the 1670s. This coincided with the rise of Amsterdam and London as important world trading centres (*pages 132–33*), and with a permanent shift in Europe's economic centre of gravity from the Mediterranean to the North Sea/Baltic zone – a shift reflected in population trends (*graph 1* and *map 1*).

THE RISE OF HOLLAND

The 17th century, often described as Holland's "golden age", was also the period of England's "apprenticeship" to the Dutch Republic. In the wake of the Dutch revolt against Spain in 1572 and also after the revocation of the Edict of Nantes by the French crown in 1685 (*pages 154–55*), Protestant refugees were welcomed in the towns of southern England and the northern Netherlands. Bringing with them their expertise in new industries and industrial processes, including brewing, papermaking, the manufacture of glass and ceramics, and silk weaving, they made a significant impact on the English economy. In an increasingly scientific age, the Dutch capacity for visualization was highly valued, showing itself in a range of skills associated with the "art of describing": mapmaking, engraving, drawing, painting and the making of scientific instruments. Dutch engineers were active in promoting drainage and embankment works in countries throughout Europe (*map 2*).

By the early 18th century an international division of labour was emerging, shaped as much by government policy as by market forces. In France and England especially, new forms of economic nationalism had emerged during the

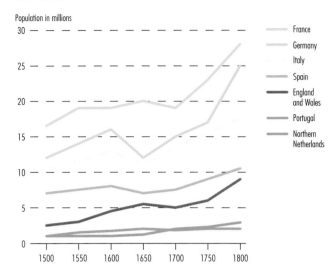

Population in millions

Legend: France, Germany, Italy, Spain, England and Wales, Portugal, Northern Netherlands

▲ While the populations of Spain and the Italian and German states declined sharply during the period 1600–50, those of England and the Dutch Republic continued to grow. From around 1650 populations in southern Europe and Germany began to increase, while overall numbers in England and the Netherlands stagnated.

1660s and 1670s, embodied in policies designed to promote overseas and colonial trade, and industrial diversification, at the expense of competitors. Anglo-Dutch and Anglo-French rivalry was sharpened by the imposition of protectionist import duties and restrictions on the export of raw materials, and above all by the English Navigation Acts of 1651 and 1660 which sought to wrest the colonial carrying trade from the Dutch. By the early 1670s the Dutch economic miracle was over, and English merchants would soon displace the Dutch as the dynamic force behind European and world trade (*graph 2*).

ANGLO-DUTCH COMPETITION

Anglo-Dutch competition was evident in many fields, including the North Sea herring fisheries, woollen textile manufacture, textile dyeing and finishing, and by the 18th century, sugar refining, tobacco processing and linen bleaching. These activities all involved processing and as such were fields in which the Dutch excelled by virtue of their success in controlling the markets for finished products. English industry, on the other hand, was more deeply embedded in the domestic manufacturing economy, and relied on the labour of rural households.

Trade rivalry and industrial competition created an international climate in which warfare became endemic, from the Anglo-Dutch wars of 1652, 1665–67 and 1672–74, to the intermittent Anglo-French struggles of 1689–1815. Military expenditure by the British state multiplied fivefold between the 1690s and the Napoleonic Wars, and provided a huge stimulus to the industrial and construction sectors. Shipbuilding, the metallurgical and arms industries, civil engineering and the building and supply of naval dockyards stimulated employment, investment and innovation through increased public spending.

As the Scottish political economist Adam Smith realized, the Anglo-French wars of the 18th century represented a struggle for economic supremacy as much as for political power in Europe, India and North America. France was a late starter in the race for colonial trade and territory, but made remarkable progress during the middle decades of the 18th century, especially in the West Indies (*graph 2*). Nevertheless, British domination of the Atlantic economy was secure by the end of the Seven Years War (1756–63). On the eve of the American War of Independence (1775–83) British imports from the West Indies and the American mainland colonies far exceeded those from either the North Sea or Mediterranean zones, and the lion's share of British manufactured exports went across the Atlantic.

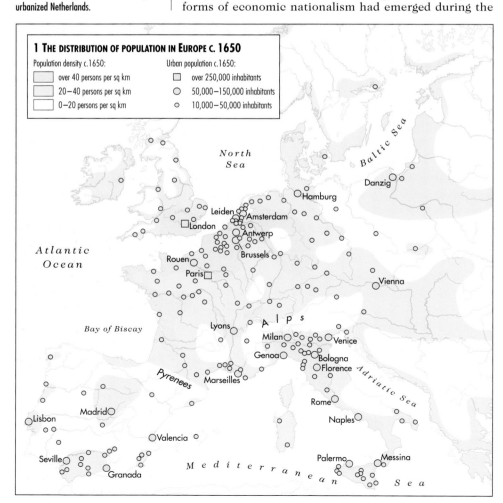

1 THE DISTRIBUTION OF POPULATION IN EUROPE c. 1650

Population density c.1650:
- over 40 persons per sq km
- 20–40 persons per sq km
- 0–20 persons per sq km

Urban population c.1650:
- over 250,000 inhabitants
- 50,000–150,000 inhabitants
- 10,000–50,000 inhabitants

North Sea · Baltic Sea · Atlantic Ocean · Bay of Biscay · Mediterranean Sea · Adriatic Sea · Alps · Pyrenees

Hamburg · Danzig · Leiden · Amsterdam · London · Antwerp · Brussels · Rouen · Paris · Vienna · Lyons · Milan · Venice · Genoa · Bologna · Florence · Marseilles · Rome · Naples · Lisbon · Madrid · Valencia · Seville · Granada · Palermo · Messina

In the last resort, however, the European economies were dependent on their natural resources and the legacy of political history. This was especially true in the case of agricultural and primary production, and the extent to which nations and regions were able to commercialize these sectors. Whereas the Dutch chose to develop a compact and specialized agricultural sector and to depend on large-scale food imports, the English chose agricultural self-sufficiency, protectionism and, after 1689, the manipulation of food prices in the interests of producers by means of subsidized exports. French peasant agriculture, on the other hand, constrained by labour-intensive farming methods and a host of geographical, political and institutional limitations, was strongly resistant to commercialization. Above all, it was on the basis of plentiful energy sources that Britain was able to surge forward towards industrialization. The availability of coal released British producers from dependence on organic materials such as timber and charcoal at a time when Dutch peat supplies were becoming exhausted. In short, the Dutch Republic faced the limitations of a city-state underpinned by merchant capital – just as Britain was emerging as a strong nation-state, with a developing industrial base.

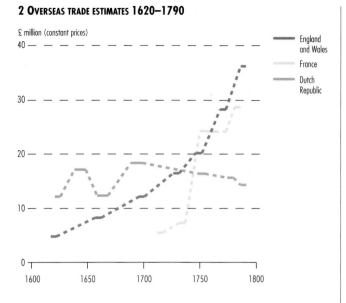

2 OVERSEAS TRADE ESTIMATES 1620–1790

£ million (constant prices)

- England and Wales
- France
- Dutch Republic

◀ In the two centuries before 1800 English overseas trade expanded steadily while that of the Dutch Republic stagnated. France's overseas trade accelerated more rapidly than England's in the 18th century, showing a fivefold increase during the period from 1716 to 1788 – double the increase registered for England at this time.

◀ In the period 1650–1750 there were several highly commercialized centres of production in western Europe, but rural industry, particularly the processing of textile fibres, was to be found throughout Britain and northwest Europe. Woollen cloth, linens, fustians and silk were the main textiles produced. Coalmining was concentrated in England and Scotland, from where coal was exported to nearby Europe.

2 THE ATLANTIC ECONOMIES 1650–1750

Areas of textile production:
- woollens
- linen
- fustian
- silk
- Area of metal production
- Manufacture of iron wares
- Coalmining
- Coal trade
- Area drained by Dutch engineers in the 17th century
- Main herring fishing grounds
- Principal port

▲ During Holland's "golden age" in the 17th century, Dutch merchants – such as the one on the right in this painting – were to be found throughout the world, from the Baltic to the Americas and Asia. However, from the 1650s their dominant role in European and world trade was increasingly threatened by the English.

◀ EUROPE 1350–1500 *pages 106–7* ◀ THE INDUSTRIAL REVOLUTION IN BRITAIN 1750–1850 *pages 168–69*

THE RISE OF EUROPEAN COMMERCIAL EMPIRES 1600–1800

T he geographical discoveries by Europeans in the late 15th and early 16th centuries gave Europe access to many new sources of wealth: land, precious metals and new products such as coffee and tobacco. However, in the rush to exploit all these, the rivalry between the European states produced a world divided into commercial empires. In the short term the discoveries probably acted as a drain on European commercial and financial resources, particularly those of Spain and Portugal. The profits from the silver mines of Spanish America and the Portuguese spice trade were substantial for those directly involved, but while the outflow of precious metals from the Americas may have quickened economic activity in Europe, it also intensified the inflationary pressures that were already present.

Overall, the growth of transoceanic trade (*map 1*) made little impact on the European economy before the 1550s, and it has been suggested that it was not until the late 17th century that commercial and industrial profits from European trade with Asia and the Americas became visible and significant, initiating a commercial revolution. By this time the benefits resulting from Iberian overseas trade and investment had become more widely diffused across Europe, accruing principally to the Dutch Republic, followed closely by England and, later, France.

▼ In the 17th and 18th centuries the countries of northwest Europe were at the centre of an expanding world economy, often able to trade on terms that were heavily in their favour. In many of the colonized parts of the Americas and Asia the production of a narrow range of primary products for export markets was encouraged, thus planting the seeds of future economic dependency and backwardness.

NEW COMMERCIAL ORGANIZATIONS

Whereas Spain and Portugal relied on the formation of government agencies to promote colonial and commercial enterprise, the newer colonial states adapted existing forms of corporate organization to serve new purposes. In this respect, the English and Dutch East India Companies (formed in 1600 and 1602 respectively) can be seen as forerunners of the modern multinational corporations. Owned by shareholders, managed by boards of directors and employing accountants and other salaried workers, these independent companies wielded great political power at home and abroad. Their efficiency and the impact of their monopoly powers have been questioned, but they undoubtedly played an important role in the expansion and integration of the global economy.

Trade in the Far East was enmeshed with politics and diplomacy, and required powerful trading bodies to act on behalf of states. However, this was not the case in the colonies of North America and the Caribbean where, with the exception of the Dutch West India Company (1621–1791), trade was conducted mainly by private, unincorporated merchants. Such merchants operated through social networks that were formed on the basis of religious, family and other personal ties. Before 1700 the bulk of

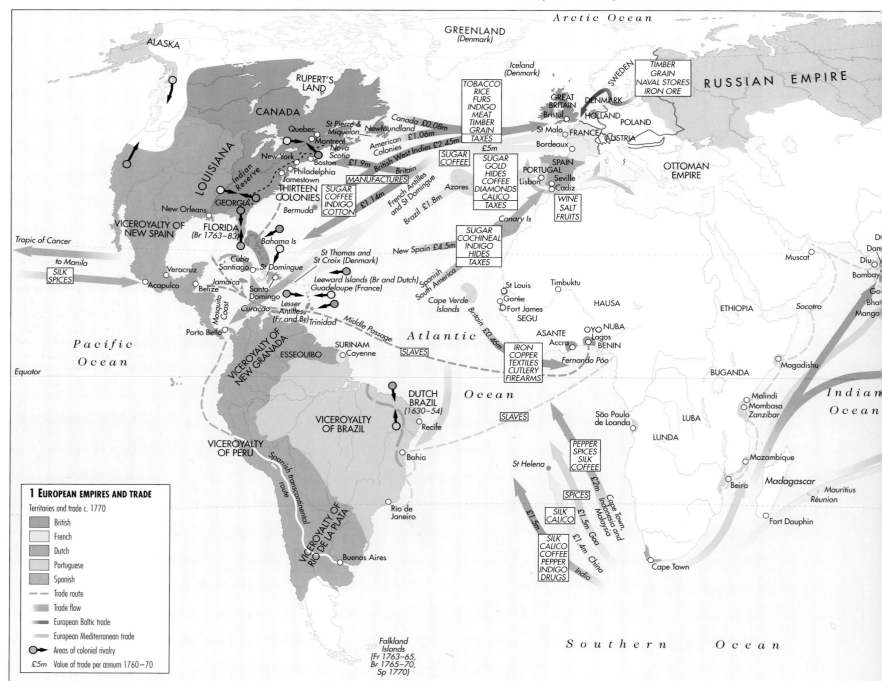

transatlantic commerce was conducted by British merchants operating through colonial agents, but local merchants increased their share of trade from the early years of the 18th century, especially in the northern colonies. Although institutionalized monopoly powers were not necessary for the development of trade with North America, the English Navigation Act of 1651 (prohibiting imports to England from outside Europe unless carried in English ships) effectively established a national monopoly which played an important role in undermining Dutch competition during the following century.

As the world economy expanded the Americas, Europe and the Far East became inextricably linked through trade, shipping and bullion flows. Silver bullion from the mines of Central and South America enabled the northern European economies to buy goods from Asia and the Baltic (*map 2*). Imports from the Baltic region, such as timber for shipbuilding, iron ore and naval stores, contributed to the further expansion of long-distance trade, while the flow of Asian imports – silk, calico, spices and drugs – brought consumer goods to Europe and North America. It was not until the second half of the 18th century that the amount of silver bullion exported to Asia fell sharply, compensated for by rising exports of British manufactured goods.

THE STIMULUS TO COMMERCIAL EXPANSION

A major stimulus behind the commercial revolution of the 17th century was an increase in consumer demand. In spite of demographic stagnation in Europe, towns and cities continued to expand (*pages 132–33*), and as they did so new patterns of consumption and social behaviour evolved. Contributing to the diversification of consumption habits was the arrival of new and exotic commodities such as spices, tobacco, tea, coffee, sugar, tropical fruit, dyestuffs and Asian textiles. Such commodities resulted in, for example, the development of coffee houses, more fashionable clothing and household furnishings, and new domestic rituals such as tea-drinking. Maize and potatoes helped to feed Europe's growing population in the 18th century, without competing with home-produced foodstuffs. New industries such as sugar refining, tobacco processing, cotton manufacture and textile printing developed as a result of long-distance trade and colonial development.

However, despite the benefits of trade with Asia and the Americas, economic growth in Europe depended mainly on trade within Europe itself, and on improvements in domestic agriculture and manufacturing. Long-distance trade was expensive, not always profitable, and did not contribute a great deal to capital formation within those countries which were at the core of the world economy. Competition between the European states – and the consequent need to defend, administer and control colonial territories – involved increased public expenditure and more complex government administration. Furthermore, the growing European demand for imported products resulted in balance of payments problems for the countries involved, to which there were two obvious solutions: to increase the volume of re-exported goods, and to provide shipping services. In this sense, the commercial revolution generated its own momentum.

GOVERNMENT INVOLVEMENT IN COMMERCE

The countries that gained most from this economic expansion were nation-states such as France and England, which were capable of developing the machinery of strong central government alongside aggressive mercantilist policies. Mercantilism aimed to increase employment through the encouragement of overseas trade, especially the import of essential raw materials, while protecting home industry by the imposition of high import duties. In comparison with the English and French variants, Dutch mercantilism remained weak and incidental, particularly in the colonial field. The decentralized federal structure of the United Provinces, together with the deeply entrenched interests of its merchants overseas, inhibited the kind of aggressive unity that was partly behind the increasing power of its larger neighbours – France and England.

▲ Coffee houses were representative of the new social habits that evolved in Europe in the 17th and 18th centuries as a result of the import from Asia and the Americas of commodities then regarded as exotic.

▼ Silver from the mines of Central and South America reached Europe via Spain and Portugal, where it entered the arteries of world trade. The Dutch, who were the dominant commercial power in Europe, operated as Europe's bankers in circulating coin and bullion, using it to purchase goods from three principal areas: the Baltic, the Middle East and East Asia.

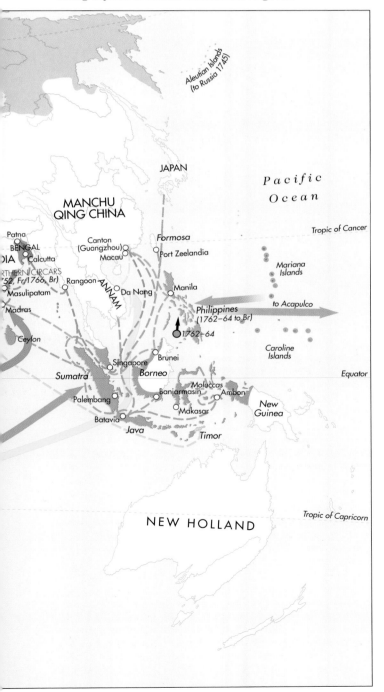

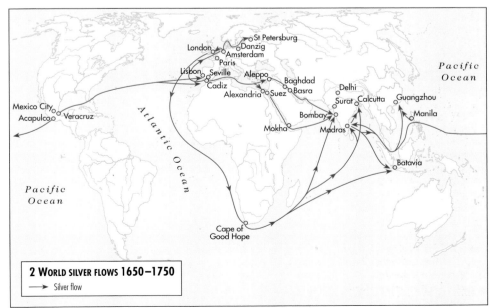

2 WORLD SILVER FLOWS 1650–1750
→ Silver flow

● THE EUROPEAN DISCOVERY OF THE WORLD 1450–1600 *pages 116–17* ● WORLD TRADE AND EMPIRES 1870–1914 *pages 208–9*

EUROPEAN URBANIZATION
1500–1800

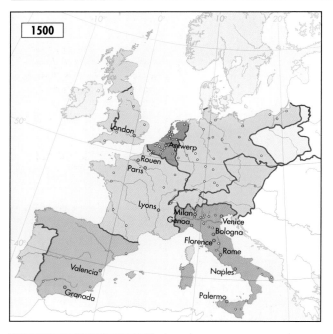

1500

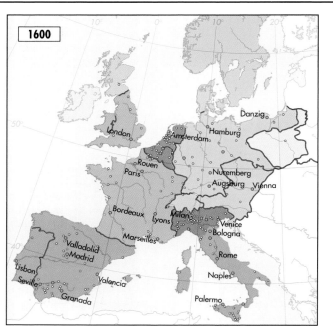

1600

◀ The process of urbanization in Europe involved three overlapping phases. In the first of these, from 1500 to around 1650, there was general growth of towns and cities of all sizes. In the second phase, between 1650 and 1750, a few large cities – most notably London, Paris and Amsterdam – expanded rapidly, while in the third phase there was an increase in the size and number of smaller cities and a relative levelling off in the growth of larger cities. In the 16th century the most urbanized regions in Europe – defined by the percentage of the total population resident in towns and cities – were the northern and southern Netherlands, and Italy. From the early 17th century, however, urban growth subsided in the last two regions while cities in the northern Netherlands expanded rapidly, in common with those of England and Scotland. By comparison, only moderate urbanization took place in France.

1–3 EUROPEAN URBANIZATION 1500–1700

Percentage of population living in towns, by region:

0–1%	15–20%
1–5%	20–25%
5–10%	25–30%
10–15%	over 30%

Town with population of:
- 8,000–40,000
- 40,000–200,000
- 200,000–400,000
- over 400,000

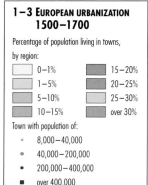

▲ In the mid-18th century the Monument – a column erected to commemorate the Great Fire of London of 1666 – was surrounded by spacious brick and stone buildings that were a great improvement on the wooden structures that had stood in their place before the Fire. There were, however, many features of London that continued to pose a threat to the health and safety of its citizens, including the streets that were often rutted dirt tracks strewn with mounds of rubbish. The standard of sanitation was very poor and was to be the cause of many outbreaks of cholera and typhus throughout the 18th and 19th centuries.

By the early 16th century a European-centred world economy was emerging, characterized not only by the rise of transoceanic trade but also by new and distinctive patterns of urban growth in Europe itself. Between 1500 and 1800 the towns and cities of Europe came to form a single urban system, involving the integration of regional trading networks and the commercialization of predominantly rural economies.

In 1500 the most urbanized regions in Europe were Italy and the Netherlands, but from the early 17th century the potential for urban growth began to move steadily northwards, with the northern Netherlands becoming the most urbanized area while rates of urban growth in Italy and the southern Netherlands subsided (maps 1–4). The Dutch Republic (the northern Netherlands) approached a ceiling in the mid-17th century because in the preceding century there had been no increase in the number of smaller centres from which cities could develop. England, by contrast, contained hundreds of market towns and industrial villages capable of expansion. By the early 19th century the rate of urban growth in Britain had reached that attained by the Dutch a century earlier, but at a much higher level of population. Between 1680 and 1820 the population of England and Wales grew by 133 per cent, while that of the Dutch Republic increased by only 8 per cent. In both countries, however, a single dominating commercial centre had emerged by 1700.

THE GROWTH OF LONDON AND AMSTERDAM

London's meteoric growth (map 5) overshadowed that of all its rivals, including Paris (graph). In 1600 about 5 per cent of the English population lived in London; by 1700 this proportion had reached 10 per cent, much higher than in other European capital cities apart from Amsterdam, which contained 8 per cent of the Dutch population. Paris, by comparison, contained only 2.5 per cent of the French people. The exceptional position of London may account for the rapid development of the English economy in the late 17th and 18th centuries, at a time when London was absorbing half the natural increase of the entire population.

This rapid expansion led to problems of overcrowding and insanitary conditions, bringing disease and high death rates. It was therefore only through substantial migration from the countryside that London and other large cities could continue to grow. A more healthy environment for Londoners only began to evolve with the replacement of timber by brick as a building material, and the introduction of building regulations after the Great Fire of London in 1666. In Amsterdam, efforts to create a more carefully planned city intensified after 1613, when construction of the spacious outer girdle of canals began.

1700

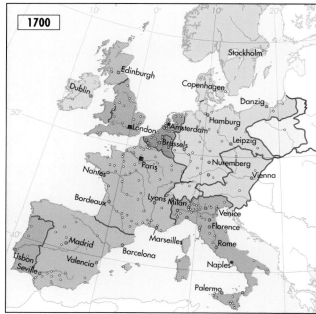

THE CHANGING ROLE OF CITIES

From the 14th to the 19th centuries the European economy was dominated by a sequence of leading mercantile cities: Venice, followed by Antwerp, Genoa, Amsterdam, and finally London. However, these cities were gradually overtaken by nation states in the deployment of commercial wealth, capital and military power. In Germany towns and cities lost their autonomy as princes absorbed them into petty feudal states, while in Italy the towns themselves became city states. The Dutch Republic, forged in the struggle against Spanish centralization in the late 16th century, emerged as something of a hybrid, a federation of city states dominated by Amsterdam as first among equals. As Europe's commercial and financial centre of gravity shifted from Amsterdam to London in the early 18th century, a strong territorial state and an integrated national economy provided the resources for a new type of commercial metropolis, the modern "world city".

In the advanced pre-industrial economies of Europe, dominant cities acted as centres of innovation in many fields, especially in the luxury trades, textile finishing, scientific instrument making, printing, and the fine and decorative arts. Since the 12th century, when universities had begun to take over the educational role of the monasteries, European cities had played a key role in the dissemination of knowledge. To their traditional educational function was added, from the later 17th century, a growing

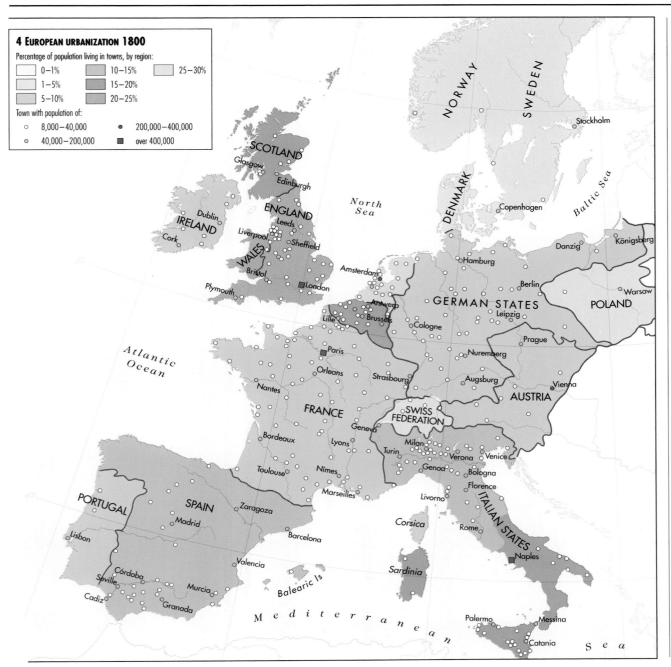

4 EUROPEAN URBANIZATION 1800

Percentage of population living in towns, by region:

- 0–1%
- 1–5%
- 5–10%
- 10–15%
- 15–20%
- 20–25%
- 25–30%

Town with population of:
- ○ 8,000–40,000
- ● 200,000–400,000
- ○ 40,000–200,000
- ■ over 400,000

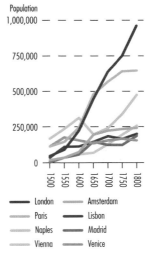

◄ In the period 1750–1850 the majority of large cities grew at much the same rate as the population as a whole, while smaller centres experienced a much higher rate of growth. The notable exception to this rule was London, whose meteoric growth continued unabated.

THE GROWTH OF EUROPEAN CITIES 1500–1800

Population

— London
— Paris
— Naples
— Vienna
— Amsterdam
— Lisbon
— Madrid
— Venice

▼ The population of London expanded from about 120,000 in 1550 to 575,000 by 1700. This latter figure represented 10 per cent of the English population, a uniquely high proportion in comparison with other European capital cities at the time.

public sphere of political debate, scientific discourse, and literary and aesthetic criticism. Newspapers first made their appearance in London in the 1620s, and by the 1690s they were carrying regular advertisements for a wide range of goods and commercial ventures, including books, medicines, lotteries, real estate and auction sales. Amsterdam led the way in the circulation and analysis of commercial information, as informal business correspondence was transformed into printed lists of commodity prices from 1613 onwards.

NEW URBAN CENTRES

As population levels rose in Europe after 1750 a new pattern of urban growth began to unfold. Expansion was no longer confined to the larger cities; indeed, it was the growth of small cities and the emergence of new urban centres which lay behind an overall increase in the pace of urbanization. There are two possible explanations for this, both arising from the overall growth in population. First, there was an increased demand for food, which in turn stimulated the rural sector and the expansion of regional marketing and administrative centres. Second, the clustering of rural producers in and around industrial villages during the preceding century had created the basis for several new manufacturing centres that were now able to emerge in response to growing markets.

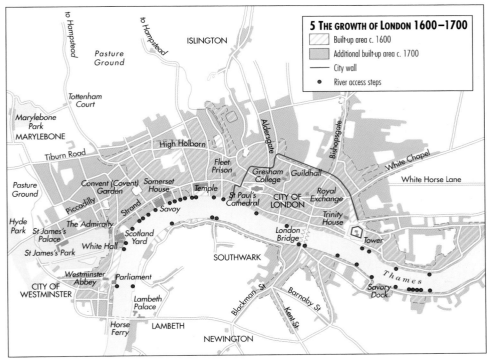

5 THE GROWTH OF LONDON 1600–1700

- ▨ Built-up area c. 1600
- ▨ Additional built-up area c. 1700
- — City wall
- ● River access steps

◄ URBAN COMMUNITIES IN WESTERN EUROPE 1000–1500 *pages 102–3* ► WORLD POPULATION GROWTH AND URBANIZATION 1800–1914 *pages 210–11*

THE DEVELOPMENT OF SCIENCE AND TECHNOLOGY IN EUROPE 1500–1770

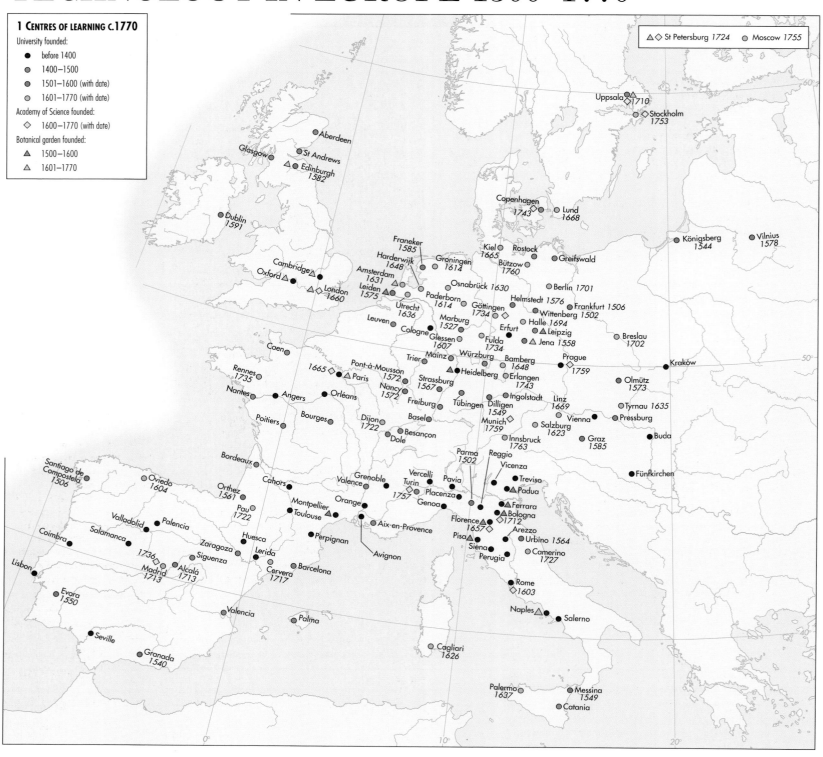

1 CENTRES OF LEARNING c.1770

University founded:
- ● before 1400
- ● 1400–1500
- ● 1501–1600 (with date)
- ○ 1601–1770 (with date)

Academy of Science founded:
- ◇ 1600–1770 (with date)

Botanical garden founded:
- △ 1500–1600
- △ 1601–1770

△◇ St Petersburg *1724*   ○ Moscow *1755*

▲ From the mid-16th century botanical gardens were established in many university towns, and in the following century academies of science added a new dimension to the range of institutions which promoted learning. The most important of these were the Roman Accademia dei Lincei (1603), the Accademia del Cimento in Florence (1657), the Royal Society of London (1660) and the Académie Royale des Sciences in Paris (1665).

Between the early 16th and mid-18th centuries there was a remarkable growth both in the understanding of the natural world and in the capacity to exploit it. In 1500 the study of mathematics was well established in major universities across Europe (*map 1*) and by the end of the 16th century it was a central discipline in both Protestant and Catholic centres of learning. The idea that the world should be represented geometrically formed a central strand of the Renaissance and was especially influential in the development of perspective representation by Italian painters and architects. The research of a number of people – including Nicolaus Copernicus (in Kraków), Johannes Kepler (in Tübingen and Prague), Galileo Galilei (in Padua and Florence) and Isaac Newton (in Cambridge) – suggested that God's Creation had been made according to a mathematical blueprint. England was briefly predominant in the field of natural philosophy following the publication of Newton's *Principia Mathematica* in 1687, but in the 18th century cities as far apart as Basel, St Petersburg, and Paris became centres of European scientific creativity.

CENTRES OF LEARNING

The works of Aristotle formed the basis of the university curriculum until the end of the 17th century, when Cartesian and then Newtonian doctrines began to take hold in most of Europe. A number of factors were involved in bringing about this shift: new discoveries, as well as a more critical attitude to ancient texts, progressively weakened the credibility of Aristotelian styles of explanation, while the development of print and paper production meant that information was available to unprecedentedly large numbers of people, particularly the new urban elites. Moreover, with the exception of Newton's research at Cambridge, innovation in the exact sciences ceased to be university-based after the late 16th century. Instead, the princely courts in Germany and Italy became the major centres of creative work, while the Roman Accademia dei Lincei at the start of the 17th century was the first of a number of academies, both metropolitan and provincial, which promoted learning in natural philosophy and astronomy (*map 1*). Little of note could have been achieved without networks of correspondence which connected individuals in all the major European cities, the most significant being those organized in the 17th century by Marin Mersenne, Samuel Hartlib and Henry Oldenburg. Many of these letters were printed in philosophical journals – the *Journal des Savants* and the *Philosophical Transactions* – which were established in the 1660s.

THE DEVELOPMENT OF BOTANY

From the late 15th century European voyages to the Americas, Africa and Asia (*pages 116–17*) provided novel and extraordinary facts which greatly supplemented and even contradicted the existing Classical texts. Botany was galvanized by information and samples pouring in from places outside Europe. From the Americas came maize, potatoes, runner beans, pineapples and sunflowers, and by 1585 peppers from South America were being cultivated in Italy, Castile and Moravia. New drug plants included guaiacum, Chinese root and sarsaparilla. Botany was practised at universities with strengths in medicine, and botanical gardens were set up to cultivate rare and exotic plants (*map 1*). Books such as Leonard Fuchs's *De Historia Stirpium*, published in 1542, pioneered naturalistic depictions of plants, and the number of plants recorded in such books expanded from less than a thousand in 1500 to the 6,000 recorded in Gaspard Bauhin's *Pinax* of 1623.

SCIENTIFIC INSTRUMENTS

Throughout the 17th and early 18th centuries systematic observation and the use of experimentation and the microscope accelerated the development of botanical and zoological knowledge across Europe. At the same time the development of the telescope revolutionized the study of astronomy, with major new astronomical discoveries made by scholars in London, Danzig, The Hague and Rome.

Research into the existence and nature of a vacuum linked developments in natural philosophy to those in technology. A vacuum was impossible in the Aristotelian system, but in the 1640s experimenters in France argued that the space at the top of a tube inverted in a bowl of mercury was void of matter. At about the same time Otto von Guericke of Magdeburg began trials with the evacuation of air from a copper surrounding. His ideas were taken up by Robert Boyle and Robert Hooke in Oxford, who constructed an air-pump with a glass receiver in 1659. The Dutchman Christiaan Huygens supervised the construction of a pump at the Académie Royale in Paris in 1665, and a number of instrument makers sold different sorts of pumps in Paris in the 1670s. London, Paris, Leipzig and Leiden all became particularly influential centres of pump construction in the 18th century, while London alone became the most important general site of instrument manufacture (*map 2*).

INDUSTRIAL TECHNOLOGY

There were also momentous developments in the area of industrial technology. As pits were dug deeper and deeper to extract coal and minerals such as tin and lead, steam engines emerged as a response to the need to rid mines of water. At the start of the 17th century a number of people considered the possibility of using steam to raise water, either for clearing mines or for producing fountains and cascades for aristocratic gardens. It is no coincidence that a pioneer of air-pump design, Denis Papin, was also extremely influential in the early history of the steam engine. Having worked on air-pumps with Boyle and Huygens in the 1670s, he wrote an article in 1690 describing how steam could raise a piston which would then be allowed to fall due to atmospheric pressure.

Papin's article may well have influenced Thomas Savery, who produced the first workable apparatus for raising water by fire at the end of the 1690s. Savery was the latest in a line of engine constructors based around London, and although his machine was practical in limited situations, it was of no help in deep mines and suffered repeatedly from boiler explosions.

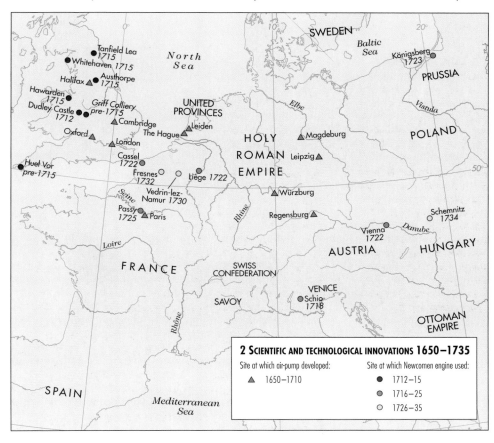

It was the Englishman Thomas Newcomen's piston-driven atmospheric engine which would transform industry in the period before James Watt's innovations revolutionized the design of steam engines towards the end of the 18th century. Newcomen's first working engine was installed in Staffordshire in 1712 (*map 2*). The design of Newcomen's engine was a closely guarded secret, and for the first 15 years no machine outside Britain was made to work without the support and maintenance of a British engineer. The success of the Newtonian system and the domination enjoyed by the British in the art of engine design throughout the 18th century are indicative of the geographical shift in innovative science and technology which had drifted northwards from Italy at the end of the 16th century.

◀ Thomas Newcomen's engine consisted of a cylinder fitted with a piston, which was attached to a counterweighted rocking beam. This, in turn, was connected to a pumping rod. Steam created in the cylinder forced the piston up; cold water was then used to condense the steam, creating a vacuum in the cylinder. Atmospheric pressure subsequently caused the piston to move down, so raising the other end of the rocking beam and lifting the pumping rod.

▼ From the 1650s the air-pump was developed in a number of European cities and by the 1670s air-pumps were on sale in Paris. The Musschenbroek brothers then developed another centre of production in Leiden, which became the most important supplier of air-pumps, telescopes and microscopes in Europe.

The first Newcomen engine was installed in 1712 at Dudley Castle in Staffordshire and the design was quickly taken up by coalfields and other mining operations across the north of England, although the engine's appetite for fuel was colossal. Its running costs were, however, a major obstacle to its diffusion across Europe.

2 SCIENTIFIC AND TECHNOLOGICAL INNOVATIONS 1650–1735

Site at which air-pump developed:
▲ 1650–1710

Site at which Newcomen engine used:
● 1712–15
◉ 1716–25
○ 1726–35

AFRICA
1500–1800

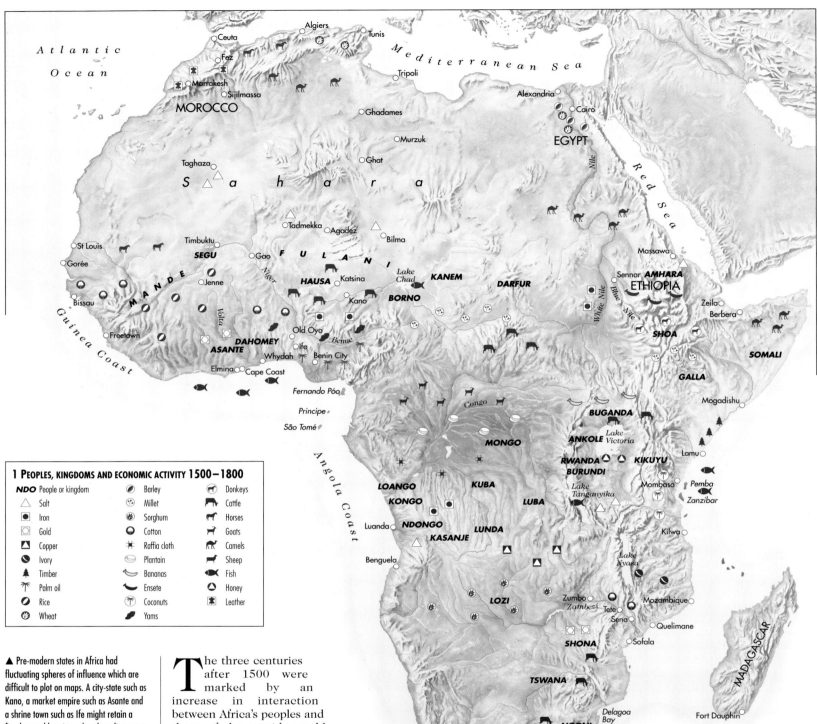

1 PEOPLES, KINGDOMS AND ECONOMIC ACTIVITY 1500–1800

NDO People or kingdom		
△ Salt	🍃 Barley	🐴 Donkeys
◉ Iron	🌼 Millet	🐂 Cattle
▣ Gold	🌿 Sorghum	🐎 Horses
▲ Copper	◯ Cotton	🐐 Goats
◐ Ivory	✳ Raffia cloth	🐪 Camels
🌲 Timber	🍲 Plantain	🐏 Sheep
🌴 Palm oil	🍌 Bananas	🐟 Fish
◑ Rice	🌾 Ensete	🐝 Honey
🌀 Wheat	🌴 Coconuts	🏺 Leather
	🌱 Yams	

▲ Pre-modern states in Africa had fluctuating spheres of influence which are difficult to plot on maps. A city-state such as Kano, a market empire such as Asante and a shrine town such as Ife might retain a fixed central location – but the ruling courts of the Amhara of Ethiopia, or the Mande of Mali, or the Lunda of Congo regularly moved from place to place in the manner of medieval European royalty. Specialists in animal husbandry such as the Fulani of West Africa, the Somali of East Africa or the Tswana of South Africa became even more mobile than the rulers of farming communities as they sought out the best ecological opportunities for grazing their camels and cattle. In contrast to this, fishermen and miners had fixed settlements and defended their economic assets.

The three centuries after 1500 were marked by an increase in interaction between Africa's peoples and those of the outside world, though this increase should not be exaggerated. On the east coast there was no radical change in the pattern of cultural and commercial exchange that had existed since the time of the Roman Empire, but Indians and Europeans encouraged the further exploitation of East Africa's copper mines, mangrove forests, elephant herds, gold deposits and shore-line fisheries (*map 1*). Foreigners also exploited opportunities to recruit voluntary, and more especially involuntary, migrant labour to serve as ships' crews and pearl divers, as household slaves and concubines, or as field hands in the coconut groves and date plantations of the Middle East.

The central interior of Africa was only indirectly affected by the globalization of Africa's external relations before 1800. Local merchants and kingdoms fought over salt quarries, iron mines and fishing lakes. Africa's ongoing agricultural revolution took a new leap forward when traditional grains such as millet and sorghum were supplemented by the slow diffusion of tropical grains from the Americas such

as flour maize and flint maize, while the traditional crops of root yam and vegetable banana were augmented by new carbohydrates processed from cassava.

THE INFLUENCE OF ISLAM AND CHRISTIANITY
In the northern third of tropical Africa, Islam slowly percolated along the ever-changing dust tracks of the Sahara, up the cataracts of the Nile and down the sailing routes of the Red Sea to bring new spiritual energy, theological ideas, commercial codes of practice, jurisprudence, the Arabic alphabet and mosque-based scholarship to the towns of Africa. Perambulating scholars settled in Timbuktu and Kano, where local holy men synthesized their own customs with those of Mediterranean Islam. Islamic art and architecture spread too – as seen in the great minarets of the Niger Valley, regularly coated in river clay, and the palaces of the Swahili east coast, which were built of carved coral.

In western Africa, Christianity was the vehicle for religious change and adaptation. In the Kongo kingdom, one faction seized power in 1506 with the help of foreign priests who subsequently built chapels and schools, created a small bureaucracy and archive, and developed powerful Christian rituals to match local ones. A hundred years later the Papacy sent Capuchin friars to Kongo and the surrounding principalities with a view to spreading the new religion into the provincial and rural areas. Rustic traditionalists proved more resistant to religious change than ambitious townsmen, however, and Christianity created factionalism, discord and eventually a civil war.

TRADE AND COLONIZATION

The impact of European merchants on the Atlantic seaboard of Africa was older, and initially more pervasive, than that of Christianity. Much merchant activity was carried out at open beaches off which 200-tonne sailing vessels anchored; on lagoons where canoes plied, carrying merchandise and slaves; and in creeks where timber vessels that were no longer seaworthy were permanently anchored as floating storehouses. On the Gold Coast (*map 2*) the pattern of trade was different, with around 40 gold-trading fortresses being built by European trading nations. Among the greatest of these castle-warehouses was Cape Coast Castle, the headquarters of the English. Its installations were matched by the fortifications and slave-trading houses of the French on the island of Gorée and, in the south, the Portuguese fortress at Luanda, which was to become Africa's largest slave-exporting harbour on the Atlantic Ocean.

During the 16th and 17th centuries three attempts at colonization of parts of Africa were made by foreigners. The Ottomans spread through North Africa during the early 16th century, capturing cities from Cairo to Algiers and creating an empire which only began to break up when Napoleon attacked Egypt in 1798. The next great colonizing episode was the Portuguese attempt to gain and retain commercial dominance on both the western and eastern flanks of Africa after 1570. Unlike the Ottomans, the Portuguese were unable to conquer significant parts of the mainland, though they attempted to do so in both Morocco and Ethiopia. They did, however, create creole communities on the islands and in a few fortress towns, notably along the Zambezi River. The part of Africa most vulnerable to foreign attack proved to be Angola, where Portuguese merchants became *conquistadores* in the Spanish-American style. The third episode of early colonization was carried out by the Dutch, who between 1637 and 1652 captured three strategic points – the gold-trading castle of Elmina, the slave harbour of Luanda and the prospective military base at Cape Town. Although the Portuguese were able to recover Luanda in 1648 and resume their conquest of Angola, the Dutch influence there proved pervasive. At Cape Town the creolized Dutch remained a distinctive segment of the population after the British captured the city in 1806.

The African response to the European opening of the Atlantic to long-distance shipping was to build their markets, their cities and their royal capitals away from the coast and beyond the range of direct foreign interference. In Angola, where European armies penetrated 300 kilometres (200 miles) inland, the greatest of the African trading empires built the royal compounds of Lunda beyond the reach of the *conquistadores*. In Asante, by contrast, the resistance to invasion was so effective that a royal city with permanent palaces could be safely established at a strategic crossroads little more than 150 kilometres (100 miles) from the coast. The Asante Empire was able to absorb several older kingdoms which had been brokers between the coast and the interior. The empire of Oyo partially eclipsed the ancient trading city of Benin and absorbed the powerful shrine city of Ife; a brash new trading state was created in Dahomey and attracted Latin American and European merchants anxious to buy prisoners of war in exchange for firearms and gunpowder as well as textiles and luxuries.

CONSEQUENCES OF THE SLAVE TRADE

The period 1500–1800 saw an enormous increase in the scale of the American, Mediterranean and Asian purchase of slaves. In some areas, such as Angola, the consequence was a demographic haemorrhage as thousands of people were sold abroad each year, thereby undermining the capacity of communities to renew themselves. In Guinea the slave trade caused such acute social malaise that small communities became dominated by secret societies which manipulated a rising fear of witchcraft. In the Niger Basin whole communities were devastated by raids which caused death, famine and disease on a spiralling scale. In contrast to this, some successful broker kingdoms built up their agrarian economies with new crops and preserved their population by refusing to sell young women captives abroad.

In the long term, however, the effects of the slave trade were to entrench violence as a way of life and create a damaging intellectual climate which presumed that white people were superior to black people. The decolonizing of the minds of both the perpetrators and the victims of the slave trade was to be a slow process, further delayed by the colonial interlude which affected Africa during the first half of the 20th century.

▲ When the Portuguese first arrived in Benin City in 1486 they found a sophisticated and wealthy kingdom. Royal patronage was the basis for the production of elaborate sculptures and artefacts, and the demand for copper and brass for this work formed the basis for early trade with the Portuguese. This 16th-century ivory carving, probably intended for the European market, shows a Portuguese soldier engaged in the slave trade.

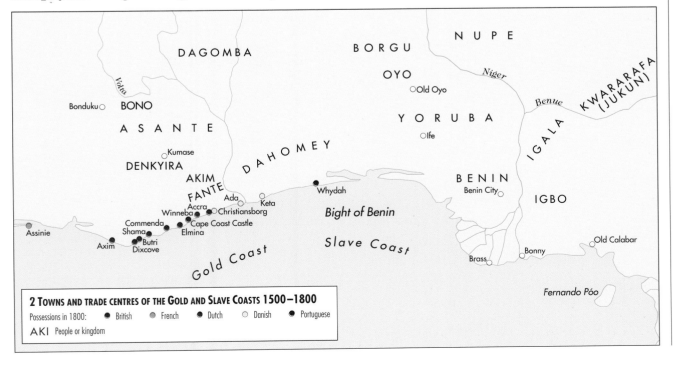

2 TOWNS AND TRADE CENTRES OF THE GOLD AND SLAVE COASTS 1500–1800
Possessions in 1800: ● British ● French ● Dutch ○ Danish ● Portuguese
AKI People or kingdom

◀ The Gold Coast and the Slave Coast were the most intensively exploited parts of the African seaboard. Here Europeans built fortified castle-warehouses to protect their chests of gold and stocks of textiles from plunder and to serve as warehouses, cantonments, slave-pens and well-appointed residences for European governors.

MING AND MANCHU QING CHINA
1368–1800

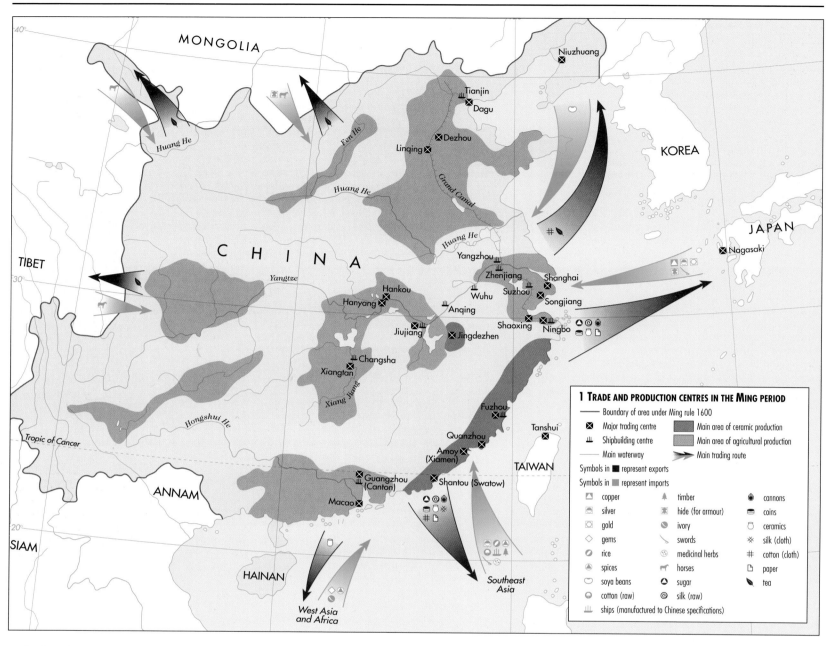

1 TRADE AND PRODUCTION CENTRES IN THE MING PERIOD

- —— Boundary of area under Ming rule 1600
- ⊠ Major trading centre
- ⛟ Shipbuilding centre
- —— Main waterway
- ▬▬▶ Main trading route
- ▨ Main area of ceramic production
- ▨ Main area of agricultural production

Symbols in ■ represent exports
Symbols in ▨ represent imports

copper	timber	cannons
silver	hide (for armour)	coins
gold	ivory	ceramics
gems	swords	silk (cloth)
rice	medicinal herbs	cotton (cloth)
spices	horses	paper
soya beans	sugar	tea
cotton (raw)	silk (raw)	
ships (manufactured to Chinese specifications)		

In 1368 the Mongols, who had ruled China since 1271, were ousted by a peasants' revolt, the leader of which crowned himself Emperor Taizu and founded the Ming dynasty. The Ming period (1368–1644) marked a renaissance in China's cultural, political and economic strength. Administrative systems for running the empire dating from 221 BC were resumed, the imperial examinations for applicants to the civil service were reinstated, and there was a national census and land registration for the purposes of taxation. The Spiritual School (*xinxue*), based on the tradition of the Ideologist School of Confucianism (*lixue*) was established, supporting the need for social order according to the "Will of Heaven". It was to remain popular throughout the Ming and subsequent Qing period.

DEVELOPMENTS IN AGRICULTURE

An agricultural system based on small freeholds was rebuilt, and initially attempts were made by Emperor Taizu to control the tax burden on the poorer farmers. During the second half of the Ming period, however, ownership of land became increasingly concentrated in the hands of a few. This led to the introduction of dual ownership, under which a freeholder could offer land for permanent lease. Sharecropping – a system by which a proportion of the crops produced by the leaseholder is handed over in rent – was also common.

There were significant technological improvements in Chinese agriculture. From the second half of the 16th

century new crops were adopted from the outside world, including the potato and sweet potato, maize, sugar beet, tomato, kidney bean, mango, papaya, agave, pineapple, chilli and tobacco; several improved species, such as the American peanut and cotton, were also introduced. This resulted in an agricultural revolution, with an increase in the use of marginal land and, as a consequence, in agricultural production. China's landscape and the Chinese diet were both dramatically altered. The publication of the *Complete Treatise on Agricultural Administration* in around 1625 also had a major impact. Its author, Xu Guangqi, was the *de facto* Prime Minister, and he enthusiastically promoted the new crops and Western technology for water control. As a result, the Chinese economy was able to survive the increasingly frequent natural disasters of the second half of the Ming period.

TRADE AND EXPANSION OF INFLUENCE

Ming China was active in domestic and foreign trade. Trading guilds were well established in commercial centres and long-distance trade in staple products flourished (*map 1*). China was essentially open to foreign trade, as is evident from the outflow of ceramics and silk, and the inflow of silver that enabled China to adopt its first silver standard. A large number of Chinese settled in Southeast Asia, along the maritime trading routes. In addition, European Christian missionaries in China introduced Western technology. Some, such as Matteo Ricci in the

16th century, were appointed to high positions in the Imperial Court.

Chinese influence was extended by the state-sponsored voyages of the early 15th century, led by Admiral Zheng He. The admiral and his fleet crossed the South China Sea, the Indian Ocean and Arabian Sea, visiting among other places Sumatra, Calicut, Zufar and Mogadishu (*map 2*). The armada – consisting of 27,800 mariners on 200 ships – was well equipped with charts and compasses, and its captains were knowledgeable about meteorological and hydrological conditions. Its voyages, which represent the most spectacular episode in Chinese maritime history, helped to consolidate China's sphere of influence in Asia.

Western powers presented little threat during the Ming period. In 1622–24 the imperial navy twice defeated invading Dutch fleets: off China's south coast, at Macau and Amoy, and off the Pescadore Islands near Taiwan. Only Japanese pirates generally caused concern on the coasts. The real danger to the empire came from the Tatar and Manchu invasions on the northern and northwestern frontiers, and in 1449 Emperor Zhu Qizhen was captured while fighting the invaders. Between 1368 and 1620, 18 major construction projects were carried out to overhaul the 6,700 kilometres (4,200 miles) of the Great Wall (*map 3*).

THE DECLINE OF THE MING DYNASTY
The military strength of the empire gradually faded, and internal rebellions broke out every year from 1522. There was a decline in the efficiency of the Ming government, partly due to interference in the process of government by court eunuchs, but also because rampant tax evasion threw the government into financial difficulties. In response, around 1573 a "one-whip method" of taxation was introduced, intended to lower administrative costs by reducing the number of different taxes levied, and to spread the tax burden more fairly. This reform was short-lived, however, and financial and socio-economic crises were to haunt the Ming dynasty until its downfall.

The Ming dynasty ended in 1644 with the suicide of Emperor Zhu Yiujian following the fall of Beijing to rebels. Officials of the Ming government enlisted the aid of the Manchus – a hitherto nomadic people from beyond the Great Wall who had adopted the Chinese culture – to help them drive the rebels from Beijing. However, once in control of the capital the Manchus refused to leave, and the rule of the Manchu Qing dynasty (1644–1911) began. (A Ming exile government survived in Taiwan until 1683 in the form of a city state with a large fleet and an extensive trading network in East and Southeast Asia.)

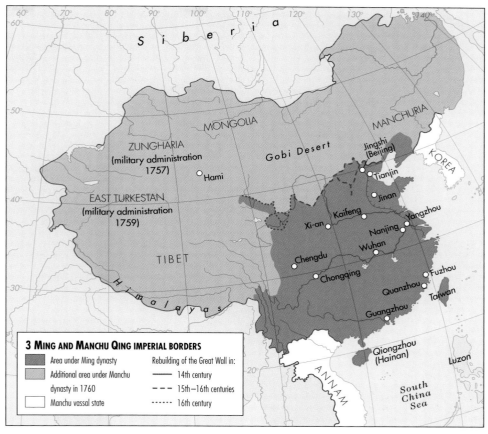

3 MING AND MANCHU QING IMPERIAL BORDERS

- Area under Ming dynasty
- Additional area under Manchu dynasty in 1760
- Manchu vassal state

Rebuilding of the Great Wall in:
- —— 14th century
- – – – 15th–16th centuries
- ······ 16th century

EARLY MANCHU QING RULE
The legitimacy of the Manchu Qing dynasty was always in question, and perhaps as a consequence it made few innovations; its language, state machinery, legal framework and economic policies were all inherited from the Ming. The early Qing can, however, be credited with maintaining a long internal peace and with expanding the Chinese empire to its greatest extent ever, by joining the Manchu territory in Manchuria and Siberia to China, consolidating military control over the part of Turkestan known as the "New Territory", and developing a political link with Tibet (*map 3*). As a result, the population of the Chinese Empire reportedly tripled from around 143 million in 1740 to over 423 million in 1846. From 1800 onwards, however, the Qing dynasty was increasingly under threat from internal uprisings – caused by famine and a corrupt government – and from aggressive Western powers.

▲ Under the Manchu dynasty the Chinese Empire, already extensive, trebled in size. However, with the exception of Manchuria, the territory gained was neither highly populated nor particularly fertile. Although the vassal states of Korea and Annam provided the empire with only a small income, they did form buffer zones against potential invaders.

◀ Zheng He's fleets, which numbered 200 ships, sailed on a series of voyages across the Indian Ocean as far as Arabia and the east coast of Africa, and throughout the islands of Southeast Asia. The ships returned laden with goods and exotic plants, as well as prisoners of war (including the King of Ceylon). Zheng's fleets used force on three occasions: in Sumatra in 1404, in Ceylon (Sri Lanka) in 1410, and in Sumatra in 1413, mainly against Chinese pirates.

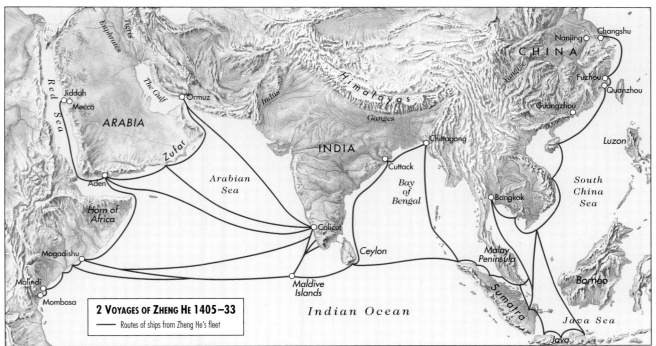

2 VOYAGES OF ZHENG HE 1405–33
—— Routes of ships from Zheng He's fleet

◆ THE MONGOL EMPIRE 1206–1405 *pages 98–99* ◆ LATE MANCHU QING CHINA 1800–1911 *pages 198–99*

TOKUGAWA JAPAN
1603–1867

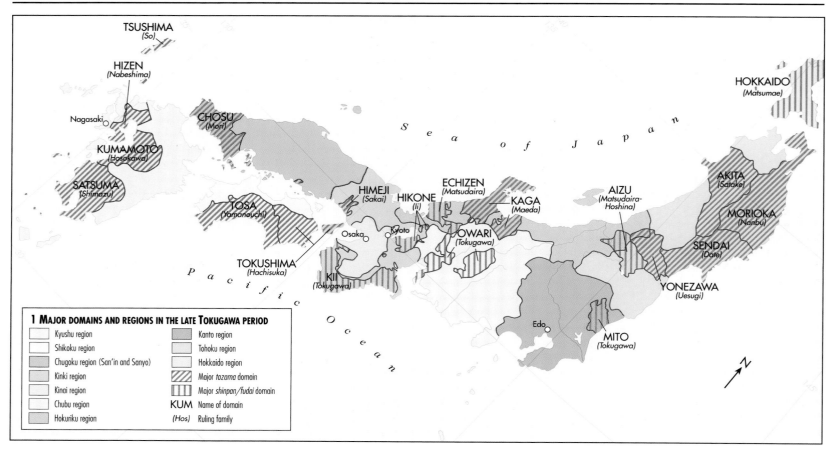

1 MAJOR DOMAINS AND REGIONS IN THE LATE TOKUGAWA PERIOD

- Kyushu region
- Shikoku region
- Chugoku region (San'in and Sanyo)
- Kinki region
- Kinai region
- Chubu region
- Hokuriku region
- Kanto region
- Tohoku region
- Hokkaido region
- Major *tozama* domain
- Major *shinpan/fudai* domain
- KUM Name of domain
- (Hos) Ruling family

▲ Throughout the Tokugawa period Japan remained divided into a largely stable number of domains, with the Tokugawa and related families (*shinpan*) together controlling over 25 per cent of the land. However, people generally identified themselves with a particular region rather than a domain, and economic and social developments occurred on a regional basis.

▲ Tokugawa Ieyasu was responsible for the establishment of the Tokugawa Shogunate in 1603. The shogunate achieved peace throughout the islands of Japan for two and a half centuries – but only through the imposition of strict controls on all classes of society and a policy of isolation from the rest of the world.

In 1603, after many decades of civil war, Japan came under a new structure of military government headed by the Tokugawa family. The emperor, resident in Kyoto, no longer had any real political power, although the Tokugawa administration, called the Shogunate or *Bakufu*, ruled in his name. It discharged some of the functions of a national government but a degree of decentralization persisted, with the country divided into domains, each ruled by a semi-autonomous *daimyo* (lord). Former enemies of the regime became *tozama* (outside) lords, while those deemed friendly were denoted *fudai* and were given important government posts. *Fudai* domains, along with those of collateral branches of the Tokugawa family (*shinpan*), were concentrated in the centre of the country (*map 1*). The shogunate had no power to tax within any of the domains, or, in general, to intervene in the political control of these private fiefdoms. Its only income came from lands directly owned by the Tokugawa and related (collateral) families, including, for example, the Ii and Matsudaira.

In an attempt to ensure their continued dominance, the Tokugawa implemented controls over individual lords and the population in general. Contacts with countries outside Japan were restricted to a minimum, giving rise to a period of national seclusion, or "isolation". All *daimyo* had to visit the shogunal capital, Edo, regularly, and leave their families there as hostages. They were compelled to engage in public works to restrict their finances, and public disorder within domains could incur heavy penalties. A strict hereditary caste system headed by the ruling *samurai* (warrior) caste, followed in descending order by farmers, artisans and merchants, was enforced. The economy was based on rice, with the size and wealth of the various domains measured in terms of the rice crop. The *daimyo* paid their warrior retainers stipends measured in rice, and the warrior caste as a whole marketed any surplus not required for consumption to purchase other necessities and luxuries.

URBANIZATION AND ECONOMIC GROWTH
Although the influence of the Tokugawa over the *daimyo* progressively weakened, the ruling structure remained broadly unchanged until the fall of the shogunate in 1867. However, the very success of the regime in achieving political and social stability stimulated changes which were ultimately to contribute to its downfall. Removal of the likelihood that output would be plundered or destroyed encouraged both farmers and artisans to increase production, while peace made the transport of raw materials and finished products easier (*map 2*).

By the end of the Tokugawa period a growing proportion of the population resided in towns of over 5,000 people, and in some areas this proportion reached over 30 per cent (*map 3*). The need for the ruling caste to transform their rice income into cash stimulated the rise of powerful merchant families, many based in the city of Osaka. These merchant houses accumulated great wealth, despite their low social status, and a growing proportion of the population engaged in educational and cultural pursuits.

Agricultural output increased with the aid of improved techniques and land reclamation, and the majority of peasants ceased to be simple subsistence rice producers, becoming involved, along with artisans, in the supply of handicrafts and other goods. The population, after growing in the first half of the Tokugawa period, stabilized. The latter years saw the rise of manufacturing activities outside the towns, the development of local specialities and the emergence of what has been termed "proto-industrialization". It is generally agreed that these economic developments were a significant factor in supporting Japan's subsequent process of industrialization.

SOCIAL CHANGE AND UNREST
The scale of economic growth and change in the 17th and 18th centuries put pressure on the old system, with the authorities becoming powerless to control the expanding commercial interests and networks. Social status and wealth no longer went hand in hand, and the *daimyo* and their followers found themselves in debt to rich merchants who were nominally at the bottom of the social hierarchy. The distinctions between castes became blurred as individuals ceased to confine themselves to their prescribed occupations; the *samurai*, in particular, now had little reason to demonstrate their military role, instead becoming bureaucrats, scholars and, increasingly, anything that would make ends meet. New economic structures, such as landlordism,

INCIDENCE OF PEASANT UPRISINGS IN THE TOKUGAWA PERIOD

▼ Peasant uprisings peaked in the 1830s — an era of famine — when unrest not only involved greater numbers than ever before but also spread to embrace whole regions. Rioting occurred both in towns and in the countryside, culminating in a major uprising in Osaka in 1837.

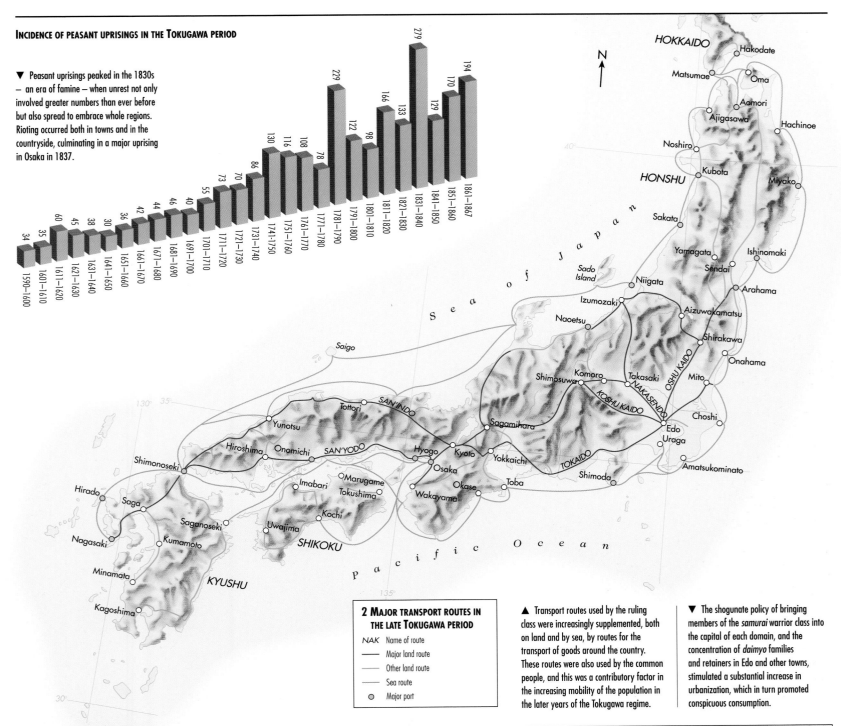

2 MAJOR TRANSPORT ROUTES IN THE LATE TOKUGAWA PERIOD

NAK Name of route
— Major land route
— Other land route
— Sea route
○ Major port

▲ Transport routes used by the ruling class were increasingly supplemented, both on land and by sea, by routes for the transport of goods around the country. These routes were also used by the common people, and this was a contributory factor in the increasing mobility of the population in the later years of the Tokugawa regime.

▼ The shogunate policy of bringing members of the *samurai* warrior class into the capital of each domain, and the concentration of *daimyo* families and retainers in Edo and other towns, stimulated a substantial increase in urbanization, which in turn promoted conspicuous consumption.

threatened to undermine the traditional tribute relationship between peasant and warrior. Above all, the benefits of growth were not evenly spread. Not only did the ruling caste lose out through their dependence on relatively fixed rice prices at a time of inflation, but the lower strata of agricultural workers and urban residents proved highly vulnerable to crop failures, market manipulation and arbitrary exactions by some of their rulers. Local unrest, often violent, became an increasingly frequent occurrence, particularly from the late 18th century (*bar chart*).

The ultimate failure of the ruling caste in many areas – particularly those controlled by the shogunate and its closest followers – to cope adequately with the effects of all these pressures fundamentally weakened the system, rendering it vulnerable to political and military opposition from within, and Western threats from without. When, after 1853, Western countries managed to breach Japan's seclusionist policy, their presence further weakened the integrity of an already shaky system, and contributed to growing internal conflicts. In 1867 these resulted in the downfall of the Tokugawa and the establishment by its enemies of a new regime, nominally headed by the emperor, the following year.

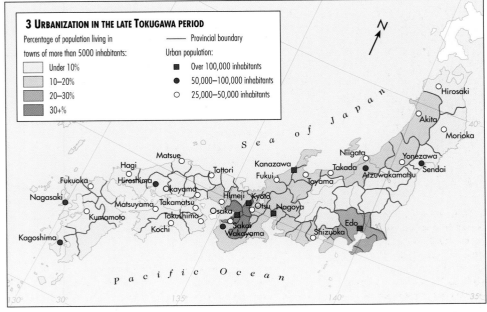

3 URBANIZATION IN THE LATE TOKUGAWA PERIOD

Percentage of population living in towns of more than 5000 inhabitants:
- Under 10%
- 10–20%
- 20–30%
- 30+%

— Provincial boundary

Urban population:
- ■ Over 100,000 inhabitants
- ● 50,000–100,000 inhabitants
- ○ 25,000–50,000 inhabitants

◀ EAST ASIA 907–1600 *pages 86–87* ▶ THE MODERNIZATION OF JAPAN 1867–1937 *pages 200–1*

THE OTTOMAN AND SAFAVID EMPIRES 1500–1683

▼ The Ottoman Empire, already substantial in 1500, continued to expand in the 16th and 17th centuries, though not without setbacks, such as its defeat in the naval Battle of Lepanto in 1571. Its decline can be dated from 1683 when Ottoman troops were forced to retreat after failing in their attempt to take Vienna.

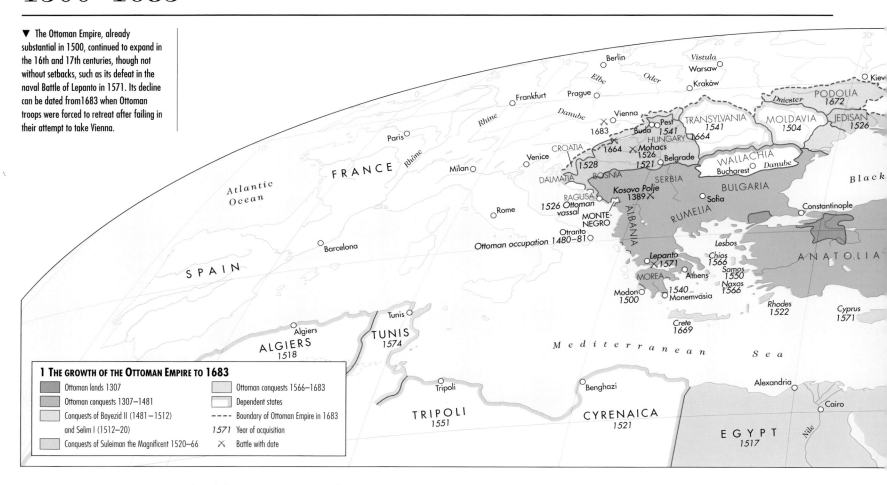

1 THE GROWTH OF THE OTTOMAN EMPIRE TO 1683

- Ottoman lands 1307
- Ottoman conquests 1307–1481
- Conquests of Bayezid II (1481–1512) and Selim I (1512–20)
- Conquests of Suleiman the Magnificent 1520–66
- Ottoman conquests 1566–1683
- Dependent states
- - - - Boundary of Ottoman Empire in 1683
- *1571* Year of acquisition
- ✕ Battle with date

The Ottoman and Safavid states represented twin peaks of Islamic political and cultural achievement, and each handed down a powerful and complex legacy to the modern Islamic world. From the mid-15th century to 1683 the Ottoman Empire was also one of the most successful and militarily effective states of all time. Its sultan, whom Western contemporaries called "The Grand Signior", was regarded with immense respect throughout Christendom. Ottoman power was based on gunnery, the maintenance of a navy and an effective system of military recruitment and training. Originally, the Ottoman Janissary regiments were maintained by the *devshirme* – the "gathering" of child slave recruits from the margins of the empire, who eventually were able to leave military service as free Muslims. However, by the 17th century local, Muslim-born recruits were beginning to dominate the army.

The Ottoman state displayed a high level of religious tolerance for the substantial proportion of the empire's subjects who were not Ottoman Turks or even Muslims. Members of minority communities became senior Ottoman commanders and administrators; indeed, the Orthodox Greek community was probably richer and more numerous than that of the ruling Ottoman Turks.

The Ottoman economy was based on an agricultural society which supported a system of military and religious fiefdoms. A vital adjunct to this peasant world was provided by the empire's most notable and outward-looking communities – the Greeks, Armenians, Syrians and Sephardi Jews who dominated many of the empire's cities and towns.

Territorial expansion was intrinsic to Ottoman power (*map 1*). As late as the 17th century there was no sign that policy-makers in Constantinople believed that Ottoman territorial authority had reached saturation point or achieved natural frontiers. Yet this was, in effect, the case. The Ottoman threat to Italy faded and Vienna – the "Red Apple of the West" in Ottoman military folklore – remained a prize that eluded the sultans. The defeat of the last great Ottoman expedition to Vienna in 1683 marked the beginning of the empire's long decline.

THE SAFAVID STATE

The Safavids made their mark by nurturing the culture that defines modern Iran. The founder of the Safavid dynasty was Shah Ismail I (r. 1501–24), who re-established a central government amid the political chaos into which Persia had fallen in the aftermath of the age of Timur-leng. Ismail's partisans were the Qizilbash – red-capped Turcoman devotees of the Safawi religious brotherhood. The shah welded the Qizilbash into a political force by

▼ The area east of the Euphrates was the subject of much dispute between the Ottomans and Safavids in the 16th and early 17th centuries, until a boundary between the two empires was finally agreed with the Peace of Zuhab in 1639.

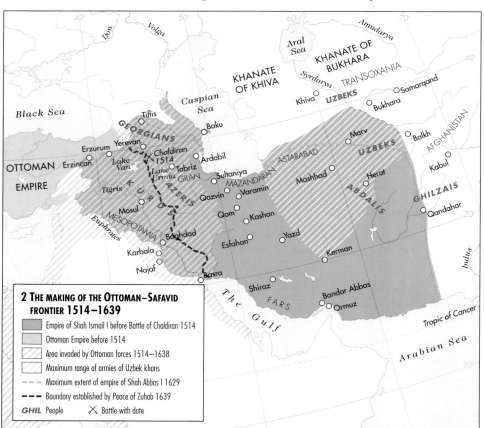

2 THE MAKING OF THE OTTOMAN–SAFAVID FRONTIER 1514–1639

- Empire of Shah Ismail I before Battle of Chaldiran 1514
- Ottoman Empire before 1514
- Area invaded by Ottoman forces 1514–1638
- Maximum range of armies of Uzbek khans
- - - - Maximum extent of empire of Shah Abbas I 1629
- - - - Boundary established by Peace of Zuhab 1639
- **GHIL** People
- ✕ Battle with date

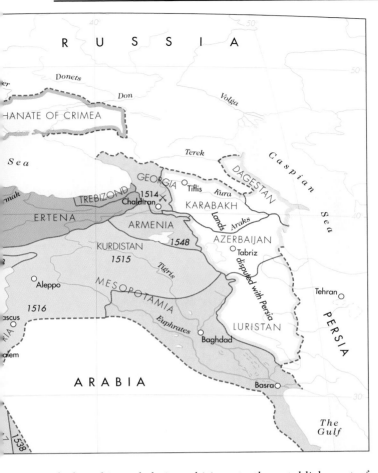

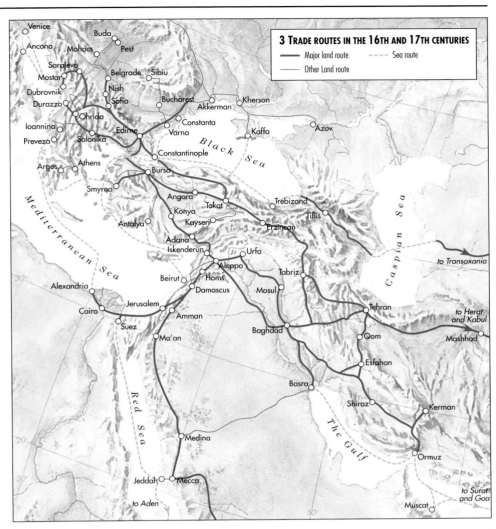

linking his and their ambitions to the establishment of "Twelve Shiism" as the religion of the Persian state. In the wider Islamic world, this nostalgic Shiite tradition was increasingly a marginal or sectarian faith, regarded by the Sunni majority as heretical. In Safavid Persia, Shiism became the defining national creed, providing the Safavids with an ideological focus. Unfortunately, it also exacerbated enmities between Persia and its Sunni Muslim neighbours and rivals, the Ottomans to the west and the Uzbek raiders from Transoxania (*map 2*).

Safavid shahs – most notably Abbas I (r. 1587–1629) – were deliberate propagandists of Shiite culture. They were patrons of representational art, usually in miniature, and undertook a magnificent building programme of religious architecture, palaces and public works. The greatest splendours survive in Abbas I's capital, Esfahan.

THE FORGING OF A FRONTIER

The Ottoman Turks inherited from their Byzantine predecessors a determination to keep the Black Sea dependent on Constantinople, free from control by Central Asian rulers. When Shah Ismail and his Qizilbash forces began to infiltrate eastern Anatolia from Tabriz in the early 16th century, they provoked a massive Ottoman military response. The armies of Sultan Selim the Grim were in the forefront of contemporary military capacity, and the Ottoman artillery gained a dramatic victory over the lightly-armed Persians at Chaldiran in 1514.

The Battle of Chaldiran appears to have shifted the centre of gravity of the Persian Empire to the east, but it was not a final encounter. It led to more than 120 years of intermittent Ottoman–Safavid conflict over land occupied by Azeris, Kurds and Mesopotamian Arabs (*map 2*). (By diverting Ottoman attention from the Balkans, this conflict relieved western Europe of some of the military pressure to which it had been exposed since the Ottoman elimination of the Byzantine Empire in 1453.) The standard pattern in this long conflict was one of an Ottoman offensive countered by Persian "scorched earth" and guerrilla tactics. Shah Abbas I was briefly able to set the Safavid forces on the offensive and reconstitute most of the empire

once ruled by his predecessor Ismail, but the eventual settlement, enshrined in the lasting Peace of Zuhab in 1639, favoured the Ottomans. The frontier had no logic in terms of language, ethnicity or culture. It divided rather than defined communities, splitting Sunni from Sunni and Shiite from Shiite, but it formed the basis for the frontier between the Ottoman and Persian empires and survived as the Iraq–Iran border. The Safavid Empire continued until the invasion of its lands by the Ghilzai Afghans in 1722 heralded the demise of the dynasty in 1736.

THE WORLD OF MERCHANTS AND CARAVANS

The Ottoman and Safavid states governed lands that had been in contact with a wider world since antiquity. The empires were crossed by commercial and pilgrimage routes and contained gateways by land and sea which linked the Mediterranean and Levantine worlds to the Indian subcontinent, Southeast Asia and China (*map 3*).

Many Ottoman and Safavid traders were also Muslim pilgrims undertaking journeys to Mecca. However, a good proportion of the traders and migrants from the Islamic empires were not Muslims but members of Christian and Jewish minority groups operating in partnership with Europeans, many of whom were based in Constantinople, Smyrna, Aleppo and Alexandria – the empire's "windows to the West". Safavid contacts with the Western world were tenuous and bedevilled by the difficulties of the Persian terrain, but during the 16th century European adventurers did make their way to Esfahan and back. At the same time, the powers of western Europe began to establish their own sea routes to the East (*pages 118–19*), thus threatening to wrest control of Eurasian trade from the Muslims. However, although in 1515 the Portuguese captured Ormuz, a Gulf market for horses and spices, they lost it again to the Safavids in 1622. Thereafter, the old trade in spices and silk – and a new trade in tea – continued to be serviced by caravan routes into the 18th century.

▲ The territory ruled by the Ottomans and Safavids was criss-crossed by land and sea routes used by merchants and pilgrims alike. Sea travel was risky but could be relatively straightforward on Mediterranean short hops or in regions governed by the alternating monsoon winds. Overland traffic was arduous and slow but continued to play an important role in trade with Asia until well into the 18th century.

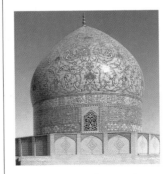

▲ The dome of the Madrasa-yi Madar-i Shah mosque is among the many splendours of Safavid architecture built in the 17th century in Esfahan, the capital of Abbas I.

INDIA UNDER THE MUGHALS
1526–1765

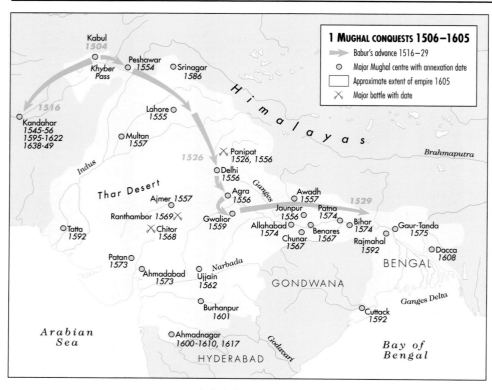

1 MUGHAL CONQUESTS 1506–1605

→ Babur's advance 1516–29
○ Major Mughal centre with annexation date
▢ Approximate extent of empire 1605
✕ Major battle with date

▲ On the death of Babur in 1530 the Mughal Empire was little more than an area in northern India under military occupation. During the reign of Akbar, between 1556 and 1605, it was much expanded and became a centrally governed state.

▼ The artisan industries of India – especially those manufacturing cotton textiles – were at first stimulated by the arrival of the Europeans in the 16th century. As a result, India became the workshop of the world known to Europeans.

The Mughal Empire was founded in 1526 by Babur, Sultan of Kabul. Babur was of Turkic origin and traced his ancestry back to Timur-leng (Tamerlane) and to Chinggis Khan, the Mongol Emperor of China. His advance from Kabul was at the expense of Afghan warlords who themselves had spread into the plains of India, conquering the Sultanate of Delhi and establishing the Lodi dynasty. Babur defeated Ibrahim Lodi at the Battle of Panipat in 1526 and then, until his death in 1530, progressively extended his sway across the Ganges Valley as far east as the borders of Bengal (*map 1*).

CONSOLIDATION UNDER AKBAR

Babur's successor, Humayan (r. 1530–56), faced a resurgence of Afghan power and, between 1540 and 1555, was driven into exile while the empire was ruled by Sher Shah and his sons. In 1555 Humayan retook Delhi to restore the Timurid monarchy, and when he died the following year the succession passed to his son Akbar (r. 1556–1605). Having driven the Mughals' enemies from Delhi, Akbar used his long reign both to expand the empire and, even more significantly, to consolidate and transform it, converting a rulership founded on warrior nomadism into one based on centralized government.

The state which Akbar constructed had a number of key features. At the top he built a "service" nobility of *mansabdars* who provided administration across the empire. Many *mansabdars* were immigrants from elsewhere in the Islamic world, whose loyalty was owed exclusively to the emperor himself. Beneath them, Akbar incorporated the Hindu Rajput chieftains who ruled over lower castes and commoners. These chieftains possessed local power bases which were notionally independent of Mughal authority, but their status and security were enhanced by membership of an imperial aristocracy. To facilitate their incorporation, Akbar – who was fascinated by all religions – also promoted a cultural style which crossed strict religious boundaries. Beneath the *mansabdari*-Rajput elite, the empire rested on the labour of millions of peasants and artisans from whom large revenues were extracted.

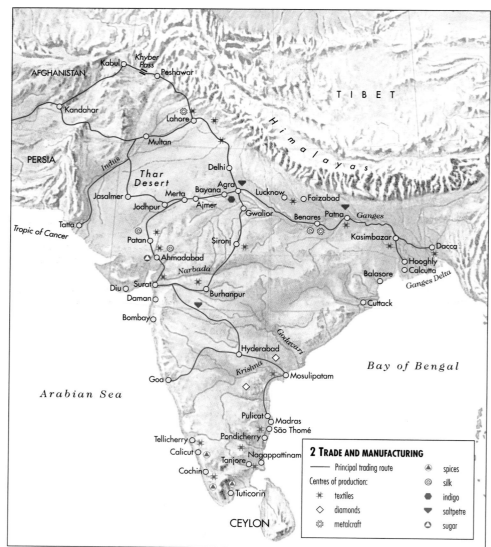

2 TRADE AND MANUFACTURING

— Principal trading route
Centres of production:
✳ textiles
◇ diamonds
⚙ metalcraft
▲ spices
◉ silk
⬣ indigo
▼ saltpetre
△ sugar

4 AN EMPIRE IN DECLINE
▢ Mughal territory c. 1765
▨ Maratha territory c. 1680
▨ Maratha territory c. 1750
▢ Areas autonomous under nawabs c. 1765
▢ Areas autonomous under independent chiefs c. 1765

▲ Following the death of Aurangzeb in 1707 many regional states competed for power, and the roles which the Europeans were acquiring in trading and banking became increasingly significant. Frequently the regional states depended on European commercial agencies – such as the British East India Company – which, as a result, moved more directly into the political foreground during the 18th century.

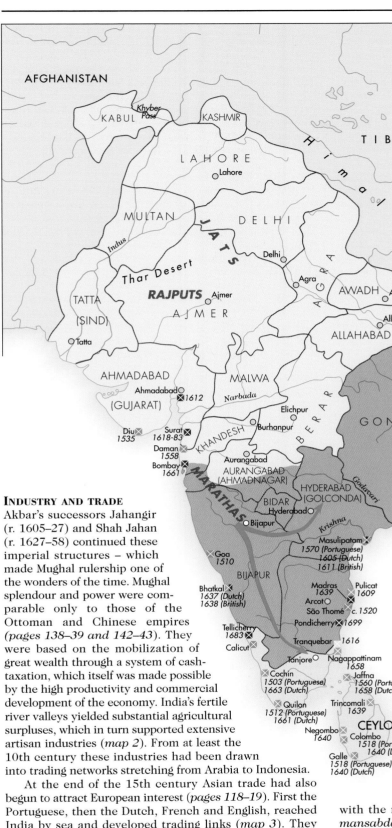

The map legend:

3 EXPANSION AND ENCROACHMENTS 1605–1707

- Approximate extent of Mughal Empire 1605
- Additional area claimed by Aurangzeb 1707
- DELH The 21 provinces (subahs) 1707
- ○ Provincial (subah) headquarters 1707
- —— Boundary of province
- → Advance of Marathas
- **MA** People in rebellion against the empire c. 1700

Major European trading posts:
- ⊠ Portuguese
- ⊠ British
- ⊠ Danish
- ⊠ French
- ⊠ Dutch

◄ Aurangzeb attempted to establish Mughal power in southern India. However, in doing so he came up against foes – in particular, the Marathas – whom he could do little to contain. The Marathas introduced new forms of warfare, based on guerrilla tactics, which defied Mughal armed might. Also, as chieftains risen from the peasantry – rather than imposed on top of it – Maratha leaders spurned the kinds of inducements which had made the Rajputs susceptible to imperial influence. From the 1680s Maratha armies broke through the Mughal cordon meant to contain them, and ravaged far and wide. The Europeans, who had established trading posts around the coast, were mere observers of events at this time.

INDUSTRY AND TRADE

Akbar's successors Jahangir (r. 1605–27) and Shah Jahan (r. 1627–58) continued these imperial structures – which made Mughal rulership one of the wonders of the time. Mughal splendour and power were comparable only to those of the Ottoman and Chinese empires (*pages 138–39 and 142–43*). They were based on the mobilization of great wealth through a system of cash-taxation, which itself was made possible by the high productivity and commercial development of the economy. India's fertile river valleys yielded substantial agricultural surpluses, which in turn supported extensive artisan industries (*map 2*). From at least the 10th century these industries had been drawn into trading networks stretching from Arabia to Indonesia.

At the end of the 15th century Asian trade had also begun to attract European interest (*pages 118–19*). First the Portuguese, then the Dutch, French and English, reached India by sea and developed trading links (*map 3*). They brought with them huge quantities of gold and silver taken from the Americas, further stimulating the Indian economy.

However, the European presence also spelled danger – although its character did not become fully apparent until the 18th century. At that point, and most notably after the death of the Emperor Aurangzeb (r. 1658–1707), Mughal power went into precipitate decline (*map 4*). The empire was unable to respond to invasions from abroad or to rebellions at home. Even the *mansabdari* elite turned against it, as governors (or *nawabs*) declared themselves independent and sought to establish their own kingdoms. Although the emperorship retained a symbolic significance throughout the rest of the century (and was not formally abolished until 1857), the real substance of Mughal power was weakening even by 1730.

THE EMPIRE'S COLLAPSE

Many different explanations have been put forward for the sudden collapse of so mighty and established an empire. Nearly all of these have rooted the problem in Aurangzeb's reign. He sought to expand Mughal power southwards, taking virtually the whole of the subcontinent under imperial rule. However, in doing so he became involved in protracted conflict against opponents whom he could neither defeat nor incorporate.

Aurangzeb's long wars in the south proved extremely costly. They stretched the finances of the empire and promoted changes in its internal structures. He increased the weight of taxation, which fomented revolt in other provinces. Frustrated by the Hindu Marathas, he became increasingly intolerant in his religious practices – threatening the Hindu-Muslim accord which had marked Akbar's empire. To cope with the rising pressures, Aurangzeb also expanded the *mansabdari* elite in ways which reduced the representation of Muslim immigrants and thus increased that of local Indian powers. The empire which he bequeathed to his successors in 1707 was already deeply strained.

Yet there may have been other causes of Mughal decline, which point to the growing influence of a wider world. Rapid commercial expansion in the 17th century, when an ever-growing number of trading posts was established, both altered the political geography of India and changed the social balance between military and economic power. Commerce was based on overseas trade and most enriched the maritime provinces. It also strengthened the position of mercantile groups and the gentry classes. The Mughal Empire, founded by warrior descendants of the "Mongol Horde" and centred on cities in India's heartland, was singularly ill-equipped to manage such developments.

▲ The Mughals are renowned for their architectural achievements, the most famous of which is the Taj Mahal, built between 1632 and 1648 by Shah Jahan. Painting also flourished, particularly during the reign of Jahangir, shown here looking at a portrait of Akbar, his father.

● THE MUSLIM WORLD 1000–1400 *pages 88–89* ● THE BRITISH IN INDIA 1608–1920 *pages 194–95*

EUROPEAN STATES
1500–1600

Maps of 16th-century Europe are often deceptive in that they appear to suggest that the western countries – France, Spain and England – and the eastern countries – Poland and Russia – were consolidated and centralized, while sandwiched between them many tiny entities were grouped together to form the Holy Roman Empire (*map 1*). In fact, all the European states were highly decentralized and regionalized in 1500. France (*map 2*) actually saw an increase in devolution during the 16th century as many provinces escaped central control in the French Wars of Religion (1562–98).

Spain consisted largely of a union of the kingdoms of Castile and Aragon, with Castile itself made up of a number of component kingdoms. In 1512 Ferdinand of Aragon added to this by annexing the kingdom of Navarre, though not the portion of it north of the Pyrenees. Stability in Spain rested on the willingness of the government (centred at Madrid from the 1560s) not to touch the immunities and privileges of these kingdoms, another of which was added to the Spanish Habsburg realm in 1580 when King Philip II of Spain also became King of Portugal.

Poland was divided up into counties and governorships dominated by the nobility, and was formally made up of two realms, the kingdom of Poland and the vast Grand Duchy of Lithuania. Agreements reached between 1569 and 1572 turned the kingdom into an elective monarchy in which the power of the king was limited by a *diet* made up of senators and delegates.

The Russian Empire came into being as a multi-ethnic empire only after the coronation of Ivan IV in 1547. It was created through the conquest of the Tatar khanates of Kazan and Astrakhan in the 1550s and expansion across the Urals into Siberia from the 1580s (*pages 148–49*). Though often ruled brutally, it hardly consisted of a centralized realm and, indeed, for a decade of Ivan's reign (1564–74) it was deliberately divided by the tsar into a personal domain, in which his word was law, and the rest of the country, in which the *boyars* (nobles) ruled.

THE HOLY ROMAN EMPIRE
By the 16th century the jurisdiction of the Holy Roman Empire was, in reality, confined to the territory north of the Alps. The Italian section continued formally as part of the Empire, with its rulers nominally invested as Imperial Vassals, but as time went on this had less and less meaning. The Swiss Confederation gained exemption from imperial duties in 1499 and was formally released from imperial jurisdiction in 1648.

In 1500 and 1512 the rest of the Empire was organized in Imperial Circles for purposes of raising taxes and administering justice. The Netherlands was formed as the Burgundian Circle, the northern provinces of which were formally recognized as independent of the Empire in 1648. As a result of the Lutheran Reformation (*pages 154–55*), many of the ecclesiastical territories were secularized after 1520. The basic constitution of the Empire (the Golden

▼ Frontiers in Europe changed considerably between 1500 and 1560. In 1500 the border between France and the Holy Roman Empire, for example, was that defined by the Treaty of Verdun in 843, with the addition to France of Dauphiné in 1349 and Provence in 1481. The treaties of Madrid (1526) and Cambrai (1529) fundamentally modified the border in the north by transferring Flanders and Artois from France to the Empire.

1 EUROPE c. 1560
— Boundary of the Holy Roman Empire
- Austrian Habsburg territories
- Spanish Habsburg territories
- Ottoman Empire
- Tributary to the Ottoman Empire
- Venetian territories
- Major German secular states

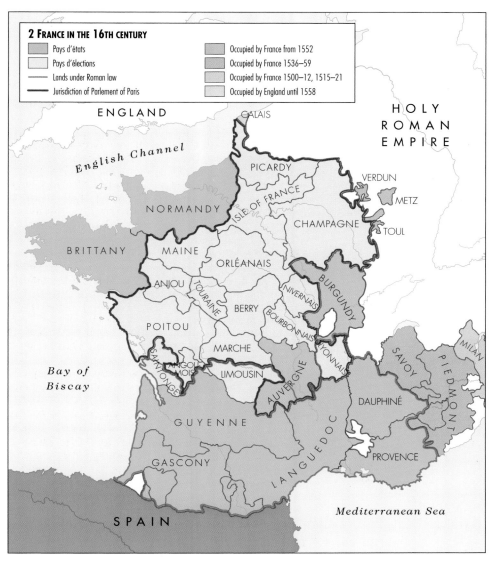

2 FRANCE IN THE 16TH CENTURY

- Pays d'états
- Pays d'élections
- Lands under Roman law
- Jurisdiction of Parlement of Paris
- Occupied by France from 1552
- Occupied by France 1536–59
- Occupied by France 1500–12, 1515–21
- Occupied by England until 1558

◄ France was composed of provinces, some of which were under centralized control (*pays d'élections*) while others raised local taxes through regional assemblies (*pays d'états*). Law differed widely between regions, the main distinction being between the Roman-based law of the south and the customary law of the north.

Bull of 1356, which defined the princes who had the right to elect the Emperor), was modified by the Treaty of Augsburg of 1555 to accommodate these changes, granting princes and cities the right to be Lutheran and recognizing the secularization of church property up to 1552.

EUROPEAN DYNASTIES

Most European states were to some extent dynastic – they were regarded as a family inheritance. The collection of lands under the rule of the King of Spain in the second half of the century (Portugal, Castile, Navarre, Catalonia, Naples and Sicily) was the product of dynastic inheritance under the Habsburg Charles V, Holy Roman Emperor from 1519 to 1558 (*pages 152–53*). In the British Isles, King Henry VIII of England claimed the throne of Ireland in 1541, and in 1603 King James VI of Scotland inherited the English throne, thus uniting all three kingdoms under one monarch.

In central Europe at the beginning of the 16th century, one branch of the Jagiellon dynasty of Poland ruled over Poland–Lithuania while another ruled over Bohemia and Hungary. Hungary, one of the largest kingdoms of the late Middle Ages, was a union of Hungary itself (with power devolved to powerful regional magnates), Croatia and parts of Bosnia. After King Lajos II of Hungary was overwhelmed by the Ottomans at the Battle of Mohacs in 1526 (*pages 142–43*), much of his inheritance passed to the Habsburgs through his sister's marriage to Ferdinand I, the brother of Emperor Charles V.

From the 1540s the borderland between this eastern Habsburg territory and the Ottoman Empire was marked by a number of territories: Hungarian Transylvania (Erdely), Moldavia and Wallachia were ruled by local princes as tributaries of the sultan, whose direct rule extended to Buda and the central region of Hungary. In the north the Union of Kalmar of 1397, which had brought together Denmark, Norway and Sweden–Finland under the same monarch, was broken in 1523 with the secession of Sweden–Finland under Gustav I Vasa (*pages 150–51*).

DYNASTIC WARS

The ruling dynasties of Europe were all closely related to each other, though this did not prevent the fighting of wars. Often described as "Wars of Magnificence", these were pursued for glory and the vindication of dynastic title, and were considered more admirable than "common wars" fought for the annexation of territory or other forms of gain. An example of this occurred in Italy (*map 3*) where the House of France and the Spanish House of Aragon – whose rights were inherited by the Habsburg Charles V – both laid claim to Naples in the south and to Lombardy and the duchy of Milan in the north. In the latter, the richest part of Italy, the struggle was more than one of inheritance. Francis I of France gained control of Milan in 1500, lost it in 1512 and reconquered it in 1515, but Charles V had to oppose this if his power in Italy were not to crumble. War began in 1521 (the French evacuated Milan in 1522), and lasted intermittently in the peninsula until the Treaty of Cateau-Cambrésis in 1559. Signed by representatives of Henry II of France and Philip II of Spain, this treaty had the effect of liquidating French ambitions in Italy while maintaining French acquisitions in Lorraine – Metz, Toul and Verdun (*map 2*). This established a new international order which was to survive with modifications until the Treaty of Westphalia in 1648.

PARMA

Until 1512 part of Milan
1512–15 held by Papal States
1515–21 held by France
1521–45 held by Papal States
1545 granted out by the Pope as a duchy to his son, founder of the Farnese dynasty

3 ITALY 1500–59

- Boundary of the Holy Roman Empire
- Under Spanish control from given date
- Papal States 1500
- Under Papal occupation at given date
- Este lands
- Occupied by Venice 1499–1509
- Occupied by Venice 1503–30
- Associate of Swiss Confederation from given date
- Absorbed by Florence

◄ In the period of intermittent war between France and the Habsburgs from 1521 to 1559, France occupied the territory of Savoy–Piedmont (1536–59) as a gateway across the Alps into Italy. Despite the disaster of the sack of Rome in 1527 by troops of Charles V, papal authority over Romagna was strengthened, with the Venetians agreeing to evacuate Ravenna in 1530. Parma was acquired from Milan by Pope Julius II in 1512 and granted out as a duchy by Pope Paul III to his son Pierluigi Farnese in 1545.

◐ EUROPE 1350–1500 *pages 106–7* ◑ REVOLUTION AND STABILITY IN EUROPE 1600–1785 *pages 156–57*

THE EXPANSION OF RUSSIA 1462–1795

The expansion of Russian rule into Europe and Asia was a process of exploration and discovery comparable with the contemporaneous exploration of the oceanic world by western European peoples. It was, however, also the creation of a highly autocratic land empire. In the mid-15th century the Russian state of Muscovy was just one of many small principalities in northern Europe which paid tribute to the Tatars; by the end of the 18th century it was at the heart of an empire that stretched from the Baltic Sea to the Bering Strait.

▶ Grand Duke Ivan III extended his territory by annexing the neighbouring principalities of Novgorod in 1478, Tver in 1485 and Viatka in 1489. In 1494 he pushed westwards into Poland–Lithuania, occupying Viazma and the towns of the upper Oka basin. Ivan's son, Vasili III, continued with this policy of aggressive expansion, taking Smolensk, Chernigov, Pskov and Riazan.

▼ As part of the process of expansion, *ostrogs* (fortified trading posts) were established at strategic points. An *ostrog* was founded at Tomsk in 1604 and by 1607 Turuchansk on the Yenisei River had been reached. The river became the frontier of the empire in 1619, with another string of *ostrogs* being established along it.

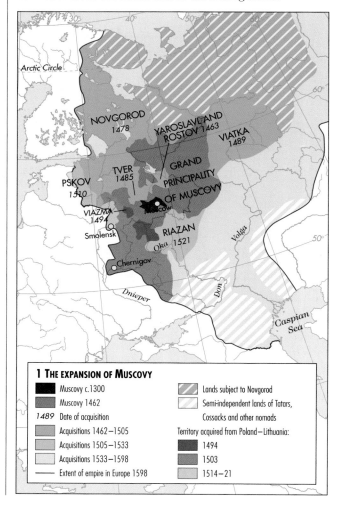

1 THE EXPANSION OF MUSCOVY

- ■ Muscovy c.1300
- ■ Muscovy 1462
- *1489* Date of acquisition
- Acquisitions 1462–1505
- Acquisitions 1505–1533
- Acquisitions 1533–1598
- — Extent of empire in Europe 1598
- ⧄ Lands subject to Novgorod
- ⧄ Semi-independent lands of Tatars, Cossacks and other nomads
- Territory acquired from Poland–Lithuania:
  - ■ 1494
  - ■ 1503
  - ■ 1514–21

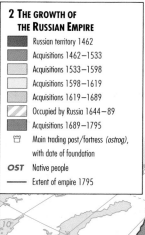

2 THE GROWTH OF THE RUSSIAN EMPIRE

- ■ Russian territory 1462
- ■ Acquisitions 1462–1533
- Acquisitions 1533–1598
- Acquisitions 1598–1619
- Acquisitions 1619–1689
- ⧄ Occupied by Russia 1644–89
- ■ Acquisitions 1689–1795
- ⌂ Main trading post/fortress (*ostrog*), with date of foundation
- **OST** Native people
- — Extent of empire 1795

The process of expansion began after Muscovy had freed itself from Tatar domination in the 1450s. Grand Duke Ivan III (r. 1462–1505) and later his son, Vasili III (r. 1505–33) set about extending his territory by annexing neighbouring regions (*map 1*). Ivan IV became the next grand duke in 1533 at the age of three, and during his minority the *boyars* (nobles) vied with each other for control of the state. No further territorial expansion took place until after he was formally crowned as the first "tsar" (emperor) in 1547. However, in 1552 a successful campaign was launched against the Tatar stronghold of Kazan, and this was followed by the seizure of Astrakhan on the Caspian Sea in 1556. Russian territory now extended the entire length of the Volga, bisecting Tatar domains and dominating the peoples of the northern Caucasus and eastern Caspian.

EXPANSION INTO ASIA

In the east the foundation in 1560 of a fortified post at Perm on the River Kama brought the Muscovites to within easy reach of the Urals, where trading in furs promised to be a great source of wealth. From 1578 the Stroganovs, a family of merchants who had been granted a vast tract of unexplored land by the tsar, took the lead in exploration and settlement beyond the Urals. Their allies in this process were the Cossacks, descendants of peasants who had fled from worsening economic conditions in Russia to become fighting guards of the frontier. The Khanate of Sibir was conquered in 1581, and the colonists founded *ostrogs* – fortified trading posts – along the Irtysh and Ob rivers, controlling the lower reaches of both by 1592 (*map 2*).

Expansion continued to be rapid in the 17th century. The Lena River was reached in 1632, the Indigirka in 1639 and the Kolyma in 1644. The explorer Dhezhnev reached the Bering Strait in 1648 and Khabarov got to the Amur River in 1649. The Khamchatka Peninsula was entered by Russian explorers in 1679. These territorial advances took place largely at the expense of the indigenous, often nomadic, peoples who were powerless in the face of Russian imperialism. Any resistance was effectively suppressed by punitive expeditions from the *ostrogs*.

RUSSIAN AMBITIONS IN THE WEST

In the west, Russian ambitions were more circumscribed. In 1558, in an attempt to take land around the Baltic, Ivan IV became embroiled in a devastating war of 25 years which ruined both Livonia and Estonia and left the Russian armies prostrate. By the end of his reign all Ivan's western conquests

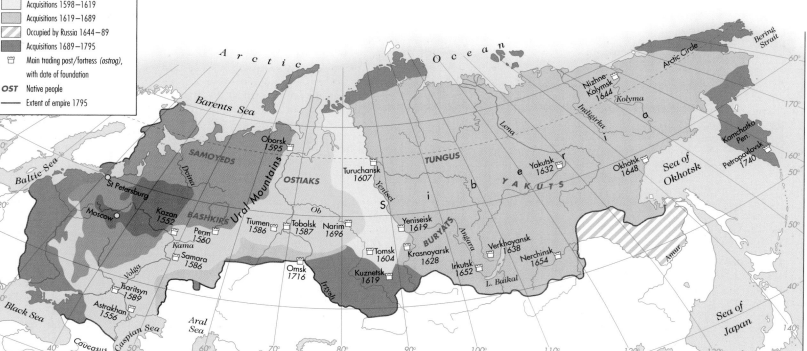

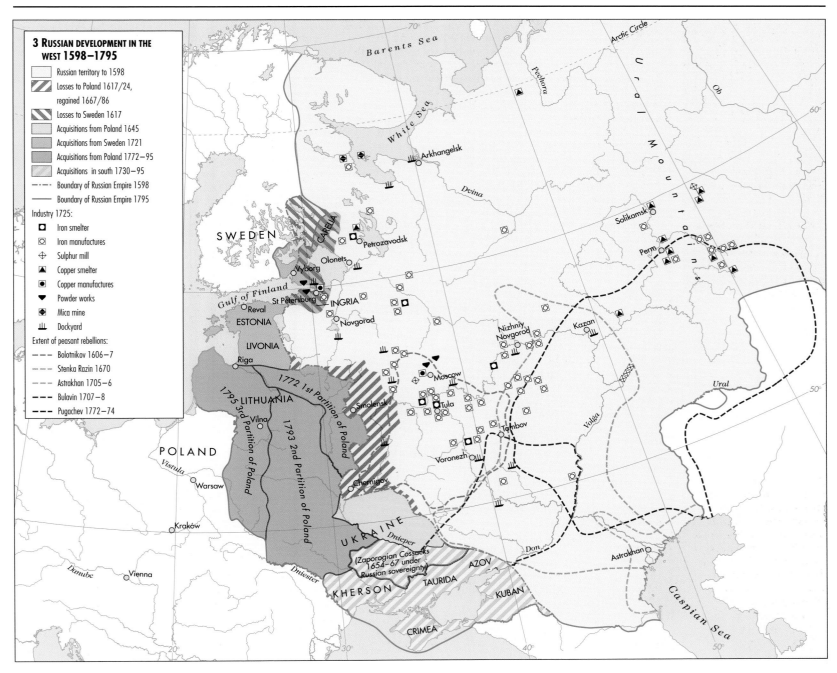

3 RUSSIAN DEVELOPMENT IN THE WEST 1598–1795

- Russian territory to 1598
- Losses to Poland 1617/24, regained 1667/86
- Losses to Sweden 1617
- Acquisitions from Poland 1645
- Acquisitions from Sweden 1721
- Acquisitions from Poland 1772–95
- Acquisitions in south 1730–95
- Boundary of Russian Empire 1598
- Boundary of Russian Empire 1795

Industry 1725:
- ▢ Iron smelter
- ◎ Iron manufactures
- ✛ Sulphur mill
- ▲ Copper smelter
- ◉ Copper manufactures
- ▼ Powder works
- ✦ Mica mine
- ⊞ Dockyard

Extent of peasant rebellions:
- Bolotnikov 1606–7
- Stenka Razin 1670
- Astrakhan 1705–6
- Bulavin 1707–8
- Pugachev 1772–74

had been lost. His death in 1584 unleashed a generation of instability culminating in the "Time of Troubles", a period of political and social upheaval and foreign occupation that was not settled until a national revolt led to the installation of a new dynasty, the Romanovs, in 1613.

At this time Russia's main western enemy was Poland, which took advantage of Russia's internal problems to take back Smolensk and Chernigov in 1618. Another threat was the growing power of Sweden (pages 150–51), which acquired Ingria and Carelia from Russia in 1617. Russia, however, was able to take advantage of the Swedish invasion of Poland in the 1650s to conclude a treaty with the Ukrainian Cossacks and detach them from Poland. Between 1667 and 1689 Russia also regained Smolensk and Chernigov from Poland.

PETER THE GREAT
By the beginning of Peter the Great's reign (1689–1725), Russia had tripled its territory in a century. In Siberia, consolidation was now the order of the day, but in the west, Russia faced the military power of Sweden under Charles XII. As a consequence, the Great Northern War broke out in 1700. Sweden was defeated by Russia in the Battle of Poltava in 1709 (pages 150–51), and the outcome, formalized in 1721, was the acquisition from Sweden of Estonia and Livonia, and the return to Russia of Ingria and Carelia.

The coastal fortresses of Vyborg, Reval and Riga had fallen into Russian hands, and Peter had been able to found the new Baltic port of St Petersburg in 1703 (map 3).

Acquiring a port on the Baltic was one element of Peter's ambitious plans to overhaul the state and "Europeanize" Russia. So, too, was the construction of a navy and the acquisition of a port on the Black Sea. He achieved the latter when he captured Azov in 1696, but he lost it again in 1711 during the Great Northern War. It was not regained until the reign of Anna in 1739. Thereafter, the conquest of the land surrounding the Sea of Azov (Kuban, Crimea and Taurida) had to wait until the 1780s, during the reign of Catherine II (1762–96).

WESTERNIZATION AND THE ECONOMY
In order to compete with other western powers, Russia needed to industrialize. A few ironworks had been set up by foreigners in the 1630s in the Tula and Moscow regions, but Russia remained an overwhelmingly peasant society and lagged far behind western Europe. Peter the Great operated an essentially mercantilist policy, patronizing certain commercial interests in order to encourage export trade. As a result there was rapid growth of both mining and the armaments industry (map 3), but this "forced industrialization", impressive as it seemed at the time, had little impact on the living standards of the peasants.

▲ During the reign of Peter the Great the number of industrial plants increased from about 20 to around 200. Many of these produced armaments, while others were mining and metallurgical plants in the Urals. However, conditions for the vast majority of Russian people – oppressed by both landlords and the state – continued to deteriorate, leading to massive peasant rebellions which periodically convulsed Russia in the 17th and 18th centuries.

◀ THE MONGOL EMPIRE 1206–1405 pages 98–99 ▶ RUSSIAN TERRITORIAL AND ECONOMIC EXPANSION 1795–1914 pages 180–81

SWEDEN, POLAND AND THE BALTIC 1500–1795

▲ Under King Gustav II Adolf (r. 1611–32), Sweden became a major power in the Baltic region. As well as modernizing the army, Gustav introduced a number of constitutional, legal and educational reforms before being killed in battle during the Thirty Years War.

At the beginning of the 16th century the Baltic region was still dominated by power blocks which had been in place for over a hundred years. In Scandinavia the Union of Kalmar, dating from 1397, joined together Denmark, Norway and Sweden–Finland in a loosely governed monarchy centred at Copenhagen. All round the southern Baltic the alliance of free Hanseatic cities, such as Danzig and Lübeck, controlled trade. In the east, the Order of the Teutonic Knights still ruled over a region that included East Prussia, Estonia, Livonia and Courland (*map 1*). The largest country was Poland–Lithuania, created in 1386 when the ruler of the vast Grand Duchy of Lithuania came to the Polish throne.

The Baltic, however, stood on the verge of great changes. Economically, it was already in the process of becoming a major supplier of raw materials to the increasingly urban capitalist society of northwestern Europe. Poland was becoming a major supplier of grain, while furs and hemp from Novgorod and Muscovy, and timber and ores from Sweden, were already major elements in European trade and production. Consequently control of the ports, tolls and waterways to western Europe was an increasingly important factor in the politics of the Baltic region.

A NEW ORDER IN THE BALTIC

In 1521 a Swedish nobleman, Gustav Vasa, led a successful revolution in Stockholm against the Danish king, thus ending the Kalmar Union. Gustav Vasa became king in 1523, beginning a new period of Swedish independence and nationhood. The civil wars which followed in Denmark and Sweden re-established the power of the aristocracy and limited that of the monarchy.

In the 1520s the Reformation (*pages 154–55*) hastened the disintegration of the lands of the Teutonic Order, while in Estonia, Livonia and Courland the Order became fragmented, leading eventually to civil war in 1556–57. The Livonian lands now became a prime object of competition between Poland, Muscovy (Russia), Sweden and Denmark. During the resulting war, the emergence of Sweden as a real power in the Baltic region was confirmed when the Hanseatic port of Reval placed itself under Swedish protection in 1560 (*map 1*). Thereafter, the maintenance of this foothold in Estonia became a

major determinant of Swedish policy – though Denmark, the most powerful state in the region, opposed Swedish pretensions. In 1582 a treaty between Poland and Russia left most of Livonia in Polish hands, and in 1595 Sweden made good its hold on Estonia by signing the Treaty of Teusino with Russia.

At the beginning of the 17th century Denmark was still the leading Baltic power, with control of the Sound – the only deep-water access to the Baltic. As a result of a war with Sweden in 1611–13, it succeeded in expelling the Swedes from their only port on the North Sea (Älvsborg) and gaining trading access to Livonia. However, military intervention in northern Germany in 1625–29 was a disastrous failure and a severe blow to Danish power.

THE RISE AND DECLINE OF SWEDEN

From 1603 Poland and Sweden fought for control of the great Baltic trading centres such as Riga, Dorpat and Reval. King Gustav II Adolf (r.1611–32) of Sweden succeeded in capturing Riga in 1621 and the whole of Livonia by 1625, and the following year he occupied most of the ports along the Prussian coast. The war was only ended by the Truce of Altmark in 1629, allowing Sweden to continue to milk the revenues of the Prussian ports.

By 1630 Sweden was a force to be reckoned with in European politics. Having modernized his armies, King Gustav II Adolf went to war in Germany to counter the threat to Sweden's security posed by the Habsburgs (*pages 152–53*). With his epic march through Germany in 1630–32, Sweden temporarily became the military arbiter of Europe and, despite setbacks in 1634–36, emerged in 1648 as one of the victors of the Thirty Years War (*map 2*).

Sweden's growing ascendancy over Denmark was recognized in 1645 by the Treaty of Brömsebro, which gave

3 SWEDEN IN 1721
- Swedish territory

(Map shows: Stockholm, Kexholm, St Petersburg, Riga, Copenhagen)

▲ The Great Northern War of 1700–21, involving Sweden, Russia and Denmark at different times, finally exhausted Swedish military strength. Treaties in 1719–20 handed Bremen and Verden to Hanover and Stettin to Prussia, and in 1721 the Treaty of Nystadt conceded the loss of Livonia, Estonia and Ingria to Russia. The overseas bases for Sweden's Baltic empire were thus cut away.

1 SWEDISH EXPANSION IN THE 16TH AND 17TH CENTURIES
- Sweden 1560
- Swedish acquisitions 1560–1660
- Swedish colonization in Finland
- Swedish occupation of Russia 1613
- Denmark–Norway
- Occupied first by Denmark, then Sweden 1645
- Seas and lakes frozen in winter
- Personal royal union with date
- Principal trade route
- ○ Hanseatic port 1500
- ⊙ Iron mining
- ▲ Copper mining
- ⊞ Silver mining
- ◎ Gold mining

► In the 16th century Sweden was a small country of just over a million people. However, with the aid of its natural resources, it built a Baltic empire, reaching the summit of its power between 1621 and 1660.

150

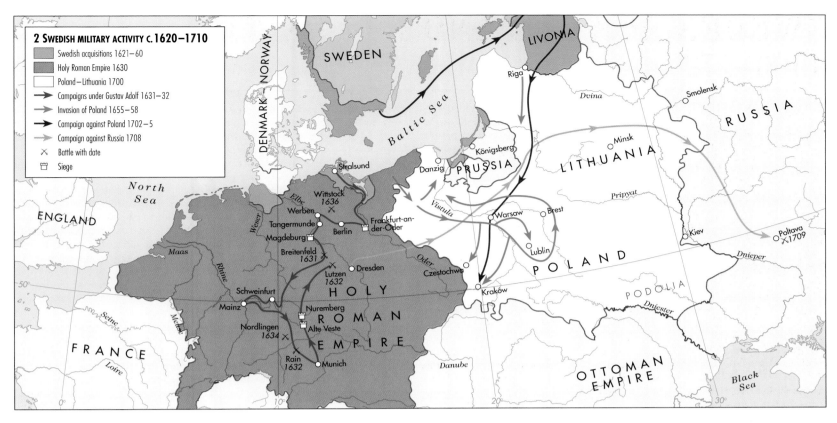

2 SWEDISH MILITARY ACTIVITY c. 1620–1710

Swedish acquisitions 1621–60
Holy Roman Empire 1630
Poland–Lithuania 1700
Campaigns under Gustav Adolf 1631–32
Invasion of Poland 1655–58
Campaign against Poland 1702–5
Campaign against Russia 1708
Battle with date
Siege

Sweden Jämtland and Härjedalen as well as a 20-year lease on Halland and freedom of passage through the Sound. Denmark also conceded Bremen and Verden, confirmed in the Treaty of Westphalia (1648) which also transferred western Pomerania to Sweden. These treaties, however, did not entirely settle the issue of predominance. Sweden still needed to assure its control of the Prussian ports, and in 1655 King Charles X mounted an invasion of Poland that led to its virtual collapse. He then moved against the Danes and in 1658 forced them to abandon their provinces on the Swedish mainland – Bohuslän, Halland, Skåne and Blekinge – as well as Trondheim in Norway (returned in 1660).

The year 1660 marks in some ways the summit of Swedish imperial power based on a military system, both at land and sea, that made Sweden the envy of Europe. There were, however, a number of factors that threatened to weaken Sweden. The population was only a little over a million, and the constitution was liable to sudden fluctuations between limited and absolute monarchy. The possessions in northern Germany were extremely vulnerable and often lost during wars, only to be retained by diplomatic manoeuvres.

The culmination of this was the Great Northern War of 1700–21 and the Battle of Poltava in 1709 between Charles XII and Peter I of Russia (*map 2*). The Treaty of Nystadt in 1721 marked the end of Sweden's hegemony over the Baltic, with the loss of Livonia and Estonia to Russia as well as part of western Pomerania to Prussia (*map 3*).

THE DISINTEGRATION OF POLAND

To the south, Swedish military adventurism was a key factor, along with Russian ambitions (*pages 148–49*), in the disintegration of the Polish state (*map 4*). Poland never recovered from the Swedish occupation of 1655–58, and in 1667 it lost the eastern Ukraine and Smolensk to Russia. Thereafter, Poland became increasingly a plaything of surrounding powers. It was a major theatre of the Great Northern War of 1700–21, and by 1717 Peter the Great of Russia had turned it into a Russian protectorate. When a faction of the Polish nobility began to challenge this from the 1760s, the protectorate ceased to serve a useful purpose and Poland was divided up between Russia, Prussia and Austria in a series of partitions from 1772 to 1795 (*map 5*).

▲ Swedish military power was based on a national standing army established after 1544 by Gustav I. This was supplemented by mercenaries when a larger force was needed for foreign conquest. In the early 17th century the army was further reformed by Gustav II Adolf, paving the way for Swedish success in the Thirty Years War (1618–48) and beyond.

▼ After a brief period as a Russian protectorate, Poland was carved up in the course of three partitions in 1773, 1793 and 1795 between Russia, Austria and Prussia.

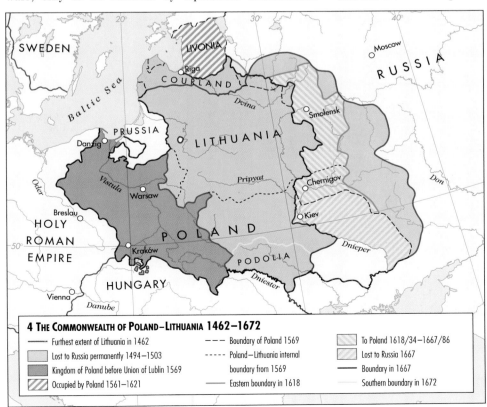

4 THE COMMONWEALTH OF POLAND–LITHUANIA 1462–1672

Furthest extent of Lithuania in 1462
Lost to Russia permanently 1494–1503
Kingdom of Poland before Union of Lublin 1569
Occupied by Poland 1561–1621
Boundary of Poland 1569
Poland–Lithuania internal boundary from 1569
Eastern boundary in 1618
To Poland 1618/34–1667/86
Lost to Russia 1667
Boundary in 1667
Southern boundary in 1672

5 PARTITIONS OF POLAND 1772–95

First Partition 1772	Second Partition 1793	Third Partition 1795
To Prussia	To Prussia	To Prussia
To Russia	To Russia	To Russia
To Austria	Boundary of Poland 1793	To Austria
Boundary of Poland 1772		

◄ Poland–Lithuania first lost ground to Muscovy (Russia) between 1503 and 1521. In 1561, however, Poland gained control of the Courland territory of the Livonian Order and in 1618 regained part of the Smolensk region. Following Swedish invasions in the 1650s and renewed war with Russia, this territory was lost again in 1667.

THE HABSBURG EMPIRE
1490–1700

In 1490 the Habsburg dynasty was just one of a number of ancient dynasties – among them the Valois of France, the Trastamaras of Castile and Aragon and the Jagiellons of Poland, Bohemia and Hungary – that were in the process of creating major princely states. Remarkably, by the 1520s the Habsburgs had accumulated under Emperor Charles V the largest conglomeration of territories and rights since the age of Charlemagne in the 9th century (*map 1*). The military and diplomatic system needed to rule and defend them in the emperor's name was formidable by the standards of the age. Yet in some ways it is a misnomer to talk about a Habsburg "empire" at this time, for Charles ruled his many territories largely through rights of inheritance and they all maintained their separate constitutions.

THE EXTENT OF HABSBURG TERRITORIES

Charles was the grandson of Maximilian I of the House of Habsburg, which had ruled over domains centred on Austria since the 13th century. Holy Roman Emperor from 1493 to 1519, Maximilian gained control, through marriage, of what was left of the territories of the extremely wealthy Valois dukes of Burgundy. In 1506 Charles inherited these territories from his father, Philip the Handsome, and in the course of his reign he made a number of additions (*map 2*). In 1516 he inherited through his mother, Juana, daughter of Isabella of Castile (d.1504) and Ferdinand of Aragon, Spanish territories

that included Majorca, Sicily and Naples. Milan was added to his territories in Italy through conquest in 1522. An alliance was formed with the Genoese Republic in 1528; the defeat of French expeditions to Milan and Naples (1528–29) and the overthrow of the French-backed Florentine Republic in 1530 sealed Habsburg predominance in Italy. Thereafter, French challenges – the occupation of Piedmont in 1536–59 and invasions in 1544 and 1556–57 – proved transitory.

In 1519 Charles was elected Holy Roman Emperor, a role which brought formal prestige as the first prince of Christendom but little more. The King of France, in any case, regarded himself as the equivalent of the emperor in his own kingdom and recognized no superior. Charles ruled more directly as Archduke of the Netherlands and of Austria. Control of the eastern Habsburg lands centred on Vienna was devolved to his brother Ferdinand, who was elected heir to the imperial throne in 1531. Charles's hopes of maintaining his prerogatives as emperor were undermined by the determination of several German princes to defy him over the ban placed on Martin Luther, who had provoked the first serious challenge to the Catholic Church at the Imperial Diet at Worms in 1521 (*pages 154–55*).

In both the Mediterranean and central Europe Charles directly confronted the power of the Ottoman Empire. The Ottomans had occupied Rhodes in 1522 and went on to defeat the Hungarian army in 1526. The Austrian territories

▼ The Habsburg Emperor Charles V presided over a vast collection of territories and faced formidable enemies – Valois France, the Ottoman Empire and various alliances of German princes. In 1556, after Charles's abdication, the empire was divided in two, with Ferdinand I ruling the Austrian domains and Philip II inheriting his father's Spanish lands.

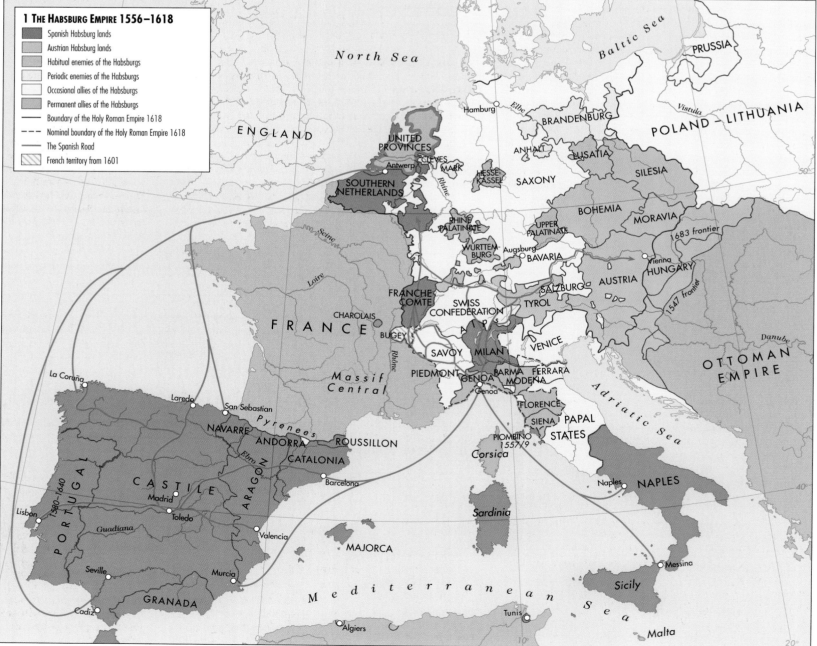

1 THE HABSBURG EMPIRE 1556–1618

- Spanish Habsburg lands
- Austrian Habsburg lands
- Habitual enemies of the Habsburgs
- Periodic enemies of the Habsburgs
- Occasional allies of the Habsburgs
- Permanent allies of the Habsburgs
- —— Boundary of the Holy Roman Empire 1618
- ---- Nominal boundary of the Holy Roman Empire 1618
- —— The Spanish Road
- ▨ French territory from 1601

were therefore in the front line, and Vienna withstood a major siege in 1529 and a threat of one in 1532. The Ottoman threat was only held at bay by the combined dynastic and imperial power of the Habsburgs. In the western Mediterranean Charles sought, through the conquest of Tunis in 1535 and the disastrous expedition against Algiers in 1541, to build on the footholds already acquired in coastal North Africa.

DIVISION OF THE EMPIRE

Charles reached the height of his power at the Battle of Mühlberg in 1547, when he managed to crush the forces of the Protestant rulers of Hesse and Saxony (*pages 154–55*). He then tried to reverse many of the religious and political developments in Germany since the 1520s, but his position quickly began to crumble. In 1552 the rebellion of the League of Princes in Germany allied to Henry II of France forced him to accept that the inheritance was too large to be ruled by one man and that, as a family and dynastic concern, it had to be shared. Consequently, on his abdication in 1556 the empire was divided between his son, Philip II, who inherited the Spanish possessions, and his brother, Ferdinand, who inherited the Austrian domains.

THE EMPIRE IN CENTRAL EUROPE

As Charles's deputy in Germany, Ferdinand I had consolidated the Habsburg family's position as central European dynasts. When King Louis II of Hungary was killed at Mohacs in 1526 (*pages 142–3*) Ferdinand was elected to the Bohemian and Hungarian thrones by the magnates, who saw him as the best guarantor of their safety against the Ottoman Turks. However, Ferdinand was opposed by one Hungarian magnate – Jan Zapolya of Transylvania, who was backed by the Turks – and all that he could salvage of Hungary were the territories of "Royal Hungary" (the west of modern Hungary and modern Slovakia). By the late 16th century these territories were elective monarchies, with large and powerful Protestant nobilities, whose independence Ferdinand II (King of Bohemia from 1617 and of Hungary from 1618, and Holy Roman Emperor 1619–37) became determined to crush, while at the same time reversing the decline in imperial power within Germany.

As a result of the Thirty Years War (1618–48) the Habsburg territories in central Europe were welded into a much more coherent dynastic empire, though the opposition of the princes of the Empire had undermined ambitions in Germany by 1635. With the weakening of the Ottoman Turks in the 17th century, the dynasty was able to begin the piecemeal reconquest of Hungary (*map 3*). Largely complete by the end of the century, this established the Habsburgs as the major dynastic power of central Europe.

THE SPANISH EMPIRE

In the west the Spanish branch of the dynasty descended from Philip II (r. 1556–98) continued the trend which was clear from the middle of Charles V's reign: the development of a Spanish empire that was dependent on the wealth arising from the Castilian conquest of the New World and on the deployment of military power and diplomatic alliances in Europe. Power was transmitted along a series of military routes leading from Spain to the Low Countries known collectively as the "Spanish Road" (*map 1*), and was challenged in the late 16th century by rebels in the Low Countries and by England. Ultimately, Spain proved unable to maintain its control of the northern provinces of the Netherlands and agreed a temporary truce in 1609.

The axis of power between Madrid and Vienna remained vital to the Spanish system and was reinforced as the Habsburgs in central Europe came under pressure from rebellious nobles and Protestants. The axis was reaffirmed in 1615 and Spanish troops were deployed in central Europe and the Rhineland from 1619, while war was renewed with

the Dutch in 1621. The last phase of the Spanish military system in western Europe showed that it was remarkably resilient in the face of massive setbacks such as the rebellions in Portugal and Catalonia in 1640 and the defeats in the Low Countries by France at Lens in 1643 and Rocroi in 1648 (*pages 158–59*). Nevertheless, the Treaty of Westphalia in 1648 forced the recognition of the independence of the United Provinces, and the Peace of the Pyrenees with France in 1659 registered a serious shift in the balance of power towards France. For the rest of the 17th century, Spain and its dependencies were constantly on the defensive. They were certainly not in a position to aid the Austrian Habsburgs, who had to contend with the last great advance of the Ottomans (*map 3*). This reached its most western limit in 1683 but would continue to pose a threat well into the following century.

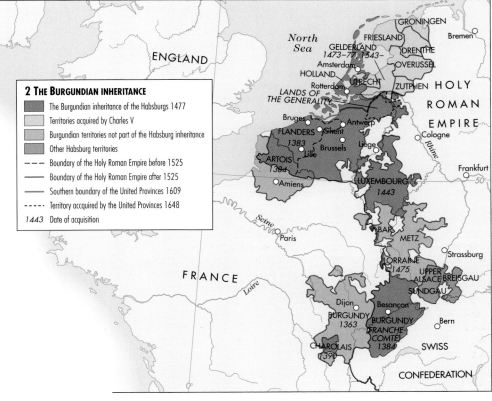

2 THE BURGUNDIAN INHERITANCE

- The Burgundian inheritance of the Habsburgs 1477
- Territories acquired by Charles V
- Burgundian territories not part of the Habsburg inheritance
- Other Habsburg territories
- --- Boundary of the Holy Roman Empire before 1525
- —— Boundary of the Holy Roman Empire after 1525
- —— Southern boundary of the United Provinces 1609
- ---- Territory accquired by the United Provinces 1648
- *1443* Date of acquisition

▲ The lands which Charles V inherited in 1506 consisted of most of the provinces of the Netherlands and the free county of Burgundy, but not the duchy of Burgundy, which had been confiscated by Louis XI of France in 1477. In the course of his reign Charles annexed Gelderland, Groningen, Friesland and the bishopric of Utrecht. His successor, Philip II, faced serious opposition from the nobility from 1565 and a full-scale revolt in Holland from 1572. This led to the formal repudiation of Philip in 1581 by what were to become the seven United Provinces of the Netherlands.

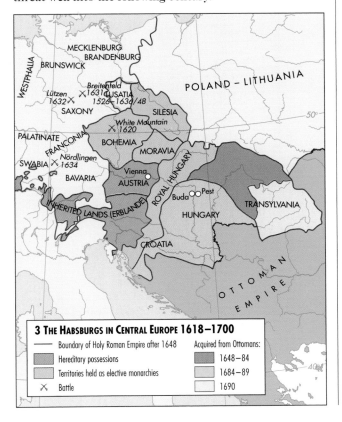

3 THE HABSBURGS IN CENTRAL EUROPE 1618–1700

- —— Boundary of Holy Roman Empire after 1648
- Hereditary possessions
- Territories held as elective monarchies
- ✕ Battle

Acquired from Ottomans:
- 1648–84
- 1684–89
- 1690

◄ During the 16th and 17th centuries the Austrian Habsburgs extended their territory across Hungary and along the Danube as far east as Transylvania. However, in 1682 the Ottomans claimed Hungary as a vassal state and sent an army of 200,000 men to advance on Vienna. The subsequent two-month siege of the city in 1683 was only lifted when a Polish army attacked the Ottoman forces and sent them into retreat. The Habsburgs eventually regained Hungary from the Ottomans under the Treaty of Carlowitz in 1699.

THE REFORMATION AND COUNTER-REFORMATION IN EUROPE 1517–1648

▼ Protestantism took a number of forms across Europe. In Germany and Scandinavia local secular rulers promoted the establishment of new churches, mostly along Lutheran lines. In the Netherlands, Calvinism became politically predominant during the later 16th century, while in England the Anglican Church under Elizabeth I was Calvinist with an episcopal government. Further east, Calvinism was adopted in Transylvania (in Hungary) – and in Poland so many nobles became Protestant that special provisions for their toleration had to be agreed in 1569–71.

The Reformation is commonly associated with an outraged response to the corruption of the Church in the late 15th and early 16th centuries. In fact, the corruption of the Church had come under attack before. What was new at this time was the emergence of a powerful force of religious revivalism which swept across Europe and sought an increased role for the laity in religious life.

THE IMPACT OF LUTHERANISM

The Protestant Reformation is traditionally dated from 31 October 1517 when Martin Luther's *Ninety-Five Theses* against indulgences (documents sold by the Church which were widely thought to remit the punishments of purgatory) were posted on the door of the castle church at Wittenburg in Saxony. Luther's *Theses* provoked a hostile reaction from the upper hierarchy of the Church. Moreover, the circulation of printed copies of the *Theses* and other writings meant that they received the attention of a wider public than might otherwise have been the case. His attack on financial abuses within the Church, and his emphasis on the spiritual nature of Christianity and the teachings of the gospel, found support among a broad range of the laity.

Before 1517 reform of the Church had been seen as a legitimate objective; now Luther's call for "reformation" was regarded as a fundamental threat to both the Church and the Holy Roman Empire. Luther was excommunicated in 1521 after denying the primacy of the Pope, and later that year he was placed under an imperial ban.

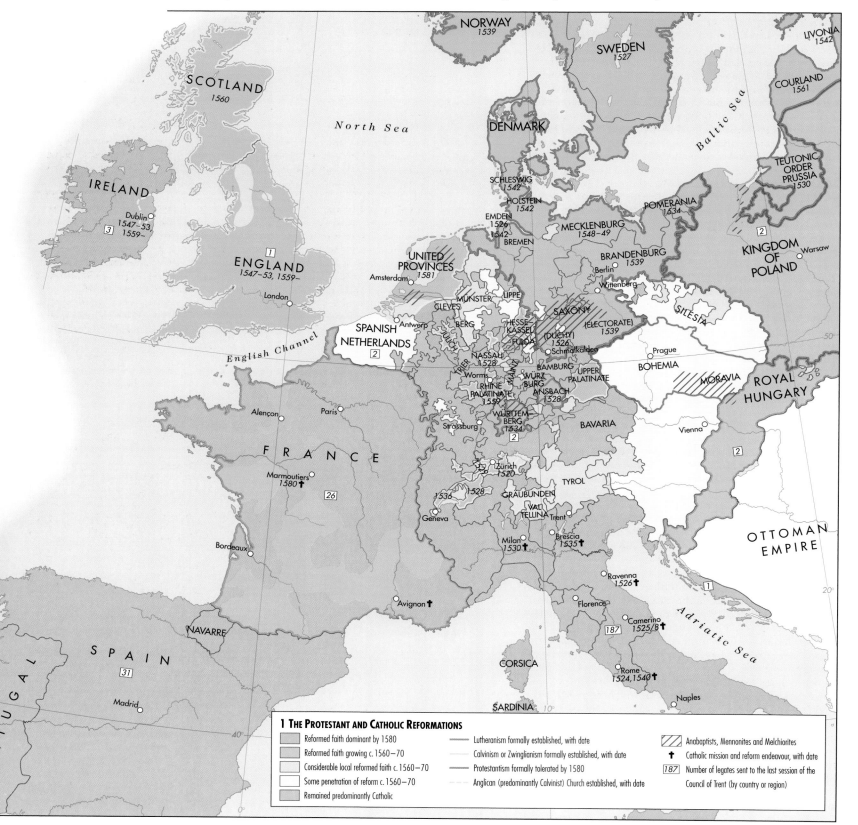

1 THE PROTESTANT AND CATHOLIC REFORMATIONS

Reformed faith dominant by 1580	
Reformed faith growing c. 1560–70	
Considerable local reformed faith c. 1560–70	
Some penetration of reform c. 1560–70	
Remained predominantly Catholic	

— Lutheranism formally established, with date
— Calvinism or Zwinglianism formally established, with date
— Protestantism formally tolerated by 1580
--- Anglican (predominantly Calvinist) Church established, with date

▨ Anabaptists, Mennonites and Melchiorites
✝ Catholic mission and reform endeavour, with date
187 Number of legates sent to the last session of the Council of Trent (by country or region)

A number of German princes broke with Rome and adopted Lutheranism, gaining stronger political control over the Church in their own territories as a result. This was met with fierce opposition from Charles V at the Augsburg Reichstag in 1530, and in response a League of Protestant estates – including Hesse, Saxony, Wurttemberg, the Palatinate and several imperial cities – was formed at Schmalkalden, thus splitting the Empire into two warring camps. It was not until 1555 that Charles V was finally forced to concede the Peace of Augsburg, granting full rights to the secular estates of the Empire to adopt Lutheran reform.

RADICAL REFORMATION

The reform movement spread rapidly (*map 1*) but for many it was the ideas of local reformers that mattered most. By the end of the 1520s a split between the Lutheran Reformation and the radical (or Reformed) churches was clear. Thomas Müntzer encouraged a more radical view that was to culminate in the "Kingdom of Zion" of the Anabaptists at Münster, while in Zürich Huldreich Zwingli led a reformation which differed from Lutheranism over, among other things, the sacrament of Communion.

Protestantism in Switzerland received a blow with the death of Zwingli in battle in 1531, but it was ultimately revived by Calvin, a humanist and lawyer born in northern France. Calvin, who controlled the Genevan church by 1541 (*map 2*), gave the French-speaking world a coherent and incisive doctrine as well as an effective organization. He proved to be the most significant influence on the emergence of the Reformation in France from the 1540s onwards, when he sent out a network of preachers to the main French cities. By 1557 an underground church was in existence and in 1559 it declared itself openly.

THE COUNTER REFORMATION

In Spain and Italy, where Spanish power posed a significant block to Protestantism, the internal reform of the Catholic Church was pushed forward by the foundation of many new religious orders devoted to charitable and evangelical work in the lay world, as well as by the militant Society of Jesus (Jesuits) founded by Ignatius Loyola in 1534.

Within the Catholic Church as whole, the establishment of the means to resist Protestantism was a priority. The three sessions of the General Council of Trent held between 1545 and 1563 restated theological doctrine in a way which precluded reunion with Protestants, and a series of decrees aimed at reforming the clergy and church organization was issued. Although the pronouncements of the Council of Trent were not immediately translated into action, the Council signalled that the Catholic Church was to become an evangelical movement, seeking to win converts both among heretics in Europe and the "pagans" of the overseas world. Crucial in this process was the growing identification between the Catholic Church and absolute monarchs, who had the power, through patronage, to win back disaffected nobles to the Roman Catholic faith.

In France, although the Jesuits were at first not allowed to preach, a resurgence of Catholic piety and fundamentalism eventually put a limit to any further expansion of Protestantism. When Catherine de Medici (the Queen Mother) ordered the liquidation of the Protestant leadership on the eve of St Bartholomew's Day 1572, mass fanaticism led to the massacre of 10–12,000 Protestants throughout the country (*map 3*). The ensuing factional chaos enabled Protestants to extract from the French crown a lasting guarantee of religious toleration in the Edict of Nantes (1598), but this in effect confirmed their minority status. When their guaranteed strongholds (*places de sureté*) were removed by the Crown in the 1620s, they were reduced to a position of sufferance. In 1685 the Edict was revoked and around 200,000 Protestants (Huguenots) were forced to convert to Catholicism or flee the country.

In the Netherlands a Calvinist minority seized power in Holland and Zeeland in 1572 but had to fight a bitter and prolonged war with Spain which was to last until 1648. In Germany the Peace of Augsburg (1555) began to break down. Some princes converted to Calvinism in defiance of the Peace, and the spread of Catholic evangelism (and Protestant fears of Catholic acts of revenge) created enormous tension in the Holy Roman Empire, culminating in the start of the Thirty Years War in 1618 (*pages 158–59*).

By the end of the war in 1648, when the Treaty of Westphalia recognized a new order in Europe, Roman Catholicism had been re-established in France, Poland, Hungary and Bohemia. However, there was no return to religious war and, to some extent, religious pluralism was reluctantly accepted between, if not within, states.

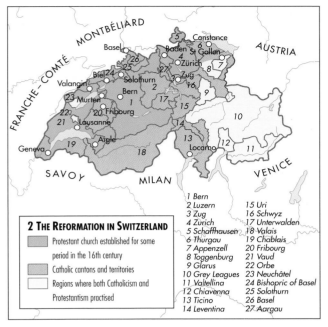

2 THE REFORMATION IN SWITZERLAND

- Protestant church established for some period in the 16th century
- Catholic cantons and territories
- Regions where both Catholicism and Protestantism practised

1 Bern	15 Uri
2 Luzern	16 Schwyz
3 Zug	17 Unterwalden
4 Zürich	18 Valais
5 Schaffhausen	19 Chablais
6 Thurgau	20 Fribourg
7 Appenzell	21 Vaud
8 Toggenburg	22 Orbe
9 Glarus	23 Neuchâtel
10 Grey Leagues	24 Bishopric of Basel
11 Valtellina	25 Solothurn
12 Chiavenna	26 Basel
13 Ticino	27 Aargau
14 Leventina	

◄ Switzerland was a major powerhouse of the Protestant Reformation but was intensely divided. The inner "forest" cantons were hostile to Zwingli and feared the power of Zürich where he was based. After his death in 1531 Bern took up the military leadership of Protestantism, giving its protection to Geneva which, although not technically part of the Swiss Confederation, was to become the centre of Swiss Protestant doctrine.

▼ French Protestantism was overwhelmingly urban. Crucial to its survival, however, was the support of a very large minority of the nobility. Its greatest concentration was eventually in a "crescent" stretching from Dauphiné in the east to Poitou in the west. This was largely a result of the course of the French Wars of Religion (1562–98) which rendered life precarious for Protestants north of the Loire, especially after the St Bartholomew's Day Massacre in August 1572.

3 THE REFORMATION AND RELIGIOUS CONFLICT IN FRANCE

- Protestant church established for some period in the 16th century
- Site of Catholic massacre of Protestants August 1572
- Protestant Academy
- Court for judging cases between Catholics and Protestants (from 1576)
- Place de sûreté

● EUROPE 1350–1500 *pages 106–7* ● REVOLUTION AND STABILITY IN EUROPE 1600–1785 *pages 156–57*

REVOLUTION AND STABILITY IN EUROPE 1600–1785

▲ The trial and execution of Charles I of England, Scotland and Ireland (*top*) in January 1649 was followed by the abolition of the monarchy and the declaration of a republic. Oliver Cromwell (*bottom*) came to prominence as a military leader during the Civil War of 1642–48 between supporters of the king and of Parliament. When parliamentary government failed in 1653 he became Lord Protector and proceeded to rule England until his death in 1658.

▶ It has been suggested that a general crisis in the 17th century, in which wars and revolts broke out across Europe, reflected global factors – in particular, a deterioration in climate that led to famine, mass migrations and a halt in population growth. It is in fact the case that there were plague epidemics in Europe and China in the 1640s as well as parallel political upheavals.

In the 17th century the major states of Europe were embroiled in the long conflict in central Europe known as the Thirty Years War (*pages 158–59*), which combined dynastic and strategic conflict with religious struggles, the latter breaking out both within and between states. The growth of armies and of military technology in this period (*pages 158–59*) could only be achieved through an increase in taxation that was so large as to challenge the basis on which states had been governed since the late 15th century.

REBELLION AND CIVIL WAR

When Spain intervened in Germany on behalf of the Austrian Habsburg emperor in 1619, and then renewed its conflict with the Dutch in 1621, it became committed to massive military expenditure which devastated its finances. In Castile, which had undergone a loss of population since the 1590s, the monarchy found the burden increasingly difficult to bear. Unable to solve the problem by concluding peace, the government restructured the tax system so that the hitherto privileged regions of Portugal, Aragon, Catalonia and Naples bore a greater share of the tax burden. This caused a national uprising in Portugal in 1640, followed by rebellions in Catalonia (1640–53) and in Naples (1647–8) (*map 1*). All this nearly brought down the Spanish state.

In France – governed by Cardinal Richelieu from 1624 – the steadily increasing tax burden was accompanied by an increase in royal tax officialdom at the expense of the local machinery of voting taxes through representative assemblies. In addition to the massive increases in direct taxes from 1635 (when France formally entered the war against Spain and the Habsburgs) and the spread of a whole range of indirect revenues such as those on salt (the *gabelle*), the direct costs of billeting and supplying the army were borne by the civil population with increasing reluctance. From around 1630 numerous local revolts broke out, often supported by regional notables resentful at the infringements of their privileges by the Crown. In 1636–37 the Crown was faced by a large-scale rebellion in the southwest which brought together under the name of *Croquants* many peasant communities outraged by army taxes. In lower Normandy in 1639 the *Nu-Pieds* rebelled against the extension of the full salt tax regime to that area.

Cardinal Mazarin succeeded Richelieu as Chief Minister in 1643 and continued the same policies of high taxes and prolonged war against Spain, even after the Treaty of Westphalia in 1648. By then the Crown faced not only a discontented peasantry but also opposition from within the royal

bureaucracy over the suspension of salaries, and a nobility unhappy with the exercise of power by the Chief Minister. The result was a confused period of civil war known as the *Frondes*, which paralysed French policy until 1653.

CRISIS ACROSS EUROPE

In Britain the attempts of Charles I to impose his religious policies on the Scots exposed the weakness at the core of the Stuart monarchy. Charles attempted to govern and raise revenues without Parliament throughout the 1630s, but he was confronted by a tax-payers' revolt and by the fact that he could not raise an army without some form of parliamentary grant. The summoning of Parliament in 1640 triggered a sequence of events that imposed shackles on the king's powers and then provoked him to try a military solution. The resulting civil war (1642–48) led to the king's execution and the proclamation of a republic in 1649. Opposition in Ireland and Scotland was crushed in 1649–50 by the New Model Army under Oliver Cromwell. In 1653 the republic was replaced by a military dictatorship, with Cromwell as "Lord Protector".

During the same period, in the United Provinces of the Netherlands (formed in 1579 after the Protestant Prince William I of Orange led a revolt against Spanish Catholic rule), an attempt to impose quasi-royal rule under William II of Orange collapsed and the Orangist Party was purged from positions of power by the oligarchic States Party. There were also struggles for power in Sweden, and in the 1620s and 1630s large-scale peasant revolts broke out in the Alpine territories of the Austrian Habsburgs. Further east, Cossack rebellions flared up in the Polish Commonwealth in the 1640s and 1650s and in Russia in the 1670s.

Not surprisingly, some contemporaries saw a pattern in all this. The English preacher Jeremiah Whittaker declared in 1643 that "these are days of shaking and this shaking is universal". Some modern historians have discerned a systematic "general crisis" in which the political upheavals of the mid-17th century were a symptom of profound economic transformation. In contrast, the trend throughout Europe after 1660 was towards political stability.

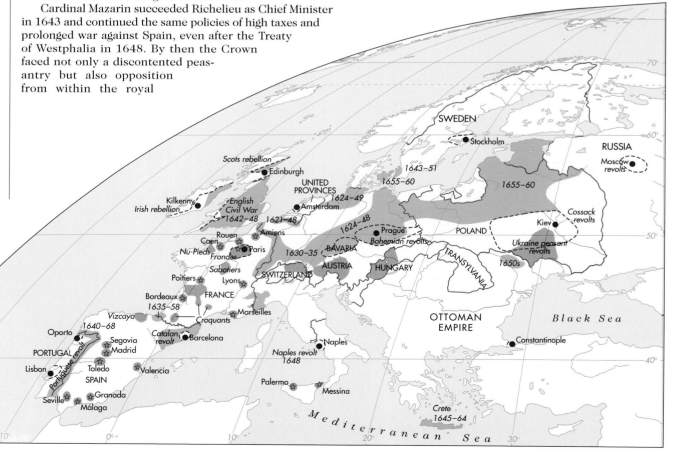

1 WARS AND REVOLTS IN EUROPE 1618–1680

Areas affected by:
- war
- popular revolt
- --- political revolt
- ☆ Centre of popular revolt
- ● Centre of political revolt

THE ESTABLISHMENT OF STABILITY

Peasant revolts continued in France until the 1670s. However, despite the continuation of severe economic problems and the massive growth of armed forces to enable the annexation of territory (*map 2*), these revolts did not seriously threaten the state. After Mazarin's death in 1661 Louis XIV assumed personal rule, which deflected the discontent of the nobility and assuaged the conflicts between government, officialdom and the courts. Thereafter he ruled as absolute monarch with the aid of a centralized bureaucratic government – a pattern which was to continue until 1789. Without any significant opposition, Louis was able to impose religious uniformity in 1685.

The doctrine of "absolute power", though not new, became the keynote for many rulers eager to imitate the splendours of Louis' court at Versailles. In east-central Europe the Hohenzollerns – rulers of Brandenburg and Prussia – gradually increased their power after the Elector Frederick William I came to an agreement with the nobility, under which his military powers were extended in return for the reinforcement of their controls over their tenantry. By the middle of the 18th century the power of the Prussian state (*map 3*) equalled that of the Habsburgs in Vienna, who were themselves building an empire in the Danubian region (*pages 152–53*).

CONCERT OF EUROPE

Elsewhere in Europe the defeat of the monarchy led to the emergence of oligarchic parliamentary systems – Britain from 1689, the United Provinces from 1702, Sweden from 1721. In Spain, the regime of the Bourbon dynasty, confirmed by the Treaty of Utrecht of 1713, imposed a centralized government on the French model. Thus, although major wars continued to be endemic and commercial rivalry both in Europe and overseas was fierce, governments were far more securely anchored than in the earlier 17th century. Religious uniformity, while still formally insisted on, was in practice no longer so vital. A Europe in which one or other dynastic state (Spain in the 16th century, France in the 17th century) threatened to dominate the rest had been replaced by a "concert of Europe" of roughly balanced powers that was to last until the revolutionary period in the 1790s.

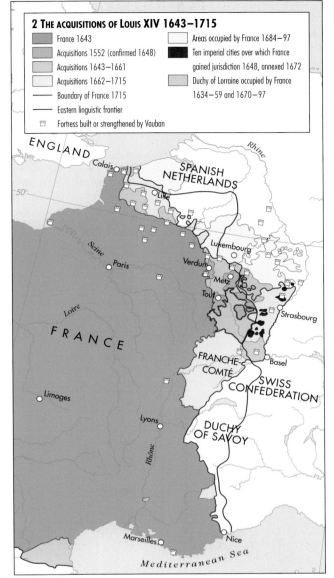

2 THE ACQUISITIONS OF LOUIS XIV 1643–1715

- France 1643
- Acquisitions 1552 (confirmed 1648)
- Acquisitions 1643–1661
- Acquisitions 1662–1715
- Areas occupied by France 1684–97
- Ten imperial cities over which France gained jurisdiction 1648, annexed 1672
- Duchy of Lorraine occupied by France 1634–59 and 1670–97
- Boundary of France 1715
- Eastern linguistic frontier
- Fortress built or strengthened by Vauban

◀ From the 1660s Louis XIV built on acquisitions made under Cardinal Richelieu to expand French territory at the expense of the Holy Roman Empire. The high point of his achievements came in 1684 when his acquisition of Luxembourg during a war with Spain and the Empire was confirmed by the Treaty of Regensburg. From 1685 the threat he posed to other powers led to a series of alliances being formed against him. Eventually, the Treaty of Utrecht (1713) placed limits on French expansion.

▼ The duchy of Prussia, founded in 1525 out of the remaining lands of the Teutonic Knights, passed to the Hohenzollern electors of Brandenburg in 1618. Under Elector Frederick William I (1640–88), Brandenburg–Prussia did well out of the Peace of Westphalia in 1648 and the Northern War (1655–60) to extend its territories. His successors continued the process of expansion until Frederick the Great (1740–86) put the seal on the emergence of Prussia as a great power by his successful annexation of Silesia in the War of the Austrian Succession (1740–48).

3 THE EXPANSION OF PRUSSIA 1618–1795

- Brandenburg–Prussia 1618

Acquisitions:
- to 1688
- 1688–1713
- 1713–1740
- 1740–1786

1772 Year of acquisition or period of possession
— Boundary of the Holy Roman Empire 1786
✕ Battle with date

THE DEVELOPMENT OF WARFARE IN EUROPE 1450–1750

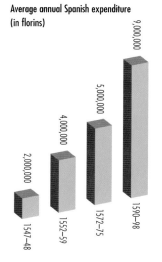

▲ By the late 16th century the military expenditure of the Spanish monarchy had placed a severe burden on Castile. Philip II's armies were periodically left without pay, resulting in nine major mutinies in the army of Flanders between 1570 and 1607.

▼ The development of frontiers was accompanied by the construction of linear networks of fortifications, for example in northern France and in Russia. The Habsburgs established a militarily governed frontier zone in Hungary and Croatia, in which soldiers (often Serbs) were settled in villages for defence against the Ottomans.

Between the 15th and 18th centuries European warfare was massively transformed in scale and complexity, and this had a powerful impact on both state and society. It has been argued that the transformation amounted to a "military revolution" led by the Swedes and the Dutch in the decades around 1600. However, this view underestimates the role of France and Spain, and the process of military change is now seen as one that was evolutionary rather than revolutionary.

ARMIES AND THE STATE

The driving force behind military change was the development of a highly competitive state structure, both regionally (as in 15th-century Italy) and across Europe. Countries which had not invested in major military reorganization by the 17th century – such as Poland – were seriously disadvantaged, but in those countries where military expenditure was high the impact was felt at all levels of society. Governments needed to be able to mobilize resources for war on a large scale, and this led to many western European states becoming "machines built for the battlefield", their essential purpose being to raise, provision and deploy armies in the pursuit of their ruler's strategic objectives. In going to war, European rulers in the 16th and 17th centuries were primarily concerned with safeguarding the interests of their dynasties, as in the case of the Italian and Habsburg-Valois Wars in the 1520s to 1550s (*pages 146–47, 152–53*), although at times religious and commercial

concerns also played a role. In addition there were several civil wars involving a degree of ideological or religious dispute, such as the French Wars of Religion (1562–98) and the English Civil War (1642–48).

ARTILLERY AND SIEGE WARFARE

Changes in warfare were made possible by a number of crucial technical innovations. First, the growing sophistication of artillery in the 15th century altered the terms of war in favour of attack. In mid-15th-century France, more effective, smaller-calibre bronze cannons replaced the existing, unreliable wrought-iron version. One of the most widely noted features of Charles VIII's invasion of Italy in 1494 was his deployment of the formidable French royal artillery. Bronze, however, was expensive, and the next important development was the manufacture of reliable cast-iron guns in England during the 1540s. Cast-iron guns were three or four times cheaper than their bronze equivalents, and the traditional cannon foundries of Europe were unable to compete until the next century.

The earliest cannons were huge and unwieldy, best suited for sieges. The major powers – Italy, France and Spain – therefore embarked on highly expensive programmes of refortification to render fortresses and cities impregnable to artillery bombardment. By the late 16th century, high and relatively thin walls and towers had given way to earthwork constructions consisting of ditches and ramparts which were to dominate the landscape of many

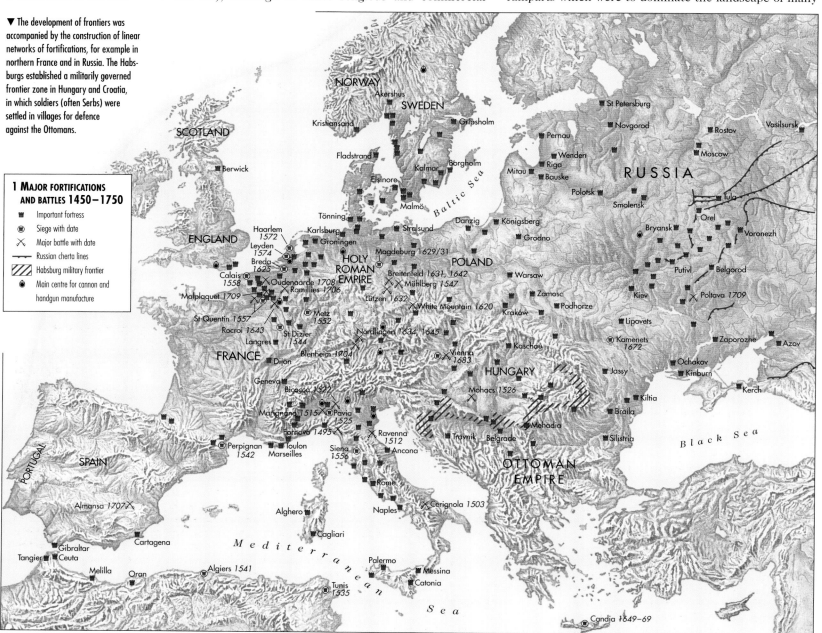

1 MAJOR FORTIFICATIONS AND BATTLES 1450–1750
- ▪ Important fortress
- ◉ Siege with date
- ✕ Major battle with date
- — Russian cherta lines
- ▨ Habsburg military frontier
- ◉ Main centre for cannon and handgun manufacture

2 THE THIRTY YEARS WAR 1618–48
→ Military intervention
✕ Battle with date
● Town sacked or plundered
— Boundary of the Holy Roman Empire 1618
Theatres of war:
--- Bohemian War 1618–20 and 1621–23
--- Lower Saxon–Danish and Polish–Swedish War 1625–29
--- Swedish War 1630–34
--- Franco-Swedish War 1635–48

◄ The Thirty Years War was in fact a complex of wars which combined dynastic and strategic conflict with religious struggles, the latter breaking out both within and between states. Germany became a battleground in which all the military powers developed and tested their strength; the armies frequently plundered towns, villages and farms for supplies, adding to the devastation. Each phase of the war saw a widening area of operations. The Holy Roman Emperor's power was at its height in 1629 but thereafter began to collapse. Foreign intervention prolonged the war from 1635 to 1648.

THE COMPOSITION OF ARMIES

Spanish Army of Flanders 1575

Spanish Army of Flanders 1640

▢ Walloons   ▢ Cavalry
▢ Spaniards   ▢ Italians
▢ Germans   ▢ British
              ▢ Burgundians

French Royal Army 1552

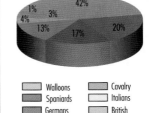

French Royal Army 1562–69

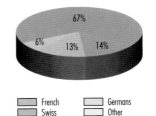

▢ French   ▢ Germans
▢ Swiss   ▢ Other

▲ During the 16th century foreign mercenaries frequently outnumbered national subjects in the armies of the kings of France and Spain. Gradually the Italians, who had been the great soldiers of fortune in the 15th century, were supplanted first by the Swiss and then by Germans from the Rhineland and Westphalia. English mercenaries served in the Netherlands in the later 16th century, and Scots were particularly active in Germany during the Thirty Years War.

European cities until they were dismantled in the 19th century. This rendered warfare much more static, with campaigns centring on great siege operations; some of the major battles of the period – Pavia (1525), St-Quentin (1557), Nördlingen (1634), Rocroi (1643) and Vienna (1683) were linked to such sieges (*map 1*). As a consequence of these developments in siege warfare, wars of rapid movement of the kind embarked upon by the English in 14th-century France became unthinkable.

CHANGES ON THE BATTLEFIELD
Artillery had its place on the battlefield, but because of difficulties in using it tactically, it was slow to gain dominance. A further agent for change was the application of a diversity of armaments, formations and tactics: heavily armed cavalry gradually gave way to massed ranks of pikemen and, from the early 16th century onwards, archers began to be replaced by infantry armed with handguns. At Ravenna (1512), Marignano (1515) and Bicocca (1522), field artillery and handguns inflicted severe casualties on pike squares. To combat this, large mixed infantry formations were used, armed partly with pikes and partly with muskets.

Despite these developments, the heavy cavalry did not disappear; in fact cavalry in general was overhauled to make

use of firearms, most notably among the German *reiters*. Commanders now sought to organize infantry and cavalry more effectively. However, it was still difficult to manoeuvre large groups of men on the battlefield, especially since the main battles consisted of vast squares of infantrymen. The necessity of increasing the rate of fire of handguns led to the development by the Dutch armies in the 1590s of "volley fire", in which the infantry was laid out in long lines, firing rank after rank. The development of the "countermarch" – a combination of volley fire, advancing ranks and cavalry charging with their swords drawn – gave the Swedish king Gustav Adolf's armies the crucial edge in the 1630s, for example in the Battle of Breitenfeld in 1631 (*map 2*).

All these changes meant that battles took place over larger areas and involved greater numbers of soldiers. In 1525, at Pavia, the French king's army of 28,000 men was defeated by a Habsburg army of 20,000; at Breitenfeld Gustav Adolf had 41,000 against 31,000 Habsburg troops; in 1709, at Malplaquet, a French army of 76,000 faced an Allied army of 105,000. While the maximum number sustainable for a whole campaign in the mid-16th century seems to have been about 50,000, by 1700 the number was around 200,000 and by 1710 France, for example, could sustain a total military establishment of 310,000 men.

◖ EUROPE 1350–1500 *pages 106–7* ◖ REVOLUTIONARY FRANCE AND NAPOLEONIC EUROPE 1789–1815 *pages 166–67*

THE AGE OF REVOLUTIONS

Between 1770 and the outbreak of the First World War in 1914 a succession of revolutions, industrial as well as political, brought widespread material progress and social change. These developments were international in character although their global impact was unevenly distributed. They had a common origin in the unparalleled expansion of European influence – economic, political, demographic and cultural – throughout the world.

In this period most of the Americas, Africa and Australasia, together with much of Asia, became dominated either by European states, or by peoples of European culture and descent. This process, which slowly but surely transformed the character of global civilization and forged the modern world, was based largely on Europe's economic and technological ascendancy. By the mid-18th century European commercial primacy was already established, but its lead in manufacturing was apparent only in some areas, such as armaments, ships and books, and it lagged behind Asia in a few fields, such as porcelain and textile manufacture.

In the later 18th and early 19th centuries there was a new wave of economic growth and development, first in Britain and then in northwest Europe. This involved the concentration and mechanization of manufacturing in factories, and the use of coal to generate steam power – changes which, while not entirely replacing domestic production or more traditional energy sources, revolutionized production, initially of textiles and iron and subsequently of other industries. Later

known as the "Industrial Revolution", the changes led to such a rapid increase in manufacturing that by the middle of the 19th century Britain was described as "the workshop of the world".

THE SPREAD OF INDUSTRIALIZATION

During the 19th century, industrialization spread first to northwestern Europe and the eastern states of the United States, and then further afield. This led to an enormous increase in world trade (which trebled between 1870 and 1914) and in mass manufacturing. By 1900 both the United States and Germany surpassed Britain in some areas of production, such as that of iron and steel. Despite this, Britain remained the leading international trader and investor, with London the centre of the world capital market and of the international gold standard. Britain was also the most urbanized society in the world, with only a tiny minority of its population directly working in agriculture.

Elsewhere, the majority of the population – even in developed countries such as the United States and France – still lived and worked in rural areas, much as their forebears had done. Global trade,

▶ In the mid-19th century Britain was the world's leading industrial nation, although the process of industrialization was gathering momentum in continental Europe and the United States. Britain's leading position was demonstrated by the Great Exhibition, which opened in London in 1851 and contained over 7,000 British and as many foreign exhibits divided into four main categories: raw materials, machinery, manufactures and fine arts. It was housed in a specially built iron and glass exhibition hall (the "Crystal Palace") which was itself a fine example of British engineering skills.

industrialization and urbanization were still relatively undeveloped in 1914, yet Western innovations had already transformed many aspects of life throughout the world. Steam power provided energy not only for factories but also for railways and ocean-going ships, which, along with the telegraph and later the telephone, dramatically reduced the time and cost of long-distance transport and communications.

POLITICAL REVOLUTIONS

In the political sphere the American Revolution of 1775–83, which ended British rule over the Thirteen Colonies, was followed by the French Revolution, which began in 1789 and signalled a new era in the "Old World". Tom Paine, an influential transatlantic radical wrote in 1791: "It is an age of Revolutions in which everything may be looked for." His optimism was premature, however, for the French Revolution failed in both its Jacobin and Napoleonic forms and was followed, after 1815, by a period of reaction in Europe, led by the autocratic rulers of Russia, Austria and Prussia.

This did not, however, prevent the growth of Liberalism in Europe, which led to revolutions in France and Belgium in 1830 and to reforms in other countries such as Britain. In 1848 there were further revolutions in France and Germany which, although not entirely successful, led to the democratization of political institutions in western Europe. By the early 20th century all European states, including Russia, had representative assemblies, most of which were elected by a wide adult male suffrage. Women were still generally excluded from the franchise, but this restriction was being challenged and undermined by campaigners in Europe and North America. In the United States and the British dominions most white men and some women could vote, but not the non-European ethnic groups.

In most of the world non-democratic forms of government prevailed (*map 1*). In both the Middle and the Far East, dynastic rulers with autocratic powers flourished until the second decade of the 20th century. In the Asian, African and Caribbean colonies of the European powers, the native inhabitants were generally not allowed any direct voice in government. Even in Europe, democracy developed under the cloak of a much older and more absolutist political tradition: hereditary monarchy. France was the only major European power to become a republic before 1917. Bismarck – the dominant political figure in late 19th-century Europe – remained Chancellor of Germany only as long as he retained the support of the kaiser. The importance of hereditary dynasties in the European state system was illustrated when the murder of the Austrian archduke, Franz Ferdinand, at Sarajevo in 1914 precipitated the First World War.

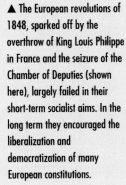

▲ The European revolutions of 1848, sparked off by the overthrow of King Louis Philippe in France and the seizure of the Chamber of Deputies (shown here), largely failed in their short-term socialist aims. In the long term they encouraged the liberalization and democratization of many European constitutions.

◄ All independent countries in the Americas embraced republicanism during the 19th century, although the franchise was usually extremely limited and elections were often suspended. By 1914 much of Europe was ruled by elected governments, although outside France and Portugal monarchs still acted as heads of state. The extent to which they actually exercised power varied from country to country, as did the proportion of citizens entitled to vote. Those areas of Asia and Africa not under European control or influence were ruled by autocratic monarchs.

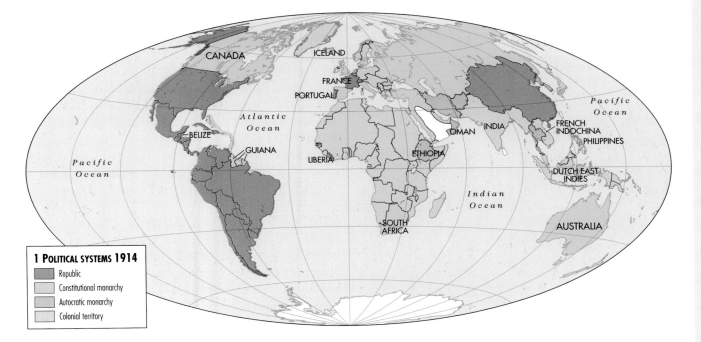

1 POLITICAL SYSTEMS 1914
- Republic
- Constitutional monarchy
- Autocratic monarchy
- Colonial territory

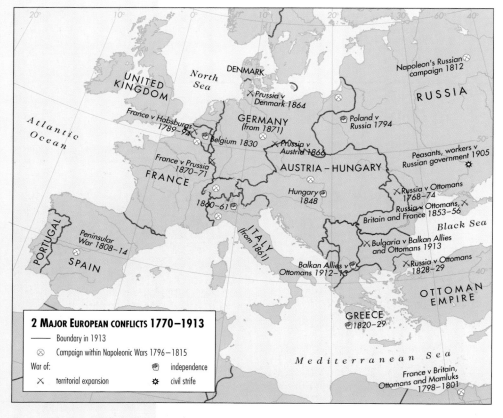

▲ Military conflicts within Europe in this period were caused largely by the territorial ambitions of the French, the Russians and the Prussians. Smaller conflicts arose as Belgium, Greece, Hungary, Italy and, at the very end of the period, the Balkan states, fought off colonial rule and established their independence.

▼ The American Civil War was the bloodiest conflict in American history. The unsuccessful attempt by the outnumbered Confederates to storm the Unionists during the Battle of Gettysburg in July 1863 is generally considered to be the turning point of the war.

MILITARY CONFLICTS

In the 19th century Europe was the most powerful region in the world both in economic and military terms, but it was seldom united either at the national or the international level. The growth of nationalist sentiment encouraged the emergence of "nation-states" such as Germany and Italy, but several great powers – Russia, Austria and the United Kingdom – were composed of different ethnic groups whose antipathies to each other were increased by the growth of nationalist feeling. Nationalism and territorial ambition led many European countries to attack one another. There were numerous wars in western Europe as well as in the unstable region of the Balkans (*map 2*).

The Franco-Prussian war of 1870–71 generated not only hundreds of thousands of casualties but also the Paris Commune, in which socialists briefly seized power. The late 19th century saw the emergence of new ideologies of egalitarianism and class conflict – Marxism, syndicalism and anarchism – which rejected liberal democracy and favoured "direct action" such as industrial strikes and assassination.

Europe was a divided continent long before the First World War (1914–18) exacerbated its problems. This was apparent even on other continents, where many wars in the late 18th and 19th centuries were fought between European powers (*map 3*). France and Spain, for example, helped the American colonists gain their independence from Britain, and Britain captured many French, Spanish and Dutch colonies during its struggle with Napoleon.

RESISTANCE TO IMPERIAL RULE

The period 1770–1914 has been described as the "Age of European Imperialism" because it was characterized by a rapid expansion in European influence over the rest of the world. However, at no time between 1770 and 1914 was most of the world under direct European control. In the Americas European colonial rule was confined to the periphery, while in the Middle East and Asia important indigenous states survived despite the expansion of European influence. The extensive Manchu Qing Empire remained largely intact until the second decade of the 20th century.

Japan acquired a maritime empire and rapidly developed its manufactures and foreign trade with the help of Western technology. Other Asian rulers, such as the shahs of Persia and the kings of Siam, kept their independence by playing off European rivals against each other. Even in India – regarded by the British as the most valuable part of their empire – control of about half the subcontinent was shared with native maharajahs. In Africa most of the interior remained beyond direct European control until the late 19th century. Furthermore, some native African states inflicted defeat on European armies – as the Zulus did at Isandhlwana in 1879, the Mahdists at Khartoum in 1885 and the Ethiopians at Adowa in 1896.

Most European colonies were of minor economic importance to their mother countries, although there were some notable exceptions. Few colonies outside North America attracted large numbers of European settlers, except Australia, where the initial settlements were established with the aid of transported convicts. Very few Europeans settled in equatorial Africa or Asia, and even India attracted only a few thousand long-term British residents.

CHANGES IN POPULATION

In the 19th century the distribution of the world's population changed considerably. Although Asia remained far more populous than any other continent, the population of Europe increased rapidly, while that of North America exploded – largely as a result of European migration. The

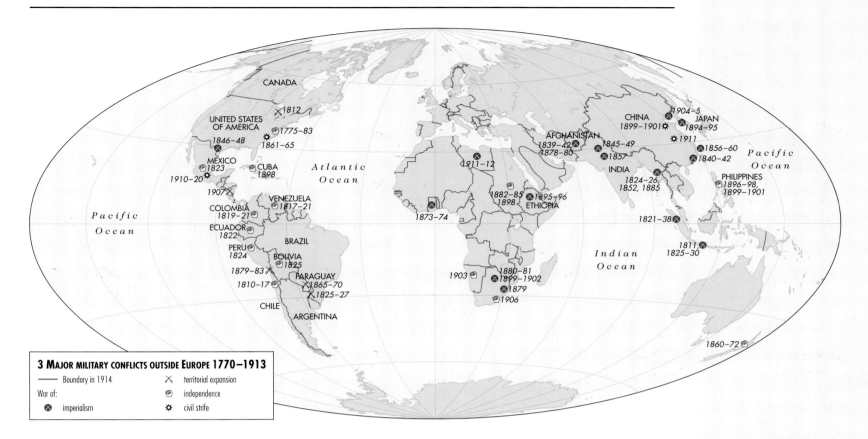

3 MAJOR MILITARY CONFLICTS OUTSIDE EUROPE 1770–1913

——— Boundary in 1914 ✕ territorial expansion

War of: 📖 independence

⊗ imperialism ✿ civil strife

expansion of the European empires in Africa and Asia facilitated both Asian and European migration, while the African slave trade continued to Brazil and Cuba until the late 19th century.

The great majority of people who left Europe – more than 30 million over the period – migrated to the United States. Americans, although they often retained some aspects of their European heritage, were proud that they had left the restrictions and conflicts of the "Old World" for the opportunities and advantages of the "New World" and supported the isolationist policy of the US government. The combination of a low tax burden with rapid westward expansion and industrialization gave the majority of white Americans a very high standard of living. By the late 19th century the United States was the richest nation in the world, although its military power and international status were still relatively undeveloped.

CROSS-CULTURAL INFLUENCES

The worldwide success of the European peoples encouraged them to believe in their own superiority, but it also exposed them to other cultures which subtly altered their own civilization. Japanese art, for example, inspired French and Dutch painters and British designers, while Hinduism prompted the fashionable cult of theosophy. In North America, popular music was influenced by African-American blues and jazz.

In Latin America Roman Catholicism became the main religion of the native peoples, but was obliged to make compromises with local practices and beliefs. Outside the Americas European Christianity had little success in converting other

ethnic and religious groups. Islam, for example, remained dominant in the Middle East and much of South and Southeast Asia, while Hinduism remained the religion of the majority in India. The Chinese and the Japanese largely remained loyal to their traditional religions, despite much missionary activity by the Christian churches, which was often prompted by deep divisions between the Protestant and Roman Catholic churches.

Throughout the period the vast majority of the world's ethnic groups remained attached to their own indigenous traditions and had little knowledge of other languages or cultures. Even in 1914 European influence on the world was still limited and undeveloped in many respects. The largest European transcontinental empires – those of Britain and France – did not reach their apogee until after the First World War, and European cultural influence only reached its zenith in the later 20th century, by which time it had been subsumed in a wider "Westernization" of the world.

◀ Many of the wars outside Europe were fought by European powers, or by people of European origin. In Latin America, for example, there was a sequence of wars of liberation, as the Spanish colonial elites staged successful revolutions against rule from Spain.

◀ One effect of the increased contact between Europe and the countries of Asia during the 19th century was an exchange of cultural influences. The landscape woodcuts of Katsushika Hokusai, such as this view of Mount Fuji from Nakahara — one of a series entitled *Thirty-Six Views of Mount Fuji* (1826–33) — are recognized as having influenced the work of Van Gogh and other European artists.

THE AMERICAN REVOLUTION 1775-83

▲ The Declaration of Independence was drafted by Thomas Jefferson (*right*), with the assistance of Benjamin Franklin (*left*) and John Adams (*centre*), and adopted by the Continental Congress on 4 July 1776.

▼ In 1763 Britain antagonized the American colonists by unilaterally deciding to maintain a standing army in North America to protect its newly acquired assets, and by prohibiting white settlement to the west of an imposed Proclamation Line.

The American Revolution or War of Independence gave birth to a new nation, the United States of America. It involved two simultaneous struggles: a military conflict with Britain, which was largely resolved by 1781, and a political conflict within America itself over whether to demand complete independence from Britain and, if so, how the resulting new nation should be structured.

Prior to the outbreak of war in 1775, the territory that became the United States comprised thirteen separate British colonies, each with its own distinct burgeoning culture, institutions and economy (*map 1*). Before 1763 the colonists, with their own colonial legislatures, had enjoyed a large measure of self-government, except in overseas trade, and had rarely objected to their membership of the British Empire. Changes to British policy after 1763 gradually destroyed this arrangement and created a sense of common grievance among the colonies.

CAUSES FOR GRIEVANCE

The spoils of the Seven Years War (1756-63) greatly enlarged the territory of British North America and established British dominance over the continent (*map 2*). In order to police this vast area and to reduce substantial wartime debt, the British government took steps to manage its North American empire more effectively. Customs officers were ordered to enforce long-standing laws regulating colonial shipping (Navigation Acts, 1650-96), and a series of measures was passed by the British parliament which for the first time taxed the colonists directly (Sugar Act, 1764; Quartering Acts, 1765; Stamp Act, 1765). Having no representation in the British parliament, the colonists viewed these measures as a deliberate attempt to bypass the colonial assemblies, and they responded by boycotting British goods. Although most of these taxes were repealed in 1770, Committees of Correspondence were organized throughout the Thirteen Colonies to publicize American grievances.

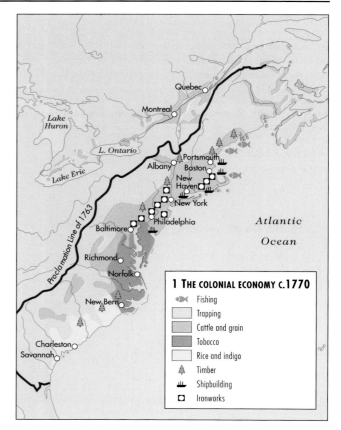

1 THE COLONIAL ECONOMY C.1770
- Fishing
- Trapping
- Cattle and grain
- Tobacco
- Rice and indigo
- Timber
- Shipbuilding
- Ironworks

▲ Between 1700 and 1770 the economic structure of the American colonies became increasingly diversified and sophisticated as the population increased sixfold to some 1,500,000. Manufacturing developed on a significant scale and there was a dramatic growth of trade, not only with the mother country and the British West Indies but also – illegally – with the French West Indies and continental Europe.

In response to the Tea Act of 1773, a symbolic "tea party" was held when protestors dumped incoming tea into Boston harbour rather than pay another "unjust" tax. The situation worsened when the boundaries of the now-British colony of Quebec were extended to the territory north of the Ohio River (Quebec Act, 1774). Feeling the need to enforce its authority, Britain passed the Coercive Acts of 1774 (the "Intolerable Acts"), which closed Boston harbour and imposed a form of martial law. Meeting in Philadelphia in 1774, the First Continental Congress asserted the right to "no taxation without representation" and, although still hoping that an amicable settlement could be reached with Britain, denounced these new British laws as violations of American rights. When Britain made it clear that the colonies must either submit to its rule or be crushed (the Restraining Act, 1775), the movement for full American independence began. War broke out when British troops clashed with the colonial militia at Lexington and Concord in April 1775.

At the start of the war, the American cause seemed precarious. The colonists were deeply divided about what they were fighting for and faced the full might of the British Empire. Britain had the greatest navy and the best-equipped army in the world, although the small size of the British army in the American colonies – composed of regular soldiers, American loyalists, Hessian mercenaries and Native American tribes, especially the Six Nations and the Cherokee – is evidence that Britain did not initially take the American threat seriously. The Americans, however, with militiamen and volunteers, had more than enough manpower to defend themselves, and in most battles they outnumbered British troops. Much of the fighting, especially in the south, took the form of guerrilla warfare, at which American militiamen, aided by the civilian population, were much more adept than the British regular troops. They had the advantage of fighting on their own territory and, unlike the British, had easy access to supplies. By the war's end America had also won the support of Britain's enemies – France, Spain and Holland.

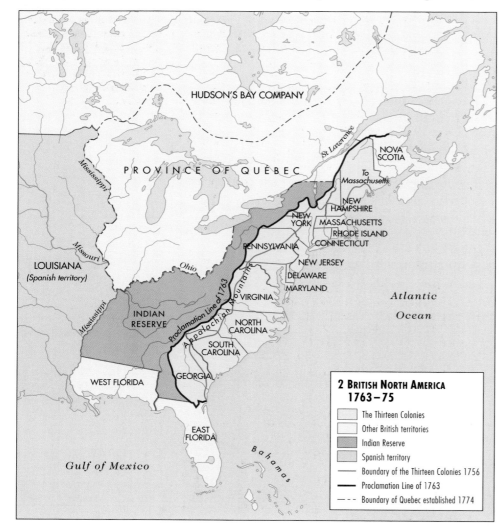

2 BRITISH NORTH AMERICA 1763-75
- The Thirteen Colonies
- Other British territories
- Indian Reserve
- Spanish territory
- —— Boundary of the Thirteen Colonies 1756
- —— Proclamation Line of 1763
- --- Boundary of Quebec established 1774

THE DECLARATION OF INDEPENDENCE

On 4 July 1776 the Second Continental Congress adopted Thomas Jefferson's Declaration of Independence. This document furnished the moral and philosophical justification for the rebellion, arguing that governments are formed in order to secure the "self-evident" truth of the right of each individual to "life, liberty and the pursuit of happiness" and that their power is derived from the consent of those they govern. Grounded in the notion that "all men are created equal", the Declaration asserted the colonists' independence from Britain and effectively cut all ties with the mother country.

3 THE AMERICAN WAR OF INDEPENDENCE 1775–83

- ⊗ British victory
- ⊗ American victory
- ✕ No decisive victory
- ➤ British troop or naval movements
- ➤ American troop or naval movements
- ⇕ French naval presence
- —— Proclamation Line of 1763

▲ The battlefronts of the American War of Independence stretched from Quebec in the north to Florida in the south, and from the Atlantic coast as far west as what is now southwestern Illinois. The dense American forest and wilderness had a crucial impact on the movement of troops, and the proximity of almost all the battlefields to either the sea or a river indicates the still-primitive nature of overland communication.

PHASES IN THE FIGHTING

The fighting took place in three distinct phases. The first phase (1775–76) was mainly located in New England but culminated in the American failure to capture Quebec in December 1775, thus enabling the British to retain Canada. The middle phase (1776–79) was fought mainly in the mid-Atlantic region. The American victory at Saratoga (October 1777) proved to be a major turning point in the war as it galvanized France into entering the war on America's side, contributing badly needed financial aid and its powerful navy and troops. The final phase took place in the south and west (1778–81). Naval warfare now assumed greater importance, with French/American and British ships fighting for control of the Atlantic Ocean and Caribbean Sea. Spain declared war against Britain in June 1779, followed by Holland in 1780. In September 1781 the French fleet drove the British navy from Chesapeake Bay, preparing the way for the British surrender at Yorktown (October 1781), the last major battle of the war.

Occasional fighting continued for over a year, but a new British cabinet decided to open peace negotiations. The Treaty of Paris (September 1783) recognized the new republic and established generous boundaries from the Atlantic Coast to the Mississippi, and from the Great Lakes and Canada to the 31st parallel in the south. The Revolution was not accepted by all Americans (about one-third remained loyal to Britain), and up to 100,000 colonists fled the country to form the core of English-speaking Canada (*pages 188–89*). The ideas expressed in the Declaration of Independence were enshrined in the American Constitution of 1789, which legally established the federal republic and was subsequently used as an inspiration for other liberation movements, most notably in France.

◀ COLONIZATION OF NORTH AMERICA 1600–1763 *pages 124–25*   ▶ WESTWARD EXPANSION OF THE UNITED STATES 1783–1910 *pages 182–83*

REVOLUTIONARY FRANCE AND NAPOLEONIC EUROPE 1789–1815

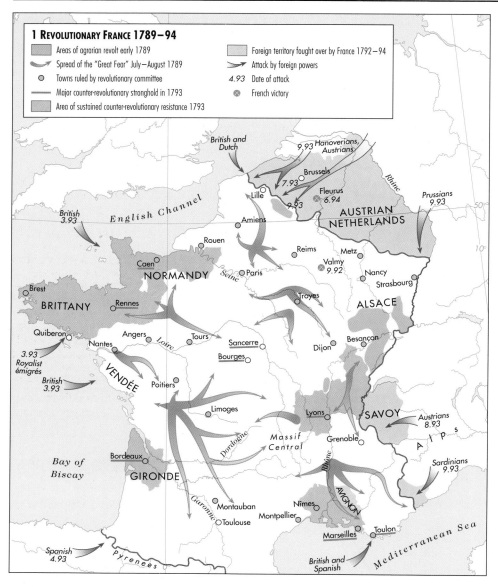

1 REVOLUTIONARY FRANCE 1789–94

- Areas of agrarian revolt early 1789
- Spread of the "Great Fear" July–August 1789
- Towns ruled by revolutionary committee
- Major counter-revolutionary stronghold in 1793
- Area of sustained counter-revolutionary resistance 1793
- Foreign territory fought over by France 1792–94
- Attack by foreign powers
- 4.93 Date of attack
- French victory

▲ The French Revolution did not occur simultaneously throughout the country, but spread out into the countryside from urban centres. Some areas remained stubbornly resistant to revolutionary rule, but by the mid-1790s even these were brought under the control of central government. The crowned heads of Europe feared the spread of revolutionary fervour into their own countries, and were thus anxious to quell the revolutionary French. However, the Austrians were eventually defeated at Fleurus, while the Prussians were repulsed in Alsace, as were the Sardinians in Savoy, the Spanish in the south, and the British on the Vendée coast and the Mediterranean. Avignon (a papal state) was incorporated into France in 1791.

The French Revolution of 1789 represented a major turning point in the history of continental Europe, for it marked the beginning of the demise of absolutist monarchies and their replacement by nation states in which the middle classes held political power. It arose partly from attempts by King Louis XVI to overcome a mounting financial crisis by summoning the Estates-General, a body of elected representatives which had not met since 1614. He thus aroused hopes of reform among the Third Estate (the bourgeoisie or middle classes) – hopes that could only be fulfilled by an attack on the judicial and financial privileges of the First and Second Estates (the aristocracy and clergy). While the king prevaricated, the First and Second Estates refused to surrender any of their privileges, and on 17 June 1789 the Third Estate proclaimed itself a National Assembly.

Riots had broken out in many parts of France early in 1789 (*map 1*) in response to a disastrous harvest in 1788 that had reduced many peasants and industrial workers to starvation. When the people of Paris stormed the Bastille prison – symbol of royal absolutism – on 4 July 1789, an enormous wave of popular unrest swept the country, and in what was known as the "Great Fear" the property of the aristocracy was looted or seized. The National Assembly reacted by abolishing the tax privileges of the aristocracy and clergy and promulgating the "Declaration of the Rights of Man and of the Citizen", in which the main principles of bourgeois democracy – liberty, equality, property rights and freedom of speech – were enunciated. Other reforms followed, including the replacement of the provinces of France by a centralized state divided into 84 departments.

Powerless to stop these changes, the king tried, unsuccessfully, to flee the country in June 1791, thus provoking

anti-royalist attacks. Tension between the moderates and anti-royalists grew as French royalist armies, backed by Austria and Prussia, gathered on France's borders. In April 1792 war was declared on Austria, and in September the Prussians invaded northeastern France, but were repulsed at Valmy (*map 1*). A new National Convention, elected by universal male suffrage, declared France a republic.

THE TERROR

Louis XVI was put on trial and executed in January 1793. Anti-revolutionary uprisings, the presence on French soil of enemy armies and continuing economic problems, led to a sense of national emergency. The Assembly appointed a Committee of Public Safety, dominated by the extremist Jacobins and led by Robespierre. A reign of terror began, with the aim of imposing revolutionary principles by force, and more than 40,000 people (70 per cent of them from the peasantry or labouring classes) were executed as "enemies of the Revolution".

In order to combat the foreign threat, the Committee of Public Safety introduced conscription. During 1794 the French proved successful against the invading forces of the First Coalition (*map 3*), and victory at Fleurus in June left them in control of the Austrian Netherlands. In July the moderate faction ousted Robespierre, who went to the guillotine. Executive power was then vested in a Directory of five members, and a five-year period of moderation set in.

THE RISE OF NAPOLEON

The Directory made peace with Prussia, the Netherlands and Spain, but launched an offensive against Austria in Italy, headed by a young general, Napoleon Bonaparte (*map 2*). He was brilliantly successful during 1796, forcing Austria out of the war, but then led an unsuccessful expedition to Egypt to try and cut Britain's communications with its Indian empire. Meanwhile, the Directory had become profoundly unpopular with all sections of the population, and was overthrown by Napoleon on his return to France in October 1799. In 1800, following the first-ever plebiscite, from which he gained overwhelming support, he was confirmed as First Consul of France – a position that gave him supreme authority. He proceeded to introduce a number of

2 NAPOLEONIC EUROPE 1796–1815

- France 1792

Area under direct French rule 1792–1815:
- for more than 10 years
- 5–10 years
- less than 5 years
- Satellite regimes in 1810
- Area within which departmental administration introduced

French victory ⊗ with dates, French defeat ⊗ with dates in wars against:
- First Coalition 1796–97
- Second Coalition 1798–99
- Third Coalition 1805
- Fourth Coalition 1806–7
- Fifth Coalition 1809–15 (including Peninsular War 1808–14, Austrian War 1809, Russian Campaign 1812–13, War of Liberation 1813, Campaign of 1814–15)
- Egyptian Campaign 1798–1801

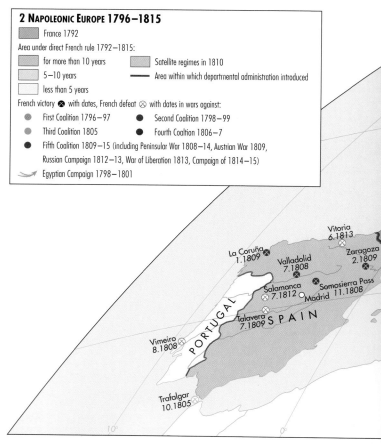

measures to create a centralized administrative structure, including the founding of the Bank of France in 1800. Between 1801 and 1804 a body of laws was created, known as the Napoleonic Civil Code, which embodied many of the fundamental principles of the French Revolution and was subsequently imposed in countries conquered by Napoleon's armies. In 1801 he signed a concordat with the Pope, thus helping to ensure that he received the Pope's approval when he declared himself emperor in 1804.

MILITARY CAMPAIGNS

By the end of 1800 France had once again defeated Austrian forces in northern Italy and by February 1801 it had made peace with all its opponents except Britain. The following year it signed the Treaty of Amiens with Britain, but the resulting period of peace was not to last long, and in 1805 Austria, Russia and Sweden joined Britain to form the Third Coalition (*map 3*). In October the French fleet was completely destroyed by the British in the Battle of Trafalgar, but by the end of the year Napoleon's armies had inflicted heavy defeats on the Austrians and Russians at Ulm and Austerlitz respectively. They then moved on through the German states, defeating the Prussians in October 1806. Following his defeat of the Russians at Friedland in June 1807, Napoleon persuaded the tsar to join forces with France to defeat Britain, which once again was isolated as Napoleon's sole effective opponent.

WAR AGAINST THE FIFTH COALITION

In 1808 Charles IV of Spain was forced to abdicate in favour of Napoleon's brother Joseph. The Spanish revolted and the British sent a supporting army to the Iberian Peninsula (*map 2*). Elsewhere in Europe the economic hardships

resulting from the French military presence tended to make Napoleon's rule unpopular with his subject nations. The imposition of the Napoleonic Civil Code in countries annexed by France, while potentially beneficial to the citizens of Europe, still represented an unwelcome domination by the French. It also caused disquiet among Napoleon's allies, the Russians, who in 1810 broke with France, eventually joining Britain and Portugal in the Fifth Coalition.

In 1812 Napoleon attempted his most ambitious annexation of territory yet, launching an invasion of Russia. Although he reached Moscow in September, he found it deserted and, with insufficient supplies to feed his army, he was forced to retreat. In Spain the British and Portuguese armies finally overcame the French, chasing them back onto French soil. At the same time the Prussians, Austrians and other subject states seized the opportunity to rebel against French rule. The Fifth Coalition armies took Paris in March 1814, Napoleon abdicated and was exiled to the island of Elba, and Louis XVIII ascended the French throne.

A year later, while the Coalition members were negotiating the reshaping of Europe at the Congress of Vienna, Napoleon escaped and raised an army as he marched north through France. Following defeat at Waterloo in 1815, he was sent into permanent exile on St Helena. The reconvened Congress of Vienna deprived France of all the territory it had acquired since 1792. It could not, however, prevent the spread of revolutionary and Napoleonic ideas in Europe, as the maintenance or adoption of the Napoleonic Civil Code in a number of countries after 1815 testified.

▼ Napoleon's armies waged war across Europe in his attempt to impose French rule and the Civil Code throughout the continent. The turning point in his fortunes came in 1812 when, with an army already fighting in Spain, he embarked on an invasion of Russia. French supply lines were stretched too far to support the army through the Russian winter, and the troops were forced to retreat, with most of the survivors deserting. Napoleon was eventually captured in 1814 on French soil by the armies of the Fifth Coalition, and imprisoned on the island of Elba. The final battle occurred following his escape, when a revived French army was defeated at Waterloo, in Belgium, on 18 June 1815.

From 1793 onwards the rulers of the European states formed various alliances in an attempt to counter the threat from France. Britain was a common member, with other countries joining when it became expedient to do so. Russia also joined all five coalitions, although from 1807 to 1810 it was allied to France. Spain, a member of the First Coalition, became a French ally and then puppet state from 1796 until the Spanish people rose up in protest in 1808 and precipitated the Peninsular War.

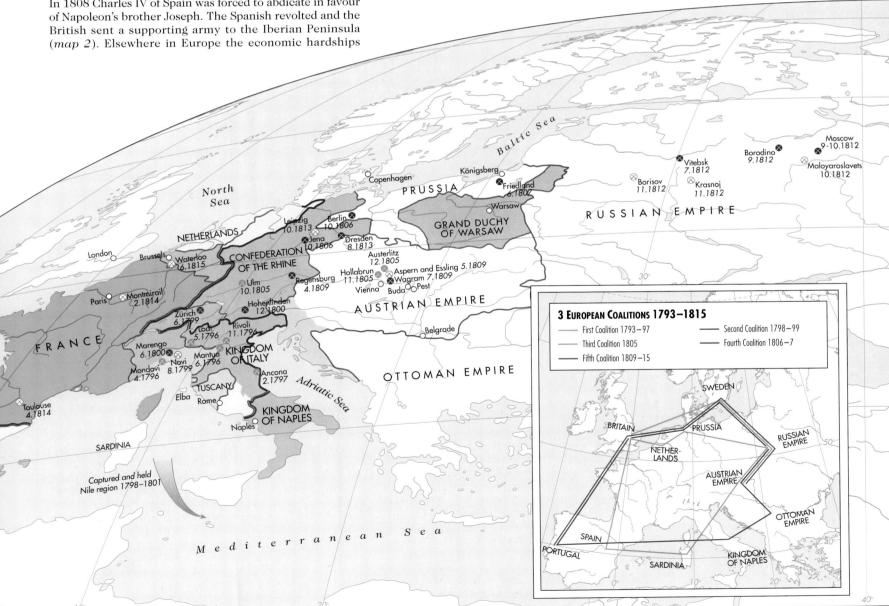

3 EUROPEAN COALITIONS 1793–1815
- First Coalition 1793–97
- Second Coalition 1798–99
- Third Coalition 1805
- Fourth Coalition 1806–7
- Fifth Coalition 1809–15

◀ REVOLUTION AND STABILITY IN EUROPE 1600–1785 *pages 156–57* ▶ REVOLUTION AND REACTION IN EUROPE 1815–49 *pages 172–73*

THE INDUSTRIAL REVOLUTION IN BRITAIN
1750–1850

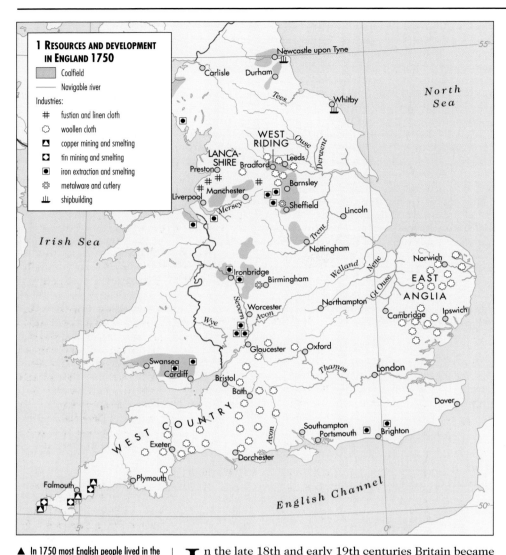

1 RESOURCES AND DEVELOPMENT IN ENGLAND 1750

- Coalfield
- Navigable river

Industries:
- # fustian and linen cloth
- ○ woollen cloth
- ▲ copper mining and smelting
- ◨ tin mining and smelting
- ◉ iron extraction and smelting
- ⚙ metalware and cutlery
- ⛵ shipbuilding

▲ In 1750 most English people lived in the countryside but many worked in the well-established local industries as well as on the land. The largest centre of manufacturing was London, whose products included silk, gin, soap, glass and furniture. Its population had increased from an estimated 120,000 to 675,000 between 1550 and 1750, and the resultant demand encouraged developments in agriculture, industry and transport. Around 650,000 tonnes of coal was shipped to London from Newcastle each year – a trade that employed 15,000 people by 1750.

1 PERCENTAGE OF LAND ENCLOSED IN ENGLAND 1500–1914

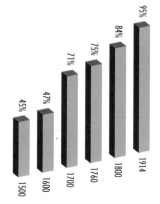

- 1500: 45%
- 1600: 47%
- 1700: 71%
- 1760: 75%
- 1800: 84%
- 1914: 95%

▲ In 1760, 75 per cent of the agricultural land in England was already enclosed and agricultural productivity had been improving for 200 years.

In the late 18th and early 19th centuries Britain became the world's leading industrial nation in a process of economic growth and change that is regarded as the world's first industrial revolution. In some respects, however, the process was of an evolutionary nature, with change occurring at different speeds in different sectors of the economy.

There were a number of reasons why the process of industrialization first occurred in Britain rather than any other country in Europe. In 1750 Britain had a well-developed and specialized economy, substantial overseas trade and an average per capita national income that was one of the highest in Europe. Domestic textile industries, iron smelting and the manufacturing of iron goods were well-established (*map 1*). The country was also fortunate in its natural resources, among them fertile land on which a productive agricultural sector had been able to develop. Early enclosure of fields (*bar chart 1*), together with crop improvements and livestock breeding, meant that British agriculture could feed a rapidly increasing urban workforce. Supplies of coal – fundamental to the nature of Britain's industrialization – were widespread and plentiful, and the development of a national market in coal was facilitated by coastal trade. Navigable rivers provided initial internal transport, while faster-flowing rivers supplied water power for industry and corn-milling.

The British government also played a very important role in establishing the conditions under which industry could thrive. Britain was free from the internal customs barriers and river tolls which stifled trade in Europe, while laws protected the textile and iron industries from foreign competition. Private property rights and a stable currency stimulated economic development, as did the stability provided by a strong state in which warfare, taxation and the public debt were managed by sophisticated bureaucracies. Shipping and trade were protected by Britain's naval

supremacy, which also helped to secure trading privileges and build up a worldwide colonial empire obliged to conduct trade using British ships.

Rapid economic progress was further encouraged by Britain's success in war, in particular the war of 1793–1815 against France (*pages 166–67*), during which Britain remained free from invasion and escaped the economic dislocation engendered by war on the continent of Europe. The war created a demand for armaments, ships and uniforms, which in turn stimulated Britain's shipbuilding, iron-smelting, engineering and textile industries.

THE TEXTILE INDUSTRY

In 1750 a variety of textiles – silk, linen, fustian (a mixture of linen and cotton) and, in particular, wool – had long been produced in Britain. The West Riding of Yorkshire, the West Country and East Anglia were centres of the woollen industry, while the fustian industry had developed in Lancashire (*map 1*). The skilled workforce employed in both industries was largely home-based and organized by merchants who thus built up capital and entrepreneurial skills. Such skills were used to great effect in the second half of the 18th century, when the cotton industry developed rapidly. Technological change allowed Lancashire to produce and sell cotton cloth more cheaply than India, where production depended on low-paid labour. Inventions such as Arkwright's water frame and Watt's steam-powered rotative engine transformed cotton spinning in the last decade of the 18th century into a factory-based, urban industry. This led to an unprecedented rise in productivity and production. Lancashire became the centre of the world's cotton manufacturing industry (*map 2*) and exported cotton cloth throughout the world. The woollen industry continued to be of importance, especially in the West Riding of Yorkshire, where mechanization was introduced and British wool was supplemented by merino wool imported from Australia.

IRON, COAL AND TRANSPORT

Innovation in iron production in the 18th century facilitated smelting, and later refining, using coke instead of charcoal. Steam power, fuelled by plentiful coal supplies, began to replace man, horse and water power, encouraging the development of the factory system and rapid urbanization near to coalfields. These developments were self-sustaining, for while steam engines increased the demand for coal and iron, better steam-driven pumps and rotary winding equipment facilitated deeper coalmines.

Transport developed in response to the economic changes. Canals were constructed to carry heavy and bulky goods, and roads were improved by turnpike trusts, opening up the national market for goods. The combination of colliery waggonways and the steam engine led to the piecemeal development of a rail network from 1825 onwards which by 1850 linked the major urban centres. It also encouraged further industrialization by generating a huge fresh demand for coal, iron, steel, engineering and investment (*map 3*).

THE CONSEQUENCES OF THE INDUSTRIAL REVOLUTION

The economic and social effects of industrialization were complex and wide-ranging. Between 1750 and 1850 the population of England almost trebled. By 1850 more than half the population lived in towns or cities, compared with only 25 per cent in 1800 (*bar chart 2*). Eleven per cent lived in London, which remained the largest manufacturing centre, and more than 60 towns and cities had over 20,000 inhabitants. Such a process of rapid urbanization was unprecedented and unplanned. Crowded and insanitary living conditions meant that urban death rates were considerably higher than those in rural areas. At the same time, the development of the factory system generated issues of discipline, as some workers resented capitalist control of work processes and the replacement of traditional skills by machines. There were outbreaks of machine

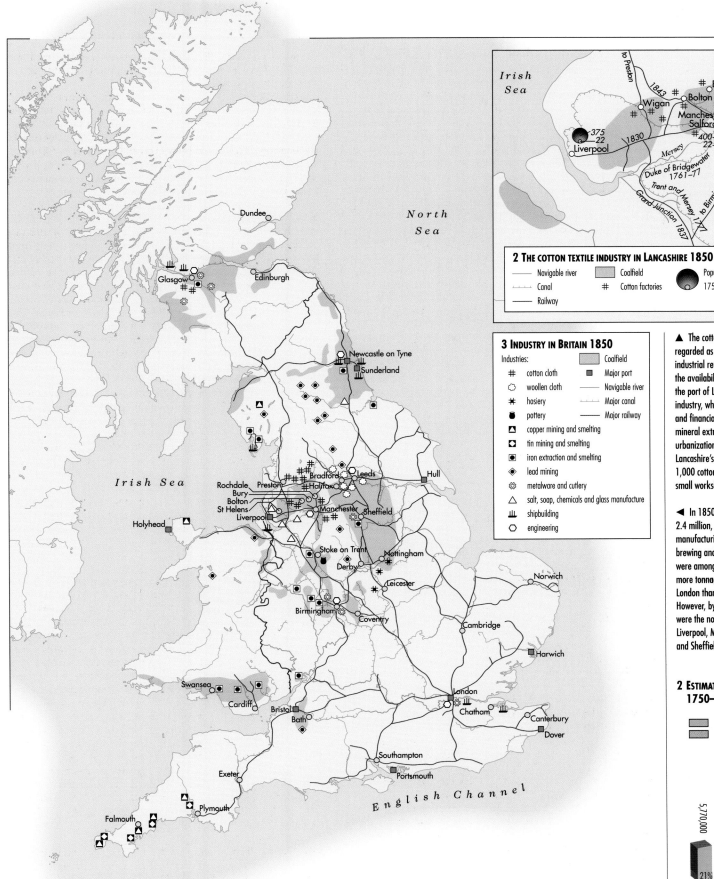

2 THE COTTON TEXTILE INDUSTRY IN LANCASHIRE 1850

—— Navigable river	▨ Coalfield	● Population of city in thousands for
⊥⊥ Canal	# Cotton factories	1750 (inner) and 1850 (outer)
—— Railway		

Irish Sea

North Sea

3 INDUSTRY IN BRITAIN 1850

Industries: ▨ Coalfield
cotton cloth   ■ Major port
○ woollen cloth   —— Navigable river
✳ hosiery   ⊥⊥ Major canal
▮ pottery   —— Major railway
△ copper mining and smelting
⬙ tin mining and smelting
◉ iron extraction and smelting
◈ lead mining
⚙ metalware and cutlery
△ salt, soap, chemicals and glass manufacture
⫿⫿ shipbuilding
⬡ engineering

Irish Sea

English Channel

▲ The cotton mills of Lancashire are often regarded as being at the centre of Britain's industrial revolution. A long textile tradition, the availability of coal and the presence of the port of Liverpool encouraged the cotton industry, which in turn promoted commercial and financial institutions, trade, transport, mineral extraction, engineering and urbanization. By 1830 one third of Lancashire's population worked in around 1,000 cotton factories and numerous small workshops.

◀ In 1850 London, with a population of 2.4 million, was still the predominant manufacturing centre in Britain. London's brewing and refining industries in particular were among the largest in the country, and more tonnage passed through the port of London than any other port in Britain. However, by 1850 the fastest-growing cities were the northern industrial centres of Liverpool, Manchester, Birmingham, Leeds and Sheffield.

2 ESTIMATED POPULATION OF ENGLAND 1750–1851

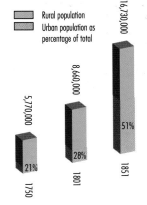

▨ Rural population
▨ Urban population as percentage of total

5,770,000 1750 21%
8,669,000 1801 28%
16,730,000 1851 51%

▲ As the population of England increased, its geographical distribution shifted in favour of the developing industrial regions. In 1750 Middlesex, Lancashire, the West Riding and Devon, the most populated counties, shared 10 per cent of the total English population. By 1851 the four most industrialized counties – Lancashire, West Riding, Staffordshire and Warwickshire – contained nearly a quarter of the English population.

---

breaking, especially in times of trade depression. Moral debates were prompted by the employment of women and children as cheap labour.

Even as late as 1850, however, when British manufactured goods were traded all over the world, many areas of Britain remained rural. In some regions industries had actually declined, among them wool production in the West Country and iron manufacture around Ironbridge (*maps 1 and 2*). The vast majority of the industrial working population was employed in retailing and warehousing, workshops and small enterprises rather than in factories. Capital and technology had become less involved with agriculture and more involved with industry, especially

manufacturing, and with trade and construction related to industry. Yet agriculture was still the largest single occupation and most of Britain's food was still home-produced.

By 1850 Britain was no longer the only country to have undergone an industrial revolution. Similar changes had begun to occur in continental Europe (*pages 170–71*), sometimes with the aid of British machinery, entrepreneurial and financial skills. British industrial workers had also taken their skills to the Continent. In the second half of the century a considerable number were to emigrate to the United States, where the process of industrialization (*pages 186–87*) was eventually to lead to Britain losing its position as the world's greatest industrial power.

◔ SCIENCE AND TECHNOLOGY IN EUROPE 1500–1770 *pages 134–35*　◗ THE INDUSTRIALIZATION OF EUROPE 1830–1914 *pages 170–71*

THE INDUSTRIALIZATION OF EUROPE
1830–1914

▲ By the outbreak of the First World War Germany's industrial development had outstripped that of all other European countries, giving it an economic and political confidence which is reflected in this striking advertisement of 1914.

The industrialization of Europe is considered to have started in the 1830s, some decades after the beginning of the Industrial Revolution in Britain in the late 18th century. Much debate has centred on whether British industrialization "spilled over" into Europe (and if so, to what extent), or whether European countries accumulated their own technological and manufacturing knowledge. There is no question that there were substantial flows of skilled labour, entrepreneurs, capital and technology from Britain, and later from France and Germany, to the less industrialized parts of Europe. However, although the basic model of industrialization remained British, each country developed its own national characteristics. Substitutes were found for the particular resources that Britain possessed but which other countries lacked, more organized banking systems supplied finance to accelerate growth, and more aware governments supplied the ideologies and incentives to motivate growth. As a result, industrialization in the countries of continental Europe was more state-driven and more revolutionary in character than in Britain. The culmination of this model was the abrupt industrialization of the USSR under the Soviet system from 1917 onwards.

REGIONAL DEVELOPMENT

In the first half of the 19th century many of Europe's modern nation-states were yet to come into existence.

Germany and Italy were still fragmented into small political entities, while at the other extreme lay dynastic empires that spanned several nationalities, such as the Habsburg Austrian Empire, the tsarist Russian Empire (which included Poland), and the Ottoman Empire (which included much of the Balkans). The process of industrialization often took place in the context of shifting political allegiances and the forging of national identities. Political alliances and wars, such as the Franco-Prussian War of 1870–71, introduced border changes that were often somewhat haphazard in economic terms. On the other hand, some of the German states used economic unification – initially in the form of a customs union (Zollverein) in 1834 – as a step towards political union in 1871 (*pages 176–77*).

Industry in its early stages was predominantly confined to a number of rather circumscribed regions. Some, such as the region just west of Kraków and a large area of northern Europe, cut across national boundaries (*map 1*). The existence of coal and iron was the most important criterion for determining the speed at which regions developed, but locally available resources were also important, especially the supply of skills in textile regions. Some of the emerging industrial regions subsequently faded, such as the areas around Le Havre, Leipzig and Dresden, while some new ones emerged, such as that bordering the Ruhr in Germany. In general, industrialization can be said to have come to

1 THE GROWTH OF INDUSTRY AND RAILWAYS

— International boundary 1871

Major railway lines constructed:
— by 1848
— 1848–70
--- 1870–1914

Industry c. 1870:
- coalmining
- iron working
- textile production

Industry c. 1914:
- ◻ steel
- ○ engineering
- ⚓ shipbuilding
- ◇ chemicals
- ⚡ electrical industry

▶ The development of the European rail network followed the 19th-century pattern of industrialization, starting in northern France, Belgium, the Netherlands and northern Germany, and spreading to Spain, Italy and Austria–Hungary as the century progressed. The availability of resources such as coal and iron ore largely determined the sites for the development of new heavy industries, but elsewhere long-standing home-based manufacture of textiles was transformed into factory-based manufacture, by the use water-power if coal was not readily available.

regions rather than to nations. Even at the beginning of the 21st century, much industrial activity in Europe is dominated by regional "clusters" of activity, rather than by a general spread of industrialization to all corners.

DEVELOPMENT OF INDUSTRY

The pattern of European industrialization (starting in northwest Europe and moving northwards, southwards and eastwards) tends to support the idea that it was based on that of Britain. It is certainly beyond doubt that the technological advances developed in Britain, for example in textile machinery and steam engines, did not need to be re-invented. However, the technology often needed to be modified to suit local conditions. For example, the type of steam engine most popular in Britain (developed by James Watt) consumed too much coal for its use to be worthwhile in regions where coal was more expensive than in Britain. As a result, water-wheels and the more efficient water turbines were often used to power machinery in France and Italy. Similarly, in the textile industry it was found that machinery developed for the manufacture of woollen and cotton cloth in Britain was not as suitable for the finer textiles of France and Spain.

The scattering of industrial areas encouraged the growth of railway systems, to facilitate the delivery of raw materials to manufacturers and the distribution of manufactured goods to customers. The first track was laid in northern Europe in the 1840s, and the network had reached all corners of Europe by 1870 (*map 1*). In countries such as Spain and Italy the railway was envisaged as the catalyst that would set in motion the process of industrialization, but in these countries, which were among the last to industrialize, the building of railway lines had little appreciable effect. In general, railways were successful at connecting already industrializing areas, rather than fostering the growth of new areas.

THE SPEED AND IMPACT OF INDUSTRIALIZATION

The impact of new industries and new technologies can be gauged from the levels of industrialization achieved, measured in terms of the volume of industrial production per person (*maps 2 and 3*). In 1830 the figure for Britain was more than twice as high as in any other European nation except Belgium, and even as late as 1913 Britain remained ahead, although it was rapidly being caught by Switzerland, Belgium and, of course, Germany, whose steel production had by this time outstripped that of Britain (*pages 216–17*). Indeed, while Britain had a 13.6 per cent share of the world industrial output in 1913, Germany, with its much larger population, had 14.8 per cent, and was thus second only to the United States in terms of its industrial might.

The most obvious effect of industrialization was on economic growth and on the living standards of the populations of the industrialized countries. While industrialization had developed first in countries whose societies were relatively

egalitarian, such as Belgium and France, it often had the effect of widening social inequalities for some years. The national income per head, the most common indicator of overall prosperity and growth, rose throughout Europe (*graph*), but its steepest increase was in northern Europe, where industrialization took its strongest hold. So, despite the squalor and misery of industrial regions and cities, it seems that industrializing nations as a whole, and certain sectors in particular, enjoyed long-term economic benefits.

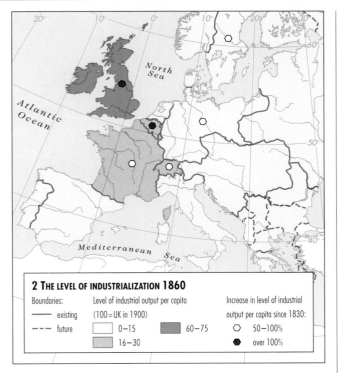

2 THE LEVEL OF INDUSTRIALIZATION 1860

Boundaries:
— existing
--- future

Level of industrial output per capita (100=UK in 1900)
☐ 0–15
☐ 16–30
☐ 60–75

Increase in level of industrial output per capita since 1830:
○ 50–100%
● over 100%

◄ Britain, with its head start, steamed ahead of the rest of Europe in terms of industrial output per capita in the first half of the 19th century, but Belgium, with readily available sources of coal and iron ore, also experienced an increase in output of more than 100 per cent. Elsewhere in northern Europe, and in Switzerland, industrialization made considerable headway, although the intense industrialization of northern France and Germany is not reflected in the per capita figures of those countries, since the majority of the population was still engaged in agricultural production.

▼ Countries underwent their main periods of industrialization at different times. Belgium experienced a spurt early on and then again at the turn of the century, while others, in particular the Scandinavian countries, were relatively late developers. Germany also started comparatively slowly but increased the volume of its industrial production per person by 240 per cent between 1880 and 1913.

▼ The degree of industrialization in Europe is clearly reflected in the growth of countries' Gross National Product (GNP). The nations of northern Europe (including Denmark) pulled away from the rest of Europe in terms of their national wealth.

The Scandinavian countries of Norway, Sweden and Finland all had a lower GNP per capita than those of southern Europe in 1830, but had outstripped them by 1910 as a consequence of a period of intense industrialization late in the 19th century.

RELATIVE GROWTH IN GNP PER CAPITA ACROSS EUROPE 1830–1910

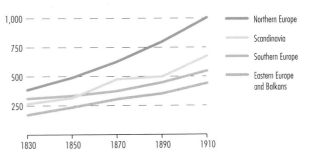

— Northern Europe
— Scandinavia
— Southern Europe
— Eastern Europe and Balkans

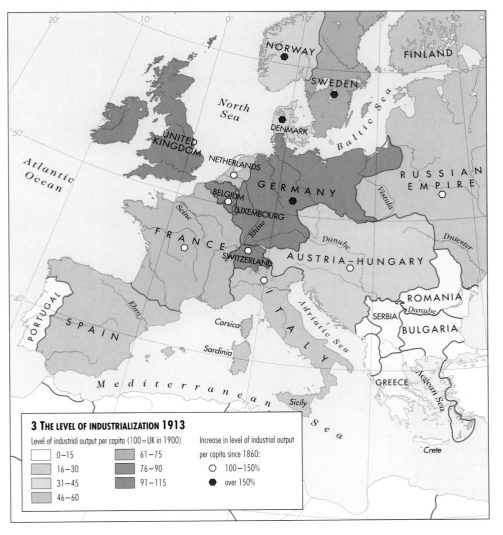

3 THE LEVEL OF INDUSTRIALIZATION 1913

Level of industrial output per capita (100=UK in 1900)
☐ 0–15
☐ 16–30
☐ 31–45
☐ 46–60
☐ 61–75
☐ 76–90
☐ 91–115

Increase in level of industrial output per capita since 1860:
○ 100–150%
● over 150%

REVOLUTION AND REACTION IN EUROPE 1815–49

► The Congress of Vienna resulted in several major boundary changes. France had its borders returned to those of 1792, Poland was divided once again and 39 German-speaking states were organized into the German Confederation, dominated by Prussia, which was given half of Saxony. Austria lost its possessions in northwest Europe to the Dutch in the newly created United Netherlands, but was given much of northern Italy by way of compensation.

▼ During the 1820s and early 1830s rebellions broke out across Europe, with liberals calling for an end to absolute monarchy in Spain and Portugal and in the Italian peninsula. The Greeks, with the help of the French, British and Russians, drove the Ottomans from Morea. The Russians also intervened to crush rebellion in Poland in 1830, having defeated their own Decembrist Revolution in 1825. The French brought about a degree of constitutional reform following the replacement of Charles X by Louis Philippe in 1830, and Belgium achieved independence from the United Netherlands the same year.

1 Treaty settlements in Europe 1814–15

- ▨ Area of expansion
- ── Border of German Confederation 1815
- Newly created states/confederations:
- ▨ United Netherlands 1815–30
- ▨ Grand Duchy of Luxembourg 1815
- ▨ Union of Norway and Sweden 1815–1905
- ▨ Republic of Kraków 1815–46
- ▨ Switzerland 1815

Following their initial victory over Napoleon in 1814, the major European powers met at the Congress of Vienna (1814–15) to decide on the future political map of Europe. The Congress was dominated by three principles: territorial compensation for the victors, the restoration and affirmation of the ruling royal dynasties, and the achievement of a balance of power between the major European states. As a result of their deliberations the German Confederation was formed, replacing the Holy Roman Empire (*map 1*). Elsewhere, national boundaries were redrawn, often with little regard to ethnic groupings, thus planting the seeds of nationalist tensions.

There was a shared conviction that the spread of republican and revolutionary movements must be prevented. In September 1815 Russia, Austria and Prussia formed a "Holy Alliance", agreeing to guarantee all existing boundaries and governments and to uphold the principles of Christianity throughout Europe. The alliance was subsequently joined by the other major European powers – with the exception of Britain, the Pope and, not surprisingly, the Ottoman sultan – and over the next 40 years there were several occasions when the autocratic rulers of Europe took military action to suppress uprisings in states other than their own.

REVOLUTIONARY ACTIVITY IN THE SOUTH
In 1820 there was an explosion of revolutionary activity in Spain. Following the defeat of Napoleon, a liberal constitution had been introduced in 1812, but this had been annulled by King Ferdinand VII on his return from exile in 1815. In 1820 his authority was challenged by an army revolt, supported by riots across Spain (*map 2*), with the result that the liberal constitution was re-established.

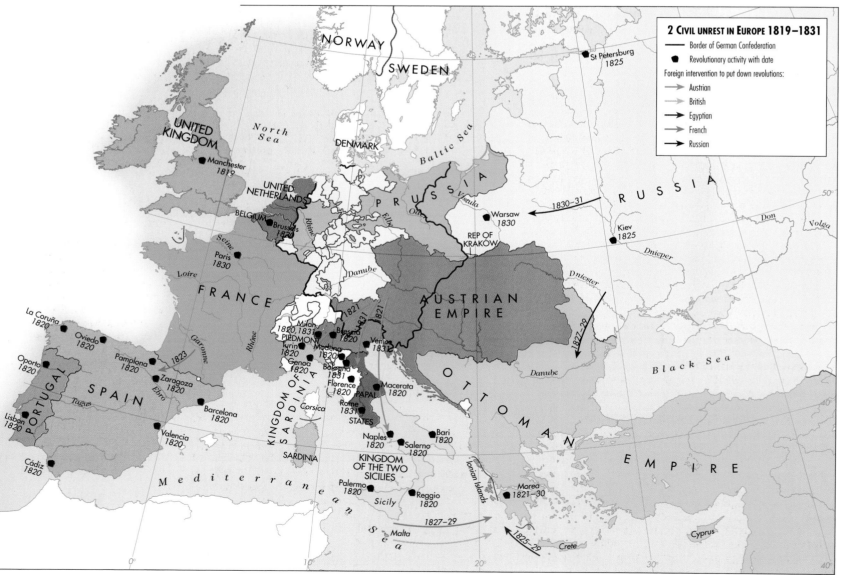

2 Civil unrest in Europe 1819–1831

- ── Border of German Confederation
- ● Revolutionary activity with date
- Foreign intervention to put down revolutions:
- → Austrian
- → British
- → Egyptian
- → French
- → Russian

Insurrections in Naples, Piedmont and Portugal in the summer of 1820 also attempted to introduce constitutional forms of government, and initially met with some success. However, Tsar Alexander I of Russia persuaded the Austrians and Prussians to support him in threatening military intervention, and in March 1821 Austria sent an army to crush the revolts in Piedmont and Naples. In December 1825 Russia faced revolutionary action on its own soil when a group of military officers tried unsuccessfully to prevent the accession to the tsardom of Nicholas I, preferring his more liberal-minded brother. The following year the continuing instability in Portugal prompted the British to intervene, in this instance with the intention of aiding the preservation of its constitutional government.

In Greece a revolution broke out in 1821 with the aim of shaking off Ottoman rule and uniting the whole of the ancient Hellenic state under a liberal constitution. The Ottomans enlisted support from the Egyptian viceroy Muhammad Ali, whose troops seized a large area of the country by 1826, when Russia, France and Britain intervened to defeat the Muslim forces. However, the London Protocol of 1830, which proclaimed Greek independence, fell far short of the aspirations of the revolutionaries in that it only established a Greek monarchy in southern Greece, under the joint protection of the European powers (*map 3*).

UNREST IN THE NORTH

By 1830 revolutionary passions were rising in France. King Charles X dissolved an unco-operative Chamber of Deputies and called an election, but when an equally anti-royal Chamber resulted, he called fresh elections with a restricted electorate. Demonstrations in Paris during July forced him to abdicate in favour of Louis Philippe, whose right to call elections was removed. His reign, known as the "July Monarchy", saw insurrections as industrial workers and members of the lower middle class, influenced by socialist and utopian ideas, demanded an increased share of political power, including the vote.

Nationalist resentment at decisions taken at the Congress of Vienna led to insurrection in both Belgium and Poland in the 1830s. In Belgium, which had been given to the United Netherlands in 1815, riots broke out in 1830 and independence was declared in October. In the kingdom of Poland, an area around Warsaw that had been given to the Russian tsar, a revolt by Polish nationalists resulted in a brief period of independence before the Russians crushed the movement in 1831, and subsequently attempted to destroy Polish identity in a campaign of "Russification".

Britain also experienced a degree of social unrest. A mass protest in Manchester in 1819 was crushed and 11 people were killed by troops in what became known as the "Peterloo Massacre". Inequalities in the electoral system provoked a strong movement for reform, which resulted in the Great Reform Bill of 1832. This expanded the electorate by 50 per cent and ensured representation from the newly developed industrial centres. Further calls were made by the Chartists for universal suffrage, with petitions presented to Parliament in 1838 and again in 1848.

THE REVOLUTIONS OF 1848

By 1848 many of the European countries were suffering from an economic crisis; the failure of the potato and grain crops in 1845–46 was reflected in the price of food. There was political discontent at different social levels: peasants demanded total abolition of the feudal system, industrial workers sought improvements in their working conditions, and middle-class professionals wanted increased political rights. In Italy and Germany there were growing movements for unification and independence (*pages 176–77*).

Revolutionary agitation began in Paris in February 1848, forcing the abdication of Louis Philippe and the establishment of the Second Republic. It then spread across central Europe (*map 3*). The Habsburg Empire, faced with demands for a separate Hungarian government, as well as demonstrations on the streets of Vienna, initially gave in to the

demands of the Hungarian nationalists and granted them a separate constitution. This, however, was annulled some months later, leading to a declaration of independence by Hungary. The Austrian response was to quell the revolt in 1849 with the help of Russian forces (*pages 174–75*).

Discontent in Austria spilled over into the southern states of the German Confederation, and liberals in Berlin demanded a more constitutional government. As a result, the first National Parliament of the German Confederation was summoned in May 1848.

FROM REVOLUTION TO REACTION

In June 1848 struggles between the moderate and the radical republicans culminated in three days of rioting on the streets of Paris. In crushing the rioters the more conservative factions gained control, a trend that was repeated in Prussia, where royal power was reaffirmed. The second half of 1848 was marked by waves of reaction that spread from one city to another. The restoration of Austrian control over Hungary was achieved partly by playing off against each other the different ethnic groups within the empire. However, despite the suppression of the 1848 revolutionaries, most of the reforms they had proposed were carried out in the second half of the century, and at least some of the nationalist movements were successful.

▲ Rebellions broke out across Europe during 1848, inspired by the success of the French in abolishing their monarchy in February. The Habsburgs faced rebellions in Hungary and in the Italian cities of Milan and Venice, which were supported by Piedmont. Although the revolutions in Italy, Germany and Hungary were all defeated, the liberal constitutions, unification and independence they were seeking did eventually come about.

THE HABSBURG EMPIRE: EXPANSION AND DECLINE 1700–1918

▶ During the 18th century the Habsburg Empire took every opportunity to expand its territory at the expense of its neighbours. As a result of the War of the Spanish Succession, the Habsburgs gained territory in the Netherlands and Italy. They fared less well in the east, however, where territory taken from the Ottoman Empire in 1718 was regained by the Ottomans in 1739.

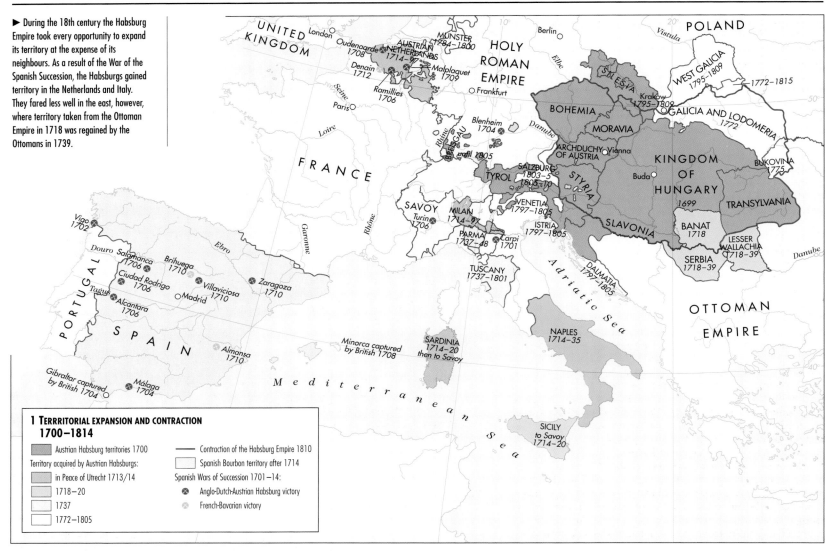

1 TERRITORIAL EXPANSION AND CONTRACTION 1700–1814

- Austrian Habsburg territories 1700
- Territory acquired by Austrian Habsburgs:
  - in Peace of Utrecht 1713/14
  - 1718–20
  - 1737
  - 1772–1805
- —— Contraction of the Habsburg Empire 1810
- Spanish Bourbon territory after 1714
- Spanish Wars of Succession 1701–14:
  - ⊗ Anglo-Dutch-Austrian Habsburg victory
  - ⊗ French-Bavarian victory

The Spanish Habsburg dynasty ended in 1700 with the death of Charles II. King Louis XIV of France supported the claim to the Spanish throne of Philip, Duke of Anjou, who was his infant grandson and the great-nephew of Charles. The British and Dutch, fearing French domination, supported the claim of the Austrian Archduke Charles, and the War of the Spanish Succession (1701–14) ensued (*map 1*). The outcome, formalized in the Peace of Utrecht (1713/14), was a compromise under which Philip attained the Spanish throne on condition that he renounced any claim to France, and the Austrians gained control of territory in Italy and the Netherlands.

During the 18th century the Austrian Habsburgs were the major dynastic power in central Europe. They were threatened, however, when on the death of Charles VI of Austria in 1740 other crowned heads of Europe refused to recognize his daughter Maria Theresa as his successor. In the resulting War of the Austrian Succession (1740–48), Bavaria, France, Spain, Sardinia, Prussia and Saxony joined forces against Austria, the Netherlands and Britain in an unsuccessful attempt to oust Maria Theresa.

REFORM OF THE MONARCHY

During her long reign (1740–80) Maria Theresa embarked on transforming the diverse Habsburg dominions into a centralized nation state, and initiated many progressive reforms in the spheres of education, law and the Church. Her minister, Hagwitz, put the Habsburg finances on a more stable footing, and these reforms reduced the rivalry between ethnic Germans and Czechs. When Joseph II succeeded his mother in 1780, he was able to build on her centralizing policies, and although his most radical reform – that of the tax system – was abolished by his successor, Leopold II, before it was given a chance to work, Joseph is generally considered to have been a strong and enlightened monarch.

In the years immediately after the French Revolution of 1789, and during the period of Napoleon's leadership, the Habsburg Empire became involved in a succession of wars against France (*pages 166–67*), as a result of which it temporarily lost much of Austria, as well as territories in northern Italy and along the Adriatic. Under the peace settlement negotiated at the Congress of Vienna in 1815, the Habsburgs renounced their claim to the Netherlands in exchange for areas in northern Italy (*map 2*).

Austria was by this time largely under the control of Foreign Minister Metternich, who used his influence to persuade the other major European powers to assist Austria in crushing revolts in Spain, Naples and Piedmont. His own methods involved the limited use of secret police and the partial censorship of universities and freemasons.

THE REVOLUTIONS OF 1848–49

The years 1848 and 1849 saw a succession of largely unsuccessful uprisings against the absolutist rule of the Habsburg monarchy (*pages 172–73*). Although reforms of the legal and administrative systems (known as the "April Laws") were set to take effect in Hungary later that year, they did not apply to the rest of the Habsburg territories.

The unrest started in Vienna in March 1848 (as a result of which Metternich was dismissed) and spread to Prague, Venice and Milan. A Constituent Assembly was summoned to revise the constitution, but its only lasting action was to abolish serfdom. By the autumn the unrest had reached Hungary as a number of ethnic groups within the empire (*map 3*) made bids for greater national rights and freedoms. In December the ineffectual Ferdinand I abdicated in favour of his nephew, Francis Joseph. Not feeling bound by the April Laws, Francis Joseph annulled the Hungarian constitution, causing the Hungarian leader Louis Kossuth to declare a republic. With the help of the Russians (who

▲ During her 40-year reign Empress Maria Theresa centralized control of the Habsburg territories through improved administrative systems, and won popular support with her social reforms.

feared the spread of revolutionary fervour), and the Serbs, Croats and Romanians (who all feared Hungarian domination), the Austrian army succeeded in crushing the revolt in 1849 (*map 4*).

From 1849 onwards an even more strongly centralized system of government was established. Trade and commerce were encouraged by fiscal reforms, and the railway network expanded. Coupled with peasant emancipation – for which landowners had been partially compensated by the government – these measures led to a trebling of the national debt over ten years. Higher taxes and a national loan raised from wealthier citizens led to discontent among the Hungarian nobles, who wished to see the restoration of the April Laws. In 1859 war in the Italian provinces forced the Austrians to cede Lombardy (*map 2*).

CRISIS AND CHANGE

Several factors combined in the 1860s to create a period of crisis for the Habsburg Empire. It was becoming clear that Prussia, under Bismarck, presented an increasing threat, but Austria was unable to keep pace with military developments because of the insistence of the international banks that it balance its budget. Unrest in Hungary was presenting a threat to the monarchy, and also making it difficult to collect taxes and recruit for the army. A centralized government was unacceptable to the Hungarian nobility, but provincial government would be unworkable because of ethnic conflict. Austria was forced to reach a constitutional settlement with Hungary in 1867, forming the Dual Monarchy of Austria–Hungary. Although Francis Joseph was crowned head of both, and there were joint ministries for finance, foreign policy and military affairs, each nation had an independent constitution and legislature.

Encouraged by the constitutional change of 1867, many of the ethnic groups within the Dual Monarchy became increasingly vocal in their demands for the right to promote their language and culture, if not for outright autonomy. In Hungary, although other languages were not actually repressed, a knowledge of Hungarian was necessary for anyone with middle-class aspirations. Croatia was granted partial autonomy within Hungary in 1878, but continued to be dominated by its larger partner. There were also demands for greater autonomy from the Czechs in Austria, which were resisted by the German-speaking majority.

▼ Throughout the 19th century the ethnic minorities within the Habsburg, and subsequently the Austro-Hungarian, Empire did not generally seek independence. Instead they sought to gain greater local autonomy within a reformed monarchy.

THE RISE OF SERB NATIONALISM

Bosnia, predominantly inhabited by impoverished peasants, was administered by the Austro-Hungarian Empire under terms agreed at the Congress of Berlin in 1878. It was annexed in 1908 in order to protect Habsburg trade routes to and from the Dalmatian coast. The resulting incorporation of a large number of Serbs into the empire was actively opposed by Serbian nationalists and was to contribute to the outbreak of the First World War in 1914. Following the defeat of the Austro-Hungarians in the war, the Treaty of Saint-Germain (1919) broke up the empire, granting autonomy to its constituent nations and reducing Austria and Hungary to less than a quarter of their former area.

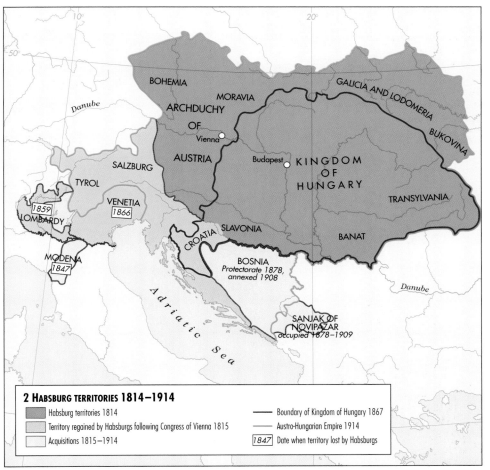

▲ In 1815 the Austrian Habsburgs regained territory they had gained and then lost during the Napoleonic Wars. However, they were forced to give it up in the mid-19th century during the process of Italian unification, and in 1867 were persuaded to grant Hungary equal status to that of Austria.

◀ The unrest in Hungary in 1848 and 1849 was largely an expression of Magyar nationalism, and as such was opposed by those from minority ethnic groups, in particular the Croats. In 1849, with Louis Kossuth appointed president of an independent republic of Hungary, the Austrians accepted Russian assistance, offered in the spirit of the Holy Alliance, and the rebels were eventually crushed at the Battle of Timisoara.

THE UNIFICATION OF ITALY AND OF GERMANY 1815–71

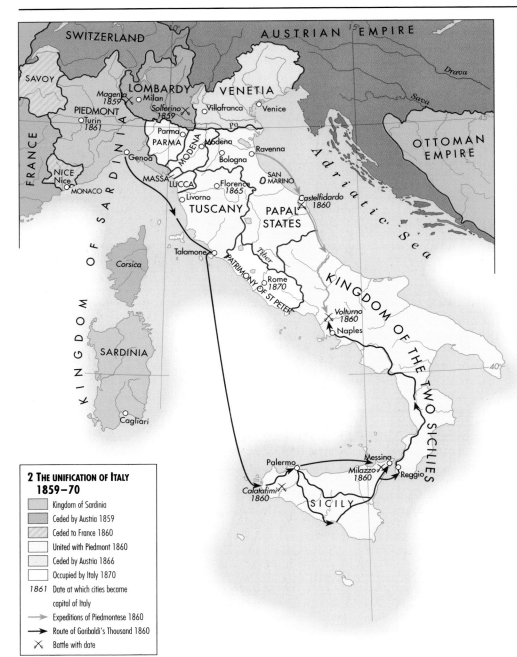

2 THE UNIFICATION OF ITALY 1859–70

- Kingdom of Sardinia
- Ceded by Austria 1859
- Ceded to France 1860
- United with Piedmont 1860
- Ceded by Austria 1866
- Occupied by Italy 1870
- *1861* Date at which cities became capital of Italy
- → Expeditions of Piedmontese 1860
- → Route of Garibaldi's Thousand 1860
- ✕ Battle with date

▲ In 1859, following a war waged by Piedmont and France against the Austrian Habsburgs, Lombardy was liberated from Austrian rule. The autocratic rulers of Florence, Parma and Modena were also overthrown and provisional governments set up under Piedmontese authority. France was granted Savoy and Nice by Piedmont.

In May 1861 Garibaldi answered requests for support from Sicilian revolutionaries and landed an army in western Sicily. He proceeded to rout the Neapolitan army in a series of battles and to proclaim himself ruler of the Kingdom of the Two Sicilies. The Piedmontese, anxious to unify the whole of Italy, despatched an army southwards to take the Papal States, and Garibaldi was persuaded to hand over his authority in the south to King Victor Emmanuel II.

Venetia was ceded by Austria to Italy, following Austria's defeat of 1866 at the hands of the Prussians, whom Italy had supported. Rome and its surrounding territory was seized by Italy in 1870.

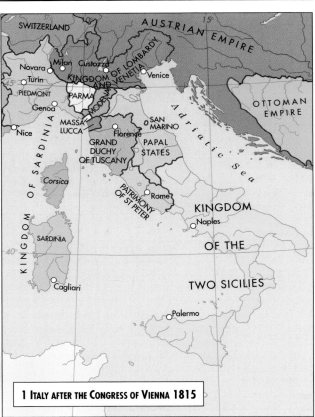

1 ITALY AFTER THE CONGRESS OF VIENNA 1815

▲ The Congress of Vienna in 1814–15 restored boundaries within Italy that had been lost under Napoleon's rule. It also restored members of the conservative Austrian Habsburg dynasty to power in Modena, Parma and Tuscany.

Among the most important developments in 19th-century Europe was the unification of Italy and Germany as nation-states – a process that fundamentally altered the balance of power in the continent. Although nationalist feeling had been stimulated by the French Revolution of 1789, and was originally associated with liberal ideas, unification was actually the result of diplomacy, war and the efforts of conservative elites rather than of popular action. German unification was promoted by Prussia, the most powerful German state, in order to protect its own domestic political stability; in Italy, Piedmont played this role for similar reasons.

ATTEMPTS TO UNIFY ITALY

The Napoleonic Wars (*pages 166–67*) had a dramatic effect on Italy. Napoleon redrew boundaries and introduced French political and legal ideas. At the Congress of Vienna in 1814–15 the major European powers attempted to reverse these changes by restoring deposed leaders, including members of the Habsburg dynasty, and giving conservative Austria effective control of Lombardy and Venetia in northern Italy (*map 1*). These developments were a major setback for Italian nationalists, who sought to remove foreign interference and unite Italy. The movement for national unification, or Risorgimento, continued to grow, despite the suppression of revolts in the 1820s and early 1830s (*pages 172–73*). A major figure in this movement was the idealist Giuseppe Mazzini, who hoped the people would overthrow their existing rulers, both Italian and foreign.

In 1848 a wave of revolutionary fervour swept the cities of Europe – including those in Italy, where the rebels attempted to dispense with Austrian domination and to persuade local rulers to introduce constitutions. King Charles Albert of the kingdom of Sardinia hoped to defuse the revolutions by expelling the Austrians from Lombardy and Venetia, but military defeats at Custozza and Novara forced him to abdicate in 1849 in favour of his son Victor Emmanuel II. In Rome, Venice and Florence republics were briefly established, but France intervened to restore Pope Pius IX to power and the Austrians reconquered Lombardy and restored the conservative rulers of central Italy.

THE RISE OF PIEDMONT

Moderate nationalists concluded that the best hope for Italian unification lay with Piedmont, which was economically advanced and had introduced a relatively liberal constitution. The Piedmontese prime minister, Count Camillo di Cavour, had already decided that foreign help would be needed to remove Austrian influence and achieve unification, and reached a secret agreement with Napoleon III of France at Plombières in 1858. Accordingly, when Cavour embarked on a war with Austria in 1859 France supported him; Austria was defeated and forced to cede Lombardy to Piedmont (*map 2*).

Piedmont's subsequent role in uniting Italy was partly a response to the actions of Giuseppe Garibaldi, one of the radicals who had created the Roman Republic in 1848. In 1860 Garibaldi led an expedition of republican "Red Shirts" (also known as Garibaldi's Thousand) through the Kingdom of the Two Sicilies, whose conservative ruler he defeated (*map 2*). Piedmont, anxious to preserve its constitutional monarchy, sent a force to annex the Papal States. Garibaldi then transferred the territory he had conquered to the Piedmontese king, who became head of the unified kingdom of Italy proclaimed in 1861. The remaining territories of Venetia and the Patrimony of St Peter were annexed during the subsequent ten years.

THE GERMAN CONFEDERATION

Before the Napoleonic Wars Germany consisted of over 300 states, loosely bound in the Holy Roman Empire. In 1806 Napoleon dissolved the empire, replacing it with a new Confederation of the Rhine comprising states in southern and western Germany, but excluding Austria and Prussia. The Confederation became a French satellite; its constitution was modelled on that of France and it adopted the Napoleonic legal code. It was dissolved after the defeat of the French at the Battle of Leipzig in 1813 (*pages 166–68*).

The German Confederation, created as a result of the Congress of Vienna in 1814–15, included 39 states, the largest and most powerful being Austria and Prussia (*map 3*). A *diet* (parliament), presided over by Austria, was established at Frankfurt, but plans to create a federal army and achieve constitutional harmony among the states failed.

As in other parts of Europe, 1848 saw a wave of revolutionary activity in Germany (*pages 172–73*). Following unrest in Berlin, the Prussian king, Frederick William IV, introduced constitutional reforms and seemed sympathetic towards German unification. Middle-class German nationalists established a parliament at Frankfurt which drew up a constitution for a future German Empire. However, they were divided over whether to pursue a "Greater Germany", to include Catholic Austria, or a smaller grouping, dominated by Protestant Prussia. The parliament fell apart in July 1849 and by the end of the year the old order had been restored in both Germany and the Austrian Empire.

Although Austria and Prussia tried to co-operate during the 1850s, Prussia was already outstripping Austria in economic terms (*pages 170–71*). In 1834 Prussia had established a Customs Union (Zollverein) that bound the economies of the north German states closely, while excluding Austria (*map 4*). Industrialization made Prussia the richest German state, and increased its military power relative to that of Austria.

▼ German unification can be seen as the annexation by Prussia of the smaller states of the Confederation. Following Prussia's display of military strength in France in 1870–71 the southern states acceded to Prussian demands for a unified Germany.

THE EXPANSION OF PRUSSIA

The leading role in German unification was played by Otto von Bismarck, the Prussian Chancellor between 1862 and 1871. Bismarck, who had come to see Austrian and Prussian interests as incompatible, sought to secure Prussian influence over northern and central Germany, and to weaken Austria's position. He hoped that success in foreign affairs would enable him to control Prussia's liberals. In 1864 Austria and Prussia jointly ousted Denmark from control of the duchies of Schleswig and Holstein, but the two powers increasingly competed for control of the German Confederation. When Bismarck engineered a war with Austria in 1866 (Seven Weeks War), most German states supported Austria. Prussia, however, enjoyed advantages in military technology and defeated Austria quickly, signalling the end of the German Confederation and making German unification under Prussian leadership more likely.

In 1867 Bismarck secured the creation of a North German Confederation (*map 4*). Each member state retained some autonomy, but the Prussian king, William I, became the Confederation's president, responsible for defence and foreign policy. Although the south German states were apprehensive about Prussian domination, Bismarck used their fear of the territorial ambitions of Napoleon III of France to persuade them to ally with Prussia. Bismarck needed to neutralize France if he was to achieve German unification on his terms, and he therefore provoked a war over the succession to the Spanish throne. In the resulting Franco-Prussian War (1870–71) France was decisively defeated, losing the largely German-speaking areas of Alsace and Lorraine to Prussia.

In January 1871, in the Hall of Mirrors at Versailles, the German Empire was declared, merging the south German states with the North German Confederation. The new empire had a federal constitution, leaving each state with some powers, but the Prussian king became emperor and most government posts were put into Prussian hands. With well-developed industrial regions in the north and east (*pages 170–71*), a united Germany represented a powerful new economic force in Europe.

▲ During 1870–71 the Prussians, under Kaiser William I and Chancellor Bismarck, defeated the French army and laid siege to Paris. This display of strength convinced the southern German states to join with the North German Confederation in a unified Germany – dominated by Prussia.

▼ The German Confederation was established following the end of the Napoleonic Wars in 1815. It comprised 39 German-speaking states, by far the largest of which was Prussia, and included states under the control of the Habsburg Empire.

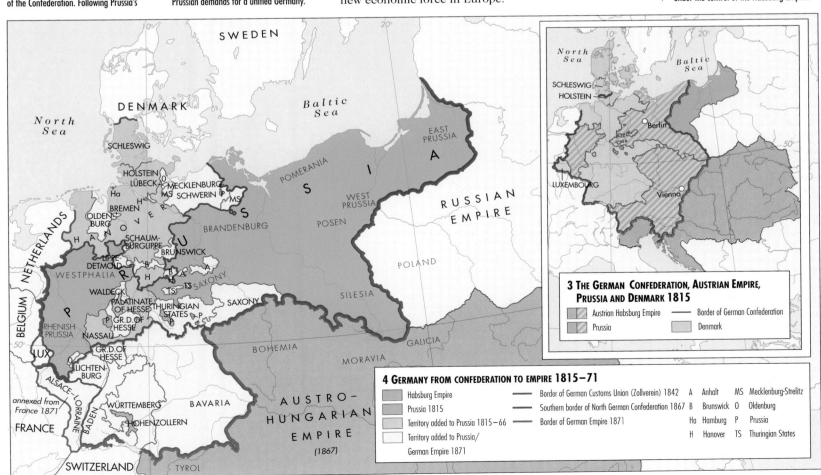

3 THE GERMAN CONFEDERATION, AUSTRIAN EMPIRE, PRUSSIA AND DENMARK 1815

▨ Austrian Habsburg Empire	— Border of German Confederation
▨ Prussia	▨ Denmark

4 GERMANY FROM CONFEDERATION TO EMPIRE 1815–71

Habsburg Empire	— Border of German Customs Union (Zollverein) 1842	A Anhalt · MS Mecklenburg-Strelitz
Prussia 1815	— Southern border of North German Confederation 1867	B Brunswick · O Oldenburg
Territory added to Prussia 1815–66	— Border of German Empire 1871	Ha Hamburg · P Prussia
Territory added to Prussia/German Empire 1871	(1867)	H Hanover · TS Thuringian States

● REVOLUTIONARY FRANCE AND NAPOLEONIC EUROPE 1789–1815 *pages 166–67* ● THE BUILD-UP TO THE FIRST WORLD WAR 1871–1914 *pages 216–17*

THE DECLINE OF THE OTTOMAN EMPIRE 1683–1923

The decline of the Ottoman Empire is often said to date from the massive defeat of the Ottomans outside Vienna in 1683, but despite the territorial losses resulting from the subsequent Treaty of Karlowitz in 1699, the 18th-century Ottoman state remained the biggest political entity in Europe and western Asia (*map 1*). Although the effectiveness of the empire's prestige troops, the Janissaries, was weakened by increasing internal unrest, Ottoman forces were able to hold Serbia. They also got the better of their old Renaissance opponent, Venice, by recovering the Morea in 1718 (*map 2*).

During the 18th century the major European states became more of a threat to the Ottomans. There were large-scale Russian encroachments around the Black Sea in the later part of the century, and in 1798 a French army under Napoleon Bonaparte made a devastating, if shortlived, surprise attack on Egypt, the empire's richest Muslim province. It was clear that the weaponry and the military capacity of the European states were moving ahead of those of their Islamic counterparts. At the same time, Europe's ideological conflicts reverberated among the Ottoman Empire's Christian subjects, encouraging bids for separatism and liberty which usually had Russian backing. Whole communities in the Caucasus switched their allegiance from the Ottoman (and Persian) states to the Russian Empire, and disaffection spread among the prosperous and previously co-operative Greeks of the empire's heartlands. In 1821 the western Greeks struck out for independence, and by 1832 they had won a mini-state (*map 1*).

THE SLIDE INTO DEPENDENCY

The Ottoman state responded to its losses with a programme of expensive remilitarization, as well as political and economic reform and development, funded precariously from what were now seriously reduced revenues. The strategy for survival was to replace the empire's traditional patchwork of cultural and religious communities with a new model Ottoman society in which there was one legal system, one citizen status and one tax rating for all. This was progressive, liberal 19th-century policy, but it attacked vested interests in the provinces and among the Muslim clergy.

The reform movement engendered a limited revival of international confidence in the Ottomans. During the Crimean War of 1853–56, British and French armies fought to defend Ottoman interests against Russian military escalation in exchange for an Ottoman commitment to equality of status for its Muslim and non-Muslim subjects. This was a deal the Ottoman state was unable to honour; twenty years after the Crimean campaign, the Ottoman authorities were still employing ill-disciplined troops to contain unruly Balkan Christians, provoking an international outcry and eventually the resumption of full-scale war with Russia. Under the agreement reached at the Congress of Berlin in 1878, the region's political map was redrawn (*maps 1 and 3*). "Turkey in Europe" became a much-reduced presence.

▼ Between 1699 and 1739 the Ottomans lost large areas in the Balkans, although they regained the Morea from Venice in 1718, and Serbia and Wallachia from the Austrian Habsburgs in 1739.

2 RETREAT IN THE BALKANS 1699–1739

Territorial gains under Treaty of Karlowitz 1699 by:
- Austria
- Poland
- Venice

Territorial gains under Treaty of Passarowitz 1718 by:
- Austria 1718
- Republic of Ragusa
- Ottoman Empire

Territorial gains under Treaties of Belgrade and Constantinople 1739 by:
- Ottoman Empire
- Russia
- Territory held by Ottomans 1739

▼ The Ottoman Empire reached its furthest extent in the mid-17th century, but when its troops failed to take Vienna in 1683 European powers took advantage of their disarray and seized territory in central Europe. The subsequent disintegration of the empire took place over the next 240 years.

The British took control of Egypt in 1882, and the Middle Eastern territories were lost as a result of an Arab uprising during the First World War.

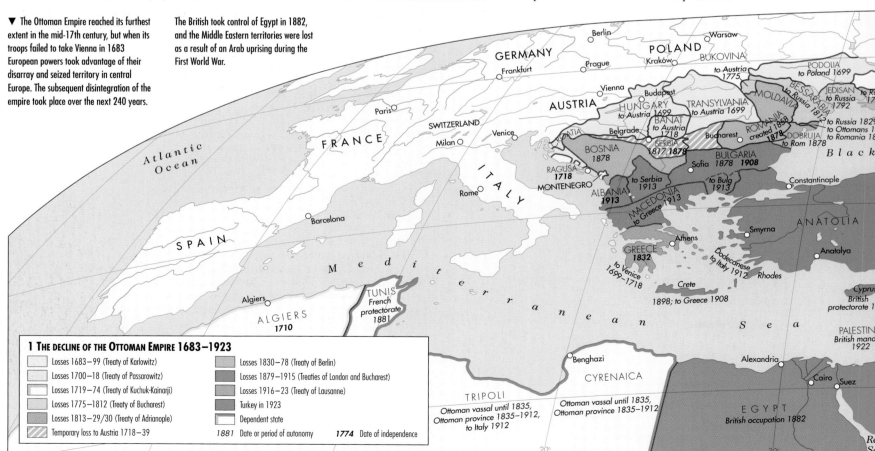

1 THE DECLINE OF THE OTTOMAN EMPIRE 1683–1923
- Losses 1683–99 (Treaty of Karlowitz)
- Losses 1700–18 (Treaty of Passarowitz)
- Losses 1719–74 (Treaty of Kuchuk-Kainarji)
- Losses 1775–1812 (Treaty of Bucharest)
- Losses 1813–29/30 (Treaty of Adrianople)
- Temporary loss to Austria 1718–39
- Losses 1830–78 (Treaty of Berlin)
- Losses 1879–1915 (Treaties of London and Bucharest)
- Losses 1916–23 (Treaty of Lausanne)
- Turkey in 1923
- Dependent state
- *1881* Date or period of autonomy
- *1774* Date of independence

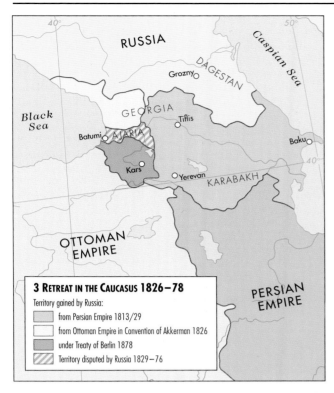

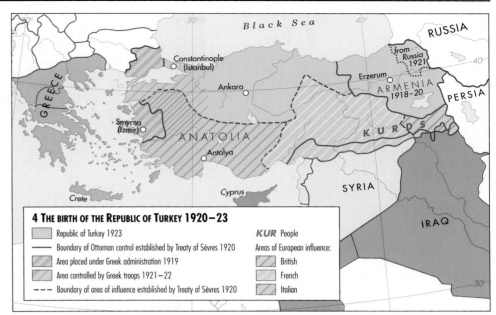

3 RETREAT IN THE CAUCASUS 1826–78

Territory gained by Russia:

- from Persian Empire 1813/29
- from Ottoman Empire in Convention of Akkerman 1826
- under Treaty of Berlin 1878
- Territory disputed by Russia 1829–76

4 THE BIRTH OF THE REPUBLIC OF TURKEY 1920–23

- Republic of Turkey 1923
- Boundary of Ottoman control established by Treaty of Sèvres 1920
- Area placed under Greek administration 1919
- Area controlled by Greek troops 1921–22
- Boundary of area of influence established by Treaty of Sèvres 1920
- **KUR** People

Areas of European influence:
- British
- French
- Italian

▲ Following Russia's defeat of the Ottomans in 1878, the Treaty of Berlin awarded an area of the Caucasus to Russia. This land was returned in 1921 by Bolshevik Russia to those fighting for the establishment of the Turkish Republic.

THE RISE OF THE "YOUNG TURKS"

The new sultan, Abdul Hamid II, swiftly shelved the constitution he had adopted as the price of survival in 1876. He ruled in the tradition of the Ottoman dynasty – as a despot. His empire had two faces: a westward-facing and cosmopolitan Constantinople, run by European-educated officials who might also be slave-owners, governing a society that faced east. The empire's political geography was now predominantly Middle Eastern, and Abdul Hamid was keen to exploit

his status as caliph (senior ruler in the Islamic world) which gave Ottoman agents access to Muslim communities worldwide, including those living under the British Raj.

Pan-Islamic policies met widespread, if covert, criticism from those within the Ottoman elite who would have preferred a state with a nationalist Turkish identity to one with a more diffuse Ottoman or Islamic facade. The empire's fault lines were exposed by a new political force: the Committee of Union and Progress (CUP), a successful, originally conspiratorial, pressure group dominated by Turkish nationalist army officers, commonly nicknamed the "Young Turks". The CUP was committed to the retention of "Turkey in Europe" and relatively dismissive of the empire's Middle Eastern provinces and peoples. In 1908 they forced the sultan to renew the long-suspended constitution of 1876, and the following year deposed him in favour of his more pliant brother.

The CUP set out with democratic ideals but found that these were incompatible with the empire's ethnic divisions. Showpiece general elections served chiefly to demonstrate the voting power of the minorities, particularly the Arabs. CUP administration survived only by becoming increasingly dictatorial, particularly when it faced a new round of territorial losses. It was in an attempt to remedy this situation that the leader of the CUP, Enver Pasha, with German military assistance, took the Ottoman Empire to war in 1914.

Between 1914 and 1916 the empire survived a series of Allied invasions (*pages 218–19*). Casualties were immense and the loyalty of the empire's minority populations was suspect, with thousands of Christian Armenians massacred for their pro-Russian sympathies. Apathy and disaffection among the empire's Arab Muslims was even more dangerous. In 1916 the Hashemi "sharif", governor of Mecca, raised a desert army which, allied with the British, successfully detached all remaining Arab provinces from Turkish control.

THE BIRTH OF THE NEW TURKEY

Post-war schemes for dismembering the empire and reducing the Ottoman sultanate to puppet status were built into the Treaty of Sèvres (1920), which the sultan's administration in Constantinople meekly accepted, thereby losing any last shred of credibility. An alternative Turkish nationalist government was set up at Ankara, led by Mustafa Kemal, later named "Atatürk" (Father of the Turks). By 1923 the Ankara regime had won diplomatic and military recognition from all its former antagonists, including the Greeks, who had been defeated by Kemal's forces in 1922.

The Sèvres agreement was replaced by the more generous Treaty of Lausanne (1923), which legitimized Ankara's right to govern an independent Turkish Republic in a region broadly corresponding to modern Turkey. The Ottoman sultanate was abolished by the treaty and the archaic caliphate followed it into extinction in 1924.

▲ The Treaty of Sèvres (1920) stripped the Ottomans of the remains of their empire, and divided Anatolia into European "spheres of influence", leaving only a small portion to be directly ruled by the sultan. The Greeks, who saw the Turkish defeat as an opportunity to claim territory in western Anatolia, had a substantial Greek population, had dispatched troops to Smyrna in 1919. Between 1920 and 1922 their troops established a firm grip on the region. During this time, however, Turkish nationalists became increasingly organized under the leadership of Mustafa Kemal, and in August 1922 a Turkish nationalist army attacked the Greek forces and drove them from Anatolia in disarray. The other European powers, recognizing the overwhelming Turkish support for Kemal, withdrew, and the Republic of Turkey was founded in 1923.

▲ As President of Turkey (1923–38), Mustafa Kemal ("Atatürk") instigated a series of reforms that created a modern secular state from the remains of the Ottoman Empire.

◀ THE OTTOMAN AND SAFAVID EMPIRES 1500–1683 *pages 142–43* ▶ THE MIDDLE EAST SINCE 1945 *pages 260–61*

RUSSIAN TERRITORIAL AND ECONOMIC EXPANSION 1795–1914

▼ Between 1795 and 1914 Russia sought to expand its territory in all possible directions but met with resistance from Austria, Britain and France when it threatened their interests in the Balkans in the 1850s. Expansion to the south and east was intermittent up until the 1880s, when it was halted by British power and by internal financial difficulties. To the east, the Russian Empire extended even onto the continent of North America, as far as northern California, until Alaska was sold to the Americans for $7.2 million in 1867. To the southeast, Russia continued to exert its influence in Manchuria and Mongolia in the early years of the 20th century, despite its defeat at the hands of the Japanese in 1905.

During the 19th century Russia continued a process of territorial expansion that had begun in the 1460s but which was now largely confined to Asia. Victory over Napoleon Bonaparte in 1815 brought the acquisition of the western part of Poland ("Congress Poland") and confirmation of earlier gains in Finland in 1809 and Bessarabia in 1812 (*map 1*). However, this marked the end of expansion to the west and in fact Romania soon cut its ties with Russia and in 1883 made an alliance with Germany and Austria. In the southwest the Transcaucasian territories were acquired between 1801 and 1830 and the route to them finally secured by the conquest of Chechenia – completed in 1859 – and Cherkessia in 1864.

In Central Asia, Russia seized large areas, often moving in where there was a political vacuum it could fill and perhaps resources it could exploit (although it failed to actually exploit them until the 1920s). The conquests began

in the 1820s and accelerated from 1853 onwards. In 1885, however, Russian troops clashed with Afghan forces at Pendjeh and came up against another imperialist power, Britain, which sent a stern warning that Afghanistan was not for the taking.

In the mid-19th century Russia also turned its attention to the eastern end of Asia, acquiring the regions north and south of the Amur River. This enabled it to establish Vladivostok – the vital warm-water port that gave year-round maritime access to the Far East. The Trans-Siberian Railway – built between 1891 and 1904 – linked Vladivostok to Moscow, and brought the potential for trade with the Far East. It tempted Russian policymakers to take over Manchuria in order to provide a more direct route to the coast, despite warnings from economic pressure groups that they should be concentrating on expanding internal markets in Siberia. The dream of eastern expansion reached both its apogee and its catastrophe in the Russo-Japanese War of 1904–5, which resulted in a humiliating defeat for Russia. The limits of the empire were thus finally set.

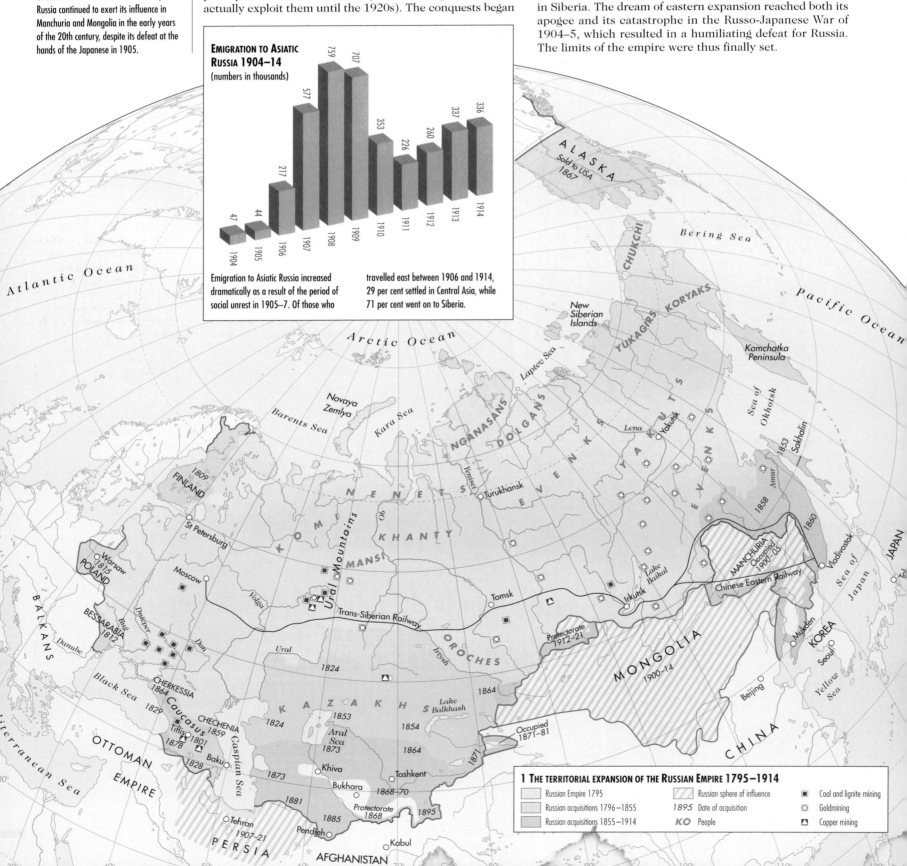

EMIGRATION TO ASIATIC RUSSIA 1904–14
(numbers in thousands)

Year	Thousands
1904	47
1905	44
1906	217
1907	577
1908	759
1909	707
1910	353
1911	226
1912	260
1913	337
1914	336

Emigration to Asiatic Russia increased dramatically as a result of the period of social unrest in 1905–7. Of those who travelled east between 1906 and 1914, 29 per cent settled in Central Asia, while 71 per cent went on to Siberia.

1 THE TERRITORIAL EXPANSION OF THE RUSSIAN EMPIRE 1795–1914

☐ Russian Empire 1795	▨ Russian sphere of influence	■ Coal and lignite mining
☐ Russian acquisitions 1796–1855	*1895* Date of acquisition	◎ Goldmining
☐ Russian acquisitions 1855–1914	**KO** People	◮ Copper mining

ECONOMIC DEVELOPMENT

The economic development of the Russian Empire (*map 2*) was continuous throughout the 19th century and into the 20th century, but four periods can be distinguished. First there was slow and steady growth from 1800 to 1885, interrupted by setbacks in the 1860s when the iron industry in the Urals was adversely affected by the emancipation of the serfs. (Many who had been forced to work in the mines fled from the region on being freed.) Then, from 1885 to 1900, there was rapid government-induced growth, with a one-sided emphasis on railway building and heavy industry. Economic stagnation, prolonged by the effects of the revolution of 1905–7 (*map 3*), constituted the third period. The final period, from 1908 to 1914, was a time of renewed economic growth on a broader front.

It was during this last period that the big rush to emigrate to Siberia began, stimulated by the government itself, with the intention of solving the problem of land shortage in European Russia that had contributed greatly to the rural disturbances of 1905–7. Emigration to Siberia increased rapidly (*graph*) and the population of Siberia rose from 5.7 million in 1897 to 8.2 million in 1910. Settlement was concentrated along the Trans-Siberian Railway, which provided a link back to the west for a developing capitalist agriculture and the gold, copper and coal mines.

THE 1905 REVOLUTION

Russia's economy expanded in the 1890s with little attention to infrastructure and a complete refusal to link economic with political changes. This created tremendous tensions in the Russian social fabric, which were exacerbated by the government's repressive measures and its attempts at a gigantic foreign-policy diversion. "What we need to stem the revolutionary tide," said the reactionary, anti-Semitic Minister of the Interior Plehve in 1903, "is a small, victorious war". However, the result of the Russo-Japanese War of 1904–5 was precisely the opposite: the "revolutionary tide" nearly swept away the whole tsarist system. Only the loyalty of parts of the imperial army at the decisive moment, in December 1905, saved the situation for Nicholas II.

The revolution of 1905 (or, more accurately, 1905–7) started under liberal slogans, and indeed the demand for representative popular government on the Western model was a common denominator throughout. It developed, however, into something much more threatening than a mere change of political regime. The workers who went on strike in 1905 set up councils, or "soviets", in every major city of the Russian Empire (*map 3*). These institutions acted as local organs of power, initally side by side with the old authorities, and in some cases led armed revolts that aimed at the complete overthrow of the imperial government. They were to resurface in 1917, with a decisive impact on Russian and world history.

The revolution of 1905 was not simply an urban movement of Russian workers and intellectuals. Agriculture had been neglected by the state in its drive for industrialization, and since the emancipation of the serfs in 1861 it had experienced either stagnation or a slight improvement, interrupted by the dreadful famine of 1891. It is hardly surprising that the peasants lost patience. The peasant revolts of 1905–7 were the first large-scale risings since the 18th century, and they forced the government into an abrupt change of policy (the Stolypin Reforms of 1906–10). This was, however, ultimately ineffectual, since the government carefully side-stepped the peasants' major grievance: the issue of gentry landholding. The peasant movement would revive with a vengeance in 1917 (*pages 222–23*).

The non-Russian nationalities also revolted in 1905, demanding autonomy or independence, depending on their level of social and national maturity. These demands would also resurface in 1917, leading to the complete disintegration of the Russian Empire, although the formation of the Soviet Union in 1922 delayed the establishment of independent national states on the territory of the former Russian Empire for nearly 70 years.

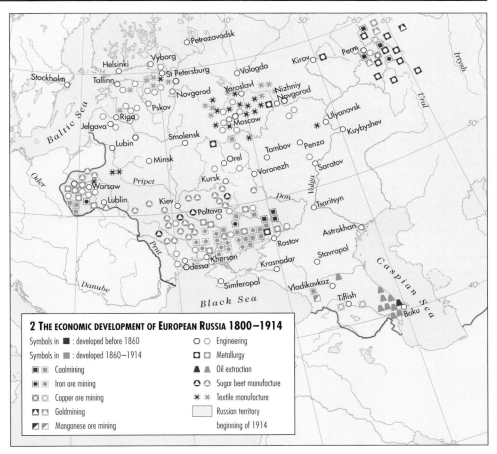

2 THE ECONOMIC DEVELOPMENT OF EUROPEAN RUSSIA 1800–1914

Symbols in ■ : developed before 1860
Symbols in ▨ : developed 1860–1914

◼▢ Coalmining
◉◎ Iron ore mining
◒◓ Copper ore mining
▲△ Goldmining
◪◩ Manganese ore mining

◯ ○ Engineering
◻ ▢ Metallurgy
▲ △ Oil extraction
⬠ ⬡ Sugar beet manufacture
✳ ✲ Textile manufacture

▭ Russian territory beginning of 1914

▲ Industrial expansion occurred mainly in engineering, metalworking and mining, with the development of engineering around Moscow and oil extraction around Baku particularly noticeable. Overall, the period 1800–1914 saw a clear shift in the centre of economic gravity from the Urals to the Ukraine and Poland.

▼ During the years of revolution, 1905–7, urban revolt was widespread across European Russia, with strikes and armed uprisings. In some cities workers organized themselves into soviets. Revolts also took place in large cities in Siberia and Central Asia, where there was a substantial Russian or Ukrainian population. Rural revolt, on the other hand, was most intense in the Ukraine and to the south of Moscow, in provinces where land was held in common by the peasants and redivided every 20 years according to family size. This led to a strongly developed sense of community, making the peasants sympathetic to socialist revolutionary agitators.

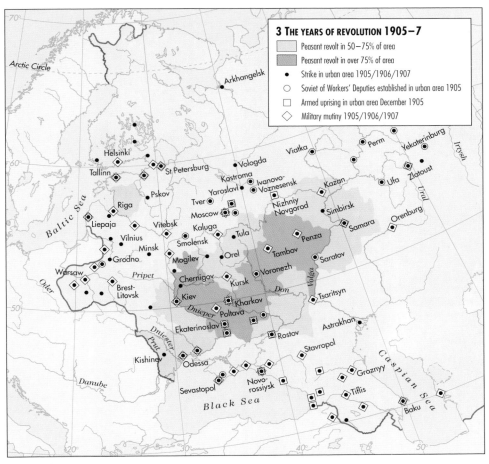

3 THE YEARS OF REVOLUTION 1905–7

▨ Peasant revolt in 50–75% of area
▨ Peasant revolt in over 75% of area
● Strike in urban area 1905/1906/1907
◯ Soviet of Workers' Deputies established in urban area 1905
▢ Armed uprising in urban area December 1905
◇ Military mutiny 1905/1906/1907

◀ THE EXPANSION OF RUSSIA 1462–1795 *pages 148–49* ▶ THE RUSSIAN REVOLUTION 1917–39 *pages 222-23*

THE WESTWARD EXPANSION OF THE UNITED STATES 1783–1910

▲ The expedition of Meriwether Lewis and William Clark in 1804–6 succeeded in its quest to find a route from the Mississippi to the Pacific. Like so many pioneering journeys in the West, it relied heavily on the local knowledge of Native Americans. Sacajawea (pictured here with Clark) – a Shoshone woman who had lived with the Mandan – was particularly valuable to the venture as a translator.

Throughout the 19th century American pioneers moved inexorably westwards across the Appalachian Mountains in search of good farmland and new opportunities. Either through diplomacy, conquest or purchase, millions of acres of new territory came under United States control to form the transcontinental nation that we recognize today. This enormous landmass was swiftly occupied by settlers, and as these new areas gained large populations they were admitted to the Union as states.

In 1783 the new nation extended from the Atlantic coast westwards as far as the Mississippi River (*map 1*). Its territory was subsequently enlarged in two great expansionist movements. Firstly, with great astuteness, Thomas Jefferson bought a great swathe of the Midwest from France in 1803 for a meagre $15 million. The "Louisiana Purchase", as it was known, instantly doubled the size of the United States. West Florida was annexed in 1813, while under the Adams–Onis Treaty of 1819, Spain ceded all of East Florida to the United States and gave up its claim to territory north of the 42nd parallel in the Pacific northwest.

The second wave of expansion involved the acquisition of Texas, Oregon and California. In 1835 American settlers in Texas staged a successful revolt against Mexican rule, winning the Battle of San Jacinto in 1836, and the Republic of Texas was born. The Mexican War (1846–48) between the United States and its weaker southern neighbour resulted in the Treaty of Guadalupe Hidalgo (1848), which gave the United States not only California but a huge region in the southwest (*map 1*).

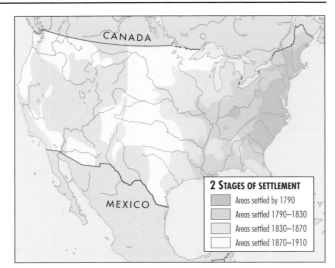

▲ Settlement took place in a number of stages, often as a result of the displacement of people from areas within the United States caused by political and economic developments. Many European economic migrants also became American pioneers.

the West. Zebulon Pike (1804–7) explored the sources of the Mississippi and visited Colorado and New Mexico, while Stephen H. Long (1817–23) investigated lands near the Red and Arkansas rivers. As well as these government agents, traders and fur trappers, such as Jedediah Smith, travelled extensively between the Missouri and the Pacific coast. It was they who opened the Santa Fe Trail between New Mexico and Missouri in 1821, while "mountain men", hunting in the Rockies in the 1820s, spread word of the riches to be found there.

WESTWARD MIGRATION

The American people flowed west in several distinct migration waves (*map 2*). The War of 1812 against Britain led to many people overcoming their fear of opposition from Native Americans and travelling westwards to find new agricultural land. Thousands of newcomers established small farms in what was known as the "Old Northwest" (now part of the Midwest). Most of the first settlers were southerners who had been displaced by the growth of the plantation system with its slave labour force. By 1830 their settlements filled southern Indiana and Illinois and were overrunning Missouri. In the following decade newcomers from the northeast settled around the Great Lakes, and by 1840 almost all the Old Northwest had been carved into states. Many pioneers had also moved into the newly acquired territory of Florida and into the land bordering the Gulf of Mexico. Most settlers here came from the southeast, looking for fields where they could grow cotton. Small farmers had been followed by large-scale planters, who brought slaves to the region – the majority from the eastern states. Once settlers had occupied the entire area, pioneers began to push beyond the Mississippi.

Many Americans believed in "manifest destiny", the idea that America was destined by God and by history to expand its boundaries over the whole of North America. After 1843, each spring, eager adventurers gathered at Independence, Missouri to organize wagon trains to travel the overland Oregon Trail across the Great Plains (*map 3*). This early trickle of settlement was hugely accelerated by the discovery of gold in California in 1848. When gold fever swept the nation, more than 100,000 "Forty-Niners" poured into California. Although relatively few found gold, many stayed on as farmers and shopkeepers.

Utah was settled not by profit-seeking adventurers but by Mormons searching for an isolated site where they could freely worship without persecution. The journey of the Mormons to the shores of Great Salt Lake in 1847 was one of the best-organized migrations in history.

Much of the West remained unsettled even after the frontier reached the Pacific Ocean. During the Civil War (1861–65) pioneers settled in the region between the Rocky

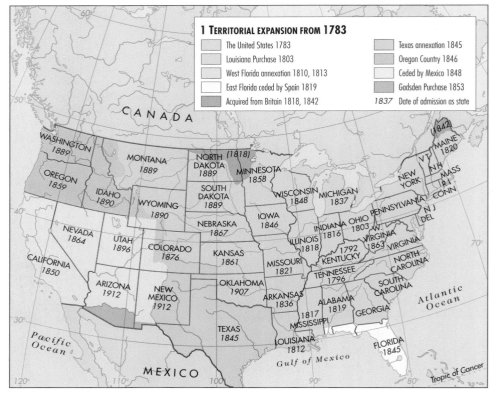

▲ The United States expanded westwards to the Pacific by a series of financial deals, negotiated settlements and forcible annexations. As each new territory was colonized by American settlers and a viable government formed, it became eligible for admission to the Union as a state and entitled to representation in Congress.

For many years, Britain had contested America's claims to the Oregon Country. Its Hudson's Bay Company controlled the region but, in the face of growing American immigration in the west of the region, Britain surrendered most of the area south of the 49th parallel to the United States in the Oregon Treaty of 1846 (*map 2*). With the Gadsden Purchase in 1853, the United States owned all the territory of its present states except for Alaska (purchased from Russia in 1867) and Hawaii (annexed in 1898).

EXPLORERS OF THE WEST

At the beginning of the 19th century part of the impetus to venture west came from the desire to increase trade – not only with the Native Americans but also with Asia. Reports from the expedition of Lewis and Clark (1804–6) (*map 3*) provided valuable information about the natural wealth of

Mountains and the Sierra Nevada, and after the war ranchers and farmers occupied the Great Plains west of the Mississippi. Cattle ranching on the open ranges involved driving herds over long distances along recognized trails (*map 3*), from the pasture lands to the railhead and on to market. However, the "cattle kingdom" was short-lived. The pastures became exhausted, and the Homestead Act of 1862 encouraged farmers to move from the east onto free or low-cost land. The settlers enclosed the pasture lands, barring the roving cattle herds. This settlement was greatly facilitated by the new east–west railroads (*pages 186–87*).

THE NATIVE AMERICANS

As the pioneers moved westwards they ruthlessly took over land from Native Americans and fighting often broke out (*map 4*). The US government sent in support for the settlers and federal troops won most encounters of the so-called Indian Wars (1861–68, 1875–90). Settlement of the West largely brought an end to the traditional way of life of the Native Americans. Farmers occupied and fenced in much of the land, and white settlers moving west slaughtered buffalo herds on which many Native Americans depended for their survival. At the same time, the federal government pushed more and more Native Americans onto reservations.

In the short period of one century, the United States expanded from being an infant rural nation confined to the Atlantic coast to a transcontinental powerhouse, with a large rural and industrial population. This territorial expansion occurred at a phenomenal speed and settlement proceeded rapidly, despite formidable physical and human obstacles. Having established its own internal empire from the Atlantic to the Pacific, the USA was now in a position to challenge European supremacy on the world stage.

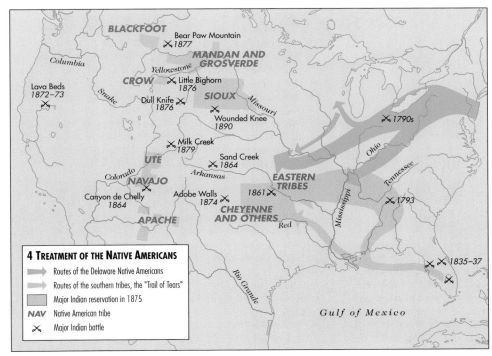

4 TREATMENT OF THE NATIVE AMERICANS
- Routes of the Delaware Native Americans
- Routes of the southern tribes, the "Trail of Tears"
- Major Indian reservation in 1875
- **NAV** Native American tribe
- ✕ Major Indian battle

▼ In 1806 a government-funded expedition, led by Lewis and Clark, established a route between the Mississippi River and the west coast. Alternative overland routes were established by pioneers seeking land or gold, and by surveyors looking for railroad routes.

▲ During the 18th century the Delaware Native Americans made a slow westward migration and in 1830 the Indian Removal Act also forced the southern tribes westward. Demands by white settlers for more land led to the establishment of Indian reservations and a series of bloody conflicts.

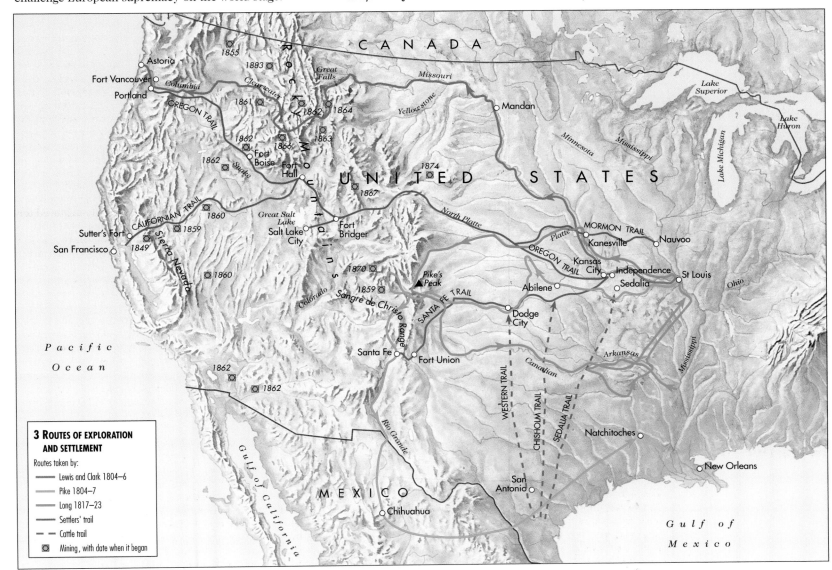

3 ROUTES OF EXPLORATION AND SETTLEMENT

Routes taken by:
- Lewis and Clark 1804–6
- Pike 1804–7
- Long 1817–23
- Settlers' trail
- Cattle trail
- ◎ Mining, with date when it began

◗ THE AMERICAN REVOLUTION 1775–83 *pages 164–65* ◗ THE INDUSTRIAL GROWTH OF THE UNITED STATES 1790–1900 *pages 186–87*

THE AMERICAN CIVIL WAR
1861–65

CASUALTIES OF THE CIVIL WAR

275,175
360,222
1,564,803

UNION: Total area represents 2,200,200

125,000
417,000
258,000

CONFEDERATE: Total area represents 800,000

- Killed
- Wounded
- Unharmed soldiers

▲ The Union was able to muster many more troops than the Confederacy, and suffered a smaller proportion of casualties. Overall, 20 per cent of soldiers in the Civil War died – the majority of them as a result of disease.

▼ Although it was the issue of slavery that prompted the Southern states to secede from the Union, the situation was not clearcut, with four of the Union states – Delaware, Maryland, Missouri and Kentucky – permitting slavery. Kansas joined the Union as a free state in 1861.

The American Civil War was fought between the Northern states (the Union), who wished to maintain the United States of America as one nation, and the Southern states (the Confederacy), who had seceded to form their own nation. The causes of the war included the long-standing disagreements over slavery and its expansion into the new territories, as well as conflicts over economic disparities between North and South and the division of power between the federal government and individual states.

Although slavery had been a marginal issue in the founding of the Republic, abolitionists began to attack this Southern institution in the early 19th century. Following the Missouri Compromise of 1820, which forbade slavery in the Louisiana Purchase (*pages 182–83*) north of 36° 36′, many thought that slavery would gradually die out as the tobacco industry declined. After 1830, however, the opening up of virgin lands in the Deep South to the cotton economy (*map 1*), coupled with the ever-increasing demand of European textile mills for raw cotton, suddenly enhanced the value of slave labour.

THE SECTIONAL DIVIDE
American politics began to divide according to sectional interests, focusing on the status of slavery in the new western territories. The Compromise of 1850 forbade slavery in California (*map 2*), while the Kansas–Nebraska Act of 1854 opened up these two territories to slavery – leading to much violence in Kansas.

Against this background, the Republican Party was formed to prevent further expansion of slavery, although in the controversial Dred Scott decision in 1857 the Supreme Court ruled that Congress could not exclude slavery from the territories.

The issue of slavery came to the forefront during the presidential election of 1860. The Republican candidate, Abraham Lincoln, was hostile to slavery and opposed its extension to new territories, although he had pledged not to interfere with it where it already existed. Following his election as President in 1860, however, South Carolina immediately seceded from the Union, a decision followed by Georgia, Florida, Alabama, Mississippi, Louisiana and Texas. These seven states formed the independent Confederate States of America early in 1861 and they would be joined by four more (Virginia, North Carolina, Tennessee and Arkansas) once war was declared.

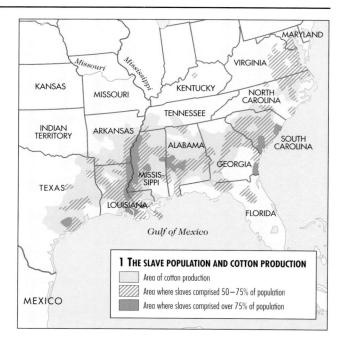

1 THE SLAVE POPULATION AND COTTON PRODUCTION
- Area of cotton production
- Area where slaves comprised 50–75% of population
- Area where slaves comprised over 75% of population

▲ The census of 1860 revealed that there were nearly four million slaves in the southern United States, the majority of whom were agricultural workers. They were considered vital to the profitability of cotton production, which had expanded to meet an increased demand from the rapidly industrializing countries of western Europe.

THE OUTBREAK OF WAR
War broke out on 12 April 1861 when Southern forces opened fire on federal-owned Fort Sumter. Arguing that secession was illegal and that the Union must be preserved, Lincoln took this as a declaration of war. Given the South's dependence on European imports, the strategy of the North was to starve the South into submission by encirclement and blockade (*map 3*).

The Confederacy won some early victories in 1861–62, successfully repelling Union attempts to capture their capital at Richmond, Virginia. The Union was forced (in particular by the defeat at the First Battle of Bull Run in July 1861) to disband its militia in favour of a new army of 500,000 volunteers. As the war progressed, however, both sides were forced to introduce conscription to raise troops.

While the Union cause seemed imperilled in the east, in the southwest Union forces were successful in their attempt to seize control of the Mississippi, culminating in the capture of New Orleans, the largest city and most important port in the Confederacy. The Confederate attempt to invade Maryland in September 1862 was thwarted at the Battle of Antietam. This encouraged President Lincoln to sign the Emancipation Proclamation on 1 January 1863, which freed all slaves in the Confederacy. Although it did not apply to Union states in which slavery was still permitted (*map 2*), it nevertheless gave the conflict a new moral purpose: to preserve the Union and abolish slavery. Freedom for the slaves took place gradually as the Union armies moved southwards, and the Proclamation helped break down the opposition to recruitment of African-American soldiers. By the war's end, 186,000 of them had served in Union armies, albeit in segregated regiments under the command of white officers and at vastly reduced levels of pay.

As the war progressed, the Union's greater manpower and superior economic and industrial resources began to prevail. The Union victory at Gettysburg, Pennsylvania in July 1863 proved to be the major turning point. The Confederacy was never strong enough again to undertake another major offensive. The next day the Confederate garrison of Vicksburg, Louisiana, which had been besieged by the Unionists since mid-May, surrendered. Not only had the Confederacy suffered huge and irreplaceable losses in the east, but it was also now split in two, with Union troops controlling the Mississippi. The second half of 1863 saw further decisive battles in the west in the Tennessee campaign, with the Confederate forces being driven back into Georgia.

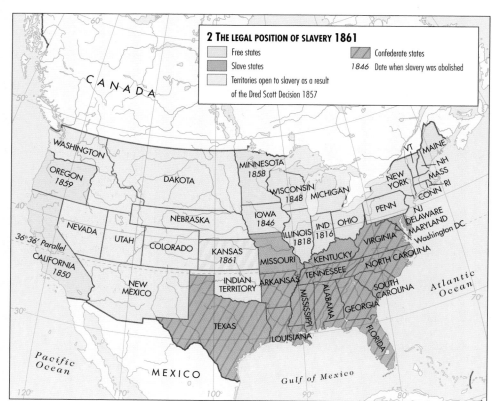

2 THE LEGAL POSITION OF SLAVERY 1861
- Free states
- Slave states
- Territories open to slavery as a result of the Dred Scott Decision 1857
- Confederate states
- 1846 Date when slavery was abolished

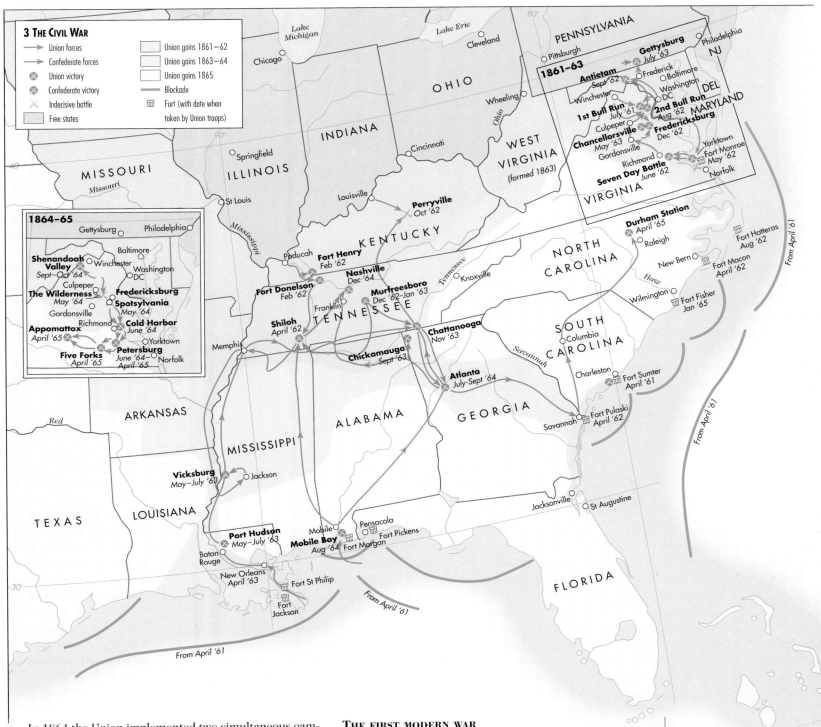

3 THE CIVIL WAR

- → Union forces
- → Confederate forces
- ⊗ Union victory
- ⊗ Confederate victory
- ✕ Indecisive battle
- Free states
- Union gains 1861–62
- Union gains 1863–64
- Union gains 1865
- Blockade
- ⛫ Fort (with date when taken by Union troops)

In 1864 the Union implemented two simultaneous campaigns. The first, centred on Virginia, saw some of the fiercest fighting of the war (*map 3 inset*), with no real victory for either side, although this war of attrition gradually depleted the human and material resources of the Confederates. In the second Union campaign Atlanta was captured, followed by General Sherman's "scorched earth" march through Georgia to Savannah and then north through the Carolinas, which caused much devastation and famine in its wake. Wilmington, the Confederate's last remaining seaport, was effectively closed down at the beginning of 1865 as a result of the Union naval blockade of Southern ports. At the outset of the war, the Confederacy had believed that the demand from Britain and France for cotton would force them to enter the war on its behalf. As the war progressed, however, the two countries decided not to risk intervention for a losing cause.

The Confederate General Robert E. Lee was forced to evacuate Petersburg and Richmond, and surrender to General Ulysses S. Grant at Appomattox Court House on 9 April 1865, effectively ending the war. By the end of May, the last Confederate forces had laid down their arms.

THE FIRST MODERN WAR

In many ways the Civil War was the first "modern" war. It was fought by mass citizen volunteer and conscript armies, rather than by professional soldiers. Railroads played a crucial role in the movement of troops and raw materials, while telegraphs were used for military communication as well as for virtually immediate Press reporting. The war also saw the first use of rudimentary iron-clad battleships, machine-guns, trench systems and dugouts.

The Civil War was fought at the cost of enormous loss of life (*pie charts*), but it had the ultimate effect of preserving the United States of America as one nation by settling the dispute over the division of power between the federal government and individual states in favour of the former. It also effectively ended the institution of slavery, although it did little to resolve the problem of race relations, which reached a climax a century later (*pages 240–41*). Furthermore, as the final decades of the 19th century were to reveal, the Civil War brought many economic benefits to the North, under whose leadership the United States had developed, by the end of the century, into the world's greatest industrial power.

▲ Most of the fighting in the Civil War took place on Southern territory, with the Confederates adopting defensive tactics on familiar terrain, and the Union side forced to maintain lengthy supply lines. The Union side devised the "Anaconda Plan", by which they first encircled the Southern states by land and sea, and then split them up by seizing control of the Mississippi River in the spring of 1863 and marching through Georgia in the winter of 1864–65.

◀ WESTWARD EXPANSION OF THE UNITED STATES 1783–1910 *pages 182–83*   ▶ INDUSTRIAL GROWTH OF THE UNITED STATES 1790–1900 *pages 186–87*

THE INDUSTRIAL GROWTH OF THE UNITED STATES 1790–1900

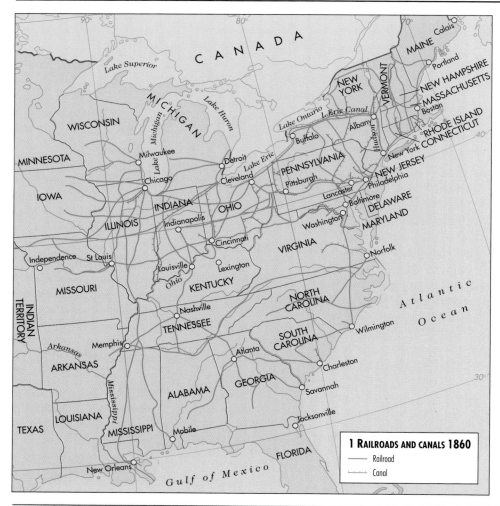

1 RAILROADS AND CANALS 1860
- Railroad
- Canal

During the course of the 19th century the United States was transformed from a simple agrarian republic into a modern industrial nation. This process of industrialization occurred in two main phases. In the first, from 1800 to the Civil War (1861–65), developments in transportation and manufacturing, and an increase in population, resulted in a capitalist commercial economy. In the second phase a dramatic acceleration in the rate of change after 1865 led to the creation of the modern American industrial superpower.

EARLY INDUSTRIALIZATION

Changes in transportation provided the main catalyst for industrialization: improved national communication created larger markets and greatly facilitated the movement of goods, services and people. The earliest manifestation of this development was the laying down of hard-surfaced roads, known as turnpikes, mainly in New England and the mid-Atlantic states. During the "Turnpike Era" (1790–1820) more than 3,200 kilometres (2,000 miles) of road were constructed, the earliest being the Lancaster Pike (1794) between Philadelphia and Lancaster, Pennsylvania. The most famous turnpike, the government-financed National Road, had crossed the Appalachian Mountains from Maryland to Virginia by 1818 and reached Illinois by 1838. These roads provided an early stimulus to economic development and westward expansion.

The turnpikes were followed by advances in river and lake transportation. The first of the commercially successful steamboats started operating on the Hudson River in 1807, but these ships became more widely used further west, travelling up and down the Ohio and Mississippi rivers and

◄ The development of canal and railroad systems, coupled with the navigation of rivers by steamboats, enabled a two-way trade flow whereby raw materials from the west and south were transported to the east and returned as manufactured goods.

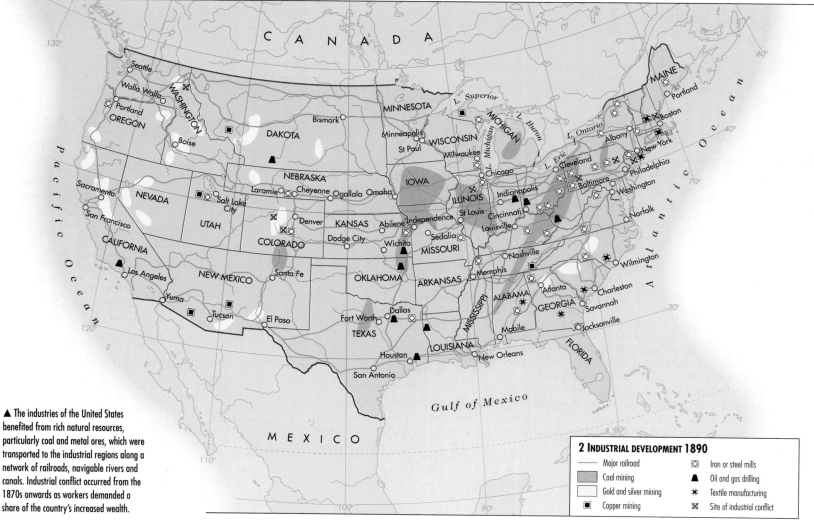

▲ The industries of the United States benefited from rich natural resources, particularly coal and metal ores, which were transported to the industrial regions along a network of railroads, navigable rivers and canals. Industrial conflict occurred from the 1870s onwards as workers demanded a share of the country's increased wealth.

2 INDUSTRIAL DEVELOPMENT 1890
- Major railroad
- Coal mining
- Gold and silver mining
- Copper mining
- Iron or steel mills
- Oil and gas drilling
- Textile manufacturing
- Site of industrial conflict

their tributaries. The steamboats stimulated the agricultural economies of the Midwest and the south by providing quick access to markets for their produce at greatly reduced prices, and enabled manufacturers in the east to send their finished goods westwards.

The first half of the 19th century also witnessed wide-scale building of canals. In 1816 there were only 160 kilometres (100 miles) of canal; by 1840 this figure had risen to 5,321 kilometres (3,326 miles) (*map 1*). The Erie Canal was completed in 1825, connecting Albany, New York to Buffalo on Lake Erie, thereby giving New York City direct access to the growing markets of Ohio and the Midwest via the Great Lakes, and to the Mississippi via the Ohio River.

The first railroad was opened between Baltimore (which funded the project) and Ohio in 1830. Other cities followed Baltimore's example, and, with the markets of Ohio, Indiana and Illinois in mind, 5,324 kilometres (3,328 miles) of track had been laid by 1840 – a figure which trebled over the next ten years. In the 1860s federal land grants encouraged railroad building to link together all parts of the nation and enable the quick and inexpensive movement of goods and people over great distances (*map 1*).

The introduction of the telegraph in 1837 further enhanced the speed of communication. By 1861 there were 80,000 kilometres (50,000 miles) of telegraph cable in the United States, connecting New York on the Atlantic with San Francisco on the Pacific coast.

DEVELOPMENTS IN MANUFACTURING

Alongside developments in transportation, the early 19th century also saw the transition from craftwork in homes and in small shops to larger-scale manufacturing with machines. Domestic US manufacturing began to flourish when imports were scarce during the War of 1812 against Britain. The textile industry spearheaded these developments, with Francis Lowell founding, in 1813, the first mill in North America that combined all the operations of converting raw cotton into finished cloth under one roof: a "factory" system based on machine technology. These early forms of manufacturing were concentrated in the east and mainly processed the products of American farms and forests.

A primary factor in the industrial growth of the United States was an abundance of raw materials (*map 2*). In addition, the country benefited from a large and expanding labour force, which also provided a vast domestic market for industrial goods. By 1860 its population had reached 31.5 million, exceeding that of Britain.

INDUSTRIALIZATION AFTER THE CIVIL WAR

In 1860 American industry was still largely undeveloped. Most industrial operations were small in scale, hand-crafting remained widespread and there was insufficient capital for business expansion. This situation changed fundamentally after the Civil War (*pages 184–85*), with the rapid development of new technologies and production processes. Machines replaced hand-crafting as the main means of manufacturing, and US productive capacity increased at a rapid and unprecedented rate. Industrial growth was chiefly centred on the north, while the south largely remained an agricultural region.

More than 25 million immigrants entered the United States between 1870 and 1916 (*bar chart*). Mass immigration, coupled with natural growth, caused the population to more than double between 1870 and 1910 to reach 92 million. In the new industrialized nation great cities and an urban culture flourished (*map 3*).

In the late 19th century mass industrialization was stimulated by a surge in technological innovation and improved factory production methods, enabling goods to be produced faster, in greater quantity and thus more cheaply than ever before. The typewriter was introduced in 1867, followed by the cash register and the adding machine. Electricity was first used as a power source in the 1870s, while international telegraph cables and the invention of the telephone assisted communication in the latter part of the century.

Railroad-building likewise increased at a dramatic rate, providing a great stimulus to coal and steel production and rivalling the steamboat and canal barge as a means of transportation. By the 1880s a nationwide network of railroads enabled goods to be distributed quickly and cheaply throughout the country, often over great distances from the point of production (*map 2*).

The highly profitable railroads provided the model for the development of the modern corporations that financed and directed this great industrial expansion. In order to eliminate cut-throat competition between companies and to encourage capital investment for further expansion and greater efficiency, enterprises were increasingly consolidated into large-scale units, often monopolies, owned by limited liability shareholders. The federal government helped to create an entrepreneurial climate in which business and trade could flourish without undue hindrance.

As a result of these developments the United States was transformed, by the end of the 19th century, from an essentially agrarian economy into a country in which half of its now culturally diversified population lived in its ever-growing cities. It had replaced Britain as the world's leading industrial power, and was thus set to dominate the global economy in the 20th century.

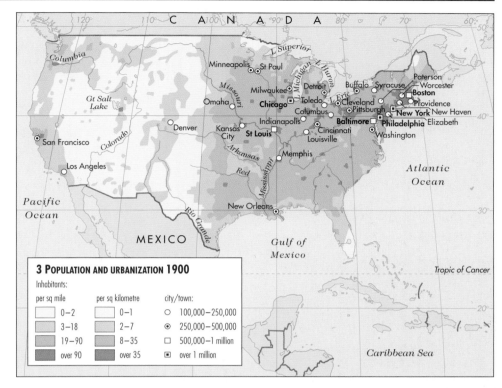

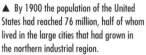

▲ By 1900 the population of the United States had reached 76 million, half of whom lived in the large cities that had grown in the northern industrial region.

▼ The pattern of migration to the United States was influenced partly by political and economic developments in Europe. Before the 1890s most immigrants came from northern and western Europe, in particular from Ireland following the Potato Famine in the 1840s, and from Germany. By 1900 the majority of migrants were from central and eastern Europe, Russia and Italy.

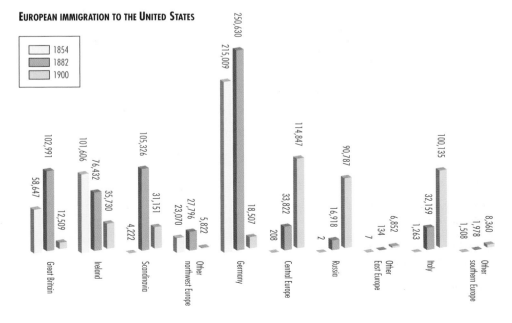

EUROPEAN IMMIGRATION TO THE UNITED STATES

THE DEVELOPMENT OF CANADA
1763–1914

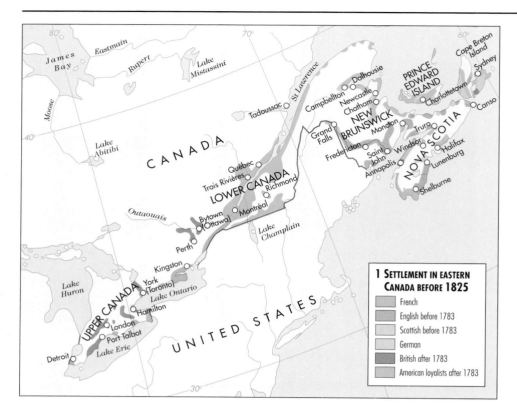

1 SETTLEMENT IN EASTERN CANADA BEFORE 1825

- French
- English before 1783
- Scottish before 1783
- German
- British after 1783
- American loyalists after 1783

▲ Since the 17th century French-speaking Canadians had largely settled along the St Lawrence River. However, in the 1770s and 1780s American Loyalists, escaping from the newly formed United States, migrated to the southwestern part of the old province of Quebec and to the British colony of Nova Scotia, necessitating the creation of another colony, New Brunswick.

▲ In 1792 Alexander Mackenzie led an expedition from Lake Athabasca to find an outlet to the Pacific Ocean. The explorers braved the rapids of the Peace and Fraser rivers before emerging on the west coast of North America at Bella Coola the following year.

▶ Expansion west into the prairies and along the west coast during the 19th century was preceded by journeys of exploration, which were often undertaken by fur traders. The completion of the Canadian Pacific Railroad in 1885 provided a huge boost to trade across Canada, and numerous settlements developed along its route.

During the 18th century territorial rivalry between the French and British in North America gradually increased, coming to a head in the Seven Years War of 1756–63. Although the British initially suffered defeats, their troops rapidly gained the upper hand after the appointment of General Wolfe in 1757 and by 1760 they had effectively defeated the French. France surrendered Canada to Britain in the Treaty of Paris in 1763, and Britain found itself in the unprecedented situation of having a colony with a large white population of approximately 6,500, who were non-English-speaking and Roman Catholic. The British parliament passed the Quebec Act in 1774, which greatly enlarged the territory of Quebec (*pages 164–65*), guaranteed freedom of religion to French Canadians (at a time when Roman Catholic subjects in Britain were effectively excluded from political participation), and recognized the validity of French civil law. These measures succeeded in securing the loyalty of the Canadians at a time of increasing discontent in the British colonies elsewhere in America. During the American Revolution (1775–83) (*pages 164–65*) attempts by the Thirteen Colonies first to secure Canadian support, and then to invade the region, failed.

The creation of the United States of America had significant repercussions for Canada. It not only defined the Canadian–American border (with Britain giving up all land south of the Great Lakes) but also fundamentally altered the composition of Canada's population. Between 40,000 and 60,000 Americans who remained loyal to the British crown flooded into Canada during and after the war, creating the basis for Canada's English-speaking population (*map 1*).

THE CONSTITUTIONAL ACT OF 1791
The loss of the Thirteen Colonies encouraged Britain to tighten its rule over its remaining North American possessions. Acknowledging the bicultural nature of the Canadian population and the loyalists' desire for some form of representative government, the Constitutional Act of 1791 divided Quebec into two self-ruling parts – English-speaking Upper Canada (now Ontario) and French-speaking, largely Catholic, Lower Canada (now Québec) – dominated by a British governor and an appointed legislative council. There were also significant English-speaking pockets in Lower Canada, most notably the dominant merchant class in Montreal and farmers in the eastern townships. Canadian independence was further secured when repeated American invasions were repelled in the War of 1812.

WESTWARD EXPANSION
Canada's survival as an independent country ultimately depended on population growth and economic development. In the east, internal communications were improved in the first half of the 19th century through the construction of roads and canals. Canada's western Pacific regions had been opened up in the last decades of the 18th century by explorers such as Alexander Mackenzie (*map 2*), Simon Fraser and David Thompson, with fur traders and the British Hudson's Bay Company (which also controlled vast tracts in the northeast of the country) following swiftly behind. In the central region, south of Lake Winnipeg, settlement was encouraged by the Scottish philanthropist Lord Selkirk, who set up the Red River colony for Scottish

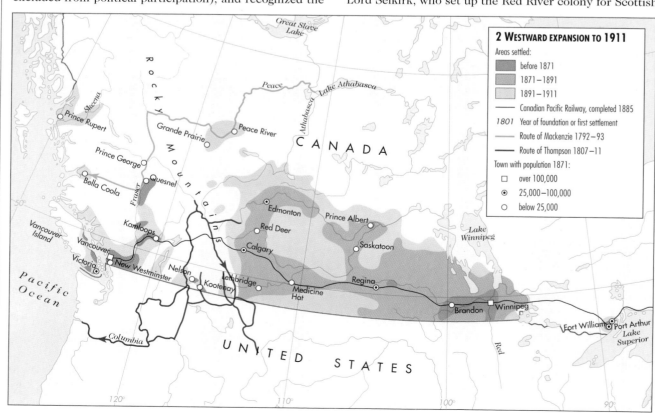

2 WESTWARD EXPANSION TO 1911

Areas settled:
- before 1871
- 1871–1891
- 1891–1911

— Canadian Pacific Railway, completed 1885
1801 Year of foundation or first settlement
— Route of Mackenzie 1792–93
— Route of Thompson 1807–11

Town with population 1871:
- ☐ over 100,000
- ◉ 25,000–100,000
- ○ below 25,000

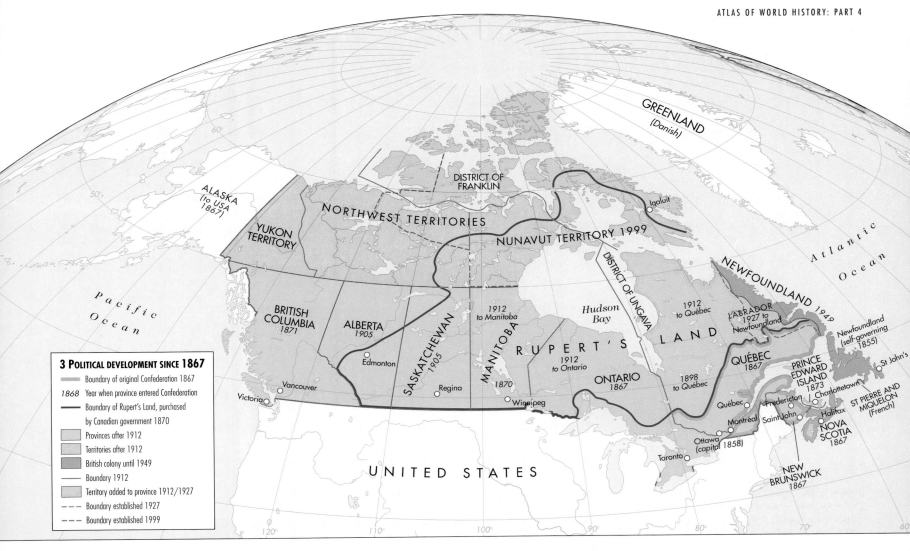

3 POLITICAL DEVELOPMENT SINCE 1867

- Boundary of original Confederation 1867
- *1868* Year when province entered Confederation
- Boundary of Rupert's Land, purchased by Canadian government 1870
- Provinces after 1912
- Territories after 1912
- British colony until 1949
- Boundary 1912
- Territory added to province 1912/1927
- Boundary established 1927
- Boundary established 1999

immigrants in 1812. Two British colonies were founded on the Pacific coast: Vancouver Island (1849) and British Columbia (1858), which united in 1866.

FROM UNION TO CONFEDERATION

Canadian discontent with oligarchic rule led to two short rebellions in both Upper and Lower Canada in 1837 and 1838, forcing Britain to reassess how best to keep Canada within the empire and how to unite the French and English Canadians. The resulting Act of Union of 1840 combined Upper and Lower Canada into the new Province of Canada and by 1848 Canadians had gained a degree of self-government. Under this system, however, both Canada West and Canada East (formerly Upper and Lower Canada respectively) had equal representation in the province's legislative assembly. This did little to ensure national unity and encouraged political stalemate; further problems arose after 1850, when the population of Canada West exceeded that of Canada East, with the former unsuccessfully demanding representation by population.

During the 1850s and 1860s calls grew to dissolve this ineffectual union and to replace it with some form of federal government by which each part of Canada could control its own affairs while a central government protected national defence and common interests. Constitutional change was also spurred on by external events. Britain increasingly wanted Canadians to shoulder the burden of their own defence, while Canada felt increasingly threatened by fears of an anti-British American invasion during the American Civil War (1861–65) and by the reality of raids across its borders in the 1860s by Fenians (Irish Americans demanding Irish independence from Britain). After conferences in Charlottetown and Québec (1864), the British North America Act was signed by Queen Victoria in 1867.

This act created the largely self-governing federation or Dominion of Canada under the British crown, with a constitution based on the British parliamentary system. It initially comprised only four provinces (*map 3*), with a population of 3.5 million people, only 100,000 of whom lived west of the Great Lakes. The driving ambition of the "Fathers of the Confederation" was to unite all of the

remaining British colonies in North America in order to achieve the economic and social development necessary for a viable nation, especially in the face of ongoing American expansionism.

In 1870 the government vastly extended Canadian territory by purchasing Rupert's Land from the Hudson's Bay Company (*map 3*); while the company retained its trading station and forts, it gave up its monopoly of the area which had long been difficult to enforce. The province of Manitoba was created in the same year, following the Red River Rebellion by settlers of mixed French and Native American ancestry, led by the *métis* Louis Riel. In 1871 British Columbia joined as Canada's sixth province after the promise of a transcontinental railroad (completed in 1885) linking it to eastern Canada (*map 2*). Similar financial incentives enabled Prince Edward Island to become the seventh province in 1873, although Newfoundland remained a proud self-governing colony until 1949.

Realizing that population growth was necessary for national survival, the Canadian government actively promoted immigration from the British Isles and the United States and, towards the end of the century, from central and eastern Europe; this once more changed the cultural and ethnic mix of Canada's population. The new settlers moved primarily to unoccupied lands on the prairies (*map 2*), which enabled the provinces of Alberta and Saskatchewan to be created in 1905. In 1912 the remaining parts of the former Hudson's Bay Company lands were added to Québec, Ontario and Manitoba.

TENSIONS BETWEEN THE BRITISH AND FRENCH

The position of French Canadians as a cultural minority within the Confederation led to ongoing tension, exacerbated by Canada's decision to send volunteer troops to fight for the British Empire in the Boer War (1899–1902). The situation reached crisis point when, in 1917, the Canadian parliament introduced conscription. Ironically, the fact that 55,000 Canadians lost their lives fighting for the empire in the First World War led ultimately to the transformation of Canada into a fully independent sovereign nation under the Statute of Westminster in 1931.

▲ Between the establishment of the original four provinces of the Dominion of Canada in 1867 and the outbreak of the First World War in 1914, the political map of Canada changed dramatically. As the population grew in the newly settled territories, provinces were created and federated to the central government in Ottawa. In 1912 Manitoba and Ontario were greatly enlarged to the north, with the annexation of land from the Northwest Territories. Further boundary changes occurred in 1927, when the colony of Newfoundland was enlarged at the expense of Québec, and in 1999, when the Nunavut Territory – administered by its majority Inuit inhabitants – was created.

◀ THE COLONIZATION OF NORTH AMERICA AND THE CARIBBEAN 1600–1763 *pages 124–25*

INDEPENDENCE IN LATIN AMERICA AND THE CARIBBEAN 1780–1830

In 1800 (*map 1*) few people, either in Europe or the Americas, could have anticipated that 25 years later all of Spain's mainland American colonies would be independent republics. Several colonial rebellions had occurred during the late 18th century, but they had all been defeated, and should not be interpreted as antecedents of independence. The most significant of these uprisings, in Peru, was interesting for what it revealed about the fundamental allegiances of Spanish American creoles (those of Spanish descent, born in the colonies). In 1780 a creole revolt against Spanish tax increases was superseded by an anti-Spanish rebellion among the American Indians, led by Tupac Amaru. The small minority of creoles hastily jettisoned their own protest in favour of helping the colonial authorities to suppress this revivalist Inca movement – at the cost of 100,000 lives, most of them Indian.

CREOLE ALLEGIANCE

The creoles' fear of the African, Indian and mixed-race peoples, who made up approximately 80 per cent of Spanish America's population in the late 18th century, meant that many of them looked to Spain to defend their dominant social and economic position. This rationale was strengthened after a slave revolt in the French Caribbean colony of Saint Domingue in 1791 led to the founding, in 1804, of Haiti, the first African–Caribbean republic in the Americas. Most creoles calculated that their interests ultimately depended on Spain, despite an expanding list of grievances against the mother country. It was not until Napoleon invaded Spain in 1808, and installed Joseph Bonaparte in place of the Bourbon King Ferdinand, that some creoles began to reconsider their options. They were presented with three main choices: to support Joseph Bonaparte; to declare allegiance to the provisional Spanish

▼ In 1800 the majority of Latin America was under Spanish control, administered by viceroys and captains-general. The Portuguese were still in control of Brazil and the British ruled in Guiana, where they had temporarily expanded to take over the adjacent Dutch territory (now Surinam). The French had taken control of Santo Domingo from the Spanish but were to lose it in 1809. They had already lost the colony of Saint Domingue in 1804, when it became independent Haiti. The Spanish territory was rich in minerals and included Potosí, the silver-mining capital of the world, although its resources were by now on the verge of being exhausted.

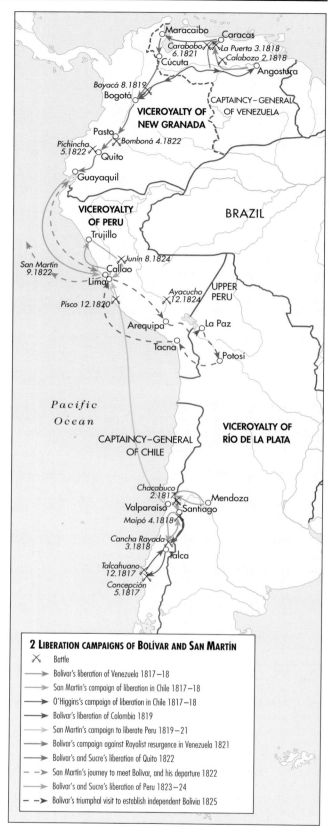

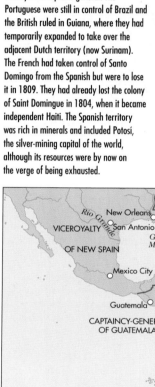

1 LATIN AMERICA AND THE CARIBBEAN 1800

Colonial power:
- Spain
- Portugal
- Britain
- France

2 LIBERATION CAMPAIGNS OF BOLÍVAR AND SAN MARTÍN

- ✕ Battle
- → Bolívar's liberation of Venezuela 1817–18
- → San Martín's campaign of liberation in Chile 1817–18
- → O'Higgins's campaign of liberation in Chile 1817–18
- → Bolívar's liberation of Colombia 1819
- → San Martín's campaign to liberate Peru 1819–21
- → Bolívar's campaign against Royalist resurgence in Venezuela 1821
- → Bolívar's and Sucre's liberation of Quito 1822
- --→ San Martín's journey to meet Bolívar, and his departure 1822
- → Bolívar's and Sucre's liberation of Peru 1823–24
- --→ Bolívar's triumphal visit to establish independent Bolivia 1825

▲ Venezuelan-born Simón Bolívar was involved in two failed insurrections before his successful campaigns against the Spanish in New Granada in 1817–22, resulting in the creation of a new Republic of Gran Colombia.

During this time José de San Martín, aided by Bernardo O'Higgins, had been liberating Chile. Leaving O'Higgins behind as president of the new state of Chile, San Martín travelled north to take Lima and to attempt to liberate what was to become Peru. In 1822 he was forced to seek help from Bolívar, and in September 1822 retired from command. Bolívar subsequently completed the liberation of Peru at the Battle of Junín. In this he was aided by Antonio José de Sucre, who went on to win the final battle against the Spanish at Ayacucho in 1824.

The following year Bolívar made a triumphal visit to the region, during which he established the independent republic of Bolivia, which was named in honour of the "Great Liberator". Bolívar himself returned to Colombia but was unable to hold together the republic he had created, and in 1830 (the year of his death) it broke up into the three modern-day states of Venezuela, Colombia and Ecuador.

authorities that rapidly developed in resistance to French rule in the name of Ferdinand; or to establish autonomous ruling authorities. It was the third option that was adopted by most creoles, even though they took care to emphasize that this was a temporary measure until Ferdinand regained the Spanish throne.

Creoles were, however, dissatisfied with Spanish rule on two main counts: commercial monopoly and political exclusion, both of which stemmed from attempts in the second half of the 18th century by the Bourbon kings to extract more revenue from the colonies. Spain's commercial monopoly had been tightened up, and Spanish Americans were unable to exploit legally what they perceived as lucrative trading opportunities in the British and US markets. Taxes had been increased and collection vigilantly enforced. A new system of colonial administration had been introduced that interfered with well-established informal mechanisms for allocating power and resources within Spanish American societies. Bourbon absolutism aimed to strengthen the position of *peninsulares* (Spaniards born in Spain) at the expense of Spanish Americans. By the end of the 18th century, creoles accounted for a far smaller proportion of the upper levels of the colonial bureaucracy than in 1750.

INDEPENDENCE FROM SPAIN

During the first two decades of the 19th century there was a gradually developing sense among elite creoles in Spanish America that their interests might best be served by self-government. This redefinition of their position was enhanced by an incipient sense of national identity that had been developing within creole communities throughout the 18th century – an idea of being distinct not only from Spaniards but also from each other. The political ideas of the French Enlightenment, although probably less influential in the development of independence movements than was once thought, were certainly of importance to some of their leaders, notably the Venezuelan, Simón Bolívar.

During the 1810s, as Spain oscillated between reformist liberalism and absolutism, Spanish Americans first declared, and then fought for, their independence (*map 2*). Nevertheless, the battles between republicans and royalists remained fairly evenly balanced until events in Spain during 1820–21 provided the final catalyst to the creation of a political consensus among creoles that was needed to secure independence. Once it had become clear that Spanish liberalism, which returned to power in 1821, was bent on restoring the pre-1808 relationship between Spain and the American colonies, commitment to independence became widespread throughout Spanish America – with the exception of Peru, where memories of the Tupac Amaru rebellion remained vivid. Peru was eventually liberated in 1824 by Bolívar's troops, after the retreat of the Spanish had been initiated by an invasion from the south led by the Argentine José de San Martín. By 1826 the last royalist troops had been expelled from South America, and Spain's empire in the Americas was reduced to Puerto Rico and Cuba (*map 3*).

INDEPENDENCE FROM PORTUGAL

Brazil's independence was partly the result of colonial grievances, although less severe than those felt by Spanish Americans. However, in overall terms, it was even more attributable to events in Europe than was the decolonization of Spanish America. The Portuguese monarchy implemented milder versions of the Bourbon reforms in the late 18th century, but in general the local elite played a far greater role in governing Brazil than their counterparts in Spain's colonies. The main event which triggered an increasing awareness of Brazil's distinct identity was the Portuguese Prince Regent's establishment of his court in Rio de Janeiro in 1808, after he had fled from Napoleon's invasion of

Portugal. This represented a shift in political power from Portugal to Brazil which was to prove irreversible. When the French were ousted from Portugal in 1814, the Prince Regent chose to stay in Brazil, which was raised to the status of a kingdom equal to that of Portugal. As King John, landowners resented his bowing to British pressure to end the slave trade, while merchants were unhappy about increasing British penetration of the Brazilian market, but these issues were causes of disaffection rather than rebellion. It was attempts by the Portuguese government in 1821 to return Brazil to its pre-1808 colonial status that was the main cause of its declaration of independence in 1822 under Pedro I – the region's only constitutional monarchy.

Brazil was unique in that it won its independence largely without the damaging consequences of civil war and economic collapse that occurred elsewhere in the region. In Spanish America mineral production plummeted to less than a quarter of its level before its independence struggles, industrial output declined by two-thirds, and agriculture by half. Socially, independence brought relatively little change. The corporate institutions of Spanish colonialism remained intact, the Church remained strong, and militarism was strengthened. Creoles simply took over the property abandoned by fleeing Spaniards and established themselves as a new oligarchy, which regarded the masses with at least as much disdain as their Spanish predecessors had done.

▼ In a remarkably short space of time, from 1818 to 1825, the Spanish were ousted from Central and South America, leaving only the strongholds of Cuba and Puerto Rico in the Caribbean. The ruler of Brazil, Dom Pedro, had declared its independence from Portugal in 1822, crowning himself emperor. A successful revolt in the southern area of the country resulted in an independent Uruguay in 1828.

3 LATIN AMERICA AND THE CARIBBEAN 1830

- Independent country
- 1818 Date of independence

Colonial power:
- Britain
- Netherlands
- Denmark
- France
- Spain

UNITED STATES

Gulf of Mexico

MEXICO 1821

Havana

CUBA 1902

SANTO DOMINGO 1821

PUERTO RICO

HAITI 1804

Caribbean Sea

Mexico

UNITED PROVINCES OF CENTRAL AMERICA 1823

Panama

Caracas

VENEZUELA 1821

Bogotá

NEW GRANADA 1831

GUIANA

Pacific Ocean

Quito

GRAN COLOMBIA 1822

Negro

Amazon

ECUADOR 1830

Purus

Madeira

Tapajós

Trujillo

PERU 1824

Lima

EMPIRE OF BRAZIL 1822

São Francisco

Tocantins

Salvador

La Paz

BOLIVIA 1825

Potosí

PARAGUAY 1811

Rio de Janeiro

Asunción

CHILE 1818

Santiago

ARGENTINE CONFEDERATION 1816

Paraná

Uruguay

URUGUAY 1828

Buenos Aires

Montevideo

Salado

Colorado

PATAGONIA

▲ Simón Bolívar was instrumental in the liberation from Spanish rule of much of South America. However, he failed in his attempts to hold together the Republic of Gran Colombia, and died disillusioned.

◖ THE COLONIZATION OF CENTRAL AND SOUTH AMERICA 1500–1700 *pages 122–23*   ◗ LATIN AMERICA AND THE CARIBBEAN 1830–1914 *pages 192–93*

LATIN AMERICA AND THE CARIBBEAN
POST-INDEPENDENCE 1830–1914

The newly independent republics of Spanish America faced formidable challenges of reconstruction in the years following their wars of independence. The first problem was territorial consolidation. Their boundaries were roughly based on colonial administrative divisions, but none was clearly defined, and nearly all Spanish-American countries went to war to defend territory at some point during the 19th century (*map 1*). The only nation on the continent that consistently expanded its territory at the expense of its neighbours was Brazil.

FOREIGN INTERVENTION

Foreign powers were active in the region throughout this period, and acted as a significant constraint on the ability of the new states to consolidate their sovereignty. Spain was too weak to do much beyond defending its remaining colonial possessions, but it fought two wars over Cuban independence (1868–78 and 1895–98) before US military intervention in 1898 led to the Spanish-American War and the secession of Cuba and Puerto Rico to the United States. Following a three-year military occupation Cuba was declared an independent republic, albeit with a clause in its constitution (the "Platt Amendment") stipulating the right of the USA to intervene in its internal affairs. Mexico, which achieved independence in 1821 following a civil war, subsequently lost large amounts of territory to the USA. It was briefly ruled by the Austro-Hungarian, Maximilian von Habsburg, as emperor (1864–67), supported by French troops. Britain had colonies in Guiana and British Honduras, and consolidated its commercial and financial dominance throughout most of the region, especially in Brazil and Argentina.

ECONOMIC DEVELOPMENTS

Throughout the 19th century Latin American economies remained dependent on the export of raw materials (*maps 1, 2 and 3*), continuing patterns of production established in colonial times. Although there has been considerable debate about the wisdom of this policy, in practice they had little choice. The colonial powers had left behind scant basis for the creation of self-sufficient economies, and the independent states simply did not have the resources necessary for such development. Attempts were made to encourage industrialization in Mexico, Colombia and Brazil in the 1830s and 1840s, but they all succumbed to competition from European imports.

The export of primary products brought considerable wealth to Latin America, especially once the development of steamships and railways in the 1860s had modernized transportation. In the last quarter of the 19th century Latin American economies were able to benefit from the overall expansion in the world economy fuelled by European and US demands for raw materials and markets for their manufactured goods (*pages 208–9*). At the time it made economic sense for Latin America to exploit its comparative advantage in the world market as a supplier of raw materials. Although this strategy later proved to be flawed, it did result in rapid economic growth and a wave of prosperity among Latin American elites in what became known as "la belle époque" of Latin American development (c. 1880–1914). On the eve of the First World War, the region was producing 18 per cent of the world's cereals, 38 per cent of its sugar and 62 per cent of its coffee, cocoa and tea.

1 SOUTH AMERICA 1830–1914

— Confederation of Peru and Bolivia 1836–39

1885 Date slavery abolished

— Railways 1910

Primary products:

- ▲ timber
- ◌ rubber
- ♦ cocoa
- ⬭ coffee
- ⤸ bananas
- ⬭ sugar
- ◉ cereals
- ⬤ cotton
- ◉ tobacco
- ◣ copper
- ◪ manganese
- ♦ tin
- ▣ silver
- ✛ nitrates
- ✦ guano
- ꭓ sheep
- ꬷ cattle

▲ In the years following independence most countries became involved in wars over their boundaries. Argentina lost the Falkland Islands to the British in 1833, but secured Patagonia in 1881. Both Peru and Bolivia lost out to Chile in the War of the Pacific in 1879, surrendering territory rich in nitrates and, in Bolivia's case, an outlet to the sea.

ELITIST POLITICS

Politics in 19th-century Latin America was entirely an elite affair, with electoral contests typically involving at most ten per cent of the population and dominated by rivalry between liberals and conservatives. Most of the republics had adopted liberal constitutions based on that of the United States, but these were to prove an inadequate blueprint for the authoritarian reality of Latin American politics.

The major challenge in most countries was to consolidate central state authority over remote and often rebellious areas. Until well into the 1850s local leaders, known as *caudillos*, raised armies to fight for their interests, holding sway over their followers by a combination of charisma, blandishment and brutality. In these circumstances, many liberal statesmen found themselves obliged to pursue distinctly illiberal policies. As the century wore on, Latin American liberalism, which came to power in most Latin American countries during the 1850s and 1860s, took on an increasingly conservative cast. One distinctive legacy of liberalism was an appreciable reduction in the wealth of the Catholic Church, particularly in Mexico, although liberals did not succeed in diminishing the religious devotion of the majority of the populations.

SOCIAL CHANGES

Conditions barely improved for the Latin American masses. Indeed, American Indians had good reason to feel that their plight had been less onerous under colonial rule, when they had at least enjoyed a degree of protection from the Spanish crown against encroachments on their communal lands. The attempts of liberal governments to turn Indian peasants into smallholders by forcibly redistributing their lands left most Indians worse off, particularly those in Mexico.

Slavery was abolished in Central America as early as 1824 (*map 3*), and in the Spanish South American republics during the 1850s (*map 1*), but it continued in Portuguese-dominated Brazil, where a weak emperor was reluctant to antagonize the powerful plantation owners. Brazil did not pass legislation to end the trade in slaves until 1850 and it took until 1888 – the year before Brazil declared itself a republic – for slavery itself to be abolished. Even in conditions of allegedly "free" labour, however, the lack of alternative work meant that many former slaves had little choice but to join a floating rural proletariat, subject to seasonal work in exchange for pitiful wages.

During the middle part of the 19th century the populations of most Latin American countries more than doubled (*bar chart*), and by the end of the century Latin America's integration into the world economy was beginning to bring about changes in the socio-economic structure which independence had not. Urbanization, industrialization and their consequences continued from the 1880s onwards. The late 19th century saw the emergence of a middle class based

on professionals and state bureaucrats. Trade unions among the working classes – most of which were organized by European immigrants to Argentina or Brazil – first became active during this period, and public education programmes were initiated in the larger countries. It was not until after the First World War, however, that the political consequences of all these socio-economic changes were to manifest themselves.

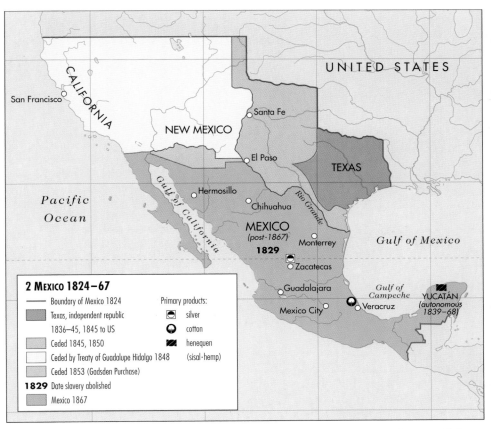

2 MEXICO 1824–67
- Boundary of Mexico 1824
- Texas, independent republic 1836–45, 1845 to US
- Ceded 1845, 1850
- Ceded by Treaty of Guadalupe Hidalgo 1848
- Ceded 1853 (Gadsden Purchase)
- **1829** Date slavery abolished
- Mexico 1867

Primary products:
- silver
- cotton
- henequen (sisal-hemp)

▲ Mexico was substantially reduced in size during the mid-19th century. It lost Texas to an independence movement in 1836 and California, New Mexico and Arizona after being defeated in the 1846–48 war with the United States. (Mexicans rarely need reminding that the California Gold Rush began in 1849.) Further territory was ceded in 1850 and again in 1853, as a result of the Gadsden Purchase.

▼ Most of Central America and the larger Caribbean islands had gained independence by 1910. The smaller islands remained European colonies, while the United States retained control of Puerto Rico.

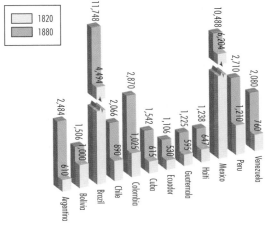

LATIN AMERICAN POPULATION IN 1820 AND 1880 (in thousands)

- 1820
- 1880

▲ The 19th century saw large population increases in most Latin American countries. Many countries experienced a doubling of their numbers between 1820 and 1880, while the population in the economically successful Argentina quadrupled.

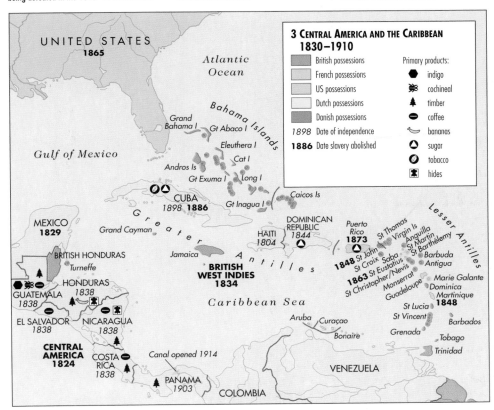

3 CENTRAL AMERICA AND THE CARIBBEAN 1830–1910
- British possessions
- French possessions
- US possessions
- Dutch possessions
- Danish possessions
- **1898** Date of independence
- **1886** Date slavery abolished

Primary products:
- indigo
- cochineal
- timber
- coffee
- bananas
- sugar
- tobacco
- hides

◀ INDEPENDENCE IN LATIN AMERICA AND THE CARIBBEAN 1780–1830 *pages 190–91* ▶ LATIN AMERICA 1914–45 *pages 226–27*

THE BRITISH IN INDIA
1608–1920

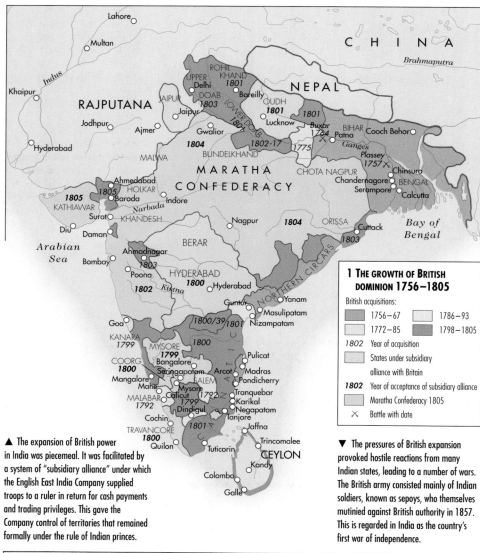

▲ The expansion of British power in India was piecemeal. It was facilitated by a system of "subsidiary alliance" under which the English East India Company supplied troops to a ruler in return for cash payments and trading privileges. This gave the Company control of territories that remained formally under the rule of Indian princes.

1 THE GROWTH OF BRITISH DOMINION 1756–1805

British acquisitions:
- 1756–67
- 1772–85
- 1786–93
- 1798–1805
- *1802* Year of acquisition
- States under subsidiary alliance with Britain
- *1802* Year of acceptance of subsidiary alliance
- Maratha Confederacy 1805
- ✕ Battle with date

▼ The pressures of British expansion provoked hostile reactions from many Indian states, leading to a number of wars. The British army consisted mainly of Indian soldiers, known as sepoys, who themselves mutinied against British authority in 1857. This is regarded in India as the country's first war of independence.

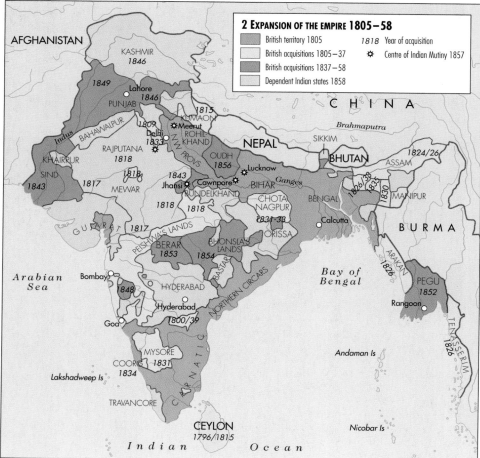

2 EXPANSION OF THE EMPIRE 1805–58

- British territory 1805
- British acquisitions 1805–37
- British acquisitions 1837–58
- Dependent Indian states 1858
- *1818* Year of acquisition
- ✿ Centre of Indian Mutiny 1857

An English East India Company fleet first reached India in 1608 and, over the course of the next century, the Company developed its trade steadily around the coasts of the subcontinent. It quickly established trading posts, known as "factories", starting at Surat in 1619 and followed by Madras in 1634, Bombay in 1674 and Calcutta in 1690.

Although originally entering the "Indies" trade in pursuit of spices, the Company made most of its fortune from cotton textiles, whose manufacture was highly developed in India. However, until the second quarter of the 18th century, there was little to suggest that the British presence in India heralded an empire. Europeans in general were economically outweighed by indigenous trading and banking groups and were politically subordinate to the great Mughal Empire (*pages 144–45*).

The turning point, which was to lead to British supremacy in India, came only in the mid-18th century when the Mughal Empire began to break up into warring regional states, whose needs for funds and armaments provided opportunities for the Europeans to exploit. Another factor was the growing importance of the English East India Company's lucrative trade eastwards towards China, which enhanced its importance in the Indian economy, especially in Bengal.

BRITISH–FRENCH RIVALRY

Conflicts between the European powers started to spill over into Asia, with the French and British beginning a struggle for supremacy that was not finally resolved until the end of the Napoleonic Wars in 1815. In southern India from 1746 the British and French backed rival claimants to the Nawabi of Arcot. In the course of their conflict Robert Clive, who rose from a clerkship to command the English East India Company's armies and govern Bengal, introduced new techniques of warfare borrowed from Europe. These not only prevailed against the French but opened up new possibilities of power in the Indian subcontinent.

In 1756 Siraj-ud-Daula, the Nawab of Bengal, reacted to the growing pretensions of the British by sacking their "factory" at Calcutta and consigning some of their officers to the infamous "Black Hole". Clive's forces moved north in response and defeated Siraj-ud-Daula's army at Plassey in 1757 (*map 1*). This created an opportunity for the conversion of the Company's economic influence in Bengal into political power; the defeat of the residual armies of the Mughal emperor at Buxar in 1764 completed this process.

However, it was to take another 50 years for the British to extend their dominion beyond Bengal, and a further 100 years for the limits of their territorial expansion to be established. First, they faced rivalry from other expanding Indian states which had also adopted the new styles of warfare, most notably Tipu Sultan's Mysore (defeated in 1799) and the Maratha Confederacy (defeated in 1818). It was not until the annexation of Punjab in 1849 that the last threat to the Company's hegemony was extinguished (*map 2*). Even after this, the process of acquisition was continued: smaller states that had once been "subsidiary" allies were gobbled up and Baluchistan and Burma were brought under British control, in 1876 and 1886 respectively, as a means of securing unstable borders (*map 3*). Nor was political stability within the empire in India achieved with any greater ease. Most notably, in 1857 the "Great Mutiny" of Indian soldiers in the Bengal army saw the British lose control of the central Ganges Valley and face rebellion in the heartland of their empire.

EFFECTS OF BRITISH RULE

The carrying forward of the imperial project in the face of so many problems was a reflection of the importance attached to India by the British. In the course of the 19th century it became "the jewel in the crown" of the British Empire, to which it was formally annexed in 1858 when the English East India Company was dissolved. Although there was little "white" settlement and most of its economy

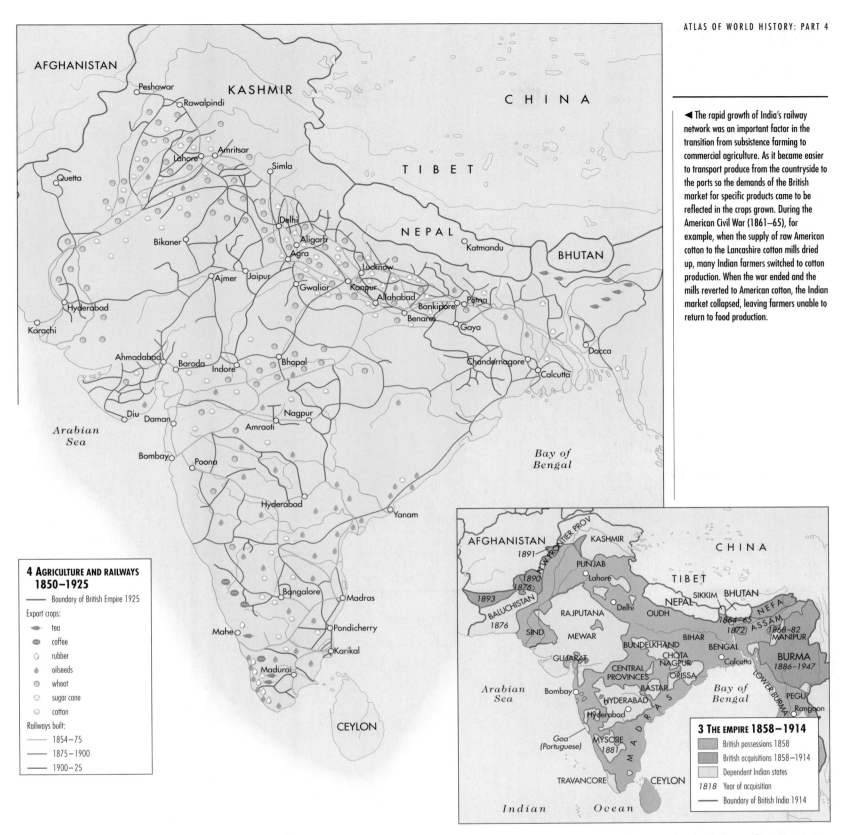

4 AGRICULTURE AND RAILWAYS 1850–1925

——— Boundary of British Empire 1925

Export crops:
- tea
- coffee
- rubber
- oilseeds
- wheat
- sugar cane
- cotton

Railways built:
- 1854–75
- 1875–1900
- 1900–25

3 THE EMPIRE 1858–1914
- British possessions 1858
- British acquisitions 1858–1914
- Dependent Indian states
- *1818* Year of acquisition
- ——— Boundary of British India 1914

and key social institutions remained in indigenous hands, India was manipulated to yield singular advantages to Britain. Its most significant role was to supply a large army which was extensively used for imperial defence around the world. In addition, India became a captive market for the products of Britain's industrial revolution, a major exporter of agricultural commodities and an important area for the investment of British capital, especially in the rapidly expanding railway network (*map 4*).

What effects British rule had on India remains a controversial question. The agricultural economy grew, with expanding foreign trade and British capital providing the rudiments of a modern transport infrastructure. However, the once-great textile industry declined and few other industries rose to take its place. Ambiguity also marked British social policy. A strong imperative, especially from the 1840s onwards, was to "civilize" India along Western lines, introducing "scientific" education, a competitive market economy and Christian ethics. However, a conservative view held by some in the British administration in India warned against disturbing "native" custom. After the

Mutiny, such conservative counsels won out and were reinforced by a deepening British racism, which denied equal rights to Indian subjects of the British monarch.

The reactions of Indian society to British rule were extremely mixed. Some groups mounted a ferocious defence of their traditional rights, but others responded positively to what they regarded as modernizing trends, especially taking up Western education. For such groups, the racism of the late-Victorian British and their turning away from earlier liberal ideals proved disappointing and frustrating. An Indian National Congress had been formed in 1885 to advance the cause of Indians within the empire. However, by the early 1900s it had already begun to reject the politics of loyalism and to express more fundamental objections. As the shadow of the First World War fell across the Indian landscape, the British Empire, which had succeeded in bringing India into the 19th century, was fast losing its claims to lead it through the 20th. In 1920–22, shortly after the war, Mahatma Gandhi launched the first of the mass civil disobedience campaigns which signalled the beginning of the end of British rule in India.

▲ As the frontiers of Britain's empire in India slowly stabilized, over a third of the subcontinent remained governed by Indian rulers, although the British used trade and defence agreements to exert their influence over these areas.

◄ INDIA UNDER THE MUGHALS 1526–1765 *pages 144–45* ► SOUTH ASIA SINCE 1920 *pages 248–49*

SOUTHEAST ASIA IN THE AGE OF IMPERIALISM 1790–1914

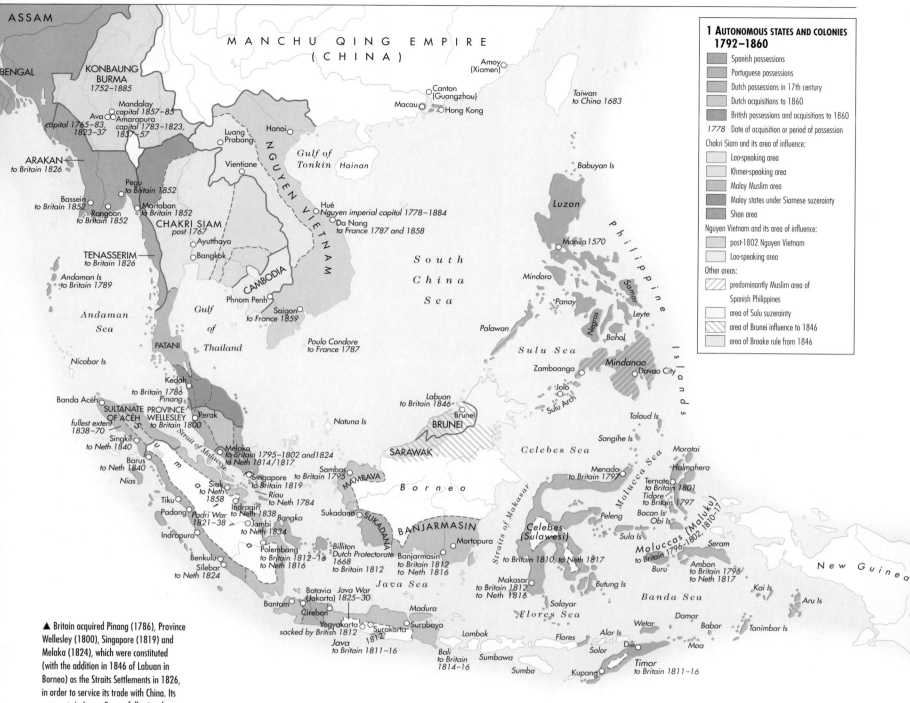

1 AUTONOMOUS STATES AND COLONIES 1792–1860

- Spanish possessions
- Portuguese possessions
- Dutch possessions in 17th century
- Dutch acquisitions to 1860
- British possessions and acquisitions to 1860
- *1778* Date of acquisition or period of possession

Chakri Siam and its area of influence:
- Lao-speaking area
- Khmer-speaking area
- Malay Muslim area
- Malay states under Siamese suzerainty
- Shan area

Nguyen Vietnam and its area of influence:
- post-1802 Nguyen Vietnam
- Lao-speaking area

Other areas:
- predominantly Muslim area of Spanish Philippines
- area of Sulu suzerainty
- area of Brunei influence to 1846
- area of Brooke rule from 1846

▲ Britain acquired Pinang (1786), Province Wellesley (1800), Singapore (1819) and Melaka (1824), which were constituted (with the addition in 1846 of Labuan in Borneo) as the Straits Settlements in 1826, in order to service its trade with China. Its conquests in Lower Burma following the Second Anglo-Burmese War (1852) – including Pegu and the seaports of Martaban, Bassein and Rangoon – were designed to protect India's eastern frontier. Meanwhile, the victory of the Dutch over Wahhabi-influenced Muslim reformers in western Sumatra in the Padri War (1821–38) enabled them to undertake limited expansion along the east and west coasts of Sumatra. Dutch authority was established in Jambi (1834), Indragiri (1838), Singkil and Barus (1839–40), but attempts to move further north were thwarted by the combination of the resurgent power of the Sultanate of Acèh and the influence of the mainly British and Chinese merchants in the Straits Settlements.

The outbreak of the Revolutionary and Napoleonic Wars in Europe in April 1792 marked the beginning of a more intense European imperial involvement with Southeast Asia – an involvement which reached its peak between 1870 and 1914. By then nearly the whole of Southeast Asia was under European rule, the major exception being Chakri-ruled Siam (modern Thailand).

BRITISH, DUTCH AND SPANISH COLONIALISM

Britain's emergence as the leading commercial and seaborne power in the region was confirmed after 1795 when its naval forces, operating from Madras and Pinang in the Strait of Malacca, captured Dutch East India Company possessions throughout the Indonesian archipelago. By 1815 Britain controlled Java and the Spice Islands (Moluccas), and was soon to establish itself in Singapore (1819) and in Arakan and Tenasserim in Lower Burma following the First Anglo-Burmese War (1824–26) (*map 1*). Although Java was handed back to Holland in 1816, Dutch power in Indonesia remained totally dependent on British naval supremacy until the Second World War.

Commercially and militarily Britain owed much to India. British India (*pages 194–95*) provided the troops for its colonial conquests in Southeast Asia, and Bengal opium was the mainstay of Britain's lucrative trade with China (*pages 198–99*). Between 1762, when the English East India Company was granted a permanent trading post in Canton (Guangzhou), and the 1820s, when Assam tea production began, total Bengal opium exports increased 1,500 per cent from 1,400 to 20,000 chests per annum, and exports of Chinese tea tripled from 7,000 to over 20,000 tonnes. Britain's interest in Southeast Asia in this period was driven by its need to find trade goods saleable in Canton in exchange for tea, and by its desire to protect its sea lanes.

Elsewhere, before the 1860s, European expansion was slow. Dutch control of fertile Java was only consolidated following the bitterly fought Java War (1825–30), and Dutch finances only improved following the introduction of the "Cultivation System" (1830–70). This required Javanese peasants to grow cash crops (mainly sugar, coffee and indigo) for sale at very low prices to the colonial government. By 1877 this had produced 832 million guilders for the Dutch home treasury, which represented over 30 per cent of Dutch state revenues. In the Philippines, Spanish power was checked in Muslim-dominated Mindanao and Sulu by the strength of the local sultans, while on the main island of Luzon, the seat of Spanish colonial authority since the late 16th century, the emergence of an educated mixed-race – Filipino-Spanish-Chinese – elite, known as the *ilustrados* ("the enlightened ones"), began to challenge the political predominance of the Iberian-born friars and the Madrid-appointed colonial administrators.

SOUTHEAST ASIAN RESISTANCE

The existence of newly established dynasties and kingdoms, especially in mainland Southeast Asia, complicated the task of the European colonialists. From the mid-18th century onwards Burma, Siam and Vietnam had all experienced extensive political renewal under the leadership of new dynasties. This encompassed a revitalization of Theravada Buddhism and Confucianism; the subjugation of minority populations to new state-sponsored forms of culture, religion, language and governance; the development of Chinese-run revenue farms and commercial monopolies; and the limited acquisition of Western military technology.

The principal reason for the British annexation of Lower Burma between the 1820s and 1850s was to check the expansionist policies of a succession of Konbaung monarchs. French involvement in Indochina, which began with the capture of Da Nang in 1858, was spurred by the anti-Catholic pogroms initiated by the Vietnamese emperor Minh-mang (r. 1820–41) and his successors.

The political and cultural self-confidence of the Southeast Asian rulers went hand in hand with rapid economic and demographic growth. After a century of stagnation, the exports of Southeast Asia's three key commodities (pepper, coffee and sugar) increased by 4.7 per cent per year between 1780 and 1820, with Acèh alone accounting for over half the world's supply of pepper – 9,000 tonnes – by 1824. In the same period the region's population more than doubled to over nine million. This meant that when the Europeans began to move in force against the indigenous states of Southeast Asia after 1850, they encountered fierce resistance. It took the Dutch 30 years (1873–1903) to overcome Acèhnese resistance, and when the British eventually moved into Upper Burma in

November 1885 and overthrew the Konbaung monarchy, it required another five years of sustained operations to "pacify" the remaining guerrilla fighters.

In the Philippines the energies unleashed by the emergence of indigenous resistance movements proved too much for the incumbent colonial administration. Two years (1896–98) of armed struggle by the *ilustrado*-led Filipino revolutionaries brought the Spanish administration to its knees and facilitated the intervention of the United States, which acquired the Philippines from Spain in the Treaty of Paris (December 1898). However, three more years were to pass before the military forces of the Philippine Republic were finally subdued in a series of bitter campaigns which required the deployment of over 60,000 American troops.

NATIONALIST MOVEMENTS

Apart from the Chakri monarchs in Siam (whose power lasted until 1932) none of the Southeast Asian dynasties survived the height of Western imperialism intact (*map 2*). Instead, new Western-educated elites emerged to take their place, eventually demanding political rights and recognition of what they saw as legitimate nationalist aspirations.

Between 1906 and 1908 the foundation of the Young Men's Buddhist Association in Rangoon and the "Beautiful Endeavour" (*Boedi Oetomo*) organization of Javanese medical students in Batavia (Jakarta) led to the development of more radical forms of nationalism. In Vietnam this took the form of the anti-French agitation of the "Confucian scholar activists", such as Phan Chu Trinh and Pham Boi Chau, both of whom advocated the use of violence against the colonial state. Meanwhile, Japan's victory over tsarist Russia in 1904–5 (*pages 200–1*) had given the lie to the myth of Western superiority. The fact that Western colonial authority rested for the most part on very small numbers of troops and armed police – 42,000 for a population of 62 million in the case of the Dutch in Indonesia – made it vulnerable both to external attack and internal subversion. The rise of Japanese militarism during this period and the emergence of increasingly well-organized Southeast Asian nationalist movements sounded its death knell.

▲ Prince Dipanagara (1785–1855), leader of the Javanese forces against the Dutch in the Java War (1825–30), attempted to restore Javanese control of the island and to enhance the role of Islam. Widely revered as a Javanese "Just King", he ended his days in exile in Celebes (Sulawesi).

▼ The heyday of Western imperialism in Southeast Asia was brief, but it left a problematic legacy. The introduction by the colonialists of Western-style bureaucracies, education, capitalist means of production and communications systems – especially the telegraph (which was introduced into Southeast Asia in 1870–71), railways and steamships – led to the demise of older monarchical forms of authority and the rise of Western-educated, nationalist elites.

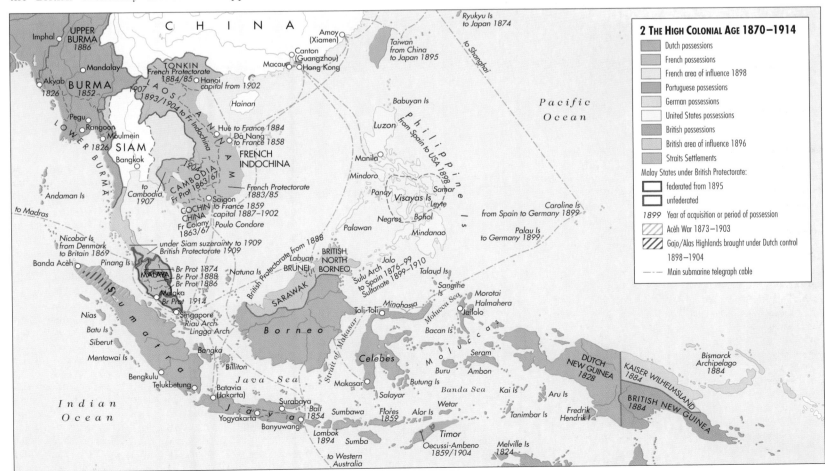

2 THE HIGH COLONIAL AGE 1870–1914
- Dutch possessions
- French possessions
- French area of influence 1898
- Portuguese possessions
- German possessions
- United States possessions
- British possessions
- British area of influence 1896
- Straits Settlements

Malay States under British Protectorate:
- federated from 1895
- unfederated

1899 Year of acquisition or period of possession
- Acèh War 1873–1903
- Gajo/Alas Highlands brought under Dutch control 1898–1904
- Main submarine telegraph cable

◀ EUROPEANS IN ASIA 1500–1790 *pages 118–19* ▶ SOUTHEAST ASIA SINCE 1920 *pages 250–51*

LATE MANCHU QING CHINA 1800–1911

► The First Opium War was the British response to attempts by the Qing rulers to restrict trade to the government-monitored custom houses of Canton (Guangzhou), and to ban the damaging import of opium. British gunships bombarded Chinese ports along the full length of its coast in 1840 and again in 1841–42, even venturing up the Yangtze to Nanjing, until the Chinese agreed peace terms which allowed for the opening up of "treaty ports" (*map 2*). Not satisfied with the outcome, however, the British joined forces with the French in 1856 to exact further concessions in the Second Opium War. China was defeated again by the French in 1885, and lost control of Korea to the Japanese in 1895.

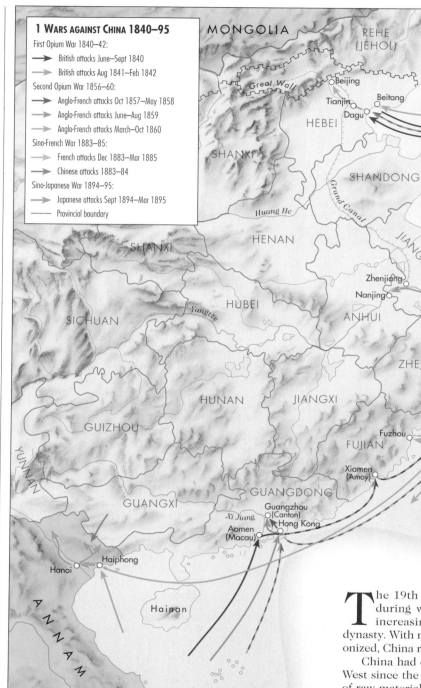

1 WARS AGAINST CHINA 1840–95

First Opium War 1840–42:
→ British attacks June–Sept 1840
→ British attacks Aug 1841–Feb 1842

Second Opium War 1856–60:
→ Anglo-French attacks Oct 1857–May 1858
→ Anglo-French attacks June–Aug 1859
→ Anglo-French attacks March–Oct 1860

Sino-French War 1883–85:
→ French attacks Dec 1883–Mar 1885
→ Chinese attacks 1883–84

Sino-Japanese War 1894–95:
→ Japanese attacks Sept 1894–Mar 1895
— Provincial boundary

CHINA'S TRADE DEFICIT WITH INDIA

Three-year average, in millions of pounds:
— total value of imports from India
— total value of opium imported
— total value of exports to India

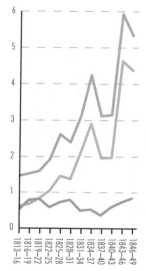

▲ Throughout the period 1800–37 the total value of imports from the English East India Company increased steadily, while Chinese exports remained fairly static. Opium imports grew during this period, leading the Chinese to impose restrictions and the British to use force in order to protect their market. Following the defeat of China in the Opium War of 1840–42, the value of opium imported more than doubled.

► During the Sino-Japanese War of 1894–95 the Chinese defenders were easily overcome by the more modern weaponry of the invading Japanese. As a result of its defeat, China was forced to cede the island of Taiwan to Japan.

The 19th century was a turbulent period for China, during which the Western powers posed an ever-increasing threat to the sovereignty of the Manchu dynasty. With most of South and Southeast Asia already colonized, China represented the final target in the Asian world.

China had enjoyed sizeable surpluses in trade with the West since the 17th century, exporting increasing amounts of raw materials – in particular tea, sugar and raw silk – in the face of growing competition from Japan and India. However, it had also become economically dependent on the West, as it had few precious metals and needed the inflow of silver from foreign trade to facilitate the expansion of its internal trade. In 1760 the Manchu Qing government had restricted the activities of foreign traders to just four ports, thus facilitating the collection of duties from these traders. By the late 18th century this had led to a system under which Canton (Guangzhou) was the sole port for foreign trade and all activities had to go through the government-monitored chartered trading houses (*cohung*). Westerners attempted, but failed, to persuade the Qing government to reform its restrictive policies, and it became clear that such policies could not be shaken off by peaceful means as long as Qing sovereignty remained intact.

THE OPIUM WARS

Western traders soon found ways to get around the *cohung* system, and smuggling was widely practised. More significantly, the British discovered an ideal commodity to sell in China: opium. In the China–India–Britain trade triangle, China's tea exports were no longer offset by silver bullion but by opium, and from the beginning of the 19th century a balance of trade rapidly developed in favour of the English

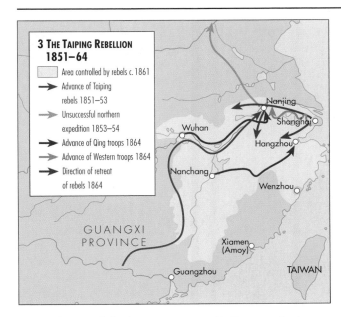

3 THE TAIPING REBELLION 1851–64

- Area controlled by rebels c. 1861
- → Advance of Taiping rebels 1851–53
- → Unsuccessful northern expedition 1853–54
- → Advance of Qing troops 1864
- → Advance of Western troops 1864
- → Direction of retreat of rebels 1864

▲ During the Taiping Rebellion the Qing lost control of much of China's most fertile region, resulting in a 70 per cent drop in tax revenues. The Qing army was largely unsuccessful against the rebels, which were only crushed with the aid of Western troops.

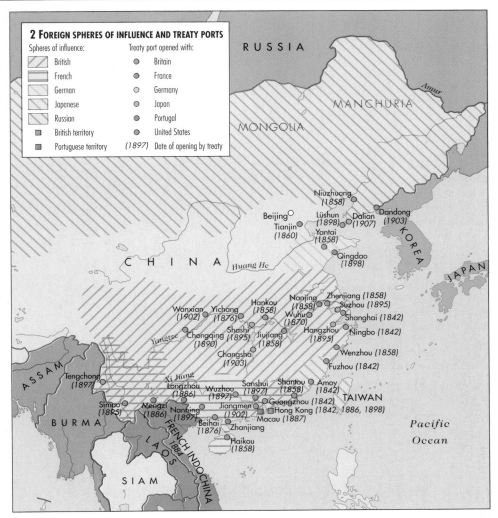

2 FOREIGN SPHERES OF INFLUENCE AND TREATY PORTS

Spheres of influence:
- British
- French
- German
- Japanese
- Russian
- ■ British territory
- ■ Portuguese territory

Treaty port opened with:
- ● Britain
- ● France
- ● Germany
- ● Japan
- ● Portugal
- ● United States
- (1897) Date of opening by treaty

East India Company (*graph*). China's hard-earned silver began to flow out in large quantities, causing severe deflation in the economy. The Manchu Qing, who did not want to see the resulting loss of tax revenue, responded by imposing a total ban on the opium trade. This triggered the invasion, in 1840, of British gunships, against which the Qing armed forces proved to be no match. The First Opium War (*map 1*) came to an end in 1842 when, under the Treaty of Nanjing, the victorious British secured the lifting of the ban on the opium trade and the opening up to trade of the "treaty ports" (*map 2*). The state monopoly was over.

The events of 1840 heralded the end of China as a world power in the 19th century. British and French allied forces extracted further concessions from China in the Second Opium War in 1856–60 (*map 1*), while the Russians annexed around 1 million square kilometres (386,000 square miles) of Chinese Siberia north of the River Amur, and further territory in Turkestan. Furthermore, China's control over its "vassal states" in Southeast Asia was weakened when Annam became a French colony after the Sino-French War in 1883–85, and China was forced to relinquish control of Korea after the Sino-Japanese War in 1894–95 (*map 1*).

These successive military and diplomatic defeats cost the Chinese Empire dearly in terms of growing trade deficits and of mounting foreign debts, mainly incurred by war reparations. China was forced to adopt what amounted to a free-trade policy. By the end of the 19th century a series of treaties had resulted in the country being largely divided up by the foreign powers (*map 2*). Although China remained technically independent, its sovereignty was ruthlessly violated – a situation that led to the anti-foreign, anti-Christian Boxer Rebellion of 1899–1901.

INTERNAL STRIFE

Partly as a result of the numerous concessions made to the foreign powers, there was an upsurge in nationalism and in the widespread antipathy to the Qing rulers, who originated from Manchuria and were therefore not considered "Chinese". In the struggle for their own survival, the Qing rulers leaned increasingly towards the West, relying on Western troops, for example, to help suppress the Taiping Rebellion (*map 3*). However, while employing the support of the West delayed the demise of the Manchu Qing government for half a century, in the long term it proved a fatal strategy. In 1911 the Nationalists, who until then had been only loosely organized, rose up in armed rebellion (*map 4*). The revolution began in Hankou on 10 October 1911, and although the Qing troops recaptured the city on 27 November, the movement to secure independence had by

▲ By the end of the 19th century China was effectively "carved up", with all its major ports and trading centres allocated by treaty to one or other of the major Western powers. In order to ensure a constant supply of goods for trading, the Western powers also exercised their influence over large areas of the Chinese hinterland. In addition, Britain was granted a lease on the territory of Hong Kong and the Portuguese gained the territory of Macau.

◄ The 1911 revolution started with the Nationalists seizing control of Hankou on 10 October. Similar uprisings in most of the major cities then followed rapidly. Only in the northeast, and in the province to the southwest of Beijing, were rebellions successfully put down by Qing troops. Following the truce of 18 December, Emperor Xuantong abdicated, and control of Beijing passed to General Yuan Shikai. The Nationalists subsequently established their capital in Nanjing.

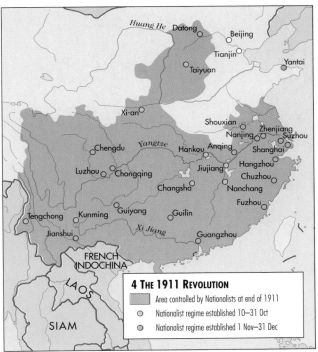

4 THE 1911 REVOLUTION

- Area controlled by Nationalists at end of 1911
- ● Nationalist regime established 10–31 Oct
- ● Nationalist regime established 1 Nov–31 Dec

this time already spread across southeast and central China. Bowing to pressure from the Western powers, whose trading interests were likely to be disrupted by civil war, the Qing emperor signed a truce with the rebels on 18 December, which stipulated his abdication and the elevation of his general, Yuan Shikai, to the position of President. The independent provinces recognized Nanjing as their new capital, and elected the Nationalist leader Sun Yat-sen as provisional President on 1 January 1912, although he stepped down on 14 February in favour of Yuan Shikai.

◄ MING AND MANCHU QING CHINA 1368–1800 *pages 138–39* ► THE REPUBLIC OF CHINA 1911–49 *pages 224–25*

THE MODERNIZATION OF JAPAN 1867–1937

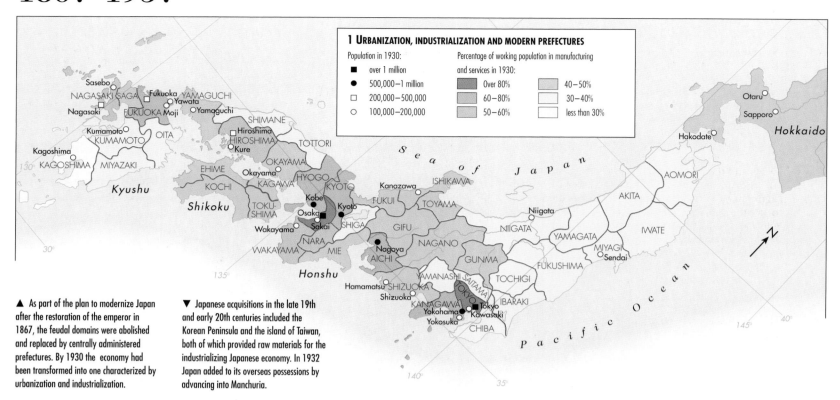

1 URBANIZATION, INDUSTRIALIZATION AND MODERN PREFECTURES

Population in 1930:
- ■ over 1 million
- ● 500,000–1 million
- □ 200,000–500,000
- ○ 100,000–200,000

Percentage of working population in manufacturing and services in 1930:
- Over 80%
- 60–80%
- 50–60%
- 40–50%
- 30–40%
- less than 30%

▲ As part of the plan to modernize Japan after the restoration of the emperor in 1867, the feudal domains were abolished and replaced by centrally administered prefectures. By 1930 the economy had been transformed into one characterized by urbanization and industrialization.

▼ Japanese acquisitions in the late 19th and early 20th centuries included the Korean Peninsula and the island of Taiwan, both of which provided raw materials for the industrializing Japanese economy. In 1932 Japan added to its overseas possessions by advancing into Manchuria.

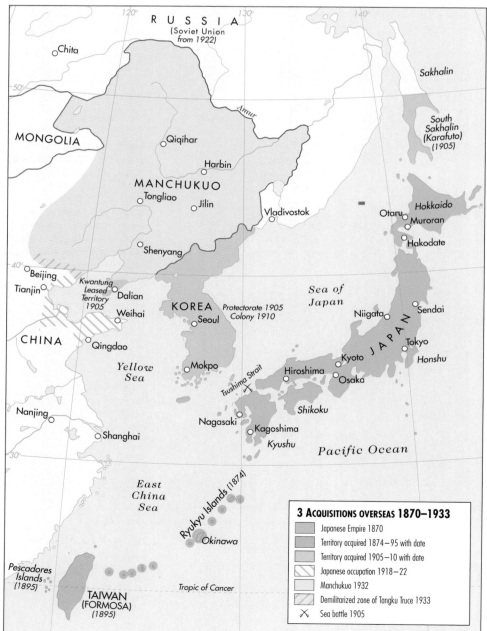

3 ACQUISITIONS OVERSEAS 1870–1933
- Japanese Empire 1870
- Territory acquired 1874–95 with date
- Territory acquired 1905–10 with date
- Japanese occupation 1918–22
- Manchukuo 1932
- Demilitarized zone of Tangku Truce 1933
- ✕ Sea battle 1905

The collapse of the Tokugawa regime in 1867 initiated a period of momentous change in Japan, in which society, the economy and politics were transformed. After more than 200 years of isolation, in the 20th century Japan emerged onto the world stage as a major power.

The new leaders believed that to achieve equality with the nations of the West, Japan had to pursue an aggressive foreign policy, and for this it needed a viable and modern military capability, backed up by a modern industrial sector. It would be a mistake to exaggerate the role of the state in the transformation of Japan into a modern industrial power. However, the government played a leading role in setting the tone for change and in laying the framework within which non-government enterprises could take the initiative.

A NEW CONSTITUTION

The new government moved swiftly, rapidly disbanding the old caste hierarchy, abolishing the domains (*pages 140–41*), and ruling the country from the centre through a system of prefectures (*map 1*). All this was done in the name of the emperor, who had been the focus of the anti-Tokugawa movement. However, disagreement within the new ruling oligarchy, and problems in dismantling the social, economic and political structures of the Tokugawa government, meant that the new imperial constitution did not take effect until 1890. The constitutional structure arrived at involved maintaining a balance of power between the various elites: the emperor, the political parties within the diet (legislative assembly), the privy council, the military and the bureaucracy. This system remained in place until 1945, with different groups dominant within it at different times.

Democratic participation was limited. Universal male suffrage was not granted until 1925, women were barred from political life, and there were draconian restrictions on labour activity as well as on ideologies and organizations deemed to be potentially subversive. The concept of the "family state" was promoted, according to which the emperor – said to be descended from ancient deities – was the benevolent patriarch of the Japanese. Any criticism of the "emperor-given" constitution was regarded as treason.

Three emperors reigned under this constitution: the Meiji Emperor (r. 1867–1912), who became identified with the national push for change; the Taisho Emperor (r. 1912–26), who was mentally impaired and made no lasting impact; and the Showa Emperor (Hirohito), who took over as regent from his father in 1921, and reigned in his own right from 1926 until his death in 1989.

MODERNIZATION OF THE ECONOMY

In their efforts to compete with the West, Japan's leaders studied and imitated Western economies, borrowing ideas as they saw fit. The legal and penal systems and the military were all remodelled along Western lines. Financial and commercial infrastructures were "westernized", and transport networks were improved; railway mileage, for example, expanded rapidly (*map 2 and graph*). A system of compulsory education was implemented from the turn of the century. Agricultural output (based on rice) increased substantially, and then levelled off from the First World War (1914–18) onwards, but there was sustained growth in commercial agricultural products, especially silk cocoons.

Up to 1914 manufacturing remained largely focused on handicraft production of traditional products for the domestic market, which in turn enabled capital accumulation for the growth of larger-scale, mechanized production. By the end of the Meiji period, factory-based silk reeling and cotton spinning were both major export industries, and the first heavy industrial plants had been established. The First World War gave a major boost to manufacturing growth, and after 1918 the industrial structure was transformed. By 1930 the percentage of the population in many prefectures working on the land or in fishing had fallen substantially (*map 1*). The relative contribution of agriculture to the Gross National Product had declined dramatically. The service sector had grown, and light industry (especially textiles), while remaining crucial in exports, had been gradually overtaken by heavy industry.

During the 1920s and 1930s some industrial sectors came to be dominated by business groupings called *zaibatsu*, who controlled multiple enterprises and huge assets. Some *zaibatsu* came under fierce attack in the wake of the Depression (1929–33), when falling prices and general instability brought agricultural crisis in some areas, and increasing internal political conflict. Despite the growth of the Japanese economy in the 1930s, living standards were squeezed and the distribution of benefits was unequal.

JAPAN AND THE WORLD

One of the most pressing concerns of the new government was to rid the country of the "unequal treaties" imposed on Japan by the Western powers towards the end of the Tokugawa period. These treaties, forcing Japan to open its ports to trade with the West, had been an important contributory factor in the collapse of the Tokugawa regime. Japan eventually achieved a revision of the treaties in 1894,

and in 1902 an alliance was concluded with Britain. Relations with her neighbours were rarely harmonious, however, as Japan gradually encroached on their sovereignty (*map 3*). Conflict with China over interests in Korea brought war between the two countries in 1894–95, resulting in a Japanese victory and the acquisition of Taiwan (Formosa). Tension with Russia culminated in the war of 1904–5. Although the Japanese victory was less than clear-cut, it gave Japan a foothold in Manchuria and the freedom to annex Korea as a colony in 1910. In all its overseas territories, but particularly in Korea, Japanese rule was harsh. After the First World War (1914–18) the League of Nations mandated the former German colonies of the Caroline, Marshall and Mariana islands (except for Guam) to Japan.

Relations with China remained tense as Japan sought to obtain increasing concessions in the wake of the 1911 Revolution, and to strengthen her control of Manchuria, regarded by the Chinese as an integral part of China's territory (*pages 224–25*). In 1927 Japanese troops in Manchuria were involved in the murder of a leading warlord, and in 1931 engineered an "incident", in the wake of which the Japanese army, acting initially without the sanction of Tokyo, occupied the territory. The following year the puppet state of Manchukuo was established. Tension between Japan and China finally erupted into full-scale war in 1937.

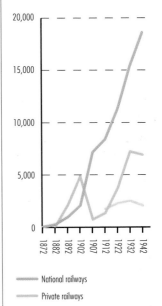

◄ In the Battle of Tsushima Strait in May 1905 (*map 3*) the Russian fleet was overwhelmed by the Japanese under the command of Admiral Heihachiro Togo. Russian losses of men and ships vastly exceeded those of the Japanese and as a result of this humiliation, and other losses on land, the Russians conceded defeat in September 1905.

GROWTH OF RAILWAY MILEAGE 1872–1942

— National railways
— Private railways
— Streetcars (trams)

▲ The nationalization of much of the railway system in 1906 more than trebled the extent of Japan's state-owned lines.

▼ The rapid development of a railway network was one feature of the dramatic changes in transport and other parts of the infrastructure that occurred from the 1870s.

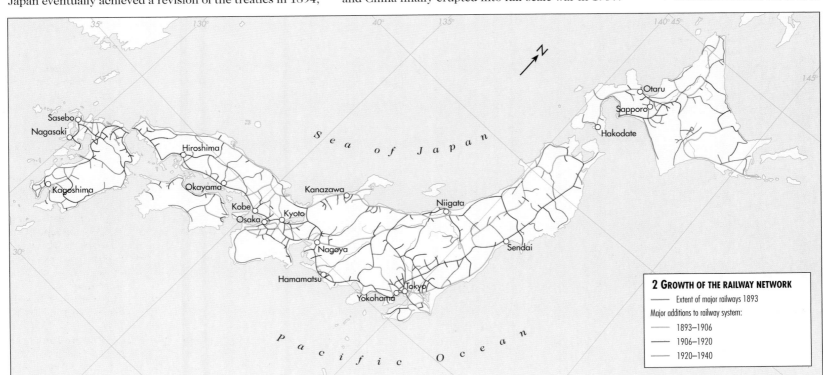

2 GROWTH OF THE RAILWAY NETWORK
— Extent of major railways 1893
Major additions to railway system:
— 1893–1906
— 1906–1920
— 1920–1940

◖ TOKUGAWA JAPAN 1603–1867 *pages 140–41* ◗ THE WAR IN ASIA 1931–45 *pages 234–35*

THE DEVELOPMENT OF AUSTRALIA AND NEW ZEALAND SINCE 1790

The history of both Australia and New Zealand long predates the arrival of Europeans in the late 18th century. Australia had been inhabited by its Aboriginal population for around 60,000 years, while New Zealand had been home to the Polynesian Maori (who called it Aotearoa) for around 1,000 years. During the 17th century Dutch explorers charted the western and northern coasts of Australia, and in 1642 Abel Tasman sighted Van Diemen's Land (later Tasmania) and followed the coastline of New Zealand (*map 1*). In 1769–70, during his first Pacific voyage, James Cook charted the coast of New Zealand and landed on the eastern coast of Australia, which he claimed for Britain.

The first British colony was founded at Port Jackson (Sydney) in January 1788, with the arrival of around 750 convicts, guarded by just over 200 marines and officers. (Over the subsequent 60 years a further 160,000 convicts would be shipped out to penal colonies established all round the eastern and southern coasts.) As the land immediately around Sydney was unsuitable for agriculture, the colony relied heavily on intermittent supplies of foodstuffs shipped out from England throughout the 1790s.

THE GROWING ECONOMY

Initially, economic activity in Australia was confined to whaling, fishing and sealing, but in the early 1820s a route was developed to the inland plains and, with access to vast expanses of pastoral land, newly arrived free settlers turned to sheep-rearing. The wool they exported to Britain became the basis of Australia's economy, and further colonies based on this trade were established over the next three decades in Tasmania, Victoria, Western Australia and Queensland.

The ever-increasing demand for pasture brought the settlers into conflict with the Aboriginal population. As well as seizing land and using violence against the Aborigines, the settlers carried with them alien diseases such as smallpox and influenza. These imported diseases had disastrous consequences for the indigenous population, whose numbers certainly declined (to an extent that can only be estimated) and would continue to do so until the 1930s (*bar chart*).

Large-scale immigration of non-convict, mainly British, settlers accelerated from the 1830s, as more agricultural territory was opened up (*map 2*). It was further encouraged by gold strikes in the 1850s. The development of overseas trade,

dependent on coastal ports, and the expansion of mining industries helped to foster an increasingly urban society. Australia's population grew dramatically from 405,000 in 1850 to 4 million by the end of the century.

The Australian colonies developed political systems based on that in Britain, and most became self-governing during the 1850s. The creation of the Commonwealth of Australia in 1901 promoted freer trade between the states within this federation and facilitated a joint approach to defence. However, one of the first measures taken by the Commonwealth was to adopt the "white Australia policy", designed to exclude non-white immigrants.

WHITE SETTLERS IN NEW ZEALAND

New Zealand was initially treated by the British as an appendage of New South Wales. It only became a separate colony following the controversial Treaty of Waitangi in 1840, which provoked decades of conflict between the white settlers and the Maori, mainly because the treaty, which gave sovereignty to Britain, was not clearly translated for the Maori chiefs who agreed it. While the Maori population declined, the settler population grew dramatically during the second half of the 19th century. Wool and gold formed the basis of the colony's economy, and with the invention of refrigerated shipping in the 1870s the export of meat became increasingly important (*map 3*). Tension over land triggered the Maori Wars of 1860 to 1872, after which large areas of Maori land were sold or confiscated by the government.

New Zealand evolved quickly to responsible government, and a central parliament, including Maori representatives, was established in 1852. By 1879 the country enjoyed almost universal male suffrage, and women obtained the vote in 1893. In 1907 New Zealand became, like Australia, a self-governing dominion within the British Empire, although its economy remained heavily dependent on British markets.

BREAKING TIES WITH BRITAIN

Until the 1950s both Australia and New Zealand retained close political ties with Britain, fighting alongside Britain in the two world wars. Britain's inability to defend the region adequately during the Second World War, however, encouraged both countries to enter into defensive arrangements with the United States, leading to the ANZUS Pact of 1951.

▼ Early exploration of Australia and New Zealand was confined to the coastline, which was explored and charted by James Cook in the 18th century and, at the beginning of the 19th century, by separate expeditions around Australia under the leadership of Matthew Flinders from Britain and the Frenchman Nicholas Baudin. In the mid-19th century explorers ventured into Australia's inhospitable interior. Without the survival techniques of the Aboriginal population many perished from lack of water (most famously, Burke and Wills). In New Zealand, however, Dieffenbach and Brunner both took Maori guides, who were largely responsible for the white men's survival.

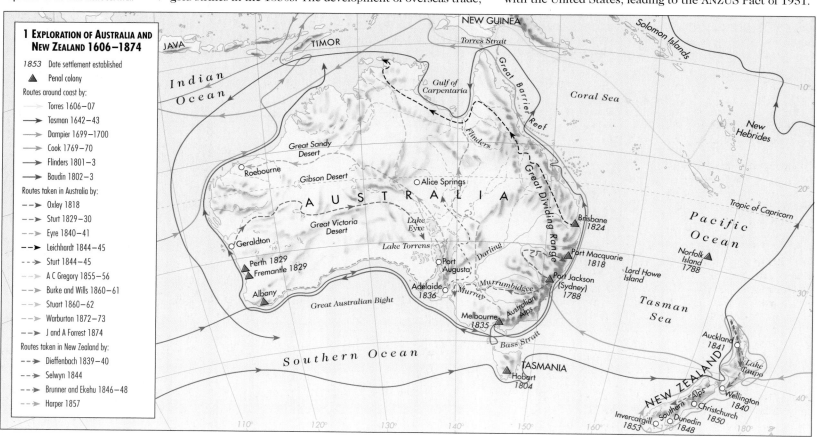

1 EXPLORATION OF AUSTRALIA AND NEW ZEALAND 1606–1874

1853 Date settlement established

▲ Penal colony

Routes around coast by:
- Torres 1606–07
- Tasman 1642–43
- Dampier 1699–1700
- Cook 1769–70
- Flinders 1801–3
- Baudin 1802–3

Routes taken in Australia by:
- Oxley 1818
- Sturt 1829–30
- Eyre 1840–41
- Leichhardt 1844–45
- Sturt 1844–45
- A C Gregory 1855–56
- Burke and Wills 1860–61
- Stuart 1860–62
- Warburton 1872–73
- J and A Forrest 1874

Routes taken in New Zealand by:
- Dieffenbach 1839–40
- Selwyn 1844
- Brunner and Ekehu 1846–48
- Harper 1857

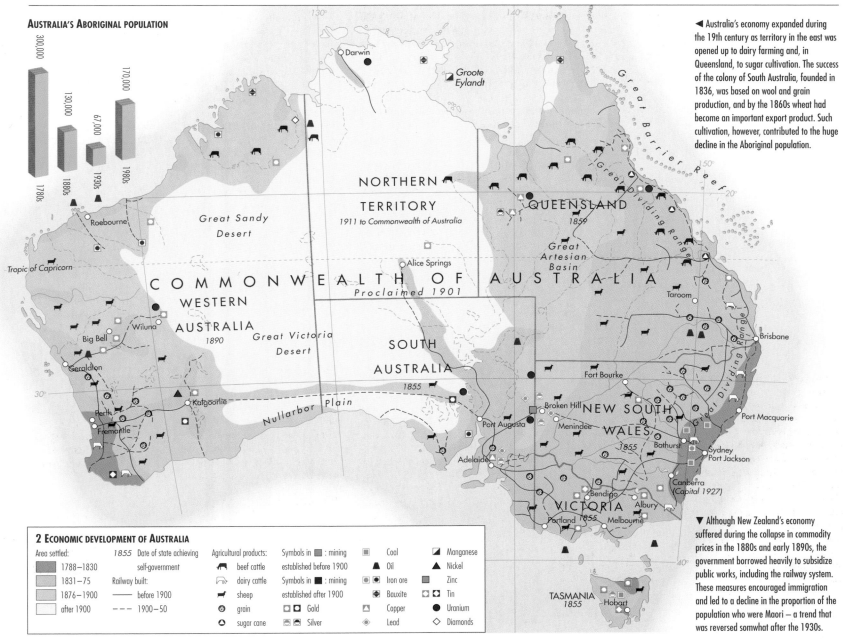

AUSTRALIA'S ABORIGINAL POPULATION

300,000
170,000
130,000
67,000

1780s
1880s
1930s
1980s

◀ Australia's economy expanded during the 19th century as territory in the east was opened up to dairy farming and, in Queensland, to sugar cultivation. The success of the colony of South Australia, founded in 1836, was based on wool and grain production, and by the 1860s wheat had become an important export product. Such cultivation, however, contributed to the huge decline in the Aboriginal population.

2 ECONOMIC DEVELOPMENT OF AUSTRALIA

Area settled:		
1788–1830		
1831–75		
1876–1900		
after 1900		

1855 Date of state achieving self-government

Railway built:
— before 1900
--- 1900–50

Agricultural products:
🐂 beef cattle
🐄 dairy cattle
🐑 sheep
grain
sugar cane

Symbols in ▢ : mining established before 1900
Symbols in ■ : mining established after 1900
▢ ▢ Gold
▢ ▢ Silver

▢ Coal
▲ Oil
◉ ◉ Iron ore
◈ Bauxite
▲ Copper
◉ Lead

⬛ Manganese
▲ Nickel
⬛ Zinc
▢ ▢ Tin
● Uranium
◇ Diamonds

▼ Although New Zealand's economy suffered during the collapse in commodity prices in the 1880s and early 1890s, the government borrowed heavily to subsidize public works, including the railway system. These measures encouraged immigration and led to a decline in the proportion of the population who were Maori – a trend that was reversed somwhat after the 1930s.

Economic ties with Britain also declined after 1945, especially once Britain joined the European Economic Community in 1973. Australia and New Zealand have increasingly focused on economic diversification and in developing ties with the United States, Japan and other countries of the "Pacific Rim" (*pages 242–43*).

MAORI AND ABORIGINAL RIGHTS

One of the most important recent political developments has been campaigns in both New Zealand and Australia to achieve fairer treatment for the Maori and Aboriginal populations. A cultural reawakening among the Maori was evident by the beginning of the 20th century (in the Ratana movement), and Maori political campaigning began in earnest in the 1920s and 1930s. Participation in the Second World War, urbanization and reviving population figures (*bar chart*) helped strengthen Maori assertiveness, and in the 1970s legislation was introduced to address grievances dating back to the Treaty of Waitangi. It took another 20 years and further protests, however, before any land was returned to the Maori, most of whom inhabit North Island.

Australia's Aborigines had begun to assert their identity and demand an end to discrimination during the 1930s, but it was not until 1967 that they won equal citizenship. In the early 1970s the federal authorities began to promote the return of land to Aboriginal communities, but although the number of Aborigines is rising, they remain the most disadvantaged sector of Australian society.

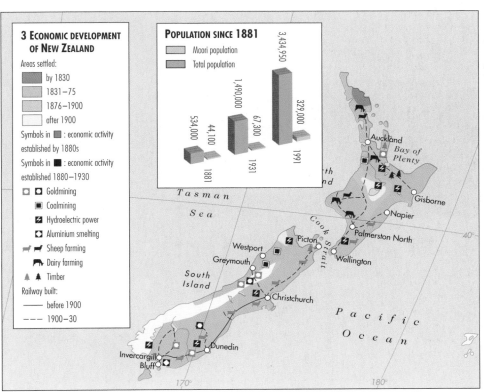

3 ECONOMIC DEVELOPMENT OF NEW ZEALAND

Areas settled:
by 1830
1831–75
1876–1900
after 1900

Symbols in ⬛ : economic activity established by 1880s
Symbols in ■ : economic activity established 1880–1930
▢ ▢ Goldmining
■ Coalmining
⬛ Hydroelectric power
▢ Aluminium smelting
🐑 Sheep farming
🐄 Dairy farming
▲ ▲ Timber

Railway built:
— before 1900
--- 1900–30

POPULATION SINCE 1881

Maori population
Total population

3,434,950
1,490,000
534,000
44,100
67,300
327,000

1881
1931
1991

AFRICA
1800–80

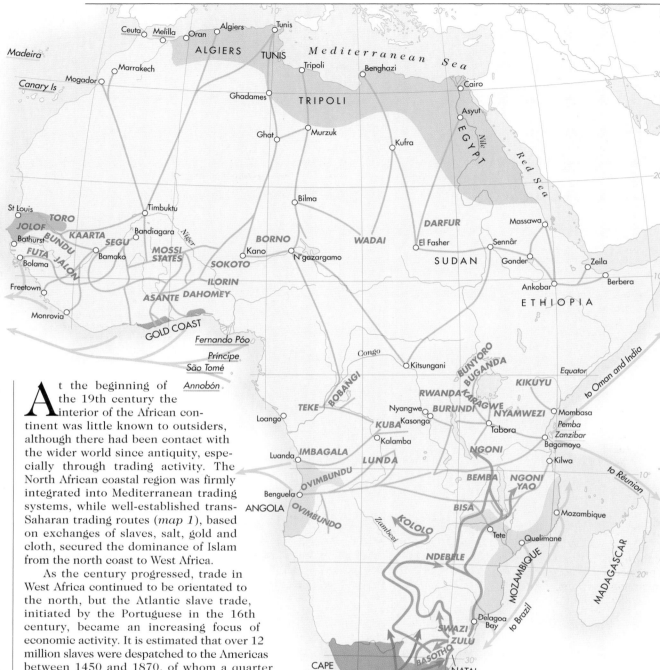

► In the mid-19th century European traders operated from bases on the coast, supplied with goods by the African trading network. In the south the dominant Zulu nation caused the dispersal of other ethnic groups throughout the region.

1 PRINCIPAL AFRICAN AND EUROPEAN TRADING ROUTES C. 1840

Areas controlled by non-African powers:

- Britain
- France
- Oman
- Portugal
- Spain
- Ottoman Empire

TEKE   African state or ethnic group

   African trade route

   Mfecane warfare and population dispersal

   Slave route

   Route of Voortrekkers 1835–40s

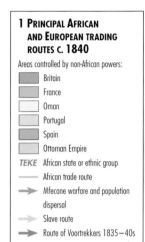

▲ The city of Timbuktu served for centuries as a trading post for trans-Saharan caravans. By the 19th century it had declined in importance but was still a focus of curiosity for Europeans, for whom travel in the region was made dangerous by Muslim antipathy to Christians. In 1853–54 the German explorer Heinrich Barth spent some time there in the course of an extensive expedition (*map 3*), and the illustration above was published in his account of his travels.

At the beginning of the 19th century the interior of the African continent was little known to outsiders, although there had been contact with the wider world since antiquity, especially through trading activity. The North African coastal region was firmly integrated into Mediterranean trading systems, while well-established trans-Saharan trading routes (*map 1*), based on exchanges of slaves, salt, gold and cloth, secured the dominance of Islam from the north coast to West Africa.

As the century progressed, trade in West Africa continued to be orientated to the north, but the Atlantic slave trade, initiated by the Portuguese in the 16th century, became an increasing focus of economic activity. It is estimated that over 12 million slaves were despatched to the Americas between 1450 and 1870, of whom a quarter were exported during the 19th century. The political, social and economic reverberations of European competition for slaves along the west and central African Atlantic coast extended far into the interior. Slaves were exchanged for firearms, metal goods, beads and other manufactured goods. With the formal abolition by Britain of the slave trade in 1807 (and despite the defiance by other European countries of this ban for many years after), ivory, rubber, palm oil, cloth, gold and agricultural products assumed ever greater importance as trading commodities.

In East Africa trading activities were somewhat less developed, as was urbanization and the formation of states. Nevertheless, Indian Ocean ports such as Mombasa, Bagamoyo, Kilwa and Quelimane were important in bringing Bantu-speaking Africans into commercial contact with Arabs, Indians and Portuguese (*map 1*). The slave trade in this region remained relatively unaffected by its formal illegality until the latter part of the 19th century.

ENCROACHMENTS BY EUROPEANS

At the start of the 19th century the European presence in Africa was largely restricted to the coastal regions of northern, western and southern Africa. The French invaded the Algerian coast in 1830 and also established a presence on the west coast. Spain had been in control of the Moroccan ports of Ceuta and Melilla since the 16th century. The Portuguese were in possession of large parts of Angola and Mozambique. In West Africa, British interests were expanding into the hinterland from the slave-trading regions of present-day Sierra Leone, Nigeria and Ghana. British influence in the region was consolidated after 1807, when the Royal Navy took on the role of enforcing an end to the slave trade and merchants extended the domain of legitimate commerce. A major area of British expansion was in southern Africa, where the Cape Colony was wrested from Dutch control in 1806. The frontiers of this settler society expanded throughout the 19th century and a second British colony, Natal, in the east of the region, was established in 1845.

AFRICAN POLITICS

Dynamic changes occurred, sometimes intensified by European contact, at other times with little reference to encroachment from the outside. In southern Africa the *mfecane* migrations, occasioned by the rise of the Zulu state

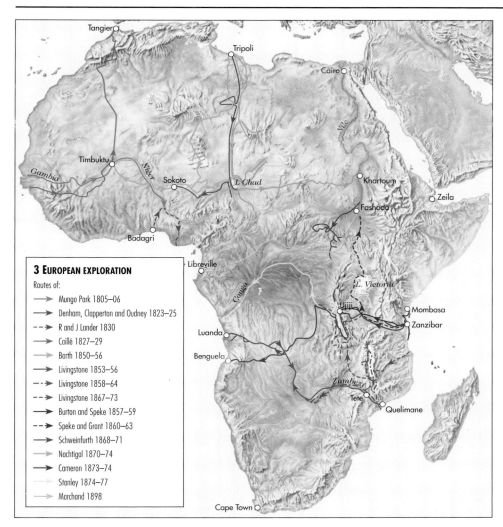

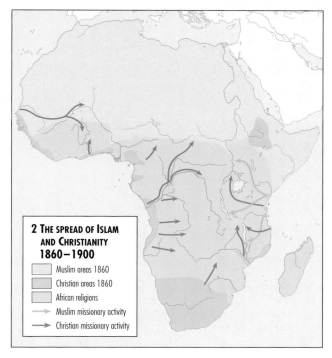

3 EUROPEAN EXPLORATION

Routes of:

→ Mungo Park 1805–06

→ Denham, Clapperton and Oudney 1823–25

--→ R and J Lander 1830

→ Caillé 1827–29

→ Barth 1850–56

→ Livingstone 1853–56

-·-→ Livingstone 1858–64

--→ Livingstone 1867–73

→ Burton and Speke 1857–59

-·-→ Speke and Grant 1860–63

→ Schweinfurth 1868–71

→ Nachtigal 1870–74

→ Cameron 1873–74

→ Stanley 1874–77

→ Marchand 1898

2 THE SPREAD OF ISLAM AND CHRISTIANITY 1860–1900

☐ Muslim areas 1860

☐ Christian areas 1860

☐ African religions

→ Muslim missionary activity

→ Christian missionary activity

◄ The first European "explorers" in Africa were those that ventured into regions in West Africa already well known to Berber traders, but hitherto considered too dangerous for Christians. From the mid-19th century onwards Europeans made expeditions into central Africa. Their motives were mixed. David Livingstone summed them up as: "Christianity, commerce and civilization", but the pursuit of scientific knowledge also played a part.

▲ During the 19th century the two main religions – Christianity and Islam – competed for domination of the African interior. The Muslim religion spread south from North Africa (although the Coptic Christians held out in Ethiopia) and inland from Arab trading bases in East Africa. The Christian churches sent out missionaries from European colonies in the south, east and west of the continent, with the Catholics and Protestants vying for converts.

during the 1820s, caused a massive dispersal of population throughout the region and resulted in the emergence of several new polities or nations, such as those of the Kololo, the Ndebele, the Swazi and the Ngoni (*map 1*). This political turbulence was exacerbated by the arrival in the southern African interior from the 1830s onwards of migrant Boer Voortrekkers, attempting to escape control by British colonists. They sought to establish independent states, largely in territory depopulated as a result of the *mfecane,* although they came into conflict with the Zulu in Natal, most spectacularly at the Battle of Blood River in 1838. Many moved on again when the British annexed the republic of Natal in 1845.

In West Africa the advance of Islam, associated with the Fulani *jihad* of 1804, resulted in the disintegration of long-established kingdoms, such as the Yoruba empire of Oyo and the Bambara state of Segu, though the Fulani were resisted in Borno. By the 1860s the Fulani caliphate of Sokoto was pre-eminent in the region, having absorbed much of Hausaland into its aegis.

In Egypt the autocratic modernization strategy adopted by Muhammad Ali in the early decades of the century transformed this province of the Ottoman Empire into an independent state in all but name; Egyptian authority was extended southwards and the Sudan was invaded in 1820–22 in order to secure the upper Nile and find a more reliable source of slaves.

Around Lake Victoria in East Africa, the kingdoms of Buganda, Bunyoro and Karagwe were linked by the trading activities of the Nyamwezi to the Swahili- and Arab-dominated coastal region, extending outwards from Zanzibar. To the north, in Ethiopia, the ancient Christian state centred on Axum was fragmented and in disarray until the mid-19th century. Thereafter, under the leadership of John IV and Menelik II, the Ethiopian Empire underwent consolidation and expansion; Ethiopia has the distinction of being the only African state to have successfully resisted 19th-century European colonial occupation.

RIVAL RELIGIONS

The creation and expansion of new states and societies, whether originating from within Africa or from external forces, were accompanied by cultural change and accommodation. Religion was a key aspect of such change (*map 2*). In North and West Africa, conquest and the spread of Islam were closely associated, although one did not presuppose the other. Christianity had been present in North Africa from the 2nd century and, though checked by the rise of Islam, had become firmly established in Coptic Ethiopia. Efforts to convert other parts of Africa to Christianity had been led by the Portuguese from the 15th century. It was in the 19th century, however, that intense Catholic and Protestant proselytization occurred; some, indeed, see missionaries as crucial precursors of European colonialism. Christianity did not, however, replace indigenous African religious traditions in any simple manner. Adaptation and coexistence was more the norm and, in many instances, African forms of Christianity emerged that would later serve as an important ideology in mobilizing resistance to European colonialism.

EUROPEAN EXPLORERS

Along with trading and missionary activity, explorers played an important role in "opening up" Africa to Europe (*map 3*). At the start of the 19th century the interior of Africa was barely known to the outside world. Expeditions, whether motivated by scientific and geographic curiosity or the search for natural resources and wealth, attracted considerable popular interest in Europe; the exploits of travellers and explorers were celebrated both in terms of individual achievement and as sources of national pride. Among the best-known 19th-century expeditions were those that explored the sources of the Nile, the Congo, the Zambezi and the Niger. The exploration and mapping of Africa proved of considerable importance to the drawing of colonial boundaries in the late 19th century.

◄ AFRICA 1500–1800 *pages 136–37* ◄ THE PARTITION OF AFRICA 1880–1939 *pages 206–7*

THE PARTITION OF AFRICA
1880–1939

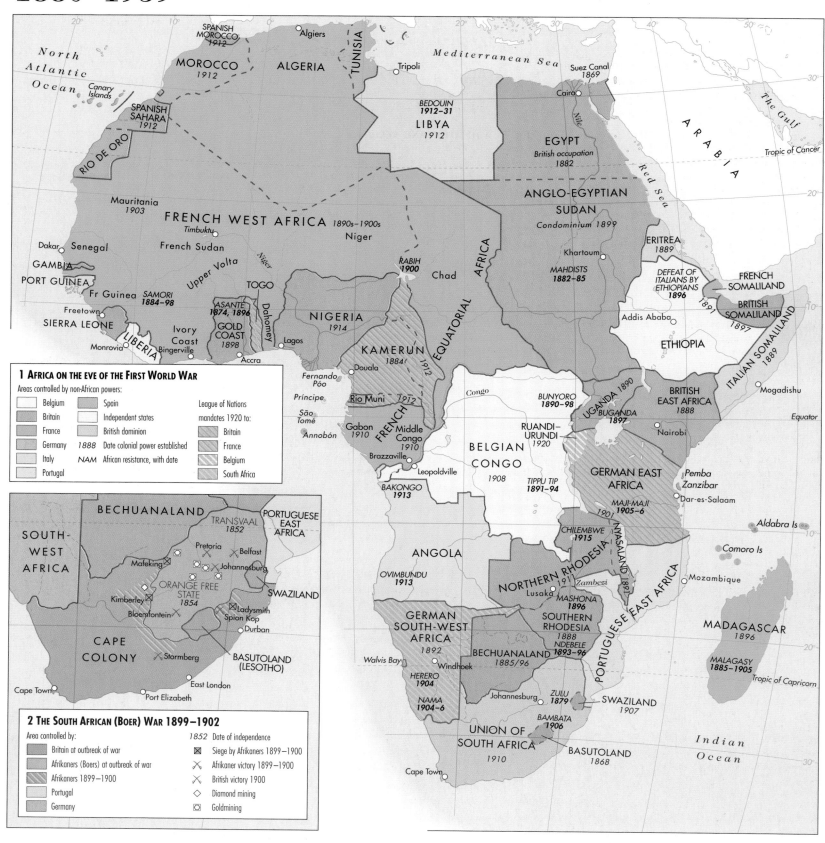

1 AFRICA ON THE EVE OF THE FIRST WORLD WAR

Areas controlled by non-African powers:

Belgium	Spain	League of Nations
Britain	Independent states	mandates 1920 to:
France	British dominion	Britain
Germany — 1888 Date colonial power established	France
Italy — NAM African resistance, with date	Belgium
Portugal		South Africa

2 THE SOUTH AFRICAN (BOER) WAR 1899–1902

Area controlled by:

	1852 Date of independence
Britain at outbreak of war	⊠ Siege by Afrikaners 1899–1900
Afrikaners (Boers) at outbreak of war	✕ Afrikaner victory 1899–1900
Afrikaners 1899–1900	✕ British victory 1900
Portugal	◇ Diamond mining
Germany	⊡ Goldmining

▲ The South African (Boer) War of 1899–1902 was one of the longest and costliest in British imperial history. In the initial phase the Afrikaners secured notable victories, but in 1900 their main towns were captured by the British. General Kitchener finally defeated them by burning their farmsteads and imprisoning civilians in concentration camps. In the Peace of Vereeniging (May 1902) the Afrikaners lost their independence. In 1910, however, the Union of South Africa gained independence under the leadership of the Afrikaner general Louis Botha.

Between 1880 and 1914 the whole of Africa was partitioned between rival European powers, leaving only Liberia and Ethiopia independent of foreign rule (*map 1*). The speed of the process was bewildering, even more so when one considers that most of the African landmass and its peoples were parcelled out in a mere ten years after 1880. European competition for formal possession of Africa was accompanied by intense nationalist flag-waving and expressions of racial arrogance, contributing in no small manner to the tensions that resulted in the outbreak of the First World War.

Many explanations have been given for the partition of Africa. Some lay particular stress on economic factors: the attractiveness of Africa both as a source of raw materials

▲ The partition of Africa was formalized at the Berlin Conference of 1884–85, attended by all the major European nations. It was agreed that a nation that was firmly established on a stretch of coast had the right to claim sovereignty over the associated hinterland on which its trade depended for the supply of goods.

and as a virtually untapped market for finished goods during Europe's "second" industrial revolution. Others view the partition of Africa in terms of intra-European nationalist rivalry, emphasizing the prestige associated with possession of foreign territory and the ambitions of individual statesmen and diplomats. Another explanation relates to geopolitical concerns, in particular the strategic designs of military and naval planners seeking to preserve lines of communication, such as the route to India through the Suez

Canal (opened 1869) and around the Cape. A variant of this theory emphasizes conditions on the ground, claiming that European powers were sucked further and further into Africa as a result of local colonial crises and trading opportunities. Technological advances (including the telegraph), as well as more effective protection against disease, facilitated the "scramble for Africa".

One of the first examples of colonists fighting for freedom from European domination occurred following the discovery of diamonds and gold in territory controlled by Afrikaner farmers (descendants of Dutch settlers, known to the British as "Boers"). Prospectors of all nationalities flooded into the region, and Britain was concerned about a possible alliance between the Afrikaners and the Germans to the west. In October 1899 the Afrikaners took pre-emptive action, besieging British troops massing on their borders (map 2). British reinforcements won several major battles, but the Afrikaners then adopted guerrilla tactics which were eventually overcome by the ruthless approach of General Kitchener.

RELATIONS BETWEEN AFRICANS AND EUROPEANS

The partition of Africa cannot be satisfactorily understood without taking into account the dynamics of African societies themselves. In some instances colonial expansion was made possible by indigenous leaders who sought to enrol Europeans as convenient allies in the struggle to establish supremacy over traditional enemies. Trading and commercial opportunities encouraged certain groups of Africans to cement ties with Europeans. Some African leaders proved adept at manipulating relationships with European powers to their own advantage, at least in the short term; elsewhere, land or mineral concessions were made to Europeans in the hope that full-scale occupation could be averted.

In a number of celebrated instances (map 1), Africans resisted the initial European colonial advance, or rose in rebellion soon after. Common informal means of resistance included non-payment of taxes, avoidance of labour demands, migration, or membership of secret religious societies. Usually, Africans sought some sort of accommodation with the advancing Europeans in order to avoid outright confrontation. Appearances are therefore deceptive: although the map indicates European possession of virtually all of Africa by 1914, in many areas control was notional. Portuguese control of Mozambique and Angola was especially tenuous. In non-settler societies and beyond major towns and centres, many Africans were more or less able to ignore the European presence and get on with their own lives.

LABOUR MARKETS AND TRADE

Perhaps the surest measure of the intensity of colonial rule is the extent to which Africa was integrated into the world economy (map 3). In southern Africa, the discovery and exploitation of diamonds and gold created huge demands for African labour. Migrant workers came from as far afield as Mozambique, Northern Rhodesia and Nyasaland. Demands for agricultural labour threatened the viability of independent African cultivators in the region, although in some areas – as in the case of cocoa production in the Gold Coast and Nigeria, for example – colonial systems relied on indigenous peasant cultivators, who were frequently able to prosper from their participation in export markets. Forced labour was widely used by agricultural concession companies in Mozambique and Angola, and by the rubber plantations of the Belgian Congo.

COMMUNICATION INFRASTRUCTURE

Railway networks werre built that linked coastal ports to the hinterland and served as a major stimulus to trade and commodity production. Railways proved particularly important for the development of mining as well as for commercial agriculture. They were also vital for the supply of labour and were crucial for the economic development of the region.

After the initial phase of railway construction, road-building programmes, especially in the inter-war years,

brought some of the most remote areas into direct contact with the colonial economy. The arrival of trucks stimulated the re-emergence of an African merchant class, particularly in West Africa. Rapid urbanization, a remarkable feature of the colonial era, was stimulated by the development of transport links and of internal and external trade.

EDUCATION AND RELIGION

In much of colonial Africa the spread of education was closely linked to religious change. Christianity in particular underwent exponential growth. The spread of Western education, building on earlier missionary endeavours, tended to be geared to the requirements of colonial regimes – providing skilled workers, clerks and petty officials. Many Africans eagerly embraced education, often as a means of social advancement. Thus, the spread of literacy opened up new horizons and possibilities that could not easily be controlled by the colonial powers. It is striking that many of the early African nationalists were the products of mission education – men who became politicized when the opportunities opened up by their education were denied them by the inequalities inherent in colonial rule.

Education and Christianity were not, however, universally welcomed by Africans. While offering social mobility to many, these agencies also threatened the power of traditional elites. Frequently, forms of Christianity evolved which combined African belief systems and traditions with Western ones. The Bible also offered fertile ground for reinterpretation in ways that challenged European rule.

Colonialism was the source of great and profound changes: economic, political, social, cultural and demographic. Significant and wide-ranging as these changes were, however, innovations were seldom imposed on a blank slate. Rather, colonial institutions were built on existing structures and moulded according to circumstances. Far from capitulating to alien rule, many African societies showed great resilience and adaptability in surviving it.

▼ The export of raw materials from Africa affected agriculture and labour markets throughout the continent. Although mining operations and large plantations were controlled by colonists, small-scale peasant production did survive in many places and benefited from export markets. Railways were crucial to economic development, in particular for the transportation of mineral ores. Their effect, however, was mixed: because they tended to disturb more traditional forms of transport, the areas they bypassed often suffered economically.

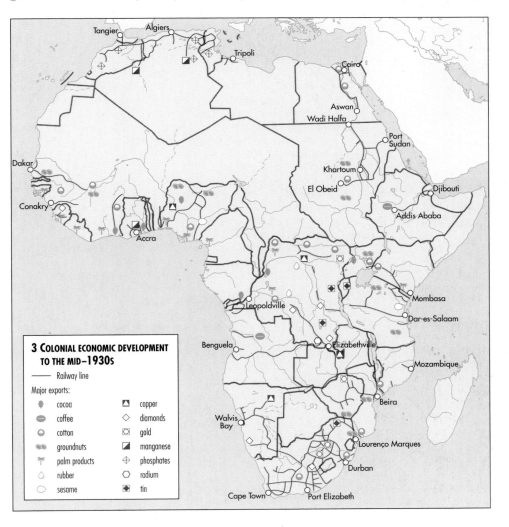

3 COLONIAL ECONOMIC DEVELOPMENT TO THE MID-1930s

—— Railway line

Major exports:

cocoa		copper	
coffee		diamonds	
cotton		gold	
groundnuts		manganese	
palm products		phosphates	
rubber		radium	
sesame		tin	

◀ AFRICA 1800–80 *pages 204–5*   ▶ AFRICA SINCE 1939 *pages 256–57*

WORLD TRADE AND EMPIRES
1870–1914

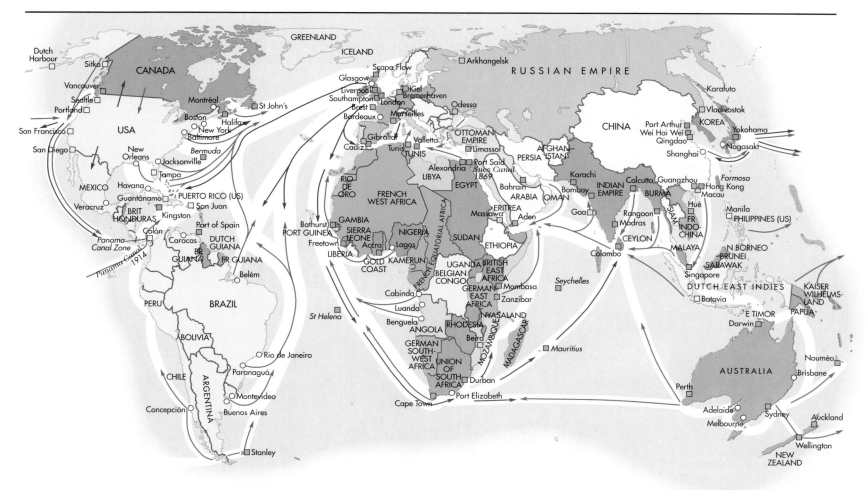

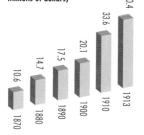

▲ The strengthening of colonial rule was linked to a number of economic and political factors, including the need for raw materials to supply rapidly industrializing economies and the desire to find new markets for manufactured goods.

The late 19th century witnessed dramatic changes, not only in the world economy but also in the relationship between the manufacturing countries and those regions of the world from which raw materials were obtained. The volume of international trade more than trebled between 1870 and 1914 (*bar chart 1*) alongside large-scale industrialization in Europe and the United States, and the spread of colonial rule, particularly in Asia and Africa. By 1913 Britain had been replaced by the United States as the world's leading manufacturing nation, but it still handled more trade than any other country (*bar chart 2*). London remained the world's leading financial centre through its operation of the international gold standard, which defined the value of the major currencies and so facilitated trade.

TRANSPORT AND COMMUNICATIONS
The enormous expansion of international trade was greatly helped by technological developments, especially in transport and communications. Sailing ships gave way to larger and faster steam vessels, which required coaling stations strategically placed around the globe (*map 1*), and merchant shipping fleets expanded to cope with the increased volume of trade. Voyages between continents were facilitated by the opening of the Suez Canal (1869) and the Panama Canal (1914). Railways also helped to increase trading activities, notably in North America and Asiatic Russia. The electric telegraph network made business transactions between continents easier (*map 2*). These technological developments also encouraged massive migrations, including that of 30 million Europeans who emigrated to North America during the 19th and early 20th centuries.

The creation of wealth in the industrialized countries led to growing interest in investing some of that wealth in the developing countries. By financing railway building or mining development in these areas, industrial economies helped to increase imports of food and raw materials, and to create larger export markets for their manufactured goods. Britain, France, other European countries and later the United States made substantial overseas investments

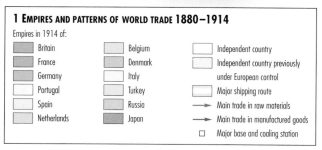

1 EMPIRES AND PATTERNS OF WORLD TRADE 1880–1914
Empires in 1914 of:

■ Britain	■ Belgium	□ Independent country
■ France	■ Denmark	□ Independent country previously
■ Germany	□ Italy	under European control
□ Portugal	□ Turkey	□ Major shipping route
□ Spain	■ Russia	→ Main trade in raw materials
■ Netherlands	■ Japan	→ Main trade in manufactured goods
		□ Major base and coaling station

(*map 2 and pie chart*), and were anxious to safeguard these from political instability and from rivals.

FACTORS INFLUENCING IMPERIAL EXPANSION
In the late 19th century the world economy was becoming more integrated, with different regions increasingly dependent on one another. Inevitably, competition between states intensified, spilling over into the political sphere. Britain's early lead as the first industrial power was linked, by many observers, to the expansion of the British Empire from the late 18th century onwards, above all in India. Other countries tried to emulate Britain by building up empires of their own. As business conditions worsened in the 1870s and 1880s, a growing number of countries also sought to protect their home markets, imposing tariffs to limit the influx of foreign goods. The attraction of untapped markets in Africa and Asia intensified as a result.

Political factors in Europe also contributed to the growth of imperialism. National prestige was always a major consideration, but it became even more so as international rivalries heightened (*pages 216–17*). The newly formed countries of Germany and Italy, as well as the declining state of Portugal, saw the acquisition of colonies as a way of asserting their status as world powers. Overseas expansion also helped to divert attention from the domestic social problems created by industrialization and population growth. Further motivation was provided by Christian missionaries, who were effective in lobbying governments to defend their activities overseas.

1 THE GROWTH OF WORLD TRADE
(exports plus imports in millions of dollars)

▲ There was a particularly sharp increase in world trade between 1900 and 1910, with the build-up of armaments by Britain and Germany — and the associated demand for raw materials — a contributory factor.

Political and economic changes taking place within non-European societies created important opportunities for the European powers to increase their influence. Local "elites" – groups who became wealthy through trade and collaboration with European powers – often facilitated the colonization of an area. Territory was sometimes acquired in order to protect existing colonial interests from rivals, or because it was particularly valuable for strategic, rather than economic, reasons. Often, however, the colonizing powers found that in order to support a limited initial claim it became necessary to expand inland from coastal bases and establish further trade links.

Although no single factor can explain the growth of imperialism in this period, the results were nevertheless far-reaching, as evidenced by the "scramble" for overseas territories in the 1880s and 1890s. By 1914 nearly all of Africa had been divided up between the European powers – chiefly Britain, France and Germany – which had also extended their control of Southeast Asia and the Pacific. China, also highly prized by the Western powers because of the enormous potential market it represented, escaped formal partition only because the Western powers could not devise a means of dividing it that was acceptable to all of them. Even here, however, European influence was strengthened following victory for Britain and France in the "Opium Wars" of 1840–42 and 1856–60 and the opening of "treaty ports" (*pages 198–99*).

The European powers were not alone in their enthusiasm for overseas expansion. After defeating Spain in the war of 1898, the United States inherited many of the former Spanish colonies, notably the Philippines and Puerto Rico. Japan, too, lacking economic resources to fuel its rapid modernization, increasingly looked to China and Korea. It was the Europeans, however, who gained most from this phase of imperialism. By 1914 the British Empire covered a fifth of the world (*map 1*) and included a quarter of the world's population, while the second-largest empire, that of France, had expanded by over 10 million square kilometres (4 million square miles) since 1870.

Although this phase of activity generated great tension among the colonial powers, aggravating their already existing mutual suspicions and feelings of insecurity, it was accomplished without direct conflict between them. (The partition of Africa, for example, was largely the result of diplomatic negotiation at the Berlin Conference of 1884–85.) The actual process of laying effective claim to territories was, however, often accompanied by extreme violence against indigenous populations, in campaigns of so-called colonial "pacification".

THE CONSEQUENCES OF COLONIAL RULE

Imperial control had far-reaching consequences for the new colonies. Their economies became more dependent on, and more vulnerable to, fluctuations in international trade. Transport and other infrastructures tended to be developed to meet the needs of colonial, rather than local, needs. Artificial colonial boundaries frequently included different ethnic or linguistic groups, sowing the seeds of future divisions. Initially, the social and cultural impact of colonial rule was limited, but Western education, medicine and religion eventually led to a devaluing of indigenous cultures. Although the colonial powers lacked the resources to employ force on a routine basis, they maintained their dominance of a region by repeated assertions of their superiority, alliances with local interest groups and occasional displays of firepower.

FOREIGN INVESTMENT IN 1914
(in millions of dollars)

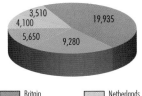

3,510
4,100
5,650
9,280
19,935

Britain	Netherlands
France	USA
Germany	

▲ European overseas investment was considerable. Its aim was to ensure a continuing supply of raw materials and to stimulate new markets for finished products. The United States, which was less reliant on overseas trade, made a comparatively small investment given the size of its manufacturing output.

◄ In 1913 the United Kingdom was still the largest trading economy, with Germany second. The United States was by this time the world's leading manufacturer, but with its rich supplies of raw materials and enormous internal market it had less need for external trade.

▼ By 1914 an extensive intercontinental telegraph network facilitated the conduct of overseas business and enabled stock markets to communicate with each other. European nations not only invested in their colonial possessions in Africa and Asia, but also in projects in North and South America and in other European countries.

2 THE VALUE OF FOREIGN TRADE 1913
(exports plus imports in millions of dollars)

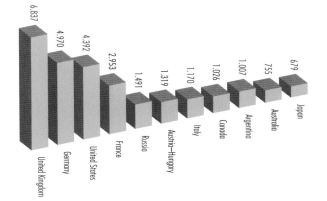

6,837	United Kingdom
4,970	Germany
4,392	United States
2,953	France
1,491	Russia
1,319	Austria-Hungary
1,170	Italy
1,026	Canada
1,007	Argentina
755	Australia
679	Japan

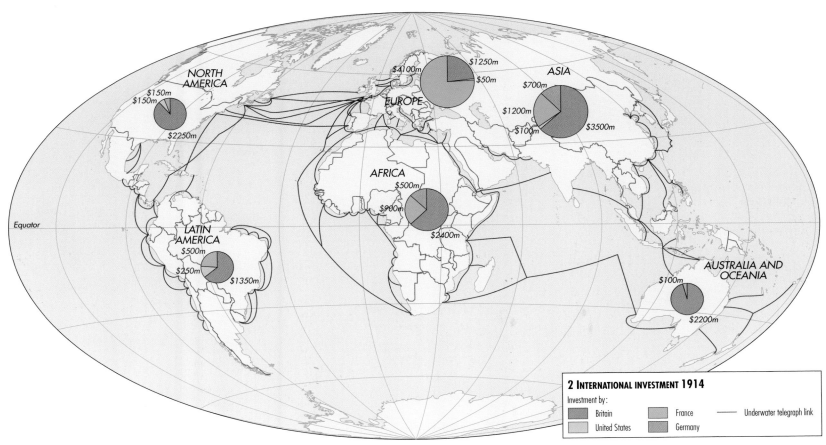

2 INTERNATIONAL INVESTMENT 1914

Investment by:
Britain	France
United States	Germany

— Underwater telegraph link

◖ THE RISE OF EUROPEAN COMMERCIAL EMPIRES 1600–1800 *pages 130–31* ◗ THE BREAKDOWN OF EMPIRES SINCE 1945 *pages 246–47*

WORLD POPULATION GROWTH AND URBANIZATION 1800–1914

▼ Population growth in the 18th and 19th centuries was unevenly distributed. Europe's population trebled, with Britain experiencing a near fourfold increase. The United States saw the most spectacular growth, caused by settlers flooding into the country, although the number of Native Americans, already decimated by war and foreign diseases, continued to decline.

High population growth around the world was matched by the development of large conurbations. In 1800 there were some 40 cities in the world with a population of between 100,000 and 500,000, of which nearly half were in Asia. By 1900 many of these had more than doubled in size and new cities had sprung up in the United States. There were now about 80 cities with a population of between 250,000 and 500,000, but only just over a fifth of these were to be found in Asia.

It is estimated that between 1500 and 1800 the world's population more than doubled, from 425 to 900 million. Then, from around 1800 the rate of increase began to accelerate so that the world's population almost doubled in just 100 years, reaching over 1,600 million in 1900. This dramatic increase was unequally distributed around the world (*map 1*). In some regions it was caused by a a higher birth rate, in others by a decline in the death rate, but in most cases it was due to a combination of the two.

FACTORS CONTRIBUTING TO POPULATION INCREASE

The birth and death rates in each country were affected by a range of socio-economic factors. One of the main ones was the increasing supply of food, which reduced the number of people dying from malnutrition, and improved people's overall health, causing them to live longer. The Agricultural Revolution in 18th-century Europe had led to the use of more efficient farming techniques, which in turn had increased food production. The expansion of the international economy and improvements in transport also contributed to improved food supplies by enabling large

quantities of cheap food to be transported from North America and elsewhere to Europe.

Industrialization was another major factor in the population growth of the 19th century. Although initially it created a new urban poverty, in most industrial countries the living standards of the working classes rose from the mid-19th century onwards as new employment opportunities became available. Medical advances made childbirth less dangerous, and the increasing use of vaccination helped prevent major epidemics. While in western Europe the use of birth control led to a drop in the birth rate from the 1880s onwards, at the same time birth rates in Asia began to rise.

INTER-CONTINENTAL MIGRATION

One consequence of the rise in population was an unprecedented intercontinental migration of people (*map 2*). Although it is usual to distinguish between "voluntary" migrants – including those seeking improved economic prospects – and "involuntary" migrants – such as those ensnared in the slave trade – for many individuals the motives for emigrating were mixed. They might involve both

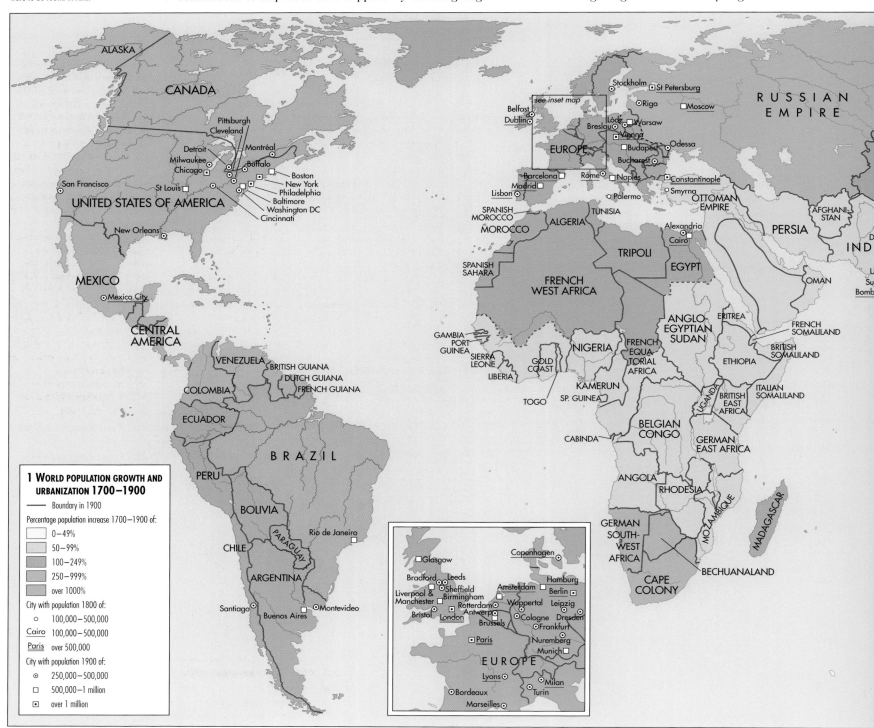

1 WORLD POPULATION GROWTH AND URBANIZATION 1700–1900

— Boundary in 1900

Percentage population increase 1700–1900 of:
- 0–49%
- 50–99%
- 100–249%
- 250–999%
- over 1000%

City with population 1800 of:
- ○ 100,000–500,000
- Cairo 100,000–500,000
- Paris over 500,000

City with population 1900 of:
- ⊙ 250,000–500,000
- □ 500,000–1 million
- ⊡ over 1 million

"push" factors, such as poverty at home, and "pull" factors, such as the availability of work in the country of destination. Between the 1880s and the outbreak of the First World War in 1914 around 900,000 people entered the United States alone each year, the majority settling in the industrializing north and east of the country (*pages 186–87*). Before the 1890s most of these migrants came from northern and western Europe, but subsequently the majority came from central and southern Europe. Europeans were particularly mobile during this period, settling not only in the United States but also in Latin America, Canada, Australasia, South Africa and Siberia.

Migration on this unprecedented scale was facilitated by the revolution in transport, which substantially reduced the cost of transatlantic travel, and by the investment of European capital overseas, which created opportunities for railway building and economic development. Chinese migrants settled in Southeast Asia, Australia and the United States, to work in mines and plantations or to build railways. Pressure on resources in Japan also led many of its citizens to emigrate to Manchuria and the Americas.

INCREASING URRBANIZATION

In addition to witnessing a large increase in overall population levels, the period 1800–1914 saw an increasing concentration of the world's population in cities (*map 1*). This was due both to population growth and, especially in Europe and the United States, to the development of new industries in the towns. At the same time, technological change in agriculture, particularly in Europe, led to a contraction in the demand for labour in rural areas.

At the beginning of the 19th century the country with the most rapid rate of urbanization was Britain, with 20 per cent of the population of England, Scotland and Wales living in towns of over 10,000 people (as against 10 per cent for Europe as a whole). By 1900 around 80 per cent of Britain's population lived in towns of over 10,000 people, and London's population had increased to over 5 million. However, despite the fact that by 1900 many large cities had developed around the world, the majority of people still lived in rural areas.

Urban infrastructures were often unable to meet the new demands being made on them, leading to inadequate housing stock, water supplies and sewage disposal. Such conditions were a factor in the cholera epidemics that affected many European and North American cities from the 1840s to the 1860s. As a result, measures to improve public health were introduced in the 1850s, and the last major European outbreak of cholera was in Hamburg in 1892. Improvements in transport, especially in the railway system, encouraged the building of suburbs, which greatly eased the problem of urban overcrowding.

◄ Rapid industrialization gave rise to urban growth that was frequently uncontrolled and unplanned. The overcrowded housing that resulted often led to squalor and disease.

▼ As the wider world became known to Europeans, many of them left their native countries in search of a better life for themselves and their families. The earliest of these European migrations was to the Americas. Around 30 million people left Europe between 1815 and 1914 bound for the United States, driven across the Atlantic by rising unemployment at home in times of economic depression and, in the case of one million Irish emigrants, the disastrous potato famine of the mid-1840s.

Sometimes migrants left Europe in order to avoid persecution of various forms, as was the case with the Russian Jews, who from the 1880s were the target of officially encouraged pogroms. Later European settlers headed for South Africa and beyond, to Australia and New Zealand. Elsewhere in the world millions of Chinese and Japanese migrated in search of work, the majority to Southeast Asia but a sizeable number to the west coast of North America.

The slave trade caused a massive involuntary migration of Africans to the Americas and also to Arabia.

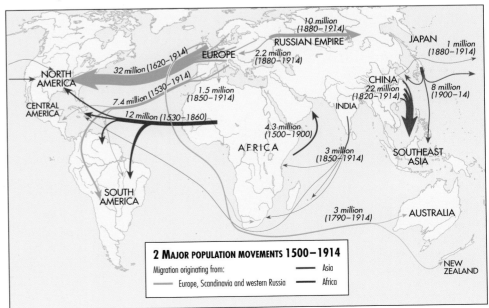

2 MAJOR POPULATION MOVEMENTS 1500–1914

Migration originating from:
— Europe, Scandinavia and western Russia
— Asia
— Africa

● EUROPEAN URBANIZATION 1500–1800 *pages 132–33* ● CHANGES IN POPULATION SINCE 1945 *pages 274–75*

THE 20ᵀᴴ CENTURY

The 20th century is often portrayed as a time of barbarism, when increasingly powerful weapons killed on an enormous scale, oppressive dictatorships flourished and national, ethnic and religious conflicts raged. Yet it was also a time when people lived longer, were healthier and more literate, enjoyed greater participation in politics and had far easier access to information, transport and communication networks than ever before.

► The two world wars were responsible for perhaps more than 80 million deaths. The First World War was essentially a European territorial dispute which, because of extensive European empires, spread as far afield as Africa and Southeast Asia. The Second World War also started as a European conflict, but spread to the Pacific when Japan seized territory. In the inter-war period disputes broke out over territory in South America and East Asia, but elsewhere the reluctance of the colonial powers to become embroiled in territorial disputes maintained an uneasy peace.

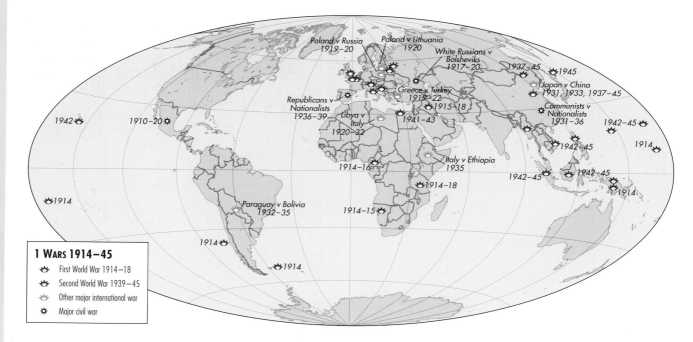

1 WARS 1914–45
- ✿ First World War 1914–18
- ✿ Second World War 1939–45
- ✿ Other major international war
- ✷ Major civil war

▼ The devastating Japanese attack on the US fleet in Pearl Harbor, Hawaii, on 7 December 1941 marked the point at which the Second World War became a truly global conflict.

The world in 1900 was dominated by the nation-states of Europe, of which the most powerful were Britain, France, Russia, Austria–Hungary and Germany. The country with the greatest industrial output in 1900 was the United States, which for the first half of the century chose to remain outside the struggle for supremacy between the European nations. Power, however, increasingly shifted away from Europe. The colonial empires which underpinned it disintegrated and the

United States became the leading world power in the second half of the century.

The first half of the century was dominated by the Russian Revolution of 1917 and the two world wars. The wars resulted in unprecedented numbers of casualties. Eight and a half million people died fighting in the First World War of 1914–18, with perhaps up to 13 million civilians dying from the effects of war. During the Second World War as many as 60 million people are believed to have died, a quarter of whom were killed in Asia and the Pacific (*map 1*). Of the total number of casualties in the Second World War it is estimated that half were civilians. The scale of the killing was largely due to the increasingly lethal power of weaponry. This reached so terrifying a peak with the invention and use of the atomic bomb at the end of the Second World War that thereafter the major powers sought to prevent local conflicts from escalating into major international wars.

THE COLD WAR
After 1945 there was no reduction in bitter international conflict, but it took a new form. The war in Europe was fought by an alliance of the communist Soviet Union with the capitalist states of Europe and the United States against the fascist regimes in Germany and Italy. Following the defeat

of fascism, the United States and Soviet Union emerged as bitterly opposed superpowers with the resources to develop huge arsenals of nuclear weapons. From 1947 a "Cold War" developed between them and their allies, in the course of which they gave support to opposing sides in conflicts in, for example, Korea, Vietnam, Angola and the Middle East, while the two superpowers remained formally at peace. The collapse of communism in Eastern Europe and the Soviet Union in 1989–91 brought the Cold War to an end.

LOCAL CONFLICTS
While there was no global war in the second half of the century, there were many local wars (*map 2*), which were waged with increasing technological expertise and precision. Some were wars of independence from colonial powers, most of which had given up their empires by 1970. Other conflicts, such as the Korean War (1950–53) and Vietnam War (1959–75), were struggles for national control between communists and non-communists, each side backed by one of the superpowers. The United Nations, established in 1945 with the aim of stabilizing international relations, failed to bring about world peace, but helped to avert or negotiate the end of some conflicts.

Some of the most persistent campaigns of violence during the 20th century were conducted by powerful governments against people of the same nation but of another political persuasion, social class, ethnic group or religious belief. In the Soviet Union under Stalin (1929–53) tens of millions of people were sent to their deaths in forced-labour camps. In Argentina and Chile in the

1970s thousands of political opponents of the government simply "disappeared", while in Cambodia in 1975–79, Pol Pot's brutal experiment in social restructuring resulted in the death of over one million people.

"Ethnic cleansing" was a term first used to describe events in the Balkans in the 1990s, but it is a concept that regularly scarred the 20th century. The Ottoman Turks deported an estimated 1.75 million Armenians from eastern Anatolia during the First World War. In Europe under the Nazis, between the mid-1930s and 1945, six million Jews, along with other minority groups, died in concentration and death camps.

◄ The opening of the gates in the Berlin Wall – symbol of the post-1945 East-West division of Europe and of the Cold War – heralded the end of communism in Europe. Mass demonstrations and political pressure from the Soviet president, Mikhail Gorbachev, forced the East German government to announce the relaxation of border restrictions. On the night of 9 November 1989 thousands of East Berliners flooded through the border to the West, many of them taking the opportunity of demonstrating their contempt for the East German authorities by climbing on, and breaking down, the Berlin Wall..

▼ As European colonial control was largely destroyed between 1945 and 1970, new nation-states were created. One result was an increase in localized wars, largely arising from boundary disputes, and in civil wars caused by conflicts between different ethnic groups or between those with conflicting religious or political beliefs. An estimated 25–30 million people died in these wars, two-thirds of whom were civilians.

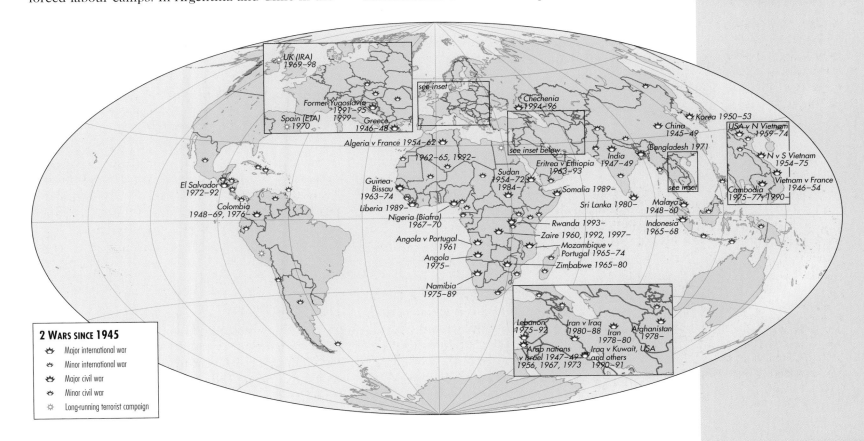

2 WARS SINCE 1945
- 🐾 Major international war
- 🐾 Minor international war
- 🐾 Major civil war
- 🐾 Minor civil war
- ☼ Long-running terrorist campaign

▲ Voting in government elections, which at the beginning of the 20th century was the prerogative of only a small proportion of the world's population, is now considered a fundamental civil right for both men and women. Democracy reached South Africa in April 1994, when the black population was allowed to vote in state elections for the first time.

HEALTH AND WEALTH

During the 20th century enormous improvements in social and economic conditions took place, although the improvements were not evenly distributed around the world. Those countries in Europe, North America and Asia that had gone through a process of industrialization in the previous century reaped the benefits, especially in the more stable economic environment of the years between 1945 and the early 1970s, when there was a general improvement in the standard of living for the majority of their citizens. In other countries, most notably those in Southeast Asia, rapid industrialization took place from the 1970s.

Advances in medical technology transformed the lives of people in, for example, Europe, North America and Japan, but were by no means widely available outside the most affluent nations. The dramatic decline in infant mortality rates and increased life expectancy in many countries during the second half of the 20th century can largely be ascribed to improved living standards, of which better medical care was just one part.

The world's population doubled between 1940 and 2000 (to reach six billion), with 90 per cent of the total growth in the 1990s taking place in the non-industrialized regions of the world. Population increases were often accompanied by rapid urbanization, frequently unplanned and unsupported by improvements in the urban infrastructure. Such rapid demographic change caused increasing social pressures, which could lead to social instability and conflict.

The supply of food and water became an overtly political issue during the later 20th century. Political and environmental factors resulted in periods of famine in some regions of the world, notably sub-Saharan Africa, while in Western Europe and North America improvements in agricultural technology and subsidies led to gluts of

certain foods, which were then stored to prevent falling prices. By the end of the century the increasing demand for water was threatening to lead to conflicts as, for example, the damming or diversion of a river by one country caused water shortages in others.

THE WORLD ECONOMY

The First World War profoundly changed European politics and society and destabilized the European-dominated world economic system. This led to reduced levels of trade and high unemployment – problems which reached crisis point in the Great Depression of 1929–33 and were still there at the outset of the Second World War in 1939.

Following the war, international agreements and institutions were established to prevent further crises and to stabilize and expand world trade. Partly in consequence, the period from the late 1940s until the early 1970s was an economic "golden age" for the industrialized countries. This economic boom came to an end when oil prices soared in the 1970s. Both rich and poor countries suffered the consequences as unemployment rose to levels comparable with those of the inter-war years. Many developing countries were encouraged

▼ During the 20th century a growing number of women became actively involved in politics. Their role was largely confined to the grassroots level, with the number of women holding government posts remaining low. However, as with this woman speaking out against the detention of political prisoners in Indonesia in 1995, they often found a voice in protest politics.

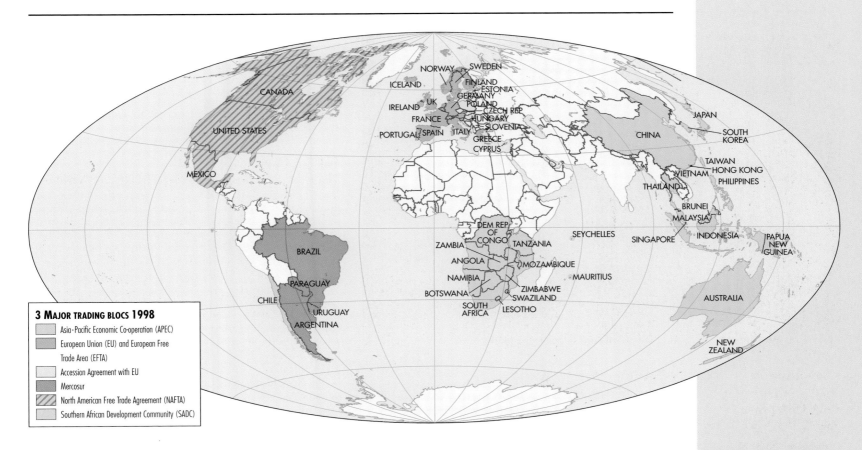

3 Major trading blocs 1998

Asia-Pacific Economic Co-operation (APEC)

European Union (EU) and European Free Trade Area (EFTA)

Accession Agreement with EU

Mercosur

North American Free Trade Agreement (NAFTA)

Southern African Development Community (SADC)

to take out huge loans, the repayment of which had a detrimental effect on their subsequent economic and social development.

THE SPREAD OF DEMOCRACY

Although at the beginning of the 20th century a number of countries had elected governments, in none of these was there universal suffrage – the right of every adult citizen to vote. A few countries had granted the vote to a high proportion of adult men, but only New Zealand had extended the vote to women. As the century progressed, representative democracy and universal suffrage spread to all continents, although it was frequently fragile as, for example, when military rulers seized control in some Latin American countries in the 1970s, or in several African countries in the 1980s and 1990s. Authoritarian communist governments, which had ruled in the Soviet Union for over 70 years and in Eastern Europe for over 40 years, collapsed in 1989–91, bringing democratic institutions to over 400 million people. At the end of the century, however, the fifth of the world's population who lived in the People's Republic of China (established by the Communist Party in 1949 after a long civil war), together with citizens of many Middle Eastern countries, still did not enjoy full political rights.

GLOBALIZATION AND NATIONALISM

The defining feature of the final decades of the 20th century was considered by some to be the process of "globalization", with multinational corporations moving their operations around the world in accordance with their needs, and individuals

travelling and communicating with one another across frontiers with unprecedented ease. However, it was questioned whether what was occurring was globalization or the "Americanization" of developing economies and of many aspects of international culture. Others stressed the significance of the new regional economic groupings which had emerged in the second half of the century (*map 3*).

A feature as strong as globalization or regionalization was nationalism – expressed both by established nations attempting to avoid domination by superpowers, and by groups within nation-states who felt oppressed and excluded on religious or ethnic grounds.

▲ Since the Second World War there has been a worldwide trend towards the creation of trading blocs between neighbouring states and erstwhile enemies.

▼ Skyscrapers have become an increasingly dominant feature of American cities since the end of the 19th century, symbolizing the enormous wealth of the United States and its position as the world's most powerful nation.

THE BUILD-UP TO THE FIRST WORLD WAR
1871–1914

After the defeat of Napoleon in 1815 Europe underwent a period of domestic transformation and upheaval that permanently altered its make-up. New nation states such as Italy were created, while the great multi-ethnic empires of the Ottomans and Austria–Hungary began to weaken. For much of the 19th century a balance of power existed in which no single European nation was strong enough to dominate, or attempt to dominate, the whole continent. This balance could not, however, endure for ever.

THE RISE OF GERMANY

The great European powers that had fought the Napoleonic Wars – Britain, Prussia, Russia, Austria and France – were growing at different rates. The most startling change occurred in the centre of Europe. Prussia, which had been the smallest of the great powers, had by 1871 been replaced by a formidable, dynamic Germany, which single-handedly defeated the Austrian Empire in 1866 and then France in 1871 (resulting in the annexation of Alsace and Lorraine) (*map 1*). The rise of Germany effectively altered the continent-wide balance of power.

The Industrial Revolution had changed the basis of national strength, making a country's production of coal, iron and steel, and the sophistication of its weaponry, even more important than the size of its population. Between 1871 and 1913 Germany moved from being the second strongest to being the leading industrial power in Europe (*bar charts*) – an economic strength that from 1890 was combined with a

▲ In an attempt to isolate France the newly unified Germany made alliances with Austria–Hungary, forming a huge power bloc in central Europe. These alliances also included Germany's arch-rival Russia (1881) and Italy (1882).

► The system of alliances between the countries of Europe in 1914 ensured that when Austria threatened Serbia following the assassination of Archduke Ferdinand, all the major European powers rapidly became involved.

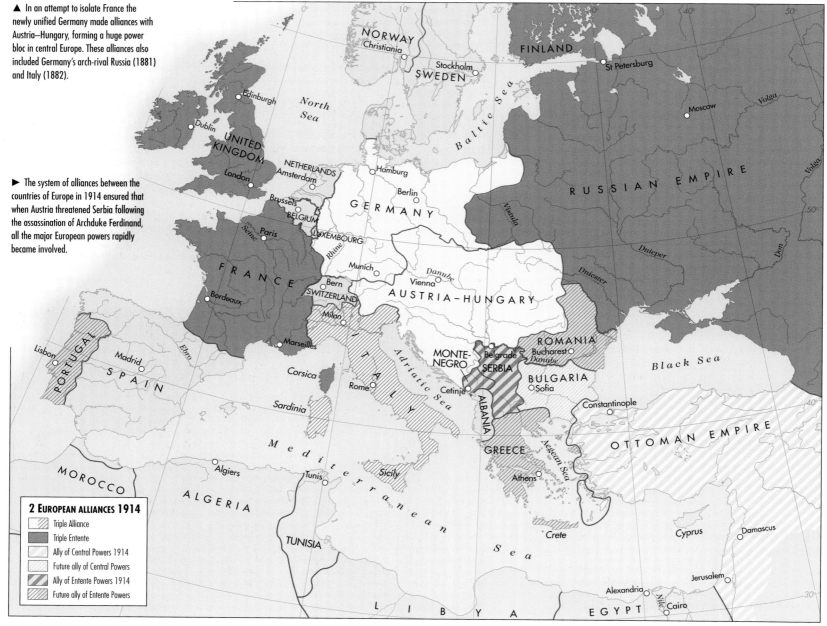

confrontational and heavy-handed foreign policy. In 1881 the German Chancellor, Count Otto von Bismarck, had concluded an alliance with Russia and Austria–Hungary, known as the "Three Emperors' Alliance" – a move intended to keep France isolated. To counterbalance this alliance with Russia (a country that might more realistically be seen as a threat), he also entered into a "Triple Alliance" with Austria–Hungary and Italy in 1882 (*map 1*). After Bismarck's fall in 1890, however, German foreign policy became increasingly concerned with the desire for expansion, both in Europe and further afield, in Africa and Southeast Asia. The Germans felt that unless they acquired a large and profitable empire they would eventually be left behind by their giant rivals: Russia, the British Empire and the United States.

THE DOUBLE ENTENTE

Meanwhile, France, which had been alternately fearful and resentful of German strength since the loss of Alsace and Lorraine in 1871, broke out of its isolation in 1894 by making an alliance with Russia. Neither country was a match for Germany on its own. France had neither sufficient population base nor industrial resources, while Russia, still relatively undeveloped industrially, could not properly utilize its enormous population and resources, as was demonstrated in the Russo-Japanese War of 1904–5 (*pages 200–1*).

The Franco-Russian alliance (the "Double Entente") was a first step towards the creation of an anti-German coalition, but if Germany's growing power was to be effectively opposed, Britain had to be included. For much of the 19th century Britain had tried to distance itself from European affairs – a policy sometimes termed "splendid isolation". With a massive and growing global empire and the world's first industrialized economy, Britain saw little profit in actively intervening on the Continent. At the end of the century, however, its isolation seemed considerably less palatable as its economic dominance disappeared with the industrialization of other European countries and the United States. Meanwhile, the criticisms levelled at its role in the South African (Boer) War (1899–1902) (*pages 206–7*) showed that much of Europe (and a sizeable proportion of the British people) resented its imperial domination.

THE TRIPLE ENTENTE

It was by no means certain that Britain would side with the Franco-Russian alliance. France and Russia had been considered Britain's greatest enemies during most of the 19th century, and in 1901 the British and German governments discussed signing an alliance of their own. However, as German power continued to grow, Britain signed an entente with France in 1904 and with Russia in 1907. Neither of these agreements was in fact a formal pledge of British mili-tary support for France and Russia in the event of a German attack, but Britain's resolve was hardened by the growth of the German navy; urged on by Admiral Alfred von Tirpitz, the Germans had, since 1898, been building up their naval strength, and by 1909 it seemed possible that they could achieve naval supremacy. Since naval supremacy had always been one of the cardinal elements of British policy, the British government, led by its very anti-German Foreign Secretary Sir Edward Grey, reacted by dramatically increasing production of British battleships. The subsequent naval construction race, won by the British, increased the rivalry between the countries and made it more likely that Britain would intervene if Germany went to war with France and Russia.

THE BALKANS

This still did not mean that war was inevitable. For the first part of 1914 Europe seemed peaceful. The issue that broke this calm was a crisis in the Balkans (*map 3*), an area of southeastern Europe that had been under Ottoman rule for centuries (*pages 178–79*). During the second half of the 19th century Serbia, Bulgaria, Bosnia-Herzegovina and Albania all agitated for independence. Austria–Hungary and Russia both coveted these areas, and in 1908 Austria annexed Bosnia into its empire. Russia was forced to accept this arrangement

3 THE BALKAN WARS 1912–13

- – – Border of country or province 1912
- —— Border of country 1914
- ▢ Austro-Hungarian Empire 1878
- ▨ Administered by Austria–Hungary from 1878

Territory gaining independence from Ottoman Empire:
- 1830–1908
- 1912–13
- Ottoman Empire 1914
- *1878* Date of independence from Ottoman Empire

because of German support for Austria. Bosnia was a multi-ethnic area populated by Croats, Serbs and Muslims of Turkish and Slavic descent. Serbian nationalists opposed Austrian rule in Bosnia, seeking to include the region in a larger Serbian national state. When Archduke Franz Ferdinand, heir to the Austrian throne, visited Sarajevo, the capital of Bosnia, in June 1914, he and his wife were assassinated by a Serbian nationalist. Austria's response was to set about crushing Serbian nationalism permanently. The Russians opposed Austrian attempts to dominate Serbia, while Germany promised to support any move the Austrians made. When the Russians duly mobilized their entire armed forces, the Germans and then the French called up their armies. As military goals became central to each nation's policies, the outbreak of the First World War became inevitable.

▲ In October 1912 Montenegro, Greece, Serbia and Bulgaria declared war on the Ottoman Empire. As a result, the Ottomans relinquished almost all their lands in southeast Europe in 1913, to the advantage of the victorious states. A second war then erupted between Bulgaria and Serbia over territory in Macedonia — a war which Serbia won, supported by Montenegro, Romania and the Ottoman Empire. These two Balkan Wars, in creating a militarily strong and ambitious Serbia, inflamed existing tensions between Serbia (supported by Russia) and Austria–Hungary and thus contributed to the outbreak of the First World War.

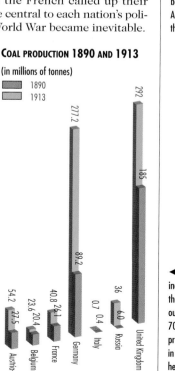

STEEL PRODUCTION 1890 AND 1913
(in thousands of tonnes)
- ▮ 1890
- ▮ 1913

Austria-Hungary: 516 / 2,611
Belgium: 221 / 2,403
France: 683 / 4,687
Germany: 2,135 / 17,609
Italy: 108 / 934
Russia: 378 / 4,918
United Kingdom: 3,636 / 7,787

COAL PRODUCTION 1890 AND 1913
(in millions of tonnes)
- ▮ 1890
- ▮ 1913

Austria-Hungary: 27.5 / 54.2
Belgium: 20.4 / 23.6
France: 26.1 / 40.8
Germany: 89.2 / 277.2
Italy: 0.4 / 0.7
Russia: 6.0 / 36
United Kingdom: 185 / 292

◀ Between 1890 and 1913 all the major industrialized nations of Europe increased their production of steel, but Germany outstripped them all with a massive 700 per cent increase. Coal, vital to the process of industrialization, was also mined in increasing quantities. This development of heavy industry was a necessary precondition for the manufacture of modern weapons, notably battleships.

THE FIRST WORLD WAR
1914–18

3 TRENCH WARFARE: BATTLE OF THE SOMME 1916

⠿ Buildings

→ First British attack, noon 1 July 1916

✳ British troops holding out in German trenches

→ Second British attack, evening 1 July 1916

0   500 m

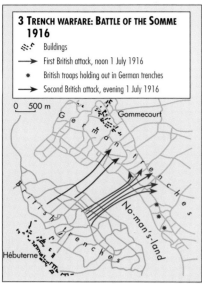

▲ After the Germans' initial attack had been repulsed by the Entente Powers, both sides dug an extensive network of trenches, often only a few hundred metres apart. Modern artillery and machine-guns made these trenches easy to defend and difficult to attack. On the first day of the Battle of the Somme, 1 July 1916, when the British attempted to break through German lines, 20,000 British troops lost their lives, with 1,000 killed in two attacks on the short sector between Hébuterne and Gommecourt alone.

▶ While the outcome of the First World War was finally decided on the Western Front, fighting took place in many areas of Europe and the rest of the world. On the Eastern Front the Russians, after some initial success, were forced back by an army equipped with modern weaponry for which they were no match. The Italians became bogged down in a small area of northeast Italy, but were finally driven back following the Battle of Caporetto in October 1917. Troops of the Ottoman Empire became involved in fierce fighting with those of the British Empire in the Tigris Valley. The Arabs assisted the Entente Powers by staging a revolt against the Ottomans, eventually driving them northwards as far as Damascus.

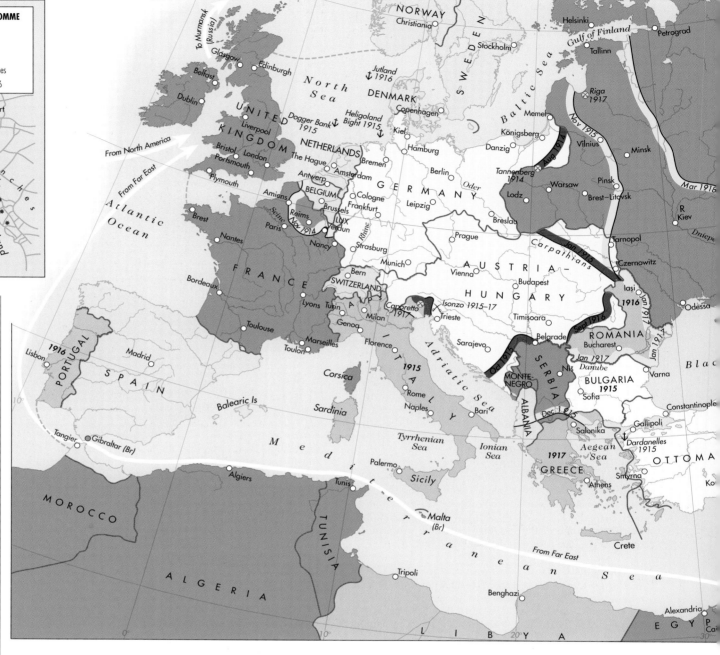

On 1 August 1914 the German army crossed the Belgian border and the First World War began. The armies of the Triple Entente (Britain, France and Russia) implemented plans drawn up in preparation for any German aggression. The French "Plan 17" called for a lightning invasion of Alsace–Lorraine on Germany's western border, and the Russians began the task of assembling their massive army and launching it against Germany's eastern frontier (*map 1*). The Germans had devised their famous "Schlieffen Plan", according to which the German army would move through Belgium into France, sweeping around Paris and encircling the French army (*map 2*) before the slower-moving Russians could muster their forces on the Germans' Eastern Front.

If executed properly the Schlieffen Plan might have resulted in a German victory in 1914, but although the German army made quick progress through Belgium, their Chief of General Staff, von Moltke, became increasingly concerned about Russian strength and transferred troops away from France to the Eastern Front. The Germans therefore had to turn south sooner than intended, allowing the French army to throw all available troops against their exposed flank on the Marne River (*map 2*). This "miracle" of the Marne was the first crucial turning point of the war.

The Schlieffen Plan was a political, as well as military, failure for the Germans. By invading Belgium, the Germans had ignored long-standing treaties guaranteeing that country's neutrality, and convinced the British of the need to enter the war. Germany thus found itself hemmed in on two sides by the Entente Powers, with only the support of Austria–Hungary, and later Turkey and Bulgaria.

THE WESTERN FRONT

Stalemate quickly ensued on the Western Front, as the Germans, British and French built long lines of trenches stretching from the Swiss border, through northern France to the English Channel. Long-range artillery pieces, accurate rifles and, most importantly, machine-guns gave the defenders a crucial advantage over the attacking forces. Industrialization and a well-developed railway system (*pages 170–71*) also meant that more ammunition and other vital supplies were available than ever before and that large armies could be transported from area to area as the situation dictated. For the next three years the Western Front was a brutal killing field (*bar chart*). The destructive nature of modern warfare was particularly demonstrated in 1916 when the Franco-German struggle over Verdun and the British offensives on the Somme led to the slaughter of 1.7 million men (*map 3*). The following year the French offensives against the retrenched German position on the Siegfried/Hindenburg line caused such heavy French casualties that there was mutiny among French troops.

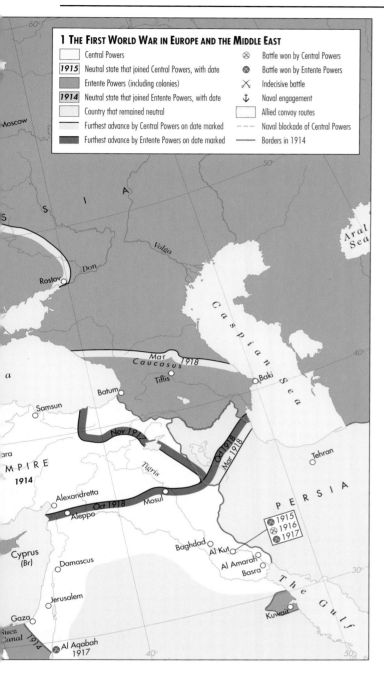

1 THE FIRST WORLD WAR IN EUROPE AND THE MIDDLE EAST

Central Powers	⊗ Battle won by Central Powers
1915 Neutral state that joined Central Powers, with date	⊗ Battle won by Entente Powers
Entente Powers (including colonies)	✕ Indecisive battle
1914 Neutral state that joined Entente Powers, with date	↓ Naval engagement
Country that remained neutral	Allied convoy routes
Furthest advance by Central Powers on date marked	Naval blockade of Central Powers
Furthest advance by Entente Powers on date marked	Borders in 1914

2 THE WESTERN FRONT

Central Power	
Entente Power	
Neutral country	
Furthest advance by Central Powers 1914	
Area of trench warfare	
Siegfried/Hindenburg Line March 1917	
Furthest advance by Central Powers 1918	
Armistice line 11 November 1918	
⊗ Battle won by Central Powers	
⊗ Battle won by Entente Powers	
✕ Indecisive battle	
Arras Name and year of battle	
[1914] Marne offensive	
The Schlieffen Plan	

FIGHTING AROUND THE WORLD

The picture on other fronts was more fluid, but just as bloody. On the Eastern Front a large Russian army was heavily defeated at the Battle of Tannenberg in August 1914 (*map 1*), and although the Russians saw limited success in 1915, ultimately their large, but poorly organized, forces were pushed back. The Germans made deep advances into European Russia in 1916, and by 1917 the morale of the Russian army and of its people was beginning to crack. The ensuing Russian Revolution and the triumph of the Bolsheviks led to Russia signing an armistice agreement with Germany at the end of 1917 (*pages 222–23*).

In the Middle East fighting also moved back and forth over a considerable area. Initially, the Entente Powers fared badly, with British, Australian, New Zealand and French soldiers being pinned down and forced to withdraw from the Gallipoli Peninsula during 1915 and early 1916, and a British Empire force from India surrendering to the Ottomans at Al Kut in April 1916. Soon, however, the tide began to turn. An Arab uprising against Ottoman rule in the summer of 1916 pushed the Ottomans out of much of the Arabian Peninsula, and in December 1917 the British captured Jerusalem. Despite these victories, the events in the Middle East had no decisive influence on the outcome of the First World War, which could really only be decided on the battlefields of Europe.

In Africa fighting broke out in all German colonies, but was most protracted in German East Africa where, in 1916, British, South African and Portuguese forces combined under General Smuts to counter the German forces.

In 1915 the Italian government, a signatory of the Triple Alliance (*pages 216–17*), joined the Entente Powers, following promises of Austrian territory. In the next two years hundreds of thousands of Italians were slaughtered before an Austrian–German force inflicted defeat on the Italian army at the Battle of Caporetto in October 1917.

THE ENTRY OF THE UNITED STATES

By 1917 the fortunes of the Entente Powers within Europe were at a low ebb, and a German victory seemed a distinct possibility. A disastrous German foreign and strategic policy was, however, to throw away their chance of victory.

It had been assumed by both sides before the war began that large fleets of battleships would engage in a decisive battle for naval supremacy. As it turned out, neither the Germans nor the British were willing to expose their surface fleets unduly, and only one large sea battle took place: the Battle of Jutland in 1916. It was a rather confused affair, with the Germans inflicting the greatest damage but being forced back to port. In the end it changed very little.

In preference to surface fighting, the Germans turned early in the war to submarine warfare as a means of cutting off vital imports to Britain. By sinking merchant ships without warning, however, the Germans inflamed US opinion. At first, after the sinking of the liner SS *Lusitania* in 1915, the Germans backed off, but in February 1917, in a dangerous gamble, they renewed their unrestricted submarine warfare around the British Isles. They were hoping to knock Britain out of the war before the United States could intervene – a rash gamble that failed when the Americans declared war on Germany on 6 April 1917.

THE FINAL PUSH

Following the signing of the Treaty of Brest-Litovsk with the Russians on 3 March 1918, the Germans were able to concentrate their resources on the Western Front. Between March and July 1918 the German army hurled itself against the French and British lines, making significant breakthroughs and advancing further than at any time since 1914. German resources were not, however, sufficient to finish the job. As US troops and supplies flooded into Europe, the German advance petered out, and the German army began to crumble in the face of a counteroffensive. Unable to increase their supply of men and weapons, the Germans realized that they had lost the war. They approached the Entente Powers for peace terms – and at 11.00 am on 11 November 1918 the fighting ceased.

▲ The original German "Schlieffen Plan" to encircle Paris from the northwest would almost certainly have resulted in a rapid victory. Instead, the German army was forced to retreat following the successful Marne offensive by the French, and the two sides dug themselves in for a war of attrition that was to last four years. In March 1917, anticipating the Nivelle offensive by the Entente Powers, the Germans withdrew to the Siegfried/Hindenburg Line. A German offensive in 1918 was initially successful, but their much smaller army was overstretched, while the Entente Powers were now reinforced by US troops. The Germans were driven back until, in November 1918, they were forced to request a truce.

▼ The two sides were unevenly matched in terms of the number of men they mobilized. The proportion of casualties (which includes those wounded, killed, reported missing in battle or dying from disease, and prisoners of war) was also uneven, with the Entente Powers suffering a casualty rate of 52 per cent against that of 67 per cent for the Central Powers.

TROOPS AND CASUALTIES

Entente Powers 42,188,810

Central Powers 22,850,00

22,089,709

15,404,477

☐ Total number of troops
☐ Number of casualties

OUTCOMES OF THE FIRST WORLD WAR 1918–29

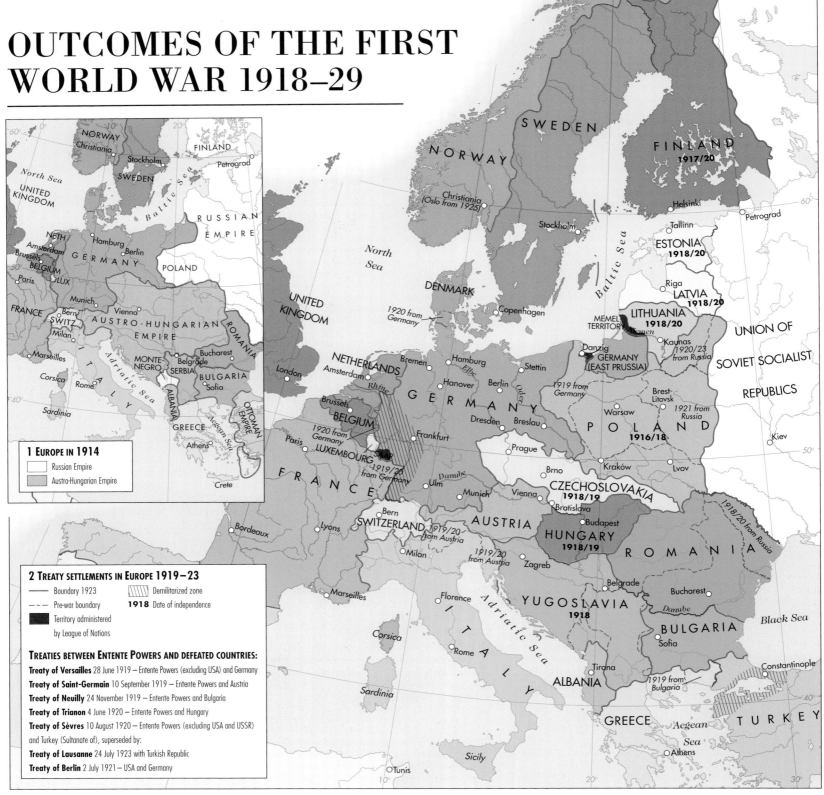

1 EUROPE IN 1914

- Russian Empire
- Austro-Hungarian Empire

2 TREATY SETTLEMENTS IN EUROPE 1919–23

—— Boundary 1923	⫽⫽ Demilitarized zone
- - - Pre-war boundary	**1918** Date of independence
■ Territory administered by League of Nations	

TREATIES BETWEEN ENTENTE POWERS AND DEFEATED COUNTRIES:

Treaty of Versailles 28 June 1919 – Entente Powers (excluding USA) and Germany

Treaty of Saint-Germain 10 September 1919 – Entente Powers and Austria

Treaty of Neuilly 24 November 1919 – Entente Powers and Bulgaria

Treaty of Trianon 4 June 1920 – Entente Powers and Hungary

Treaty of Sèvres 10 August 1920 – Entente Powers (excluding USA and USSR) and Turkey (Sultanate of), superseded by:

Treaty of Lausanne 24 July 1923 with Turkish Republic

Treaty of Berlin 2 July 1921 – USA and Germany

▲ As a result of the Paris Peace Conference of 1919 the Austro-Hungarian Empire was dismantled. Most of it was formed into small nation-states, including the new state of Czechoslovakia. In the south, however, several ethnically distinct regions were amalgamated with previously independent states to form Yugoslavia, under the domination of Serbia. Germany lost territory in the east to the recreated Poland, while a demilitarized area was established along Germany's border with France. The newly formed Union of Soviet Socialist Republics, threatened by anti-revolutionary forces, was in no position to resist moves to carve up territory on its western borders.

The First World War changed the map of Europe and the Middle East for ever. Centuries-old empires (*map 1*) were destroyed and new national states were created. The most important event in establishing the new Europe was the Paris Peace Conference (January–June 1919), which resulted in the Treaty of Versailles. The conference was called by the victorious Entente Powers after Germany had asked for an armistice in November 1918. Most of the countries involved in the war were represented in some way, but the decision-making power was held by the delegations of the "Big Three": the British, led by Prime Minister David Lloyd George, the French, led by Premier Georges Clemenceau, and the United States, led by President Woodrow Wilson.

The negotiations were delicate and often stormy. In a desire to destroy German power, the French called for the division and disarmament of Germany and for such huge reparations that the German economy would have been crippled for decades. The Americans, on the other hand, sought to establish a stable Europe and a new League of Nations to guarantee global security. They believed that the peace should be based on President Wilson's famous "Fourteen Points" and should be as magnanimous as possible. The British were stuck in the middle: they wished to

see a reduction in German power, but were wary of weakening the Germans so much that they would be completely under French domination or unable to trade. (Germany had been Britain's main European pre-war trading partner.)

THE TREATY OF VERSAILLES

The Treaty of Versailles, when signed in June 1919, represented a compromise between these different positions. The provinces of Alsace and Lorraine were given to France, while a large slice of eastern Germany was given to the re-established Polish state (*map 2*). The German city of Danzig, which was surrounded by countryside populated by Poles, was made a "Free City". Germany was also subjected to humiliating internal restrictions: the Rhineland, Germany's industrial heartland, was to be demilitarized (leaving it open to the threat of French invasion), while the German air force was ordered to disband, the army reduced to 100,000 men and the navy limited to a small number of warships. The treaty also stripped Germany of its imperial possessions in Africa and the Pacific, but since this empire had added little to German national strength, its loss did little to weaken it.

For all of its losses, Germany fared much better than its closest ally, Austria–Hungary. This multi-ethnic empire was broken up by the Treaty of Saint-Germain (1919) into a host of smaller national states (*map 2*): Poland, Czechoslovakia, Romania, Yugoslavia, Austria and Hungary. Italy, which had entered the war in 1915 because of the promise of booty from Austria–Hungary, was rewarded with a sizeable chunk of new territory.

RUSSIAN TERRITORIAL LOSSES

The greatest territorial losses of any country in Europe were those suffered by Russia, which had, under the tsar, been allied to France and Britain, but lost the war against Germany on the Eastern Front. After the Bolshevik revolution of 1917 and the ensuing Russian Civil War (*pages 222–23*), the Soviet regime found itself incapable of holding on to much of its empire in Europe. Finland and the Baltic states of Estonia, Latvia and Lithuania soon won their independence, while the province of Bessarabia was added to Romania (*map 2*). The greatest loss of Russian territory was to the newly created Poland, which gained further territory as a result of a brief war with Russia in 1921.

As a result of the Paris Peace Conference, nine new states (including Austria and Hungary) were constructed from various parts of Germany, Austria–Hungary and Russia. Whether or not this was a good thing for the European balance of power remained to be seen. Both Germany and the Soviet Union were eager to regain much of the territory they had given up against their will. In southeast Europe, meanwhile, a variety of different nationalities that had been held in check by Austria–Hungary were now exposed to a whole new set of tensions.

THE LEAGUE OF NATIONS

The Versailles treaty also called for the establishment of a League of Nations, an idea championed by President Wilson of the United States. Unfortunately, the American public was not persuaded of its necessity, and after a bitter debate in the Senate the United States decided to stay out of the League and refused to ratify the Treaty. The British and the French had been unable to master German might without American aid, and despite its losses Germany retained the potential to dominate Europe – demonstrated by the recovery in its industrial output during the 1920s.

THE DISMANTLING OF THE OTTOMAN EMPIRE

The First World War finally broke up the Ottoman Empire but still left much of the Middle East in limbo. Most of the region was assigned to British or French control (*map 3*)

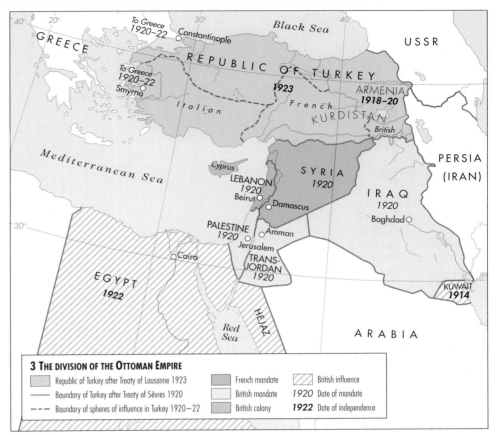

3 THE DIVISION OF THE OTTOMAN EMPIRE

Republic of Turkey after Treaty of Lausanne 1923	French mandate	British influence
Boundary of Turkey after Treaty of Sèvres 1920	British mandate	1920 Date of mandate
Boundary of spheres of influence in Turkey 1920–22	British colony	*1922* Date of independence

▲ The Treaty of Sèvres (1920) divided the defeated Ottoman Empire into British and French mandates in the Middle East, intended as temporary administrations leading eventually to independence. Kuwait, nominally independent, remained strongly influenced by Britain, as was Egypt. Large areas of Turkey were placed under European control, until Turkish resistance forced the withdrawal of all foreigners and led to the founding of the Republic of Turkey in 1923.

4 POST-WAR ALLIANCES

	French alliances
	German alliance
	Polish alliance
	Italian alliance
	Little Entente 1920/21

▲ In the 1920s France, anxious to isolate Germany within Europe, created a series of alliances with some of the newly created eastern European states. The most significant alliance of the 1920s was the "Little Entente", intended to provide mutual protection to the boundaries of its signatories, and a united foreign policy.

under League of Nations mandates. Even areas that gained nominal independence – Egypt and the new Arab kingdoms – were heavily reliant on Britain for their defence and development. The one state that grew in strength during the immediate post-war period was, surprisingly, Turkey. Shorn of its imperial burdens, the Turks, led by Atatürk, countered an invasion attempt by Greece in 1922, brutally quelled Armenian nationalists sympathetic to the Greeks, drove out the British and French and established the Turkish Republic in 1923 (*pages 178–79*).

THE LONG-TERM OUTCOMES OF THE PEACE

The Versailles treaty has been harshly criticized and, indeed, has been seen as one of the fundamental causes of the Second World War. In 1923, in response to Germany's inability to pay war reparations, the French moved their army into the Rhineland. The German mark collapsed in value and by 1924 Germany was gripped in a cycle of hyperinflation that saw some people taking home their pay packets in wheelbarrows. By the late 1920s, however, Europe seemed to be on the way to establishing a new equilibrium; the economies of all the major European countries had recovered and were experiencing strong growth.

The French saw the new eastern European states as a potential future bulwark against Germany and were eager to knit them into a defensive alliance system (*map 4*). For a while the strategy seemed quite successful, as eastern Europe developed a new stability. Czechoslovakia evolved into a democracy, Poland became a nation-state capable of defeating the Soviet Union and establishing friendly relations with its neighbours, while Yugoslavia seemed able to accommodate a multi-ethnic population. Perhaps if the prosperity of the 1920s had continued for longer, eastern Europe might have become stable enough to survive German and Russian attempts to take back their lost lands.

The Great Depression that started in 1929, and affected the economy of every country in Europe to some extent, brought to an end Europe's brief period of co-operation and recovery. This financial crisis served as the catalyst for the rise to power of the German Nazi party (*pages 230–31*), which swept aside the settlement laid out in the Versailles treaty and ended attempts to find peaceful solutions to Europe's complex problems.

▲ President Woodrow Wilson of the United States arrived at the Paris Peace Conference advocating a liberal approach to world affairs, including an end to colonial rule and the setting up of a League of Nations to maintain world peace. While the other victorious powers forced him to compromise on some of his aims, the League of Nations was included in the Treaty of Versailles. To Wilson's disappointment, however, the United States Senate rejected American involvement in such an organization and refused to ratify the treaty.

◗ THE FIRST WORLD WAR 1914–18 *pages 218–19* ◗ THE GREAT DEPRESSION 1929–33 *pages 228–29*

THE RUSSIAN REVOLUTION 1917–39

▲ In the period immediately after the Bolshevik Revolution of 1917 Lenin (*left*) and Stalin (*right*) worked closely together, and in 1922 Stalin was appointed Secretary-General of the Communist Party, while Lenin remained head of the government. Shortly before his death, however, Lenin made it clear that he did not regard Stalin as a suitable successor – information that Stalin ignored and repressed in his drive to become leader of the Soviet Union.

The Russian Revolution – one of the formative events of the 20th century – was precipitated by pressures arising from the hardships experienced during the First World War. A popular uprising in March 1917 led to the abdication of Tsar Nicholas II and the creation of a liberal Provisional Government, which was soon forced to share power with the socialist Petrograd Soviet of Workers' and Soldiers' Deputies. As the revolution spread, soviets sprang up in many cities, peasants seized land from the gentry and soldiers deserted. A dual system of government developed, with the soviets largely controlling those leaders who took their authority from the Provisional Government.

During the subsequent months the ideological rift between the two bodies widened, with the Provisional Government delaying the setting up of a Constituent Assembly (which was to decide on major economic and political policies), concentrating instead on a continued war effort. The Petrograd Soviet, meanwhile, came increasingly under the influence of the Bolshevik movement, led by Lenin, which secured popular urban support with its slogans "peace, bread and land" and "all power to the soviets". In November 1917 the Bolsheviks carried out a successful coup, seizing control of the Winter Palace, seat of the Provisional Government. Lenin then set about establishing a dictatorship of the proletariat and a one-party system.

CIVIL WAR

The new Bolshevik government arranged an armistice with the Central Powers in December 1917, formalized in the Treaty of Brest-Litovsk in March 1918. Under the terms of the treaty Russia relinquished control of its western territories. Anger at these losses and at the closure of the recently elected Constituent Assembly fuelled opposition to the retitled Communist (Bolshevik) Party. Civil war broke out, during which anti-communist "White" armies and foreign interventionists opposed the Red Army, led by Leon Trotsky (*map 1*). The Red Army was initially pushed back, but its military superiority over the comparatively disunited White armies enabled it to regain control of Central Asia, the Caucasus and Ukraine, although territory was lost in the war with Poland in 1920. This war did not spread the revolution into Europe, as Lenin had hoped it would. Outside Russia proletarian support for communism was limited (*map 2*) and when the Soviet Union was founded in 1922 it was confined to the territories of the old empire.

In order to back up the efforts of the Red Army, Lenin took rapid steps to impose nationalization and centralization in a process known as "war communism". However, revolts by peasants in the spring of 1921 forced him to introduce the New Economic Policy (NEP), based on concessions to the peasantry and a semi-market economy. Although the

▶ After sweeping away the Provisional Government in November 1917 the Bolsheviks faced widespread opposition both within and outside Russia. The Treaty of Brest-Litovsk in March 1918 ended the war with Germany but led to a civil war in which the Entente Powers initially supported the "Whites" (anti-Bolsheviks) against the "Reds" (the Bolsheviks). Admiral Kolchak formed an Eastern Front in Siberia and in 1919 advanced beyond the Volga. In the south, resistance was led by Denikin but he was brought to a halt short of Orel. In the north, Yudenich led his troops to the suburbs of Petrograd, but was then driven back. Wrangel, taking over what was left of Denikin's forces, defended the area around Sevastopol for some time but was finally forced to withdraw in November 1920. Meanwhile, the Poles were attempting to gain as much as they could of Lithuania, White Russia (Byelorussia) and Ukraine. They got as far as Kiev but then had to withdraw as the Red Army advanced in turn towards Warsaw. When the Poles regained the initiative Lenin decided to sue for peace and, under the Treaty of Riga in October 1920, 10 million Ukrainians and Russians were assigned to Polish rule. By the end of the year military operations were over and the communist (Bolshevik) government was in control of what was left of Russia.

1 REVOLUTION AND CIVIL WAR IN RUSSIA

- – – – Boundary of the Russian Empire 1914
- ——— German occupation line March 1918
- ✿ Centre of great Bolshevik activity
- ⟶ White Russian and interventionist attacks

Interventionists:
- **C** Canadian
- **F** French
- **G** Greek
- **B** British
- **US** American

- ——— Boundary of area controlled by Bolsheviks August 1918
- ▨ Area controlled by Bolsheviks October 1919
- ——— Polish advance into Russia May 1920
- ——— Russian advance into Poland August 1920
- – – – USSR–Polish boundary established October 1920 by Treaty of Riga
- ——— Other international boundaries 1922
- ▢ Areas lost to Russia 1914–21
- ▨ Soviet Union 1922

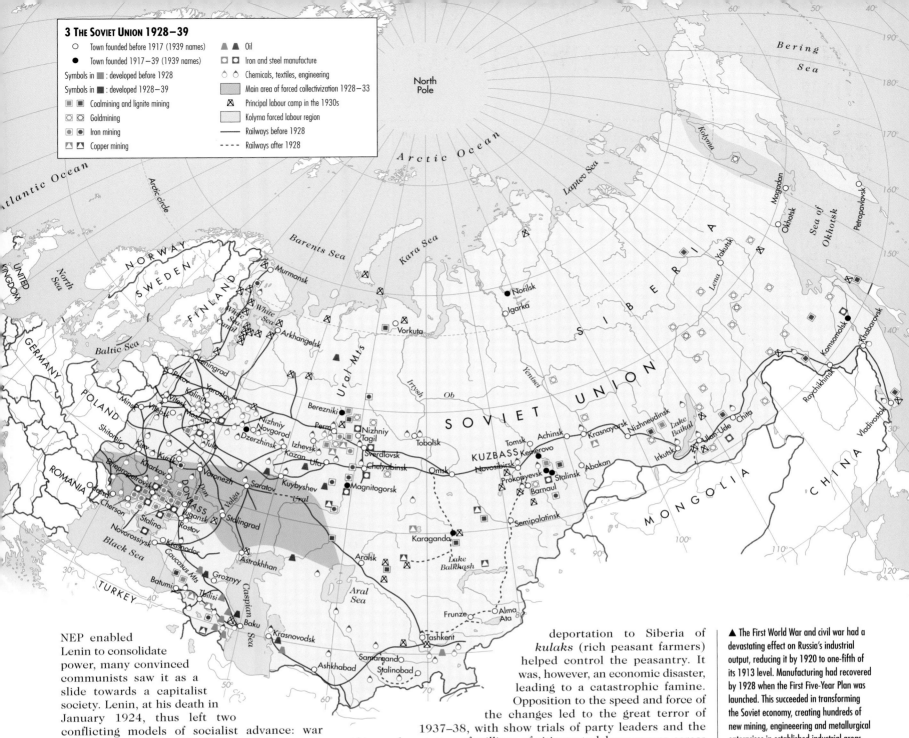

3 The Soviet Union 1928–39

○ Town founded before 1917 (1939 names)
● Town founded 1917–39 (1939 names)
Symbols in ■ : developed before 1928
Symbols in ■ : developed 1928–39
▣ ▣ Coalmining and lignite mining
◎ ◎ Goldmining
◉ ◉ Iron mining
⬔ ⬔ Copper mining
▲ ▲ Oil
▢ ▢ Iron and steel manufacture
◇ ◇ Chemicals, textiles, engineering
▨ Main area of forced collectivization 1928–33
⊠ Principal labour camp in the 1930s
▨ Kolyma forced labour region
—— Railways before 1928
- - - Railways after 1928

NEP enabled Lenin to consolidate power, many convinced communists saw it as a slide towards a capitalist society. Lenin, at his death in January 1924, thus left two conflicting models of socialist advance: war communism and the NEP. The struggle for power among his closest followers was to be fought out partly on the issue of which policy should be taken as the true Leninist line.

STALIN'S RISE TO POWER

The struggle was won by Stalin, who outmanoeuvred rivals such as Trotsky and Bukharin. Faced with foreign hostility, and convinced that the revolution should achieve an industrial, proletarian society, Stalin launched his drive to catch up with the West in ten years with a return to the centralization and utopianism of the civil war years. The First Five Year Plan was adopted in 1928, its aims being to develop heavy industry, which had been devastated during the civil war, and collectivize agriculture. Industrial advance was indeed impressive, although at the cost of enormous waste, inefficiency and suffering, as wildly over-optimistic targets for output were set. The population of the big cities nearly doubled between 1928 and 1933, and the urban infrastructure could not keep pace. Targets concentrated on heavy industry, and although they were not met, the economy was transformed. In the Urals, the Donbass and Kuzbass coalfields, the Volga area and Siberia, huge new metallurgical enterprises were developed (*map 3*). Magnitogorsk, the Turksib railway (between Tashkent and Semipalatinsk), the Dneprostroi hydro-electric complex and the White Sea Canal all date from this time. They were also all built partially with prison camp labour, for the First Five Year Plan saw a vast expansion of the concentration camps of the civil war. The secret police were deeply involved in the economy. The forcible establishment of collective farms, with the

deportation to Siberia of *kulaks* (rich peasant farmers) helped control the peasantry. It was, however, an economic disaster, leading to a catastrophic famine. Opposition to the speed and force of the changes led to the great terror of 1937–38, with show trials of party leaders and the deportation of millions of citizens to labour camps across the country. The scale of the famine, the horrors of collectivization, and the extent of the terror were not revealed to the Soviet public until the late 1980s. In 1939 the Stalin cult of personality was at its height and, to many sympathisers in Europe, this was indeed a brave new world.

▲ The First World War and civil war had a devastating effect on Russia's industrial output, reducing it by 1920 to one-fifth of its 1913 level. Manufacturing had recovered by 1928 when the First Five-Year Plan was launched. This succeeded in transforming the Soviet economy, creating hundreds of new mining, engineering and metallurgical enterprises in established industrial areas, and new factories in the empty lands of the non-Russian republics.

◄ The Bolsheviks assumed that their revolution would spark off revolutions across Europe, and in 1918–19 it looked for a while as if this would happen. A soviet republic in Hungary, led by Bela Kun, survived five months in 1919, and others in Bavaria and Slovakia lasted four and three weeks respectively. The Spartakist uprising under Rosa Luxemburg in Berlin in January 1919 was crushed by the new Weimar Republic and further insurrections in German towns were unsuccessful. Strikes spread across Europe from northern Italy to the Baltic, but the European revolution the Bolsheviks hoped for failed to materialize.

2 REVOLUTIONARY ACTIVITY IN EUROPE 1919–23

🏛 Centre of revolutionary activity
—— Boundary 1923

◄ RUSSIAN TERRITORIAL AND ECONOMIC EXPANSION 1795–1914 *pages 180–81*   ► THE SOVIET UNION AND EASTERN EUROPE 1945–89 *pages 236–37*

THE REPUBLIC OF CHINA 1911–49

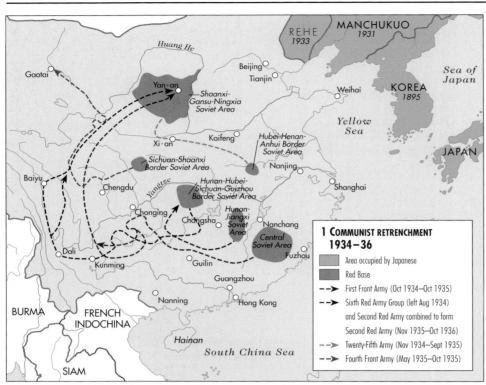

▲ From 1934 to1936 the Communists organized a series of retrenchments in the face of Kuomintang attacks. From their southern bases they embarked on lengthy journeys to the north, by way of the mountainous west. The most famous – known as "the Long March" – was that undertaken by the First Front Army, led by Mao Zedong. The casualty and drop-out rate on the marches was high: of 300,000 soldiers who set out, only 30,000 arrived in Yan-an. The Fourth Army (led by a political rival of Mao) was denied access to Yan-an and sent away to remote Gaotai, where it suffered heavy losses after confronting some well-equipped Kuomintang troops. Meanwhile, the Japanese, with the help of their Manchu collaborators, were firmly in control of Manchuria (which they renamed Manchukuo) and were poised to launch a full-scale invasion and occupation of the rest of China in 1937.

The Revolution of 1911, which had seen the overthrow of the last Manchu Qing emperor and the establishment of the first Republic, failed to solve any of China's economic or social problems (*pages 198–99*). The most important and urgent goals for the new government were the unification and defence of the country, but they were not easily achieved. The presidential term of the revolutionary leader Sun Yat-sen lasted for barely six weeks after his inauguration in January 1912, and in December 1915 President Yuan Shikai attempted to restore the monarchy by crowning himself emperor. The attempt was a failure, as was that made by General Zhang Xun and the dethroned Qing Emperor Xuantong in 1917. Both attempts, however, provided opportunities for local warlords to re-establish their power at the expense of central government. Over the next 30 years, although a fragile equilibrium existed between the various warlords and other interest groups, the Chinese Republic was in virtual anarchy.

CIVIL WAR

The first North–South War broke out in 1917 and resulted in a chain reaction that led to full-scale civil war and the establishment of a number of governing regimes across the country. To challenge the authority of the northern warlords, Sun Yat-sen formed his own southern governments in Guangzhou in 1917, 1921 and 1923. He also set about creating a united Nationalist Party (Kuomintang) and forging links with the still very small Communist Party, which was growing under the control of the Comintern (an international communist organization founded in Moscow in 1919). In 1924 Sun Yat-sen was invited to Beijing to discuss the possible unification of China, but he died there in March 1925 without concluding an agreement, and the second North–South War began the following year.

The Kuomintang was nominally unified at the end of 1928 under the leadership of Chiang Kai-shek, and gradually gained control of strategic regions. It was not, however, until the end of 1930 that real unification of the party was achieved through the military defeat by Chiang of a rival faction. For Chiang and the Kuomintang the next main task was to deal with the Communists, who now had an effective command structure and were armed. They were also entrenched in their main "Red Bases" in rural areas in the south and had considerable influence over the urban population (*map 1*).

Despite the fact that both the Kuomintang and Communists had a nationalist goal, they were more often

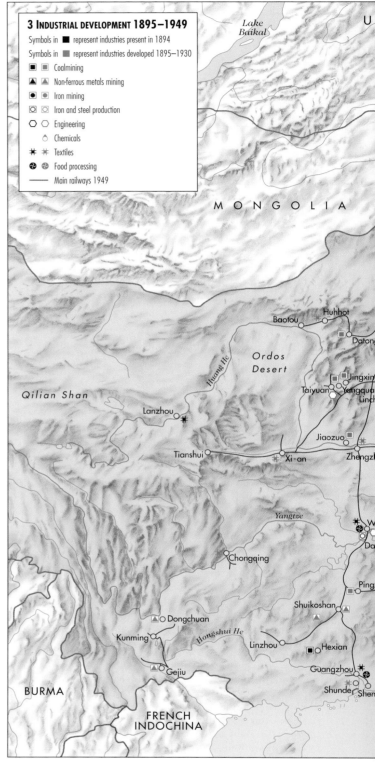

enemies than allies, competing for the same power. Most early Communists were also radical nationalists, and many had been heavily involved in the activities of the Kuomintang under Sun Yat-sen, making them doubly threatening to Chiang's regime. Consequently, immediately after the unification of the Kuomintang, Chiang launched five military campaigns to encircle and suppress the Communists in a rural area of Jiangxi province, where the communist "Central Soviet Area" was located. In October 1934 he finally succeeded in overpowering the Communists, forcing them to abandon their Jiangxi base and, under the leadership of Mao Zedong, embark on the gruelling Long March to the north. During 1935 Chiang's army was equally successful in expelling units of the Red Army from other Red Bases in the central region of the country, so that by 1936 the Communists who had survived the journey were confined to an area in the province of Shaanxi around the city of Yan-an.

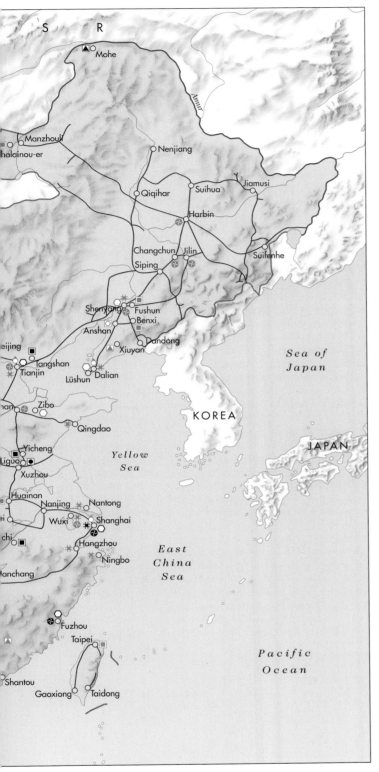

The Communists, from their stronghold in Yan-an, turned their attentions to fighting the Japanese. They proved themselves a dynamic and efficient political and military force, and took the opportunity to play the nationalist card and thus rebuild their popularity. By contrast, Chiang's concentration on suppressing his domestic rivals was by now out of tune with the wishes of the general populace – so much so that in December 1936 two of Chiang's top military commanders mutinied in order to shift Chiang's attention to fighting the Japanese. This became known as the "Xi-an Incident", and resulted in the first example of co-operation between the Kuomintang and Communists since the death of Sun Yat-sen. In January 1941, however, the Kuomintang troops ambushed and annihilated the main force of the Communist-controlled New Fourth Army, thus demonstrating just how fragile this co-operation was.

The war against the Japanese (1937–45) created opportunities for communist propaganda, recruitment and military training which proved to be invaluable when the civil war between the Kuomintang and Communists was resumed immediately after the Japanese surrender. This time the Communists were unbeatable: in their three main military campaigns in the second half of 1948, the Kuomintang were finally overpowered (*map 2*). The Communists gained control of the mainland, the Kuomintang fled to Taiwan, and the People's Republic was established in October 1949. Putting the unification of China before the defence of China had cost the Kuomintang dearly.

ECONOMIC EXPANSION

During the period between the 1911 Revolution and the birth of the People's Republic of China in 1949, the Chinese economy struggled to survive the civil wars, the Japanese occupation of large areas of the country and the mismanagement of the Kuomintang. Some indigenous industrial growth did occur along the coast and main waterways (*map 3*). This was largely due to the impact of the First World War (1914–18) and the Great Depression (1929–33), when the industrial powers relaxed their grip on the Chinese market, creating opportunities for local businesses to become established. Furthermore, while the Western gold standard collapsed during the Depression (*pages 228–29*) – resulting in severe financial crises in the West – China, which had its own silver standard, remained largely unaffected.

▲ Sun Yat-sen trained as a doctor in the early 1890s, but he subsequently turned his attention to revolutionary activity and was exiled between 1896 and 1911 before becoming the first President of the Republic of China in 1912.

▼ In 1945, at the end of the Second World War, the Communists (backed by Soviet troops) were the first to move into areas previously colonized by the Japanese. They quickly established a strong foothold in the northeast (both militarily and in terms of popular support) from which to launch their offensive against the Kuomintang, who had spent much of the previous eight years in the southwest. Fierce fighting ensued for three years, with only a temporary truce in 1946. Despite US backing, Chiang Kai-shek and the Kuomintang forces were eventually forced to retreat to Taiwan.

▲ Despite the political and economic turmoil of the first half of the 20th century, China still developed a railway network. Together with the country's system of navigable rivers, the railways provided transport for the manufactured goods and metal ores produced by the Chinese businesses that thrived as foreign firms, hit by the Great Depression of 1929–33, failed or withdrew from China.

JAPANESE AGGRESSION

Chiang's strategy was similar to that of any new ruler: to eliminate political and military competitors and reunite the country. During the 1930s, however, his aims were largely frustrated by domestic and international conditions. In particular, as Japan developed its imperialist policy towards mainland East Asia, successive Japanese governments turned their attention on a weak and fragmented China. From 1894 to 1944 they launched a series of invasions: on Beijing in 1900, Shandong in 1914, Manchuria in 1931 and Rehe in 1933, followed by a full-scale assault on east and southeast China from 1937 to 1944 (*pages 234–35*).

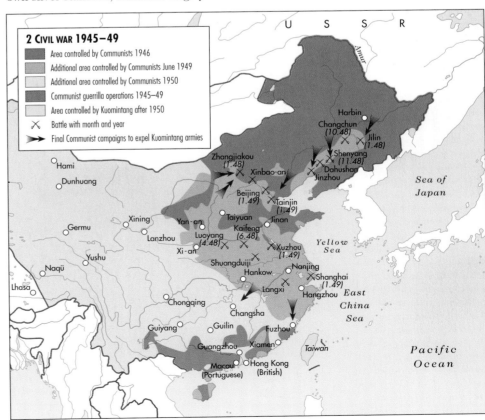

2 CIVIL WAR 1945–49

- Area controlled by Communists 1946
- Additional area controlled by Communists June 1949
- Additional area controlled by Communists 1950
- Communist guerrilla operations 1945–49
- Area controlled by Kuomintang after 1950
- ✕ Battle with month and year
- ➤ Final Communist campaigns to expel Kuomintang armies

LATIN AMERICA
1914–45

The first half of the 20th century saw many major changes in the economic and social structure of the countries of Latin America. Export-led growth based on the production of primary products (mostly minerals or agricultural goods), which had resulted in appreciable economic expansion before 1914, was shown to be severely flawed. At the same time the oligarchies whose socio-political dominance had been well-nigh absolute for most of the 19th century found their control of the state challenged by an emerging middle class. Meanwhile the majority of the population, who had previously been excluded from participation in the state, began to feature in both cultural and political debates. Finally, the dominant imperial power of the 19th century – Britain – was displaced by the United States.

VULNERABLE ECONOMIES
The problems underlying Latin America's dependence on the production of raw materials were initially felt as a result of the dislocation of world trade during the First World War (1914–18). Latin America, which at this stage relied largely on foreign banks for supplies of credit and on foreign shipping for transporting its goods, found itself isolated from international finance and trade. Production fell, imports (including food) were in short supply, and there was a high level of mass unrest. The disadvantages of export-led growth became increasingly clear: Latin American economies, especially the smaller ones, found themselves over-reliant on one or two products, the prices of which were vulnerable to fluctuations in the weather, the emergence of new centres of production or substitute products and raw materials.

Economic growth tended to follow a "boom–bust" cycle, which made it difficult for countries to plan ahead or allocate resources rationally. The Wall Street Crash of 1929 and the ensuing Great Depression (*pages 228-29*) led to the collapse of the world market on which Latin America had relied for its exports. In the 1930s Latin American countries could do little more than try to defend themselves against the effects of the Depression. However, a consensus began to develop – at least in the more advanced economies (Argentina, Brazil, Chile and Mexico) where a limited industrial base oriented towards the internal market had already evolved – that Latin America needed to adopt an economic strategy of urgent industrialization.

POLITICAL CHANGE
The early 20th century saw the first active participation by the Latin American middle classes in political life. These disparate groupings of professionals, small business owners, bureaucrats and industrialists lacked the economic power their counterparts in 19th-century Europe enjoyed as a result of the leading role they played in industrialization. Even so, governments that reflected the expanding political role of the middle classes came to power in most of the leading countries during this period, for example in Argentina (1916), Chile (1920), Peru (1919) and Mexico (1920). Their challenge to oligarchic power was incomplete and compromised – except in Mexico, which in 1910–20 experienced the world's first major social revolution of the 20th century. The outcome was to consolidate the political and economic dominance of a bourgeoisie committed to capitalist modernization. The revolution destroyed the political position of the oligarchy, and their economic strength was eroded over the next two decades by means of a programme of agrarian reform that redistributed large landed estates.

In all the major Latin American countries during the early decades of the 20th century, the issue of how to incorporate the majority of the population into national life began to be debated. Immigration and internal migration meant that the poor were becoming increasingly visible in the rapidly expanding towns and cities (*map 1*). Intellectuals and politicians, in particular those from the middle classes, became increasingly aware of the political importance of the poorer sections of society. National identities based on "the people" were proposed: images of American Indians and *gauchos* (Argentine cowboys) were celebrated as national archetypes. This did not necessarily mean that the poor themselves were treated any better, although measures were taken in Mexico to improve the lot of the Indians.

INCREASING US INFLUENCE
The Spanish-American War of 1898, which had resulted in the ejection of Spain from Latin America by the United States, signalled the rise of the United States as an imperial power in the region (*map 2*). Although Washington was reluctant to adopt a 19th-century style of colonialism (only Puerto Rico was governed as a colony), the United States consolidated its dominance in both trade and investment

3 LATIN AMERICA IN THE FIRST WORLD WAR
- Declared war on Germany
- Broke diplomatic relations with Germany
- Remained neutral
- European colonies

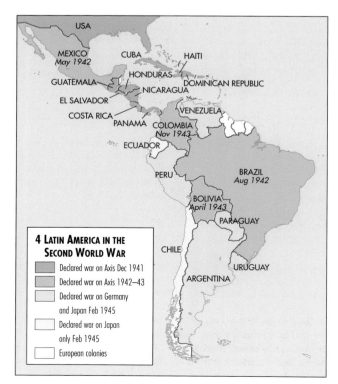

4 LATIN AMERICA IN THE SECOND WORLD WAR
- Declared war on Axis Dec 1941
- Declared war on Axis 1942–43
- Declared war on Germany and Japan Feb 1945
- Declared war on Japan only Feb 1945
- European colonies

in raw material production, especially minerals. By the end of the 1920s it had effectively displaced the European powers from Mexico, Central America and the Caribbean.

During the Second World War the US administration ensured the production of raw materials necessary to the Allied war effort by means of Lend-Lease aid agreements. Consequently, by 1945 the United States had also secured hegemony in South America. Increasing US dominance in Latin America during this period is reflected in the fact that, whereas many Latin American states had remained neutral in the First World War (*map 3*), most followed the United States into the Second World War after the Japanese bombing of Pearl Harbor in December 1941 (*map 4*). By this stage it was apparent to the governments of Latin America that only the United States could launch an effective defence of the western hemisphere.

THE RISE OF THE MILITARY

One final change that occurred during this period, which was to have a major effect on Latin American politics after the Second World War, was the rise of the military. With the consolidation of central state control in most countries during the late 19th century, the armed forces had begun a process of professionalization, mostly with the help of European advisers, which by the 1920s had given them a strong sense of corporate identity. Military coups took place in Argentina, Brazil and Peru in 1930. At this stage the military was content to intervene only briefly in the political process, but it was increasingly acquiring the conviction – subsequently to prove so detrimental to the maintenance of democracy in Latin America – that it alone was the institution which could best serve the national interest.

▼ At the beginning of the 20th century the United States professed itself reluctant to become a colonial power along the lines of some European countries in Africa and Asia. However, it was anxious to protect its own economic interests in the Caribbean and Central America. The "Platt Amendment", a clause in the Cuban Constitution of 1901 and in the treaty of 1903 between the United States and Cuba, entitled the United States to intervene in Cuban internal affairs – a right it exercised on more than one occasion. Elsewhere, it moved swiftly to repress regimes it felt might jeopardize favourable trading arrangements.

2 US INFLUENCE IN MEXICO, CENTRAL AMERICA AND THE CARIBBEAN

- British possessions
- French possessions
- US possessions
- Dutch possessions
- Danish possessions
- ★ Direct military action by USA
- ☆ Economic and political intervention by USA

1 US military occupations 1898–1902 and 1906–9
2 US military interventions 1907, 1911 and 1924
3 US military presence 1912–25 and 1927–33
4 US military occupation of Veracruz 1914
5 Panama Canal opened in zone under US sovereignty 1914
6 US military intervention and occupation 1915–34
7 General Pershing invades from the north 1916–17
8 US military intervention and occupation 1916–24
9 Bryan-Chamorro Treaty secures US option on alternative canal route 1916
10 US Marines involved in quashing strike action 1917–22
11 St Thomas, St John and St Croix bought by US from Denmark in 1917 and renamed Virgin Islands of the United States
12 USA protects oil concessions 1918–20 by non-recognition of Tinoco regime
13 General Crowder sent to supervise Cuban political process 1919–22

1 INCREASING URBAN POPULATION 1920–50

Urban population as percentage of total in 1950:
- below 30%
- 30%–50%
- 50%–70%
- over 70%

Approximate population of largest city (in thousands):
- c. 1920
- early 1930s
- 1950

Cities with population in 1950 of:
- ○ ○ 250,000–500,000 (and capital cities below 250,000)
- ◎ ○ 500,000–1,000,000
- ▢ ▢ over 1,000,000

Green symbols relate to the bar charts

▲ During the period 1920–50 the capital cities of all Latin American countries increased in size by between 100 and 300 per cent. Rapid urbanization was caused in part by the large number of European immigrants, but also by the movement of people from rural areas into the cities. By 1950 over 50 per cent of the populations of countries such as Uruguay, Argentina, Chile and Venezuela lived in urban areas.

◀ LATIN AMERICA AND THE CARIBBEAN POST-INDEPENDENCE 1830–1914 *pages 192–93*   ▶ LATIN AMERICA SINCE 1945 *pages 258–59*

THE GREAT DEPRESSION 1929–33

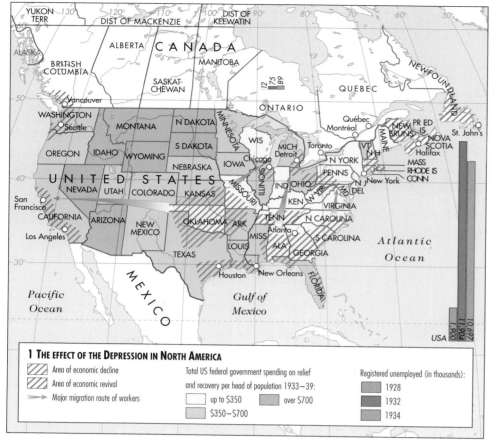

1 THE EFFECT OF THE DEPRESSION IN NORTH AMERICA

- ▨ Area of economic decline
- ▨ Area of economic revival
- ➤ Major migration route of workers

Total US federal government spending on relief and recovery per head of population 1933–39:
- ☐ up to $350
- ☐ $350–$700
- ☐ over $700

Registered unemployed (in thousands):
- 1928
- 1932
- 1934

USA 10,697 / 11,904 / 1,900

▲ The Depression caused industrial production in the United States and Canada to decline by over 30 per cent, leading to massive unemployment, particularly in the United States. People migrated in search of work, some of which was a direct result of US-government spending under the New Deal launched in 1933. It is noticeable, however, that the highest expenditure per capita was not necessarily in those areas most depressed, such as the Deep South, but in areas where the Democrat government was most anxious to win political support at the next election.

▼ Every country in Europe experienced a drop in industrial production during the Depression, with the northeast being worst hit. In Germany dissatisfaction with the high unemployment rate provided a platform on which Hitler and the Nazi Party came to power in 1933.

The Great Depression of 1929–33 was the most severe economic crisis of modern times. Millions of people lost their jobs, and many farmers and businesses were bankrupted. Industrialized nations and those supplying primary products (food and raw materials) were all affected in one way or another. In Germany and the United States industrial output fell by about 50 per cent, and between 25 and 33 per cent of the industrial labour force was unemployed.

The Depression was eventually to cause a complete turn-around in economic theory and government policy. In the 1920s governments and business people largely believed, as they had since the 19th century, that prosperity resulted from the least possible government intervention in the domestic economy, from open international economic relations with little trade discrimination, and from currencies that were fixed in value and readily convertible. Few people would continue to believe this in the 1930s.

THE MAIN AREAS OF DEPRESSION

The US economy had experienced rapid economic growth and financial excess in the late 1920s, and initially the economic downturn was seen as simply part of the boom–bust–boom cycle. Unexpectedly, however, output continued to fall for three and a half years, by which time half of the population was in desperate circumstances (*map 1*). It also became clear that there had been serious over-production in agriculture, leading to falling prices and a rising debt among farmers. At the same time there was a major banking crisis, including the "Wall Street Crash" in October 1929. The situation was aggravated by serious policy mistakes of the Federal Reserve Board, which led to a fall in money supply and further contraction of the economy.

The economic situation in Germany (*map 2*) was made worse by the enormous debt with which the country had been burdened following the First World War. It had been forced to borrow heavily in order to pay "reparations" to the victorious European powers, as demanded by the Treaty of Versailles (1919) (*pages 220–21*), and also to pay for industrial reconstruction. When the American economy fell into depression, US banks recalled their loans, causing the German banking system to collapse.

Countries that were dependent on the export of primary products, such as those in Latin America, were already suffering a depression in the late 1920s. More efficient farming methods and technological changes meant that the supply of agricultural products was rising faster than demand, and prices were falling as a consequence. Initially, the governments of the producer countries stockpiled their products, but this depended on loans from the USA and Europe. When these were recalled, the stockpiles were released onto the market, causing prices to collapse and the income of the primary-producing countries to fall drastically (*map 3*).

NEW INTERVENTIONIST POLICIES

The Depression spread rapidly around the world because the responses made by governments were flawed. When faced with falling export earnings they overreacted and severely increased tariffs on imports, thus further reducing trade. Moreover, since deflation was the only policy supported by

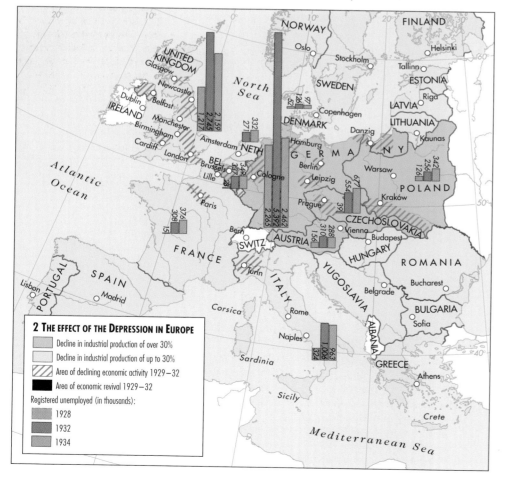

2 THE EFFECT OF THE DEPRESSION IN EUROPE
- ☐ Decline in industrial production of over 30%
- ▨ Decline in industrial production of up to 30%
- ■ Area of declining economic activity 1929–32
- ☐ Area of economic revival 1929–32

Registered unemployed (in thousands):
- 1928
- 1932
- 1934

PERCENTAGE OF INDUSTRIAL WORKERS UNEMPLOYED IN 1933

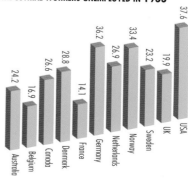

Country	%
Australia	24.2
Belgium	16.9
Canada	26.6
Denmark	28.8
France	14.1
Germany	36.2
Netherlands	26.9
Norway	33.4
Sweden	23.2
UK	19.9
USA	37.6

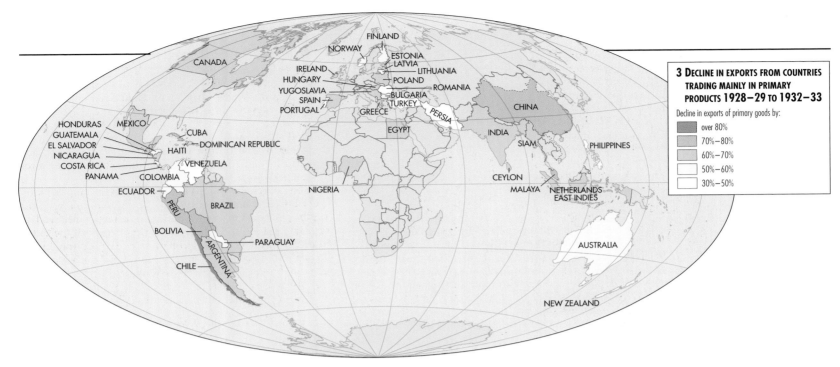

3 DECLINE IN EXPORTS FROM COUNTRIES TRADING MAINLY IN PRIMARY PRODUCTS 1928–29 TO 1932–33

Decline in exports of primary goods by:

over 80%
70%–80%
60%–70%
50%–60%
30%–50%

economic theory at the time, the initial response of every government was to cut their spending. As a result consumer demand fell even further.

Deflationary policies were critically linked to exchange rates. Under the Gold Standard, which linked currencies to the value of gold, governments were committed to maintaining fixed exchange rates. However, during the Depression they were forced to keep interest rates high to persuade banks to buy and hold their currency. Since prices were falling, interest-rate repayments rose in real terms, making it too expensive for both businesses and individuals to borrow.

The First World War had led to such political mistrust that international action to halt the Depression was impossible to achieve. In 1931 banks in the United States started to withdraw funds from Europe, leading to the selling of European currencies and the collapse of many European banks. At this point governments either introduced exchange control (as in Germany) or devalued the currency (as in Britain) to stop further runs. As a consequence of this action the gold standard collapsed (*map 4*).

POLITICAL IMPLICATIONS

The Depression had profound political implications. In countries such as Germany and Japan, reaction to the Depression brought about the rise to power of militarist governments who adopted the aggressive foreign policies that led to the Second World War. In countries such as the United States and Britain, government intervention ultimately resulted in the creation of welfare systems and the managed economies of the period following the Second World War.

In the United States Roosevelt became President in 1933 and promised a "New Deal" under which the government would intervene to reduce unemployment by work-creation schemes such as street cleaning and the painting of post offices. Both agriculture and industry were supported by policies (which turned out to be mistaken) to restrict output and increase prices. The most durable legacy of the New Deal was the great public works projects such as the Hoover Dam and the introduction by the Tennessee Valley Authority of flood control, electric power, fertilizer, and even education to a depressed agricultural region in the south.

The New Deal was not, in the main, an early example of economic management, and it did not lead to rapid recovery. Income per capita was no higher in 1939 than in 1929, although the government's welfare and public works policies did benefit many of the most needy people. The big growth in the US economy was, in fact, due to rearmament.

In Germany Hitler adopted policies that were more interventionist, developing a massive work-creation scheme that had largely eradicated unemployment by 1936. In the same year rearmament, paid for by government borrowing, started in earnest. In order to keep down inflation, consumption was restricted by rationing and trade controls. By 1939 the Germans' Gross National Product was 50 per cent higher than in 1929 – an increase due mainly to the manufacture of armaments and machinery.

THE COLLAPSE OF WORLD TRADE

The German case is an extreme example of what happened virtually everywhere in the 1930s. The international economy broke up into trading blocs determined by political allegiances and the currency in which they traded. Trade between the blocs was limited, with world trade in 1939 still below its 1929 level. Although the global economy did eventually recover from the Depression, it was at considerable cost to international economic relations and to political stability.

4 COUNTRIES ON THE GOLD STANDARD 1929–34

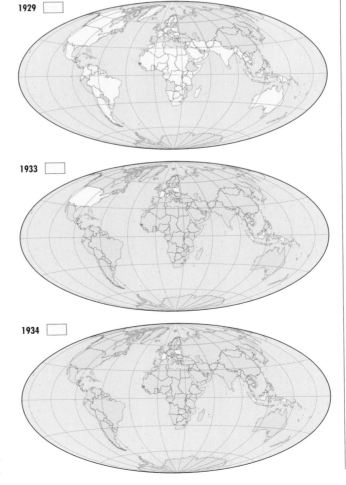

1929

1933

1934

▲ Countries around the world that supplied raw materials for the factories of the industrialized nations were hit by the drop in production during the Depression. Chile, for example, saw its exports drop by over 80 per cent, and India and Brazil suffered a fall of over 60 per cent.

◄ The gold standard linked currencies to the value of gold, and was supported by almost every country in the world. From 1931, however, countries began to leave the standard, leading to its total collapse by 1936. Although at the time this was seen as a disaster, it actually presented opportunities for recovery in many countries, allowing governments to intervene to create economic growth.

◄ OUTCOMES OF THE FIRST WORLD WAR 1918–39 *pages 220–21*   ► THE RISE OF FASCISM 1921–39 *pages 230–31*

THE RISE OF FASCISM
1921–39

In the years between the two world wars, a political and socio-cultural phenomenon known as fascism arose in Europe. Its exact form varied from country to country, but it was most commonly characterized by chauvinistic nationalism coupled with expansionist tendencies, anti-communism and a ruthless repression of all groups presumed dissident, a mass party with a charismatic leader who rose to power through legitimate elections, and a dependence on alliances with industrial, agrarian, military and bureaucratic elites.

FASCISM IN ITALY

Fascism first gained prominence in Italy, where the National Fascist Party (PFI) was founded by Mussolini in 1921. Mussolini possessed a talent for arousing enthusiasm and giving a sense of power and direction to a society in crisis. Through coercion, indoctrination and the creation of the cult of himself as "Il Duce" (the leader), he was able to balance the different interests of his supporters. His nationalist rhetoric attracted war veterans, while his promise to deal with the threat of revolutionary socialism won the support of the lower middle classes and a proportion of the peasantry. Some workers saw the fascist syndicates as an appealing alternative to socialist unions, while landowners and industrialists made large donations to fascist groups because they battered peasant and labour organizations into submission. Most importantly, the political establishment tolerated fascism and helped pave the way for Mussolini's rise to power; with the much celebrated "March on Rome" in 1922, Mussolini, now Prime Minister, signalled the beginning of a new era.

Mussolini's foreign policy wavered between aggression and conciliation. In 1923, two weeks after capitulating to

▲ Benito Mussolini started his political life as a socialist and was imprisoned for his opposition to Italy's expansionist activities in Libya in 1911–12. By the 1920s, however, he had changed his views and used his considerable rhetorical powers to whip up popular support for his fascist policies of nationalism, anti-socialism and state control of industry and the economy.

▶ The Treaty of Versailles of 1919 assigned the disputed Saar region to League of Nations protection, and denied Germany military access to the Rhineland, the region of western Germany bordering France. However, a plebiscite in Saarland in 1935 produced 90 per cent support for German rule, and in 1936 Hitler ordered troops into the Rhineland as a gesture of defiance.

In March 1938 the German Anschluss (annexation) of Austria was achieved with support from Austrian fascists, and in October, following the Munich Pact (drawn up by Britain, France, Germany and Italy), Germany took over all regions of Czechoslovakia with a population more than 50 per cent German. The Czech government (by then under a dictatorship) ceded the rest of Bohemia–Moravia in March 1939, with Slovakia becoming a German puppet state. On 1 September the Germans began their attack on Poland, and the British and French declared war. They did not, however, send troops to aid Poland, which, attacked from the east by the Soviet Union and heavily outgunned, was forced to surrender.

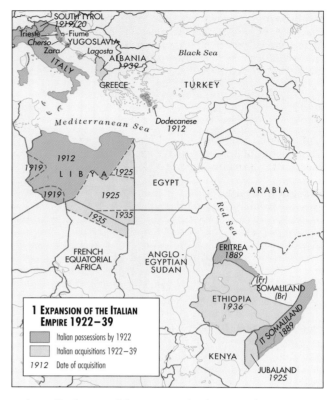

1 EXPANSION OF THE ITALIAN EMPIRE 1922–39
- Italian possessions by 1922
- Italian acquisitions 1922–39
- 1912 Date of acquisition

▲ As part of his plan to revive Italian national pride, Mussolini sought to create an Italian empire comparable to those of Britain and France. He not only expanded Italy's Libyan territory, but in 1935 launched a successful assault on Ethiopia. He also extended Italy's territories on the eastern Adriatic coast.

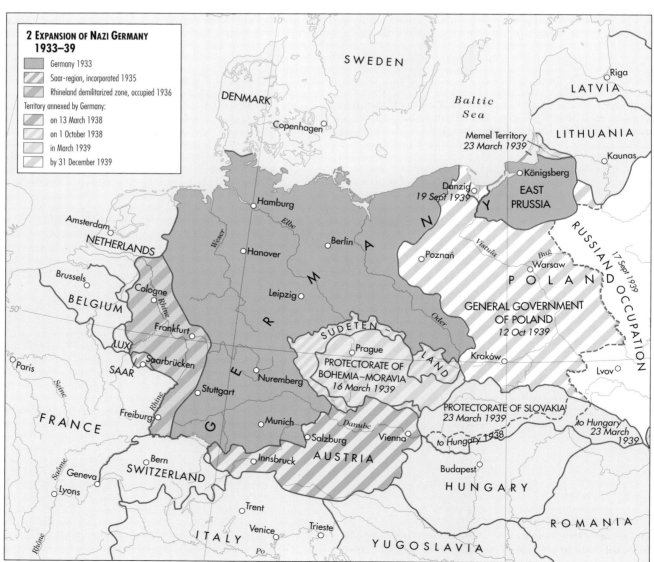

2 EXPANSION OF NAZI GERMANY 1933–39
- Germany 1933
- Saar-region, incorporated 1935
- Rhineland demilitarized zone, occupied 1936
- Territory annexed by Germany:
  - on 13 March 1938
  - on 1 October 1938
  - in March 1939
  - by 31 December 1939

the British over the "Corfu incident", he occupied Fiume (*map 1*), before concluding a treaty of friendship with Yugoslavia in a failed attempt to break the "Little Entente" (*pages 220–21*). In 1935 Italy formed an accord with France and joined in condemnation of German rearmament before invading Ethiopia in October 1935, thereby alienating itself from both Britain and France. A rapprochement with Germany was inevitable, and in 1936 the "Rome–Berlin Axis" was formed. Italy joined Germany in assisting the Nationalists in the Spanish Civil War, further alienating itself from the rest of Europe, and in May 1939 signed the "Pact of Steel" with Germany. In April 1939 it attacked Albania.

FASCISM IN GERMANY

Hitler's rise to power in 1933 can be seen partly as a product of the harshness of the Treaty of Versailles (1919), which placed an economic noose round the neck of the Weimar Republic. The Great Depression in the early 1930s (*pages 228–29*) weakened the Republic further, while Hitler's National Socialist German Workers' Party (the "Nazis") was increasing its support. In 1932 it became the largest single party and Hitler was appointed Chancellor in January 1933.

Hitler's absolute belief in the superiority of the "Aryan race" led to a series of legislative measures (1933–38) aimed at excluding Jews from German government and society, culminating in a programme of extermination: the "Final Solution" (*pages 232–33*). The regime's emphasis on ideological conformity led to heavy censorship, while the Nazis mobilized the German youth to provide a new base of mass support. The first phase of Hitler's economic plans aimed to reduce the level of unemployment, while in the second phase Germany was intended to achieve self-sufficiency both in industry and agriculture, a goal by no means realized.

Hitler's foreign policy was, however, more successful (*map 2*). With the backing of an army that had been increased to more than twice the size allowed by the Treaty of Versailles, he managed to end German isolation in Europe through the Anglo-German Naval Pact of 1935 and to remilitarize the Rhineland in 1936. In 1938 Austria was virtually incorporated into the Reich, as was the German-populated Sudetenland – an act accepted by Britain and France with the signing of the Munich Agreement in September 1938. Further gains took place in March 1939,

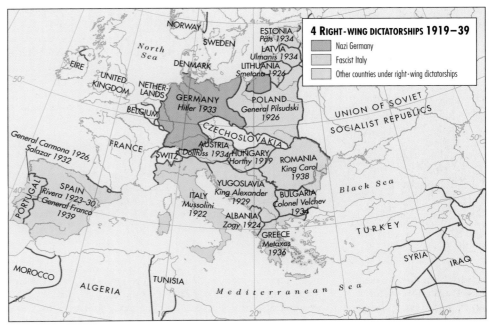

4 RIGHT-WING DICTATORSHIPS 1919–39

- Nazi Germany
- Fascist Italy
- Other countries under right-wing dictatorships

▲ During the 1920s and 1930s right-wing dictatorial regimes were established across Europe and the Iberian Peninsula. However, many dictators, such as Horthy in Hungary and King Carol of Romania, regarded fascist organizations as a threat to their rule. Even in Spain, under General Franco's regime, the influence of the fascist Falangists was replaced by the traditional bastions of order: army, Church and monarchy.

and the signing of the Pact of Steel with Italy in May 1939 was followed by the Non-Aggression Pact with the Soviet Union in August. Confident that Britain would not intervene, Hitler invaded Poland on 1 September 1939. The Second World War had begun.

THE SPANISH CIVIL WAR

The Spanish Civil War (1936–39) arose following the collapse in 1930 of Miguel Primo de Rivera's seven-year dictatorship, and the three-year rule of the left-wing Prime Minister Azaña, whose egalitarian reforms provoked bitter opposition on the part of the Establishment. In 1933 Azaña's government was succeeded by a series of centre-right coalition governments, which dismantled his reforms and resulted in social unrest. By the time of the 1936 elections Spain was polarized into two political camps, each consisting of a broad alliance: the Popular Front (Republicans) – made up of socialists, communists, liberals and anarchists – and the National Front (Nationalists) – comprising monarchists, conservatives and a confederation of Catholics. The Popular Front won the elections and Azaña formed a new government, intending to reintroduce all his earlier reforms. The army resolved to take action against the Republic. General Franco, previously exiled by Azaña to the Canaries, invaded Spain from Morocco and laid siege to Madrid in November 1936 (*map 3*). He was supported in his campaign by the fascist Falange, a party founded in 1933 by de Rivera.

The conflict attracted international interest, with Italy and Germany supporting the Nationalists and the Soviet Union the Republicans. The German bombing of the Basque town of Guernica caused an international outcry, but neither Britain nor France was prepared to confront Hitler over his assistance to Franco. When the Soviet Union decided to end its assistance to the Republicans, a Nationalist victory was assured. By spring 1939 Franco's government was recognized by most of Europe, and Spain entered an era of ruthless repression.

RIGHT-WING DICTATORSHIPS

In the 1920s and 1930s a number of right-wing dictatorships were established in Europe, both in agrarian and industrialized societies (*map 4*). They were undoubtedly influenced in their rhetoric and practice by the German and Italian models, but were also shaped by each country's indigenous features. Many of these dictators were uncharismatic figures, who actually regarded fascist movements and organizations as a threat to their rule. Only the Nazi dictatorship, with its aggressive expansionism, racism, and nationalist and militarist ideology, represented the full expression of fascism.

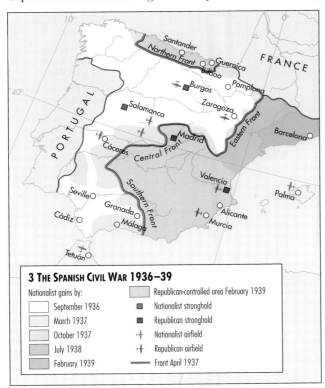

3 THE SPANISH CIVIL WAR 1936–39

Nationalist gains by:
- September 1936
- March 1937
- October 1937
- July 1938
- February 1939

- Republican-controlled area February 1939
- Nationalist stronghold
- Republican stronghold
- ✠ Nationalist airfield
- ✠ Republican airfield
- —— Front April 1937

▲ During the Civil War Spain became a battleground for fascist Germany and Italy (which backed the Nationalists) and the communist Soviet Union (which backed the Republicans). Semi-fascist Portugal allowed German supply lines across its territory.

◀ OUTCOMES OF THE FIRST WORLD WAR 1918–29 *pages 220–21* ▶ THE SECOND WORLD WAR IN EUROPE 1939–45 *pages 232–33*

THE SECOND WORLD WAR IN EUROPE 1939–45

1 MILITARY CAMPAIGNS IN EUROPE 1939–45

- Maximum extent of territory under Axis control
- Territory that remained under Soviet control
- Area under Allied control by Dec 1942
- Neutral
- → Axis advance with date
- → Allied advance with date
- ✛ German aid raid
- ✛ Allied air raid
- ▌ Soviet advance by date shown
- ▌ Western Allied advance by date shown
- — Axis advance by date shown
- ▲– Oil well and pipeline

▲ During the Second World War almost the whole of Europe came under Axis control. After Germany's invasion of western Europe, and its attempts to bomb Britain into submission, for three years the war was concentrated on the Eastern Front, with German troops sweeping across the western Soviet Union. During 1942, however, they became bogged down, with losses in the north outweighing gains in the south. In February 1943 the Soviet Union broke the siege of Stalingrad and the Germans were forced to retreat. At the same time, their forces in North Africa were also fleeing to the safety of Italy. The Germans fought a strong rearguard action, however – in the east, in Italy and, from June 1944, in western Europe, with the Allied troops eventually meeting up just west of Berlin in May 1945.

The war in Europe (1 September 1939 – 7 May 1945) was not one war but many. It began as a struggle for supremacy in Europe, but soon engulfed North Africa, the Atlantic and the Soviet Union. In December 1941, with Japan's attack on Pearl Harbor and Germany's declaration of war against the United States (*pages 234–35*), the conflict became truly global.

The French and British decision to contest Hitler's bid for European hegemony, after his invasion of Poland, took the Nazi leader by surprise. The practical implications were, however, limited. Belated rearmament meant that France and Britain could do little to prevent Germany and the Soviet Union dismantling Poland under the German-Soviet Non-Aggression Pact of 23 August 1939. Nevertheless, the Allies – at this stage, Britain, France and the Polish government in exile – were confident that Hitler could be forced by economic pressure into compromise. The initial seven-month period of calm, known as the "Phoney War", thus favoured the Allies, but a spate of spectacular military operations in the spring and summer of 1940 saw first Denmark and Norway fall to the Germans, then Belgium and the Netherlands (*map 1*). France was brought to its knees in six weeks. Puppet regimes, or direct rule from Germany, were imposed on the occupied territories, while an area of France, plus its overseas empire and fleet, was allowed to form the "Vichy" regime under Marshal Pétain (*map 2*).

During the next year Berlin consolidated and extended its political influence and control. Hitler's fascist partner, Mussolini, brought Italy into the war on 10 June, and the "Axis" was further strengthened with the signing of the Tripartite Pact between Germany, Italy and Japan on 27 September. The Balkan states soon became German satellites (*map 2*), and the remaining neutrals were forced to grant substantial economic concessions. Berlin, however, failed to achieve its strategic objectives. Against expectations, Britain refused to sue for peace and withstood the Blitz over the autumn of 1940. Unable to mount an invasion of Britain, the German foreign ministry and navy embarked on an "indirect strategy" against Britain.

Germany's submarine fleet was given the task of severing Britain's tenuous communications with the neutral United States. However, although the U-boats cut deep into Britain's reserves and posed a danger until the early summer of 1943, the indirect strategy failed to meet German expectations. Moreover, Italian efforts in 1940–41 to carve out a Mediterranean empire complicated rather than complemented Germany's war plans. Britain's maritime and imperial resources allowed it to inflict a series of humiliating setbacks on Italian forces in Egypt and Greece. Hitler was compelled to come to the aid of his ally and was drawn into campaigns of little strategic importance and marginal economic benefit, which ultimately delayed his invasion of the Soviet Union by several weeks.

THE EASTERN FRONT

On 22 June 1941 Hitler began his attack on the Soviet Union (long regarded as the Nazis' principal ideological opponent, despite the 1939 pact). As well as massive military casualties, over three million Soviet prisoners of war were deliberately killed, through starvation or overwork,

3 CENTRAL EUROPE 1945

Future Soviet bloc countries	Areas controlled by USSR
Western allies and countries liberated by them	Meeting of Soviet and Western forces
	International boundary 1945
Neutral countries	Functioning boundary of 1945
Germany	German boundary at end of 1939
Areas controlled by Western allies	City divided into four occupation zones

▲ During the final months of the war a race took place between the Western Allies and the Soviet Union for control of German territory. The two armies eventually met west of the German capital Berlin and the Austrian capital Vienna. They agreed to divide these symbolically important cities into zones of occupation, with the Soviet Union controlling the surrounding territories and thus holding the upper hand.

and millions of civilians were enslaved in German farms and factories, where many of them died. By the time winter set in, German forces had reached the suburbs of Moscow, encircled Leningrad and controlled huge swathes of Soviet territory (*map 1*).

The Soviet Union was ill-prepared to meet the German onslaught. As military resistance crumbled, industrial plant was relocated away from the advancing German forces. Aid was forthcoming from Britain and the United States, and although it was not critical, it did cover important shortfalls in transportation and communications. On learning that Japan had decided against attacking the Soviet Union in the east, Stalin transferred troops from Siberia to meet the German attacks in 1941. Better prepared for the harsh climatic conditions, the Soviet forces counterattacked the following spring, and while Germany made impressive gains in the south, in an effort to control the Soviet Union's oil resources, the retaking of Stalingrad by the Soviets in February 1943 marked a turning point. Soviet success at the massive tank battle of Kursk in July began Germany's long retreat westwards, which ended when Berlin fell to Soviet forces two years later. In terms of the number of casualties suffered and of the resources expended, the Second World War in Europe was predominantly a struggle between the Soviet Union and Germany.

THE "FINAL SOLUTION"

The war against the Soviet Union allowed Hitler to set in train the second component of his racial war: the elimination of European Jewry and those considered "defective". During 1942 death camps were erected in the occupied territories to exterminate Jews, gypsies, homosexuals and other "racial enemies" (*map 2*). By the end of the war some six million Jews, along with hundreds of thousands of other victims, had been gassed in the death camps, or starved, executed or worked to death in concentration camps. Of those that survived the camps, many died as they were forced to march away from the advancing Allies.

THE DEMAND FOR A SECOND FRONT

Given the enormity of the struggle facing the Soviet Union, Stalin demanded immediate support from his western allies. In practical terms, however, there was little that could be done. Until late 1943 the contribution of Britain's strategic bombing offensive was meagre, and was maintained largely to placate Soviet demands for a second front. In November 1942, however, Anglo-American forces landed in French Morocco and Algeria and, in conjunction with British forces in Egypt, drove the Axis back to Tunisia (*map 1*). After five months of fighting, the two Allied pincers met outside Tunis and finally ejected Axis forces from North Africa by mid-May 1943.

Against the wishes of the Soviet Union and the United States, both of whom favoured landings in northern France, Britain insisted on mounting landings in Sicily and Italy. While these campaigns knocked Italy out of the war, they failed to provide a strategic breakthrough into central Europe. Competing strategic priorities and the U-boat menace to the Atlantic convoys meant that it was only in June 1944 that the Western Allies felt sufficiently confident to create a second front by landing troops in Normandy.

German defences did not, however, crumble. Despite the Allies' massive economic, military, intelligence and technical superiority, dogged German resistance forced the Allies to fight every step of the way. In the face of inevitable defeat, an opposition cabal tried to assassinate Hitler in July 1944, but was quickly crushed. Indeed, only in the Balkans and France did armed resistance to German domination meet with any real success. Nazi Germany had to be ground down by aerial bombardment and huge land offensives.

The political consequences of the total defeat of Germany were enormous. Mutual suspicions between the Allies quickly emerged as thoughts turned to the post-war world and the division of the spoils (*map 3*). Culturally, the war dealt a blow to western European civilization and confidence from which it has struggled to recover. Though it began, and was largely fought, in Europe, the Second World War spelt the end of European influence across the globe.

▲ Despite the non-aggression pact with the Soviet Union, signed by Foreign Minister von Ribbentrop in August 1939, Nazi Germany still regarded the communist Soviet Union as its natural enemy, and launched an attack in the summer of 1941. This poster offered the German people the stark choice of "Victory or Bolshevism".

▼ Nazi Germany retained control in its conquered territories by installing puppet governments in the Balkans and its own administrations in Poland and the western Soviet Union. Italian and German troops jointly occupied Greece until the Italian surrender in 1943. Concentration and death camps were constructed, to which "undesirables", and in particular Jews, were transported from across Europe.

2 GERMANY'S 'NEW ORDER' IN EUROPE NOVEMBER 1942

German Reich	Countries occupied by Axis
Territory under German administration	Vichy-governed France
Territory under German occupation	Unconquered territory of USSR
Italy and annexed/administered territories	Territory of Allied Powers
Countries co-operating with Axis	Neutral countries
	International boundary
	German border 1937
	Concentration camp
	Major death camp

THE RISE OF FASCISM 1921–39 *pages 230–31*   THE COLD WAR 1947–91 *pages 244–45*

THE WAR IN ASIA 1931–45

The war in Asia can be seen as a series of conflicts that eventually escalated, with the Japanese attack on Pearl Harbor and Southeast Asia in December 1941, into a single element within a larger global conflagration. It began in September 1931 when the Japanese army set about seizing Manchuria as a first step in Japan's construction of an economically self-sufficient bloc under its control. By 1933 the conquest of Manchuria was complete and for the next four years there was relative peace in East Asia.

THE SINO-JAPANESE WAR

In 1937 an incident outside Beijing rapidly developed into a full-scale war between Japan and China (*map 1*). The Japanese forces proved to be superior in battle to their Chinese counterparts and by the end of 1938 Japan had seized large areas of China and had forced Chiang Kai-shek's government to retreat to Chongqing. However, despite the scale of the defeat, the Chinese refused to surrender, a fact which Japan blamed on Western support.

◄ Fierce fighting took place following the Japanese invasion of China in 1937, but despite a series of defeats, the Chinese refused to surrender.

▼ The rate of the Japanese advance in Southeast Asia and the Pacific took the Allied forces by surprise. Dutch, British and US territories fell like dominoes until Japan over-stretched itself in the Battle of Midway in June 1942. French Indochina, under the Vichy government, was sympathetic to Japan, as was Thailand. Japan ruled over its new territories with an iron fist and engaged in atrocities against both native populations and European prisoners of war.

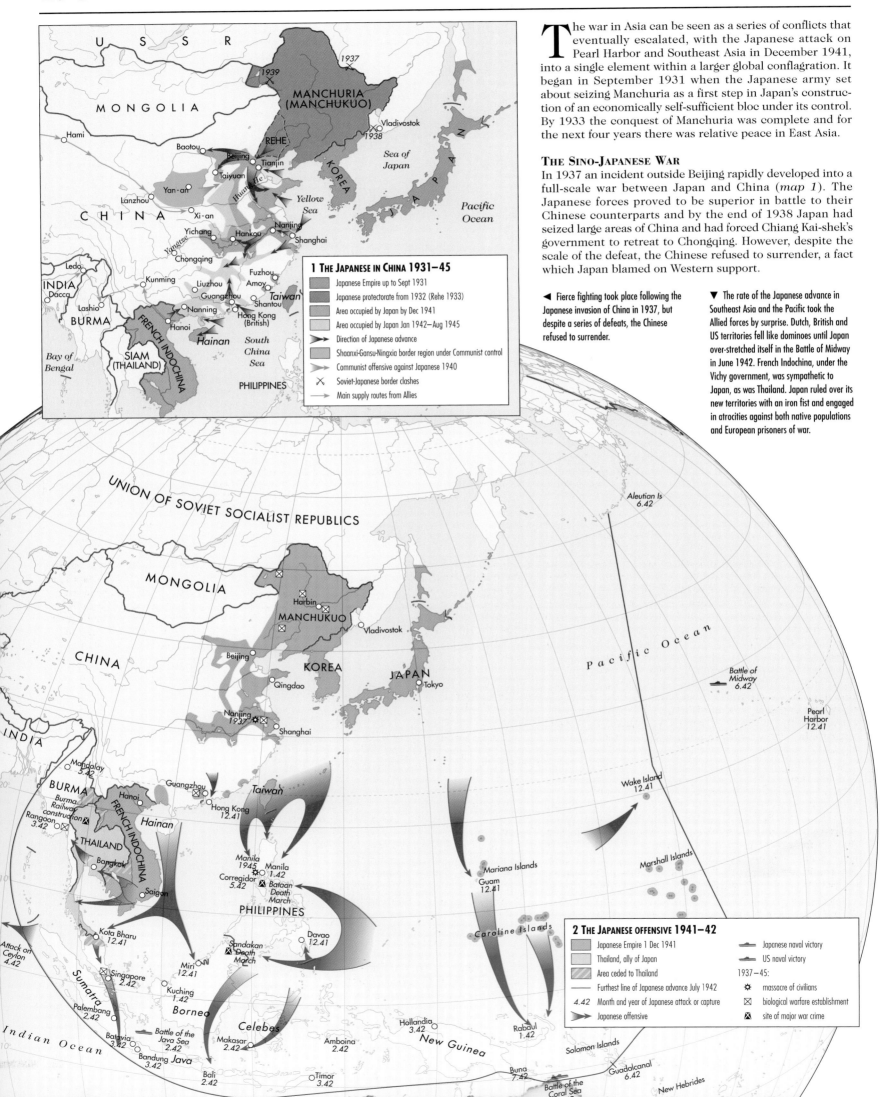

1 THE JAPANESE IN CHINA 1931–45
- Japanese Empire up to Sept 1931
- Japanese protectorate from 1932 (Rehe 1933)
- Area occupied by Japan by Dec 1941
- Area occupied by Japan Jan 1942–Aug 1945
- ⇒ Direction of Japanese advance
- Shaanxi-Gansu-Ningxia border region under Communist control
- ⇒ Communist offensive against Japanese 1940
- ✕ Soviet-Japanese border clashes
- → Main supply routes from Allies

2 THE JAPANESE OFFENSIVE 1941–42
- Japanese Empire 1 Dec 1941
- Thailand, ally of Japan
- Area ceded to Thailand
- —— Furthest line of Japanese advance July 1942
- 4.42 Month and year of Japanese attack or capture
- ⇒ Japanese offensive
- Japanese naval victory
- US naval victory
- 1937–45:
- ✿ massacre of civilians
- ⊠ biological warfare establishment
- ⊠ site of major war crime

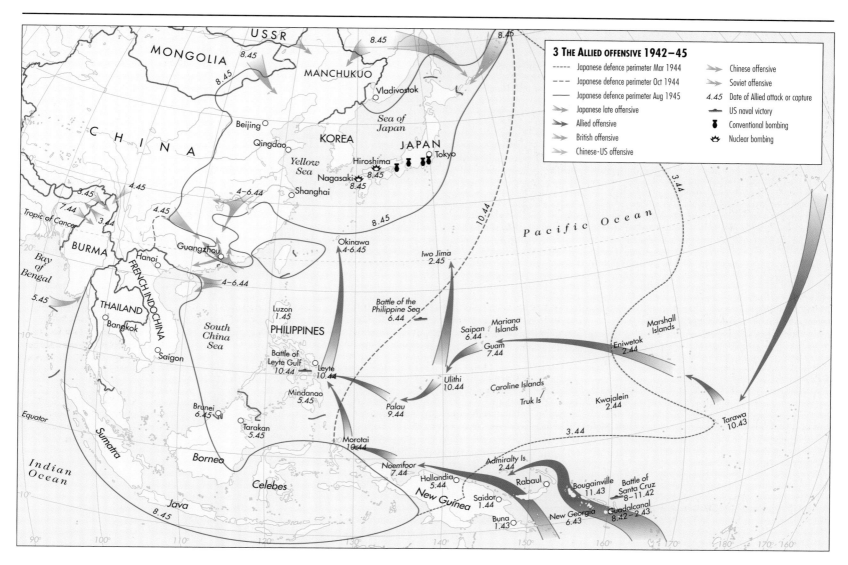

3 THE ALLIED OFFENSIVE 1942-45
- - - - - - Japanese defence perimeter Mar 1944
- - - - Japanese defence perimeter Oct 1944
——— Japanese defence perimeter Aug 1945
⟶ Japanese late offensive
⟹ Allied offensive
⟹ British offensive
⟹ Chinese-US offensive
⟹ Chinese offensive
⟹ Soviet offensive
4.45 Date of Allied attack or capture
⟶ US naval victory
▮ Conventional bombing
☢ Nuclear bombing

Japan's answer to this problem was to try to use the war in Europe to its own advantage. In the summer of 1940, following the German offensive into western Europe (*pages 232–33*), Japan sought, through diplomatic means, greater access to the raw materials of the Dutch East Indies, French Indochina and Thailand. At the same time, in an effort to deter the United States from intervening in East Asia, it signed the Tripartite Pact with Germany and Italy. In response, the United States and Britain introduced a policy of economic sanctions, culminating, in July 1941, in an embargo on oil exports to Japan. Faced with complete economic collapse or war with the Allies, the Japanese chose the latter and on 7 December 1941 launched a rapid offensive into the western Pacific and Southeast Asia in the hope of establishing an impenetrable defensive perimeter.

THE PACIFIC WAR
The speed and effectiveness of the Japanese attack, symbolized most notably by the assault on Pearl Harbor, took the US, British and Dutch forces by surprise and led to a series of humiliating defeats for the Western Allies in the first six months of the war. In February 1942 the British fortress at Singapore surrendered and by May the last US garrison in the Philippines had capitulated (*map 2*). Japan's victories led it to portray itself as the "liberator" of Asia from European imperialism. During the course of the war nominally independent states were established in Burma and the Philippines, and Japan's ally Thailand was allowed to annex areas of Indochina, Burma and Malaya. In reality, however, Japan ruled over its newly conquered territories with an iron fist and engaged in atrocities against the native population and European civilian detainees and prisoners of war.

The euphoria of victory was shortlived. In June 1942 Japan suffered its first major reverse when its naval expedition to seize the island of Midway ended in disaster with the loss of four aircraft carriers. From this point Japan was on the defensive and was out-manoeuvred strategically by the United States, which, through its "island-hopping" campaign in the western Pacific, was able to isolate the major Japanese bases such as Truk and Rabaul (*map 3*). In addition, Japan's war effort was undermined by the fact that it lacked the resources to replace its losses, with US submarines cutting the supply routes to Japan.

By 1945 it was clear that Japan was on the retreat, but the Americans feared that it would still cost many more lives to bring about its defeat. This was confirmed when the invasion of Okinawa in the spring of that year led to 10,000 American casualties. At first it was hoped that conventional bombing of Japanese cities and Soviet entry into the war in Asia would persuade Japan to capitulate, but by the summer hopes had turned to the use of the newly developed atomic bomb. The dropping of atomic bombs in early August on Hiroshima and Nagasaki – which resulted in the death of 140,000 people – and the Soviet invasion of Manchuria, proved to be the final blows for Japan, and on 15 August Emperor Hirohito announced the country's surrender.

Although Japan's attempt to carve out an empire had been defeated, the region did not return to the pre-war status quo. In Southeast Asia the war helped to inspire the rise of indigenous nationalism, which in turn laid the seeds for the wars of national liberation that were to continue into the 1970s (*pages 250–51*). In China the ineffectiveness of Chiang Kai-shek's regime and its dismal war record led many to look to the Chinese Communist Party as an alternative government and civil war soon erupted (*pages 254–55*). For the United States the war demonstrated the importance of the western Pacific to its national security and led to a permanent commitment of American forces to the region. Japan, meanwhile, eschewed militarism and sought economic expansion by peaceful means.

▲ It took the Allies more than three years to regain territory that had fallen to Japan over a six-month period. Indeed, when Japan surrendered on 15 August 1945, following the dropping of atomic bombs on Hiroshima and Nagasaki, its troops still occupied a large part of Southeast Asia.

◖ THE MODERNIZATION OF JAPAN 1867–1937 *pages 200–1* ◗ JAPAN SINCE 1945 *pages 252–53*

THE SOVIET UNION AND EASTERN EUROPE 1945–89

▲ Nikita Khrushchev emerged victorious from the struggle for power that followed Stalin's death in 1953, and went on to denounce Stalin's "reign of terror". He was deposed by conservative elements within the party in 1964 and his grandiose agricultural schemes and confrontational foreign policy, which had led the world to the brink of nuclear catastrophe during the Cuban Missile Crisis of 1962, was subsequently criticized.

▼ The 15 constituent republics of the Soviet Union were formed in the 1920s and 1930s, largely along ethnic lines. They were dominated by the Russian Federation, by far the largest and wealthiest of the republics. Russia was itself divided for administrative purposes into regions that had various degrees of local autonomy.

The Soviet Union emerged from the Second World War victorious, but devastated by the loss of 26 million people. Despite territorial gains in the west (*map 1*) there was a severe shortage of labour, aggravated by the deportation to Siberia or Central Asia of returning prisoners of war, intellectuals from the newly gained territories and whole nations accused of collaboration with the Germans (including the Volga Germans, Crimean Tatars and Chechen-Ingush). The post-war Soviet Union consisted of 15 soviet republics, some of which also contained autonomous republics, regions and national areas (*map 2*).

After 1945 Stalin sought to re-establish control of the Soviet Union. Collective farms that had been destroyed during the war were reinstated, efforts were made to develop heavy industry, and the government returned to the use of terror as a way of controlling the population. Stalinism was extended wholesale to Eastern Europe, and by 1948 communist parties were in full control throughout the region (*map 1*). The economic development of the Eastern bloc was regulated from 1949 onwards by the Council for Mutual Economic Assistance (COMECON) and defence aims were unified in 1955 with the signing of the Warsaw Pact. Only Yugoslavia, where Tito had come to power independently of the Red Army, developed a non-Stalinist form of communism.

KHRUSHCHEV AND BREZHNEV

Stalin died in March 1953 and by 1956, following a secret speech criticizing Stalin, Khrushchev had triumphed over his rivals. Political prisoners were released from the labour camps, and fresh emphasis was placed on the importance of agriculture, housing and the production of consumer goods. In order to achieve this economic change of direction at least partial decentralization was considered necessary. At the same time, Khrushchev poured money into nuclear and space research: the *Sputnik* satellite was launched in 1957, and in 1961 Yuri Gagarin made the first manned space flight.

The results of this new approach were mixed. Increased liberalization led to dissident movements in Russia and revolts across Eastern Europe. In 1956 both Poland and Hungary rose against Soviet rule. In Poland the Communist Party, under Gomulka, persuaded Khrushchev that a reformed communism would not threaten party control, but Hungary, which wanted to leave the Warsaw Pact, was invaded. Khrushchev improved relations with Yugoslavia, but his policies led to a split with China by 1960. Despite Khrushchev's successful visit to the United States in September 1959, relations with the West were soured by

1 COMMUNIST EASTERN EUROPE 1945–89
- Territory annexed by Soviet Union 1939–40
- Territory annexed by Soviet Union 1945–47
- Independent communist state after 1948
- Allied with China after 1961
- Territory under Russian occupation 1945
- Iron Curtain
- Soviet intervention 1956
- Soviet intervention 1968
- Strong Soviet pressure 1956, 1980
- COMECON member 1949–91
- Warsaw Pact member 1955–91
- City divided into zones of occupation

▲ In 1948 communist parties, supported by the Soviet Union, were in control in Eastern Europe, and from then on communication between East and West was limited.

Yugoslavia refused to align itself with the Soviet Union, Albania broke its economic ties in 1961, and from 1968 Romania developed a degree of independence.

the shooting down of a US reconnaissance plane over the Soviet Union in 1960, the building of the Berlin Wall in 1961, and the siting of Soviet nuclear missiles in Cuba in 1962 (*pages 242–43*).

Khrushchev was ousted by the Politburo in 1964, but economic reforms continued under Brezhnev and Kosygin until the invasion, in 1968, of Czechoslovakia, where Dubček threatened the Communist Party's monopoly on power. The Soviet Union then settled into a period characterized by a return to a centralized economy, with quotas that enforced quantity rather than quality. With the growing competition in armaments and space technology, and the Soviet Union's intervention on the side of the socialists in the Afghan Civil War, the Cold War intensified.

ECONOMIC DEVELOPMENT

The post-war period saw a whole series of grandiose plans for scientific management of the economy. Although Stalin's plan for the "Transformation of Nature", through windbreaks and shelter belts across the Ukraine, was shelved in 1953, Khrushchev's "Virgin Lands" scheme to grow maize across northern Kazakh SSR (*map 3*) was implemented. The resulting soil erosion ruined 40,000 square kilometres (15,440 square miles) of land and forced the Soviet Union to import grain. His scheme of the early 1960s for supranational economic sectors across Eastern Europe, with the north concentrating on industry and the south on agriculture and raw materials, failed due to

2 THE UNION OF SOVIET SOCIALIST REPUBLICS IN THE 1970s
- MOL Soviet Socialist Republic
- Autonomous Soviet Socialist Republic
- Autonomous Oblast
- National Okrug
- Soviet armed intervention

Romanian nationalism and caused Albania to establish closer links with China. A plan in 1971 for a giant computer grid to manage the whole Soviet economy was never implemented, and neither was the scheme to build a canal system that would have reversed the flow of several Siberian rivers in order to irrigate Central Asia.

Since 1917 "progress" had been envisaged as smoking factory chimneys and increased industrial production. However, Soviet economic growth rates of 5–6 per cent in the 1960s dropped to 2.7 per cent in 1976–80, and to 0 per cent in the early 1980s. Defence costs, the Afghan War and support for the countries of Eastern Europe were more than the economy could sustain. Rising expectations and a widespread black market led to labour unrest. Subsidies on food and housing took up large parts of the budget, and poor-quality consumer goods left people with little on which to spend their wages, resulting in money being put into private savings instead of back into the economy.

There were, however, successes in military and space technology, and in drilling for oil and natural gas, although exploitation of the Eastern bloc's rich mineral resources led to serious pollution – both in industrial areas and in previously untouched landscapes (*map 3*). The dangers inherent in using poorly built and inadequately managed nuclear power to generate electricity were brought home to the world by the explosion at the nuclear power plant at Chernobyl in 1986, although a larger, but unreported, nuclear accident had already occurred in 1957 at the test site "Chelyabinsk 40" in the Urals.

In Eastern Europe economic decline also set in from the mid-1970s onwards. As loans from Western banks became harder to arrange, and the Soviet Union ended its subsidized oil exports in the mid-1980s, wages

in Poland fell by 17 per cent in the period 1980–86. In Yugoslavia wages fell by 24 per cent over the same period. Declining living standards, environmental issues, pollution and related health concerns heightened demands for a release from Soviet domination.

MIKHAIL GORBACHEV

When Gorbachev came to power in 1985 it was clear that the economy needed radical reform and that the cost to the environment and to people's health had been catastrophic. Pipelines were leaking oil into the permafrost across northern Russia, and most of Russia's major rivers were polluted, in particular the Yenisei estuary around Norilsk. Grand projects, such as the building of the Baikal–Amur railway, had enabled the development of further mining enterprises, but in so doing had contributed to the destruction of the fragile ecosystem of Siberia. Damage to Lake Baikal from industrial effluent was an issue on which a growing green lobby focused, as was the drying-up of the Aral Sea, which lost 75 per cent of its volume and 50 per cent of its area between 1960 and 1989 due to overuse of its tributaries for irrigation.

Gorbachev's policies of *glasnost* (openness), *perestroika* (restructuring) and democratization initiated reforms that were to lead to the withdrawal of Soviet troops from Afghanistan in 1989, and to the ending of Soviet control of Eastern Europe.

▼ Heavy industry was central to the development of the Soviet economy, but caused severe soil and water pollution in many areas. Even the empty wastes of northern Russia were exploited for the valuable coal, oil and metals found there.

3 THE ECONOMY OF THE SOVIET UNION AND EASTERN EUROPE 1948–89

- —— Major railway
- Main industrial region
- Area of severe environmental damage
- Virgin Lands territory
- Nuclear power station
- ▲ Oil extraction
- ◆ Oil refining
- ■ Coalmining
- Ferrous ores mining
- Copper mining
- Lead and zinc mining
- Gold mining
- Silver mining
- Aluminium mining
- Nickel mining
- Tin mining

◐ RUSSIAN REVOLUTION 1917–39 *pages 222–23* ◐ FORMER SOVIET REPUBLICS SINCE 1989 *pages 262–63* ◐ EASTERN EUROPE SINCE 1989 *pages 264–65*

WESTERN EUROPE SINCE 1945

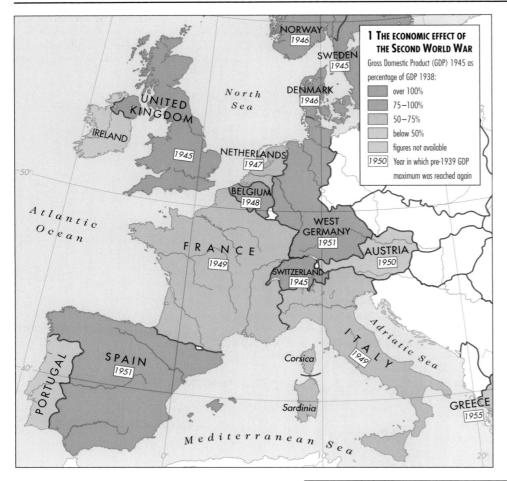

1 THE ECONOMIC EFFECT OF THE SECOND WORLD WAR

Gross Domestic Product (GDP) 1945 as percentage of GDP 1938:

- over 100%
- 75–100%
- 50–75%
- below 50%
- figures not available

1950 Year in which pre-1939 GDP maximum was reached again

In the five decades after the end of the Second World War Western Europeans experienced an unprecedented increase in material prosperity. This was the outcome of almost uninterrupted economic growth which, by the end of the 20th century, had led to average per capita incomes more than three and a half times as high as in 1950, with the income gap between "rich" and "poor" countries within Western Europe much smaller than in the immediate post-war years. This rise in the material standard of living was associated with the increasing integration and interdependence of the European economies and their reliance on economic links with the rest of the world, underpinned by a profound structural transformation in which the relative importance of the agricultural sector declined. It was also associated with increasing political integration.

PROBLEMS OF POST-WAR ECONOMIC RECONSTRUCTION

At least 40 million people died throughout Europe during the Second World War and there was extensive damage to factories, housing, transport and communications systems. In 1945 Western European countries were faced with implementing the transition from war to peace, reconstructing industries and re-establishing international trade and payments. The length of time it took for pre-war output levels to be restored largely corresponded to the amount of damage inflicted on individual economies by the war (*map 1*).

The immediate post-war period saw severe food shortages and a large number of displaced people. Economic

▼ The European Economic Community (EEC) was set up by the Treaty of Rome in 1957 and was renamed the European Community (EC) in 1967. As a first step towards stabilizing European currencies, the European Monetary System came into force in 1979. The Treaty of European Union was signed at Maastricht in February 1992, and the single European currency system (Euro) was launched on 1 January 1999.

▲ Those countries that experienced land fighting ended the war in 1945 with real GDP levels below those of 1938, while those that had not been subject to land fighting came out of the war with real incomes above their pre-war levels (the United Kingdom and neutral Spain, Sweden and Switzerland).

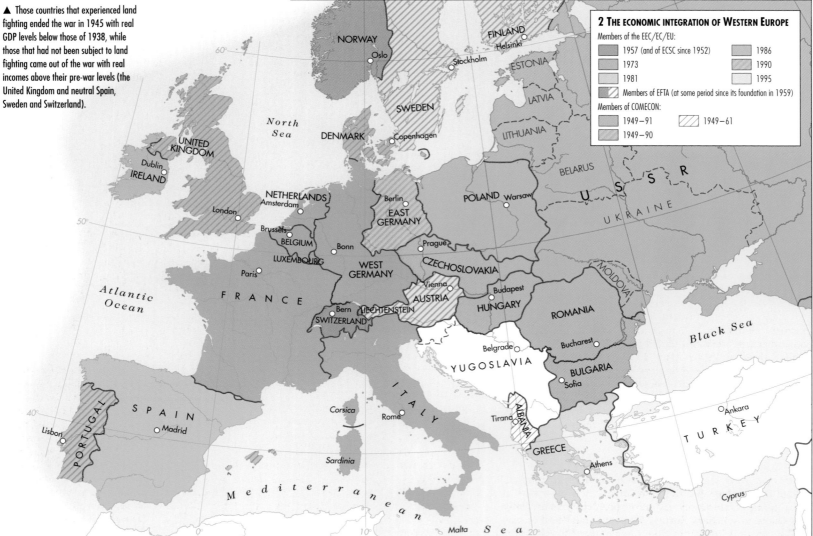

2 THE ECONOMIC INTEGRATION OF WESTERN EUROPE

Members of the EEC/EC/EU:

- 1957 (and of ECSC since 1952)
- 1973
- 1981
- 1986
- 1990
- 1995

Members of EFTA (at some period since its foundation in 1959)

Members of COMECON:

- 1949–91
- 1949–90
- 1949–61

recovery was soon got under way. A major constraint, however, was that Western Europe relied heavily on imports, especially from the United States, but had neither the currency reserves nor export dollar earnings to pay for them. To preserve their foreign currency reserves, European governments restricted imports from neighbouring countries, resulting in a low level of intra-European trade. In order to combat these problems and build Europe into a strong trading partner for the future, the United States announced the European Recovery Program (ERP or Marshall Plan). From 1948 to 1951 ERP funds enabled the countries of Western Europe to continue importing goods from the United States, and thus helped speed up the process of economic recovery. In return the United States put pressure on Western Europe to build and maintain constitutional democracy as a bulwark against the spread of communism and the revival of fascism.

Perhaps the most significant contribution of the ERP was the revitalization of intra-European trade through its support, in 1950, of the European Payments Union (EPU). This restored limited convertibility between European currencies while allowing member countries to maintain controls on imports from the dollar area. By 1958 the EPU had fulfilled its role, but the rapid expansion in trade had resulted in the increasing integration of the European economies – a process that many sought to take further.

EUROPEAN INTEGRATION
Early French post-war plans for reconstruction and modernization called for the expansion of the national steel industry, while relying on unrestricted access to coal from the German Ruhr area. The French foreign minister, Robert Schuman, suggested in 1950 the formation of a common market for coal and steel. The "Benelux" countries, West Germany and Italy accepted the French invitation and negotiated the Treaty of Paris which, in 1951, created the European Coal and Steel Community (ECSC). Its success encouraged member states to push economic integration further to create a customs union and common market – the European Economic Community (EEC) – which began operation in 1958. This increased the liberalization of internal trade and provided access to a larger market, while offering a protective shield against non-members; it also enabled the implementation of common policies. The EEC grew, via the European Community (EC), into the European Union (EU) of 15 countries in 1995 (*map 2*).

In 1959 the United Kingdom, which at that point had not signed up to the EEC, founded the European Free Trade Association (EFTA), and was joined initially by six other countries (*map 2*). Unlike the EEC/EC/EU, with its supranational institutional arrangements, EFTA was intergovernmental in nature. Yet with many of its members eventually joining the economically and politically more powerful Community, EFTA gradually lost its significance.

ECONOMIC GROWTH IN POST-WAR EUROPE
Between 1950 and the mid-1990s all of Western Europe experienced an increase in material prosperity (*bar chart*), despite variations in the rates of economic growth between countries. Moreover, by 1994 the gap in per capita income between the poorest and the richest economies was much smaller than in 1950. After 1973 practically all these economies experienced a slow-down in growth whose extent, however, differed between countries.

Western Europe's post-war growth was closely associated with changes in the employment structure that saw a large-scale shift of resources out of agriculture and industry, especially into services (*map 3*).

POST-WAR POLITICS
Closer economic integration was accompanied by gradual, though incomplete, political convergence. Institutions of parliamentary democracy had never previously been firmly established in southern Europe. The army-backed dictatorship of General Franco in Spain lasted until his death in

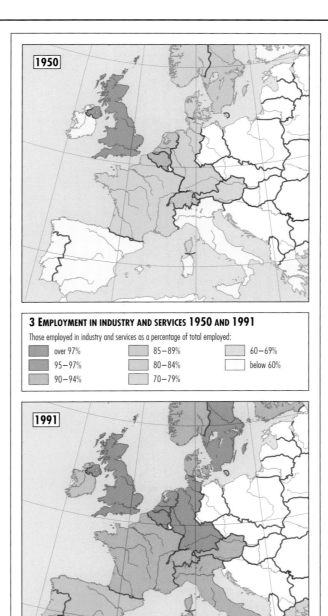

3 EMPLOYMENT IN INDUSTRY AND SERVICES 1950 AND 1991

Those employed in industry and services as a percentage of total employed:

over 97%	85–89%	60–69%
95–97%	80–84%	below 60%
90–94%	70–79%	

◀ During the second half of the 20th century employment patterns changed across Europe with the decline of the agricultural sector and the rise, in particular, of service industries.

1975, but was followed by the restoration of the monarchy of King Juan Carlos, and free elections in 1977. Greece experienced a bitter civil war, a military coup in 1967, and seven years of dictatorship that gave way to a democratic system only in 1974. Democracy did not come to Portugal until 1985. Elsewhere in Western Europe democratic systems did not escape problems. Post-war France went through frequent changes of government until stability was achieved under Charles de Gaulle in the 1950s. Italy not only had many short-lived governments throughout the second half of the 20th century but endured a serious crisis of corruption at all levels of government in the 1990s.

The 1960s saw short-lived left-wing activism, especially in Italy and Germany. In Germany the environmentalist Green movement had limited electoral success in the 1970s. The challenge to constitutional democracy in the 1980s and 1990s came from extreme right-wing, essentially racist, movements, which were most successful electorally in France and Italy. Through most of the period from 1945 to the end of the century, power swung like a pendulum, or was shared, between moderate social democratic or Labour parties and moderate conservative parties. This was the case under voting systems based on proportional representation that encouraged negotiation between political groupings and, as in Britain, a "first-past-the-post" adversarial system that encouraged competition between them.

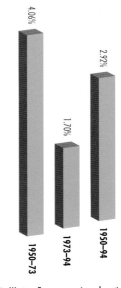

AVERAGE ANNUAL GROWTH OF GDP PER CAPITA THROUGHOUT WESTERN EUROPE

4.06% 1950–73
1.70% 1973–94
2.92% 1950–94

▲ Western Europe experienced particularly rapid economic growth from 1950 until the early 1970s. The large productivity gap separating Europe and the United States in the late 1940s was rapidly reduced, and repair to war-damaged economies and changes in economic policy also created growth. The price of raw materials remained low and there was little competition from the Asian economies. From the early 1970s onwards, however, although the Western European economies continued to grow, they did so at a much slower rate.

◀ THE SECOND WORLD WAR IN EUROPE 1939–45 *pages 232–33*

THE UNITED STATES SINCE 1900

▼ After the Second World War people began to migrate from the industrialized northeast and Midwest to the Pacific region, where high-technology industries were being developed. By the end of the century California was not only the most populous state but also an international economic powerhouse.

DISTRIBUTION OF POPULATION IN 1900

TOTAL POPULATION: 58,024,000

1941

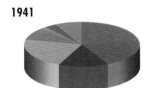

TOTAL POPULATION: 131,595,000

1996

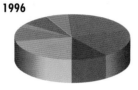

TOTAL POPULATION: 265,285,000

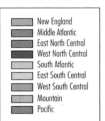

New England
Middle Atlantic
East North Central
West North Central
South Atlantic
East South Central
West South Central
Mountain
Pacific

Since 1900 there have been many dramatic changes in the nature of society in the United States. In 1900 the population was around 76 million, of whom more than half lived in either the northeast or Midwest (*pie charts*). Over 87 per cent were white and just over 10 per cent were African American. The life expectancy of an American born in 1900 was 47 years, and only 4 per cent of the population was over 65 years old. A large percentage still lived on farms, and in the years before the motorcar the railroads served as the lifeblood of the nation.

Over the coming decades great social, racial, technological and economic changes were to create a very different country. By the end of the century there were more than 270 million Americans. They were more racially diverse, more spread out (*map 1*), lived longer (76 years on average), were older (nearly 13 per cent were over 65) and generally richer (with an average Gross National Product per capita over five times that of the world average).

During the 20th century huge numbers of Americans migrated to the west and southwest in search of new jobs and greater opportunities. This mobility of labour helped the USA to remain a more flexible and productive economic power than other countries and was part of a realignment in the economy which saw the percentage employed in services increase from 40 to 76 per cent between 1920 and 1998. Meanwhile, employment in agriculture fell from 25 to 5 per cent and in industry from 35 to 19 per cent.

IMMIGRATION AND CIVIL RIGHTS

Immigration to the USA reached a peak in the early years of the 20th century, but from the 1920s onwards a more restrictive approach was adopted. A quota system was introduced for each nationality, based on the percentage of the existing US population of that nationality. This enabled northern European immigrants to be favoured at the expense of those from other regions of the world.

In 1965 the quota system was replaced by a permitted annual total of immigrants. There was an increase in the number of Hispanic Americans (people originally from Latin America, Cuba and Puerto Rico) in US society. By the end of the century they made up over 10 per cent of the population and were the fastest-growing group in the country. The size of other ethnic groups also increased dramatically, in particular those from Japan, the Philippines,

South and Southeast Asia. The Native American population also grew in the last decades of the century, although less dramatically: at the end of the 20th century they made up around 1 per cent of the population.

In 1900 African Americans were politically and socially marginalized, the majority living on farms in the Deep South (*map 2*) where their parents or grandparents – if not they themselves – had been slaves. While they were supposedly guaranteed equal rights by the constitution, most southern states, politically dominated by whites, enforced segregation. In many places they were discouraged from voting by poll taxes, literacy tests and other intimidatory tactics.

The industrial boom of the early 20th century, coupled with two world wars, created a need for factory workers in the northeast and Midwest. Many African Americans migrated there to find work and established neighbourhoods, with their own traditions and cultures, in cities such as New York, Detroit and Chicago. Their political power was still curtailed and, with the famous exception of Henry Ford's automobile plants, African Americans were usually given less prestigious and lower-paid jobs than whites.

The Civil Rights movement began in the 1950s with pressure both from above and below. In 1954 the famous Supreme Court decision Brown v Board of Education attacked the notion of state segregation. In the 1950s African Americans protested against enforced segregation and in Montgomery, Alabama they forced the town authorities to let them sit with whites on town buses (*map 3*). Subsequently, not only the South but the USA as a whole was forced to confront the issue of racial inequality. The 1960s were particularly turbulent, with legal victories for equality being won in the face of continuing racism.

POLITICAL DEVELOPMENTS

These social changes acted as a catalyst for some important political changes in the USA. At the beginning of the 20th century the country's two major political parties, the Republicans and the Democrats, were more sectional groupings – often with competing interests – than ideological entities. The Democrats were loyally supported by the bulk of southern whites, for reasons stretching back to Republican rule during the Civil War, and were also often backed by a large number of farmers from poorer western states and different ethnic coalitions in the large cities. By

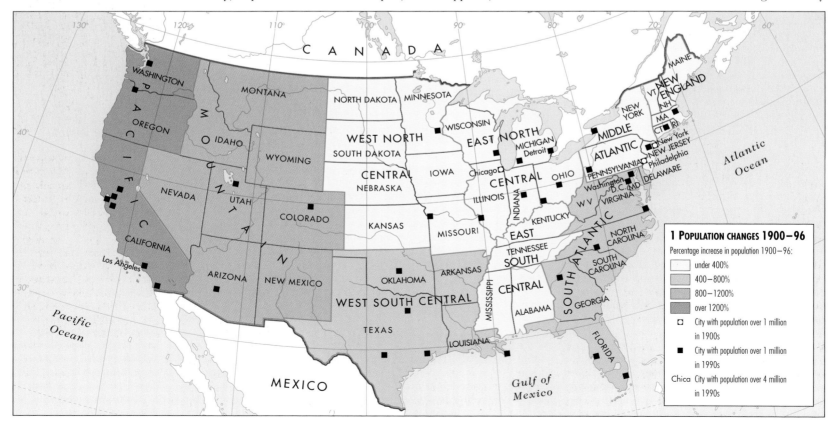

1 POPULATION CHANGES 1900–96

Percentage increase in population 1900–96:

under 400%
400–800%
800–1200%
over 1200%

▢ City with population over 1 million in 1900s

■ City with population over 1 million in 1990s

Chica City with population over 4 million in 1990s

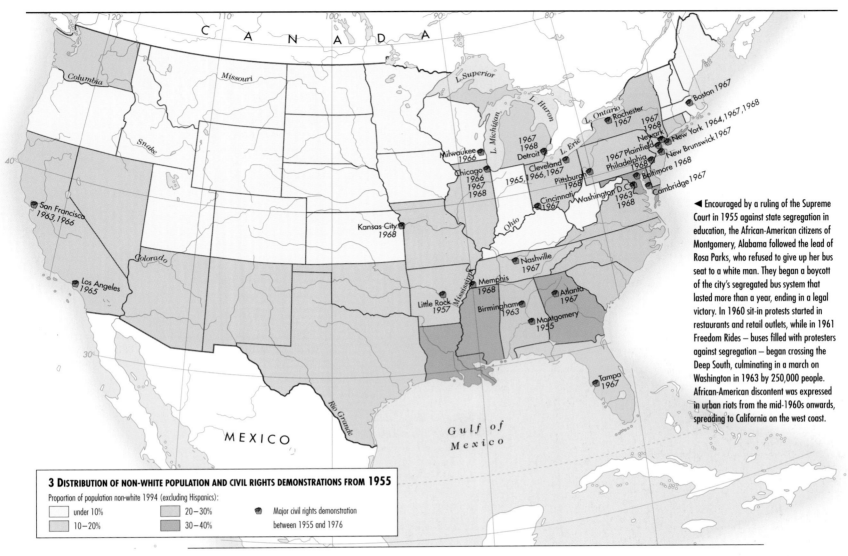

► Encouraged by a ruling of the Supreme Court in 1955 against state segregation in education, the African-American citizens of Montgomery, Alabama followed the lead of Rosa Parks, who refused to give up her bus seat to a white man. They began a boycott of the city's segregated bus system that lasted more than a year, ending in a legal victory. In 1960 sit-in protests started in restaurants and retail outlets, while in 1961 Freedom Rides – buses filled with protesters against segregation – began crossing the Deep South, culminating in a march on Washington in 1963 by 250,000 people. African-American discontent was expressed in urban riots from the mid-1960s onwards, spreading to California on the west coast.

3 DISTRIBUTION OF NON-WHITE POPULATION AND CIVIL RIGHTS DEMONSTRATIONS FROM 1955

Proportion of population non-white 1994 (excluding Hispanics):

- under 10%
- 10–20%
- 20–30%
- 30–40%
- Major civil rights demonstration between 1955 and 1976

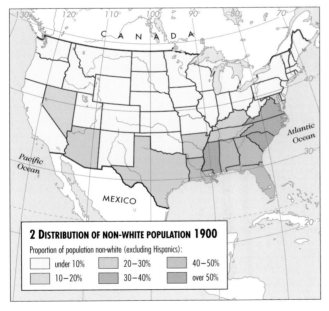

2 DISTRIBUTION OF NON-WHITE POPULATION 1900

Proportion of population non-white (excluding Hispanics):

- under 10%
- 10–20%
- 20–30%
- 30–40%
- 40–50%
- over 50%

▲ In 1900 African Americans remained concentrated in the southern states. Native Americans were scattered throughout the West, on reservations and territories to which they had been forcibly resettled in the 19th century. Hispanic Americans lived mainly in states that had been part of Mexico before 1848. By the end of the 20th century the population of many states had become more ethnically diverse (*map 3*). The non-white percentage of the population in the northeastern industrial regions, and in California, Texas and New Mexico, had increased markedly, partly as a result of internal migration, but also due to a large influx of migrant workers, many of whom were illegal immigrants. Successive US governments have placed restrictions on immigration, starting with the law of 1862 prohibiting Chinese immigration. However, illegal immigrants continue to find their way into the country, the majority crossing the border from Mexico, while others brave the dangers of the sea crossing from Cuba.

contrast, the backbone of the Republican Party was the middle-class business community and farmers in the northeast and Midwest, though the party also garnered a large part of the working-class vote. There were other, smaller, parties, including the Socialists, but they invariably performed poorly at election time.

The situation began to change significantly during the era of the Great Depression (1929–33) and the subsequent New Deal policies of Democrat President Franklin D. Roosevelt (*pages 228–29*). Previously, African Americans had, when allowed to vote, almost always supported the Republicans (the party of Abraham Lincoln), but Roosevelt's massive increases in government social spending caused both they and many working-class white voters to switch allegiance to the Democrats. As a result, the Democrats took over the Republicans' previous role as the natural party of government, and from the 1930s regularly won a majority of the seats in Congress, especially in the House of Representatives. However, during the 1980s a reverse migration of southern whites, often evangelical Christians, into the Republican Party created a situation of approximate balance. The parties have now developed more distinctive ideologies, with the Republicans on the whole supporting fewer taxes, less government regulation and smaller government welfare plans than the Democrats.

Many of the changes that have occurred since 1900 have led to an ongoing and emotional debate about what exactly it means to be "an American". The traditional idea of a "melting pot", whereby immigrants were expected to shed many of their old customs in order to become fully American, has been challenged, particularly on the Left, by the idea of a "great mosaic". Ethnic minorities are now encouraged by some to maintain their separate identities, although other factions have fought this idea, believing that it could undermine the cohesion of the American nation.

▲ The Reverend Martin Luther King started his political life as leader of the Montgomery bus boycott. His policy of passive resistance, to which he adhered in the face of criticism from more militant African-American leaders, was based on the teachings of Gandhi. He was a powerful orator, famous for his "I have a dream" speech, first delivered in 1963. Despite important legislative victories won by the civil rights movement, protests became increasingly violent in the mid-1960s – a situation that was exacerbated when Dr King was assassinated in 1968.

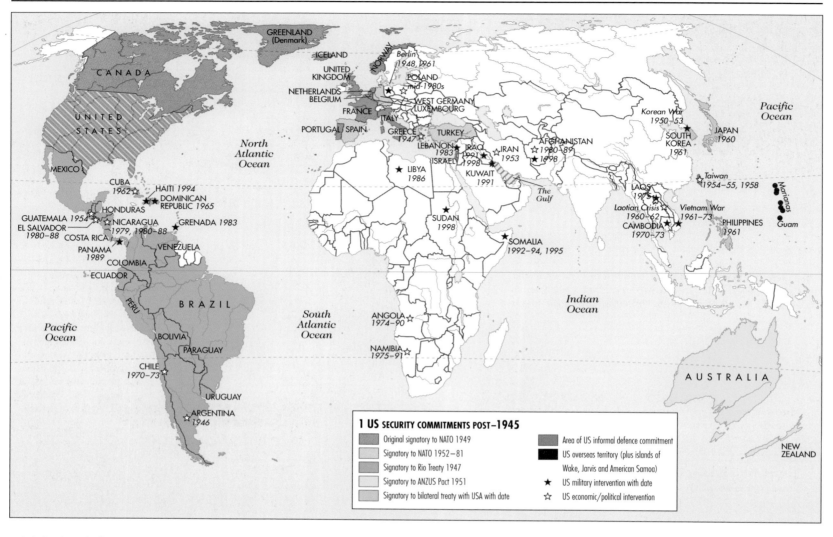

1 US SECURITY COMMITMENTS POST–1945

- Original signatory to NATO 1949
- Signatory to NATO 1952–81
- Signatory to Rio Treaty 1947
- Signatory to ANZUS Pact 1951
- Signatory to bilateral treaty with USA with date
- Area of US informal defence commitment
- US overseas territory (plus islands of Wake, Jarvis and American Samoa)
- ★ US military intervention with date
- ☆ US economic/political intervention

▲ As the United States has become more powerful economically it has extended its area of involvement beyond the American continent to Africa, Southeast Asia and Europe. Although it has sometimes considered it necessary to employ force to defend its interests, in many instances economic backing or, conversely, the threat of trade sanctions has been sufficient to achieve its objectives.

At the end of the Second World War the United States dominated the globe. It not only had the world's largest navy and air force, and its second largest army, but it also dwarfed all other national economies. With most major European and Asian countries devastated by war, it was estimated that the United States produced half of the world's goods in 1945. The question facing the United States was what it should do with its tremendous power.

Before the Second World War US foreign policy had been unpredictable. With much of the country firmly isolationist, there was no national consensus as to what part the United States should take in world affairs. Most Americans seemed content to play a dominant role in North, Central and South America (*pages 226–27*) but had little interest in intervening in conflicts elsewhere in the globe. After the Second World War many of those responsible for US foreign policy, such as President Truman and Secretary of State George Marshall, considered that a return to this pre-war position would be untenable given the growth of the power of the Soviet Union.

Although the United States and the Soviet Union had been allies during the war, this relationship had been forced on them by necessity and by no means represented a healing of the huge ideological rift between them. In the period following the end of the war the Soviets increased their domination of Eastern Europe (*pages 236–37*), and many Americans worried that if the USA withdrew its forces from Western Europe the USSR would eventually dominate the whole continent. The USA, committed to free enterprise, and hitherto dependent on Europe for a large part of its export trade, was alarmed at the prospect of communist governments restricting trade with the non-communist world. Likewise, the Soviet government, led by Stalin, was suspicious of a western hemisphere dominated by the USA, and expressed doubt that capitalism and communism could peacefully coexist for long.

THE COLD WAR YEARS

The perceived threat posed by the Soviet Union eventually proved decisive in the development of the United States into an economic and military world power. President Truman committed the USA to a policy of "containment", involving resistance to the spread of communism anywhere in the world. In 1949 the USA played a key role in the formation of the North Atlantic Treaty Organization (NATO) (*map 1*), which committed it to defending Western Europe. By this time the "Cold War" between the USA and the USSR was a reality and would continue to dominate international relations for the next four decades (*pages 244–45*).

There was a slight thaw in relations during the 1970s, when the USA (under presidents Nixon, Ford and Carter) and the USSR (under General Secretary Brezhnev) adopted a policy of "detente", whereby the two countries recognized each other as equals and tried to settle a number of outstanding differences through negotiation. However, this policy proved very controversial in the United States; many saw it as a capitulation to communism and called for less co-operation and more confrontation with the USSR. In 1980 Ronald Reagan, one of the harshest critics of detente, was elected US president. He committed his country to rolling back the "evil empire", as he described the Soviet Union, and began the largest peacetime military build-up in United States history.

Reagan and his advisers were gambling that they could bankrupt the Soviet Union without causing all-out war and without damaging the US economy. In the end the policy seemed to work. The USSR, even though it devoted a far larger proportion of its economy to military expenditure than did the USA, found it impossible to match the advanced technology of its rival. By 1989 Soviet president Mikhail Gorbachev had recognized that drastic changes were needed in order to reduce international tension and give the economy of the Soviet Union opportunities for

▲ In February 1945 Churchill, Roosevelt and Stalin met at Yalta to discuss plans for the post-war division of Europe. As the leading superpower the USA realized that its pre-war isolationist policy was no longer tenable, and that it had a major role to play in the reconstruction of Europe and in the encouragement of democratic regimes.

expansion. Gorbachev's liberalization led ultimately to the break-up of the Soviet Union in December 1991 (*pages 262–63*), as a result of which the United States lost its major adversary and the Cold War came to an end.

INTERVENTION WORLDWIDE

The policy of the United States during the Cold War was eventually successful in destroying Soviet power, but it had damaging repercussions for US international relations in some parts of the world. The USA often felt it necessary to overthrow or undermine regimes largely because they were influenced by communist ideas, while at the same time supporting manifestly corrupt and oppressive right-wing regimes considered friendly to the USA. Cuba, Guatemala, El Salvador, Nicaragua and Panama all had their governments either supported or besieged according to whether they were perceived by the US government as loyal or threatening (*map 1*). The most extreme example of US intervention was the Vietnam War. President Kennedy committed US ground troops to Vietnam in the early 1960s in an effort to "save" Vietnam and its neighbouring countries from communism (*pages 250–51*), but even with over half a million troops fighting in Vietnam the US government could not "save" a people who did not wish to be saved. During the war 60,000 US military personnel and two million Vietnamese lost their lives, with millions more Vietnamese left wounded, orphaned and homeless.

TRADING LINKS

The United States strengthened trade with its American neighbours during the second half of the 20th century, and also looked westwards to the rapidly growing Southeast and East Asian economies. This change of focus was reflected in various trade agreements: the founding of the Organization of American States (OAS) in 1948, the signing of the North American Free Trade Agreement in 1992 (effective from 1994), and the founding of the Asia–Pacific Economic Co-operation Organization in 1989 (*map 2*).

The recovery in its economic position (*bar chart*) has encouraged the USA to play a leading role in the global push towards more open trading markets, although within the USA there is a powerful lobby, including many labour organizations, that oppose the idea of tariff-free trade with other countries. Pledges by the OAS and APEC to introduced tariff-free trading in the future have also met with scepticism, both within the USA and in some of the other member states. Yet, even though no one can say with any certainty how these organizations will develop, it is clear that the role of the USA – with few national security concerns – will increasingly rely, and focus, on its economic position.

▶ The period from 1945 to the mid-1980s was tumultuous for the US economy: its position of world economic dominance was slowly whittled away as Western European countries and Japan recovered from their wartime devastation – in part with the aid of US investment. During the 1980s and 1990s, however, the US economy started to show unexpected vigour compared with the slower growth of many of its rivals.

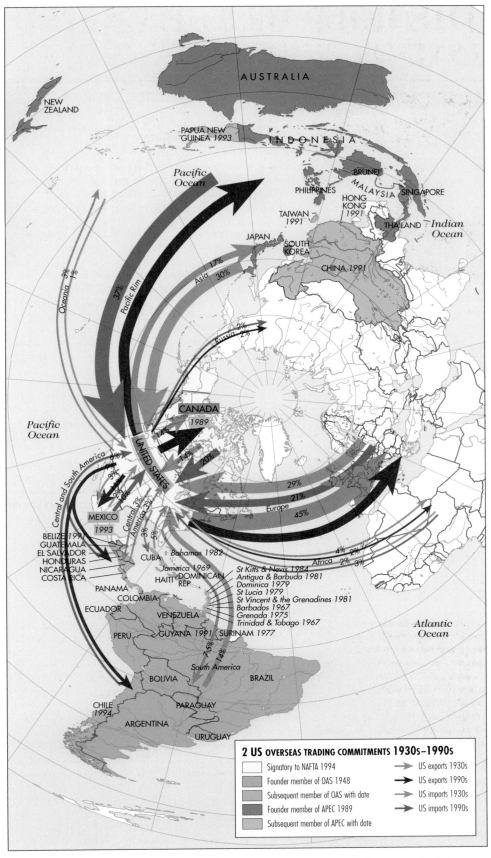

2 US OVERSEAS TRADING COMMITMENTS 1930s–1990s

Signatory to NAFTA 1994	US exports 1930s
Founder member of OAS 1948	US exports 1990s
Subsequent member of OAS with date	US imports 1930s
Founder member of APEC 1989	US imports 1990s
Subsequent member of APEC with date	

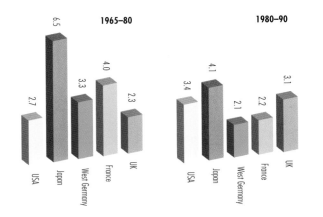

COMPARATIVE GROWTH OF NATIONAL GDP 1965–1990
(average annual percentage growth in GDP)

1965–80
USA 2.7
Japan 6.5
West Germany 3.3
France 4.0
UK 2.3

1980–90
USA 3.4
Japan 4.1
West Germany 2.1
France 2.2
UK 3.1

▲ The North American Free Trade Agreement (NAFTA), which allows for tariff-free trade between the USA, Canada and Mexico, came into effect in 1994 and has been followed by even more ambitious attempts to create wide free-trade areas. Both the Organization of American States (OAS) and the Asia-Pacific Economic Co-operation (APEC) Organization have proclaimed their intention of establishing free trade between their member states, in 2005 and 2020 respectively.

◑ THE UNITED STATES SINCE 1900 *pages 240–41*

THE COLD WAR
1947–91

▲ The phenomenal force of the nuclear bomb, which had been so effectively demonstrated in Hiroshima and Nagasaki in August 1945, dominated the Cold War years, with both sides building up huge arsenals of weapons. In 1963, in the wake of the near-disastrous Cuban Missile Crisis, the United States and the Soviet Union agreed a test-ban treaty. However, despite the Strategic Arms Limitation talks, which culminated in the signing of treaties in 1972 (SALT I) and 1979 (SALT II), and the Strategic Arms Reduction Talks (START), which opened in 1982, the destructive capacity of the two superpowers continued to grow.

▶ At the end of the Second World War Korea, previously a Japanese colony, was divided along the 38th parallel. North Korea came under the control of a communist-inspired, Soviet-backed regime, while South Korea was supported by the USA. In June 1950 North Korean troops advanced across the 38th parallel in a bid to unify the country. They had nearly gained control of the entire peninsula when United Nations (mostly US) troops landed both in the southeast of the country and at Inchon, behind North Korean lines.

The UN troops advanced almost to the border with China, which reacted to this apparent threat to its territory and launched an attack in support of the North Koreans. For the next two months the UN troops were on the defensive, but by June 1951 they had driven the Chinese and North Koreans back to a line north of the 38th parallel. Protracted negotiations followed, with a truce eventually being signed in July 1953. The war had resulted in an estimated four million casualties.

The Cold War was an ideological, political and diplomatic conflict in the years 1947–91, between the United States and its allies on the one hand and the communist bloc led by the Soviet Union on the other. Characterized by extreme tension and hostility, it had a detrimental effect on international relations in this period.

At the Yalta Conference in February 1945 the United States, the Soviet Union and Britain had agreed that free elections would be held throughout Eastern Europe. It soon became apparent, however, that the Soviet Union under Stalin intended instead to fill the political vacuum in Eastern Europe with communist governments loyal to Moscow. By 1948 the governments of Poland, East Germany, Hungary, Romania, Bulgaria and Czechoslovakia had been transformed from multiparty coalitions, as envisaged by the Yalta Declaration, to governments composed entirely of communists who adhered strictly to the ideologies, policies and practices of the government in Moscow (pages 236–37). The "Iron Curtain", dividing the communist regimes from the rest of Europe, had fully descended.

THE TRUMAN DOCTRINE

Despite these events in Europe, President Truman of the USA hoped that some form of co-operation with the USSR could continue. In February 1947, however, when the British announced that they were no longer able to provide economic and military support for the Greek and Turkish governments, the USA felt compelled to intervene. Not to do so might allow Greece, in particular, to fall to the communists, thus creating a threat to US global interests and national security. The result was the "Truman Doctrine",

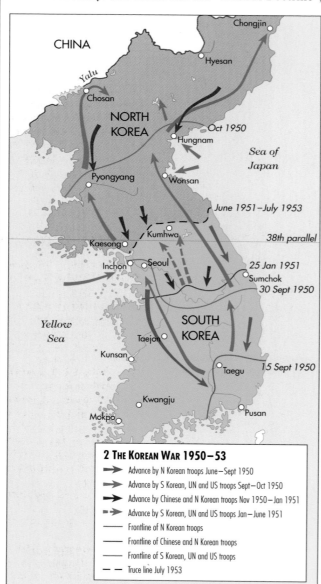

2 THE KOREAN WAR 1950–53

➤ Advance by N Korean troops June–Sept 1950
➤ Advance by S Korean, UN and US troops Sept–Oct 1950
➤ Advance by Chinese and N Korean troops Nov 1950–Jan 1951
➤ Advance by S Korean, UN and US troops Jan–June 1951
— Frontline of N Korean troops
— Frontline of Chinese and N Korean troops
— Frontline of S Korean, UN and US troops
-- Truce line July 1953

which stated that the USA would oppose any further expansion of communist territory and would provide a financial package to help Greece and Turkey defend themselves from external interference. This was followed by the Marshall Plan, which provided $13.5 billion in economic aid to the war-torn countries of Europe. It was hoped that this would combat the spread of communism across the continent, but it was only partially successful because the states in Eastern Europe refused, or were prevented by Moscow from accepting, Marshall Aid.

THE DEEPENING OF THE WAR

Following the announcement of the Truman Doctrine, the Cold War deepened (map 1) with the Berlin Blockade of 1948–49, a communist uprising in Malaya in 1948, and the formation of the People's Republic of China in 1949, when the Chinese communists, led by Mao Zedong and supported by the USSR, finally defeated the US-backed forces of Chiang Kai-shek (pages 254–55). All these crises encouraged the creation of a string of Western military alliances to deter any further expansion of communist territory, beginning with the formation of the North Atlantic Treaty Organization (NATO) in 1949.

In the same year the USSR produced its first atomic bomb, and the Cold War took on a new character. From the point of view of the NATO countries the tension was increased, while the USSR, knowing that it could match NATO in nuclear capacity, gained in confidence. In 1955 it established with other Eastern European countries a military alliance known as the Warsaw Pact. Despite, or because of, the huge arsenal of nuclear weapons stockpiled by both sides, none was ever used in warfare. Indeed, the Cold War never resulted in actual combat between US and Soviet troops, the risk of nuclear weapons becoming involved being far too high. Instead, it took on the form of an arms race – and later a space race – and the provision of economic aid and military equipment to other countries in order to gain political influence and thus strategic advantage. In some cases both sides intervened to defend their own ideology, and in a few cases one of them sent in troops.

The Korean War of 1950–53, when communist North Korea invaded South Korea, was one of the largest and bloodiest confrontations of the Cold War (map 2). It marked the beginning of over 12 years of intense global tension and rivalry between the superpowers, which culminated in the Cuban Missile Crisis of 1962 (map 3). The discovery by the USA of Soviet missiles being assembled on communist-led Cuba, within easy range of the US mainland, led to the gravest crisis of the Cold War. It almost resulted in a third world war, the tension easing only when the Soviet leader, Nikita Khrushchev, agreed to withdraw the missiles.

THE THAWING OF THE WAR

Over the next 20 years both superpowers attempted to ease tensions and "thaw" the Cold War. The resulting "detente" produced superpower summit meetings and agreements to reduce nuclear arsenals. Meanwhile, competition between the superpowers continued in Vietnam where, between 1964 and 1973, the US deployed hundreds of thousands of troops to fight communist North Vietnamese forces who were attempting to unify their country (pages 250–51).

In 1979 detente was abruptly ended when the USSR invaded Afghanistan, producing a new period of tension and hostility between the superpowers, and a fresh arms race. This lasted until 1985 when the new Soviet leader, Mikhail Gorbachev, began to de-escalate the Cold War by reviving summit meetings and arms negotiations with the USA. He also began a process of internal reform in the USSR itself and gradually relaxed the Soviet grip on Eastern Europe. This resulted in the collapse of communism throughout the Eastern European bloc following the "People's Revolutions" of 1989 and 1990 (pages 264–65), and the dissolution of the Soviet Union (pages 262–63). With the demise of the USSR and the formal dissolution of the Warsaw Pact in 1991, the Cold War came to an end.

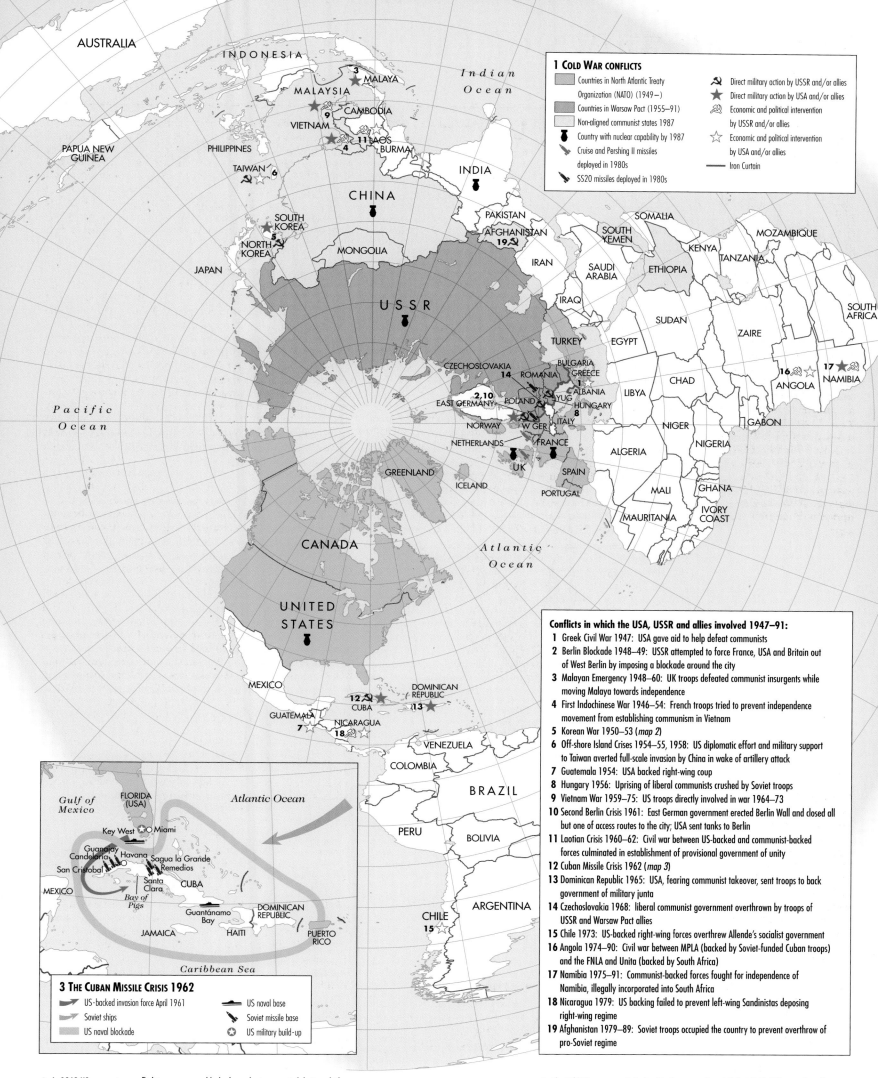

1 COLD WAR CONFLICTS

- Countries in North Atlantic Treaty Organization (NATO) (1949–)
- Countries in Warsaw Pact (1955–91)
- Non-aligned communist states 1987
- Country with nuclear capability by 1987
- Cruise and Pershing II missiles deployed in 1980s
- SS20 missiles deployed in 1980s
- Direct military action by USSR and/or allies
- Direct military action by USA and/or allies
- Economic and political intervention by USSR and/or allies
- Economic and political intervention by USA and/or allies
- Iron Curtain

Conflicts in which the USA, USSR and allies involved 1947–91:

1. **Greek Civil War 1947:** USA gave aid to help defeat communists
2. **Berlin Blockade 1948–49:** USSR attempted to force France, USA and Britain out of West Berlin by imposing a blockade around the city
3. **Malayan Emergency 1948–60:** UK troops defeated communist insurgents while moving Malaya towards independence
4. **First Indochinese War 1946–54:** French troops tried to prevent independence movement from establishing communism in Vietnam
5. **Korean War 1950–53** (map 2)
6. **Off-shore Island Crises 1954–55, 1958:** US diplomatic effort and military support to Taiwan averted full-scale invasion by China in wake of artillery attack
7. **Guatemala 1954:** USA backed right-wing coup
8. **Hungary 1956:** Uprising of liberal communists crushed by Soviet troops
9. **Vietnam War 1959–75:** US troops directly involved in war 1964–73
10. **Second Berlin Crisis 1961:** East German government erected Berlin Wall and closed all but one of access routes to the city; USA sent tanks to Berlin
11. **Laotian Crisis 1960–62:** Civil war between US-backed and communist-backed forces culminated in establishment of provisional government of unity
12. **Cuban Missile Crisis 1962** (map 3)
13. **Dominican Republic 1965:** USA, fearing communist takeover, sent troops to back government of military junta
14. **Czechoslovakia 1968:** liberal communist government overthrown by troops of USSR and Warsaw Pact allies
15. **Chile 1973:** US-backed right-wing forces overthrew Allende's socialist government
16. **Angola 1974–90:** Civil war between MPLA (backed by Soviet-funded Cuban troops) and the FNLA and Unita (backed by South Africa)
17. **Namibia 1975–91:** Communist-backed forces fought for independence of Namibia, illegally incorporated into South Africa
18. **Nicaragua 1979:** US backing failed to prevent left-wing Sandinistas deposing right-wing regime
19. **Afghanistan 1979–89:** Soviet troops occupied the country to prevent overthrow of pro-Soviet regime

3 THE CUBAN MISSILE CRISIS 1962

- US-backed invasion force April 1961
- Soviet ships
- US naval blockade
- US naval base
- Soviet missile base
- US military build-up

▲ In 1962 US reconnaissance flights detected evidence that the Soviet Union was building nuclear missile bases on Cuba, within range of the US mainland. A US naval blockade, and a tense period during which nuclear war appeared likely, eventually resulted in the USSR, under Khrushchev, agreeing to dismantle the nuclear bases.

▲ The Cold War was a period of political and economic confrontation between the two superpowers and their allies. The area of highest tension was along the "Iron Curtain" that divided Western from Eastern Europe, but the two sides' opposition to each other was played out in conflicts — some of a military nature — all over the world.

THE SECOND WORLD WAR IN EUROPE 1939–45 *pages 232–33*

THE BREAKDOWN OF EMPIRES SINCE 1945

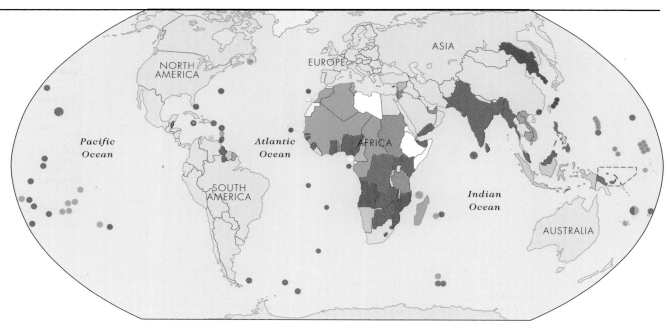

► In 1939 large areas of the world were still under colonial rule, although in India and Africa, in particular, the colonial powers depended on indigenous political rulers to administer at the local level. Immediately after the conclusion of the First World War the League of Nations established mandates according to which countries victorious in the war, such as Britain and France, undertook to administer regions that had previously been colonies of Germany or the Ottoman Empire, with eventual independence as the ultimate goal. Japan was the only country to expand its empire during the inter-war period, moving into Manchuria in 1931 as a prelude to its full-scale assault on China in 1937.

B efore the Second World War the European colonial empires seemed largely secure (*map 1*). Despite independence movements in India (*pages 248–49*) and French Indochina (*pages 250–51*), and the growth of trade unions and early political movements in Africa and the Caribbean, colonial rule was widely expected to continue well into the 21st century. Yet within 20 years of the war's end most colonies had become independent, leaving only a few outposts whose future had still to be resolved (*map 2*).

The war's corrosive effects on colonialism were initially seen most clearly in Asia. Some colonies, such as Malaya and French Indochina, experienced invasion and occupation by Japanese forces, unleashing anti-colonial nationalism which could not be reversed after the war. The African colonies, meanwhile, became vital sources of military manpower and raw materials for the Allied war effort, the mobilization of which involved economic and social change. Colonial governments were forced to depart from their traditional approach of working through local political rulers and to adopt a more interventionist approach. This laid them open to local criticism of wartime restrictions, food shortages and many other hardships – grievances that often escalated into early forms of political protest.

Paradoxically, although the war weakened most of the colonial powers, it also increased their desire to utilize colonial resources to assist their own economic recovery after the war. The colonial powers sometimes used force in the face of growing local resistance to their rule, as seen in the unsuccessful attempts by the French and Dutch to re-establish control of Indochina and Indonesia respectively, and in Britain's ultimately successful campaign to defeat a communist insurrection in Malaya.

THE INEVITABILITY OF INDEPENDENCE

Much of sub-Saharan Africa became independent between 1956 and 1962. Partly responding to the "winds of change" of African nationalism, Britain accelerated its plans for decolonization, and most of its African colonies became independent in the early 1960s (*map 2*). The major obstacle proved to be the resistance of white settlers to African

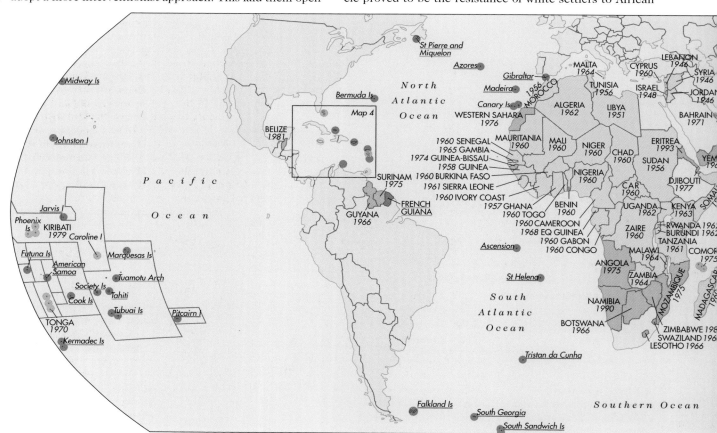

majority rule in East and Central Africa, and Zimbabwe's legal independence was delayed until 1980. By the 1970s only the Portuguese dictatorship seemed determined to retain its African colonies, fighting a series of protracted wars against guerrilla movements. The financial and human cost to Portugal was enormous, provoking a military coup and revolution in 1974, with the new government committed to rapid decolonization.

After 1945 colonialism increasingly became an international issue. Both the United States and the Soviet Union had traditionally been hostile to European colonial rule and had put pressure on their wartime allies, Britain and France, to make a commitment to reform. In the immediate post-war period the colonial powers attempted to raise the living standards of the indigenous peoples in their colonies, hoping thus to appease both local feeling within the colonies and the international community. As the Cold War intensified (pages 244–45), the superpowers competed for influence in the developing world, both in ex-colonies and in colonies soon to become independent. Moreover, the United Nations, now responsible for the territories mandated by the League of Nations, became an important forum for criticism of colonialism. Arguments for faster decolonization intensified as former colonies themselves became members of the UN.

An important factor by the early 1960s was the desire to avoid costly, and probably unwinnable, wars against colonial nationalist movements. The long and bloody Algerian War (1954–62), as a result of which France lost control of Algeria, had demonstrated the perils of opposing demands for independence. Furthermore, such conflicts risked escalating the Cold War if the communist bloc offered support to the forces fighting for independence.

Another consideration was the shifting pattern of international trade. By the late 1950s economic integration in Western Europe (pages 238–39) was giving rise to serious doubts about the likely returns from large-scale colonial investment. Moreover, as the French demonstrated, it was possible to decolonize while preserving many of the advantages, commercial and otherwise, of formal colonial rule. A major consideration influencing British and French policymakers, therefore, was the hope that their respective colonies would opt after independence to join the Commonwealth of Nations (map 3) or the French Community. The

great majority of former British colonies did choose this form of continuing association, so that decolonization seldom represented an abrupt change in relationships. Despite the effective collapse of the French Community in 1960, France has maintained close economic, diplomatic and military links with many of its former possessions.

SMALL ISLAND STATES
Decolonization posed the question of whether small island states, particularly those in the Caribbean (map 4) and the Pacific, could achieve viable independent nationhood. One solution was to group small territories together into larger political units. The Federation of the West Indies was formed in 1958 after many years of negotiation, although British Guiana and British Honduras opted not to join. However, when its larger, more prosperous members, Jamaica and Trinidad and Tobago, gained separate independence in 1962 the Federation was dissolved. Other island territories, such as Gibraltar, had originally been acquired for their strategic value, but this declined as Britain wound down its overseas defence commitments in the late 1960s and early 1970s.

The remaining European dependencies (map 2) are mostly small territories, often islands. In some cases, notably the Falkland Islands/Malvinas (claimed by Argentina) and Gibraltar (claimed by Spain), the issue of sovereignty remains unresolved. In the case of Hong Kong and Macau, the return of sovereignty to China was agreed through negotiated settlements. Some small islands, especially in the Caribbean and Pacific, have opted for a limited form of independence, retaining association with their former colonial power in matters such as defence and diplomacy, while others, including many islands in French Polynesia, have rejected offers of independence.

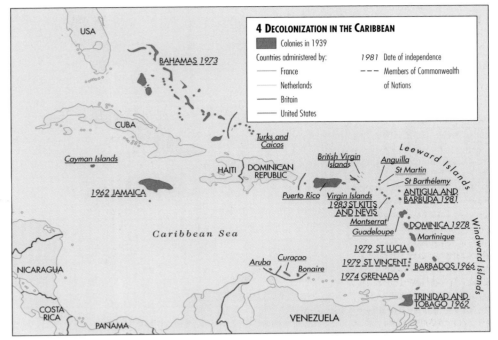

▲ Because of their small size, many of the Caribbean islands are not economically viable as independent states. Attempts to form an economic and political union, known as the Federation of the West Indies, failed when the larger ex-colonies opted out, leaving islands such as Montserrat to be administered as British dependencies. All the ex-British colonies in the Caribbean opted to join the Commonwealth of Nations on achieving their independence.

▼ The expansion of the British Commonwealth (the Commonwealth of Nations) in 1947 to include India and Pakistan enabled the organization to evolve into a multi-ethnic grouping, which nearly all Britain's former colonies decided to join. South Africa left the Commonwealth in the face of condemnation of its policy of apartheid, but rejoined in 1994. Pakistan left in 1972 in protest at the admission of Bangladesh to the Commonwealth, but rejoined in 1989. In 1997 the first countries not previously British colonies – Cameroon and Mozambique – were admitted.

SOUTH ASIA
SINCE 1920

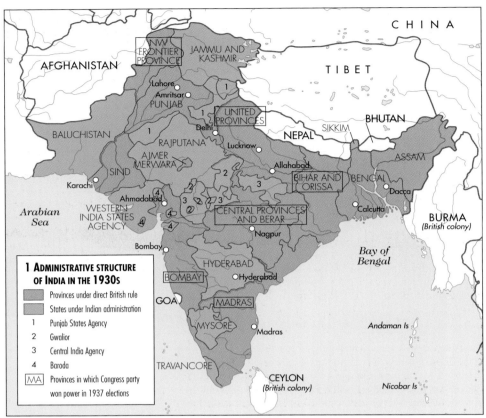

1 ADMINISTRATIVE STRUCTURE OF INDIA IN THE 1930s

Provinces under direct British rule

States under Indian administration

1 Punjab States Agency

2 Gwalior

3 Central India Agency

4 Baroda

MA Provinces in which Congress party won power in 1937 elections

▲ The administration of India in the 1930s was undertaken in some areas by the British, but in others by local Indian rulers and agencies. In the 1937 elections the Congress Party won political control in provinces across the country.

▶ India's population increased significantly in the second half of the 20th century, trebling in under 55 years. Its growth rate also accelerated, so that by the end of the century the population was increasing by 25 per cent every ten years.

1 INDIA'S POPULATION 1941–97
(in millions)

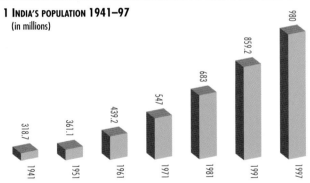

318.7 — 1941
361.1 — 1951
439.2 — 1961
547 — 1971
683 — 1981
859.2 — 1991
980 — 1997

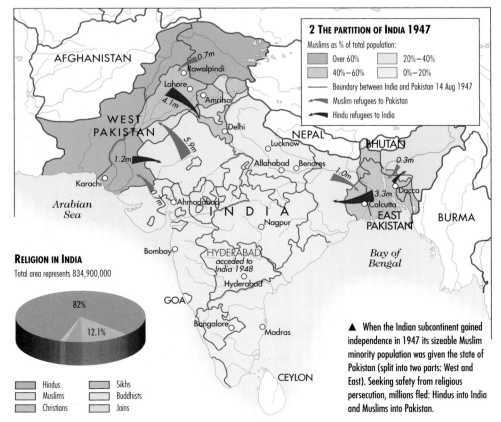

2 THE PARTITION OF INDIA 1947

Muslims as % of total population:

Over 60%   20%–40%

40%–60%   0%–20%

— Boundary between India and Pakistan 14 Aug 1947

Muslim refugees to Pakistan

Hindu refugees to India

RELIGION IN INDIA

Total area represents 834,900,000

82%

12.1%

Hindus   Sikhs

Muslims   Buddhists

Christians   Jains

▲ When the Indian subcontinent gained independence in 1947 its sizeable Muslim minority population was given the state of Pakistan (split into two parts: West and East). Seeking safety from religious persecution, millions fled: Hindus into India and Muslims into Pakistan.

During the 1920s and 1930s a struggle developed between Britain, determined to maintain control over its empire in India, and the growing force of Indian nationalism. Political reforms in 1919, which were ostensibly a step towards eventual self-government, gave elected Indians limited responsibility in provincial government, but failed to satisfy nationalists. Indian protests centred on the campaigns of non-violent civil disobedience organized by the nationalist leader, Mohandas Gandhi, seeking Indian self-rule. Gandhi, and the largely Hindu Indian National Congress Party, mobilized nationwide mass support, undermining British authority and causing alarm among India's large Muslim minority. By the late 1920s Congress was demanding complete independence. Britain's response was to combine repression (involving the detention of nationalist leaders – among them Gandhi and Jawaharlal Nehru) with constitutional reforms in 1935. These gave substantial power to autonomous provincial governments, while keeping overall control in British hands. In the 1937 provincial elections, Congress won power in a number of provinces (including the largely Muslim North West Frontier Province) (*map 1*).

The Second World War transformed the situation. In India Britain suspended talk of constitutional change until after the war and Congress ministers resigned in protest at India's involvement in the war without prior consultation. The cost of mobilizing India's economy to support the war effort was high, and was paid for by the victims of the Bengal famine of 1943 in which over one million people died. In 1942, faced by a possible Japanese invasion, Britain offered India independence after the war, in return for its wartime support. Congress replied with the massive "Quit India" protest campaign, which resulted in its leaders being imprisoned until 1945. Meanwhile, the Muslim League committed itself to forming a separate Muslim state (Pakistan).

By 1945 Britain, lacking the will or the resources to rule by force, sought to accelerate India's independence. Britain hoped to maintain Indian unity through a federal structure, but Congress insisted on a strong, centralized government, while the Muslim League demanded greater provincial autonomy. In the face of violence between the Hindu and Muslim communities, Congress agreed to the partition of India, with the creation of a separate Pakistan from the mainly Muslim western provinces and Bengal. In August 1947 India and Pakistan became independent (*map 2*), and millions of Hindu and Muslim refugees subsequently sought safety in the two new states. At least one million people died in attacks and reprisal killings carried out by one or other of the opposing religious groups. Despite the mass migration, India's population still includes a substantial proportion of Muslims (*pie chart*).

INDIA SINCE INDEPENDENCE

Since independence India has remained the world's largest democracy. During the premiership of Nehru (1947–64), his government introduced five-year plans, and controlled foreign and private enterprise, in an effort to increase agricultural and industrial production. Given India's rapidly growing population (*bar chart 1*) it was imperative to boost food production and the late 1960s saw the beginnings of a "green revolution", in which modern farming techniques were employed with some success (*bar chart 2*). Attempts were made to attack poverty and social underprivilege, although measures to emancipate women and the lower castes were seen as challenging traditional Hindu values.

In 1966 Nehru's daughter, Indira Gandhi, became prime minister. Her attempts to tackle mass poverty and encourage birth control alienated conservative opinion. She was found guilty of electoral corruption in 1975 and declared a state of emergency. Briefly imprisoned in 1978, Mrs Gandhi regained power in 1980. During the 1980s communal tensions re-emerged, with minority groups demanding greater recognition (*map 3*). Growing Sikh separatism led to Mrs Gandhi's assassination by Sikh extremists in 1984. Tensions also emerged between the central government and India's Naga, Tamil and Muslim communities.

PAKISTAN AND BANGLADESH

Pakistan began life as two ethnically distinct territories physically separated by India (*map 2*). The country faced poverty and political division, aggravated by West Pakistan's attempts to assert its dominance over East Pakistan. Whereas India was a leading force in the non-aligned movement, Pakistan aligned itself with the Western nations. While the Indian army remained non-political, Pakistan's army, which first seized power in 1958, often intervened in politics. During the 1960s the economic gap between West and East Pakistan widened. In East Pakistan separatism developed under Sheikh Mujib-ur-Rahman, whose Awami League triumphed in the 1970 elections. When West Pakistan sent troops to restore order in 1971, civil war broke out and India intervened on Mujib's behalf. Pakistan was defeated and an independent Bangladesh was created in January 1972. Continuing political instability and military interventions have since added to Bangladesh's problems of mass poverty.

Zulfikar Ali Bhutto's modernization programme in the early 1970s alienated conservative groups in Pakistan, and in 1977 he was ousted in a military coup led by General Zia-ul-Haq, who sought to make Pakistan a more Islamic state. The country returned to democracy in 1988, but faced the problems of Islamic fundamentalism and separatism.

SRI LANKA

The British colony of Ceylon contained, in addition to its majority Buddhist Sinhalese population, a large Hindu Tamil minority. When it became independent in 1948 government attempts to make Sinhalese the official language alienated the Tamil minority, who campaigned for autonomy. In 1960 Mrs Sirimavo Bandaranaike became the world's first woman prime minister. She changed the country's name to Sri Lanka in 1972 and pursued radical socialist policies. Her successor, Junius Jayawardene, reversed this trend and tried to appease the Tamil community. However, in 1983 long-standing ethnic tensions erupted into a prolonged civil war which Indian military intervention in 1987 failed to end.

TERRITORIAL DISPUTES

Since independence, South Asia has witnessed several major territorial disputes (*map 3*). Relations between India and Pakistan were soured by their rival claims to Jammu and Kashmir. Immediately after the formation of India and Pakistan, from which Kashmir initially remained independent, the new Pakistan government sent troops to lay claim to the predominantly Muslim state. The Hindu maharaja, Sir Hari Singh, immediately acceded the state to India, who sent troops in his support, forcing the Pakistanis into a partial withdrawal. The United Nations intervened and ruled in 1949 that a plebiscite should take place, but the two sides failed to reach agreement on how this should be administered. In 1965 serious fighting between India and Pakistan culminated in a Soviet-arranged truce, and in 1972 each country accepted that the dispute should be solved bilaterally. Violent protests in Kashmir for greater autonomy have, however, persisted since the 1980s.

Territorial disputes between India and China escalated after China absorbed Tibet in 1959. In October 1962 China invaded India in Arunachal Pradesh, forcing Indian troops to retreat before a ceasefire was arranged. These regional tensions have led both India and Pakistan to maintain large armies and to develop nuclear weapons, as demonstrated in a series of underground tests in May 1998.

▲ The dynastic tradition in South Asian politics has led to several women holding positions of power. Sirimavo Bandaranaike took control of the Sri Lankan Freedom Party following her husband's assassination and became the world's first woman prime minister in 1960. She served a further term during the 1970s and in 1994 was appointed for a third by her daughter Chandrika Kumaratunga, who was then serving as president.

◄ Since independence in 1947 India and Pakistan have continued to dispute control of Jammu and Kashmir. China also claims a small area of this mountainous region. Elsewhere, border disputes have occurred between India and China, and between Bhutan and China. In 1971 East Pakistan broke away from West Pakistan to form the independent state of Bangladesh, and both Pakistan and India have experienced claims for autonomy from people within their borders, among them the Baluchis in Pakistan and the Nagas in Assam.

The subcontinent's most serious separatist activity has been that of the Tamils in Sri Lanka, where fighting and bomb attacks led to the death of an estimated 30,000 people in the period 1947–99.

▼ Improvements in agricultural practices in India, known as the "green revolution", led to marked increases in productivity from the 1960s to the 1980s, with the amount of wheat harvested more than trebling.

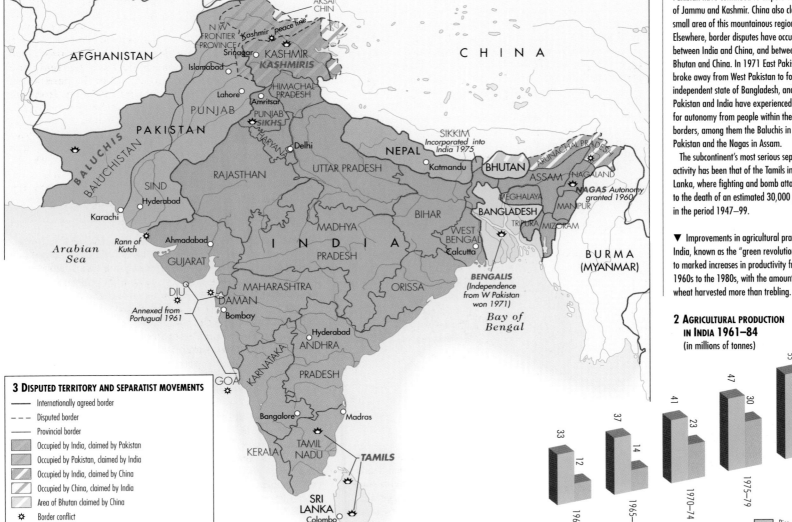

3 DISPUTED TERRITORY AND SEPARATIST MOVEMENTS

— Internationally agreed border
- - - Disputed border
— Provincial border
▨ Occupied by India, claimed by Pakistan
▨ Occupied by Pakistan, claimed by India
▨ Occupied by India, claimed by China
▨ Occupied by China, claimed by India
▨ Area of Bhutan claimed by China
✿ Border conflict
❀ People demanding autonomy

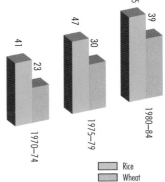

2 AGRICULTURAL PRODUCTION IN INDIA 1961–84
(in millions of tonnes)

Rice	Wheat

SOUTHEAST ASIA SINCE 1920

Indonesia 1965

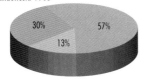

Indonesia 1985

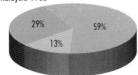

Malaysia 1965

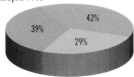

Malaysia 1985

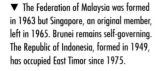

▲ As elsewhere in the world, Southeast Asia has seen a substantial increase in the number of people employed in services and industry in recent decades, at the expense of agriculture.

▼ The Federation of Malaysia was formed in 1963 but Singapore, an original member, left in 1965. Brunei remains self-governing. The Republic of Indonesia, formed in 1949, has occupied East Timor since 1975.

In 1920 Thailand was the only country in Southeast Asia that was not under Western colonial administration, although indigenous anti-colonial movements had been established in most parts of the region, even if in rudimentary form. The next 55 years were to be dominated by the struggle for self-determination – a process which differed markedly from country to country (*map 1*).

At one extreme was the peaceful transfer of power in the Philippines, which had become a colony of the United States at the conclusion of the Spanish–American War in 1898. The United States, with its strong anti-colonial tradition, was uncomfortable with its new responsibilities and moved rapidly to transfer political and administrative powers to Filipinos. In 1935 it established the Philippine Commonwealth, granting the Filipino government control of internal affairs, and promising full independence on 4 July 1946. To a large degree, the process of decolonization was driven by the colonial power itself.

At the other extreme was the turbulent situation in French Indochina and the Dutch East Indies, where anti-colonial agitation was, for much of the 1920s and 1930s, vigorously suppressed by colonial administrations. Between the two extremes was Burma, where, under pressure from the constitutional advances being made in India (*pages 248–49*), the British transferred some administrative responsibilities to the Burmese in the early 1920s.

The Western colonial presence in Southeast Asia was shattered by the Japanese military advance into the region between December 1941 and April 1942 (*pages 234–35*). The fiercely anti-Western sentiments expressed by the Japanese, and their effective destruction of the myth of white supremacy, influenced the political aspirations of the indigenous populations of the region. Following the Japanese surrender in August 1945, the Dutch and French faced severe opposition to their attempts to re-establish control over their former colonies. In the Dutch East Indies a fierce military and political battle was waged between the Dutch and the forces of the newly declared Republic of Indonesia until, towards the end of 1949, the United States – acting through the United Nations – put pressure on the Dutch to withdraw.

Burma achieved independence early in 1948, a few months after India, but was almost immediately riven by ethnic and political splits which have still to be resolved. British rule in Malaya came to an end by peaceful negotiation in 1957, although from 1948 until 1960 British and Commonwealth troops were involved in the suppression of a major communist rebellion in the country.

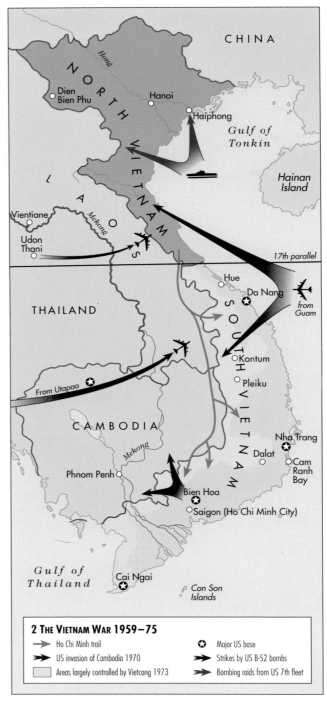

2 THE VIETNAM WAR 1959–75
- → Ho Chi Minh trail
- ⇒ US invasion of Cambodia 1970
- ▢ Areas largely controlled by Vietcong 1973
- ✪ Major US base
- ⇒ Strikes by US B-52 bombs
- ⇒ Bombing raids from US 7th fleet

▲ Vietnam's struggle for independence from the French resulted, in 1954, in the division of the country into communist North Vietnam and US-backed South Vietnam. North Vietnam attempted to overthrow the southern regime and reunify the country. The United States, anxious to prevent the spread of communism, became militarily involved in the 1960s but was eventually defeated by the Vietcong's guerrilla tactics.

THE VIETNAM WAR

In French Indochina the anti-colonial struggle was to last much longer. Open conflict between the French and the Vietminh, in effect the Indochinese Communist Party, broke out in December 1946, after negotiations to reconcile the ambitions of French colonialists and Vietnamese nationalists had failed. After a long, draining guerrilla war, the French forces were defeated at Dien Bien Phu in 1954. (The Vietnamese were the only people in Southeast Asia to achieve the withdrawal of a colonial power by military victory.) However, at the Geneva Conference which opened in May 1954, the Communists failed to secure a united Vietnam under their control. Instead, they were forced – partly by pressures imposed by China, the Soviet Union, and the United States – to accept a temporary division along the 17th parallel pending elections in 1956 (*map 2*). From 1955 a strongly anti-communist government was established in

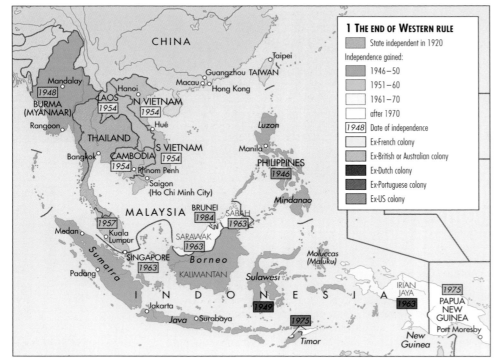

1 THE END OF WESTERN RULE
- State independent in 1920

Independence gained:
- 1946–50
- 1951–60
- 1961–70
- after 1970

- 1948 Date of independence
- Ex-French colony
- Ex-British or Australian colony
- Ex-Dutch colony
- Ex-Portuguese colony
- Ex-US colony

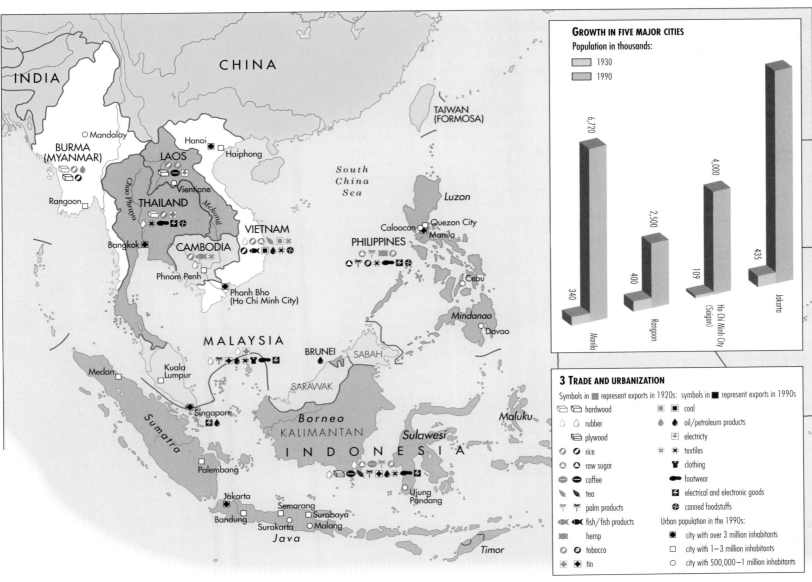

GROWTH IN FIVE MAJOR CITIES
Population in thousands:
- 1930
- 1990

6,720
4,000
2,500
435
400
340
109

Manila
Rangoon
Ho Chi Minh City (Saigon)
Jakarta

3 TRADE AND URBANIZATION

Symbols in ■ represent exports in 1920s; symbols in ■ represent exports in 1990s

hardwood		coal	
rubber		oil/petroleum products	
plywood		electricty	
rice		textiles	
raw sugar		clothing	
coffee		footwear	
tea		electrical and electronic goods	
palm products		canned foodstuffs	
fish/fish products			
hemp			
tobacco			
tin			

Urban population in the 1990s:
- ⬢ city with over 3 million inhabitants
- ☐ city with 1–3 million inhabitants
- ○ city with 500,000–1 million inhabitants

the South, under the leadership of Ngo Dinh Diem, and was soon receiving massive US economic and military support. In the late 1950s communist North Vietnam began the armed struggle to overthrow the southern regime, funnelling supplies of men and arms down the Ho Chi Minh Trail – in reality a shifting complex of jungle routes – into the South.

The United States first committed ground troops to Vietnam in 1965, although much of its military might took the form of mass bomber raids from bases in Thailand and aircraft-carriers in the South China Sea against the Ho Chi Minh Trail and urban centres in North Vietnam. In early 1968, while celebrations were underway for the lunar New Year (*Tet*), the communist Vietcong launched fierce attacks against urban centres across South Vietnam – the "Tet Offensive". However, despite some striking successes – including Vietcong fighting their way into the compound of the US Embassy in Saigon – the offensive failed to dislodge the southern regime and its ally. In 1970, in an attempt to protect its forces in the south, the United States launched an invasion into eastern Cambodia with the aim of destroying the communist sanctuaries there. It was now clear, however, that the United States could not defeat the Vietcong and, following strong domestic pressure, US forces were withdrawn from Vietnam by the end of March 1973. In April 1975 communist troops entered Saigon, the southern regime collapsed, and Vietnam was united under communist rule.

THE POST-COLONIAL ERA

The period since the mid-1960s has seen an extraordinary economic transformation in large parts of Southeast Asia. From being principally exporters of agricultural products and minerals, Thailand, Malaysia, Singapore, Indonesia and, to some degree, the Philippines, have developed a substantial industrial base, exporting finished manufactured goods – including electrical and electronic goods, clothing and footwear – to markets across the world (*map 3*). This was largely achieved through heavy investment by East Asian, European and American multinational companies, which took advantage of Southeast Asia's low wage costs. High economic growth rates were sustained over a number of decades, with a particularly rapid spurt in the late 1980s.

The industrialization of Southeast Asia was mirrored by the rapid pace of urbanization. Cities expanded rapidly (*bar chart*), with the result that a high proportion of the population now live in shanty towns surrounding the prosperous commercial centres. Rapid economic growth created fortunes for Southeast Asia's tycoons, with the large urban middle class and those living in rural areas also benefiting.

Southeast Asia's long boom was brought to a sudden halt in the middle of 1997. Beginning with the Thai baht, many of the region's major currencies came under intense speculative pressure and were forced to devalue. Stock markets plunged and banks crashed. In the wake of the financial meltdown unemployment soared and large sections of the population faced severe economic hardship. The causes of the crisis differed from economy to economy, but the overcommitment of largely unregulated banks, widespread corruption and unsustainable budget deficits by governments with over-ambitious spending plans were clearly important factors.

The economic crisis had serious political consequences in 1998. Riots in Indonesia in May led to the end of President Suharto's 30-year period in power, and in Malaysia a split in the dominant political party, coupled with popular protest against corruption, provoked a serious challenge to the prime minister, Mahathir bin Muhammad.

▲ Southeast Asia has for centuries been a provider of raw materials to Western and Japanese manufacturers. While exports of agricultural products (including hardwoods from its rapidly diminishing rainforests) continue, Malaysia, Indonesia, the Philippines and Thailand have also developed into producers of manufactured goods, in particular electrical and electronic products. As their industrial sector has expanded so have their cities, with people flooding in from agricultural regions in the hope of finding relatively well-paid employment in manufacturing and expanding service industries.

◗ SOUTHEAST ASIA IN THE AGE OF IMPERIALISM 1790–1914 *pages 196–97*

JAPAN
SINCE 1945

THE CHALLENGE OF AN AGEING POPULATION

Percentage of population aged:
- 0–14
- 15–64
- 65+

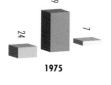

1950
35 / 60 / 5

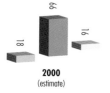

1975
24 / 69 / 7

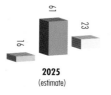

2000
(estimate)
18 / 66 / 16

2025
(estimate)
16 / 61 / 23

▲ During the 1960s Japan benefited from a youthful and rapidly growing working population, but the children of the post-war "baby boom" will eventually reach retirement age. Social and financial adjustments will be required in order to provide a decent standard of living for a large population of pensioners.

efeat in the Pacific War (1941–45) left Japan without an empire and with an industrial economy in ruins. The Allied (predominantly American) occupiers moved swiftly to incorporate democratic reforms into a revised constitution. The emperor was retained as a ceremonial figure, but power was exercised by a legislature elected by universal suffrage. The great industrial combines (*zaibatsu*) that had dominated the pre-war economy (*pages 200–1*) were broken up, labour unions were legalized, and the power of rural landlords was destroyed by wholesale land reforms that favoured small family farms.

The reforming zeal of the occupying authorities was, however, of little immediate significance to most ordinary Japanese, for whom the economic hardships of war and its aftermath were compounded by the repatriation of millions of former soldiers and colonists, and the post-war "baby boom". The failure of the economy to recover sufficiently to meet the day-to-day needs of the population soon led to revisions in economic policy, and these changes were reinforced by the political fallout from the victory of the Communists in China and the outbreak of war on the Korean Peninsula in 1950 (*pages 244–45*). By the time the United States administration ended in 1952, Japan had been redefined as a bastion of anti-communism in East Asia, and expenditure of around $3.5 billion by the United States military during the Korean War had stimulated the economy into growth.

ECONOMIC EXPANSION

Over the next two decades Japan enjoyed an extraordinary period of economic expansion. Industrial production had recovered to pre-war levels by 1955, and during the 1960s average annual growth rates exceeded 10 per cent. This success, which became a model for other Asian economies, rested on a fortuitous combination of external and internal circumstances. Japan's deficiencies in mineral resources were of little importance in an era when cheap raw materials could be acquired easily from overseas. The United States offered a ready market for manufactured exports, made more competitive by an increasingly undervalued currency. It also provided access to industrial expertise for Japanese technologists. Foreign policy focused overwhelmingly on trade promotion, although one important territorial issue was resolved with the return of Okinawa to Japanese sovereignty by the United States in 1972.

The "family state" of pre-war times was replaced by a "developmental state", in which a stable political regime under the conservative Liberal Democrats allowed major industrial groupings to re-emerge under the guiding hand of an elite bureaucracy. Large-scale movements of population from the countryside to the cities (*map 1*) guaranteed a supply of youthful and well-educated workers for Japan's factories; labour relations based on company unions and employment for life helped to

secure support for economic growth as the primary goal of the nation. A high rate of savings ensured adequate supplies of capital. As wealth accumulated, domestic demand became an increasingly important source of growth.

By the late 1960s it was apparent that such unrestrained economic expansion had environmental costs, with outbreaks of illnesses caused by industrial pollution – such as "Minamata Disease" and "Yokkaichi Asthma" – serious enough to attract international attention. Labour shortages in Japan's cities reinforced pressure for industry to relocate or raise productivity (*map 2*). Trade friction with the United States and a sharp revaluation of the yen preceded the oil crisis of 1973–74 (*pages 272–73*). Japan's vulnerability to disruption in the supply of an energy source on which it had become almost wholly dependent was exposed amid panic buying of daily essentials by the public, rapid inflation and the temporary cessation of growth.

Japan responded quickly and effectively to these challenges. Energy-intensive heavy industries were obliged to raise their efficiency and clean up their effluents or move overseas, as Japan felt the effect of competition from Korea and the other emerging industrial economies in East Asia. Small, fuel-efficient cars were suddenly in demand, and

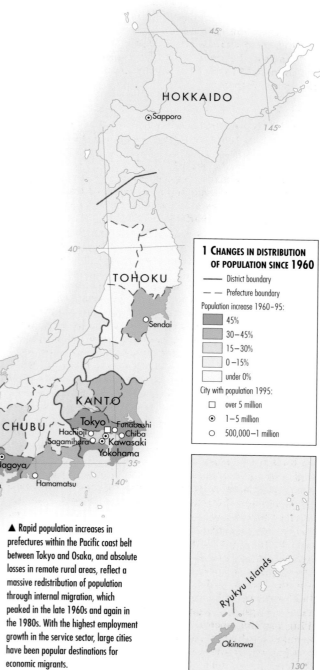

1 CHANGES IN DISTRIBUTION OF POPULATION SINCE 1960
- —— District boundary
- – – Prefecture boundary

Population increase 1960–95:
- 45%
- 30–45%
- 15–30%
- 0–15%
- under 0%

City with population 1995:
- ☐ over 5 million
- ◉ 1–5 million
- ○ 500,000–1 million

▲ Rapid population increases in prefectures within the Pacific coast belt between Tokyo and Osaka, and absolute losses in remote rural areas, reflect a massive redistribution of population through internal migration, which peaked in the late 1960s and again in the 1980s. With the highest employment growth in the service sector, large cities have been popular destinations for economic migrants.

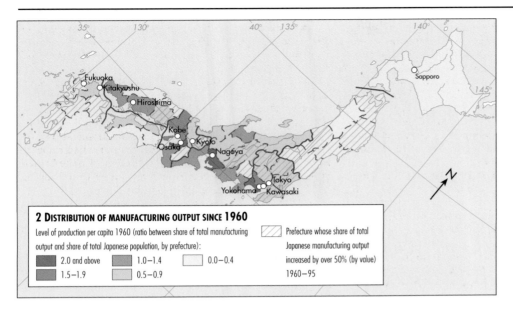

2 DISTRIBUTION OF MANUFACTURING OUTPUT SINCE 1960

Level of production per capita 1960 (ratio between share of total manufacturing output and share of total Japanese population, by prefecture):

- 2.0 and above
- 1.5–1.9
- 1.0–1.4
- 0.5–0.9
- 0.0–0.4

Prefecture whose share of total Japanese manufacturing output increased by over 50% (by value) 1960–95

◄ The major industrial regions in Japan were established before the Second World War. Investment was concentrated there in the 1960s to take advantage of the existing infrastructure. However, labour shortages, high land prices and pollution controls in large cities, plus competition from overseas, fuelled a relocation of industry within Japan to areas that had not previously proved attractive to investors.

▼ In the 1960s Japanese manufacturing was largely dominated by heavy industries such as steel production and shipbuilding. By the 1970s, however, more profitable industries, in particular vehicle manufacturing, were increasingly important. In the 1980s new industries, such as those producing semiconductors and other electronic equipment, experienced a boom and continued to expand in the 1990s.

THE CHANGING BALANCE OF INDUSTRIAL PRODUCTION

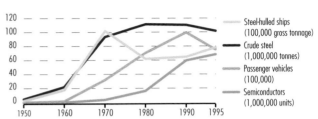

- Steel-hulled ships (100,000 gross tonnage)
- Crude steel (1,000,000 tonnes)
- Passenger vehicles (100,000)
- Semiconductors (1,000,000 units)

exports responded quickly, until the threat of protective tariffs from countries in North America and Europe on cars exported from Japan forced Japanese car manufacturers to increase their production in these regions. Industry shifted towards "knowledge-intensive" sectors such as electronics (*graph*), in which Japan established international standards and dominated world markets. Growth did slow from the heady rates of the 1960s, but still averaged over 4 per cent per annum in 1974 to 1985, and Japan was able to weather the second oil crisis of 1978.

FOREIGN RELATIONS AND TRADE

As the 1980s progressed, relations with the United States became more problematic. The cost to the United States of protecting Japan during the Cold War was high, while Japan grew ever richer on burgeoning trade surpluses. The United States became sensitive to the effect of imports from Japan on job prospects at home. It put restraints on trade in manufactured goods between the two countries, and pressure on Japan to open up its markets to US farm produce, such as rice. Japanese agriculture itself was by now heavily subsidized and plagued by inefficiencies linked to the small farms inherited from the land reforms of the 1940s. It attempted, unsuccessfully, to adapt to competition from imports by changing the crops that it produced.

The Plaza Agreement of 1985, between the United States, Japan, France, Germany and the United Kingdom, sought to resolve global trade imbalances by expanding Japan's domestic demand. The rapid appreciation of the yen was also expected to make Japanese products less competitive in international markets and to boost imports to Japan. Yet again, however, Japanese industry responded by shifting up a gear: in a flurry of direct investment in East and Southeast Asia, manufacturers sought to avoid high Japanese wages by moving production overseas (*map 3*).

This process was known as "hollowing out". It was matched by a rapid expansion in Japan's foreign aid, the aim of which was to support infrastructural improvements in neighbouring countries. This facilitated production of, and created additional demand for, Japanese products in these countries. Japan became the centre of a regional manufacturing system tied together by trade flows of raw materials, components and manufactured goods. Tokyo was transformed into one of the world's three great financial centres. Investments at home and overseas were buoyed up by low interest rates and the willingness of banks to lend against property assets, which soared in value. This speculative "bubble economy" finally burst in the early 1990s as land prices collapsed, obliging the government to shore up the ailing banking sector. The banks' problems were compounded by the subsequent economic crisis in Southeast Asia (*pages 250–51*) as loans to finance new factories in Thailand, Indonesia and elsewhere turned sour.

In the latter half of the 1990s Japan, with the world's highest life expectancy, was beginning to adjust to social changes brought about by a population in which the proportion of older people was growing (*bar charts*). Its politicians were attempting to relax bureaucratic control of domestic markets and to continue the reform of its financial systems. Such changes were a necessary counterpart to the growing climate of openness in Japan's trade and financial relations with the outside world.

The popular opposition to military participation in the Gulf War of 1991, and Japan's inability to counter the threat posed by North Korean missiles, indicated the mismatch between Japan's status as a pre-eminent global economic power and its low political and military profile. The occupation by Russia of the islands to the northeast of Japan also remained a sensitive issue at the end of the century.

▼ The "hollowing out" of the Japanese economy, which saw Japanese direct investment in Asia increase tenfold between 1985 and 1990, added a new dimension to Japan's economic ties to other countries in the region, which had previously been dominated by imports of raw materials, and exports of products manufactured in Japan.

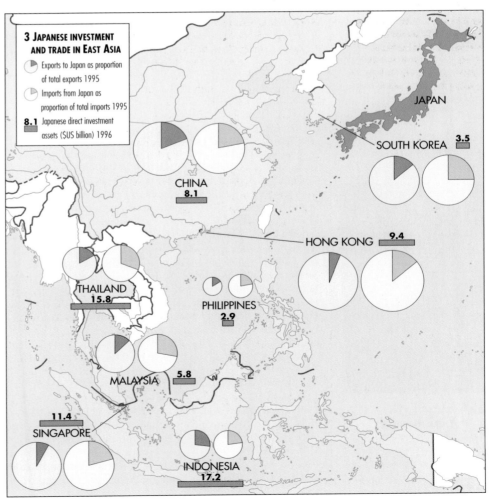

3 JAPANESE INVESTMENT AND TRADE IN EAST ASIA

- ◐ Exports to Japan as proportion of total exports 1995
- ◐ Imports from Japan as proportion of total imports 1995
- **8.1** Japanese direct investment assets ($US billion) 1996

JAPAN

SOUTH KOREA **3.5**

CHINA **8.1**

HONG KONG **9.4**

THAILAND **15.8**

PHILIPPINES **2.9**

MALAYSIA **5.8**

SINGAPORE **11.4**

INDONESIA **17.2**

● THE WAR IN ASIA 1931–45 *pages 234–35*

THE PEOPLE'S REPUBLIC OF CHINA SINCE 1949

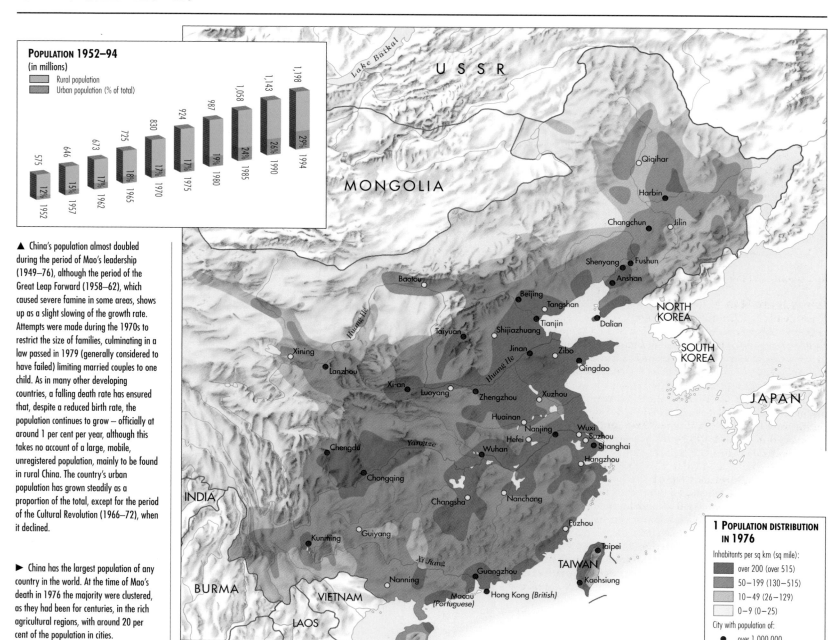

POPULATION 1952–94
(in millions)

Rural population
Urban population (% of total)

Year	Population	Urban %
1952	575	12%
1957	646	15%
1962	673	17%
1965	725	18%
1970	830	17%
1975	924	17%
1980	987	19%
1985	1,058	24%
1990	1,143	26%
1994	1,198	29%

1 POPULATION DISTRIBUTION IN 1976

Inhabitants per sq km (sq mile):

- over 200 (over 515)
- 50–199 (130–515)
- 10–49 (26–129)
- 0–9 (0–25)

City with population of:
- ● over 1,000,000
- ○ 500,000 to 1,000,000

▲ China's population almost doubled during the period of Mao's leadership (1949–76), although the period of the Great Leap Forward (1958–62), which caused severe famine in some areas, shows up as a slight slowing of the growth rate. Attempts were made during the 1970s to restrict the size of families, culminating in a law passed in 1979 (generally considered to have failed) limiting married couples to one child. As in many other developing countries, a falling death rate has ensured that, despite a reduced birth rate, the population continues to grow – officially at around 1 per cent per year, although this takes no account of a large, mobile, unregistered population, mainly to be found in rural China. The country's urban population has grown steadily as a proportion of the total, except for the period of the Cultural Revolution (1966–72), when it declined.

► China has the largest population of any country in the world. At the time of Mao's death in 1976 the majority were clustered, as they had been for centuries, in the rich agricultural regions, with around 20 per cent of the population in cities.

▲ In the mid-1960s Mao Zedong successfully reasserted control over the Communist Party by empowering Chinese youth in his Cultural Revolution. *The Little Red Book*, containing Mao's political axioms, became a symbol of revolutionary zeal, not only in China but also around the world.

The People's Republic of China was founded on 1 October 1949, following the defeat of the Japanese invaders and the unification of the country under a single government. The immediate priorities were to establish law and order, implement land reforms, balance the state budget, stabilize prices and nationalize industry. Having gained public support for these essentially nationalistic policies, from the mid-1950s onwards Chairman Mao Zedong began to introduce communist reforms. Initially, the communist programme was heavily influenced by the Soviet Union, with whom China had signed a pact in 1950. It involved wholesale rural and urban collectivization, with the assets of large property owners being taken over by the state. Those of smaller property owners were given to communes, supervised by the Communist Party. Other radical social measures were passed, including giving women equal legal status with men in terms of marriage and employment.

THE FIVE-YEAR PLANS
The main thrust of the programme was industrialization, formalized into a series of five-year plans. During the first of these (1953–58), over 100 industrial projects were set up with the help of machinery and expertise from the Soviet Union. The aim was to create an economy that did not depend on imports from capitalist countries, and the policy was initially effective in changing China's economy from one based on agriculture to one based on heavy industry.

In his second five-year plan, known as the "Great Leap Forward", Mao rejected the Soviet model and developed a specifically Chinese communism based on peasant labour. He instructed collectives to build and run small-scale iron and steel foundries. However, not only did it prove impossible to produce metal of an acceptable standard, but the scheme also took labour away from the agricultural sector. Production of food dropped as a consequence, leading to a nationwide famine that claimed tens of millions of lives (*bar chart 1*). The plan also seriously backfired in the industrial sector, with production dropping by up to 50 per cent, forcing the government to de-industrialize the economy. China's economic growth was temporarily halted.

THE CULTURAL REVOLUTION
Chairman Mao's main concern was to promote his ideology and increase his power, leading him into conflict with other, more pragmatic, members of his government, in particular President Liu Shaoqi. Mao launched his Cultural Revolution in 1966 in an attempt to revive his control over the party and society. Party officials, teachers and factory managers were among those in authority who were verbally and physically attacked, imprisoned or sent to work in labour camps. There they were joined by millions of young people, whose schools and universities had been closed. Industrial production was severely disrupted, and the economy brought near to bankruptcy during the ten-year process.

FOREIGN POLICY UNDER MAO

Immediately after the revolution of 1949, China allied itself with the Soviet Union and gave assistance to independence movements in Southeast Asia. It also provided troops to assist the North Koreans in their efforts to unify their country in 1950, and aided the Vietnamese in their battle to expel the French from Indochina in the early 1950s. From the early 1960s, however, China's relations with the Soviet Union soured, mainly due to Khrushchev's repudiation of Stalin's policies. At the same time, China also lost support among the neutral, newly independent countries of the developing world when it crushed anti-Chinese opposition in Tibet, and entered into a border dispute with India. The Cultural Revolution was a period of intense xenophobia, but in 1971 Mao, in an apparent reversal of policy, welcomed President Nixon's initiative to normalize relations with the United States. In October of that year the People's Republic of China replaced Taiwan in the United Nations and re-entered the world stage.

CHINA AFTER MAO

Mao's death in 1976 initiated a power struggle between the "Gang of Four" (which included Mao's widow) and Deng Xiaoping. Deng emerged the victor, and during his era (1978–97) pragmatism prevailed. Faced with a rapidly expanding population (*map 1 and bar chart*), economic growth became the stated priority, to be brought about by a policy of "four modernizations" (in industry, agriculture, science and technology, and the army). China's industrial output rose steadily during the 1980s, and increased dramatically during the 1990s by over 20 per cent each year. In the agricultural sector China made important gains through the reform of farming practices. Although the total land area committed to agriculture remained much the same, yields improved enormously (*map 2*).

From 1978 onwards state ownership and planning were reduced, "the market" was respected and nurtured, and property rights were gradually defined. Communes were abolished and citizens permitted to run private businesses and engage in market activities. Instead of attempting to make China self-sufficient, the new regime adopted an export-led growth strategy, copied from other newly industrialized countries.

DEMANDS FOR DEMOCRACY

As China became more open to Western economic principles and ideology during the 1980s, many people, in particular students, began also to demand modernization of the political system. Although the paramount leader Deng resisted these demands, Communist Party General Secretary Hu Yaobang was more open to change. Hu's demotion and subsequent death triggered pro-democracy demonstrations in many major cities during April 1989. Throughout May demonstrators occupied the vast Tiananmen Square in Beijing, demanding Deng's dismissal and political reform. With the world's press watching, the Chinese government held back for several weeks. However, overnight on 3–4 June the army moved in to disperse the demonstrators. Hundreds were killed and thousands were injured; arrests, imprisonments and executions followed. The international outrage that resulted soured China's relations with the outside world and briefly affected foreign investment , which had, since the 1980s, been channelled through China's "Special Economic Zones" and "open cities" (*map 3 and bar chart*).

In July 1997, shortly after Deng's death, Hong Kong was returned to Chinese rule (and designated a "Special Administrative Region"). Later that year the Chinese government decided to privatize state-owned enterprises operating at a loss – roughly 30 per cent of the state sector. With mounting unemployment from the collapse of the public sector, the trend towards a semi-capitalist society continues in uneasy contrast to the strict party control, creating a great deal of uncertainty about the political and economic future of the world's most populous nation.

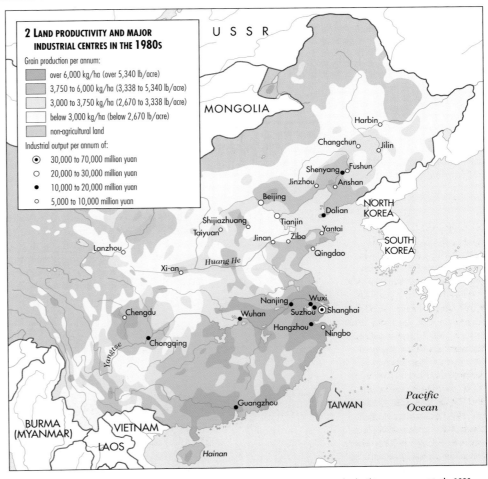

2 LAND PRODUCTIVITY AND MAJOR INDUSTRIAL CENTRES IN THE 1980s

Grain production per annum:
- over 6,000 kg/ha (over 5,340 lb/acre)
- 3,750 to 6,000 kg/ha (3,338 to 5,340 lb/acre)
- 3,000 to 3,750 kg/ha (2,670 to 3,338 lb/acre)
- below 3,000 kg/ha (below 2,670 lb/acre)
- non-agricultural land

Industrial output per annum of:
- ◉ 30,000 to 70,000 million yuan
- ○ 20,000 to 30,000 million yuan
- ● 10,000 to 20,000 million yuan
- ○ 5,000 to 10,000 million yuan

▲ The majority of industrial production in the 1980s was to be found along the Yangtze River, which was used to transport raw materials and finished goods to internal and foreign markets.

▼ Communist China represents a vast potential market to the capitalist economies. Special Economic Zones, in which a free market economy (including foreign goods and capital) could function, were established by the Chinese government in the 1980s as an experiment. They were followed by "open cities", initially along the coast but later inland, where foreign businesses have special access to the vast Chinese market.

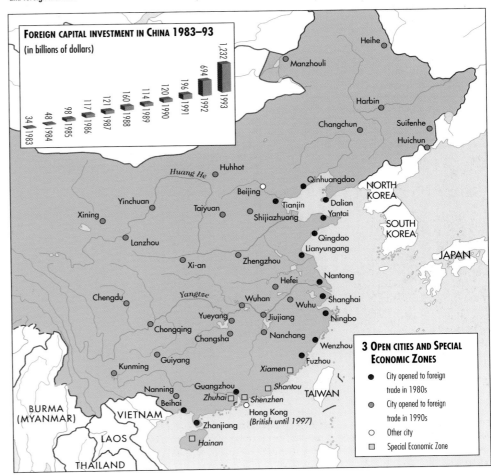

FOREIGN CAPITAL INVESTMENT IN CHINA 1983–93 (in billions of dollars)

Year	Value
1983	34
1984	48
1985	98
1986	117
1987	121
1988	160
1989	114
1990	120
1991	196
1992	694
1993	1,232

3 OPEN CITIES AND SPECIAL ECONOMIC ZONES

- ● City opened to foreign trade in 1980s
- ● City opened to foreign trade in 1990s
- ○ Other city
- ☐ Special Economic Zone

⬅ THE REPUBLIC OF CHINA 1911–49 *pages 224–25*

AFRICA
SINCE 1939

During the heyday of colonial power in Africa in the 1920s and 1930s, it looked as though European control would survive into the far distant future (*pages 206–7*). The ease with which African countries were drawn into the Second World War highlighted their status as European possessions. North Africa became a major theatre of conflict, and many African soldiers served with the Allied armies. African colonies were also used as major sources of vital raw materials and foodstuffs.

The war stimulated economic development in Africa. Industrialization and urbanization increased markedly, as did the production of foodstuffs and cash crops by African cultivators. In political terms, the refusal of the colonial powers to extend to Africa the democratic ideals for which they had fought in Europe sharpened Africans' sense of the injustice of colonialism. The independence granted to India in 1947 and other countries in Asia around this time encouraged African nationalists to press for similar political freedoms in their own continent. The rise of an educated African elite, which took advantage of new economic opportunities and skill shortages in the colonial bureaucracy,

provided a social base for the developing anti-colonial consciousness. A growing desire for independence was also fuelled by the fact that in the years immediately after the war, Britain and France relied on African raw materials, purchased at artificially depressed prices, to rebuild their shattered economies. Between 1945 and 1951 Britain made a profit of £140 million on commodity transactions with its African colonies, while injecting only £40 million in return via the Colonial Development and Welfare Acts.

THE GAINING OF INDEPENDENCE

The speed with which the process of gaining independence swept through Africa was in many ways a mirror image of the hasty 19th-century partition of Africa among the colonial powers. Libya gained independence in 1951 largely because the United Nations could not agree who should control the former Italian colony. The vast British-controlled Sudan gained independence in 1956, as did the French colony of Tunisia. It was, however, the achievement of independence by the Gold Coast as Ghana in 1957, spearheaded by the charismatic pan-Africanist leader

► With a few exceptions the boundaries of colonial Africa, hastily drawn in the "scramble for Africa", continued into modern times as the boundaries of the new independent states. Wars in southern Sudan, Zaire and the Biafran region of Nigeria all failed to establish new states. Eritrea (granted to Ethiopia by the British in 1962) finally broke away from Ethiopia after a protracted struggle. The self-proclaimed Somaliland Republic was less successful at establishing independence. Western Sahara was occupied by Morocco after being granted independence by Spain in 1976.

▼ For most states the establishment of a democratic system with multi-party elections has taken several decades, and a few have yet to achieve it. In the late 1980s and 1990s, however, the increasingly strong grassroots support for democracy was reinforced by the collapse of communism in the Soviet Union (to which many autocratic African leaders had looked for ideological inspiration) and by pressures from the International Monetary Fund and the World Bank to democratize as a condition of loan extensions.

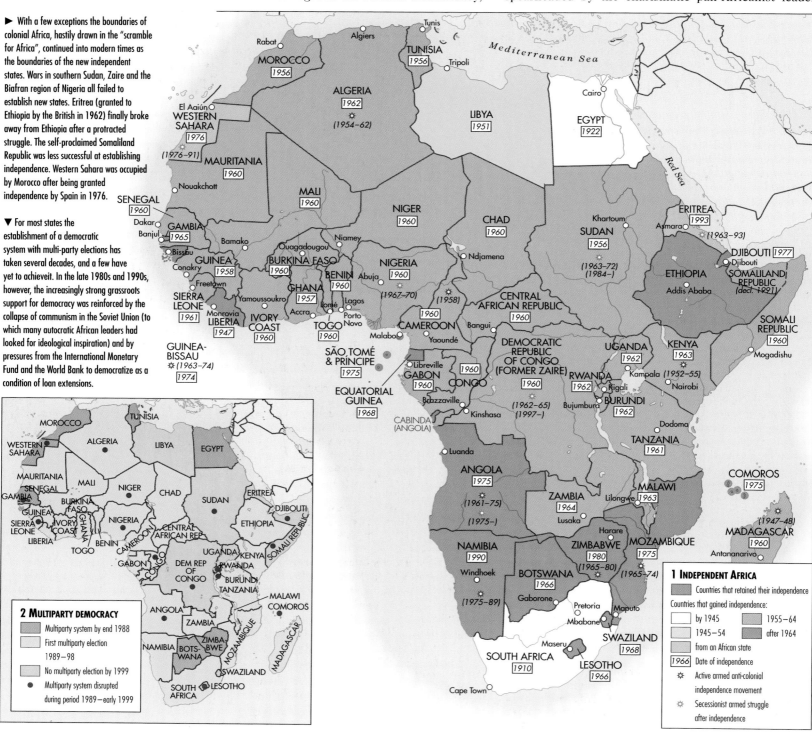

2 MULTIPARTY DEMOCRACY

- Multiparty system by end 1988
- First multiparty election 1989–98
- No multiparty election by 1999
- • Multiparty system disrupted during period 1989–early 1999

1 INDEPENDENT AFRICA

- Countries that retained their independence

Countries that gained independence:
- by 1945
- 1945–54
- 1955–64
- after 1964
- from an African state
- *1966* Date of independence
- ✿ Active armed anti-colonial independence movement
- ✾ Secessionist armed struggle after independence

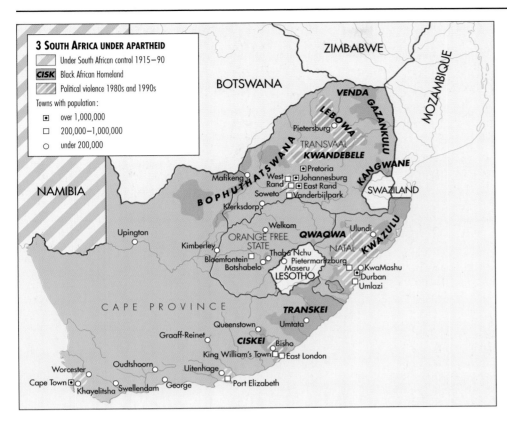

3 SOUTH AFRICA UNDER APARTHEID

- Under South African control 1915–90
- **CISK** Black African Homeland
- Political violence 1980s and 1990s

Towns with population:
- ⊡ over 1,000,000
- ▫ 200,000–1,000,000
- ○ under 200,000

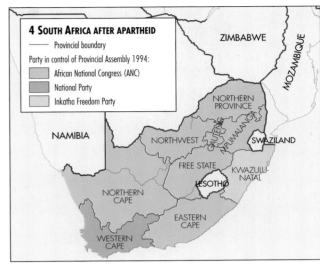

4 SOUTH AFRICA AFTER APARTHEID

- Provincial boundary

Party in control of Provincial Assembly 1994:
- African National Congress (ANC)
- National Party
- Inkatha Freedom Party

◀ Under the "apartheid" system in South Africa (1948–91) many black Africans were forced to live in "homelands" often far from the main labour markets. Violent protests, coupled with international economic pressure, eventually led to President de Klerk's announcement of the abolition of apartheid and the release from prison of the ANC leader, Nelson Mandela, in 1990.

▲ The first national elections in which black South Africans could vote were held in April 1994. Protests in Bophuthatswana (map 3) and KwaZulu Natal had threatened to disrupt them, but they passed off relatively peacefully. The African National Congress was victorious, taking 63 per cent of the vote, and Nelson Mandela was sworn in as President of South Africa in May 1994.

Kwame Nkrumah, that sparked off a wave of decolonization in sub-Saharan Africa. Ghana provided a model of relatively peaceful transition to independence, while in French-controlled Algeria and British-occupied Kenya protracted and bitter insurrection was waged by the National Liberation Front (FLN) and the Mau Mau movement respectively.

Most African colonies gained their independence in the years between 1956 and 1962 (map 1). In some instances the process was hurried and unplanned. The hastily granted independence of the Belgian Congo (Zaire, now Democratic Republic of Congo) in 1960 resulted in the attemptedsecession of the copper-rich southern region, giving rise to political instability and foreign interference that characterized the post-independence history of many African states.

Not all African countries gained independence during the first wave of national liberation. The Portuguese colonies of Angola and Mozambique finally won independence only after a coup d'etat in Lisbon in April 1974, led by General Spinola. The struggle in Guinea-Bissau (which had claimed its independence a year earlier) persuaded Spinola that the Portuguese African empire could no longer be sustained. A bitter guerrilla war was also fought in Southern Rhodesia (Zimbabwe), against a white colonial regime that had proclaimed its own independence from Britain in 1965. After Zimbabwe, where black African rule was finally achieved in 1980, the only African states still to achieve freedom for blacks were South Africa and its illegally occupied satellite, Namibia (map 3). Although Namibia won its independence in 1990, black South Africans did not vote in a national election until 1994, when Nelson Mandela (who had spent 27 years as a political prisoner) became president (map 4).

AFTER INDEPENDENCE

The upsurge of African nationalism, which brought so many countries to independence, also engendered huge optimism and unrealistic expectations of rapid economic development. All too often, however, the new governing elites were ill-prepared for office, ambitious development plans went awry, expectations of rapid industrialization were misplaced, and political instability became endemic. During the Cold War (pages 244–45) competition for influence in Africa became an important proxy for global conflict, and former colonial powers could exert great economic power. Foreign aid was often provided in the form of military training and weaponry, rather than as a stimulus to economic development.

When the Ghanaian president Nkrumah was deposed in a coup in 1966, much of the early optimism for independent Africa began to wane. The civil war that broke out when Biafra sought to secede from Nigeria in 1967 highlighted the problems of military involvement in civil affairs, and of the failure of nationalism to supersede ethnic divisions.

ECONOMIC AND SOCIAL DEVELOPMENTS

Many African countries have made solid economic and social progress since independence, with massive provision of primary and secondary schooling, and the extension of basic health facilities. Growing networks of rural clinics and the availability of cheap drugs have done much to enhance life expectancy and improve infant mortality figures, although the rapid spread of AIDS in some regions is effectively undoing many of these advances (pages 274–79).

Following independence, countries such as Ghana and Mozambique adopted the rhetoric of socialist transformation; others, such as Kenya and the Ivory Coast, proclaimed the benefits of capitalism, while Tanzania sought to disengage itself from the world economy and concentrate on autonomous development. Although none of these approaches proved particularly successful in the long run, many African countries made considerable economic progress in the 1950s and 1960s as a result of relatively high commodity prices. In Nigeria the exploitation of oil reserves provided spectacular wealth for its political elite.

Africa suffered a major economic crisis in the 1970s as a result of massive increases in oil prices (pages 272–73). Falling commodity prices and increased interest rates severely affected those economies that had been encouraged to borrow on international markets. By the mid-1980s some, such as Zambia, were so stricken by debt that they had no option but to accept "structural adjustment programmes" proposed by the International Monetary Fund, remodelling their economies on free-market principles and enforcing cuts in social provision. As a result, large parts of Africa experienced economic stagnation during the 1980s.

In the early 1990s optimism replaced the euphoria of the independence era and the gloom of the 1980s, as several civil wars ended and democratic elections were held across the continent. As the decade wore on, however, such optimism appeared ill-founded as bitter ethnic and religious disputes and civil wars broke out and the prospect of democracy and development receded in several key states.

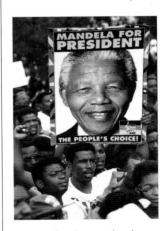

▲ A wave of popular support brought Nelson Mandela to power in the 1994 elections. Many material and social advances have been made, although expectations of rapid improvements in living conditions for the black majority population have proved somewhat over-optimistic.

◑ THE PARTITION OF AFRICA 1880–1939 pages 206–7

LATIN AMERICA SINCE 1945

MANUFACTURING AS A PERCENTAGE OF GROSS DOMESTIC PRODUCT (GDP)

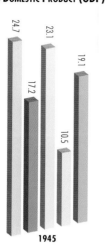

1945

24.7 23.1 17.2 10.5 19.1

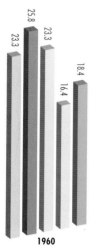

1960

25.8 23.3 23.3 16.4 18.4

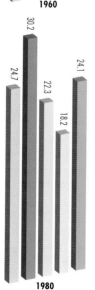

1980

30.2 24.7 22.3 18.2 24.1

Argentina	Colombia
Brazil	Mexico
Chile	

▲ The main Latin American economies have met with mixed success in their attempts to industrialize. While Brazil and Colombia managed to improve their manufacturing output in the 1950s (and Mexico produced a spurt between 1960 and 1980), output for Argentina and Chile remained static as a percentage of Gross Domestic Product.

258

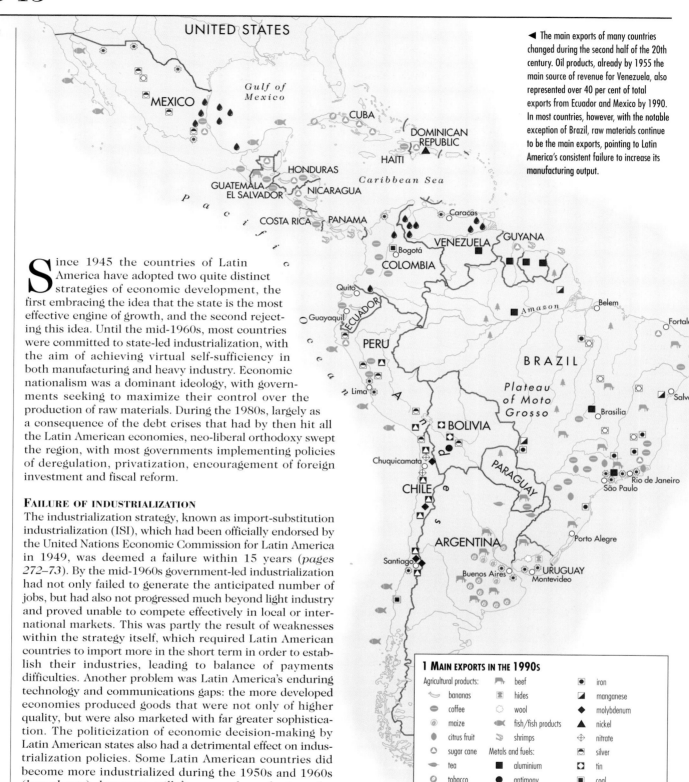

◄ The main exports of many countries changed during the second half of the 20th century. Oil products, already by 1955 the main source of revenue for Venezuela, also represented over 40 per cent of total exports from Ecuador and Mexico by 1990. In most countries, however, with the notable exception of Brazil, raw materials continue to be the main exports, pointing to Latin America's consistent failure to increase its manufacturing output.

1 MAIN EXPORTS IN THE 1990S

Agricultural products:
- beef
- bananas
- hides
- iron
- coffee
- wool
- manganese
- maize
- fish/fish products
- molybdenum
- citrus fruit
- shrimps
- nickel
- sugar cane
- Metals and fuels:
- nitrate
- tea
- aluminium
- silver
- tobacco
- antimony
- tin
- wheat
- copper
- coal
- timber
- gold
- oil/petroleum products
- manufactured goods

Since 1945 the countries of Latin America have adopted two quite distinct strategies of economic development, the first embracing the idea that the state is the most effective engine of growth, and the second rejecting this idea. Until the mid-1960s, most countries were committed to state-led industrialization, with the aim of achieving virtual self-sufficiency in both manufacturing and heavy industry. Economic nationalism was a dominant ideology, with governments seeking to maximize their control over the production of raw materials. During the 1980s, largely as a consequence of the debt crises that had by then hit all the Latin American economies, neo-liberal orthodoxy swept the region, with most governments implementing policies of deregulation, privatization, encouragement of foreign investment and fiscal reform.

FAILURE OF INDUSTRIALIZATION

The industrialization strategy, known as import-substitution industrialization (ISI), which had been officially endorsed by the United Nations Economic Commission for Latin America in 1949, was deemed a failure within 15 years (*pages 272–73*). By the mid-1960s government-led industrialization had not only failed to generate the anticipated number of jobs, but had also not progressed much beyond light industry and proved unable to compete effectively in local or international markets. This was partly the result of weaknesses within the strategy itself, which required Latin American countries to import more in the short term in order to establish their industries, leading to balance of payments difficulties. Another problem was Latin America's enduring technology and communications gaps: the more developed economies produced goods that were not only of higher quality, but were also marketed with far greater sophistication. The politicization of economic decision-making by Latin American states also had a detrimental effect on industrialization policies. Some Latin American countries did become more industrialized during the 1950s and 1960s (*bar charts*), but were still far more dependent on the production of raw materials (*map 1*) than had been anticipated when the policy of ISI was launched.

INTERNATIONAL DEBT CRISIS

The failure of the industrialization model was one factor contributing to the debt crises that hit Latin America in the early 1980s. The major cause, however, was the disintegration, during the 1960s, of the system of international financial regulation that had been in place since 1944. When oil price rises in 1973 led to a surplus of "petro-dollars" on the international lending markets, Latin American countries, which had never succeeded in generating internally the levels of capital needed for development, appeared to be ideal targets for loans. With economic depression and inflation in the developed economies, these loans were effectively set at very low, or even negative, interest rates. When US interest rates rose dramatically in the early 1980s, Latin American countries found themselves

unable to service their debts. As bankers hastened to call on the services of the International Monetary Fund (IMF), most debtor countries were obliged to sign stabilization agreements with the IMF as a prerequisite to the rescheduling of their debts. The aim of these agreements was to cut spending and increase exports, thereby maximizing revenue to make interest payments.

The 1980s are referred to as "the lost decade" of Latin American development; economies contracted and there was a huge net transfer of capital out of the region. In the 1990s capital investment returned to Latin America, and it is now accepted that much of the original debt will probably not be repaid. However, Latin America could continue to be burdened by interest payments well into the 21st century.

POLITICAL DEVELOPMENTS

Politically, this period saw the introduction of full suffrage throughout the region, with women granted the vote by the mid-1950s in all Latin American countries, and literacy qualifications gradually dropped, although not until as late as 1989 in the case of Brazil. However, for much of the period the democratic process was compromised at best, and completely suspended at worst. Most countries were governed by populist regimes in the 1940s and 1950s which, although elected, tended to use dictatorial methods once in power. Argentina's Juan Domingo Perón (1946–55) was the classic example. Nevertheless, populism generated a level of political activity among the masses which alarmed those in the property-owning classes to such an extent that most were prepared to support military coups in the 1960s and 1970s.

Such fears were shared by US governments, whose long-standing concerns about political stability in Latin America had acquired particular urgency because of the Cold War (*pages 244–45*). During the late 1940s and 1950s, the United States had taken care to consolidate not only its political alliances with Latin American nations (in the Organization of American States) but also its military links, with the USA supplying most of Latin America's weapons and military training (*map 2*). In these circumstances, the military coups of the 1960s and 1970s ushered in regimes influenced partly by the management techniques and

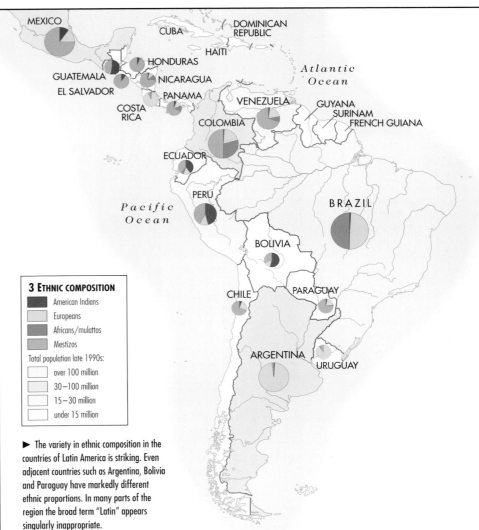

3 ETHNIC COMPOSITION
- American Indians
- Europeans
- Africans/mulattos
- Mestizos

Total population late 1990s:
- over 100 million
- 30–100 million
- 15–30 million
- under 15 million

► The variety in ethnic composition in the countries of Latin America is striking. Even adjacent countries such as Argentina, Bolivia and Paraguay have markedly different ethnic proportions. In many parts of the region the broad term "Latin" appears singularly inappropriate.

2 US INTERVENTION IN LATIN AMERICA SINCE 1945
- ★ Direct military action by USA
- ☆ Economic and political intervention by USA
- ⚒ Direct military action by USSR

1 USA attempts to thwart election of Perón (1946)
2 Popular Revolution neutralized by US economic pressure (1952)
3 CIA-organized invasion overthrows Arbenz (1954) following expropriation of United Fruit Company lands
4 Nationalist revolution (1959) and alliance with USSR (1960). USA declares economic embargo and CIA organizes failed Bay of Pigs invasion (1961). Cuban Missile Crisis (1962)
5 Covert intervention by USA against elected Marxist government of Popular Unity (1970-73)
6 Military intervention to suppress possible communist influence (1965)
7 Revolution (1979): USA funds counter-revolutionary movement (1980s)
8 Covert intervention by USA to defeat left-wing guerrillas (1980-88)
9 US invasion to restore stable government (1983)
10 US invasion to arrest President Noriega on charges of drug trafficking (1989)
11 "Negotiated" US invasion to restore democracy (1994)
12 North American Free Trade Agreement (1994)

▲ In the second half of the 20th century the United States extended its sphere of influence beyond its immediate neighbours in Central America and the Caribbean into South America. It used not only covert but also occasionally direct methods in its attempts to quash what it perceived as attempts by the Soviet Union to gain a foothold in the USA's "backyard" through communist-inspired political movements.

development economics learned either in the USA itself or at national military training schools based on the US model. The military leaders argued that only they were capable of bringing about national development and that the democratic process would have to be suspended until the country was "ready" for electoral politics. The repression for which these regimes became internationally condemned was directed initially at the Left, but gradually acquired a random nature designed to inhibit all political activity, even among moderates.

Although the military stayed in power for lengthy periods of time (Brazil 1964–85, Argentina 1976–83 and Chile 1973–89), they proved no more able than civilian politicians to achieve economic development; indeed, they presided over the debt crises (and, in many cases, their purchases of weapons contributed substantially to the debt). A process of redemocratization began in Latin America in 1980, and by 1990 there were elected governments in every country of the region apart from Cuba.

Most Latin American countries are still some distance away from being fully consolidated liberal democracies, with civilian control over the military, respect for civil rights, freedom of the press and broadly representative political parties. The process of resisting authoritarianism stimulated a wide range of grassroots organizations concerned with, for example, human rights, women's issues and neighbourhood self-help, many of which are reluctant to be recruited by formal political parties. The question of ethnic identities (*map 3*) also assumed an increasing significance, particularly in 1992, the quincentennial of the European "conquest", "discovery" or "encounter" with the Americas. (The very term used to describe Columbus's landing in 1492 is highly disputed, reflecting the intractability of the ethnic and cultural issues at stake.) There is still a potentially dangerous gap between the concerns of the people and of the government in many Latin American countries.

▲ Between 1956 and 1958 Fidel Castro led a revolutionary movement in Cuba that resulted in the overthrow of the dictator Fulgencio Batista on 1 January 1959 and the installation of Castro as president.

◄ LATIN AMERICA 1914–45 *pages 226–27*

THE MIDDLE EAST SINCE 1945

uring the Second World War calls for independence from the territories in the Middle East held as mandates by the French and British intensified. Lebanon and Syria, both promised independence by the Free French government during the war, achieved this status by 1946 (*map 1*). In the same year Britain relinquished its mandate of Jordan, but was left with the growing problem of the conflict between Jews and Arabs in Palestine.

THE NEW STATE OF ISRAEL
The issue of whether a Jewish state should be established in Palestine became a focal point of both Arab and international politics. Tensions between the growing Jewish immigrant community and the Arab inhabitants of the region, already high in the inter-war period, had been exacerbated by the influx of refugees from Nazi-occupied territories and the suggestion by the United Nations that Palestine be divided into Arab and Jewish states, with Jerusalem as an international zone (*map 2*). A civil war between Arabs and Jews from November 1947 escalated into an international war between Israel (proclaimed a state on 14 May 1948 after the British withdrawal) and the Arab countries of Egypt, Syria and Iraq, which ended in an Arab defeat and armistice agreements by July 1949.

Over 700,000 Palestinians fled to refugee camps in the West Bank (annexed by Jordan in 1950), Gaza (under Egyptian military occupation) and in other Arab countries. Further wars between Israel and its neighbours, in 1956, 1967 and 1973, resulted in the lasting Israeli occupation of the West Bank, Gaza Strip and the Golan Heights (*map 3*), although Sinai was returned to Egypt under a peace treaty signed in March 1979. In 1964 the Palestine Liberation Organization (PLO) began a guerrilla war against Israel, and in 1987 a Palestinian uprising in the West Bank and Gaza increased pressure on the Israeli government to find a negotiated solution. A breakthrough seemed to be reached with the Declaration of Principles in Oslo (September 1993), which resulted in a withdrawal of Israel from Gaza and parts of the West Bank. The process, however, was slow – marred by terrorist attacks by Palestinians and Jewish fundamentalists, and the continuing dispute over control of Jerusalem.

EVENTS IN LEBANON
The Arab–Israeli conflict had a powerful effect on neighbouring Lebanon, in which the political system was based on a balance of power between Maronite Christians, and Shiite and Sunni Muslims. The country became, in 1970, a major base for Palestinian guerrilla warfare against Israel.

▼ An estimated 70 per cent of the world's known oil reserves are located in the Middle East and North Africa, mainly on the Arabian Peninsula and in the Gulf. The resultant oil boom has facilitated the rapid modernization of the producer states. It has also contributed to the economies of the surrounding countries, partly through the wages paid to immigrant workers in Saudi Arabia and the Gulf states, and partly through the provision by the oil-rich countries of politically motivated development aid. The Organization of Petroleum Exporting Countries (OPEC), whose most powerful members are in the Middle East, attempts to ensure a minimum price for crude oil by controlling supplies.

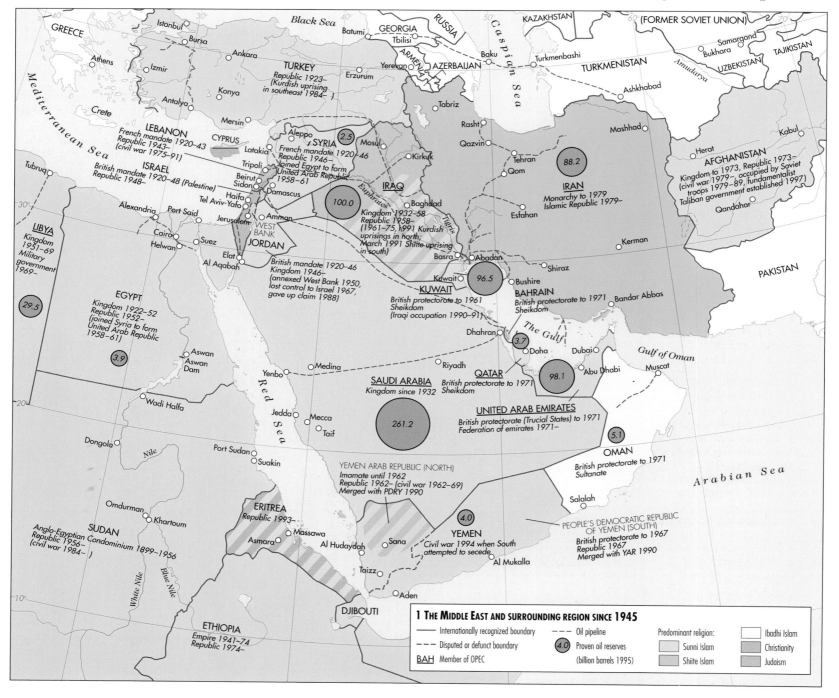

1 THE MIDDLE EAST AND SURROUNDING REGION SINCE 1945

—— Internationally recognized boundary	--- Oil pipeline
---- Disputed or defunct boundary	(4.0) Proven oil reserves
BAH Member of OPEC	(billion barrels 1995)

Predominant religion:
Sunni Islam — Shiite Islam — Ibadhi Islam — Christianity — Judaism

In 1975 internal disputes and external pressures resulted in a civil war between the Christian Phalangists, backed by the Israelis, and Lebanese Muslims, backed by the Syrians and the PLO. Although the Agreement of Ta'if (1989) prepared the ground for peace, fighting only ended in 1991, with victory for the Muslims. Israel continued to occupy the southern border areas of Lebanon, and Syria to maintain a military presence in much of the country.

SOCIALISM, NATIONALISM AND FUNDAMENTALISM

Defeat by Israel in 1949 served as a catalyst for the emergence in Egypt, Syria and Iraq of army-led, nationalist, secular regimes that advocated socialist economic reforms to improve living conditions for the countries' rapidly growing populations. Gamal Abdel Nasser of Egypt became the champion of Arab nationalism, advocating non-alignment, with some degree of co-operation with the Soviet Union, as a way of curtailing the influence of the Western powers in the Middle East. His decision to nationalize the Suez Canal in 1956 resulted in Britain and France sending in troops in a failed attempt to gain control of this vital sea-route. Egypt's anti-Western approach was opposed by Saudi Arabia, Israel and Iran, which saw Egypt's growing power as a threat. The conflict was played out in a proxy war, when Egypt and Saudi Arabia supported opposing sides in the civil war in the Yemen Arab Republic in 1962–69 (map 1).

By the 1970s most of the major industrialized countries of the world relied on oil from the Middle East – a situation that the Arab members of the Organization of Petroleum Exporting Countries (map 1) used to their advantage when they placed an oil embargo on countries who supported Israel in its 1973 war with Egypt and Syria (pages 272–73).

The tensions arising from the widening social rifts in many oil-rich states resulted in the emergence of "political Islam", which combined the religious teaching of Islam with the desire for social and political change. The Iranian revolution of 1979 under Ayatollah Khomeini, with its specifically Shiite character, encouraged other Islamic opposition movements. These erupted across the Middle East, from Egypt to Afghanistan. In Afghanistan, Islamic groups fought the Soviet intervention of 1979 before engaging in a civil war among themselves which resulted in the victorious Taliban establishing a fundamentalist government in 1997.

WARS IN THE GULF REGION

The Iranian revolution caused particular concern in neighbouring Iraq, which feared a similar rebellion from its own large population of Shiite Muslims. Both countries also included large Kurdish populations, and Iraq accused Iran of supporting an uprising of the Iraqi Kurds in 1979. The main motive for an Iraqi attack on Iran in 1980, however, was to expand into the oil-rich region on their joint border (map 4). At the end of an eight-year war in which an estimated one million people died, neither side had made significant gains. During the war Iraq received aid from most of the Arab states and, shortly before the end of the fighting, used chemical weapons against its own Kurdish population, some of whom had supported Iran.

Debts incurred by Iraq in its war against Iran, territorial claims, disputes over the price to charge for oil, and loss of prestige were all factors that contributed to Iraq's invasion of Kuwait on 2 August 1990. Ignoring international condemnation, Iraq annexed Kuwait and could not be persuaded by United Nations sanctions to withdraw. In January 1991 an international alliance led by the United States declared war on Iraq, initially concentrating on an aerial bombardment of Iraqi military installations. On 24 February ground forces moved in, and by the end of February Iraqi troops had been forced to withdraw from Kuwait. Iraq's subsequent suppression of revolts by Shiites in the south and Kurds in the north led to UN-backed "no-fly zones" for Iraqi aircraft north of the 36th and south of the 32nd parallels. Rivalries among Kurdish groups, Iraqi intervention, and repeated invasion by Turkish troops seeking to suppress the revolt in Turkish Kurdistan by eliminating camps in Iraq, reduced the Kurds

to abject poverty. The whole Iraqi population suffered from punitive economic sanctions, imposed in an attempt to force the Iraqi government to comply with UN requirements to eradicate the country's weapons of mass destruction.

Many had hoped that with the end of the Cold War (pages 244–45) the Middle East would reap the fruits of the so-called "peace dividend" and make a gradual transition to democracy. However, many of the fundamental social and economic problems, as well as the authoritarian regimes and regional ethnic and religious tensions, persist.

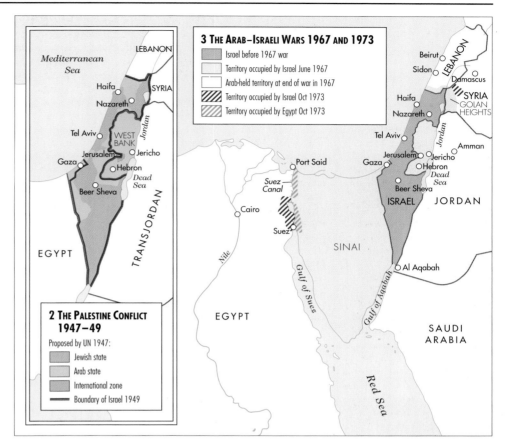

▲ The UN's proposed division of Israel was abandoned after Israeli independence in May 1948 (map 2). Israel also expanded its territory in 1967 and 1973, although the Sinai region was returned to Egypt in 1979.

▼ Iraq's desire for further oil-rich territory prompted its attacks on Iran in 1980 and on Kuwait in 1990. Despite heavy casualties, Iraq failed to make territorial gains.

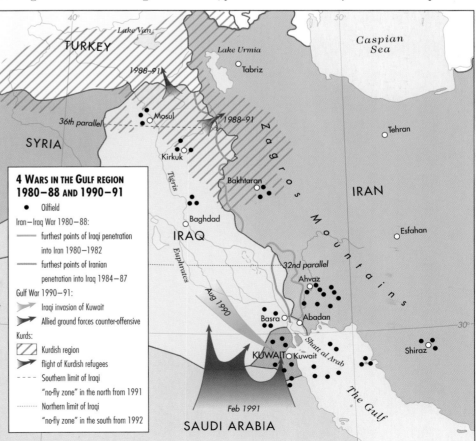

⬤ OUTCOMES OF THE FIRST WORLD WAR 1918–29 pages 220–21

THE FORMER REPUBLICS OF THE SOVIET UNION SINCE 1989

▼ The Soviet Union was formally abolished in December 1991 and the Soviet Socialist Republics became independent states. Most felt the need for some degree of continuity in defence, international relations and currency, and they eventually formed the Commonwealth of Independent States (CIS). This has, however, enjoyed limited success, with the parliaments of many of the states anxious to assert their autonomy. The Russian Federation is divided into administrative regions that are directly controlled from Moscow and constituent republics which, since 1993, have been entitled to their own constitutions.

Following the passing of discriminatory ethnic laws in many of the new states, around three million Russians returned to their native country during the 1990s. There was also movement between the new states over the same period. The descendants of Germans encouraged to settle along the Volga by Catherine the Great in the 18th century, but moved to Central Asia by Stalin in the 1940s, migrated back to Germany. Many Asians migrated to Belarus and Ukraine in the hope of finding an easy route into western Europe.

Mikhail Gorbachev became the General Secretary of the Communist Party – and as such supreme ruler of the Soviet Union – in March 1985. He appointed reformers such as Yakovlev, Rykov and Shevardnadze to positions of power, and introduced a policy of *perestroika* (economic restructuring), which attempted to introduce competition and market forces into the planned economy. Although heavy industry and collective farms remained under state control, private individuals could form co-operatives. Non-profitable firms were no longer propped up by the state, but allowed to go bankrupt. Nevertheless, economic growth continued to fall, while crime, inflation and unemployment rose. Strikes among miners in 1989 were the first sign of popular discontent at the Soviet Union's economic problems, exacerbated by the devastation caused by the explosion at the Chernobyl nuclear reactor in 1986 and the Armenian earthquake of 1988.

DEMOCRATIZATION

Gorbachev also introduced a policy of *glasnost* (openness), leading to an almost free press which, ironically, undermined his hopes of reviving support for a reformed Leninism. Democratization of the Communist Party apparatus allowed a choice of candidates in elections, followed by the participation of other parties in the Congress of People's Deputies in the summer of 1989. Finally, Article 6 of the Soviet constitution, which guaranteed the Communist Party a monopoly of power, was abolished in February 1990, and Gorbachev was appointed President of the Soviet Union. His radical approach to internal affairs was matched by his foreign policy. The withdrawal of Soviet troops from Afghanistan in 1988–89, negotiations with the United States to end the arms race, and encouragement of, or tacit support for, the countries of Eastern Europe in their bid to free themselves from Soviet domination in 1989–90 all had a tremendous

effect on world politics. However, while Gorbachev was praised abroad for his bold foreign-policy decisions, his popular support at home was waning. The economic crisis within Russia in the autumn of 1990 proved a turning point. A "500-day plan" for rapid market reform was rejected by Gorbachev, as a consequence of which reformers left the government, and under pressure from political hard-liners and military and industrial leaders, Gorbachev appointed more reactionary communists to power.

Meanwhile, Popular Fronts to support *perestroika* were formed in the republics, enabling dissidents to stand in elections in the Socialist Republics in March 1990, and leading to non-communist gains in areas such as the Ukraine and Lithuania (*map 1*). By 1989 there were conflicts between Moscow and the republics over religion, language and control of the economy, between republics and their own minorities, such as that between Georgia and South Ossetia, and between the republics of Azerbaijan and Armenia over the region of Nagorno–Karabakh (*map 2*). The Baltic States demanded outright independence but Gorbachev was desperate to keep the Soviet Union together, and force was used in Vilnius (Lithuania), as well as in Tbilisi (Georgia) and Baku (Azerbaijan). The rise of Russian nationalism allowed Boris Yeltsin, sacked by Gorbachev from the position of Mayor of Moscow in 1987, to return to politics, first as head of the Russian Supreme Soviet and then as democratically elected, anti-communist President of Russia, in June 1991.

THE BREAK-UP OF THE SOVIET UNION

Gorbachev's plan for a new Union Treaty, which recognized the independence of the Baltic States and decentralized power to the republics, sparked off a hard-line communist coup against him in August 1991 (*map 3*). Yeltsin managed to gain the support of the Russian parliament

1 THE BREAK-UP OF THE SOVIET UNION SINCE 1991
- Border of Soviet Union until 1991
- Russian Federation
- Constituent republic within Russian Federation
- Member state (with Russian Federation) of CIS
- State not member of Commonwealth of Independent States
- ✷ Area of armed conflict
- → Ethnic Russian immigrants
- → Other refugees or returnees
- → "Volga Germans" emigrating to Germany
- → Asian immigrants

▶ Ethnic tensions and rivalries in the Caucasus region, held in check by the centralized control of the Soviet Union, broke out into armed conflicts after the collapse of the Soviet Union in late 1991. Many smaller regions within the larger republics battled to achieve autonomy. Chechenia declared independence from Russia in 1991, but although Grozny and the surrounding region was extensively bombed, the Russian army failed to defeat the guerrillas and the republic achieved *de facto* independence in 1997. Georgia was also the scene of armed conflict, both for control of the republic (1991–93) and as a result of successful attempts by the regions of Ossetia and Abkhazia to assert their independence. The republics of Armenia and Azerbaijan waged a bloody war over control of Nagorno–Karabakh, which Armenia won.

2 CAUCASUS REGION 1988–98
—— International boundary
········ Boundary of constituent republic/autonomous region
Area ruled by Russian central government
Constituent republic within Russian Federation
Area that has achieved de facto autonomy
✵ Area of armed conflict with date

against the rebels, and his defiance was largely responsible for the failure of the coup. Thus Yeltsin's position was strengthened, and although Gorbachev was reinstated his power was diminished. The Ukrainian independence referendum in December 1991 made the continuation of the Soviet Union untenable, and when Yeltsin and the presidents of Ukraine and Belarus met in Minsk to create the Commonwealth of Independent States (CIS), the Soviet Union collapsed into 15 independent republics (*map 1*). Gorbachev resigned on 25 December 1991.

YELTSIN'S PRESIDENCY

Yeltsin, as President of the Russian Federation, inherited the unresolved problems of his predecessor. Although he introduced rapid market reform, including privatization, the economic decline continued. Inflation reached 245 per cent in January 1992, while industrial output slumped. Some people made huge profits but savings were wiped out, leading to real hardship among the population. The Orthodox Church gained support, as did nationalist, right-wing parties such as Zhirinovsky's Liberal Democrats. Yeltsin did not call new elections for the communist-led Supreme Soviet, now called the *duma* (parliament), but ruled by decree instead. Furthermore, he did not form his own political party, and neither did the democrats, thereby weakening the democratic system. Yeltsin's banning of the Communist Party in 1991 was declared unconstitutional, and led to its rebirth under Zyuganov. From December 1992 there was open conflict between Yeltsin and the *duma*, and Yeltsin replaced his reformist prime minister with the more conservative Viktor Chernomyrdin.

Yeltsin won public support in a referendum in April 1993, but conflict with the *duma* continued and in September it was dissolved. The political leaders within the *duma* retaliated by proclaiming Yeltsin's removal from the presidency, with the result that in October they were besieged in the parliament building. Their response was to order an attack on the Kremlin and other key buildings, leading to a three-hour battle. The army rescued Yeltsin and shelled parliament, leaving 145 dead and over 700 injured. New elections resulted once again in a majority for the Nationalists–Communists, but Yeltsin, although in ill-health, won the presidential elections of June 1996. His reformist policies failed once again to improve the economy.

A financial collapse in the summer of 1998 discredited the market reformers and brought a new conflict between Yeltsin and the *duma*. The latter rejected Yeltsin's attempt to restore Chernomyrdin as prime minister. A compromise resulted in the appointment of Primakov, who brought a degree of stability to the government.

NATIONALIST DEMANDS

Nationalism, responsible for the break-up of the Soviet Union, also threatened the Russian Federation. Autonomous republics, such as Tatarstan and Yakutia (now Sakha), demanded "sovereignty", in which their own laws would take precedence over those of Moscow. Yeltsin's Union Treaty of March 1992 compromised by granting them considerable autonomy, and finally even Tatarstan signed in February 1994. Chechenia split from Ingushetia and declared independence after the August 1991 coup. Yeltsin sent in Russian troops at the end of 1994, and following international condemnation of Russia's actions, Chechenia achieved a *de facto* independence in 1997 (*map 2*).

Conflict continued on the peripheries of the old Soviet Union. The so-called Dnestr Republic (*map 1*) rejected Moldovan rule with Russian military support, and there was conflict between Russia and Ukraine over the Crimea and over which country should control the ships of the former Soviet navy, based in the Black Sea. Newly independent republics brought in citizenship laws that discriminated against Russian residents, causing a migration of ethnic Russians into Russia (*map 1*). In Georgia, President Gamsakhurdia's extreme nationalism led to his overthrow in 1992. The new president, Shevardnadze, clamped down on civil war and joined the CIS, but lost Abkhazia when the province rebelled with Russian support (*map 2*). Azerbaijan and other oil-rich states in Central Asia attracted Western investment, but a revival of Islamic fundamentalism led to civil war in Tajikistan. At the end of the 20th century the future of the region remained uncertain, both in economic terms and in relation to democratic reform.

▲ Mikhail Gorbachev, who led the Soviet Union through a period of rapid reform in the late 1980s, was forced to resign in December 1991 when the Soviet Union broke up into its constituent republics.

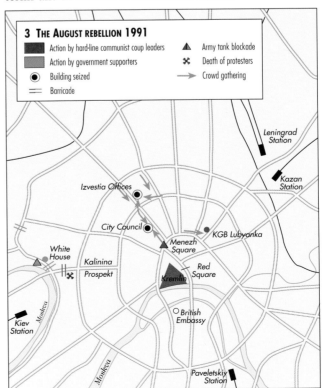

3 THE AUGUST REBELLION 1991
Action by hard-line communist coup leaders
Action by government supporters
◉ Building seized
═ Barricade
▲ Army tank blockade
✖ Death of protesters
→ Crowd gathering

◀ In August 1991 Moscow experienced street fighting unprecedented since the "October Revolution" of 1917. Hard-line communists tried to reassert the Communist Party's monopoly of power and prevent President Gorbachev's proposed Union Treaty from being signed, but the people of Moscow took to the streets in support of the government and barricaded the streets around the Russian parliament (the White House). Three of them were killed by the army, which was divided in its support. With Gorbachev a prisoner in his summer retreat in the Crimea, Boris Yeltsin, then President of Russia, eventually persuaded the army to stand firm behind Gorbachev, and thus defeated the communist rebels.

EASTERN EUROPE SINCE 1989

INCREASING ETHNIC HOMOGENEITY
1930–91

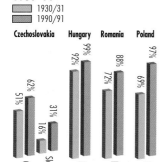

	1930/31
	1990/91

Czechoslovakia   Hungary   Romania   Poland

▲ Boundary changes, war losses, extermination, migration, expulsions and population exchanges between 1938 and 1948 significantly reduced the ethnic mix in all Eastern European countries so that there was a higher degree of ethnic homogeneity in 1991 than had been the case in 1930.

▼ The collapse of the communist regimes of Eastern Europe occurred between 1989 and 1990. In general, the "peoples' revolutions" were carried off relatively peacefully. Only in Romania, where the communist regime put up a fight, and in the former Yugoslavia did fighting break out.

Throughout the 1980s the communist regimes of Eastern European underwent a profound crisis. They experienced increasing economic difficulties as a result of inefficiency, low productivity and declining growth, compounded by the growing environmental crisis affecting, in particular, parts of East Germany, Czechoslovakia and Poland (*pages 236–37*). The unelected communist governments had always had trouble maintaining their legitimacy in the eyes of their electorates, but since the radical reforms introduced in the Soviet Union under Mikhail Gorbachev they could no longer threaten critics with the ultimate sanction of Soviet military intervention.

In the second half of 1989 all the communist regimes collapsed, although they did so in various ways (*map 1*). In the most reformist of the communist regimes – Hungary – the demise was gradual and was managed by the communist government itself. Some of its increasingly radical measures had a profound effect on other communist governments. The decision, for example, to open the borders with Austria and let thousands of East German "tourists" depart for the West forced the East German government into belated attempts to save itself by offering concessions of its own. In Poland, where the Solidarity movement challenged the hegemony of the state as early as 1979, the end of communism was negotiated and brought about by partial elections held as a result of negotiations between government and opposition. The East German and Czechoslovak regimes both collapsed as a result of public demonstrations. In Bulgaria the government fell following a coup, which overthrew Todor Zhivkov, and in Romania the end of the Ceausescu regime was brought about by a violent uprising.

POLITICAL AND ECONOMIC TRANSITION

All the post-communist countries embarked on the construction of a democratic system of government and the conversion of a centrally planned economy into one that was market-led. One of the major problems was their lack of experience of democratic government. Although some institutional and legal changes, such as a multiparty system and free elections, were introduced quite rapidly, the development of a democratic political culture proved more difficult. The bulk of the electorate still expected the state to guarantee not just security but also their well-being. Increasing inflation and declining Gross Domestic Product (*map 2*) caused most people's living standards to decline. In this economic climate former communists gained significant popular support with promises to minimize the negative consequences of economic change.

The problem was how to liberalize and privatize an economy under conditions of relative instability. Major disagreements existed between the proponents of the gradualist approach and those who advocated the "short, sharp shock treatment" involving simultaneous radical liberalization of prices and large-scale privatization. Some countries – particularly those in which former communists still held power, such as Romania and Bulgaria – adopted a slow and often inconsistent approach; others, such as Poland, adopted a radical path. Although the West provided some financial and technical help, this was not on a scale to make a significant difference, except in East Germany where, after the reunification of Germany in 1990, the transition process was financed by a massive influx of West German capital.

1 THE TRANSITION FROM COMMUNISM TO DEMOCRACY 1989–96

🐦 Location of main civil unrest

A further aim of the post-communist countries was a "return to Europe". In this respect Poland, Hungary and the Czech Republic proved more successful than countries such as Bulgaria and Romania. Not only were they in the first wave of new entrants to NATO in 1999, but were among the first group of applicants from Eastern Europe to embark on negotiations for entry with the European Union (*map 2*).

THE EFFECTS OF NATIONALISM

Developments since 1989 have largely completed the process – started in the late 19th and early 20th centuries and accelerated by the Second World War – of the creation of ethnically homogeneous states in the region (*bar chart*). In post-communist Czechoslovakia the national grievances felt by many Slovaks resurfaced and were compounded by the fact that the process of industrialization undergone by the region of Slovakia since 1948 had left it largely dependent on markets in the Soviet Union and other Eastern European countries. This placed it at a disadvantage in a country that was increasingly seeking Western European trading partners. Furthermore, while the Czechs preferred a centralized state, the Slovaks sought a loose confederation. These differences proved intractable and the Czechoslovak state broke up on 1 January 1993 into two national states: the Czech Republic and Slovakia.

In Yugoslavia the federal system developed by President Tito in the 1950s and 1970s gave some credence to national autonomy while controlling nationalist self-assertion in the constituent republics. With the decline of communist power, the economic disparities between the constituent republics and the pressure for democratization gave rise to nationalist resentments. Demands were made by Slovenia, Croatia and Macedonia for a large measure of sovereignty, and by Serb nationalists for a larger Serb state (to include parts of Slovenia, Croatia and Bosnia-Herzegovina).

The former Yugoslav Republic of Macedonia, with a Serb population of 2 per cent of the total, achieved independence peaceably in 1991. The process of independence in Slovenia, which also included a Serb population of around 2 per cent, was accomplished in 1991 with only a brief intervention by the Yugoslav (Serbian) army. In Croatia, however, the conflict that broke out in 1991, following the Croatian declaration of independence, was more violent, with the Yugoslav army fighting on behalf of a Serbian minority of around 12 per cent of the total.

The bloodiest conflict occurred in ethnically and religiously mixed Bosnia-Herzegovina, where the 1991 census showed that 31 per cent of the population were Serb, 17 per cent Croat, and 44 per cent were classified as "Bosnian Muslim" (although some of these were of no religious persuasion). An organized campaign of "ethnic cleansing" was undertaken, principally by the Serbs, with the aim of creating ethnically homogeneous regions in Bosnia as a prelude to its dismemberment and incorporation into Serbia and Croatia. The war, and the terrorist methods used against the civilian population, resulted in large-scale movements of populations (*map 3*).

In Kosovo, a region in southern Serbia where the large ethnic Albanian population sought independence, violence erupted in 1998 between the Kosovo Liberation Army and the Yugoslav army. Attempts to bring about a negotiated settlement failed and ethnic Albanians in Kosovo became the target of a Serbian campaign of ethnic cleansing, as villages were burned and people were driven from their homes. In June 1999, following a NATO campaign of air strikes, Serbia agreed to NATO troops entering Kosovo.

Significant Hungarian minorities remain in Romania and Slovakia, and the Bulgarian population is around 10 per cent Turk. There is also still a sizeable Roma population in Romania, Hungary, Slovakia and the Czech Republic, although accurate figures are difficult to come by. The Roma people are subjected to a variety of forms of discrimination, and a significant increase in violent incidents arising from anti-Roma feelings since 1989 has encouraged many to attempt to emigrate to Western Europe.

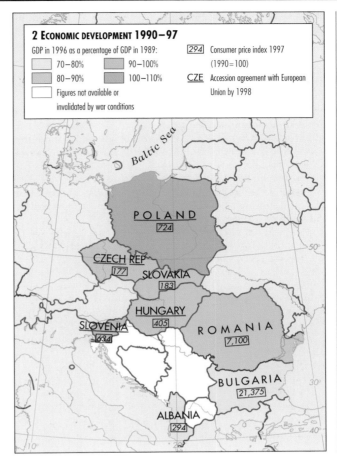

2 ECONOMIC DEVELOPMENT 1990–97

GDP in 1996 as a percentage of GDP in 1989:
- 70–80%
- 80–90%
- 90–100%
- 100–110%
- Figures not available or invalidated by war conditions

294 Consumer price index 1997 (1990=100)

CZE Accession agreement with European Union by 1998

POLAND 724
CZECH REP 177
SLOVAKIA 183
HUNGARY 405
SLOVENIA 634
ROMANIA 7,100
BULGARIA 21,375
ALBANIA 294

◄ The varied approaches taken by the elected governments over the conversion to a free-market economy yielded varying degrees of short-term success. In the mid-1990s Poland's more radical approach appeared to have paid off, although at the end of the 20th century it was still unclear as to which country would be the most successful in the long term. All the Eastern European countries were keen to join the European Union, but not all passed the EU's various entry criteria, which relate to the effectiveness of both their market economy and their democratic system.

▼ In the constituent republics of the former Yugoslavia, democratically elected governments sought independence from the Serb-dominated Yugoslav Federation. The government of Serbia, however, was anxious to defend the rights of Serbs throughout the region, and bloody conflicts ensued. Despite the Dayton Peace Agreement of 1995, which divided Bosnia-Herzegovina into a Serb Republic and a Muslim/Croat Federation, in 1998 there were still around 1.5 million refugees and displaced persons in the region as a whole (and a further quarter of a million elsewhere in Europe). In 1999 the crisis in Kosovo led to another massive movement of people as over 850,000 ethnic Albanian Kosovans fled from Yugoslavia.

3 FORMER YUGOSLAVIA 1991–99
- Boundary of former Yugoslavia 1991
- Boundary established by Dayton Peace Agreement 1995

Croatia and Bosnia-Herzegovina (June 1998):
- 274 refugees (in thousands)
- mainly Serb refugees
- mainly Croat/Bosnian Muslim refugees
- 49 displaced persons (in thousands)

Kosovo (mid-June 1999):
- 445 refugees (in thousands)
- Ethnic Albanian refugees
- 60? displaced persons (in thousands)

UNITED NATIONS PEACEKEEPING SINCE 1945

The first purpose of the United Nations, enunciated in the UN Charter, is to maintain international peace and security, and its founders originally envisaged the creation of a UN security force dedicated to doing this. When negotiations between the superpowers – the United States and Soviet Union – over the creation of such a force failed, various alternatives were suggested. "Peacekeeping" emerged as an improvised response to this failure and to developing international crises, in particular the 1948 crisis in Palestine. The term is used to describe efforts made by the United Nations to diffuse civil and regional conflicts.

In 1948 the United Nations Secretary-General, Trygve Lie, requested that the Security Council authorize the creation of the first UN ground force to police the truce in the Middle East: the United Nations Truce Supervision Organization (*map 1*). In the period 1948–56 other UN truce supervision forces were established in areas of dispute, although it was not until 1956 that a fully fledged peace-keeping force, the United Nations Emergency Force, was established by the General Assembly to police and monitor the ceasefire between Egypt and Israel. This provided the model for future operations: the creation of an impartial UN force composed of troops contributed by member countries, serving under the UN flag, interposed with the consent of the protagonists, and resorting to arms only in self-defence. In such operations, members of the peacekeeping force have acted as intermediaries, with responsibility for helping the belligerents negotiate a settlement.

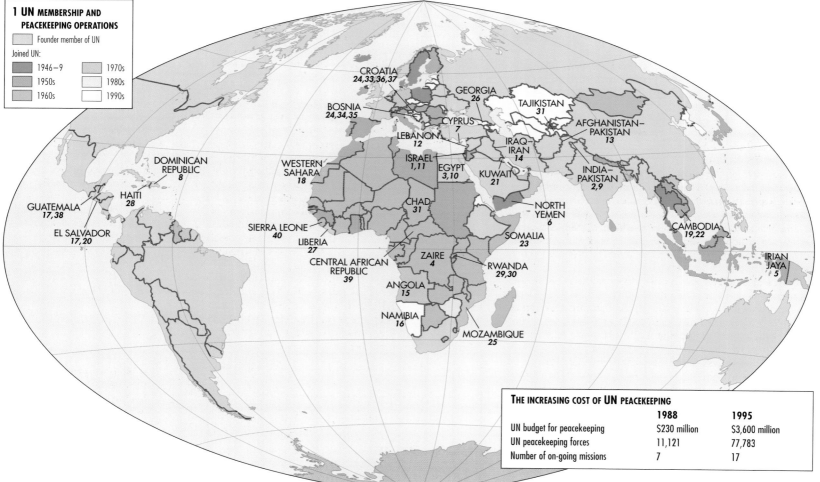

1 UN MEMBERSHIP AND PEACEKEEPING OPERATIONS

- Founder member of UN

Joined UN:
- 1946–9
- 1950s
- 1960s
- 1970s
- 1980s
- 1990s

THE INCREASING COST OF UN PEACEKEEPING		
	1988	**1995**
UN budget for peacekeeping	$230 million	$3,600 million
UN peacekeeping forces	11,121	77,783
Number of on-going missions	7	17

▲ UN peacekeeping operations around the world have included those attempting to restore or maintain peace between warring nations, such as the Iran–Iraq Military Observer Group of 1988–91, and those intervening to protect and bring aid to the civilian population in a state affected by civil war, such as the Operation in Mozambique in 1992–95. The UN budget for peacekeeping increased dramatically in the last decade of the 20th century, with more than half its peacekeeping missions being initiated during that time while other, more long-term, operations continued.

1. UN Truce Supervision Organization (UNTSO) (June 1948–)
2. UN Military Observer Group in India and Pakistan (UNMOGIP) (Jan 1949–)
3. First UN Emergency Force (UNEF I) (Nov 1956–June 1957)
4. UN Operation in the Congo (ONUC) (July 1960–June 1964)
5. UN Security Force in West New Guinea (West Irian) (UNSF) (Oct 1962–Apr 1963)
6. UN Yemen Observation Mission (UNYOM) (July 1963–Sept 1964)
7. UN Peacekeeping Force in Cyprus (UNFICYP) (Mar 1964–)
8. Mission of the Representative of the Secretary-General in the Dominican Republic (DOMREP) (May 1965–Oct 1966)
9. UN India–Pakistan Observation Mission (UNIPOM) (Sept 1965–Mar 1966)
10. Second UN Emergency Force (UNEF II) (Oct 1973–July 1979)
11. UN Disengagement Observer Force (UNDOF) (June 1974–)
12. UN Interim Force in Lebanon (UNIFIL) (Mar 1978–)
13. UN Good Offices Mission in Afghanistan & Pakistan (UNGOMAP) (Apr 1988–Mar 1990)
14. UN Iran–Iraq Military Observer Group (UNIIMOG) (Aug 1988–Feb 1991)
15. UN Angola Verification Missions (UNAVEM I, Jan 1989–June 1991), (II, June 1991–Feb 1995), (III, Feb 1995–June 1997), (MONUA) (July 1997–1999)
16. UN Transition Assistance Group (UNTAG) (Apr 1989–Mar 1990)
17. UN Observer Group in Central America (ONUCA) (Nov 1989–Jan 1992)
18. UN Mission for the Referendum in Western Sahara (MINURSO) (Sept 1991–)
19. UN Advance Mission in Cambodia (UNAMIC) (Oct 1991–Mar 1992)
20. UN Observer Mission in El Salvador (ONUSAL) (July 1991–Apr 1995)
21. UN Iraq–Kuwait Observation Missions (UNIKOM) (Apr 1991–)
22. UN Transitional Authority in Cambodia (UNTAC) (Mar 1992–Sept 1993)
23. UN Operations in Somalia (UNOSOM I, Apr 1992–Apr 1993), (UNOSOM II, May 1993–Mar 1995)
24. UN Protection Force (UNPROFOR) (Mar 1992–Dec 1995)
25. UN Operation in Mozambique (UNUMOZ) (Dec 1992–Jan 1995)
26. UN Observer Mission in Georgia (UNOMIG) (Aug 1993–)
27. UN Observer Mission in Liberia (UNOMIL) (Sept 1993–Sept 1997)
28. UN Mission in Haiti (UNMIH) (Sept 1993–June 1996), (MIPONUH) (Dec 1997–)
29. UN Observer Mission Uganda–Rwanda (UNOMUR) (Oct 1993–Sept 1994)
30. UN Assistance Mission for Rwanda (UNAMIR) (Oct 1993–Mar 1996)
31. UN Aouzou Strip Observer Group (UNASOG) (May 1994–Mar 1996)
32. UN Mission of Observers in Tajikistan (UNMOT) (Dec1994–)
33. UN Confidence Restoration Operation in Croatia (UNCRO) (Mar 1995–Jan 1996)
34. UN Preventive Deployment Force (UNPREDEP) (Mar 1995–1999)
35. UN Mission in Bosnia–Herzegovina (UNMIBH) (Dec 1995–)
36. Transitional Administration for Eastern Slavonia, Baranja, and Western Sirmium (UNTAES) (Jan 1996– Jan 1998)
37. UN Mission of Observers in Prevlaka (UNMOP) (Jan 1996–)
38. UN Human Rights Verification Mission in Guatemala (MINGUA) (Jan–May 1997)
39. UN Mission in the Central African Republic (MINURCA) (April 1998–)
40. UN Mission of Observers in Sierra Leone (UNOMSIL) (July 1998–)

Mediterranean Sea

Kyrenia

Nicosia

Famagusta

Larnaca

Paphos

Limassol

2 THE DIVISION OF CYPRUS 1974

Area controlled by Greek Cypriots

Area controlled by Turkish Cypriots

UN-patrolled buffer zone

British military base

Urban area

Ceasefire line of Cypriot National Guard

Ceasefire line of Turkish forces

▲ The island of Cyprus, only 100 kilometres (55 miles) south of Turkey but with 80 per cent of its population Greek-speaking, has been divided in two since the invasion of Turkish forces in July 1974. The UN Peacekeeping Force in Cyprus, which arrived on the island in 1964 to avert civil war, polices the "green line" between opposing Turkish and Greek Cypriot forces.

This buffer zone is 180 kilometres (112 miles) long and includes part of the northern suburbs of Nicosia. The UNFICYP investigates hundreds of violations of the ceasefire each year, and its task is to ensure that none of these escalates into full-scale war.

There are also two British military bases on the island, under an agreement made when Cyprus became independent in 1960.

"CLASSICAL" PEACEKEEPING

Following the success of UNEF I, this type of peacekeeping became a popular UN policy option. Used in cases of inter-state conflict, it is known as "first" or "classical" peacekeeping. It attempts to bring about an end to the fighting, separate the opposing forces and encourage the creation of a lasting peace. Such operations have usually included the supply of UN humanitarian assistance to the affected civilian population. From the 1960s to the late 1980s classical peacekeeping was used in the majority of peacekeeping operations, including that of the United Nations Force in Cyprus (map 2), deployed on the island in 1964 in order to separate warring Turkish and Greek Cypriot communities, and the United Nations Disengagement Observer Force, sent to supervise the Syrian Golan Heights in 1974, following the Arab–Israeli War.

All of the UN's peacekeeping efforts between 1948 and 1990 were, however, constrained by the existence of the Cold War (pages 244–45), during which the majority of conflicts were affected to some degree by rivalry between the United States and the Soviet Union, neither of whom wanted UN involvement if this compromised its own national interests.

"SECOND GENERATION" PEACEKEEPING

Since the end of the Cold War new opportunities have arisen for UN action in dealing with threats to peace, and this has stimulated an increase in the form of operation known as "second generation" peacekeeping. This occurs when the UN becomes involved in intra-state conflicts in "failed states", where governmental functions are suspended, the infrastructure is destroyed, populations are displaced and armed conflict rages. In these circumstances the UN has performed three different peacekeeping roles.

First, it has acted as a neutral force and honest broker between the warring factions, seeking to encourage the negotiation and implementation of a peace agreement and to prepare and conduct national elections as a means of furthering reconciliation and stability. This was the case with the United Nations Angola Verification Missions from 1989 onwards and the UN mission to Cambodia in 1991–95.

Second, it has interposed itself between warring parties to ensure the delivery of humanitarian aid to the war-torn population, as in the case of the United Nations Operations in Somalia in 1992–95.

Finally, "second generation" peacekeeping has been used to create a stable environment for the re-establishment of democracy, as was the purpose of the United Nations Transition Assistance Group in Namibia in 1989–90 and the United Nations Mission in Haiti in September 1993.

These "second generation" peacekeeping missions have become more common since the end of the Cold War, and have led to an increase both in the number of forces deployed and in the total expenditure on peacekeeping (table). In the case of the UN operations in Bosnia (map 3), Somalia and Rwanda, however, the UN did not have the consent of the various warring factions. Rather, the UN was forced by the international community to act in the interests of the civilian populations. The UN's hasty reaction to such demands resulted in clouded mandates, which made the implementation of peacekeeping problematic.

Peacekeeping is inherently risky, and over 1,500 peacekeepers have lost their lives since 1948. The UN's role has also at times been compromised by a failure to remain neutral, as when a large force, sent to the Congo in 1960 by the Security Council, lost its impartiality, and became involved in fighting against the Soviet-orientated, democratically elected prime minister, Patrice Lumumba. At other times failure has resulted from lack of military strength and restrictions on its freedom of action, such as when the United Nations Protection Force was unable to enforce the "Safe Areas" it had created in Bosnia in 1993 (map 3).

UN peacekeeping operations have generally worked well where the task is fairly limited and clear cut – such as the patrolling of ceasefire lines in Cyprus – but when the situation is more complex, as in Rwanda or Bosnia, the UN peacekeepers have often found themselves out of their depth. Nevertheless, peacekeeping has, in many cases, assisted in ending war and in creating the conditions in which the causes of the war can be addressed through diplomacy, and the economic and social reconstruction of a war-torn country can commence.

▲ Kofi Annan, a Ghanaian diplomat, was elected Secretary-General of the United Nations in 1996 – the first black African to hold the position. Among the international crises in which he became involved as peace-maker in the late 1990s were those arising from events in Bosnia and Iraq.

SLOVENIA

HUNGARY

Zagreb

CROATIA

VOJVODINA

Sisak

Osijek

SLAVONIA

Zrenjanin

Danube

Novi Sad

KRAJINA

Banja Luka

Brčko

Belgrade

Bihać

Sava

Maglaj

Tuzla

Vrbas

BOSNIA–

Zenica

Srebrenica

SERBIA

HERZEGOVINA

Žepa

Titovo Užice

Gornji Vakuf

Sarajevo

Split

Goražde

Neretva

Mostar

Adriatic

KOSOVO

MONTENEGRO

Pec

Dubrovnik

Sea

Podgorica

Dakovica

ALBANIA

◄ The UN became involved in Bosnia, a multi-ethnic constituent republic of Yugoslavia, in 1992, after the Yugoslav (predominantly Serbian) army invaded to prevent the formation of an independent state. Sarajevo was besieged and the UN attempted to keep the airport open to allow supplies to be flown in. In an attempt to protect the Bosnian Muslim population from attack by Bosnian Serb forces, six towns were nominated by the UN as "Safe Areas". The UN force lacked sufficient military strength, however, to implement their policy; with only limited freedom of action it was forced to withdraw from two of the areas (Žepa and Srebrnica) in the summer of 1995, leaving them to be overrun by Bosnian Serbs.

3 THE UN IN BOSNIA 1994

Area of Bosnia and Croatia controlled by Serbs

Area of Bosnia controlled by Bosnian Croats

Area controlled by Bosnian Muslims

UN Safe Area established April 1993

Boundaries of Bosnia and Croatia

Border of former Yugoslavia

HUMAN RIGHTS
SINCE 1914

In 1998 the United Nations celebrated the 50th anniversary of the Universal Declaration of Human Rights, the preamble of which asserts that the "recognition of the inherent dignity and of the equal and inalienable rights of all members of the human family is the foundation of freedom, justice and peace in the world." The Declaration, according to the General Assembly of the United Nations, was to be a "common standard of achievement for all peoples and all nations", and during the second half of the 20th century efforts were made to define, articulate and enforce the fundamental rights of all peoples of all nations.

DEFINITION OF HUMAN RIGHTS
The United Nations, chartered in 1945, was not the first body to recognize and assert basic human rights. The first ten amendments to the US Constitution, the Bill of Rights (ratified in 1791), outline what early Americans believed to be their inalienable rights. The League of Nations, the international organization established as a result of the Treaty of Versailles (1919), drew up conventions on slavery and forced labour. Yet the United Nations was the most powerful force within the field of human rights in the 20th century, and the breadth of conventions created in the first 50 years of its existence surpassed those of any prior body. They cover areas such as employment, the rights of children, refugees, development, war crimes and the eradication of hunger and malnutrition. The earliest conventions were generally concerned with civil and political rights, while more recently the UN has turned its attention to the rights of people to economic and social development and to peace and security.

ELECTIVE DEMOCRACY
In the first half of the 20th century most democratic governments (those resulting from multiparty elections) were to be found in countries in Europe and in North and South America (*map 1*), although in some of these countries sections of society were still barred from voting for reasons of ethnic origin, gender or income. After the Second World War, and in particular in the last two decades of the 20th century, elective democracy spread to the great majority of countries in the world, although the fifth of the world's population who live in the People's Republic of China were still not able to exercise full democratic rights.

It remains to be seen how the spread of democracy will affect human rights. Governments that can be voted out by their electorate are less likely to abuse their citizens (as demonstrated by the contrast between the democratic society of Chile in the 1990s, and the society under the military dictatorship of Pinochet in the preceding two decades). In countries where political opposition is not tolerated, however, governments often go to great lengths to ensure that political rivals are silenced, and human rights abuses, including a ban on the freedom of speech, imprisonment without a fair trial, torture and execution, are common.

RELIGIOUS CONFLICT
The right to practise the religion of one's choice is enshrined in a UN Declaration of 1981, yet persecution on religious grounds is still prevalent throughout the world (*map 2*). Discrimination on the basis of religion often occurs when a religious group is seen as a threat to the status quo because of demands for autonomy, although it is difficult to distinguish it from discrimination on ethnic or political grounds.

An example of an area riven by sectarian conflict is Ireland (*map 3*), where British rule and domination by Protestants was resisted by Catholic Nationalists for centuries. A guerrilla war, fought by the Irish Republican Army (IRA) against British forces from 1918, came to a temporary end in 1921 with the Anglo-Irish Treaty, under which the British agreed to a large area of Ireland (in which Catholics predominated) becoming an independent state (initially within the Commonwealth). Six of the nine northern counties of Ulster remained part of the United Kingdom, albeit with their own parliament. Although Protestants predominated in much of the north, there was still a sizeable Catholic minority, which found itself under-represented in the political system, and in the allocation of public housing and of public investment.

These factors led to the development of a Catholic civil rights movement in Northern Ireland in the late 1960s and to clashes between Protestant and Catholic paramilitary groups and civilians, as a result of which the British army was deployed in the province. The introduction of internment (imprisonment without trial) in 1971 was seen by many Catholics as a transgression of their civil and political rights and an escalation of political violence ensued. On 30 January 1972 the British army killed 13 Catholics in

▼ During the second half of the 20th century democracy was introduced to most of the countries of Africa, Central America and, following the collapse of their communist regimes in 1989–90, to the countries of Eastern Europe and Central Asia. In addition, democratic processes were reinstated in many South American countries, which experienced periods of right-wing dictatorship during the 1970s and early 1980s. However, in many countries democracy is only tenuously established, and human rights abuses continue; in Africa some of the newly democratic countries have slipped back to being one-party states, and in others there has been clear evidence of rigged elections.

The majority of the world's countries now support the International Covenant on Civil and Political Rights (ICCPR), adopted by the UN in 1966, which sets out a range of rights, including freedom of conscience, freedom from torture and slavery, and the right to demonstrate peaceably.

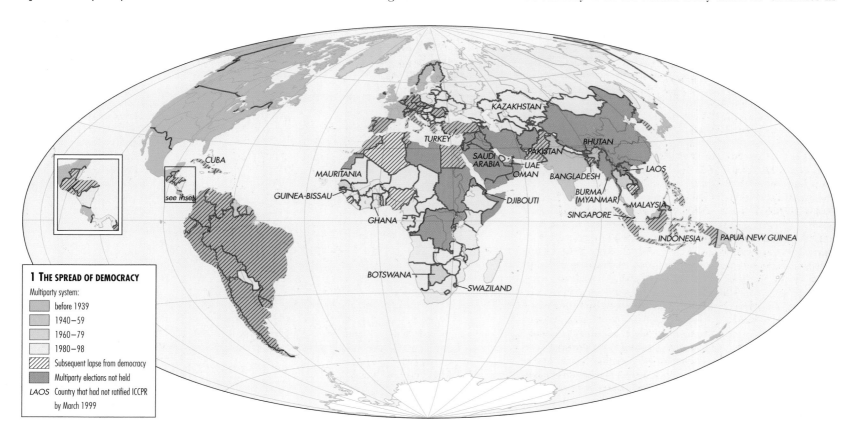

1 THE SPREAD OF DEMOCRACY
Multiparty system:
- before 1939
- 1940–59
- 1960–79
- 1980–98
- Subsequent lapse from democracy
- Multiparty elections not held
- *LAOS* Country that had not ratified ICCPR by March 1999

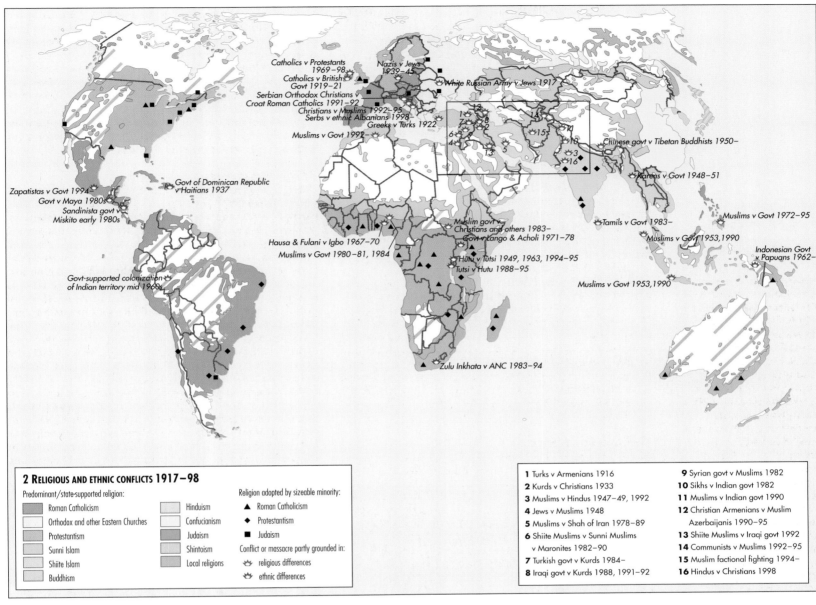

2 RELIGIOUS AND ETHNIC CONFLICTS 1917–98

Predominant/state-supported religion:

▨ Roman Catholicism	▨ Hinduism
▨ Orthodox and other Eastern Churches	▨ Confucianism
▨ Protestantism	▨ Judaism
▨ Sunni Islam	▨ Shintoism
▨ Shiite Islam	▨ Local religions
▨ Buddhism	

Religion adopted by sizeable minority:

▲ Roman Catholicism
◆ Protestantism
■ Judaism

Conflict or massacre partly grounded in:
🕊 religious differences
🕊 ethnic differences

Map labels:
- Catholics v Protestants 1969–98
- Catholics v British Govt 1919–21
- Serbian Orthodox Christians v Croat Roman Catholics 1991–92
- Christians v Muslims 1992–95
- Serbs v ethnic Albanians 1998–
- Greeks v Turks 1922
- Muslims v Govt 1992
- Nazis v Jews 1939–45
- White Russian Army v Jews 1917
- Chinese govt v Tibetan Buddhists 1950–
- Karens v Govt 1948–51
- Muslims v Govt 1972–95
- Govt of Dominican Republic v Haitians 1937
- Zapatistas v Govt 1994
- Govt v Maya 1980s
- Sandinista govt v Miskito early 1980s
- Hausa & Fulani v Igbo 1967–70
- Muslims v Govt 1980–81, 1984
- Muslim govt v Christians and others 1983–
- Govt v Lango & Acholi 1971–78
- Hutu v Tutsi 1949, 1963, 1994–95
- Tutsi v Hutu 1988–95
- Tamils v Govt 1983–
- Muslims v Govt 1953, 1990
- Indonesian Govt v Papuans 1962–
- Govt-supported colonization of Indian territory mid 1960s
- Muslims v Govt 1953, 1990
- Zulu Inkhata v ANC 1983–94

1 Turks v Armenians 1916	9 Syrian govt v Muslims 1982
2 Kurds v Christians 1933	10 Sikhs v Indian govt 1982
3 Muslims v Hindus 1947–49, 1992	11 Muslims v Indian govt 1990
4 Jews v Muslims 1948	12 Christian Armenians v Muslim Azerbaijanis 1990–95
5 Muslims v Shah of Iran 1978–89	13 Shiite Muslims v Iraqi govt 1992
6 Shiite Muslims v Sunni Muslims v Maronites 1982–90	14 Communists v Muslims 1992–95
7 Turkish govt v Kurds 1984–	15 Muslim factional fighting 1994–
8 Iraqi govt v Kurds 1988, 1991–92	16 Hindus v Christians 1998

what became known as "Bloody Sunday". In March 1972 the Northern Ireland parliament was dissolved and direct rule imposed from London. The subsequent 25 years, during which over 2,750 civilians, soldiers and RUC officers lost their lives, saw several peace proposals and peace movements gain support and then founder. On Good Friday 1998 an agreement was brokered between political representatives of the two sides, which established a Northern Ireland Assembly, with both Catholic and Protestant representation.

HUMAN RIGHTS AND REFUGEES

Between 1970 and 1995 the world's refugee population increased by over 900 per cent to 27 million people. This was partly due to wars (*map 2*), but also due to people seeking refuge from poverty, persecution and economic and environmental disasters. Refugees often end up in the poorest countries, which lack money to support their own citizens, let alone refugees. These displaced populations are a growing concern to the international community.

With so many nations still struggling to develop economically and politically, the provision of basic human rights on a world scale seems an immense task. A strong international legal foundation has been laid for the respect of human rights, and ordinary people are becoming increasingly aware of their entitlements – even if they are unable to achieve them. However, the reluctance of the international community to use economic and military sanctions against governments that abuse human rights – and the apparently ineffectual nature of these sanctions – means that worldwide transgressions of human rights are likely to continue.

▲ Religious and ethnic differences have led to intense conflict in many regions of the world, although issues such as inequality of social status, income and land distribution are frequently strong contributing factors. Demands for autonomy by minority groups, including the Bosnian Muslims and Kosovan Albanians in former Yugoslavia, and the Kurds in Iraq and Turkey, have resulted in attempts by the governments concerned to suppress entire peoples and eradicate their cultures.

3 THE DIVISION OF IRELAND 1922

Catholics as percentage of population:
- ▨ over 90%
- ▨ 70–90%
- ▨ 50–70%
- ☐ under 50%

◀ In 1922, following centuries of religious conflict, Ireland was divided in two. Catholics predominated in the Irish Free State, and also formed the majority in large rural areas of Protestant-controlled Northern Ireland, which were included in the province in order to provide it with sufficient agricultural land.

THE POSITION OF WOMEN SINCE 1914

In 1893 New Zealand became the first country to grant universal suffrage to women. Today few women anywhere in the world are excluded from political participation, and most women are able not only to vote in national and local elections, but to run for office as well (*map 1*). In some countries, such as the United States and most Western European nations, the female franchise was preceded by long fights for political equality; in other countries women were granted the right to vote partly in recognition of the contribution they made towards the struggle for independence from colonial rule.

Improving women's lives has become an international concern in the 20th century. Women's lives differ from men's in every area, including education, health and employment, in ways that have not always been readily apparent. Gender inequality means different things in different cultures, but the use of gender as a category of analysis in measuring the quality of people's lives has greatly changed perceptions of the social interactions of women and men.

THE UNITED NATIONS DECADE FOR WOMEN
The first United Nations Decade for Women took place between 1976 and 1986. During this period the UN began to compile statistics on women for regional and international comparison, in relation to such areas as maternity and reproduction, leadership and decision making, family life, economics, education and health. These statistics have served as a focus for discussions, and have helped to identify areas needing attention and improvement.

The increased desire in the 20th century to recognize the importance of women's daily lives has also led to greater scrutiny of the employment of women and the ways in which work is measured. International statistics on employment, for example, indicate the extent to which women are participating in paid employment (*map 2*), and the type of job in which they are employed. However, the 1995 United Nations Fourth World Conference on Women in Beijing stressed the importance of valuing unpaid labour. In the industrialized world work is often valued by the remuneration attached to it. Volunteer, domestic and childrearing work (unpaid labour that is most often performed by women) has been devalued and, in terms of statistics, gone unreported. Activities such as subsistence production and housework, in which a large proportion of women in developing countries are involved, are now being measured more effectively, although progress remains to be made.

Statistical information on women's lives has revealed not only that governments have invested less in females than in males, but that women provide more care to children and older people, have different access to education and employment from that of men, and usually work longer hours in and out of their homes throughout their lifetime than men. In short, women often experience a poorer quality of life than their male counterparts.

Although overall there has been a global trend towards improvement in the provision of secondary education for girls (*map 3*), this disguises the fact that within individual countries attendance at school may be affected by war or by economic difficulties. Furthermore, when assessing improvements in women's lives it is necessary to look at more than one variable. Even in countries that awarded women the vote relatively early (such as Turkey and Japan), women may still be represented in fewer than 10 per cent of administrative and managerial jobs, whereas in countries that granted women the vote relatively late (such as Switzerland, Honduras and Botswana) more than 30 per cent of women are in such employment.

One indication of women's status in society is the number who are political representatives, specifically those holding ministerial-level appointments. There have often been long periods between a country's enfranchisement of women and the election of the first woman to the national parliament. At the end of the 20th century there was still little female representation worldwide. Even in a country such as the United States, where over 50 per cent of women

▼ While women in New Zealand were fully enfranchised as early as 1893, elsewhere in the world, with the exception of a few US states (*map 4*), women had to wait until well into the 20th century before they could vote. In several European countries, including France and Switzerland, women were not given the right to vote until after the Second World War.

Native American women in Canada not able to vote in federal elections until 1950

Norwegian women over specified age and income enfranchised 1907–full franchise 1913

Women of former USSR enfranchised by Russian Revolution in 1917 but unable to exercise democratic rights until 1990s

British women between 21 and 30 years of age not allowed to vote until 1928

Vote extended only to literate women in Bolivia in 1930s– full franchise 1952

Non-white women not included in 1930s enfranchisement

Chilean women unable to vote in legislative or presidential elections until 1949

Kenyan women not given unconditional right to vote until early 1960s, although women of European origin could vote in 1919

1 WOMEN AND THE RIGHT TO VOTE
Women first enfranchised:
- pre-1914
- 1914–20
- 1921–45
- 1946–70
- 1971–
- no suffrage

Australian Aboriginal women not given full voting rights until 1967

▼ In Africa, parts of Asia and South America women are largely responsible for the agricultural work done in their community. They not only provide their families with food, but frequently produce cash crops for sale in local markets.

▶ Women make up a very small percentage of the workforce in some Muslim countries, such as Saudi Arabia. However, in several countries of Asia and southern Africa more women than men are in paid employment.

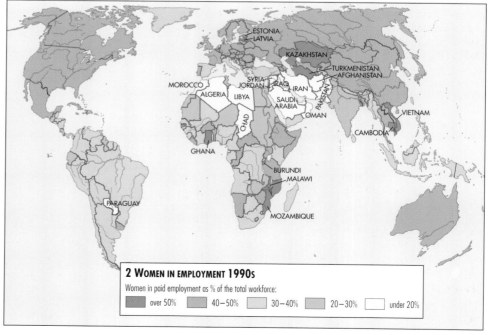

2 WOMEN IN EMPLOYMENT 1990s
Women in paid employment as % of the total workforce:
- over 50%
- 40–50%
- 30–40%
- 20–30%
- under 20%

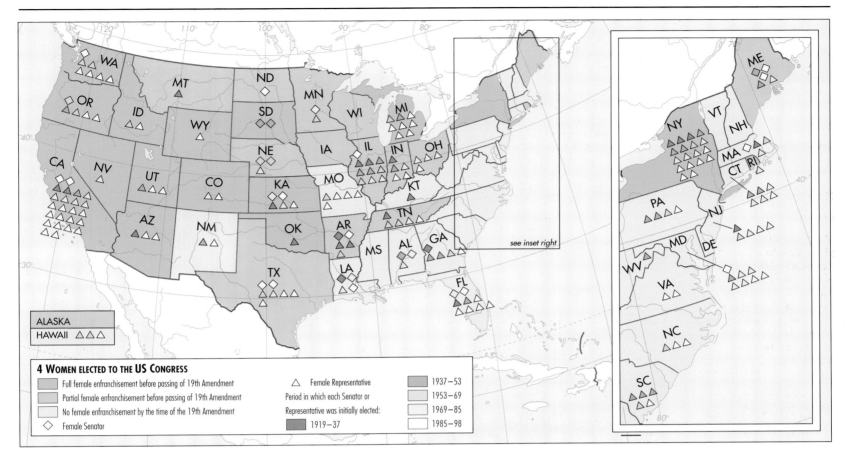

4 WOMEN ELECTED TO THE US CONGRESS

Full female enfranchisement before passing of 19th Amendment

Partial female enfranchisement before passing of 19th Amendment

No female enfranchisement by the time of the 19th Amendment

◇ Female Senator

△ Female Representative

Period in which each Senator or Representative was initially elected:

1919–37

1937–53

1953–69

1969–85

1985–98

ALASKA

HAWAII △△△

see inset right

were employed in administrative and managerial posts, relatively few women had been elected to Congress (*map 4*).

In order to understand change in women's lives it is necessary to appreciate how different aspects of women's lives are interwoven: how a girl's physical and mental development will affect the woman she will become; how a woman's status in relation to that of a man changes throughout the different phases of her life; and the difficulty in disentangling the inter-relationship between education, employment, fertility and contraception. For example, in many instances there is a clear correlation between a high female literacy rate and low birth rate (*bar chart*). There seems to be a two-way effect whereby education gives women the information and confidence to make family-planning decisions, and access to contraception gives young women the opportunity to fulfil their educational potential before starting a family.

▼ The percentage of girls receiving secondary education is a useful measure of a country's attitude to its female citizens, and the role they are expected to play in society. In many countries, although girls might receive a primary education, they are then expected to leave school and work in the home or the fields. Some cultures still consider secondary education for girls a largely wasted investment.

NON-GOVERNMENTAL ORGANIZATIONS

Many of the changes brought about in women's lives have come not from governments but from grassroots activists. Although women may be poorly represented worldwide in the traditional spheres of national politics, women have found that they can bring about change through participation in professional groups, trade unions, locally elected bodies and a growing number of non-governmental organizations (NGOs), of which there are estimated to be 30,000 worldwide. Such groups have allowed women's concerns to be voiced and supported on local, national and international levels, enabling them to build the skills necessary to exert political pressure and to collect the statistical information required to persuade governments to act.

Although disparities between the lives of men and women still exist, and progress remains to be made in the way in which men and women live and work together, the past century has witnessed vast changes in the way some men and women perceive women's roles. Women's rights have become human rights and the work of women has begun to be recognized as having no less an impact on society and the economy than that of men.

▲ The first women in the world to be given the vote were those in Wyoming in 1869, but female enfranchisement was only granted in all US states in 1920, after the passing of the 19th Amendment. Although the US Constitution did not actually prohibit women from standing for office, the first female Representative was not elected until 1917. The majority of Congresswomen have come from the eastern states and the west coast, although in 1998 Vermont, New Hampshire and Delaware were among those which had still never elected a woman.

▼ There is a strong correlation between the percentage of a country's women who are literate and its fertility rate. Women in industrialized nations, where literacy rates are much higher, have smaller families than those in non-industrialized nations, where educational provision is often fairly limited and that for girls is particularly poor.

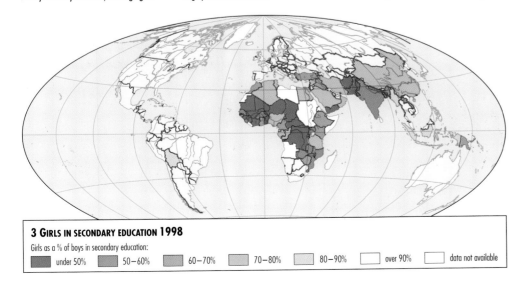

3 GIRLS IN SECONDARY EDUCATION 1998

Girls as a % of boys in secondary education:

under 50%   50–60%   60–70%   70–80%   80–90%   over 90%   data not available

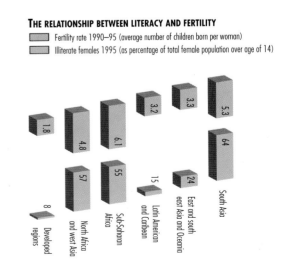

THE RELATIONSHIP BETWEEN LITERACY AND FERTILITY

Fertility rate 1990–95 (average number of children born per woman)

Illiterate females 1995 (as percentage of total female population over age of 14)

1.8 — Developed regions

4.8 / 57 — North Africa and west Asia

6.1 / 55 — Sub-Saharan Africa

3.2 / 15 — Latin American and Caribbean

3.3 / 24 — East and south east Asia and Oceania

5.3 / 64 — South Asia

8 — Developed regions

THE WORLD ECONOMY SINCE 1945

► The comparative wealth of the major economies of the world changed during the second half of the 20th century. Although the United States maintained its position as the world's wealthiest nation, countries such as Argentina, Uruguay and Mauritius, whose wealth was largely based on the export of raw materials, had slipped out of the "top 20" by 1970. The oil-producing countries of Saudi Arabia and Venezuela both featured in 1970, but were overtaken in 1990 by the newly industrialized countries of Western Europe and East Asia.

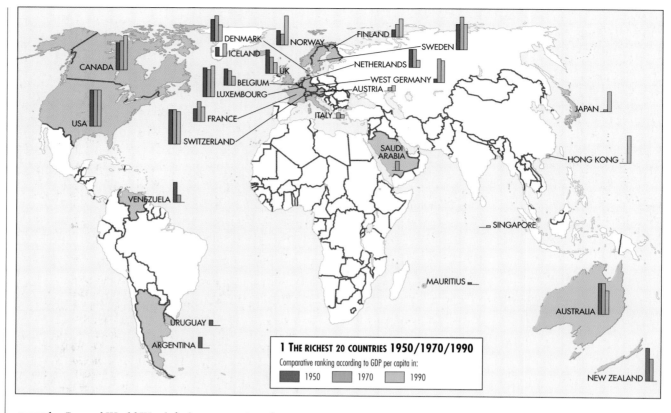

1 THE RICHEST 20 COUNTRIES 1950/1970/1990

Comparative ranking according to GDP per capita in:
- 1950
- 1970
- 1990

The Second World War left the economies of continental Europe, the Soviet Union and Japan ravaged, with manufacturing and agricultural output severely disrupted. The US economy remained strong, however, and its strength became a mainspring of recovery in Europe. The European Recovery Programme (or "Marshall Plan") provided US investment for Western European economies from 1948 to 1951 – effectively speeding up the process of economic recovery. In giving aid to Germany and Austria, as well as to the victorious Allied nations, it also engendered a more positive spirit than the one which emerged from the punitive Versailles agreement of 1919 (*pages 220–21*).

Co-operation between Europe and the United States aided recovery to the extent that by 1951 all Western European economies had at least recovered to their highest pre-war level of output (*pages 238–39*) and were entering a "golden age" of growth that was to last until the first oil crisis in 1973. Japan also received US financial support, and found its economy boosted by demand for supplies to

► The oil crisis of 1973–74 arose largely as a result of the Arab–Israeli War. The Organization of Petroleum Exporting Countries (OPEC) controls the majority of the world's oil exports and in 1973 its Arab members persuaded the organization to place an embargo on the supply of oil to those nations that supported Israel. The subsequent shortage of oil to the industrialized world severely disrupted production and oil prices soared.

support the UN troops in the Korean War (1950–53) (*pages 252–53*). New institutions, such as the International Monetary Fund (for the financial system), the World Bank (for developing countries) and the General Agreement on Tariffs and Trade (for the trading system), were designed by the United Nations Monetary and Financial Conference at Bretton Woods in 1944 in order to support the recovery.

In the decades following the Second World War the world economic situation changed markedly, with countries that were wealthy in pre-war times being overtaken by newly enriched nations (*map 1*). The United States was, and has remained, the wealthiest economy in the world, and for the early part of the post-war period it was also the major source of technological change; large US companies took their innovations abroad and invested in new plants in less advanced economies. In 1975 the total value of such multinationals' overseas stock was 4.5 per cent of world output, rising to 9.5 per cent by 1994. About a quarter of the stock is located outside the major industrialized nations, spreading new technologies to newly industrializing countries.

THE GOLDEN AGE OF GROWTH 1950–73

Between 1950 and 1973 Gross Domestic Product (GDP) per capita grew on average by 4 per cent a year in Western Europe as a whole. This growth was based on high levels of

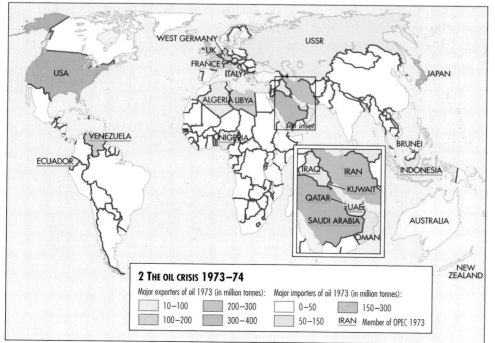

2 THE OIL CRISIS 1973–74

Major exporters of oil 1973 (in million tonnes):
- 10–100
- 100–200
- 200–300
- 300–400

Major importers of oil 1973 (in million tonnes):
- 0–50
- 50–150
- 150–300

IRAN Member of OPEC 1973

▼ During the 1970s OPEC engineered two substantial increases in the price of oil, largely through the tactic of restricting supply. The price of oil subsequently dropped again from the mid-1980s onwards as member nations ignored OPEC's limitations on exports. Fears are growing of a worldwide shortage of oil in the 21st century.

INDEX OF OIL PRICES (ADJUSTED FOR INFLATION)

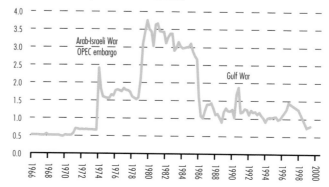

productive investment, the import of US technologies, and improvements in the quality of the workforce through education and training. In France, for example, there was a two-year rise in the average length of time spent in primary and secondary education (to 11.7 years), while in West Germany there was a rise of more than one year (to 11.6 years). The increasing integration of the European economies through the "Common Market" (EEC) also stimulated growth (*pages 238–39*). By 1970 the 20 countries with the highest GDP per capita were mainly to be found in Europe, and the world's wealth was concentrated largely in the North Atlantic.

Developments in East Asia, however, were just as remarkable, with Japan entering the "top 20" economies for the first time in 1970. Japan's output had grown by more than 9 per cent a year since 1950, driven by high investment and the rapid adoption and adaptation of US technology. The skills of the workforce had also improved rapidly, with the average length of time spent in primary and secondary education rising from 9 to 12 years.

THE SLOWING OF GROWTH RATES SINCE 1973

The golden age had been supported by low oil prices and cheap commodities, with the advanced economies becoming increasingly dependent on imported fuels as their incomes rose (*map 2*). The extent to which this made them vulnerable became all too apparent in 1973 when the Organization of Petroleum Exporting Countries (OPEC) – a cartel whose Arab members were the most powerful – placed an embargo on oil exports to the nations that supported Israel in the Arab–Israeli War. Oil suddenly became scarce and prices rose sharply (*graph*), causing major disruption in the United States and Europe.

A major slowdown in activity followed, and it took the advanced economies time to recover. They were just doing so when oil prices rose again in 1979. The richer European countries had largely caught up with the United States by this time, with the result that their growth was beginning to slow from 4 per cent per annum to a figure closer to the US level, which had dropped from 2.4 to 1.7 per cent following the 1973 oil crisis. With a post-1979 growth rate of only 1.7 per cent in Western Europe, unemployment rose sharply. In Japan growth remained high at 3 per cent, although this was well below the level of 8 per cent during the golden age.

OPENNESS AND GROWTH IN THE MODERN WORLD

Countries adopted different growth strategies after 1950. Those in Latin America, many in Africa and some in Asia – such as India – opted for a more self-sufficient approach, substituting home-produced goods for imports. The Europeans and many countries in Southeast and East Asia, on the other hand, opted for a strategy centred on openness to trade – importing and exporting a large share of their GDP (*map 3*). The open strategy made it necessary for these countries to react to external demands, and to adjust their methods of production accordingly. As the world moved, especially after 1970, beyond simple mass production towards the specialized production of high-technology products, the countries that had adopted the strategy of openness became increasingly successful.

Lessons have been learnt, and trading arrangements that remove barriers between member nations are becoming more common. The European Union, one of the oldest trading blocs (*pages 238–39*), is set to expand to include central and eastern European states. Its barriers to external trade stimulate inward investment by countries such as Japan. More recently formed regional trading blocs include the North American Free Trade Area (*pages 242–43*) and Mercosur (comprising Argentina, Brazil, Paraguay and Uruguay). The East Asians have set up an outward-looking bloc in APEC (*pages 242–43*), in an attempt to stimulate trade. However, as they learnt in the economic crisis of 1997–98, openness may aid growth, but it can leave their economies vulnerable to the vagaries of the world market.

◄ India is one of many Asian countries that have made huge economic and technological advances since 1945. However, a large proportion of its population continues to live without what are regarded as basic amenities – such as running water – in the industrialized world.

▼ A country's openness to trade is calculated by adding together the value of exports and imports (trade), and dividing the total by its Gross Domestic Product. In countries such as Argentina (with the lowest "openness" score) trade represents less than 12 per cent of its GDP, while others, of which Singapore is the prime example, import manufactured parts, assemble them into products, and export the finished goods. This has the effect of producing a ratio of trade to GDP of over 100 per cent.

In general the economies of those countries that have been open to trade (especially the smaller nations) have expanded most rapidly, as seen in the contrast between the low growth rates in some countries of South America and Africa, and the high growth rates in Southeast Asia. The western European economies have also grown rapidly because trade barriers have fallen within the region, with much of Europe becoming one large market.

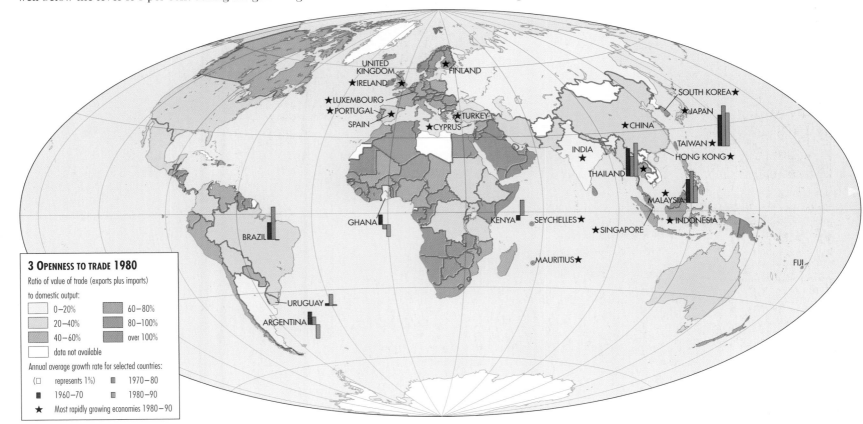

3 OPENNESS TO TRADE 1980

Ratio of value of trade (exports plus imports) to domestic output:

- 0–20%
- 20–40%
- 40–60%
- 60–80%
- 80–100%
- over 100%
- data not available

Annual average growth rate for selected countries:

- (□ represents 1%)
- 1960–70
- 1970–80
- 1980–90
- ★ Most rapidly growing economies 1980–90

◄ THE GREAT DEPRESSION 1929–33 *pages 228–29*

CHANGES IN POPULATION SINCE 1945

► Population growth is unevenly spread around the globe, with many of the more established industrial nations experiencing increases below 50 per cent since 1950. The populations of many of the newly industrialized nations, on the other hand, have increased by over 250 per cent in the same period. The Gulf states in the Middle East have seen the largest increases, mainly because of the economic expansion arising from their oil revenues.

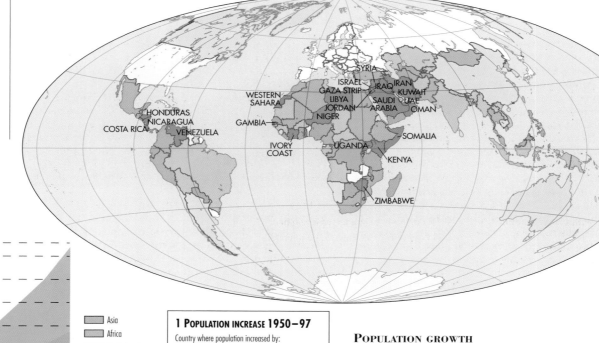

GLOBAL POPULATION
(in millions)

1 POPULATION INCREASE 1950–97
Country where population increased by:

- 0–50%
- 50–100%
- 100–150%
- 150–200%
- 200–250%
- 250–300%
- over 300%

Legend (graph): Asia, Africa, South America, North America, Europe, USSR/CIS, Asia

▲ The increase in global population has accelerated rapidly since 1950, although it is projected to slow down somewhat in the second decade of the 21st century. Over half of the world's population now lives in South, East and Southeast Asia.

The human population has more than doubled since 1940, with the total at the end of the 20th century standing at around six billion (*graph*). Despite indications that the rate of growth is slowing slightly, projections put the total population for the year 2025 as high as 8 billion. The majority of the growth since the mid-20th century has been in developing countries (*map 1*), with the increase in these regions contributing over 75 per cent of the world total growth in the 1950s, and over 90 per cent in the 1990s.

POPULATION GROWTH

The population explosion of the 20th century is not only the result of more babies being born, but also of better health care, nutrition, education and sanitary conditions, all of which have led to increased life expectancy. These conditions have aided population growth even in the face of disasters such as famines and epidemics. However, high population growth rates can also put greater pressure on public services and lead to a fall in living standards, poor nutrition, inadequate education and high unemployment.

The negative aspects of high population growth are compounded in developing regions (where over 75 per cent of the world's population lives) because of the greater incidence of poverty and economic instability. Most countries do not have the resources to support such large populations and the number of people without access to food, sanitation,

2 URBANIZATION OF THE WORLD
· City with at least 1 million inhabitants

▲ Dacca, the capital of Bangladesh, increased in size from 1.7 million people in the early 1980s to over six million by the end of the 1990s.

◄▼ Urbanization is one of the most extreme changes to have affected the world in the 20th century. In 1900 there were only a handful of cities with populations of over a million. By 2000 such settlements were scattered liberally around the globe.

THE WORLD'S LARGEST CITIES
(by millions of inhabitants)

- 1960
- mid-1990s

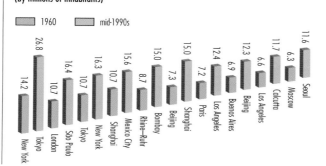

New York 14.2 / 26.8, Tokyo 10.7, London 10.7, São Paulo 16.4, Tokyo 16.3, New York 10.7, Shanghai 15.6, Mexico City 8.7, Rhine-Ruhr 15.0, Bombay 7.3, Shanghai 15.0, Paris 7.2, Buenos Aires 12.4, Beijing 12.3, Los Angeles 6.9, Calcutta 6.6, Moscow 11.7, Seoul 6.3 / 11.6

► In the 1950s there were fewer than ten cities with five million or more inhabitants, but by the mid-1990s there were over 30 cities of this size. The ten largest cities in the 1990s all had over ten million inhabitants, and the majority were to be found in the newly industrializing world.

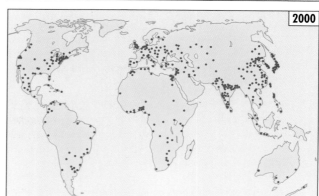

safe water and health services increases as the population grows. Furthermore, the inability of a country to provide for its citizens' basic needs affects its chances of maintaining or achieving economic and social growth. Balancing the growth of the population with the Earth's resources and society's ability to provide these basic necessities is crucial for a healthy population and continued development.

With a growth rate of 0.5 per cent per annum, the human population is set to double in 139 years; a growth rate of 1 per cent reduces that time to 69 years, 2 per cent to 35 years and 3 per cent to 23 years. Thus, what may appear as low rates of growth per annum can actually result in significant increases in population over a few generations.

Recognition of the adverse effects of our burgeoning population assisted in reducing growth rates in the 1980s and 1990s. This was achieved through a combination of improved education and the wider availability of contraceptives. However, while growth rates in developing regions have decreased, many will remain as high as 3 per cent or more in the 21st century. European countries currently reflect the lowest rates of growth (mostly below 1 per cent), with some countries – such as Bulgaria, Hungary, Romania, Latvia and Estonia – actually experiencing negative growth rates, leading to population decline. When coupled with the migration of people into cities, population decreases affect rural communities most severely.

URBANIZATION AND MIGRATION
Population growth in the developing world has been accompanied by an increasing number of people living in the cities of these regions, making urbanization a global phenomenon (map 2). Before the 20th century comparatively few people lived in cities, and the urbanization that occurred was largely the result of industrialization. Urbanization is now also a result of migration into the cities of people from agricultural areas unable to support them financially.

In the 1960s most of the world's largest cities were in industrialized countries, whereas now the majority are to be found in Central and South America, Asia and Southeast Asia (bar chart). This rapid urbanization of the world has resulted, among other things, in increasing levels of urban air pollution and waste, rapid growth in slum settlements, homelessness, insanitary water supplies and vast changes in the landscape (pages 280–81).

Populations have not only moved from rural environments to cities within their own country. During the 20th century substantial migrations took place (for economic and political reasons) across national boundaries (map 3). In many cases these migrations have resulted in significant minority cultures developing in the host nation. Many countries in the developed world now have multicultural populations, and people with racially mixed backgrounds are becoming more common.

DEMOGRAPHIC AGEING
The populations of many developed countries are getting older as a result of falling birth rates accompanied by improved health and healthcare, and the same process is predicted to occur in developing countries, assuming current improvements in life expectancy. At the beginning of the 21st century the number of people aged over 65 stands at around 390 million, but is projected to rise to 800 million by 2025, representing 10 per cent of the predicted population. Latin American and Asian countries are likely to experience increases of 300 per cent by 2025 in the number of people over 65 years old.

This demographic shift towards societies in which older people predominate can be a positive reflection of a country's health and prosperity, but it also signals the need for changes in the structure of the labour force, and for a shift away from a youth-centered culture towards one in which better health and social services are a priority. Growing and demographically changing populations have many implications for societies around the world in terms of standards of living, trends in health and ill-health, and the quality of the environment.

▼ The world's population has always been migratory to a certain extent, but the 20th century saw increased movement. This was partly as a result of economic factors but also as a result of political pressure and war. European Jews, an increasing number of whom migrated to Israel after the First World War, were forced by German Nazism to seek asylum elsewhere in Europe and in the United States in the 1930s. Most of those who did not escape were transported to death camps in eastern Europe. Stalinist policies in the Soviet Union also resulted in millions of people being forced into Siberian labour camps. Since the Second World War, major migrations have taken place in Asia and Africa as a result of war, and economic migrants from developing countries have sought work in the economies of North America, Europe and the Gulf states.

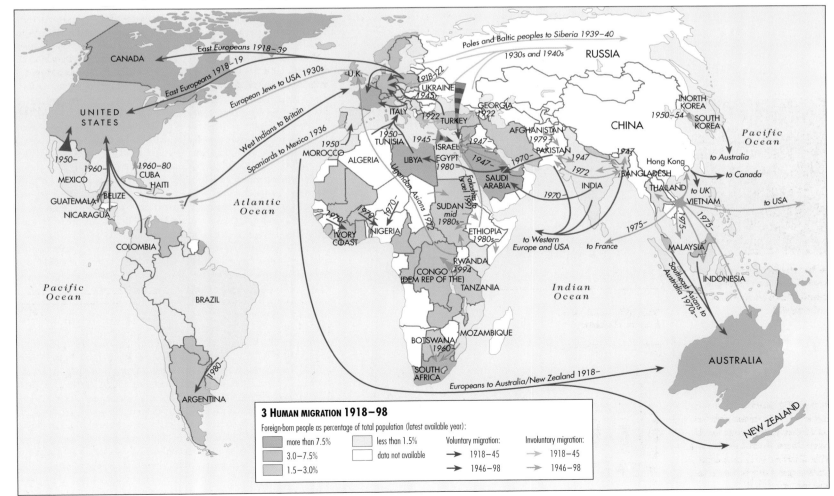

3 HUMAN MIGRATION 1918–98

Foreign-born people as percentage of total population (latest available year):

- more than 7.5%
- 3.0–7.5%
- 1.5–3.0%
- less than 1.5%
- data not available

Voluntary migration:
→ 1918–45
→ 1946–98

Involuntary migration:
→ 1918–45
→ 1946–98

PATTERNS OF HEALTH AND ILL-HEALTH SINCE 1945

▲ Child immunization programmes have been a major contributing factor in the worldwide increase in life expectancy.

▼ Spending on health care as a proportion of a country's Gross National Product largely increased during the second half of the 20th century. However, in some countries – among them the United States – this was largely due to private health schemes rather than government spending.

A worldwide increase in life expectancy during the 20th century suggested that the human population was the healthiest it had ever been, and increased health spending also gave cause for optimism (*maps 1 and 2*). However, at the end of the 20th century millions of people continued to live in poverty and had no access to adequate food, safe water or health services. New infectious diseases, such as AIDS and Hepatitis C, had spread across the world, while epidemics of older infections, such as cholera and yellow fever, had also broken out. Treatment of bacterial infections – after making huge advances with the introduction of penicillin in the 1940s – had been complicated by the evolution of drug-resistant bacteria. Health services are now widely recognized as crucial to economic development, but they are often the first to be axed by governments in the face of economic instability.

IMPROVEMENTS IN HEALTH

Better nutrition, improved access to health care and greater understanding of disease control have allowed people to live longer, healthier lives. Since the 1950s life expectancy has increased by over 50 per cent in developing regions and by 12 per cent and higher in industrial countries, to approximately 63 and 74 years respectively. Global immunization programmes have reduced the occurrence of diseases such as tuberculosis (TB) and measles, and have helped to contain the spread of many controllable diseases. Although the percentage of infants immunized against TB and measles in 1994 was as low as 20 per cent in some African countries, estimates for developing regions as a whole include rates of 70–90 per cent. These health measures have contributed substantially to a fall in infant and child mortality rates (*map 3*), and new and better vaccines are continually being developed.

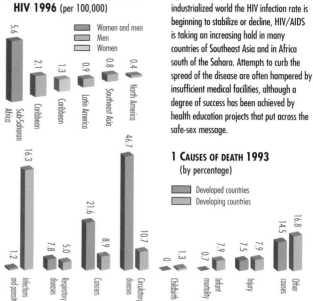

2 NUMBER OF PEOPLE INFECTED WITH HIV 1996 (per 100,000)

- Women and men
- Men
- Women

◄ While the indications are that in the industrialized world the HIV infection rate is beginning to stabilize or decline, HIV/AIDS is taking an increasing hold in many countries of Southeast Asia and in Africa south of the Sahara. Attempts to curb the spread of the disease are often hampered by insufficient medical facilities, although a degree of success has been achieved by health education projects that put across the safe-sex message.

1 CAUSES OF DEATH 1993 (by percentage)

- Developed countries
- Developing countries

▲ The marked differences in lifestyle and diet between the developed and developing world are reflected in the major causes of death. Diseases of the circulatory system and cancers, caused partly by high-fat diets, account for nearly 70 per cent of all deaths in the developed world, as against 20 per cent in the developing world.

CAUSES OF ILL-HEALTH

Improved health for some has been accompanied by greater ill-health for others, and a major cause of this has been poverty, which at the end of the 20th century affected over one billion children and adults throughout the world. Lack of funds for basic needs naturally leads to undernourishment and higher susceptibility to disease. Some of the most extreme poverty is to be found in the growing number of urban centres (*pages 274–75*), where public health systems cannot keep up with the demands placed on them by growing populations.

Both poverty and wealth can lead to ill-health. The high death rates from cancers, and heart and circulatory diseases in developed countries (*bar chart 1*) are partly due to greater life expectancy, but they are also undoubtedly related to unhealthy lifestyles. While wealthier, industrialized countries often have better education, more advanced medical technology, access to better health care and the higher incomes to pay for it, their populations as a whole also tend to have unhealthy diets, indulge in excessive drinking and smoking, and suffer from lack of exercise. The populations of industrial countries, and of large cities throughout the world, are also plagued by pollution, in particular air pollution, which is thought to be causing an alarming rise in respiratory problems such as asthma.

In developing countries, by contrast, infectious and parasitic diseases account for the majority of deaths. AIDS is one example of a modern plague. Since the 1980s health professionals have watched the disease spread worldwide, into all sectors of society, but in particular to the poorest, and estimates suggest that in the late 1990s over 33 million people were infected with the HIV virus that is believed to lead to AIDS, of whom 95 per cent lived in the developing world (*bar chart 2*). Water-borne diseases (such as cholera, typhoid, diarrhoea and guinea worm disease) are also common. In the 1990s the World Health Organization (WHO) estimated that 78 per cent of people living in developing countries still had no access to safe water. Despite world food surpluses, death from malnutrition, often caused by drought, remains a problem in many regions (*map 4*).

Shortage of water is projected to become an increasing problem in the 21st century, with populations growing in areas where there is little available. Advances in agricultural science and practice are being made in order to make the best use of limited resources, but international conflicts threaten to break out over use of river water.

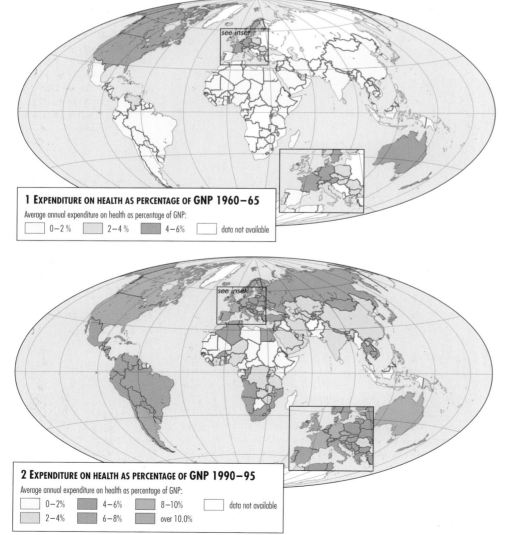

1 EXPENDITURE ON HEALTH AS PERCENTAGE OF GNP 1960–65

Average annual expenditure on health as percentage of GNP:
- 0–2 %
- 2–4 %
- 4–6%
- data not available

2 EXPENDITURE ON HEALTH AS PERCENTAGE OF GNP 1990–95

Average annual expenditure on health as percentage of GNP:
- 0–2%
- 2–4%
- 4–6%
- 6–8%
- 8–10%
- over 10.0%
- data not available

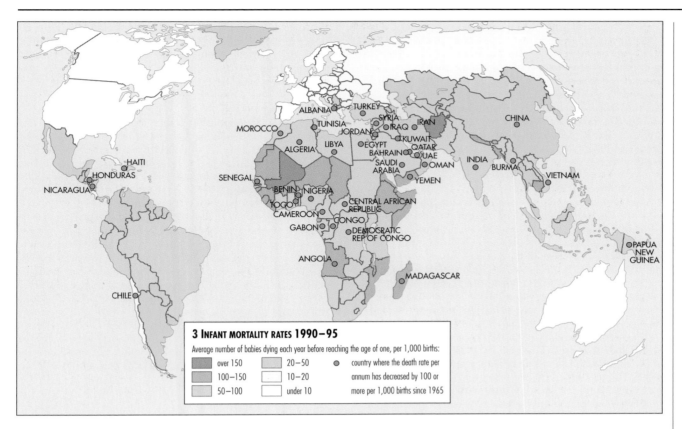

3 INFANT MORTALITY RATES 1990–95

Average number of babies dying each year before reaching the age of one, per 1,000 births:

- over 150
- 100–150
- 50–100
- 20–50
- 10–20
- under 10
- country where the death rate per annum has decreased by 100 or more per 1,000 births since 1965

◄ In the period between 1955 and 1995 the number of deaths per live births or children aged under one year decreased by 60 per cent worldwide, from an average of 148 deaths per thousand live births to 59. Most of the developed countries managed to reduce their rates by over 60 per cent between the 1960s and the late 1990s. While the developing world has, on the whole, not managed such large percentage drops, in many countries the infant death rate has been cut substantially in real terms.

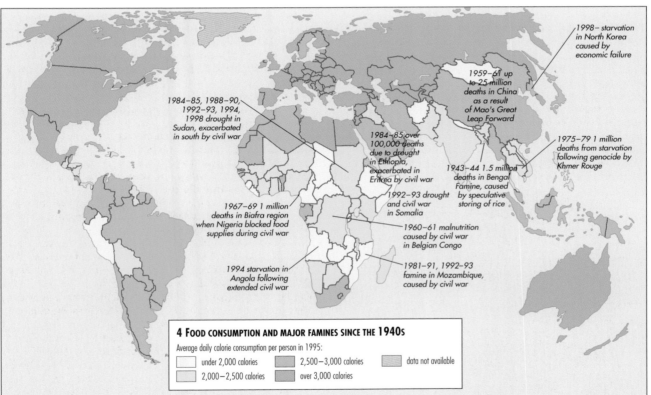

1998– starvation in North Korea caused by economic failure

1959–61 up to 25 million deaths in China as a result of Mao's Great Leap Forward

1984–85, 1988–90, 1992–93, 1994, 1998 drought in Sudan, exacerbated in south by civil war

1984–85 over 100,000 deaths due to drought in Ethiopia, exacerbated in Eritrea by civil war

1975–79 1 million deaths from starvation following genocide by Khmer Rouge

1943–44 1.5 million deaths in Bengal Famine, caused by speculative storing of rice

1967–69 1 million deaths in Biafra region when Nigeria blocked food supplies during civil war

1992–93 drought and civil war in Somalia

1960–61 malnutrition caused by civil war in Belgian Congo

1994 starvation in Angola following extended civil war

1981–91, 1992–93 famine in Mozambique, caused by civil war

4 FOOD CONSUMPTION AND MAJOR FAMINES SINCE THE 1940S

Average daily calorie consumption per person in 1995:

- under 2,000 calories
- 2,000–2,500 calories
- 2,500–3,000 calories
- over 3,000 calories
- data not available

◄ The average daily consumption of calories in the industrialized nations is nearly twice as much as in the non-industrialized nations. The five countries consuming least per head of population are Mozambique, Liberia, Ethiopia, Afghanistan and Somalia. Periods of drought in sub-Saharan Africa have severely affected agricultural production, and in many areas this has been exacerbated by war. In other parts of the world, such as China, Cambodia and North Korea, the policies of political leaders have been responsible for millions of deaths from starvation.

▼ It is estimated that around a third of the global adult population smokes. Although smoking is declining in parts of the industrialized world, in other areas, notably China (included in the figures for Western Pacific), smoking is becoming increasingly popular. The World Health Organization estimated that in the mid-1990s over 60 per cent of Chinese men smoked.

3 CONSUMPTION OF CIGARETTES

Average percentage annual change 1970–92 per adult

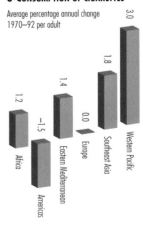

- Africa 1.2
- Americas –1.5
- Eastern Mediterranean 1.4
- Europe 0.0
- Southeast Asia 1.8
- Western Pacific 3.0

PREVENTIVE MEDICINE

The promotion of preventive health care was one of the greatest achievements of the last two decades of the 20th century. Instead of just treating illness, health promotion programmes help people take steps to improve their overall health. Widespread immunization programmes, better education and nutrition, and increased access to family planning services all contributed to reductions in the infant mortality rate during the latter part of the 20th century (map 3), particularly in parts of Africa and Asia.

One area where preventive health practices are fighting for a stronger foothold is in the use of addictive substances. Nicotine is one of the most widely used drugs; WHO statistics indicate that nearly one third of all adults smoke (and nearly half of all men), with most people starting before they reach the age of 20. Many developed countries have seen a significant decline in consumption since the 1970s (although the popularity of smoking among young people, in particular young women, gives cause for concern), but the tobacco industry continues to seek an expansion of its market. As a consequence, cigarette smoking in industrialized countries is on the increase (bar chart 3), and health departments and practitioners expect to see an upsurge in smoking-related heart disease and cancers.

Human health is possibly the most important issue facing the world in the 21st century in that it is both affected by and has an impact on environmental and demographic changes, and on social and cultural developments.

STANDARDS OF LIVING SINCE 1945

The Gross World Output (the total amount of money generated worldwide) in 1950 was $3.8 trillion. In the mid-1990s it was estimated to be $30.7 trillion. This near-tenfold increase was not, however, distributed evenly around the world. At least half of the extra wealth was created by the United States, Japan and the countries of Western Europe, where per capita incomes (the amount of money generated by a country divided by its population) grew markedly. By contrast, elsewhere in the world economic underdevelopment and high population growth rates resulted in per capita incomes actually decreasing.

WEALTH AND POVERTY

The result of this unequal growth is an increasing disparity between the national wealth of the richest and the poorest countries (*map 1*). Equally noticeable, however, is the disparity within a country between those with an income sufficient to provide a decent standard of living and the poorest members of society. The gap between rich and poor is most pronounced in the developed countries, where the average income of the poorest 20 per cent of the population may be as little as a quarter of the average per capita income (*bar chart*).

Poverty can be defined in different ways. In the United States the "poverty line" is calculated in relation to the cost of providing a nourishing diet for one person for one year. In 1996, 15 per cent of the US population was considered to be living below the poverty line, with a disproportionate number from the minority ethnic groups. In some European Union countries poverty is defined in relative terms, giving a typical figure of between 2 and 6 per cent.

THE HUMAN DEVELOPMENT INDEX

Despite the wide disparity of incomes within the industrialized countries, the majority of their populations have their most basic health and educational needs met. In many non-industrialized countries, on the other hand, free (and easy) access to doctors and schools is by no means universal. The disparity between the conditions experienced by the populations of the richest and the poorest nations of the world prompted the United Nations in 1990 to develop an index that defined and measured human development. The income of a country is one factor included, but figures for life expectancy and for literacy are also taken into account,

producing an overall score for each country. *The Human Development Report 1997*, based on figures for 1994, showed Canada at the top of the scale, scoring 0.96 out of the maximum possible score of 1, with Sierra Leone at the bottom, scoring 0.176 (*map 2*).

LIFE EXPECTANCY

The Human Development Index scores a country on the basis of the age to which a baby born in that country might be expected to live. In so doing it takes into account not only the general health of the population, but also the infant mortality rate. While the latter has improved dramatically since 1960 (*pages 276–77*), at the end of the 20th century it was still over 10 per cent in many non-industrialized countries, resulting in an average life expectancy at birth of between 40 and 50 years of age. However, those who survive the early years of life can expect to live well beyond their forties. For example, in Malawi, where the infant mortality rate is around 14 per cent, a girl who has survived until 15 years of age can expect to live, on average, until she is 62 years of age. In many countries improved health care, including vaccination, has resulted in substantial increases in life expectancy for both children and adults. Programmes to provide access to fresh water are also helping to improve the health of young and old people alike, and thus not only to improve life expectancy but also to raise the quality of people's lives.

EDUCATION AND LITERACY

In 1959 the United Nations General Assembly proclaimed that "The child is entitled to receive education, which shall be free and compulsory, at least in the elementary stages." Education became, for the first time in history, the right of young people worldwide. In 1962 the UN went further and attempted to remove barriers to education for such reasons as sex, religion, ethnic group and economic conditions. Education thus became the right of all people, but the extent to which they are given the opportunity to exercise that right remains highly variable, depending on where a person lives and whether they are male or female.

It is difficult to compare the amount of money spent on education by the different nations of the world. Expenditure on education as a percentage of Gross National Product (GNP) gives an idea of the importance a country attaches

COMPARISON OF INCOMES EARLY 1990S

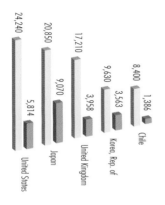

Average per capita income (in US dollars)

Per capita income of the poorest 20%

▲▼ The world's wealth is very unevenly distributed. The richest countries generate amounts of money that, when divided by the total population, produce (theoretical) per capita incomes over four times the world average; the equivalent figure for the poorest nations is one tenth of the average (*map 1*). Within most countries there is also a huge differential between the average incomes of the population as a whole and that of the poorest 20 per cent (*bar chart*).

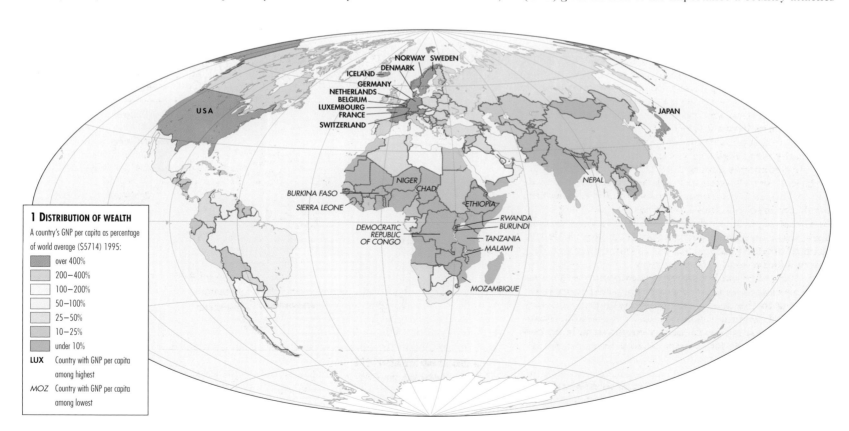

1 DISTRIBUTION OF WEALTH

A country's GNP per capita as percentage of world average (S5714) 1995:

- over 400%
- 200–400%
- 100–200%
- 50–100%
- 25–50%
- 10–25%
- under 10%

LUX Country with GNP per capita among highest

MOZ Country with GNP per capita among lowest

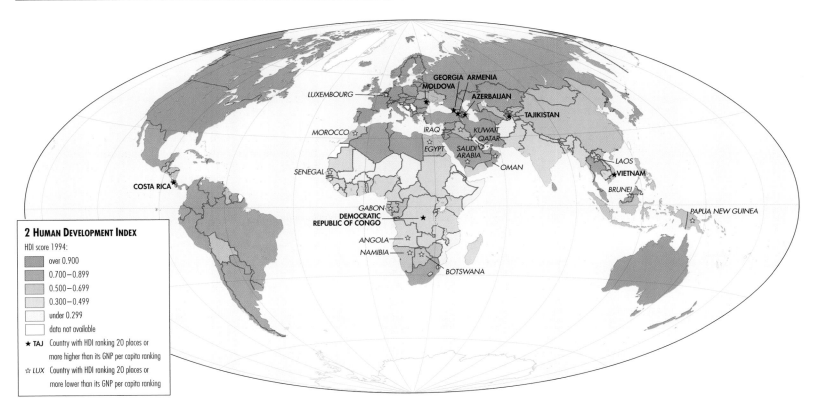

2 HUMAN DEVELOPMENT INDEX

HDI score 1994:

- over 0.900
- 0.700–0.899
- 0.500–0.699
- 0.300–0.499
- under 0.299
- data not available
- ★ TAJ Country with HDI ranking 20 places or more higher than its GNP per capita ranking
- ☆ LUX Country with HDI ranking 20 places or more lower than its GNP per capita ranking

to education; it indicates, for example, that some of the poorest nations of the world recognize how vital literacy is to their economic development and so invest a comparatively high proportion of their GNP in education (*map 3*). Their resources are meagre, however, in comparison with those available to the countries of the industrialized world.

RISING ENROLMENT IN EDUCATION

The United Nations Educational, Scientific and Cultural Organization (UNESCO) has estimated that during the second half of the 20th century student enrolment rose from 300 million to more than 1 billion. Enrolment in primary education, which begins at any time between the ages of 5 and 7 and provides the basic elements of education, increased markedly, with the result that the majority of children now receive some form of schooling.

Secondary education (enrolment at ages 10–12 years) and tertiary education (enrolment at ages 17–19 years), in institutions such as middle and high schools, vocational schools, colleges and universities, experienced an even more startling increase during the second half of the 20th century, with enrolments more than doubling. The take-up of higher education was highest in North American countries, and at its lowest in such areas as sub-Saharan Africa and China. High primary education enrolment levels did not necessarily mean high levels of post-primary education.

Many countries experienced setbacks in educational progress in the 1980s as war and decreased aid and trade led to cutbacks in government provision of free education. Enrolment in school often drops if parents have to shoulder the burden of paying for their children's education, and even where education is free, parents may keep their children at home to provide vital agricultural labour, or because they cannot afford to clothe them properly.

At the end of the 20th century education was just one of the necessities denied to many of the world's population – pointing to the need for a redistribution of monetary wealth and natural resources on a worldwide basis. However, the focus of each country continues to be on how it can best provide for its own citizens and operate in a growing global economy.

▶ In many countries half the population have not achieved basic standards of literacy. Some of the poorest nations spend over 6 per cent of their GNP on education, but this is still not enough to guarantee free access to a decent education for all.

▲ The Human Development Index scores each country according to how close it is to a target standard: an average lifespan of 85 years, universal access to education and a reasonable income for all. It also ranks the countries of the world according to both their development score and their GNP per capita. Some countries (particularly those in eastern Europe and the former Soviet Union) achieve a much higher development ranking than would be expected from their GNP per capita, while the development rankings of other, comparatively wealthy, countries (in particular many of the Arab oil states) are lower than expected.

◀ For many of the world's children an outdoor classroom is the best they can hope for at school. Many do not even have desks, while books, paper and writing equipment are all in short supply.

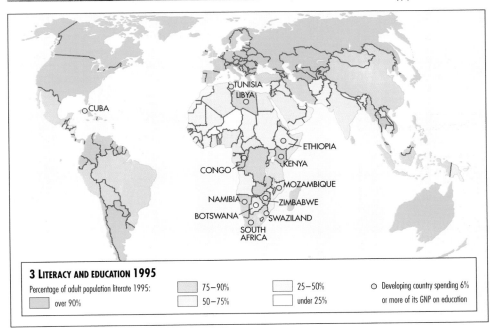

3 LITERACY AND EDUCATION 1995

Percentage of adult population literate 1995:

- over 90%
- 75–90%
- 50–75%
- 25–50%
- under 25%
- ○ Developing country spending 6% or more of its GNP on education

THE CHANGING ENVIRONMENT SINCE 1945

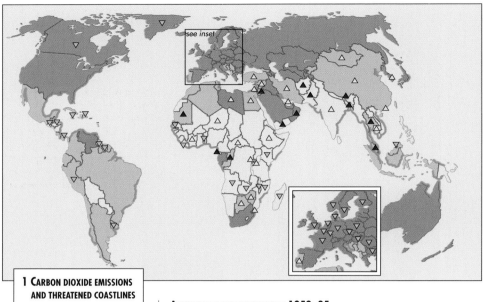

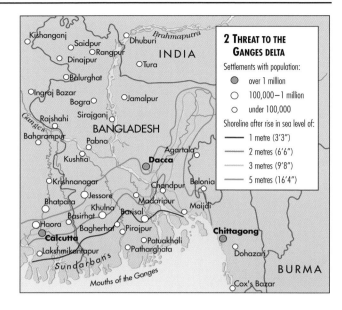

1 CARBON DIOXIDE EMISSIONS AND THREATENED COASTLINES

Emissions of CO_2 in tonnes per person per year (1992):

- over 10
- 5–10
- 1–5
- under 1

Changes in CO_2 emissions 1980–90:

- ▲ over 100% increase
- △ 50–100% increase
- ▽ reduction in emissions
- — Mainland coasts in danger of flooding from rising sea levels

▲ The emission of carbon dioxide into the atmosphere from the burning of fossil fuels is believed to increase the naturally occurring "greenhouse effect", causing a rise in the Earth's air and sea temperatures. This is likely to have far-reaching effects on the climate and possibly lead to an increase in sea level of around 50–100 centimetres (19–39 inches) in the 21st century.

▼ The world's tropical rainforests are being cut down at an ever-increasing rate. The timber trade makes an important contribution to the economies of many tropical regions, and population growth has also created demand for more farmland. Once the trees have been removed, however, the land can only be used for a short while for agricultural and grazing purposes before the topsoil becomes nutritionally depleted or eroded.

2 THREAT TO THE GANGES DELTA

Settlements with population:
- ● over 1 million
- ○ 100,000–1 million
- ○ under 100,000

Shoreline after rise in sea level of:
- — 1 metre (3'3")
- — 2 metres (6'6")
- — 3 metres (9'8")
- — 5 metres (16'4")

AVERAGE GLOBAL TEMPERATURE 1959–95

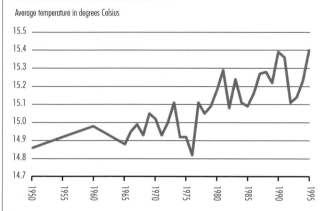

Average temperature in degrees Celsius

◄ Annual average global temperatures showed a marked increase during the second half of the 20th century, with the three hottest years of the century occurring in the 1990s.

▲ The low-lying region of the Ganges delta would be severely affected by a one-metre (three-foot) rise in sea level. One of the most densely populated regions of the world can ill afford to lose fertile land in this way.

Human activity has always had an impact on the natural environment, but the industrialization, urbanization and a rapidly increasing population of the last two centuries have had far-reaching adverse effects never before experienced. Changes in the environment range from those readily visible – such as deforestation, desertification and air pollution or smog – to less visible phenomena, such as climate change, damage to the upper ozone layer, mineral depletion, water pollution, and the extinction of plants and insects. Although these changes began to occur before the 20th century, it is only since the 1960s that they have been brought to public attention.

SUSTAINABLE DEVELOPMENT

Government policies regarding the environment, and various environmental conferences since the 1970s, including the UN Conference on Environment and Development in Rio de Janeiro in 1992 (the "Earth Summit"), have brought world leaders together to discuss the state of the environment and draw up plans of action. For cultural, economic and geographic reasons, numerous divergent views are held on the state of the environment, but it is generally agreed that some environmental monitoring and action is necessary. One of the most important concepts in environmental theory at the beginning of the 21st century is that of "sustainable development" – an approach to the use of the Earth's natural resources that does not jeopardize the well-being of future generations.

GLOBAL WARMING

Among the most widely publicized environmental problems in the 1990s was that of global warming (*graph*). A layer of carbon dioxide (CO_2) in the Earth's atmosphere traps heat from the sun's rays in a naturally occurring process known as the "greenhouse effect". Although the Earth's average temperature has always fluctuated naturally, many believe that emission of CO_2 from the burning of fossil fuels such as coal and oil are increasing the greenhouse effect and have been responsible for a rise of around 0.5° Celsius (1° Fahrenheit) during the 20th century.

Emissions of CO_2 have risen steadily since the 1950s. The larger industrial countries emit most (*map 1*), although many are now working towards curtailing, or at least stabilizing, their emissions. However, countries that have industrialized only recently are reluctant to restrict their industrial development or invest in new technology necessary to bring about a reduction. Predictions vary as to the amount by which temperatures are set to rise over the next century, and the possible effects of further global warming. It is likely, however, that global warming will cause the temperature of the world's oceans to increase and thus expand, causing flooding in low-lying areas (*map 2*).

Forests naturally absorb harmful CO_2, and deforestation also contributes to rising CO_2 levels. Rainforests have been destroyed at an increasing rate since the 1960s, with those in South America and Asia the most heavily affected (*map 3*).

The nuclear power industry has provided an alternative to the use of fossil fuels, generating 350 per cent more power worldwide in 1990 than in its early days in the 1960s. Nuclear power is not without its risks, however. The accidents at Three Mile Island in the United States in 1979 and at Chernobyl in the Ukraine in 1986, coupled with the problems associated with the disposal of nuclear waste, have led many to see the nuclear industry as one of the major threats to humans and to the environment.

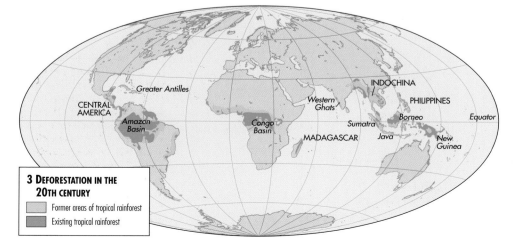

3 DEFORESTATION IN THE 20TH CENTURY

- Former areas of tropical rainforest
- Existing tropical rainforest

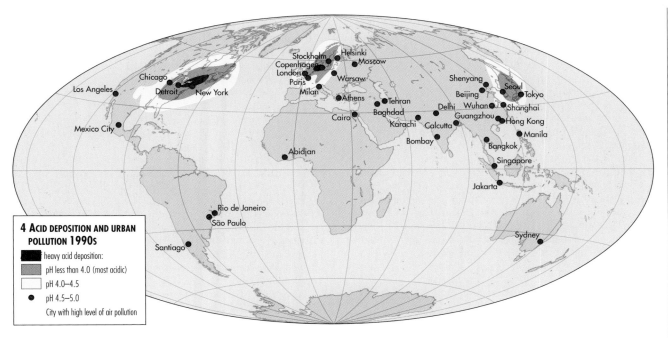

INDUSTRIALIZATION AND GLOBALIZATION

An increasing demand for electricity is made by the world's industries. While providing many benefits, such as increased wealth, employment and self-sufficiency, industrialization can also lead to an increase in air and water pollution, to changes in land use and to rapidly growing urban environments. One of the effects of industrialization has been increased emission into the atmosphere of sulphur and nitrogen. This falls back to Earth, either as dry deposits, or, combined with natural moisture, as "acid rain" (*map 4*), not only damaging trees and natural vegetation but also affecting crops and fish stocks in freshwater lakes.

Technological developments, particularly in areas such as transport and electronic communication, have helped to create a global economy in which people, products and information can move easily around the world. However, aircraft, ships, trains, passenger and heavy goods vehicles all pollute the environment, and require large-scale changes

to the landscape. They can also lead to environmental disasters, such as oil and chemical spills (*map 5*).

Oceans are particularly susceptible to environmental damage. Since the 1960s regulations have been established regarding such activities as offshore oil drilling, navigation and fisheries. The United Nations Convention on the Law of the Sea, which came into force in 1994, not only gives countries economic control over their coastal regions, but also the obligation to monitor and regulate marine pollution.

Global efforts are being made to conserve land and protect ecosystems, but preservation or protection is costly and may be hard to achieve in countries whose resources are already insufficient to meet population needs and whose economies are racing to catch up with those of the richer nations. The notion of sustainable development requires changes in the way people live their lives, and in the relative importance they assign to consumption over protection of the Earth's resources – changes that are difficult to achieve.

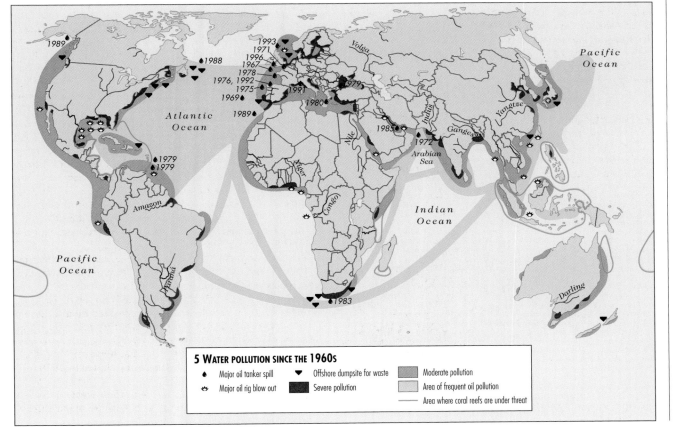

TRANSPORT AND COMMUNICATION SINCE 1945

▼ Car ownership is unevenly distributed around the world, with many families in the industrialized world owning two or more cars, and millions of people in the non-industrialized world never having the opportunity of travelling in one. In the 1960s the United States was still the largest car producer in the world, but it experienced no significant increase in output from the 1960s onwards and by the end of the 20th century had been overtaken by Japan. Of the European countries, Germany and France are in the same league as the United States, although the biggest increase in production was seen in Spain. The most remarkable development in car production was in Japan, China and Korea, with increases of over 5,000 per cent between the 1960s and the end of the 20th century.

When the American Wright brothers made the first flight in a motorized aircraft at Kitty Hawk, North Carolina in 1903 their invention was recognized as a milestone in transportation history. At the beginning of the century steamboats and trains were well-established methods of transport worldwide, and use of the recently invented telephone and car was spreading through the industrialized nations. However, the manner in which people travel and the methods by which they communicate have changed dramatically since then, and in particular since the 1980s.

High-speed trains, planes and cars, mobile phones, personal pagers, computers, electronic mail and the Internet have all contributed to an ease of travel and immediacy of communication that has created what has been termed a "global village". At the same time, in vast areas of the non-industrialized world, millions of people continue to live in real villages, excluded from, or touched only lightly by, the technological wonders of the late 20th century.

THE TRANSPORT REVOLUTION

Car ownership and production in the industrialized nations grew at an enormous rate during the 20th century. Cars were initially owned only by the well-off, but the innovation of mass-produced, and therefore relatively inexpensive, cars greatly expanded their ownership in North America and Europe during the 1920s and 1930s. Even so, in 1950 the number owned worldwide was still below 100 million, whereas 40 years later it was approaching 600 million.

Japan, in particular, saw a boom in car production and ownership from 1965 onwards, and by the end of the 20th century China had also increased its car production, from 80,000 cars a year in 1970 to around 1.5 million. Nevertheless, at the end of the 20th century the main mode of transport for millions of people, in China and elsewhere, was still a bicycle or other non-motorized vehicle. While car ownership has almost reached saturation point in many industrialized nations, with one car for fewer than five people and some cities forced to place restrictions on car use, in large areas of the world there is only one car per 1,000 people (*map 1*).

Alongside the marked increase in car ownership, air travel has also become the norm for those in the industrialized world. The total number of kilometres flown each year continues to grow (*graph*), as people venture further and further afield for reasons of business and pleasure (*map 2*).

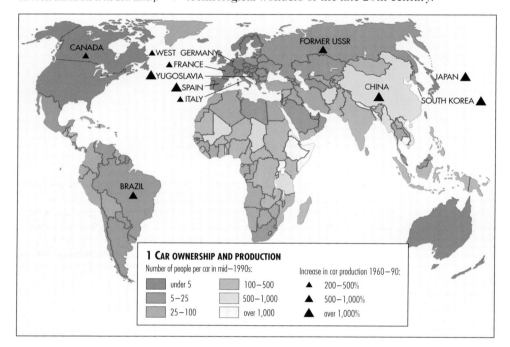

1 CAR OWNERSHIP AND PRODUCTION

Number of people per car in mid–1990s:

under 5	100–500
5–25	500–1,000
25–100	over 1,000

Increase in car production 1960–90:

▲ 200–500%
▲ 500–1,000%
▲ over 1,000%

▼ Increased vehicle ownership and a general decline in the availability of public transport led to over-stretched road systems and to more frequent traffic jams throughout the industrialized world at the end of the 20th century.

▼▶ Both the number of flights taken each year and the distances flown have increased as people have become accustomed to travelling further for recreation and business. It is now the norm for many Europeans, North Americans and Australians to fly to foreign destinations for their holidays, with the more "exotic" locations in relatively inaccessible areas becoming more and more popular. In large countries, such as the United States, Canada and Russia, people travelling to destinations within their country have increasingly turned from rail to air travel.

NUMBER OF PASSENGER KILOMETRES FLOWN 1970–95 (in millions)

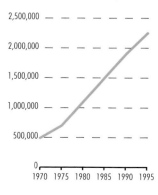

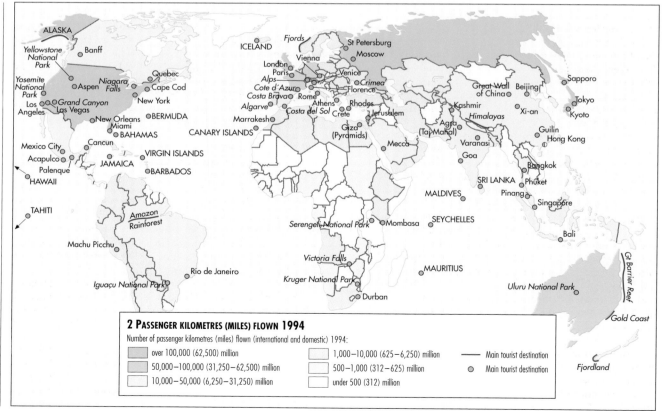

2 PASSENGER KILOMETRES (MILES) FLOWN 1994

Number of passenger kilometres (miles) flown (international and domestic) 1994:

over 100,000 (62,500) million	1,000–10,000 (625–6,250) million
50,000–100,000 (31,250–62,500) million	500–1,000 (312–625) million
10,000–50,000 (6,250–31,250) million	under 500 (312) million

— Main tourist destination
○ Main tourist destination

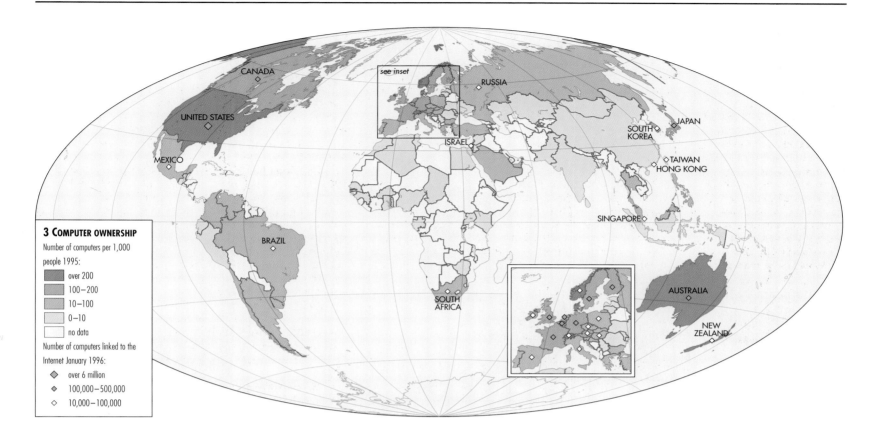

The manner in which we travel has an impact on the environment. The construction of roads, railways, waterways and airports often requires extensive changes to the landscape, and cars and trucks, aircraft, ships and trains all produce pollutants that are released into the atmosphere (*pages 280–81*). In order to reduce environmental pollution, governments, town planners and vehicle manufacturers are being urged to consider these issues when designing new transport networks and developing new models.

COMMUNICATION

At the beginning of the 20th century the quickest way to send a message across the world was by telegraph, via a network of overland and undersea cables (*pages 208–9*). The invention of the radio-telephone in 1902 and subsequent improvements in the quality of transcontinental telephone signals enabled the human voice to travel huge distances. However, the most significant advance in this sphere was the development, during the 1960s, of a network of communications satellites that allowed not only aural, but also visual, signals to be sent up into space and bounced back, greatly enhancing telephone links and enabling live television broadcasts to be made from one side of the world to the other. Several hundred active communications satellites now orbit the globe, and without these none of the major developments in communications of the late 20th century would have been possible.

Mass television ownership enables people worldwide to share programmes. American and British soap operas are shown, for example, dubbed, on Russian television. Major events, such as the football World Cup Finals, are watched simultaneously by hundreds of millions of people. For those without access to a television set, the radio often provides a link with the outside world. The BBC World Service alone had an estimated 140 million regular listeners worldwide in the late 1990s, enabling people to obtain news they might otherwise be prevented from hearing.

The most spectacular development in international communication since the 1980s has been the Internet, giving millions of people in the industrialized world almost instant access to a vast network of information, and the means to communicate with each other speedily and cheaply. It has been made possible largely through the development of the microprocessor, which enabled small personal computers to be manufactured from the mid-1970s onwards. By the mid-1980s these machines had become powerful enough for their users to be able to access the Internet, a worldwide computer networking system. First developed in the 1970s for the United States Department of Defense, it was subsequently extended to the academic community, commercial organizations and the general public. By the end of the 20th century there were more than 130 million users of the "worldwide web", created in 1994, with millions more using electronic mail (e-mail).

Technological advancement is the province of the rich nations, with, for example, almost 50 per cent of the world's personal computers to be found in the United States (*map 3*). The technological gap between rich and poor nations is an enormous challenge for those in the process of industrializing, although it may also be to their advantage if there is sufficient money to buy the latest technology. In China, for example, where until recently few households had a telephone, the old telecommunications technology, involving the laying of cables, is being bypassed in favour of the installation of radio masts for mobile telephones.

CULTURAL INTEGRITY

All forms of communication require language and there are estimated to be over 5,000 languages in use. Of these, English, Mandarin Chinese, Hindi, Spanish and French are the most widely spoken, but far more people use them as their official language than as their mother tongue (*bar chart*). Although there are, of course, benefits to a country having a common language, there are also disadvantages. There has been a sharp decline in the number of different African languages spoken, leading to a disintegration of the cultural values and traditions attached to those languages.

Cultural integrity is also challenged by developments in global communications, which have provided the most technologically advanced countries with a powerful means of spreading their ideologies and culture.

The extent to which countries can participate in the "global village" will affect their future prosperity. There is no question that modes of transportation and communication will continue to evolve at great speeds, but whether they will become universally available remains uncertain.

▲ Computer technology represents huge profits for the countries involved in producing it, and has provided businesses and individuals worldwide with enormous benefits. It has also created a widening social and economic gap between those who have access to it and those who do not.

▼ The most widely used official language in the world is English, partly as a result of British influence in the 19th century but more recently because of the domination of US culture. In countries such as China and India, where many languages are spoken, it is essential to have a single language in which official communications can be made.

THE WORLD'S MAJOR LANGUAGES 1990 (in millions of speakers)

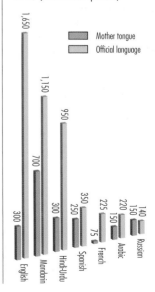

5 MILLION YEARS AGO – 1001 BC

The number(s) at the end of entries refer to the pages with relevant maps.

	ASIA	AFRICA	AMERICAS/AUSTRALASIA
5 million years ago		c. 5–2 million Evolution of *Australopithecines*, the first hominids to walk upright ◐16	
	c. 1.8 million *Homo* spreads through much of Asia, giving rise to *Homo erectus* in East Asia ◐16, 17	c. 2.5 million First members of genus *Homo* appear; diet includes scavenged meat c. 2 million Evolution of *Homo ergaster*	
	c.500,000 Evidence from cave of Zhoukoudian, in northeast China, indicates use of fire ◐16, 17 by 130,000 Neanderthals living in West Asia ◐16, 17	c. 200,000 Stone blade tools made, stone tips hafted to spears	
120,000 years ago	by 90,000 *Homo sapiens* living in West Asia; subsequently spreads through Asia ◐16, 17	c. 120,000 Emergence of earliest modern humans (*Homo sapiens*), spreading into West Asia by c. 100,000 ◐17	
			c. 60,000 *Homo sapiens* reaches Australia and New Guinea by sea ◐26 by 30,000 Colonization of Bismarck archipelago ◐26 c. 25–15,000 Extinction of Australian megafauna and colonization of Australia's desert interior
	by 25,000 *Homo sapiens* reaches northeast Asia ◐16, 17		by 18,000 Spread of *Homo sapiens* to far north of Americas across landbridge from Asia ◐16, 24
10,000 BC	c. 9000 Agricultural communities emerge in Levant and Zagros region of western Iran ◐18 c. 8000 Jericho substantial walled settlement ◐18 c. 7000 First farmers in Indo-Iranian borderlands, northern Mesopotamia and Anatolia; pottery in West Asia ◐18 c. 6000 Irrigation in northern Mesopotamia and colonization of southern Mesopotamia; farming communities in Southeast Asia, also making pottery; millet cultivated in China ◐18 c. 5000 Towns emerge in southern Mesopotamia; rice cultivated in Southeast Asia and China; sedentary hunter-gatherers and fishers in Japan	c. 8000 Lakes and fishing communities in Sahara region, making pottery by c. 7500 c. 6000 Villages in Egypt; rock art and domestic cattle in Sahara region; hunters using arrows tipped with poison ◐22 c. 5000 Farming in Nile Valley, Sahara, Ethiopia and parts of West Africa ◐22	c. 12,000–9000 BC Colonization of mainland Americas as far as southern tip ◐16, 24 c. 9500–8900 Mammoth hunters in North America; extinction of many large mammals ◐24 c. 7000 Horticulture in New Guinea by 6500 Partly agricultural communities develop in Central and South America; regional groups emerge in North America c. 6000 Tasmania cut off from mainland c. 5000 Maize cultivated in highland Mesoamerica; Caribbean islands colonized
4000 BC	by 4000 City-states developing in Mesopotamia c. 4000 Trading towns on Iranian Plateau; plough used in northern China; domestication of horse in Russia	c. 4000 Sahara becoming desiccated - farmers migrate southwards ◐23	c. 4000 Pottery made in northern South America
	c. 3500 Social divisions appearing in northern Chinese society; emergence of urban civilization in Mesopotamia; farmers settle on Indus Plains	c. 3500 Farmers in Upper Nubia c. 3400 The first walled towns in Egypt c. 3100 Unification of Egypt under Menes	
3000 BC	c. 3000 Farming spreads into Korea; more hierarchical society in north; Early Dynastic period in Mesopotamia c. 2800 Silkworms raised in China c. 2600 Emergence of Indus civilization ◐29; Royal Graves at Ur in Mesopotamia ◐18; walled towns in China ◐31 c. 2500 Farming spreads to islands of Southeast Asia c. 2300 Sargon I and Akkadian Empire in Mesopotamia; trading towns in Gulf region ◐28 2112–2004 Third Dynasty of Ur in Sumeria	c. 3000 Egyptians expand into Lower Nubia; development of kingdom of Kerma in Nubia; farmers established in East Africa 2686–2181 Old Kingdom Egypt ◐30 c. 2650 Stepped pyramid of Djoser built at Saqqara ◐30 c. 2600–2500 Pyramids built at Maidum and Giza 2055–1650 Middle Kingdom Egypt – fortresses in Lower Nubia ◐37	c. 3000 Spondylus trade in South America; Old Copper culture in eastern North America c. 3000 Small tools tradition in Australia; dingos introduced from Southeast Asia c. 2600 U-shaped shrines in South America; agricultural communities appear in northwest South America and on Peruvian coast
2000 BC	c. 2000 Pastoral nomadism in steppe zone of West Asia c. 1700–1050 Shang civilization in China ◐30 1792–1750 Babylonia dominant power in southwest Asia c. 1650–1200 Hittite Empire ◐36 c. 1300 First Assyrian Empire established ◐37 c. 1100 Phoenician city-states emerge in Levant ◐38 1050–1771 Western Zhou period in China; nomad incursions cause eastward shift of capital 1771 ◐48	c. 2000 Further spread of agriculture in sub-Saharan Africa ◐23 c. 1650 Kush powerful independent state 1550–1069 New Kingdom Egypt – expands into Nubia and Levant ◐37 1490 Queen Hatshepsut's expedition to Punt (Somalia) to obtain incense trees c. 1200–1000 "Sea Peoples" menace Egypt and southern Mediterranean ◐37	c. 2000 Cultivation of local plants in southeast North America; agricultural communities in Mesoamerica ◐25 c. 1500 Bow and arrow comes into use in Arctic; Lapita culture reaches Solomons and Melanesia – by 1300 reaches Fiji and Samoa ◐26 c. 1200 Olmecs in Central America ◐32 c. 1200–200 Rise of Chavin culture in South America ◐34

EUROPE	SCIENCE AND TECHNOLOGY	ARTS AND RELIGION	
			5 million years ago
	2.5 million First stone tools made, extending human physical capabilities		
c. 1 million Spread of hominids into Europe, giving rise to *Homo heidelbergensis* ❍17	**1.5 million** Development of variety of stone tools, including handaxes	**c. 1 million** Attractive handaxes indicate development of some aesthetic sense	
c. 500,000 Hunting with wooden spears and fire probable	**c. 400,000** Brushwood shelter constructed at Terra Amata, France – possibly earlier windbreaks and shelters exist in Africa	**c. 250,000** Red ochre possibly used as paint at Becov, Europe	
c. 400,000 Stone tools made using prepared core technique			
by 120,000 Neanderthals living in Europe		**c. 90,000** Beginning of deliberate burial at Qafzeh in West Asia	**120,000 years ago**
	c. 60,000 Seaworthy boats implied by colonization of Australia	**c. 60,000** Australian art possibly created from time of earliest colonization	
by 40,000 Colonization of Europe by *Homo sapiens* ❍17	**c. 30,000** Innovations include blade tool industries, sewn clothing, tents and mammoth bone shelters	**c. 35,000** Worldwide portable and rock art, especially in Europe, Africa and Australia	
by 30,000 Neanderthals extinct	**by 23,000** Ground stone tools in Australia	**c. 26,000** World's earliest known cremations at Lake Mungo, Australia	
	by 18,000 Bow and arrows used for hunting		
	c. 11,000 Pottery made in Japan		
	c. 10,000 Dogs domesticated in many parts of world	**c. 10,000** Beginning of long-lived tradition of rock painting in central India	**10,000 BC**
	c. 8000 Boomerang invented in Australia; copper smelting, kiln-fired pottery and irrigation in West Asia	**c. 8000** Pottery made in China; burials in North America with red ochre; skull cult in Levant; linen textiles and clay figurines in West Asia	
c. 7000 Farming groups from West Asia spread into Greece and the Balkans ❍20	**c. 7000** Pottery made in West Asia		
c. 6000 Farming communities in central and western Mediterranean; large, well-furnished Mesolithic cemeteries in northern Europe ❍20	**c. 5000** Pottery made in Amazon Basin		
c. 5500 Farmers colonize central Europe from the southeast ❍20	**c. 4500** Plough and sailing boats in Mesopotamia; copper smelted in Balkans	**c. 4300** Cotton cultivated in Mesoamerica; decorated temples in Mesopotamia; bronze produced in West Asia; "lost wax" technique copper casting in Levant	
c. 4500 Megaliths begin to be constructed in France and Spain, by 4000 in Britain and Scandinavia ❍20			
c. 4000 Farming communities established in western and northern Europe; plough, woolly sheep and milking introduced into Europe ❍20	**c. 4000** Flint and copper mining in Europe	**by 4000** Lacquerware at Hemudu, China; temple pyramids in Peru	**4000 BC**
	c. 3800 Wheeled vehicles in Mesopotamia		
		c. 3500 Temples in China; cotton textiles made on Peruvian coast; burials with wheeled vehicles in southern Russia	
	c. 3400 Potter's wheel invented in Mesopotamia	**c. 3300** First writing in Mesopotamia	
c. 3100 Construction of great passage grave at Newgrange in Ireland ❍20	**3035** Earliest surviving papyrus sheets in Egypt	**3200** Megalithic stone circles in Europe	
c. 3000 Defended walled settlements established in Mediterranean region; vines and olives cultivated	**c. 3000** Abacus invented in Babylon and China	**c. 3000** Copper and gold metallurgy widespread in Europe; glass made in Egypt	**3000 BC**
	c. 2900 Egyptian hieroglyphic script fully developed	**c. 3000–2000** Maltese temples constructed	
c. 2700 Fortified settlements established in Iberia	**c. 2600** Fully developed cuneiform script in Mesopotamia; plough and script in Indus Valley	**c. 2700** Legendary Gilgamesh king of Uruk is subject of cycle of Sumerican epic poems – world's earliest written story	
c. 2550 First stone circle constructed at Stonehenge ❍21			
c. 2500 Corded Ware and Beakers made; copper metallurgy widespread ❍21	**c. 2500** Four-wheeled war chariot and reflex composite bow in Mesopotamia; bronze working in Southeast Asia; skis in Scandinavia	**c. 2500** Woven cotton in Indus Valley	
c. 2300 Bronze working and international trade; emergence of chiefdoms; palaces in Aegean region	**c. 2300** Constellations found by Indus, Babylonian and Chinese astronomy	**c. 2100** Ziggurats erected in Mesopotamia	
c. 2000–1450 Minoan civilization on Crete ❍36	**c. 1900** Bronze casting developed in China;		**2000 BC**
c. 1800 European High Bronze Age – rich burials ❍21	**c. 1700** Minoan Linear A script invented	**c. 1500** Sculptures depicting sacrifice of war captives at Cerro Sechin, Peru	
c. 1650–1100 Mycenaean civilization ❍36	**c. 1600** Chariots introduced into China	**c. 1200** Composition of the *Vedas*, earliest of Hindu sacred texts	
	c. 1500 Goldworking in Andean region; bronze technology introduced to Korea;		
c. 1300 Use of bronze for many everyday purposes ❍21	**c. 1450** Mycenaean Linear B script in use	**1352–1336** Monotheistic worship of sun-god Aten instigated by Egyptian king Akhenaten but abandoned by successors	
c. 1250–1200 Destruction of many Mycenaean centres ❍36	**c. 1400** Writing in use in China		
c. 1200 Fall of Troy	**c. 1200** Phoenician alphabet of 22 letters invented	**1270** Great Temple at Abu Simbel constructed by Rameses II	
c. 1100 End of Mycenaean culture; start of Greek Dark Ages			

1000–101 BC

	ASIA	AFRICA	AMERICAS/AUSTRALASIA
1000 BC	**by 1000** Phoenicians major sea power in Mediterranean, establishing colonies in Africa and Europe ❍*38* **early 1st millennium** Colonization by farmers and herders of Ganges Valley in northern India **9th century** Legendary Mahabharata War in India **966–926** Solomon, son of David, King of Israel; on his death kingdom splits into Israel and Judah ❍*38* **911** Assyrian Empire established; expanded in 9th C ❍*38* **800** Megalithic tombs constructed in southern India **771–221** Eastern Zhou period in China, during which Warring States compete for power	**1000–500** Megaliths built in West Africa **9th century** Kush becomes independent kingdom **9th century onwards** Greek and Phoenician colonies established in Mediterranean, including Carthage ❍*54* **c. 860** Royal centre developed at Napata in Upper Nubia, with impressive cemetery	**by 1000** Fortified settlements on North American Pacific coast **c. 1000** Long-distance exchange networks operating in Australia; colonists reach Marshall Is and Tonga by 1000 ❍*26* **c. 1000–AD 100** Adena moundbuilder culture in eastern North America; maize farming introduced to southwest North America ❍*25* **c. 850–250** Chavin de Huantar major centre in Peru ❍*34* **c. 800** Maize cultivation in Amazon Basin; beginnings of script in Mesoamerica
750 BC	**734** Neo-Assyrian Empire conquers Israel **627–539** Babylonian Empire; takes Judah 586 ❍*39* **612** Assyria falls to Medes and Babylonians ❍*39* **550** Persian king Cyrus overthrows Median king and founds Persian Empire **547–539** Persians under Cyrus conquer Greek cities of Asia Minor and Babylonian Empire ❍*42* **550** States emerge in western Arabia, based on overland trade in incense **543–491** Bimbisara makes Magadha into foremost Indian state by conquest, diplomacy and dynastic marriage ❍*46*	**c. 750** Rise of Kingdom of Kush (Nubia) **747–656** Nubian kings of Napata rule Egypt as 25th dynasty ❍*39* **665** Assyrians take over Egypt ❍*39* **c. 600** Emergence of Meroe on Sudanese Nile; beginning of Nok culture in Niger Valley ❍*23* **526–525** Egypt falls to Persians ❍*42*	**c. 700** Harpoons used in Arctic to catch whales **c. 600** Beginning of construction of ballcourts for ritual ballgame in many parts of Mesoamerica
500 BC	**by 5th century** Power in China concentrated among seven principal states **c. 500** Wet rice agriculture, bronze casting, spinning and weaving introduced to Japan; Sinhalese colonize Sri Lanka **491–459** Parricide King Ajatashatru enlarges Magadha, India's principal state ❍*46* **420** Founding of Nabataean kingdom centred on city of Petra (Jordan) ❍*52*	**c. 500** Copper and iron working develops in West Africa, in particular by Nok culture, which also produces fine terracottas ❍*23* **480** Carthage and Sicilian Greeks in conflict **409–367** Renewed hostility between Carthaginians and Sicilian Greeks – Carthaginians control a third of Sicily	**c. 500** Chieftains emerging in northwest South America **c. 500** Zapotec inhabitants of Oaxaca Valley in Mexico; found new centre at Monte Alban; Paracas culture in Peru
400 BC	**334–323** Alexander the Great conquers southwest Asia and northwest India, dying at Babylon 323 ❍*42* **321** Chandragupta Maurya usurps throne of Magadha and creates Mauryan Empire ❍*46* **316** Shu state becomes vassal of Qin China; reforming statesman Shang Yang makes important legal and administrative changes in Qin state **323** Beginning of power struggle for control of Alexander's conquests ❍*43* **312** Seleucus I gains control of Alexander's West Asian territories and founds Seleucid Empire ❍*43*	**by 400** Carthage dominates western Mediterranean ❍*40* **c. 400** Ironworking Bantu farmers move into East Africa ❍*23* **4th century** Farming villages established in equatorial forest region **331** Alexander the Great liberates Egypt from Persian rule ❍*42* **310–305** Carthaginians again fighting Sicilian Greeks; peace in 305, with further territorial gains to Carthage **305** Ptolemy gains control of Egypt ❍*43*	**c. 400** Pachacamac in Peru emerges as important shrine and place of pilgrimage; emergence of Nazca culture in Peru, centred at Cahuachi; foundation of later major city of Tiwanaku ❍*34* **c. 400** Tres Zapotes replaces La Venta as principal Olmec centre, but Olmecs in decline; new centres emerging in Mexican highlands; stone temple platforms built at Cuicuilco, main centre in Valley of Mexico ❍*35*
300 BC	**c. 300** Rice farming spreading through western Japan – defended settlements **281** Galatian Celts settle in Asia Minor **277** State of Chu defeated by Qin (conquered 223); Qin defeats state of Zhou 260 (conquered 228) ❍*48* **272** Accession of Mauryan emperor Ashoka ❍*46* **250** Parthians secede from Seleucid Empire, founding independent state; becomes Parthian Empire c. 240 ❍*53* **221** Qin Shi Huang Di unifies China and consolidates the Great Wall of China ❍*48* **206** Liu Bang (Emperor Gao Di) defeats rivals in civil war after collapse of Qin dynasty; start of Han Empire ❍*48*	**295** Nubian capital moved south to Meroe and kingdom established **274–217** Egypt and Seleucid Empire fight for control of lucrative Levant **264–241** First Punic War between Carthage and Rome ❍*54* **237–219** Carthaginians conquer Spain **218–201** Second Punic War **202** Decisive Battle of Zama in Tunisia	
200 BC	**187** Collapse of Mauryan Empire in India; Shungas rule greatly reduced kingdom in Ganges Valley ❍*47* **185** Bactrian Indo-Greeks conquer northwest India ❍*46* **165** Xiongnu defeat Yuezhi and drive them west ❍*51* **141** Sakas (Scythians) seize control of Punjab ❍*46* **139–126** Expedition by Han envoy Zhang Qian to Yuezhi establishes Chinese contacts with west via Central Asia **138–90** Han conquer large territories in Central Asia, Yue, Manchuria and Korea ❍*48*	**c. 200** Bantu farmers and ironworkers spreading southward ❍*23* **149–146** Third Punic War **146** Carthage destroyed by Rome ❍*54* **112–106** War between Roman armies and Numidians of North Africa ends in defeat of latter	**c. 200–150** New wave of eastward colonization in Pacific reaches islands of French Polynesia ❍*26* **c. 200** Cerro Blanco established as main Moche centre in Peru ❍*34* **c. 200** Hopewell moundbuilder culture emerges in eastern USA ❍*25* **c. 200** Start of construction of Nazca Lines **c. 150** Maya construct ceremonial centres with temple platforms

EUROPE	SCIENCE AND TECHNOLOGY	ARTS AND RELIGION	
c. 900 Beginning of Etruscan civilization; Etruscans engage in international trade 8th century **◯21** **9th century** Lycurgus the Lawgiver leader of Sparta **c. 8th century** (traditionally 1100) Gades founded by Phoenicians; many independent "city-states" formed in Greece **◯41** **776** First Olympic Games held **753** Traditional date for foundation of Rome on Palatine by Romulus **◯54**	**c. 1000–700** Iron, sporadically used before in West Asia, comes into widespread use in West Asia, India, China and Europe **c. 800** Alphabet adopted in Greece; fine goldwork produced by Chavin culture in Peru; underground irrigation canals constructed in Iran and Central Asia	**10th century (probable)** Zoroaster (Zarathustra) teaches in Iran **c. 950** Solomon dedicates temple in Jerusalem **c. 900–600** La Venta becomes main Olmec ceremonial centre – colossal heads carved **c. 900–700** Geometric style of Greek art	**1000** **BC**
c. 750–550 Main period of Greek colonization of Western Mediterranean and Black Sea; trade with southern Etruria **730–710** Spartans conquer Messenia **c. 700** Archaic Greece **7th century** Etruscan League – expands into Campania **c. 660** Byzantium founded **c. 600** Massilia (Marseilles) founded; trades with Celts at hillforts **◯40** **6th century** Athens develops democracy; villages of Rome coalesce to form town **509** Rome forms republic	**704–681** Irrigation works created by Assyrian king Sennacherib – builds "Palace without a Rival" at Nineveh **c. 700** Etruscans adopt alphabetic script **7th century** Use of coinage begins in India **from 6th century** Ironworking in Meroe (Sudan) and China – iron casting and copper mine in operation in China **c. 590** Sabaeans in southern Arabia build massive irrigation dam in capital Ma'rib **before 500** Beginning of calendric writing in Central America	**c. 750** *Iliad* and *Odyssey* written down **c. 604–562** Nebuchadnezzar creates Hanging Gardens of Babylon **c. 600** Latin script appears **6th century** Lao-Tze teaches in China **599** Birth of Mahavira, founder of Jainism **587** Temple of Jerusalem destroyed **563** Birth of Siddharta Gautama (Buddha) **551** Birth of Kung Fu Tzu (Confucius) **c. 550** Zoroastrianism adoped as official Persian religion	**750** **BC**
c. 500 Rome annexes neighbouring territories, making their inhabitants Roman citizens; Etruscans expand to Po valley and Adriatic Sea **5th century** Etruscans lose most of colonies to Gauls, Latins and Greeks **499–490** Ionian revolts against Persians suppressed **490, 480–470** Greeks defeat Persian invasions **◯40** **477** Delian League founded; becomes de facto Athenian empire 488 **◯41** **431–404** Sparta victorious over Athens in Peloponnesian War **◯41**	**by 500** Brahmi syllabic script in India, (ancestral to most later Indian scripts) **c. 500** "Lost wax" technique used in casting Dong Son drums in Southeast Asia **c. 500** Persian king Darius completes canal connecting Red Sea and River Nile **5th century** First Chinese coins made of bronze in shape of miniature tools; canals built in China, partly for military supplies **c. 450** Invention of crossbow in China	**c. 500** Carved stele at Monte Alban depicts sacrificed victims; Nok terracottas of human heads **c. 480** Classical style of Greek art begins **c. 460** Gold and ivory statue of Zeus at Olympia created by sculptor Phidias **c. 450** Development of La Tene style in Celtic art **495–429** Athens' Golden Age under Pericles – great works of architecture and art created, including the Parthenon	**500** **BC**
396 Rome begins conquest of Etruria (complete c. 200) **396–386** Corinthian War: Sparta against Athens, Corinth and Persians **390** Army of Celtic migrants sweeps through Italy, sacks Rome and settles in north Italy **◯21** **371** Defeat at Leuctra ends Spartan dominance of Greece **338** Philip II of Macedonia defeats Greek city-states at Chaeronea and becomes master of Greece **◯42** **338** Murder of Philip of Macedonia and accession of his son Alexander, "the Great" **323** Rome defeats Gauls, dominates Samnites and tribes of Latin League **◯54**	**c. 400** Ironworking and production of stoneware ceramics introduced into Korea from China **4th century** First Chinese disc coins introduced, made of copper with central hole to string on belt **c. 350** Mausoleum constructed at Halicarnassus (Bodrun), Asia Minor (one of Seven Wonders of the World)	**387** Plato founds Academy in Athens **372** Birth of Mengzi (Mencius), major Confucian philosopher **331** Founding of Alexandria on Egyptian coast, subsequently the great centre of Hellenistic learning	**400** **BC**
290 Roman victory over Samnites leads to confrontation with Greek colonies in southern Italy **◯54** **279** Galatian Celts invade Greece and sack Delphi **276** Macedonia falls to family of Antigonus in power struggle following death of Alexander the Great **◯54** **270** Rome gains control of all Italy **225** Celts defeated by Romans at Telamon **217** Carthaginian general Hannibal crosses Alps, defeats Romans at Trasimene and again at Cannae (216) **◯54**	**292–280** Colossus of Helos at Rhodes (one of Seven Wonders of the World) built, but destroyed by an earthquake 227 **279** Dedication of Pharos lighthouse (one of Seven Wonders of the World) in Alexandria harbour **c. 250** Dam constructed on Ming River allows huge area of plains to be irrigated **c. 220** Three-masted ships in Greece	**c. 250** Pillars erected by Ashoka at great places in life of Buddha **246** Sri Lanka converted to Buddhism under King Devanampiya Tissa **c. 225** Erection of Nubian Lion Temple at Musawwaarat es Sofra	**300** **BC**
c. 200 Fortified centres developing in Celtic Europe **◯21** **2nd century** Roman armies progressively take over Spain **168** Romans conquer Macedonia **146** Rome in control of Italy and its islands, parts of Spain and the North African coast, Macedonia, Greece and parts of Asia Minor **◯54** **133** Attalus III of Pergamum bequeathes kingdom to Rome **133–122** Gracchi brothers redistribute land to Rome's poor; fierce opposition from establishment – both are murdered	**c. 200** Archimedes screw used to pump water for irrigation **193** Concrete invented by Romans and first used in Porticus Aemilia in Rome **c. 170** Paved roads in Rome **c. 165** Parchment (vellum) invented in Pergamum, Asia Minor **127 and 102** Emperor Wu Di repairs and enlarges Great Wall of China	**c. 200** First of Dead Sea scrolls written **166** Great Altar to Zeus and Athena constructed in Pergamum **146** Conquest of Greece by Rome results in strong Greek cultural influence on Roman arts **136** Confucianism made official ideology of Han Empire in China **113** Death of Prince Liu Sheng in China – buried in jade suit	**200** **BC**

	ASIA	AFRICA	AMERICAS/AUSTRALASIA
100 BC	**66–63** Romans conquer remains of Seleucid Empire **c. 60** Chinese control caravan cities of Central Asia **◑53** **57** Establishment of Kingdom of Silla in Korea **54** Xiongnu nomads divide into north and south branches **53** Roman army defeated by Parthians at Carrhae **c. 50** Kushans of Afghanistan begin conquering Saka lands in India **◑46** **37** Traditional date of founding of Koguryo state in Korea **33** Xiongnu submit to Han Chinese rule **◑48** **18** Traditional date of founding of Paekche state in Korea	**c. 100** Establishment of Axumite state **c. 100** Camels introduced into Sahara, facilitating trans-Saharan trade **◑23** **46** Rome founds new city on ruins of Carthage **31** Octavian defeats rival Antony and Egyptian queen, Cleopatra, in Battle of Actium; Egypt annexed by Rome	**by 100** Maya civilization using raised field systems **◑33** **1st century** Complex burial rites of Adena moundbuilders in North America **◑25** **c. 50** Destruction of Cuicuilco by volcanic eruption; Teotihuacan becoming dominant in Valley of Mexico **◑32**
0	**2–11** Massive flooding on Huang He and subsequent famine leads to downfall of reforming emperor Wang Mang in 23 **53** Tiridates founds Arsacid dynasty in kingdom of Armenia **c. 60** Kushans gain control of much of northern India, founding the Kushan Empire **◑46, 53** **66–73** Jewish Revolt against the Romans; destruction of Temple in Jerusalem and siege of Masada 72–73 **78** Kanishka accedes as ruler of Kushan Empire **91** Xiongnu defeated by Han Chinese in Mongolia **◑48**	**40** Romans annex Mauretania in northwest Africa **c. 50** Expansion of Axumite kingdom in Ethiopia **◑52**	**by 1st century** Tahiti and Micronesia colonized **◑26** **early 1st century** Emergence of Basketmaker culture in American Southwest **1st century** Development in southeast Mesoamerica interrupted by eruption of Mount Ilopango; Adena mounds have chambers and mortuary enclosures added **c. 50** Beginning of construction of Pyramid of the Sun in Teotihuacan **◑33**
100	**106** Romans annex Nabataean kingdom and establish province of Arabia **◑55** **132–135** Second Jewish revolt against Rome – failure leads to dispersal of Jews (Diaspora) **◑44** **c. 150** Chinese driven out of Central Asia by Huns **184** Revolt of Yellow Turbans in China **◑49** **190** Han emperor murdered	**162** Construction of Zaghouan aqueduct to supply Carthage with water	**c. 100** Early burials in spectacular Moche royal mound cemetery at Sipan **◑34** **c. 150** Pyramid of the Moon constructed at Teotihuacan **◑33**
200	**220** Collapse of Han Empire – replaced by Three Kingdoms; rise of Korean states **◑49** **226** Sasanians under Ardashir I (Artaxerxes) overthrow Parthians and found Sasanian Empire **◑44** **c. 240** Breakup of Kushan Empire **245** Chinese envoys travel to court of Funan, Indochina **mid-3rd century** Kushans become vassals of Sasanians **250** Yamato chief controls central Japan **280** Jin dynasty reunifies China	**3rd century** Ironworking in Lake Chad region **◑23**	**c. 200** Cahuachi abandoned as Nazca centre, but continues as place of pilgrimage and burial **◑35** **c. 290** Last burials in Moche royal mound cemetery at Sipan **◑35**
300	**c. 300** Arabian kingdoms decline as sea trade replaces overland trade and Axumite incense becomes important; Yamato chiefs extend control over much of Japan **304** Huns invade China, sack city of Luoyang 311 **◑51** **313** End of Chinese control of Korea **335** Accession of Gupta king Samudragupta **383** Chinese repulse Hun invasion of south China **c. 395** Gupta Empire reaches greatest extent **◑46** **399–414** Chinese pilgrim Fa Xian travels Silk Road to visit Buddhist shrines in India and Sri Lanka and returns via Southeast Asia **◑53**	**c. 300** Bantu farmers and ironworkers reach Natal in South Africa **◑23** **325** Meroe destroyed by neighbouring kingdom of Axum **◑52** **343** Axumite kingdom adopts Christianity on conversion of King Ezana **◑44**	**by 300** Teotihuacan Empire controlling central Mexico; expansion of Monte Alban checked by growth of Teotihuacan Empire **◑32–33** **3rd century** Pit house villages in southwest North America; conflict developing between Maya city-states **378** Tikal defeats army of Uaxactun – latter's king sacrificed; close relations between Tikal and Teotihuacan **◑35**
400	**c. 400** Silk Road towns embellished with rock-cut Buddhist cave temples **◑53** **400–439** Toba Wei nomads move into northern China **◑51** **427** Mausoleum of Emperor Nintoku built **c. 450** Ruan-Ruan nomads threaten China – replacing Xiongnu; Hunas (White Huns) invade northern India **◑51** **484** Hunas (White Huns) attack Sasanian Empire	**5th century** Lost wax copper casting in Senegal; people from Southeast Asia settle in Madagascar **c. 400** Jenne-jeno in Niger Valley flourishing town **◑23** **428–40** Vandals gain control of Roman North Africa **◑57**	**by 400** Easter I and Hawaii colonized **◑26** **c. 400** Hopewell moundbuilding ceases; Teotihuacan at height of prosperity, controlling obsidian at Kaminaljuyu; Marajoara culture developing on Marajo island in mouth of Amazon **c. 400** States of Tiwanaku and Huari states emerging in Peru **◑35** **c. 450** Nazca stop building pyramids and start building stone towers **◑34**

EUROPE	SCIENCE AND TECHNOLOGY	ARTS AND RELIGION	
89 All Italy granted Roman citizenship **73–71** Slave revolt in Italy, led by Thracian gladiator Spartacus, brutally put down **60–49** First Triumvirate in Rome **49–47** Civil war between Pompey and Caesar; Caesar assassinated 44 ●*54* **43–28** Second Triumvirate in Rome **27** Octavian takes title Augustus: start of Roman Empire **9** Germans destroy Roman legions in Teutoburg Forest	**c. 100** Agricultural improvements, including crop rotation, introduced in China **1st century** Important medical advances in India; sophisticated irrigation works in Sri Lanka **50** Technique of blowing glass developed in Levant; earliest paper in East Asia **46** Julian calendar introduced by Julius Caesar – not replaced until AD 1582 **36–30** Herod builds palace on three levels at Masada – major feat of engineering	**1st century** Classically influenced Gandaran style of Buddhist art develops in northwest India **90** Parthian establish Ctesiphon as their capital **85** Death of Chinese historian Sima Qian **c. 50** Great Stupa at Sanchi, India, constructed over earlier stupa built by Mauryan emperor Ashoka **c. 4** Birth of Jesus Christ	**100** **BC**
43 Romans begin conquest of Britain **60–1** British, led by Queen Boudicca, revolt against Roman occupation **64** Disastrous fire in Rome during reign of Emperor Nero – Christians blamed and persecuted **69** Death of Nero leads to civil war over succession, with Vespasian eventual victor **79** Eruption of Vesuvius and destruction of Pompeii and Herculaneum **82–86** Rhine–Danube established as frontier of Roman Empire in Europe ●*55*	**c. 45** Construction of artificial harbour at Ostia under Emperor Claudius **c. 60** *The Periplus of Erythraean Sea* written in Greek, detailing sea trade routes between Roman world, Africa and South Asia **70** Pliny the Elder suggests using liver of mad dog as protection against rabies **80** Official opening of Colosseum in Rome	**1st century** Buddhism spreads into Central Asia under Kushan rule; spread of Mithraism through Roman Empire **5** Construction of Shrine of Ise, Japan **c. 20** Great Stupa at Sanchi embellished and gateways constructed **c. 30** Christ crucified in Jerusalem **c. 50** Mathura style of Buddhist art developing in Ganges Valley under Kushan patronage	**0**
112–125 Wall built across northern England by Emperor Hadrian to repel Picts ●*55* **117** Roman Empire reaches maximum extent, but Armenia and Mesopotamia abandoned in same year ●*54–55* **168–175** Germanic tribes repulsed after crossing Danube **197–199** Mesopotamia again temporarily annexed by Rome	**c. 110** Invention of wheelbarrow in China **112** Sheltered hexagonal harbour basin built at Ostia, connected to Tiber by canal **118–128** Construction of Pantheon under Emperor Hadrian features 43-metre (141-foot) hemispherical concrete dome **c. 130** Invention of crude seismograph in China **c. 175** Principles of anatomy and physiology established by Greek physician Galen	**2nd century** Runic alphabet in use in northern Europe; Buddhist cave temples and monasteries constructed in India **110–113** Construction of Trajan's column carved with details of Dacian campaigns of 101–6 **132** Jang Dao-ling founds Daoism based on teachings of 6th-century Chinese philosopher Lao-Tze **c. 150** Buddhism introduced to China via Silk Road through Central Asia	**100**
212 All free inhabitants of Roman Empire granted citizenship **260** Germanic confederacy invades Roman Empire; Romans temporarily abandon Rhine-Danube frontier; Alemanni invade Italy but defeated ●*56* **260–274** Brief breakaway Gallic Empire under Postumus **267** Raid on Greece and Athens by large Germanic fleet **285** Administration of Eastern and Western Roman Empires separated	**271** Magnetic compass invented in China **275** Construction of major defensive wall around Rome **292** Earliest dated *stele* (carved stone slab) in Maya lowlands	**c. 200** Completion of *Mishnah*, the codification of Jewish law **226** Mazdaism becomes official religion of Sasanian Empire **mid-3rd century** Major cultural and religious centre of Nagarjunakonda (Vijayapuri – City of Victory) founded in India **c. 250** Persian prophet Mani propounds Manichaean doctrine **284** Diocletian persecutes Christians **297** Chinese history *Weizhi* contains earliest written account of Japan	**200**
330 Byzantium becomes capital of Eastern Roman Empire **332** Alliance between Romans and Visigoths intended to make latter buffer zone against other barbarians ●*56* **370** Huns invade eastern Europe and displace Visigoths, who invade Roman Empire ●*57, 76* **378** Visigoths defeat Romans at Hadrianople ●*57* **382** Visigoths settle in Moesia under treaty with Rome **395** Roman Empire officially split into East and West parts	**c. 300** Large votive deposit at Hjortspring in Scandinavia includes first known clinker-built boat **c. 350** Methane used for lighting in China **365** Moons of Jupiter discovered by Chinese astronomers	**312** Constantine converted to Christianity **313** Edict of Milan grants toleration of Christianity throughout empire **325** Christian doctrine established by Council of Nicaea **341** Gothic alphabet devised for Bible **c. 366** Buddhism introduced into Korea **372** Confucian school established in Koguryo Korea **381** Christianity becomes official religion of Roman Empire	**300**
406 Rhine freezes – barbarians pour into Roman Empire **407** Romans leave Britain; Burgundy established ●*57* **409** Vandals and Sueves invade Spain ●*57* **410** Visigoths invade Italy; sack Rome; found kingdom in Aquitaine 418; seize much of Spain 430 ●*57* **434–453** Huns under Attila menace Europe ●*57, 76* **455** Rome sacked by Vandals from North Africa ●*57* **488** Ostrogoths under Theodoric found kingdom in Italy	**5th century** Raised field systems for irrigation in Amazon Basin **5th century** Large stone platforms (*ahu*) first built on Easter Island **475** Invention of stirrup in China	**5th century** Hinduism revives and Buddhism declines in India; Buddhism introduced to Japan; Buddhist paintings at Ajanta **410** Nestorian Church recognized as Iranian church by Synod of Seleucia **451** Coptic Church (Egypt, Ethiopia) breaks away after Council of Chalcedon **488** Sasanian king Kavadh supports revolutionary Mazdakite sect	**400**

500–1099

	ASIA AND AUSTRALASIA	AFRICA	EUROPE
500	**535** Hunas (White Huns) invade remaining Gupta territories in India; end of Gupta dynasty ◗51	**c. 500** Bantu cattle farmers from north reach Orange River ◗23	**502** Franks defeat Alemanni and conquer southern Germany ◗74
		533 Byzantine armies under Belisarius capture North Africa from Vandals ◗66	**511** Visigoths annex Spain ◗66
			527 Accession of Justinian leads to major expansion of Byzantine Empire ◗67
	552 Nomadic Turks under Bumin win decisive victory over Ruan-ruan and establish control across Central Asia ◗51	**c. 543** Warlords of Ethiopian highlands converted to Christianity by missionaries	**534** Franks conquer Burgundian kingdom
	562 Sasanian and Turk alliance crushes Hunas	**c. 550** Kingdom of Ghana established in West Africa	**552** Byzantines conquer southern Spain
	c. 575 Sasanians begin domination of Arabia (to 628)	**570** Christian Axumite armies make unsuccessful attempt to conquer Mecca	**550–570** Avars settle in Hungary ◗74
			568 Lombards invade northern Italy
	589 Sui dynasty under Wen Ti achieves short-lived reunification of China through military conquest		**c. 580** Slavs begin to migrate south into Balkans and Greece (to c. 660) ◗70, 76
600	**618** Gao-zi founds Tang dynasty in China ◗72	**616** Sasanians conquer Egypt	
	622 Muhammad and his followers flee Mecca for Medina (the *Hejira*); start of Islamic calendar	**628** Emperor Heraclius restores Byzantine control over Egypt	**629** Visigoths drive Byzantines from southern Spain
	c. 630 Chinese establish control over much of Central Asia	**639–642** Arab armies conquer Egypt ◗68	
	630 Mecca surrenders to Muhammad's army ◗68	**c. 650** Arabian traders establish Muslim settlements on east coast of Africa	**673–678** Arab armies conduct first siege of Constantinople by land and sea ◗67
	642 Defeat by Arabs marks end of Sasanian Empire ◗68	**652** Arab rulers of Egypt agree to respect existing borders of Nubian kingdoms	**681** Onogur Huns establish kingdom of Bulgaria
	645 Taika reforms establish central government in Japan		
	661 Umayyad family emerges as ruling Arab dynasty		**687** Victory over dynastic rivals at Tertry extends power of Frankish king Pippin II
	668 Kingdom of Silla unifies most of Korea ◗73	**689** Arabs destroy Byzantine Carthage	
	670 Start of Srivijaya Empire in Sumatra ◗64		
700	**713** Arab expansion reaches upper Indus Valley ◗69	**702–711** Arab conquest of the Maghreb reaches Atlantic coast ◗68	**711–718** Moor (Arab/Berber) armies invade and conquer Visigoth Spain ◗68
	750 Abbasids defeat Umayyads at Zab and take over as ruling Islamic dynasty in Middle East	**739** Zanzibar founded by Muslim traders from southern Arabia ◗82	**732** Frankish armies led by Charles Martel reverse Muslim advance at Poitiers ◗68
	c. 750 Pallava dynasty of Gopala controls eastern India	**c. 740** Some Berber peoples in northern Sahara region revolt against Umayyad rule, form independent Islamic kingdoms	**751** Pippin III becomes king of the Franks, founds Carolingian dynasty
	751 Arabs and Turks defeat Chinese at the Talas River; Tang dynasty loses control of Central Asia ◗69, 72	**753** Arab expedition crosses Sahara, makes contact with kingdom of Ghana	**774** Charlemagne defeats Lombards, annexes north and central Italy ◗74
	755–763 An Lushan revolt weakens Tang control of China		
	762 Abbasids found city of Baghdad as new capital ◗68		**793** Vikings pillage the island monastery of Lindisfarne off northeast England
	794 Japanese capital moves to Kyoto; start of Heian period ◗73	**789** Idrisid dynasty establishes Morocco as independent Islamic kingdom	**796** Frankish armies conquer Avars
800	**c. 800** Polynesians begin to colonize New Zealand ◗26	**c. 800** Kingdom of Kanem established round eastern shore of Lake Chad ◗80	**800** Charlemagne crowned Emperor in Rome by Pope Leo III
	802 Jayavarman II establishes kingdom of Angkor ◗64	**808** Fez becomes capital of Morocco	**804** Charlemagne completes conquest of Saxony ◗74
	845 Buddhism banned in China; Confucianism restored		
	c. 850 Kingdom of Pagan founded in Burma ◗64		**843** Treaty of Verdun divides Carolingian Empire into three kingdoms
	858 Yoshifsa establishes Fujiwara family as power behind emperor in Japan		
	c. 860 First Thai polity established in Thailand ◗64		**c. 865** Bulgarians converted to Christianity
	c. 880 Tibetan Lamaic unity dissolves into local rivalries		**878–885** Alfred of Wessex defeats Danes, confining them to north and east ◗79
900	**907** Tang dynasty falls in China after 289 years ◗72	**c. 900** Hausa kingdom of Daura founded	**911** Viking leader Rollo granted dukedom of Normandy in northern France ◗79
	c. 925 Chola annex Pallava kingdom in south India	**909** Shiite Fatimid family seizes power in western Tunisia	**955** Magyars defeated by East Frankish king Otto I at Lechfeld ◗77
	936 Koryo establish control over all of Korea ◗87	**c. 920** Start of golden age of Ghana ◗80	**962** Otto I crowned Emperor by Pope in Rome ◗90
	945 Iranian Shiite Buyids capture Baghdad from Abbasids	**922** Fatimids seize Idrisid Morocco	
	960 Zhao Kuang-yin reunifies northern China and establishes Song dynasty, reigning as Emperor Taizu ◗86	**960** Falasha (black Jews) warriors under Queen Gudit sack Christian Axum ◗80	**c. 969** Miesco I establishes Christian kingdom of Poland ◗71
	971 Fatimids conquer Syria and Palestine from Abbasids	**969** Fatimids invade Egypt and establish Cairo as their capital	**986** Córdoba conquers remaining Christian kingdoms in northern Spain
	985 Cholas invade island of Sri Lanka from India		**988** Russian Vladimir I becomes Christian
	998 Mahmud seizes control in Afghanistan and eastern Iran, founds Ghaznavid dynasty, raids India ◗88		**997** First Christian kingdom in Hungary
1000			**1014** Byzantine Basil II completes reconquest of Balkan region ◗67
	1040 Seljuk Turks defeat Ghaznavids at Dandankan, invade Iran, capture Baghdad 1055 ◗88		**1028** Danes under Cnut invade Norway
	1060–1067 King Anawrahta unifies Pagan kingdom ◗64	**1054** Abdullah Ibn Yasin begins Muslim Arab conquest of West Africa	**1040–1052** Normans establish control over Byzantine southern Italy ◗96
	1071 Seljuks defeat Byzantine army at Manzikert and occupy most of Asia Minor ◗88, 96	**1061–1070** Berber chief Abu Bakr establishes Almoravid dynasty, founds city of Marrakech	**1054** The Great Schism irretrievably divides Catholic and Orthodox churches
	1096 People's Crusade massacred by Turks in Asia Minor	**1075–1077** Almoravids conquer northern Morocco and western Algeria	**1066** William of Normandy conquers England, becomes King William I
	1099 First Crusade takes Jerusalem from Fatimids ◗94		**1096** First Crusade leaves France ◗94

THE AMERICAS	SCIENCE AND TECHNOLOGY	ARTS AND HUMANITIES	
c. 500 Anasazi settlement established at Mesa Verde in southeast Colorado ❍*108*	**525** Scythian mathematician Dionysius Exiguus begins practice of dating years using birth of Jesus Christ as starting point; beginning of Christian, Common or Current Era	**529** St Benedict founds first monastery, at Monte Cassino	**500**
c. 500 Moche state reaches greatest extent along lowland coast of Peru ❍*34*		**534** Emperor Justinian publishes a Law Code; it forms the basis for medieval law in western Europe	
c. 500 Cities of Tiwanaku and Huari rise to prominence in Peruvian Andes ❍*35*	**531–537** Church of Hagia Sophia constructed in Constantinople	**c. 550** Byzantine historian Procopius publishes *History arcana (Secret History)*	
		c. 550 Italian scholar Cassiodorus collects and preserves work of Greek and Latin writers, and establishes practice of monks copying manuscripts	
	580 Suspension bridge with iron chains built, in China	**c. 590** Gregory of Tours completes *History of the Franks*	
		c. 590 Isidore, Bishop of Seville, begins work on 20-volume *Etymologies,* pioneering medieval reference book	
c. 600 Hohokam people establish town and ritual centre in southern Arizona ❍*108*	**c. 600** Earliest known windmills used to grind flour in eastern Iran	**c. 600** Pope Gregory I reforms use of plainsong in Christian services – thought to have introduced Gregorian chant	**600**
	600s Cast iron used for large ceremonial structures in China	**c. 635** Sutras of the Koran collected and distributed	
c. 650 Moche civilization displaced by city-state of Huari ❍*35*	**c. 628** Indian mathematician Brahmagupta publishes his *Brahmasphutra-siddhanta (The Opening of the Universe)*	**641** Great Library at Alexandria destroyed by fire during Arab attack	
c. 650 Teotihuacan destroyed by warfare: collapse of Teotihuacan Empire ❍ *33*	**c. 650** Chinese scholars develop technique for printing texts from engraved wooden blocks	**652** The Koran reaches its final printed form – 20 years after death of Prophet Muhammad	
	685–692 Dome of the Rock mosque (Qubbat al-Sakhrah) built in Jerusalem		
c. 700 Zapotecs abandon base of Monte Alban (founded c. 500)	**c. 700** First tea grown, in China	**700** Greek language banned from public documents throughout Arab Empire	**700**
	715 Great Mosque at Damascus completed	**726** The iconoclasm movement starts in Byzantine Empire when Leo III bans figurative images in Christian art	
	767 First printed text (a million prayers) produced in Japan by Empress Shokutu	**731** Bede's *Ecclesiastic History of the English People*	
	780 Birth Islamic mathematician al-Khwarizmi, whose *Algoritmi de numero Indorum* (825) adopts Indian 10-digit number system and positional notation	**c. 750** Anglo-Saxon poem *Beowulf* written down	
c. 750 Decline of Huari control in Andean Peru		**c. 750** *The Book of Kells* produced in Ireland	
	c. 784 First Arab paper factory opens in Baghdad – using skills learned from Chinese prisoners taken at Talas River	**780** Birth of Shankara, Indian philosopher and founder of Advaita Vedanta branch of Hinduism	
		794 Heian period sees rise of classical Japanese literature	
c. 800 Start of century of declining power and population for the Maya: the so-called "Maya Collapse" ❍*84*		**804** Death of monk Alcuin; instigated civil service training and revival of Classical learning at Charlemagne's court	**800**
c. 800 Metalworking introduced into Mesoamerica from south	**c. 825** Abbasid Caliph al-Mamun establishes House of Wisdom, a library and translation academy in Baghdad	**c.820** Reign of Abbasid Caliph Harun al-Rashid in Baghdad inspires writing of the *Thousand and One Nights*	
c. 800 Cultivation of maize, and later beans, introduced to eastern North America	**c. 840** Porcelain dishes first made in China	**824** Death in China of Han Yu, leading exponent of Neo-Confucianism	
	c. 850 Gunpowder invented in China	**843** End of iconoclasm in Byzantine Empire; images (icons) again permitted in Christian art	
c. 800 Huari abandon capital city in Andean Peru	**c. 850** First European windmills built, in Islamic Spain	**c. 860** Cyril and his brother Methodius devise Cyrillic alphabet to assist their conversion of Slavs to Christianity	
c. 850 Toltecs establish military supremacy in central Mexico ❍*85*	**c. 880** Ethanol (alcohol) distilled from wine in Arabia	**868** Earliest known printed book, the Buddhist scripture *Diamond Sutra*, produced in China	
c. 900 Chimu people establish city-state of Chan-Chan in northern Peru ❍*84*	**925** Death of al-Razi, author of *Al Hawi* – a comprehensive survey of Greek, Arab, and Indian medical knowledge	**922** Shiite Muslim al Hallaj crucified for Sufi teachings against orthodox interpretations of the Koran	**900**
c. 900 Anasazi peoples establish towns around Chaco Canyon in New Mexico	**929** Death of Islamic mathematician al-Battani, author of *On the Motion of Stars*		
c. 900 Greenland discovered and explored by Norsemen ❍*78*	**953** Arab mathematician al-Uqlidsi produces first decimal fractions	**960** Rise of Chinese temple music, using choir and orchestra, under the Northern Song	
c. 950 Toltecs build legendary capital city of Tula in central Mexico ❍*85*	**976** Arabic (Indian) numerals first used in Europe (in northern Spain)		
986 Storms drive Norse explorer Bjarni Herjolfsson to Newfoundland ❍*78*	**984** Ch'iao Wei-yo invents the canal lock	**978** Birth of Japanese woman novelist Murasaki Shikibu, author of *The Tale of Genji*	
c. 987 Toltecs seize control of Mayan city of Chichen Itza			
c. 1000 Mogollon people in southwest North America begin building semi-subterranean villages	**1003** Death of French scholar Gerbert of Aurillac, translator of Arabic texts on the abacus and astrolabe	**1033** Birth of philosopher Anselm of Canterbury, proposer of logical proof for the existence of God	**1000**
c. 1000 New World's first European settlers are Norwegians, who establish settlement on Newfoundland ❍*78*	**1037** Death of Persian doctor Ibn Sina, known in Europe as Avicenna, author of *Canon Medicinae*	**c. 1040** Modern system of musical notation, using a stave, invented by Italian monk Guido d'Arezzo	
	c. 1040 Technique for printing moveable ceramic type invented in China	**1048** Birth of Persian scientist and poet Omar Khayyam	
		1086 Domesday Book survey of Norman England	
c. 1050 Cahokia begins 200 years as largest of Mississippian chiefdoms ❍*108*	**1054** Chinese astronomers observe super-nova explosion creating the Crab nebula	**1088** First officially sanctioned university in Europe established in Bologna, Italy	

1100–1499

	ASIA AND AUSTRALASIA	AFRICA	EUROPE
1100	**1118** Order of Knights Templar established	**1117** City of Lalibela becomes capital of Christian Ethiopia ❍82	**1138** Conrad III becomes Holy Roman Emperor, founds Staufen dynasty
	1127 Jurchen nomads overrun northern China; Song dynasty retreats south ❍73		**1143–50** Almohad dynasty establishes control over Almoravid areas of southern Spain ❍92
	1148 Second Crusade ends after abortive campaigns against Turks in Palestine	**1143–45** Almohads overthrow fellow Berber dynasty of Almoravids in western North Africa	**1151** Independent Serbian kingdom established
	1156 Civil war between rival clans breaks out in Japan; end of Heian period		**1153** Frederick I becomes Holy Roman Emperor
	1173 Muhammad of Ghur overthrows Ghaznavid dynasty in Afghanistan ❍88	**1171** Saladin overthrows Fatimid dynasty in Egypt, founds Ayyubid dynasty ❍88	**1154** Plantagenet Henry II becomes King of England, later formalizes control over Ireland, Wales, and Scotland ❍93
	1185 Yoritomo Minamoto institutes Shogunate in Japan		**1167** Lombard League formed against Frederick I in Italy
	1187 Saladin's victory at Hattin leads to Arab recapture of Jerusalem ❍95		**1186** Independent Bulgarian kingdom re-established
	1189-1192 Third Crusade captures Cyprus and Acre, fails to take Jerusalem ❍95		**1198** Otto VI, member of Guelph family, becomes Holy Roman Emperor; civil war breaks out in Germany
	1192 Ghurid victory over Rajputs presages Muslim conquest of northern India		**1199** John succeeds brother Richard I as King of England
1200	**c. 1200** Khmer Empire of Angkor at its peak under Jayavarman VII ❍64	**c. 1200** Kingdom of Mwenemutapa founded, centred on Great Zimbabwe	**1204** Fourth Crusade sacks Constantinople; Latin Empire created on former Byzantine territory ❍96
	1207 Mongol warrior Chinggis Khan, invades China; over following 70 years Mongol armies conquer most of mainland Asia ❍98	**1228** Hafsids wrest control of Tunisia from Almohads ❍88	**1209** English-led crusader army suppresses Albigensian heretics (Cathars) in southern France
		1239 Ziyanid dynasty overthrows Almohads in Algeria ❍88	**1212** Almohads routed at Las Navas de Toloso by Castile
	1211 Sultanate of Delhi established ❍98	**1240** King Sundiata of Mali defeats King Samanguru of Ghana, establishes empire of Mali ❍80	**1215** King John signs Magna Carta at Runnymede
	1219 Last Minamoto Shogun killed in Japan; Hojo family in control (to 1335)		**1236** Castile captures Córdoba from Moors ❍92
	1228–1229 Sixth Crusade, under Frederick II of Germany, gains Jerusalem by treaty	**1249-50** Fifth Crusade captures Egyptian port of Damietta, fails to take Cairo ❍95	**1240** Mongols capture Moscow, destroy Kiev ❍98
	1229 Shan establish kingdom of Assam	**1250** Mamluks seize power in Egypt; end of the Fatimid dynasty	**1240** Prince of Novgorod, Alexander Nevski, defeats Swedes
	1258 Baghdad destroyed by Mongol armies; end of Abbasid rule ❍89, 98	**1269** Marrakech falls to Marinid control in Morocco; end of Almohad dynasty ❍88	**1241** Mongol forces from Golden Horde defeat Polish and German knights at Leignitz, Hungarians at Pest ❍98
	1260 Mongols defeated by Mamluks of Egypt at Ayn Jalut in Palestine ❍98	**1270** Warlord Yekuno Amlak seizes control of Ethiopia, founds Solomonid dynasty	**1242** Nevski defeats Teutonic Knights at Lake Peipus
	1279 Mongols finish conquest of southern China; start of Yuan dynasty (to 1368)		**1261** King Ottokar II of Bohemia captures Austria
			1261 Greeks, assisted by Genoese, seize Constantinople
			1283 Teutonic Knights complete conquest of Prussia
			1297 William Wallace leads Scottish revolt against English
1300	**1300** Osman I proclaims himself Sultan of the Turks; start of Ottoman Empire	**c. 1300** Kingdom of Benin founded ❍80	**1319** Sweden and Norway united under Magnus Ericsson
	1303 Sultanate of Delhi conquers Rajput fort of Chitor, last Hindu stronghold in northern India	**1316** Military expedition from Egypt establishes Muslim rule in Nubia	**1325** Ivan I becomes ruler of Grand Duchy of Moscow
		1320 King Amda Seyon extends Christian control to southern Ethiopia	**1337** Edward III of England lays claim to French throne; beginning of Hundred Years War (to 1453) ❍106
	1335 Start of Ashikaga rule in Japan ❍87	**1324** Pilgrimage of Mansa Musa I to Mecca marks high point of Mali empire	**1347** Black Death reaches Italian ports from Crimea❍105
	1353 Kingdom of Laos established		
	1367 Sultanate of Delhi massacres over 400,000 Hindu civilians at Vijayanagar	**1332** War between Christian Ethiopia and neighbouring Islamic kingdoms	**1356** "Golden Bull" changes Holy Roman Empire from monarchy to elitist federation of seven Electors
	1368 Mongols expelled from China after Red Turbans' revolt; start of Ming dynasty	**1340** Portuguese discover Canary Islands (allocated to Castile by the Pope 1344)	**1358** France rocked by rising in Paris, led by Etienne Marcel, and peasant revolts (the Jacquerie) in countryside
	1369 Thais sack Khmer capital of Angkor	**1348** Black Death devastates Egypt	**1378** Great Schism in Catholic Church (to 1417) ❍107
	1369 Timur-leng's rebel forces take Chaghatai capital of Samarqand ❍99	**1349** Moroccan traveller Ibn Battuta returns home after 25-year journey to India and China	**1381** Rebellions in Florence, England and Flanders ❍107
	1392 Yi dynasty established in Korea (to 1910); end of Koryo kingdom ❍87		**1385** Portugal defeats Castile to gain independence
	1398–1399 Timur-leng sacks Delhi ❍87	**1375** Kingdom of Songhay breaks away from empire of Mali ❍80	**1386** Marriage creates Polish-Lithuanian commonwealth
			1388–1395 Timur-leng conquers the Golden Horde ❍99
			1389 Ottomans defeat Serbs at Kosovo Polje ❍97
			1397 Treaty of Kalmar unites Denmark, Norway, Sweden
1400	**1402** Timur-leng defeats Turks at Angora		**1410** Polish-Lithuanian armies defeat Teutonic Knights
	1409 Chinese invade Vietnam	**1420** Portuguese sailors occupy island of Madeira; discover Azores 1430, Cape Verde Islands 1455	**1415** Henry V invades France, wins Agincourt, takes Paris
	1420 Beijing replaces Nanking as capital of China		**1417** Council of Constance restores single papacy in Rome
	1431 Annam gains independence from China	**1421** Zheng He's ships from Ming China visit Muslim ports of East Africa ❍139	**1419** Predominantly Czech supporters of Jan Hus rise against German rule in Bohemia; start of Hussite Wars
	1433 Chinese admiral Zheng He's 28 years of exploration to Asia and Africa ended by imperial ban on distance voyages ❍139	**c. 1440** Walled enclosure and tower are built at Great Zimbabwe ❍83	**1427** Venetians conquer Bergamo to complete Terrafirma
		1468 Songhay Empire conquers Timbuktu, occupied by desert nomads since 1433	**1434** Cosimo de' Medici becomes ruler of Florence
	1461 Trebizond conquered by Ottomans ❍97	**1471** Portuguese establish Elmina trading post on coast of modern Ghana ❍80	**1435** Spanish kingdom of Sicily united with Naples ❍152
	1467 Ten-year Onin Wars presage century of conflict between warlords in Japan	**1478** Portuguese defeat Spanish fleet to attain supremacy on West African coast	**1452** Frederick III first Habsburg Holy Roman Emperor
	1471 Annam (north Vietnam) conquers Champa (south Vietnam)	**1483** Portuguese make contact with equatorial kingdom of Kongo ❍136	**1453** Turks besiege and capture Constantinople ❍97
	1479 Vietnam conquers kingdom of Laos		**1453** End of Hundred Years War ❍106
		1488 Portuguese Bartholomew Dias sails round Cape of Good Hope ❍116	**1472** Ivan III adopts title of tsar; conquers Novgorod 1478
	1498 Portuguese explorer Vasco da Gama	**1498** Vasco da Gama visits Mombasa	**1485** Wars of the Roses in England end after 30 years; Henry VII becomes king, begins Tudor dynasty
			1492 Ferdinand and Isabella conquer Granada to unite all Spain under Christian rule ❍146

1100

THE AMERICAS

c. 1130 Rise of Chibcha culture in modern Colombia

c. 1160 Last Toltec king, Heumac, flees destruction of Tula ➲*85*

1187 Maya leader Hunac Ceel leads revolt that evicts Toltecs from Chichen Itza, establishes new capital at Mayapan ➲*85*

c. 1190 Aztecs establish small state on shore of Lake Texcoco in central Mexico

SCIENCE AND TECHNOLOGY

c. 1140 Adelard of Bath translates Euclid into Latin using Greek and Arabic texts

1150 University established in Paris

1163 External flying buttresses used for first time, in construction of Nôtre Dame

1167 Oxford University founded

1174 Construction on the unintentionally Leaning Tower in Pisa started

1187 Death of Italian scholar Gerard of Cremona, who translated works of Galen from Arabic texts captured at Toledo

ARTS AND HUMANITIES

c. 1120 Peter Abelard revives teachings of Aristotle in Paris

c. 1130 Temples at Angkor Wat in Cambodia started

1140 Birth of Japanese philosopher Eisa, whose teachings found Zen Buddhism

1154 Chartres Cathedral, southwest of Paris, marks spiritualized Gothic style of church architecture

c. 1180 Chinese philosopher Zhu Xi compiles Confucian Canon

1198 Death of Muslim philosopher Ibn Rushd, known in Europe as Averroes, author of extended commentaries on Aristotelian thought that influenced early Renaissance

1200

THE AMERICAS

c. 1200 Manco Capac establishes Inca dynasty with capital at Cuzco in Peru

c. 1200 Monks Mound constructed at Cahokia, North America (in present-day Illinois)

SCIENCE AND TECHNOLOGY

1202 Italian Leonardo Fibonacci publishes *Liber Abaci* – first European book to explain Indian numerals

c. 1230 Explosive bombs and rockets first used, by Chinese against Mongol forces

c. 1250 Gunpowder first mentioned in European manuscripts

1264 French scholar Vincent of Beauvais publishes *Speculum Maius,* combination of encyclopedia and universal history

1266 English scholar and mathematician Roger Bacon completes *Opus Maius*, advocating use of scientific experiment

c. 1270 Firearms and cannon (made from reinforced bamboo) first used in China

c. 1280 Belt-driven spinning wheel introduced to Europe from India

ARTS AND HUMANITIES

1209 Franciscans founded by Francis of Assisi

c. 1210 German minnesinger Wolfram von Eschenbach writes romance *Parzival*

c. 1220 In southern China, landscape artists Ma Yuan and Hsia Kuei emphasize mist and clouds

1220 Amiens Cathedral marks start of Rayonnant Gothic style characterized by large circular windows

1248 Moors begin building Alhambra citadel in Granada

1248 Cologne Cathedral heralds start of spread of Gothic ecclesiastical architecture across northern Europe

c. 1250 "Black Pagoda" Temple of the Sun at Kanarak in India built by King Narasimhadeva

1260 Nicolo Pisano's pulpit carved for Baptistry in Pisa reflects classical Gothic church architecture

1273 Thomas Aquinas completes *Summa Theologicae*

1285 Buoninsegna's *Rucellai Madonna* revolutionizes Byzantine style of Sienese painting

1300

THE AMERICAS

c. 1300 Chimu state in Peru expands to rival that of neighbouring Incas ➲*84*

c. 1300 Mesa Verde and other Anasazi centres in southwestern North America abandoned, probably due to extended drought ➲*108*

c. 1325 Aztecs establish Tenochtitlan, city on island in Lake Texcoco ➲*111*

c. 1350 King Mayta Capa begins expanding Inca control in Peru

c. 1350 Mixtecs control Oaxaca Valley in southern Mexico ➲*85*

c. 1370 Chimu complete conquest of coastal northern Peru ➲*84*

SCIENCE AND TECHNOLOGY

c.1300 Earliest known European spectacles manufactured in Italy

1324 Earliest known European cannon manufactured in France

1340 First European factory for making paper opens in Fabriano, central Italy

1340s Windpumps first used to drain marshes in Holland

1370 French king Charles V establishes standard time according to weight-driven mechanical clock in royal palace in Paris

c. 1377 French scholar William of Oresme writes *De Moneta*, on monetary policy

1377 Single-arch bridge with span of 72 metres (236 feet) is completed at Trezzo, northern Italy

1385 Heidelberg University established

ARTS AND HUMANITIES

1304–1309 Giotto de Bondone's frescoes in Padua represent turning point towards more realism in Italian painting

1307 Dante Alighieri begins writing *Divine Comedy*

1332 Birth of Arab philosopher Ibn Khaldun

1337 Death of Chinese dramatist Wang Shifu, author of *The Romance of the Western Chamber*

1358 Giovanni Boccaccio completes *Decameron,* record of bawdy tales told by Florentines during the Black Death

c. 1365 Flemish painters establish Bruges School

c. 1370 Japanese dramatist Kanami Motokiyo establishes classic form of Zen-influenced "No" theatre

1378 Philosopher-nun Catherine of Siena writes *Dialogo*

1380 Reformer John Wycliffe translates Bible into English

c. 1380 Geoffrey Chaucer begins *The Canterbury Tales*

c. 1399 Greek artist Theophanes paints icon *The Deeds of the Archangel Michael* for Kremlin Cathedral in Moscow

1400

THE AMERICAS

1427 Itzcoatl becomes Aztec king, begins policy of military expansion ➲*111*

1438 Pachacuti's Incas conquer Chancas

1470 Incas, under Pachacutec and Tupac Yupanqui, annex Chimu kingdom ➲*110*

c. 1490 Incas expand empire into parts of Colombia, Bolivia and Paraguay ➲*110*

1492 Columbus reaches island he names San Salvador in Bahamas ➲*120*

1494 Treaty of Tordesillas divides known New World, assigning Brazil to Portugal and the rest to Spain

1496 Santo Domingo founded as Spanish centre of government in Americas ➲*120*

1497 John Cabot first European since Vikings to land on Newfoundland ➲*116*

1499 Vespucci "discovers" Amazon ➲*116*

SCIENCE AND TECHNOLOGY

1403 Moveable metal type first used for printing, in Korea

1410 Flagship of Chinese admiral Zheng He is 130 metres (427 feet) long, with five masts and 12 decks

1419 Portuguese prince Henry establishes school of navigation at Sagres

1435 Italian architect Leon Albertini outlines mathematical laws of perspective in painting in *De Pictura*

c. 1445 German Johann Gutenberg produces first Bible using moveable metal type

1452 Hungarian armourers cast 50-tonne cannon for use by Turks in successful siege of Constantinople

1474 Government of Venice issues world's first patents protecting inventors' rights

c. 1475 Italian astronomer Paolo Toscanelli proposes reaching China by sailing west

ARTS AND HUMANITIES

1406 Construction starts on Forbidden Palace in Beijing

1410 Limbourg brothers produce their famous illustrated *Les Très Riches Heures du Duc de Berry*

1413 Czech philosopher and religious reformer John Hus writes *Exposition of Belief*; he is burned for heresy 1415

1416 *Hsing Li Ta Ch'uan,* 120-volume compilation of moral philosophy, published in China

1434 Flemish artist Jan van Eyck paints *The Betrothal of the Arnolfini*

1434 Donatello casts bronze statue *David* in Florence

1469 Birth of Nanak, Indian founder of Sikhism

1484 Anonymous Flemish painting depicts organ with chromatic keyboard – a clavichord with four octaves

1485 William Caxton prints Malory's *Le Morte d'Arthur*

1494 German poet Sebastian Brandt writes *Ship of Fools*

1495 Bosch paints *The Garden of Earthly Delights*

1498 Dürer publishes album of woodcuts *The Apocalypse*

1500–1649

ASIA AND AUSTRALASIA	AFRICA	EUROPE
1500		
1501 Safavid dynasty is founded in Iran by Shah Ismail I	**1504** Christian kingdom of Soba in Nubia conquered by Islamic forces	**1501** Portuguese establish direct sea-route to import pepper and spices from India into Europe
1502 Portuguese ships destroy the Indian port of Calicut	**1505** Portuguese sack Islamic ports of Kilwa and Mombasa and establish trading posts on the east coast of Africa	**1517** German priest Martin Luther writes his *Ninety-Five Theses*; start of the Reformation in Europe ❍*154*
1510 Portuguese conquer Goa and make it their capital in India ❍*118*	**1510** Spanish capture Tripoli	**1521** Ottoman Turks capture Belgrade ❍*142*
1514 Ottoman Turks defeat the Safavids at the Battle of Chaldiran ❍*142*	**1517** Ottomans under Sultan Selim I conquer Egypt; end of the Mamluk dynasty ❍*142*	**1521–26** First war for control of Italy is fought between France and Habsburg Spain
1515 Ottomans conquer Kurdistan ❍*143*	**1518** First direct shipments of slaves from Africa to the Americas	**1524–25** Violent peasant uprisings sweep across Germany
1516 Portuguese establish a trading post at Canton (Guangzhou) in southern China		**1525** Spain defeats France at the Battle of Pavia in Italy and captures Francis I ❍*158*
1526 Afghan warlord Babur conquers western half of Sultanate of Delhi and establishes Mughal dynasty ❍*143*	**1527** Islamic Somalis invade Ethiopia	**1526** Habsburgs gain Hungary after King Lajos II is killed in the Battle of Mohács against the Turks ❍*158*
		1527 Habsburg Emperor Charles V sacks Rome
		1529 The Ottoman Turks unsuccessfully besiege Vienna
1530		
1534 Ottomans conquer Mesopotamia ❍*143*	**1531** Portuguese send troops to aid Ethiopia against Muslim invaders	**1531** War breaks out between Protestant and Catholic cantons in Switzerland ❍*155*
	1531 Portuguese capture Sena, on the River Zambezi in East Africa	**1534** Act of Supremacy is passed in England, making Henry VIII head of the English Church; Society of Jesus (Jesuits) is founded by Ignatius Loyola
1540 Mughal dynasty in India is overthrown by Afghan warlord Sher Shah	**1541** Charles V's expedition against the Turks at Algiers fails	**1541** Ottoman Turks conquer Hungary ❍*142*
1555 Burmese invade northern Thailand	**1543** Ethiopian forces, assisted by Portuguese troops, expel Islamic invaders	**1545** Council of Trent meets to reform the Catholic Church ❍*154*
1555 Humayan restores Mughal rule in India	**1545–6** Songhay forces sack the capital of the Mali Empire	**1546–47** Charles V defeats the Protestants in the Schmalkaldic War in southern and central Germany
1555 Portuguese establish a colony at Macau in China ❍*118*		**1555** Peace of Augsburg establishes freedom of worship in Germany
1556 Akbar I becomes Mughal emperor of India		**1558** Russia invades Livonia
		1559 Philip II of Spain defeats France; Naples and the Low Countries are restored to Spanish control
1560		
1563 Chinese destroy Japanese pirates who have been raiding coastal cities		**1562** Massacre of Huguenots (French Protestants) at Vassy in France starts a series of religious civil wars
		1569 Poland and Lithuania unite under Polish control ❍*150–1*
	1571 Idris III becomes king of Kanem-Borno and establishes control of the Lake Chad region	**1570** City of Novgorod is destroyed by armies from Moscow
1573 Mughals conquer Patan and Ahmadabad in Gujarat ❍*144*	**1574** Portuguese found the city of Luanda in Angola as a base for slave raiding	**1571** Venetian and Spanish fleets defeat the Ottoman Turks at the Battle of Lepanto ❍*142*
1576 Mughals conquer Bengal	**1578** A Portuguese attempt to conquer the interior of Morocco is defeated at the Battle of Alcazar-Kabir	**1572** Thousands of Huguenots (French Protestants) are massacred on St. Bartholmew's Day ❍*155*
1581 Yermak Timofeyevich begins the Russian conquest of Siberia	**1581** Morocco begins expanding south into the western Sahara	**1576** Following Spanish sack of Antwerp, Dutch provinces unite under William of Orange
1583 Mughal emperor Akbar I proclaims toleration of all religions in India	**1589** Portuguese defeat the Turks at Mombasa in eastern Africa	**1579** Dutch republic is formed; southern Netherlands remain under Spanish control
1586 Abbas I becomes shah of Safavid Empire in Iran		**1589** Henry III of France is assassinated; Henry IV accedes to the throne; start of the Bourbon dynasty
1590		
1590 Hideoshi conquers east and north Japan, reuniting country under his rule	**1591** Invading Moroccans capture Timbuktu, crush Songhay forces at the Battle of Tondibi and destroy the city of Gao; end of the Songhay Empire	**1595** After intervening in the Livonian wars, Sweden acquires Estonia ❍*150*
1597 Iranian Safavids defeat the nomadic Uzbeks and expel them from western Afghanistan ❍*142*		**1598** Boris Godunov becomes tsar of Russia
1600 English East India Company formed	**1595** Dutch establish a trading post in Guinea on the west coast of Africa	**1598** Edict of Nantes establishes limited religious freedom in France
1602 Dutch East India Company formed	**1598** Dutch establish a small colony on the island of Mauritius	**1603** James VI of Scotland inherits the English throne as James I; start of the Stuart dynasty
1603 Tokugawa dynasty begins rule in Japan ❍*140*		**1609** A truce ends fighting in the Netherlands between Philip III of Spain and the Dutch rebels
1603 Safavids under Abbas I capture Baghdad from the Ottomans		**1613** Start of Romanov dynasty in Russia
	1612 The city-state of Timbuktu becomes independent of Morocco	**1617** Under the Peace of Stolbovo with Sweden, Russia loses access to the Baltic Sea ❍*149*
1610 Dutch establish the port of Batavia (present-day Jakarta) in Java ❍*119*		**1618** Protestant revolt in Prague begins the Bohemian War, and marks the start of the Thirty Years War ❍*159*
1620		
1623 English merchants on the island of Ambon are massacred by the Dutch	**1621** Dutch capture the West African island slave ports of Arguin and Gorée from the Portuguese	**1621** Gustaf Adolf II of Sweden conquers Riga in Livonia
1635–37 Manchurian confederation conquers southern Mongolia and Korea	**1626** French establish the colony of St. Louis at the mouth of the River Senegal	**1621** Warfare between the Dutch and Spanish is renewed
1638 Japan is closed to foreigners	**1626** French settlers and traders establish a colony on the island of Madagascar.	**1631** Swedes defeat the German Catholics at the Battle of Breitenfeld and invade southern Germany ❍*159*
1639 Peace of Zuhab establishes a permanent border between Ottoman and Safavid Empires ❍*142*	**1637** Dutch capture the fortified port of El Mina from the Portuguese ❍*137*	**1635** France enters the Thirty Years War as Sweden's ally
1644 Manchurians enter Beijing; end of the Ming dynasty in China, start of the Qing dynasty ❍*142*	**1641** Dutch capture Luanda from the Portuguese	**1640** Portugal and Catalonia revolt against Spanish rule ❍*156*
1648 Janissaries (slave soldiers) revolt and depose the Ottoman sultan Ibrahim I		**1642** English Civil War begins ❍*156*
		1648 Peace of Westphalia ends the Thirty Years War
		1648–53 *Fronde* rebellions erupt in France ❍*156*
		1649 Charles I of England is executed; a Commonwealth is established

THE AMERICAS	SCIENCE AND TECHNOLOGY	ARTS AND HUMANITIES	

1501 First African slaves are brought to the Caribbean

1508–15 Spanish conquer Puerto Rico and Cuba ●120

1513 Spanish explorer Balboa crosses the Isthmus of Panama and discovers and names the Pacific Ocean ●120

1518–21 Spanish *conquistadore* Hernán Cortés conquers Mexico, destroying the Aztec capital, Tenochtitlan ●120

c.1520 Inca king Huyana Capac conquers parts of Ecuador

1520 Ferdinand Magellan sails around Cape Horn at the southern tip of South America ●116

1522 Viceroyalty of New Spain is created and Mexico City is founded on the ruins of Tenochtitlan ●120

1530 Portuguese begin the colonization of Brazil ●122

1533 Inca king Atahualpa is killed by the Spanish, who occupy Cuzco and conquer the Inca Empire ●121

1537 Spanish colonies are established at Buenos Aires, and at Asunción on the River Paraguay ●121

1539 Spanish begin the conquest of the Maya cities in the Yucatán region of Mexico ●122

1541 French explorer Jacques Cartier attempts to establish a colony at Québec in Canada

1545 Spanish begin mining silver at Potosi in Peru ●122

1547 Spanish open silver mines at Zacatecas in Mexico ●122

1555 Dutch, English, and French sailors form the Guild of Merchant Adventurers to raid Spanish shipping routes from America

1561 Following the failure of the Pensacola settlement in South Carolina, the Spanish abandon attempts to colonize the east coast of North America

1561 Spanish treasure fleets are forced to adopt a convoy system as a defence against "pirate" attacks

1565 Portuguese found the city of Rio de Janeiro ●122

1568 Spanish destroy the fleet of the English slave trader John Hawkins at Veracruz

1578 English explorer Francis Drake sails along the west coast of North America and lays claim to California ●116

1607 The London Company establishes a colony at Jamestown, Virginia, under the leadership of John Smith

1608 French explorer Samuel de Champlain founds Québec as the capital of the colony of New France

1609 Henry Hudson explores the Hudson River

1612 Tobacco first cultivated by English settlers in Virginia

1613 Dutch set up a trading post on Manhattan Island

1619 First African slaves arrive in Virginia, and the first representative assembly is held

1620 English Protestant settlers cross the Atlantic in the *Mayflower* and establish a colony in Massachusetts

1630 Dutch capture Recife in Brazil from the Portuguese

1630 A 12-year period of intensive migration from England to Massachusetts begins

1635 French establish colonies of the islands of Martinique and Guadeloupe ●125

1638 An English colony is established at New Haven on Long Island

1643 French establish the city of Montréal in Canada

1643 English colonies form the New England Confederation

1500 Leonardo da Vinci designs an impractical, but correctly principled, helicopter

c.1505 The pocket watch is invented by German clockmaker Peter Henlein

c.1510 Polish astronomer and mathematician Nicolaus Copernicus formulates his theory that the Earth orbits the Sun

c.1520 Rifling for firearms is invented in central Europe

1522 Spanish ships returning from Magellan's voyage complete the first circumnavigation of the world ●116–17

1533 German surveyor Gemma Frisius discovers the principles of triangulation

1540 Italian Vannoccio Biringuccio publishes *Pirotechnia*, a handbook of metal smelting and casting techniques

1542 French scholar Conrad Gesner publishes his *Historia Plantarum*, the first modern work of botany

1543 Polish astronomer Nicolaus Copernicus publishes his theory of the Sun-centred universe

1543 Belgian doctor Andreas Vesalius publishes his *De Fabrica Corporis Humani*, an illustrated handbook of human anatomy based on dissection

1569 Flemish mapmaker Gerhard Mercator publishes a world navigation chart that has meridians and parallels at right-angles

1572 Italian mathematician Raffaele Bombelli introduces imaginary numbers in his *L'Algebra*

1576 Danish astronomer Tycho Brahe builds an observatory for Frederick II

1586 Dutch mathematician Simon Stevin demonstrates that objects fall at an equal rate in a vacuum, irrespective of their weights

1589 A stocking-frame knitting machine is invented by William Lee in Cambridge, England

1593 Death of doctor Li Shizen (b.1518) who compiled *The Comprehensive Pharmacopoeia* of traditional Chinese medicine

1602 Italian astronomer and scientist Galileo Galilei discovers the constancy of a swinging pendulum

1608 Dutch optician Hans Lippershey invents a refracting telescope

1614 Scottish mathematician John Napier invents logarithms

1618 Dutch scientist Snellius discovers his law of the diffraction of light

1619 German astronomer Johannes Kepler outlines the third of his three laws of planetary motion in *Harmonices Mundi*

1628 English doctor William Harvey explains the circulation of the blood

1633 Italian scientist Galileo's theory about a Sun-centred solar system is condemned by the Inquisition

1637 French philosopher and mathematician Rene Descartes introduces analytical geometry in his *La Geometrie*

1639 French mathematician Girard Desargues introduces the study of projective geometry in his *Brouillon Projet*

1643 Italian scientist Evangelista Torricelli invents the mercury barometer

---

1503 Leonardo da Vinci paints *Mona Lisa*

1511 Dutch humanist Desiderius Erasmus publishes his *In Praise of Folly*

1512 Italian artist Michaelangelo Buonarotti finishes painting the ceiling of the Sistine Chapel in Rome

1513 Italian politician Niccolo Machiavelli writes his *The Prince*

1516 English scholar Thomas More publishes his *Utopia*

1520 Martin Luther writes his *The Freedom of a Christian Man*

1525 English reformer William Tyndale starts printing English versions of the New Testament in Cologne in Germany

1533 German artist Hans Holbein the Younger paints his *The Ambassadors*

1534 Luther completes his translation of the Bible into German

1534 French humanist Francois Rabelais writes his satire *Gargantua*

1540 Holy Carpet of Ardebil is woven in northern Iran

1545 Indian architect Aliwal Khan designs the octagonal tomb of the Afghan warlord Sher Shah at Sasaram, India

1559 Index of Forbidden Books is published by the Catholic Church

1563 Dutch artist Peter Bruegel the Elder paints his *The Tower of Babel*

1563 Work begins on the monastery and palace of Escorial near Madrid, Spain, deigned by Juan de Herrara

1572 Portuguese Luis vaz de Camoes publishes his epic *The Lusiads*

1575 Italian architect Giacomo della Porta designs the church of Il Gesu in Rome; this marks the beginning of the Baroque period of European art

1580 French writer Michel Montaigne publishes the first of his *Essays*

1586 Greek-born artist El Greco paints his Burial of Count Orgasz in Toledo, Spain

1590 English dramatist Christopher Marlowe writes his play *Tamburlaine the Great*

1596 English poet Edmund Spenser completes his *The Fairie Queen*

1600 William Shakespeare writes *Hamlet*

1604 Confucian Tung-lin Academy founded in China

1605 Spanish author Miguel de Cervantes publishes the first volume of *Don Quixote de la Mancha*

1615 Bukeshohatto book of warriors' wisdom is published in Japan

1619 English architect Inigo Jones designs the Banqueting House in London

1623 Death of the Indian poet Tulsi Das (b. 1532), author of the Hindu classic *Tulsi-krit Ramayan*

1624 Dutch artist Frans Hals paints his *Laughing Cavalier*

1632 Dutch artist Rembrandt van Rijn paints his *Anatomy Lesson of Dr Tulp*

1634 Taj Mahal is built in Agra, India, as a tomb for Mumtaz Mahal, wife of Mughal emperor Shah Jahan

1642 Italian composer Claudio Monteverdi completes his opera *The Coronation of Poppea*

1500

1530

1560

1590

1620

1650–1769

	ASIA AND AUSTRALASIA	AFRICA	EUROPE
1650	**1652** Russian colonists found the city of Irkutsk in Siberia ❍*148*	**1650** Ali Bey establishes himself as hereditary ruler of Tunis	**1652** Spain is reunited when Catalonia submits to Spanish rule
	1656 Mohammud Kuprulu becomes Ottoman vizier (chief minister) and stabilizes the empire	**1652** Dutch settlers found the colony of Capetown in South Africa	**1652–54** England wins a naval war against the Dutch over shipping rights
	1662 Chinese pirate warlord Zheng Chenggong (Koxinga) expels the Dutch from Formosa (Taiwan)	**1654** French occupy the island of Réunion	**1653** Oliver Cromwell becomes Lord Protector of England
	1664 Hindu raiders sack the Mughal port of Surat in India	**1658** Dutch settlers start importing slaves to the Cape Colony	**1654** Cossacks in the Ukraine defect from Poland to Russia, starting a war
	1669 Hindu religion is prohibited throughout the Mughal Empire in India, and Hindu temples are destroyed	**1662** Portugal cedes the city of Tangier in Morocco to England	**1660** Monarchy is re-established in England with the accession of Charles II
		1662 English build a fort at the mouth of the River Gambia in West Africa	**1664** Austria defeats the Turks at St Gotthard
1670			**1670** Peasants and cossacks revolt in southern Russia ❍*156*
	1674 Hindu raider Sivaji becomes independent ruler of Maratha in India	**1677** French expel the Dutch from Senegal in West Africa	**1672** France invades Holland, William III of Orange opens the sluices to save Amsterdam
	1674 Regional rulers rebel against central control in China		**1678** Treaties of Nijmegen end the wars between France and the Dutch, Germans, and Spanish
	1675–78 Sikhs rebel against their Mughal overlords in India	**1683** Prussians build a fort on the coast of Guinea in West Africa	**1683** Ottomans besiege Vienna and are defeated by a German-Polish army at the Battle of Kahlenberg ❍*158*
	1683 Manchu Qing dynasty conquers Taiwan which comes under direct rule from China for the first time	**1684** French mount naval expeditions to suppress Islamic pirates at Algiers	**1684** Venice, Austria and Poland form a Holy Alliance against the Ottoman Turks
	1689 Treaty of Nerchinsk establishes the Russian–Chinese border in the Amur region ❍*148*	**1686** French formally annex Madagascar	**1685** Louis XIV revokes the Edict of Nantes in France
		1688 Huguenot refugees from France arrive in South Africa	**1688** Austrian armies liberate Belgrade from the Ottomans
1690	**1690** English establish a colony at Calcutta in northern India	**1697** French complete the conquest of Senegal	**1690** William III of England defeats French troops under former king James II at the Battle of the Boyne in Ireland
	1691 Mughal Empire in India reaches its greatest extent ❍*144*	**1698** Portuguese expelled from most East African ports by Omanis from southeast Arabia	**1700** Sweden is attacked by Poland allied to Denmark and Russia; start of the Great Northern War. The Swedes defeat the Russians at the Battle of Navara
	1705 A Chinese attempt to impose their candidate for dalai lama provokes risings and unrest in Tibet	**1701** Asante emerges as a powerful state in West Africa ❍*136*	**1702** Grand Alliance (England, Holland, and Austria) declares war on France and Spain, the War of the Spanish Succession begins
	1707 Death of the emperor Aurangzeb is followed by the rapid disintegration of the Mughal Empire in India	**1705** Husseinid dynasty takes control in Tunis, establishing independence from the Ottoman Empire	**1703** Peter I establishes the city of St Petersburg in Russia
	1708 Sikhs establish independent control of the Punjab region of northern India	**1709** Dutch cattle farmers in South Africa trek east across the Hottentot Holland Mountains	**1709** British defeat the French at the Battle of Malplaquet ❍*158*
			1709 Swedes in alliance with Ukrainian cossacks invade Russia but are defeated at the Battle of Poltava ❍*151*
1710	**1712** War of succession between Mughal emperor Bahadur's sons divides India	**1710** French take Mauritius from the Dutch	**1713** Treaty of Utrecht between Britain and France ends the War of the Spanish Succession
	1717 A Mongol army seizes control of the Tibetan capital, Llasa	**1714** Ahmed Bey establishes the Karamanlid dynasty as independent rulers in Tripoli, Libya	**1715** First Jacobite rebellion in Scotland is defeated by British troops
	1722 Afghan ruler, Mir Mahmud, invades Iran and makes himself shah	**1717** Dutch begin importing slaves to Cape Colony in South Africa	**1720** South Seas Company fails and creates financial panic in London
	1724 Shah Mahmud orders the massacre of the Iranian aristocracy	**1723** British Africa Company claims the Gambia region of West Africa	**1721** Defeat in the Great Northern War ends Swedish dominance in the Baltic region
	1724 Chinese establish a protectorate over Tibet	**1724** Dahomey emerges as a powerful new state in West Africa ❍*136*	**1722** Peter I makes administrative reforms in Russia and limits the traditional priviledges of the aristocracy
	1728 Dutch explorer Bering discovers the Bering Strait for Russia	**1728–29** Portuguese briefly re-occupy the East African port of Mombasa	**1725** Following the death of Peter I, his wife Catherine I becomes the first of a series of weak Russian rulers
1730	**1730** Iranian chieftain Nadir Kuli drives the Afghans from Iran and restores the Safavid dynasty	**1730** Dutch northerly expansion in South Africa reaches the River Olifants	**1733–35** France allied to Spain fights the War of the Polish Succession against Austria allied to Russia
	1736 Nadir becomes shah of Iran on the death of Abbas III; end of the Safavid dynasty	**1732** Spain recaptures Oran, Algeria	**1740** War of the Austrian Succession begins when Frederick II of Prussia invades Austrian-controlled Silesia ❍*157*
	1739 Iranians invade India, defeat a Mughal army and capture Delhi		**1744** Prussians invade Bohemia
	1742 Marathas raid British Bengal	**1744** Mazrui, the Omani governor of Mombasa, declares his independence from the Sultan of Oman	**1745** Second Jacobite rebellion breaks out in Scotland
	1745 Iranians under Nadir Shah defeat the Turks at the Battle of Kars	**1745** Asante warriors armed with muskets defeat Dagomba armoured cavalry in West Africa	**1746** British troops defeat the Jacobite Scots at Culloden and the rebellion ends
	1747 Assassination of Nadir Shah leads to a period of anarchy in Iran		**1747** Orangists restore the monarchy in Holland
			1748 End of the War of the Austrian Succession; Prussia emerges as a major European power
1750	**1752** Afghans capture Lahore in northern India from the Mughals	**1750** French establish a settlement on the island of St Marie, off Madagascar	**1755** Portuguese capital, Lisbon, is destroyed by an earthquake
	1755 Clive captures Calcutta for British	**1755** Death of Emperor Jesus II of Ethiopia marks the end of strong government; Ethiopia becomes divided between rival claimants to the throne	**1756** Prussia allies with Britain and invades Saxony; start of the Seven Years War
	1757 Clive defeats ruler of Bengal at Battle of Plassey ❍*194*		**1762** Catherine II seizes power in Russia and restores strong government
	1765–69 Chinese invade Burma and establish overlordship	**1756** City of Tunis is captured by the Algerians	**1764** Russia and Prussia form an alliance to control Poland
	1767–69 War between British troops and the Indian state of Mysore ends in truce	**1766** Ali Bey establishes himself as ruler of Egypt and declares independence from the Ottomans	**1768** An anti-Russian confederation is formed in Poland, which leads to a civil war in which Russia intervenes

THE AMERICAS	SCIENCE AND TECHNOLOGY	ARTS AND HUMANITIES	
	1650 German scientist Otto von Guericke invents a vacuum pump ❍*135*	**1651** English political philosopher Thomas Hobbes publishes his *Leviathan*	**1650**
1654 Portuguese expel the Dutch from Brazil		**c.1652** English reformer George Fox founds the Society of Friends (Quakers)	
1656 English ships capture Jamaica from Spain, provoking a war	**1657** Academia del Cimento, the first scientific research institute, is established in Florence, Italy ❍*134*	**1656** Spanish artist Diego Velazquez paints his *The Maids of Honour*	
	1657 Dutch scientist Christiaan Huygens constructs a pendulum clock	**1667** French dramatist Moliere writes his comedy *The Misanthrope*	
1663 English colony of Carolina is established	**1662** Irish scientist Robert Boyle formulates his law of gas expansion	**1667** English poet John Milton publishes his epic *Paradise Lost*	
1664 New Amsterdam surrenders to the English; the city becomes known as New York	**c.1665** English scientist Isaac Newton formulates the law of gravity	**1668** Dutch artist Jan Vermeer paints his *Astronomer*	
1673 Frenchmen Father Jacques Marquette and Louis Joliet explore the upper reaches of River Mississippi	**1673** French military engineer Sebastien de Vauban demonstrates his system for attacking fortresses at the siege of Maastricht	**1670** English architect Christopher Wren begins rebuilding 50 London churches destroyed by fire	**1670**
1674 English establish a trading post at Hudson's Bay	**1675** Royal Observatory is established at Greenwich near London	**1677** Dutch philosopher Baruch Spinoza publishes his *Ethics*	
1680 A revolt by Pueblo Native Americans drives the Spanish from New Mexico	**1678** Christiaan Huygens proposes the wave theory of light	**1681** Italian composer Arcangelo Corelli writes his *Sonate da Chiesa*	
1682 French explorer Robert de La Salle reaches the mouth of the Mississippi and claims the Louisiana Territory for France	**1680** Dutch scientist Anton van Leeuwenhoek discovers bacteria		
1682 English Quaker William Penn establishes the colony of Pennsylvania	**1687** Isaac Newton publishes his *Philosophiae Naturalis Principia Mathematica*	**1689** English composer Henry Purcell writes his opera *Dido and Aeneas*	
1683 Portuguese establish the colony of Colonia on the River Plata in Argentina		**1689** English philosopher John Locke publishes his *Two Treatises on Government*	
1696 Spanish reconquer New Mexico		**1694** Death of the Japanese haiku poet Matsuo Basho (b. 1644)	**1690**
1701 Yale College is founded in Connecticut	**1698** English engineer Thomas Savery invents a practical steam-driven water pump	**1700** English dramatist William Congreve writes his comedy *The Way of the World*	
1702 French acquire the *asciento* to supply African slaves to the Spanish American colonies	**1701** English farmer Jethro Tull invents the seed drill	**1700** Samuel Sewell publishes his anti-slavery tract, *The Selling of Joseph*, in Boston, Massachusetts	
1704 Native Americans allied to the French massacre English settlers at Deerfield in Connecticut		**1702** World's first daily newspaper, the *Daily Courant*, is published in London	
1708–9 Portuguese destroy the power of the *Paulistas* (slave-raiders) in southern Brazil in the War of the Emboabas		**1709** Italian instrument-maker Bartolommeo Cristofori invents the piano by substituting hammer action for the plucking action of the harpsicord	
1709 Large numbers of Germans from the Palatinate region begin migrating to the English colonies in North America, especially Pennsylvania	**1709** British ironworker Abraham Darby perfects a technique for producing iron in a coke-fired blast furnace		
1713 By the Treaty of Utrecht Britain gains Newfoundland and Nova Scotia from the French, and a monopoly on the *asciento* slave trade with Spanish colonies	**1711** Italian naturalist Luigi Marsigli shows that corals are animals not plants	**1710** English bishop George Berkley publishes his *Treatise Concerning the Principles of Human Knowledge*	**1710**
1715 British colonists defeat the Yamassee Native Americans in South Carolina	**1714** German scientist Gabriel Fahrenheit invents the mercury thermometer	**1715** French novelist Alain le Sage publishes his *Gil Blas*	
1715 Scots-Irish immigrants begin the settlement of the Appalacian foothills	**1712** English engineer Thomas Newcomen's first atmospheric steam engine is installed in Staffordshire, England ❍*135*	**c.1715** Japanese dramatist Chikamatsu Mozaemon writes his *Love Suicides*	
		1719 British author Daniel Defoe publishes his novel *Robinson Crusoe*	
1721 José de Antequerra leads the revolt of the *communeros* against the Spanish in Paraguay	**1721** Smallpox inoculations are carried out in Boston, Massachusetts	**1721** French philosopher Charles de Montesquieu publishes his *Persian Letters* – the first major work of the Enlightenment	
1726 British colonists from New York make a treaty with the Native American Iroquois League against the French	**1727** British biologist Stephen Hales publishes his *Vegetable Staticks*		
1730 Cherokee Native Americans acknowledge British supremacy	**1730** British navigator John Hadley invents the reflecting quadrant	**1731** French author Antoine Prevost publishes his novel *Manon Lescaut*	**1730**
1735 Spanish authorities finally suppress the revolt of the *communeros* in Paraguay	**1733** British clothworker John Kay invents the flying shuttle	**1734** French writer Voltaire publishes his *English or Philosophical Letters*	
1739 British capture the Spanish settlement of Porto Bello in Panama	**1735** Swedish scientist Carolus Linnaeus publishes his *Systema Naturae*	**1738** British religious reformer John Wesley lays the foundations of Methodism	
1739 African slaves revolt and kill white settlers at Stono River in South Carolina	**1742** French scientist Paul Malouin invents a process for galvanizing steel	**1738** Russian Imperial Ballet School is founded in St Petersburg	
1743 Hostilities between Britain and Spain develop into King George's War, the American phase of the War of the Austrian Succession	**c.1745** Leyden jar electrical capacitor is invented in Holland	**1739** Scottish philosopher David Hume publishes his *Treatise on Human Nature*	
1748 British fleet captures Port Louis in St Domingue, from the French	**1747** German scientist Andreas Marggraf invents a process for extracting sugar from sugar beet	**1746** French philosopher Denis Diderot publishes his *Philosophical Thoughts*	
1750 By the Treaty of Madrid, Spain recognizes Portuguese claims in southern and western Brazil		**1755** British writer Samuel Johnson publishes his *Dictionary*	**1750**
1754 At the Albany Congress, Benjamin Franklin proposes limited union of the British colonies against the French	**1759** British clockmaker John Harrison constructs the first marine chronometer	**1758** French philosopher Claude Helvetius publishes his atheistic book *Essays on the Mind,* which is condemned and burned	
1763 End of the Seven Years War; Britain gains Canada, Tobago, and Grenada from France and Florida from Spain, France cedes Louisiana to Spain	**1764** British engineer James Hargreaves invents the Spinning Jenny	**1762** French philosopher Jean Jacques Rousseau publishes his *Le Contrat Social*	
1765 Stamp Act places a tax on books and documents in British North America; rioting breaks out in Boston and other cities	**1765** Scottish engineer James Watt improves the steam engine by adding a separate condenser	**1766** German poet Heinrich Gerstenberg publishes the first of his *Letters on the Curiosities of Literature*	
	1769 British engineer Richard Arkwright invents a water-powered spinning frame		

1770–1849

	ASIA AND AUSTRALASIA	AFRICA	EUROPE
1770	**1770** Captain James Cook lands on Australia and charts eastern coastline ◑*202* **1770** British establish trading post at Basra, southern Iraq **1773** British merchants obtain monopoly over opium production in Bengal **1774** Warren Hastings is appointed governor of India **1775** First Maratha War breaks out in India **1782** Thai king Rama I expels Burmese and establishes Bangkok **1788** First British colony founded at Port Jackson, Australia ◑*202*	**1779** War breaks out when Xhosa try to prevent eastward expansion by Afrikaners **1787** British establish colony of Sierra Leone	**1772** Power of monarchy is restored in Sweden by Gustav III **1774** Treaty of Kuchuk Kainarji ends Russo-Turkish war ◑*178–79* **1778** France declares war on Britain, in support of Americans **1779** Spain declares war on Britain, in support of Americans, and besieges Gibraltar **1781** Joseph II of Austria abolishes serfdom **1783** Russia annexes Crimea ◑*178–79* **1789** French Revolution breaks out ◑*166*
1790	**1793** First free settlers arrive in Australia **1795** British naval forces capture Dutch possessions in Indonesian archipelago **1796** British claim Ceylon as a crown colony following expulsion of Dutch **1802** Vietnamese emperor Nguyen Anh reunites Annam under his control **1802** Treaty of Bassein gives British control over central India **1803** Second Maratha War breaks out **1809** Treaty of Amritsar establishes boundary of Sikh kingdom	**1793** Dutch authorities force Afrikaners to concede territory to Xhosa **1795** British capture Cape Colony from Dutch **1798–1801** Napoleon attacks Egypt and defeats Mamluks but French are repelled by Anglo-Ottoman forces ◑*167* **1804** Fulani *jihad* in West Africa causes break-up of some African kingdoms and establishes Sokoto caliphate ◑*204* **1805–11** Muhammad Ali establishes control of Egypt **1806** British retake control of Cape Colony from Dutch ◑*204* **1807** Slave trade is formally abolished throughout British Empire	**1792** Wars break out on France's borders; French Republic founded ◑*166* **1793** Louis XVI of France is executed **1796** France launches successful offensive against Austria in Italy ◑*167* **1799** Napoleon becomes First Consul of France **1801** Work begins on Napoleonic Civil Code **1804** Napoleon becomes Emperor of France **1805** British defeat French at Battle of Trafalgar; French defeat Austrian and Russian armies ◑*166* **1806** Napoleon dissolves Holy Roman Empire and replaces it with Confederation of the Rhine **1808** Spanish revolt leads to Peninsular War ◑*166* **1809** Fifth Coalition formed against France ◑*167*
1810	**1811** British capture Dutch Java and Sumatra from Napoleonic France **1813** British government abolishes the English East India Company's monopoly on trade with India **1816** British return territories to Dutch **1818** British finally defeat Maratha Confederacy, India ◑*194* **1819** Sikhs conquer Kashmir **1819** Stamford Raffles founds British colony of Singapore **1821–38** War breaks out between Muslim reformers and Dutch in Sumatra **1824–26** First Anglo-Burmese War breaks out **1825–30** Javanese revolt unsuccessfully against Dutch	**1815** France abolishes slave trade **1815** Afrikaner revolt is suppressed by British in southern Africa **1819** Zulu establish control in Natal causing *mfecane* wars of 1820s and 1830s and migration of Kololo, Ndebele, Swazi and others ◑*204* **1820–22** Egypt invades the Sudan **1822** Settlement of freed slaves is established in Liberia, West Africa **1824–27** First Asante War breaks out, leading to defeat of Asante in Gold Coast ◑*204*	**1812** Napoleon invades Russia and retreats ◑*167* **1813** French army driven out of Spain ◑*166–67* **1814** Napoleon exiled to Elba; Louis XVIII ascends French throne **1815** Napoleon escapes but is defeated at Battle of Waterloo; Congress of Vienna decides on European boundaries; German Confederation is formed ◑*172* **1815** Norway and Sweden are united under Swedish crown **1819** British troops massacre protestors in Peterloo, Manchester ◑*172* **1820** Revolt in Spain establishes liberal constitution; revolts occur in Naples, Piedmont and Portugal ◑*172* **1821** Greek War of Independence breaks out ◑*172* **1825** Decembrist revolution in Russia attempts to establish constitutional monarchy ◑*172* **1828–29** Russo-Turkish war ends with Treaty of Adrianople ◑*178*
1830	**1832** Muhammad Ali gains control of Syria from Ottomans **1839–42** British invade Afghanistan; Afghan revolt forces their withdrawal **1840–42** British defeat Chinese in First Opium War, opening treaty ports to British traders ◑*198–99* **1840** Treaty of Waitangi with Maori chiefs establishes British sovereignty over New Zealand **1844–48** Maoris revolt against British in New Zealand **1845–49** British annex Punjab, India in Anglo-Sikh war ◑*194*	**1834** France annexes Algeria despite resistance led by Abd al-Qadir ◑*204* **1834** Slavery is abolished throughout British Empire **1835–40** Afrikaners seek new lands in southern Africa, free from British domination ◑*204* **1838–39** Afrikaners clash with Zulus in Natal **1843–45** British establish colony in Natal; Afrikaners move north **1844** French bombard Tangier in Morocco and defeat Abd al-Qadir and Moroccan allies **1848** Afrikaners establish Orange Free State **1849** Afrikaners establish Transvaal	**1830** Charles X of France is forced to abdicate in favour of Louis Philippe **1830** Belgium gains independence ◑*172* **1830–31** Warsaw uprising is crushed by Russians ◑*172* **1832** British Great Reform Bill extends franchise **1832** Greece gains independence **1834** German Customs Union (*Zollverein*) is formed ◑*177* **1841** The Straits Convention closes the Dardanelles to non-Ottoman warships **1848** Louis Philippe of France is forced to abdicate; revolts break out in Austrian territories ◑*173* **1849** Hungarian revolt led by Louis Kossuth is crushed by Austrians and Russians ◑*175*

THE AMERICAS	SCIENCE AND TECHNOLOGY	ARTS AND HUMANITIES	
		1770 Kangra style of painting is established in northern India	**1770**
1773 Tea Act is passed by British government; "Boston Tea Party" protest is held	**1774** British chemist Joseph Priestley discovers oxygen	**1772** The final volume of Denis Diderot's *Encyclopedia* is published, despite attempts by French government and church to censor it	
1774 Quebec Act extends colony of Quebec **◯***164*			
1775 War breaks out between British and Americans **◯***164*	**1779** A cast-iron bridge is completed at Coalbrookdale in Britain	**1774** Johann Goethe's novel *The Sorrows of Young Werther* is published	
1776 American Declaration of Independence is signed	**1781** British scientist Henry Cavendish proves water to be a compound of hydrogen and oxygen	**1776** Scottish economist Adam Smith's work on political economy, *The Wealth of Nations*, is published	
1777 Americans win victory over British at Battle of Saratoga; France declares support for Americans **◯***165*	**1782** Double-action steam engine is invented in Britain by James Watt	**1776** The writings of British political propagandist Thomas Paine lend support to American colonists	
1779 Spain joins war on side of Americans	**1783** Montgolfier brothers invent first passenger-carrying hot-air balloon	**1781** French philosopher Jean-Jacques Rousseau's *Confessions* are published	
1780 Netherlands joins war on side of Americans	**1785** Edmund Cartwright invents steam-powered loom in Britain	**1786** German composer Wolfgang Amadeus Mozart's *The Marriage of Figaro* is first performed	
1780 Tupac Amaru leads unsuccessful rebellion against Spanish in Peru	**1789** Radioactive element uranium is discovered by Martin Klaproth		
1783 Under Treaty of Paris Britain recognizes American republic and cedes Florida to Spain			
1789 American constitution is ratified			
			1790
1791 Constitutional Act divides Quebec into Upper and Lower Canada **◯***188*	**1792** Gas lighting using coal gas is invented by Scottish engineer William Murdock	**1791** Louis de Saint-Just's *The Spirit of the Revolution* is published	
1791 Slave revolt in Haiti leads to founding (1804) of first non-European republic in Americas **◯***191*		**1792** British feminist Mary Wollstonecraft's *Vindication of the Rights of Women* is published	
1794 Ohio Native Americans defeated at the Battle of the Fallen Timbers	**1796** Vaccination against smallpox is tested by British doctor Edward Jenner	**1798** The *Lyrical Ballads* of William Wordsworth and Samuel Taylor Coleridge launch English Romanticism	
	1798 Lithographic method of printing is invented by Bavarian cartographer Aloys Senefelder	**1798** Thomas Malthus's *Essay on the Principle of Population* is published	
	1799 French historians in Egypt discover the tri-lingual Rosetta Stone	**1800** Library of Congress is established in Washington DC	
1803 Louisiana Purchase doubles size of USA **◯***182*	**1802** First gas lighting is installed	**1804** Ludwig van Beethoven composes his *Eroica* symphony	
1804 Expedition of Lewis and Clark finds overland route to west coast of North America **◯***183*	**1803** Morphine is extracted from opium by Charles Derosne	**1805** Joseph Turner's *The Shipwreck* introduces Romantic elements into British painting	
	1807 First steamboat service begins operation on Hudson River, USA		
1809 Shawnee Native Americans oppose westward movement of settlers in USA			
1810 USA annexes West Florida **◯***182*	**1812** Food canning process is invented in Britain by Bryan Donkin	**1810** Francis de Goya's *The Disasters of War* record the horrors of the French invasion of Spain	**1810**
1811 Paraguay proclaims independence	**1815** Miner's safety lamp is invented by British scientist Humphry Davy	**1810** The first cantos of British poet George Byron's romantic poem *Childe Harold's Pilgrimage* are published	
1812–15 United States fights Britain in War of 1812	**1815** Scottish engineer John McAdam introduces improved method of road construction	**1813** British industrialist and social reformer Robert Owen publishes his *A New View of Society*	
1816 Argentina declares independence **◯***190*			
1818 Chile declares independence **◯***190*	**1820** Electro-magnetism is discovered by Danish physicist Hans Oersted	**1814** Kurozumi Munetada revives popular Shintoism in Japan	
1819 Spain cedes East Florida to United States **◯***190*	**1821** First fossilized dinosaur skeleton is discovered by British fossil collector Mary Anning		
1821 Mexico gains independence from Spain			
1822 Republic of Gran Colombia is established **◯***190*	**1822** The first steamship is sailed from New York to Havana by Henry Eckford	**1825** Russian poet Alexander Pushkin's romantic poem *Ruslan and Ludmilla* is published	
1822 Brazil gains independence from Portugal	**1825** Stockton–Darlington railway, Britain, opens with locomotive designed by George Stephenson		
1824 Peru gains independence from Spain **◯***190*		**1828** Noah Webster's *American Dictionary of the English Language* is published	
1824 Slavery is abolished in Central America **◯***193*			
1825 Bolivia gains independence from Peru **◯***190*			
1828 Uruguay gains independence from Brazil			
1829–30 Venezuela and Ecuador break away from Gran Colombia **◯***190*			
1830 Indian Removal Act leads to relocation of Native Americans to west of Mississippi River **◯***183*	**1830** Baltimore–Ohio railroad opens in USA **◯***186*	**1830** US religious leader Joseph Smith's *The Book of Mormon* is published	**1830**
	1836 Samuel Colt invents revolver in USA	**1830** French composer Hector Berlioz writes his *Symphonie Fantastique*	
1836 Texas gains independence from Mexico **◯***190*	**1837** John Deers invents steel plough in USA	**1833** Japanese artist Katushika Hokusai completes his series of colour prints of *Thirty-six Views of Mt Fuji*	
1837–38 Rebellion breaks out against British rule among French speakers in Canada	**1837** Samuel Morse invents single-wire telegraph		
	1839 Photographic processes invented in France by Louis Daguerre and in Britain by William Fox Talbot	**1839** American writer Edgar Allan Poe's *The Fall of the House of Usher* is published	
1840 Act of Union combines Upper and Lower Canada into Province of Canada	**1839** Steam hammer is invented by Scottish engineer James Nasmyth		
	1843 SS *Great Britain* is first propeller-driven ship to cross the Atlantic		
1844 Dominican Republic becomes independent of Haiti	**1847** Anaesthetic effect of chloroform is discovered in Britain by James Simpson and John Snow	**1846** Belgian instrument-maker Adolphe Sax patents the saxophone	
1846 Oregon Treaty marks US–Canadian northwestern border		**1848** German socialists Karl Marx and Friedrich Engels issue their *Communist Manifesto*	
1848 Mexico cedes territory to the United States following defeat in war **◯***193*	**1849** Anaesthetic effect of ether is reported by Crawford Long		
1849 Gold rush occurs in California **◯***183*			

	ASIA AND AUSTRALASIA	AFRICA	EUROPE
1850	**1851** Taiping Rebellion breaks out in eastern China ●*199*	**1852** Fulani leader Al-Hadj Umar launches a war against other West African states	**1851–52** Louis Napoleon overthrows Second French Republic and establishes empire
	1854 US naval officer Matthew Perry forces Japan to allow limited foreign trade	**1853** Independent Afrikaner republic of Transvaal is recognized by British	**1853–56** Russia fights Britain, France and Ottoman Empire in Crimean War
	1856–60 British and French defeat Chinese in Second Opium War; Treaty of Tientsin opens up further treaty ports ●*198*	**1853** Ras Kasa reunifies Ethiopia and proclaims himself emperor (Theodorus II)	
	1857 Indian Mutiny breaks out ●*194*	**1854** British agree to Afrikaners establishing Orange Free State	
	1858 India becomes part of British Empire		
	1858–60 Russians annex area around Amur River ●*180*	**1859** Construction of Suez Canal begins under Ferdinand de Lesseps	**1859** Austria cedes Lombardy to Piedmont ●*176*
1860	**1860** Maori uprisings break out against British in New Zealand	**1860** Spain completes invasion of Morocco	**1860** Garibaldi takes control of Kingdom of the Two Sicilies ●*176*
		1860 German traders establish settlement in Cameroon	**1861** Unified kingdom of Italy declared
	1864 Taiping Rebellion in China is suppressed with help of Western troops ●*199*	**1861** Zanzibar gains independence from Oman	**1861** Serfdom is abolished in Russia
		1861 Fulani conquer kingdom of Segu	**1864** Germany and Austria oust Denmark from Schleswig-Holstein
	1864 A joint French, Dutch, British and US expedition destroys Japanese coastal forts		
	1867 Togugawa Shogunate in Japan is replaced by Meiji emperor		**1866** Germany defeats Austria in Seven Weeks War
	1868–70 Russians annex Tashkent ●*180*	**1867–68** British send force to Ethiopia	**1867** Dual monarchy of Austria–Hungary is formed ●*175*
		1869 Suez Canal is opened	**1867** North German Confederation is formed under Prussian domination
1870	**1871** Prefectures replace feudal structure in Japan ●*200*	**1870** Slave trader Tippu Tip establishes himself as ruler in southern Congo region	**1870–71** France is defeated in war with Germany; Paris Commune is crushed; Third French Republic is established
	1872 Maori War in New Zealand ends with British making concessions to Maori	**1871** Journalist Henry Stanley finds David Livingstone	**1871** Germany is unified under Prussian emperor ●*177*
		1871 British annex diamond-producing area of Orange Free State	
	1874 Japan acquires Ryukyu Islands ●*200*	**1874** Asante rise up against British in Ghana ●*206*	
		1874 Charles Gordon becomes governor of the Egyptian Sudan	**1878** Russia defeats Ottoman Empire; Treaty of San Stefano imposes harsh terms on Ottomans
		1879 Zulus fight British in Natal ●*206*	**1878** Congress of Berlin revises terms of Treaty of San Stefano; some areas of Balkans gain independence ●*179*
1880		**1881** Tunisia established as French protectorate	**1881** The Ottomans cede Thessaly to Greece, but keep Macedonia ●*179*
	1883–85 Chinese are defeated in Sino-French War ●*198*	**1882–85** Mahdist uprising in Sudan results in defeat of Gordon at Khartoum ●*206*	**1882** Germany, Austria and Italy form Triple Alliance ●*216*
	1885 French establish protectorate in Indochina ●*196*	**1884–85** At Berlin Conference European powers agree on the partition of Africa	
	1886 British overthrow Konbaung monarchy in Upper Burma	**1886–90** Britain and Germany divide East Africa ●*206*	**1886** Bulgaria annexes East Roumelia and defeat Serbs; Austria intervenes to prevent the invasion of Serbia
1890	**1891** Work begins on Trans-Siberian railway (completed 1901) ●*180*	**1890** Limits of German South-West Africa agreed with Britain and Portugal ●*206*	**1890** Bismarck is dismissed by William II
	1893 New Zealand women are enfranchised ●*270*		**1890** Socialists in Europe initiate May Day celebrations
	1894–95 Sino-Japanese War ends in Chinese defeat ●*198, 200*	**1896** Italians defeated by Ethiopians ●*206*	**1891** Russia and France sign a defensive entente
			1893 Corinth Canal opens in Greece
	1898 USA gain Philippines as a result of Spanish-American War ●*196*	**1896** Asante rise up against British in Gold Coast colony ●*206*	**1894** France makes an alliance with Russia ●*216*
			1895 Kiel Canal opens linking the North and Baltic seas
	1899 Boxer Rebellion breaks out against foreigners in China	**1898** British, under General Kitchener, defeat Mahdists	**1896** Olympic Games are revived in Greece
			1896 Cretans rebel against Ottoman rule
	1899 Filipinos rebel against United States	**1899** South African (Boer) War breaks out ●*206*	**1897** Greeks declare war on Ottomans and are defeated
1900	**1901** Commonwealth of Australia is formed and becomes a British dominion	**1901** Kitchener combats Afrikaner guerrilla tactics with concentration camps	**1903** Russian Social Democratic Party splits into Mensheviks and Bolsheviks
	1902 Australian women enfranchised; Immigration Restriction Act passed	**1902** Peace of Vereeniging ends South African War	**1904** Anglo-French entente settles differences over Morocco, Egypt, Suez Canal and Madagascar
	1904–5 Russo-Japanese War ends in victory for Japan; Protectorate Treaty grants Japan control of Korea ●*200*	**1904–8** Insurrection of Herero and Bambata breaks out in southern Africa ●*206*	**1905** Revolution in Russia leads to tsar establishing national assembly (*duma*) ●*181*
			1905 Norway separates from Sweden
	1907 New Zealand is granted dominion status by Britain	**1906** Crisis between Germany and France over control of Morocco is resolved by Algeciras Act	**1908** Austria annexes Bosnia-Herzegovina in the face of Serbian objections ●*175*

THE AMERICAS	SCIENCE AND TECHNOLOGY	ARTS AND HUMANITIES	
1850 Compromise forbids slavery in California but allows territories to take their own decision	**1851** Great Exhibition opens in Crystal Palace, London	**1851** US writer Walt Whitman's novel *Moby Dick* is published	**1850**
		1855 Henry Wadsworth Longfellow's poem *The Song of Hiawatha* is published	
1854 Kansas–Nebraska Act opens territories up to slavery, leading to fighting between pro- and anti-slavery factions in Kansas; Republican Party formed to oppose slavery	**1854** Link between cholera and contaminated drinking water is discovered in London by John Snow	**1856** French novelist Gustave Flaubert's *Madame Bovary* is published	
		1857 French artist Jean-Francois Millet paints his *The Gleaners*	
1857 Dred Scott decision reinforces the rights of US slave-owners	**1859** British naturalist Charles Darwin publishes book outlining theory of evolution	**1857** French poet Charles Baudelaire's *The Flowers of Evil* is published	
1858–63 Civil war in Mexico; Maximillian declared emperor	**1859** First successful oil rig is built in Pennsylvania, USA	**1859** British philosopher John Stuart Mill publishes his essay *On Liberty*	
1860–61 Southern states secede from Union; American Civil War breaks out (1861) **◐**185	**1861** First ironclad warship, HMS Warrior, is launched in Britain	**1862** Russian author and dramatist Ivan Turgenev's novel *Fathers and Sons* is published	**1860**
1863 Abraham Lincoln signs Emancipation Proclamation	**1862** Machine-gun is patented in USA	**1862** British artist Edward Burne-Jones paints his *King Cophatua and the Beggar Maid*	
1864 Cheyenne Native Americans defeated by US troops	**1863** Antiseptic surgery is introduced in Britain by Joseph Lister		
1865 Confederates surrender; President Lincoln is assassinated	**1863** First underground steam railway opens in London	**1862** French author and poet Victor Hugo's novel *Les Miserables* is published	
1865–74 Paraguay loses territory to Brazil and Argentina as a result of unsuccessful war **◐**192	**1866** First transatlantic cable is successfully laid **◐**208	**1867** German philosopher Karl Marx publishes first volume of *Das Kapital*	
1867 Canada is granted dominion status by British	**1867** Dynamite is invented by Swedish chemist Alfred Nobel	**1869** Russian novelist Leo Tolstoy publishes his *War and Peace*	
1867 Russia sells Alaska to United States **◐**180			
1868–78 Cubans fight war against Spanish			
1870 Canadian government purchases Rupert's Land; Manitoba and British Columbia join Canada (1871) **◐**189		**1871** US artist James Whistler paints his *Arrangement in Gray and Black – the Artist's Mother*	**1870**
1871 US Indian Appropriations Bill abolishes the collective right of Native American peoples		**1872** German philosopher Friedrich Nietzsche's *The Death of Tragedy* is published	
	1876 First ship with refrigerated hold carries perishable food across Atlantic	**1873** First exhibition of Impressionist paintings is held in Paris	
	1876 Telephone is invented by US scientist Alexander Graham Bell	**1876** Wagner's *Ring Cycle* receives its first performance, at Bayreuth	
	1877 Phonograph is invented in USA by Thomas Edison		
1879–83 Peru and Bolivia fight Chile in War of the Pacific **◐**192	**1878** First electric railway is demonstrated in Germany by Werner von Siemens		
1882 Chinese Exclusion Act prohibits Chinese immigration to the United States	**1881** Cuban doctor Carlos Finlay discovers that mosquitoes carry yellow fever	**1880** French artist Auguste Rodin sculpts his statue *The Thinker*	**1880**
1883 The Northern Pacific Railroad is completed **◐**186	**1884** Linotype machine is invented by US printer Ottmar Mergenthaler	**1881** US writer Henry James's novel *The Portrait of a Lady* is published	
1885 Northwest Rebellion is suppressed by Canadian troops; leader Louis Riel is executed	**1886** Four-wheeled petrol-powered automobile is built by German engineer Gottlieb Daimler	**1882** Indian author Bankim Chandra Chatterji's novel *Anandamath* is published	
1886 Slavery is abolished in Cuba	**1886** Flush toilet is invented by British engineer Thomas Crapper	**1883** Spanish architect Antonio Gaudi begins work on the Church of the Holy Family in Barcelona	
1887 Canadian Pacific Railway is completed **◐**188	**1887** Monotype machine invented by US printer Tolbert Lanston		
1888 Slavery is abolished in Brazil		**1888** Dutch artist Vincent Van Gogh paints his *Sunflowers*	
1890 US troops massacre Native Americans at the Battle of Wounded Knee **◐**183	**1890** First electric underground rail system opens in London	**1890** Bengali writer Rabindranath Tagore publishes his collection of poems *Manasai*	**1890**
1891 United States of Brazil is established	**1891** Kinetoscope movie camera/projector is patented by US scientist Thomas Edison	**1893** Norwegian artist Edward Munch paints his *The Scream*	
	1892 Vaccine against typhoid fever is discovered by Almoth Wright	**1894** French composer Claude Debussy writes *Prelude à l'après midi d'un faune*	
1895 Nicaragua, El Salvador and Honduras form the Greater Republic of Central America	**1893** Innoculation against cholera is developed by Russian bacteriologist Waldemar Haffkine	**1895** Austrian composer Gustav Mahler writes his Symphony no. 2	
1895–98 Cubans revolt against Spanish; United States backs Cubans and defeats Spain	**1896** Diagnostic X-ray photographs are first taken by US physicist Michael Pupin	**1896** Nicaraguan poet Ruben Dario's *Prosas Profanas* is published	
1896 Gold is disovered in the Yukon in Canada	**1899** Aspirin goes on sale in Europe	**1897** Russian dramatist Anton Chekhov writes *Uncle Vanya*	
1901 Cuba becomes a protectorate of United States	**1901** Marconi transmits radio signals across Atlantic	**1900** Sigmund Freud's *Interpretation of Dreams* introduces psychoanalysis	**1900**
1901 US President McKinley is assassinated and succeeded by Theodore Roosevelt; United States is granted right to Panama Canal Zone	**1903** Wright brothers make first sustained flight of powered aircraft	**1905** Expressionist movement is founded by group of German painters	
1903 Bolivia cedes territory to Brazil in exchange for access by rail and water to Atlantic **◐**192	**1905** German mathematician Albert Einstein publishes theory of relativity	**1907** Exhibition of Cubist paintings opens in Paris	
1907 War breaks out between Nicaragua and Honduras	**1908** Mass production of cars begins with Model T Ford in USA	**1908** Schoenberg develops Expressionism in music and moves towards atonalism	
1909–11 Civil war breaks out in Honduras	**1909** Blériot flies across English Channel	**1909** Marinetti publishes pamphlet on Futurism	

1910–49

	ASIA AND AUSTRALASIA		AFRICA AND MIDDLE EAST		EUROPE
1910	**1910** Japan formally annexes Korea ❍*200*		**1910** Union of South Africa is granted dominion status by Britain		**1910** Republic is declared in Portugal
	1911 Chinese Republic is proclaimed		**1911** Agadir Crisis is resolved by France granting Germany territory in Congo		**1912** First Balkan War breaks out between Ottoman Empire and Bulgaria, Serbia, Montenegro and Greece ❍*178*
	1911 Russians invade and occupy Iran				**1913** Second Balkan War breaks out ❍*178*; Serbs invade Albania
	1914 German territory in China is occupied by Japan, Australia and New Zealand	FIRST WORLD WAR	**1914** British protectorate is proclaimed in Egypt	FIRST WORLD WAR	**1914** First World War breaks out when Germany invades Belgium and northern France ❍*218*
	1916 Indian National Congress Party and All-India Muslim League make pact at Lucknow calling for Indian independence		**1914–18** Entente Powers fight Germans in Africa		**1915** Western Front is established; Italy declares war on Central Powers ❍*218*
	1916 Russia invades Armenia		**1915–17** British Empire and Ottoman troops engage in Mesopotamia ❍*218*		**1916** Easter Rebellion of Irish nationalists against British breaks out
			1916 Arabs in Hejaz rise up against Ottoman Empire		**1917** Bolshevik Revolution breaks out in Russia
					1918 Russia and Central Powers sign Treaty of Brest-Litovsk; armistice is signed, ending First World War
	1919 British troops kill 400 Indian demonstrators at Amritsar		**1919–21** Nationalist uprisings occur in Egypt		**1919** Versailles Peace Conference is convened; White Russian army attacks communist Russia ❍*222*
1920	**1920** Mohandas Gandhi institutes policy of non-co-operation with British in India		**1921** Moroccan nationalists under Abd-el Krim defeat Spanish		**1920** League of Nations is established in Geneva
	1921 First Indian Parliament meets		**1922** Referendum in Southern Rhodesia rejects joining Union of South Africa		**1920–22** Greek offensive is mounted against Turkish nationalists in Anatolia ❍*178*
	1924 First Kuomintang (Nationalist Party) Congress is held in China		**1922** League of Nations mandates come into force in Middle East ❍*220*		**1920–21** Poland and Russia are at war ❍*222*
					1922 Soviet Union (USSR) is founded
	1924 Mongol People's Republic is declared as satellite of Soviet Union		**1922** Egypt achieves nominal independence		**1922** Irish Free State is established within Commonwealth
					1922 Benito Mussolini forms fascist government in Italy
					1923 Turkey becomes a republic
	1926 Australia and New Zealand are given equality with Britain in Commonwealth		**1926** Moroccan nationalists are defeated by alliance of French and Spanish		**1926** General Strike takes place in Britain
	1927 Chinese Kuomintang government is established at Hankou		**1927** Slavery is abolished in Sierra Leone		**1927** Germany's economy collapses
	1928 Chiang Kai-shek is elected President of China				**1927** Joseph Stalin assumes control of Soviet Union
					1928 Universal female suffrage is introduced in Britain
	1929 Collapse of US stock market results in Great Depression ❍*228*		**1929** Term "apartheid" is first used to describe racial segregation in South Africa		**1929** Crash of US stock market results in Great Depression
1930	**1930** Gandhi leads march against British salt monopoly in India				
	1931 Japanese invade Manchuria ❍*234*		**1931** First trans-African railway is completed		**1931** Spain becomes a republic
	1932 Indian National Congress Party declared illegal; Gandhi arrested				**1933** Adolf Hitler becomes German Chancellor and assumes dictatorial powers
	1935 Commonwealth of Philippines is inaugurated, supervised by United States		**1935** Italy, under Mussolini, invades Ethiopia		**1935** Saarland is returned to Germany following a plebiscite ❍*230*
			1936 Italians capture Addis Ababa and annex Ethiopia ❍*230*		**1936** Germany occupies the Rhineland ❍*230*
	1937 Japanese invade China, seizing Beijing, Nanjing and Hangzhou ❍*234*		**1937** French suppress nationalist uprising in Morocco		**1936–39** Spanish fight civil war ❍*231*
	1938 Japanese seize Canton (Guangzhou)				**1938** German troops seize Austria as part of Third Reich; Munich Conference grants Germany right to Sudetenland in Czechoslovakia ❍*230*
		SECOND WORLD WAR		SECOND WORLD WAR	**1939** Germany and Soviet Union sign non-aggression pact; Germany invades Czechoslovakia and Poland; Britain and France declare war on Germany
1940	**1940** French Vichy government allows Japanese troops into Indochina		**1940** Italian forces fail to capture Allied positions in North Africa		**1940** Germany invades most of western Europe; Battle of Britain and the Blitz ❍*232*
	1941 Japanese attack Malaya and Pearl Harbor US naval base		**1941** Italian forces are expelled from East Africa; Germans counter-attack in North Africa ❍*232*		**1941** Germany invades Russia ❍*232*
	1942 Japanese invade Southeast Asia and Pacific; Allies counter-attack ❍*234*		**1942** Battle of El Alamein leads to Axis retreat in North Africa ❍*232*		**1943** Russians begin counteroffensive; Allies land in Italy; Italy surrenders and declares war on Germany ❍*232*
			1943 Allied troops take control of North Africa		**1944** Allies land troops in Normandy ❍*232*
	1945 Allies drop atomic bombs on Hiroshima and Nagasaki; Japan surrenders; civil war continues in China ❍*225*				**1945** Hitler commits suicide and Germany surrenders
	1946 Fighting breaks out in Vietnam		**1947** Nigeria achieves a form of self-government		**1945** Labour Party wins landslide in British election
					1946 First United Nations General Assembly is held, in London
	1947 Independence of India is proclaimed; the partitioning of India takes effect		**1947** UN announces plan for partition of Palestine ❍*261*		**1946** Italy becomes a republic
	1948 "Mahatma" Gandhi is assassinated in India		**1948** War breaks out in Palestine between Arabs and Jews		**1946** British government lays foundation of welfare state
					1948 Marshall Aid is granted to European countries ❍*238*
					1948–49 Soviet Union blocks road and rail links to Berlin; Western powers organize airlifts
	1949 Communists take control of China; Nationalists withdraw to Taiwan ❍*225*		**1949** Apartheid programme is introduced in South Africa		**1949** COMECON is formed by communist countries ❍*236*

THE AMERICAS	SCIENCE AND TECHNOLOGY	ARTS AND HUMANITIES	

1910 Civil war breaks out in Mexico

1911 Francisco Madero is elected president in Mexico

1913 Victoriano Huerta seizes control of Mexico City and is opposed by Emiliano Zapata, Pancho Villa and Venustiano Carranza

1914 Carranza seizes power in Mexico; US troops occupy Veracruz ◗226; peace treaty is signed by USA and Mexico

1914 Panama Canal becomes fully operational ◗226

1915 SS *Lusitania* is sunk by German U-boat

1916 United States sends expedition to Mexico ◗227

1917 United States declares war on Central Powers; sends troops to fight on Western Front ◗218

1918 Woodrow Wilson proposes his "Fourteen Points" for world peace to Congress

1911 Nuclear theory is announced by Rutherford

1911 Polish scientist Casimir Funk discovers vitamins

1913 Geiger counter is used to measure radioactivity

1914 Stainless steel is made in Germany

1914 Henry Ford introduces conveyor-belt production lines in US car factory

1916 General theory of relativity is published by Einstein

1917 Tanks are used successfully for first time, at Battle of Cambrai, by the British

1918 Alcock and Brown make first transatlantic flight

FIRST WORLD WAR

1912 Bengali writer Rabindranath Tagore's *Gitanjali, Song Offerings* is published

1912 W C Handy publishes *Memphis Blues*

1913 Russian composer Igor Stravinsky's *Rite of Spring* causes a riot at first performance

1917 Siegfried Sassoon's volume of poems, *The Old Huntsman* is published

1917 Term "surrealist" is used for the first time

1917 The Original Dixieland Jazz Band makes first phonograph jazz record

1917 Pulitzer Prize is established

1919 Bauhaus school of architecture is founded by Walter Gropius

FIRST WORLD WAR

1910

1920 In the United States Prohibition comes into force, Nineteenth Amendment gives women the vote ◗270, Senate rejects involvement in League of Nations

1921 International Conference on Naval Limitation is convened in Washington

1924 Plan for the scheduling of Germany's reparations payments to France, proposed by US Vice-President Dawes, is accepted by both sides

1926 Canada is afforded equal status to Britain within the Commonwealth at Imperial Conference

1928 War breaks out between Paraguay and Bolivia

1929 US Senate ratifies Kellogg–Briand Pact

1929 Peru settles Tacna–Arica dispute with Chile

1929 Wall Street Crash results in Great Depression

1920 First regular radio broadcast is transmitted in United States

1922 Insulin is used in treatment of diabetes

1924 Commercial production of quick-frozen foods is developed by Birdseye

1925 Electric recording and reproduction techniques are developed

1926 Liquid-fuel rockets are tested in United States

1927 First public broadcast of television programme is transmitted in Britain

1928 Zeppelin airship flies around world

1922 American author T S Eliot's *Wasteland* and Irish author James Joyce's *Ulysses* are published

1924 German composer Schoenberg's first "twelve-tone" or "serial" music is performed

1925 International Style of architecture is further developed by Le Corbusier

1925 Soviet director Sergei Eisenstein's film *Battleship Potemkin* is released

1927 Fritz Lang's film *Metropolis* presents a vision of the future

1927 Sound films are popularized by *The Jazz Singer*, starring Al Jolson

1928 Walt Disney makes first "Mickey Mouse" film in colour

1920

1930 Military coups break out in Argentina, Brazil and Peru

1932 Franklin D Roosevelt is elected President of the United States

1932 Chaco War breaks out between Bolivia and Paraguay

1933 Fulgencio Batista leads a military coup in Cuba

1933 Roosevelt embarks on New Deal; Public Works Administration is created; dollar is devalued

1934 Drought and bad farming techniques combine to form the "Dust Bowl" in the US Midwest

1935 Chaco War ends with Paraguay gaining territory from Bolivia, but Bolivia gaining access via Paraguay River to the Atlantic

1938 Mexico nationalizes US and British oil companies

1939 Battle of River Plate ends with the scuttling of the German battleship *Graf Spee*

1931 Nylon is invented by US chemist Wallace Carothers; it is not patented until 1937

1935 Fluorescent lighting is introduced by General Electric, and sodium vapour lamps are developed for street lighting

1935 Radar system is invented by British scientist Robert Watson-Watt

1936 Regular television service starts to be broadcast in Britain

1936 British mathematician Alan Turing develops mathematical theory of computing

1939 Insecticidal properties of DDT is discovered by Swiss chemist Paul Muller

1932 American writer John Steinbeck's *Grapes of Wrath is* published

1933 Anti-Nazi intellectuals and artists flee Germany after Hitler's rise to power

1935 American composer George Gershwin's *Porgy and Bess* opens

1936 BBC starts regular television service

1937 Frank Lloyd Wright's "Fallingwater" building is built in Pennsylvania

1937 Picasso's *Guernica* depicts bombing of the Spanish city by Germans

1938 Benny Goodman's swing band dominates Broadway

1930

1941 United States enters war against Germany and Japan following Japanese attack on Pearl Harbor ◗234

1942 Japanese forces invade western Aleutian Islands; people of Japanese origin in the United States are interned

1943 The United States retakes western Aleutians

1944 United Nations Monetary and Financial Conference at Bretton Woods establishes International Monetary Fund and World Bank

1946 Juan Perón is elected President of Argentina

1947 Rio Treaty for mutual defence is signed by 19 republics

1947 Truman Doctrine is announced, pledging support for European countries opposing spread of communism

1949 Newfoundland joins Canada ◗188

1949 North Atlantic Treaty is signed in Washington, establishing NATO

1940 Plutonium is discovered by US chemist Glenn Seaborg

1941 First flight of jet-powered plane, with engine designed by British engineer Frank Whittle, takes place

1943 Penicillin is first used as medicine

1944 V-2 rockets are used by Germany against UK

1945 Atomic bombs are dropped on Hiroshima and Nagasaki ◗234

1948 The LP record is invented by Hungarian-born American Peter Goldmark

1949 The Soviet Union tests its first atomic bomb

SECOND WORLD WAR

1941 Orson Welles' *Citizen Kane* is first shown

1943 The musical *Oklahoma* by Richard Rodgers and Oscar Hammerstein is first performed

1944 US musician Glenn Miller is killed in an air crash

1945 British writer George Orwell publishes his novel *Animal Farm*

1947 Italian writer Primo Levi publishes his description of a Nazi concentration camp in *If This Is a Man*

1949 French feminist Simone de Beauvoir publishes *The Second Sex*

SECOND WORLD WAR

1940

1950–74

	ASIA AND AUSTRALASIA	AFRICA AND MIDDLE EAST	EUROPE
1950	**1950** Soviet Union and China sign a co-operation treaty; China invades Tibet	**1950** Jordan annexes the West Bank and East Jerusalem **⊃261**	**1951** Economic aid provided to Europe under Marshall Plan comes to end
	1950–53 North Korean and Chinese forces fight South Korean and UN forces in Korean War **⊃244**	**1951** Libya gains full independence	**1951** West Germany is admitted to the Council of Europe
	1951 Iran nationalizes its oil industry	**1952** Mau Mau start terror campaign against British settlers in Kenya **⊃256**	**1952** European Coal and Steel Community is formed
		1952 Eritrea becomes a part of Ethiopia	**1952** Greece and Turkey join NATO
	1953 Laos gains full independence from France	**1952** King Farouk is overthrown in Egypt	**1953** Tito is appointed President of the Federal People's Republic of Yugoslavia
	1954 French forces in Vietnam surrender after defeat by the Vietminh; the Geneva Peace Treaty splits Vietnam at the 17th parallel **⊃250**	**1953** Britain establishes the Federation of Rhodesia (Northern and Southern) and Nyasaland	**1953** Nikita Khrushchev becomes First Secretary in the Soviet Union after the death of Stalin
		1954 Nigeria becomes self-governing	**1954** Greek nationalist EOKA movement carries out attacks on British troops in Cyprus
	1954 South East Asian Treaty Organization (SEATO) is established	**1954** National Liberation Front (FLN) organizes an anti-French revolt in Algeria	**1954** Italy and Yugoslavia reach agreement over control of Trieste
1955	**1956** Islamic republic of Pakistan is declared	**1956** Sudan and Morocco gain independence	**1955** Allied occupation troops withdraw from West Germany, which joins NATO
		1956 President Nasser of Egypt nationalizes Suez Canal; Israel invades Egypt; Britain and France send troops but withdraw under international pressure	**1955** Communist countries of Eastern Europe form the Warsaw Pact with the Soviet Union **⊃236**
	1957 Malayan Federation gains independence **⊃250**		**1956** British deport Archbishop Makarios from Cyprus
		1957 Israel evacuates Gaza Strip	**1956** Hungarian prime minister, Nagy, leads an anti-communist uprising that is quickly crushed by Soviet troops **⊃236**
	1958 Chinese bombard the Quemoy islands off Taiwan **⊃245**	**1957** Gold Coast and Togoland are joined to form the independent state of Ghana	
	1959 Singapore becomes an independent state	**1958** Egypt and Syria form the United Arab Republic	**1957** Saar region is returned to West Germany
	1959 Antarctica is safeguarded by international treaty	**1958** The FLN rebels declare a provisional government in Algeria	**1957** Treaty of Rome establishes the European Economic Community (EEC) or Common Market **⊃238**
	1959 Chinese troops suppress a rising in Tibet; the Dalai Lama flees		**1959** Belgium, Holland, and Luxembourg become a single economic unit – Benelux
1960	**1960** Ideological and political differences split the Soviet-Chinese alliance	**1960** South African troops kill demonstrators at Sharpeville	**1960** Soviets shoot down a US U2 spyplane and capture pilot Gary Powers
	1960 Achmad Sukarno assumes dictatorial powers in Indonesia	**1960** Eleven West African states gain independence **⊃256**	**1960** Non-EEC countries form the European Free Trade Association (EFTA) **⊃238**
	1961 India conquers the Portuguese colony of Goa **⊃249**	**1960–63** Civil war breaks outs in the Belgian Congo following independence	**1960** United States, Canada and Western European nations form the Organization for Economic Co-operation and Development (OECD)
	1962 China invades northern India, then withdraws to the disputed border **⊃249**	**1961** South Africa becomes a republic and leaves the British Commonwealth **⊃247**	**1961** Soviet authorities build a wall across the divided city of Berlin
	1963 Federation of Malaysia is formed from Malaya, Northern Borneo, Sarawak and Singapore **⊃250**	**1962** Algeria becomes independent; the French nationalist OAS organizes a revolt	**1963** Britain, the USSR, and the USA sign a nuclear-test treaty banning all but underground explosions
	1964 Indonesia invades Malaysia	**1963** Organization of African Unity (OAU) is formed in Addis Ababa	**1964** UN troops are sent to Cyprus in response to fighting between Greeks and Turks
	1964 China explodes an atom bomb	**1964** Tanganyika and Zanzibar unite to form Tanzania; Zambia, Kenya and Gambia become independent **⊃256**	**1964** Khrushchev is deposed in the Soviet Union by Leonid Brezhnev and Aleksei Kosygin
1965	**1965** United States begins a bombing campaign against North Vietnam **⊃250**	**1965** White settlers declare Southern Rhodesia independent (UDI)	
	1965 Singapore becomes an independent republic **⊃250**	**1967–70** Biafra tries to secede from Nigeria; a civil war ensues during which famine develops in Biafra **⊃256**	**1966** France withdraws from NATO
	1965 War breaks out between India and Pakistan **⊃249**		**1967** A coup by Greek colonels takes power in Athens
	1966 Mao Zedong launches the Cultural Revolution in China	**1967** Israel defeats its Arab neighbours in Six Day War **⊃261**	**1967** Nicolae Ceausescu becomes head of state in Romania
		1967 People's Democratic Republic of Yemen is established **⊃260**	**1967** Forty-six nations sign the General Agreement on Tariffs and Trade (GATT) in Geneva
	1966 Raden Suharto seizes power in Indonesia and ends conflict with Malaysia		**1967** EEC becomes the European Community (EC)
	1968 North Vietnam launches the Tet Offensive; US bombing campaign halted; US troops massacre villagers at Mai Lai	**1969** Yasser Arafat is elected chairman of the Palestine Liberation Organization	**1968** Students and workers build barricades in Paris; student leader Rudi Dutschke is shot in West Germany
		1969 Colonel Muammar al-Qadafi overthrows King Idris in Libya	**1968** Czech politician Alexander Dubcek introduces reforms; a Soviet invasion ends the "Prague Spring"
	1969 Soviet and Chinese troops clash along their border **⊃249**	**1969** Left-wing group seizes power in Sudan	**1968** Albania withdraws from the Warsaw Pact **⊃236**
			1969 Sectarian violence flares in Northern Ireland
1970	**1970** Cambodia becomes the Khmer Republic; US bombing of North Vietnam is resumed	**1970** PLO is evicted from Jordan	**1970** The Strategic Arms Limitation Treaty (SALT) talks begin in Helsinki
		1971 Idi Amin seizes power in Uganda	
	1971 East Pakistan declares independence; West Pakistan declares war; India intervenes in the fighting; the republic of Bangladesh is established	**1972** Amin expels Asians from Uganda	**1972** British impose direct rule in Northern Ireland
			1972 Palestinian terrorists kidnap and kill Israeli athletes at the Munich Olympics
	1971 China is admitted to the UN; nationalist Taiwan is expelled	**1973** Arab nations attack Israel; Egyptian forces invade across the Suez Canal but are defeated **⊃261**	**1973** Paris Peace Agreement ends US involvement in Vietnam
	1973 Last US combat troops leave Vietnam	**1973** Arab states cut oil production and cause a worldwide energy crisis	**1974** Left-wingers seize power in Portugal
	1973 Afghanistan becomes a republic after a coup		**1974** Greek nationalists stage a coup in Cyprus and declare union with Greece; Turkish troops invade and conquer half the island; the Greek junta abdicates power
		1974 Ethiopian emperor, Haile Selassie, is deposed and a republic is declared	**1974** Illegal Irish Republican Army (IRA) intensifies its bombing campaign against British targets

THE AMERICAS	SCIENCE AND TECHNOLOGY	ARTS AND HUMANITIES	
		1950 Chilean poet Pablo Neruda's epic *General Song* is published	**1950**
1951 The 22nd amendment to the US constitution limits presidents to two terms of office	**1951** US engineers build UNIVAC I, the first commercial computer	**1951** Japanese director Akira Kurosawa's film *Rashomon* is released	
1952 Dwight Eisenhower is elected US president	**1952** USA explodes hydrogen bomb	**1951** US novelist J D Salinger's *Catcher in the Rye* is published	
1952 Fulgencio Batista returns to power in Cuba through a military coup and establishes a dictatorship	**1952** US researcher Jonas Salk develops a vaccine against poliomyelitis	**1952** British artist Henry Moore creates his sculpture *King and Queen*	
	1952 British Comet aircraft makes the first jet passenger flight	**1953** British scholar Michael Ventris deciphers the Linear B script	
	1953 British and US scientists Francis Crick and James Watson discover the helical structure of DNA	**1954** Welsh author Dylan Thomas writes his verse play *Under Milk Wood*	
1954 Anti-communist crusade of Senator Joseph McCarthy reaches a climax with televised hearings of his investigation committee	**1954** Chinese and US scientists Min-Chueh Chang, Gregory Pincus and Frank Colton invent the contraceptive pill	**1954** British novelist William Golding publishes his *Lord of the Flies*	
1955 US Supreme Court rules that racial segregation in public schools must end	**1955** British engineer Christopher Cockerell invents the hovercraft	**1955** Irish dramatist and novelist Samuel Beckett writes his play *Waiting for Godot*	**1955**
1955 President Péron is forced into exile from Argentina	**1956** Heart pacemaker is invented	**1955** Indian director Satyajit Ray's film *Pather Panchali* is released	
1955 Alfredo Stroessner seizes power in Paraguay	**1956** FORTRAN computer programming language is developed	**1956** US singer Elvis Presley releases his record *Heartbreak Hotel*; Frank Sinatra releases *Songs for Swinging Lovers*	
1955 US Civil Rights activist Rosa Parks sits in a whites-only bus seat in Montgomery, Alabama ➲*241*	**1957** Soviet Union launches an artificial satellite – *Sputnik I*; a dog is carried into orbit by *Sputnik II*	**1957** Australian novelist Patrick White publishes his *Voss*	
1956 Revolutionary Fidel Castro lands in Cuba	**1957** UN forms the International Atomic Energy Commission	**1958** US economist J K Galbraith's *The Affluent Society* is published	
1957 "Papa Doc" Duvalier becomes president of Haiti	**1958** Nuclear-powered submarine USS *Nautilus* passes beneath North Pole	**1958** Russian poet and author Boris Pasternak's novel *Dr Zhivago* is published	
1957 US Civil Rights Act appoints a commission to examine African-American voting rights	**1958** Stereophonic music records go on sale in the USA		
1958 US National Aeronautics and Space Administration (NASA) is established		**1959** German author Gunther Grass's novel *The Tin Drum* is published	
1959 Castro becomes prime minister of Cuba			
1959 Alaska and Hawaii become states of the USA			
1960 US civil rights activist Martin Luther King organizes a sit-in demonstration in Greesboro, North Carolina	**1960** US scientist Theodore Maiman invents the laser	**1960** Frank Lloyd Wright's Guggenheim Museum opens in New York	**1960**
1960 United States embargoes exports to Cuba and cuts Cuban sugar quotas by 95 per cent	**1961** Soviet cosmonaut Yuri Gagarin orbits the Earth	**1960** US director Alfred Hitchcock's film *Psycho* is released	
1961 Cuban exiles in the USA attempt an invasion of Cuba at the Bay of Pigs ➲*245*	**1962** US astronaut John Glenn orbits the Earth	**1961** British pop group The Beatles play at the Cavern Club in Liverpool; US singer-songwriter Bob Dylan makes his debut in New York	
1961 US civil rights activists organize "freedom rides" on segregated buses	**1962** US telecommunications satellite Telstar is launched	**1962** US economist Milton Friedman's *Capitalism and Freedom* is published	
1962 Jamaica and Trinidad and Tobago become independent ➲*247*	**1963** US astronomer Maarten Schmidt discovers quasars	**1963** US environmentalist Rachel Carson's *Silent Spring* is published	
1962 Soviet Union attempts to install nuclear missiles in Cuba; the USA imposes a naval blockade ➲*245*	**1964** US scientist Murray Gell-Mann proposes the existence of quarks	**1964** Canadian Marshall McLuhan's *Understanding Media* is published	
1963 Martin Luther King leads a civil rights march to Washington ➲*241*	**1964** Japanese railways begin running high-speed "bullet" trains		
1963 President J F Kennedy is assassinated in Dallas, Texas			
1965 Medicare and other welfare legislation is passed in the United States	**1965** Soviet cosmonaut Alexei Leonov takes a "space walk" while in orbit	**1965** US dramatist Neil Simon writes his play *The Odd Couple*	**1965**
1965 Race riots erupt in the Watts district of Los Angeles	**1966** Soviet spaceprobe *Luna 9* makes a soft landing on the Moon	**1966** Japanese author Yukio Mishima's novel *The Sailor Who Fell from Grace with the Sea* is published	
1965 Military coup leads to widespread fighting in the Dominican Republic; US marines land and are then replaced by OAS forces	**1967** South African surgeon Christiaan Barnard performs a heart transplant	**1967** US artist Andy Warhol publishes his print of Marilyn Monroe; British artist David Hockney paints *A Bigger Splash*	
1966 African-American activists Bobby Seale and Huey Newton form the Black Panthers	**1968** US astronauts orbit the Moon in the *Apollo 8* spacecraft	**1967** The Beatles release *Sergeant Pepper's Lonely Hearts Club Band*	
1967 US Court of Appeal orders the desegregation of Southern schools	**1969** Concorde supersonic passenger aircraft makes its first flight	**1967** Colombian novelist Gabriel Garcia Marquez' *One Hundred Years of Solitude* is published	
1968 UN approves a nuclear non-proliferation treaty	**1969** US spacecraft *Apollo 11* lands on the Moon; astronaut Neil Armstrong takes the first steps	**1968** US psychologist Timothy Leary's *The Politics of Ecstasy* is published	
1968 Martin Luther King is assassinated in Memphis; Senator Robert Kennedy is assassinated in Los Angeles	**1969** US engineer Douglas Engelbart invents the computer "mouse"		
1969 War breaks out between Honduras and El Salvador			
1970 US National Guard soldiers kill four protesting students at Kent State university in Ohio	**1970** Soviet space probe *Venera VII* soft-lands on Venus	**1970** Australian feminist Germaine Greer's *The Female Eunuch* is published	**1970**
1970 Salvador Allende is elected president of Chile	**1970** Boeing 747 enters service	**1970** US musician Jimi Hendrix dies of a drug overdose; The Beatles split up to pursue separate careers	
1972 Burglars are arrested in the Democratic Party election headquarters at the Watergate hotel in Washington	**1971** US engineer Ted Hoff invents the computer microprocessor ("chip")		
1973 US national security advisor Henry Kissinger is appointed secretary of state	**1971** Pocket calculator is invented	**1973** *Small is Beautiful*, by German economist Ernst Shumacher, is published	
1973 Juan Péron returns to Argentina and becomes president	**1973** US launches the Skylab space station	**1974** Terracotta army buried with the first Qin emperor is discovered in China	
1973 General Augusto Pinochet seizes power in Chile	**1974** US paleontologist Donald Johanson discovers "Lucy" (a partial skeleton of *Australopithicus afarensis*) in Ethiopia	**1974** US author Erica Jong's novel *Fear of Flying* is published; US author Robert Pirsig's *Zen and the Art of Motorcycle Maintenance* is published	
1974 US President Nixon resigns over the Watergate scandal	**1974** Bar codes are introduced in USA		

1975–99

	ASIA AND AUSTRALASIA	AFRICA AND MIDDLE EAST	EUROPE
1975	**1975** Khmer Rouge guerrillas under Pol Pot capture Phnom Penh **1975** Vietnam War ends **1975** Communists take control of Laos **1976** Chinese leader Mao Zedong and prime minister Zhou Enlai die **1977** General Zia al-Huq seizes power in Pakistan **1978–79** Vietnamese troops invade Cambodia and expel Pol Pot **1979** Revolution breaks out in Iran; US embassy staff in Tehran are taken hostage **1979** Soviet Union invades Afghanistan	**1975** Clashes between Christian Falangists and Palestinians lead to war in Lebanon **1975** South Africa establishes Transkei as a black homeland; South African security forces kill young demonstrators in Soweto **1977** South African political activist Steve Biko is murdered in police custody **1979** Tanzanian troops invade Uganda and oust Idi Amin **1979** A conference in London ends the civil war in Rhodesia between guerrillas and the white minority government **1979** Saddam Hussein becomes President of Iraq	**1975** Helsinki accords on peace and human rights mark a major step in the process of détente between NATO and the Warsaw Pact **1975** Turkish Federated State of North Cyprus is established; UN forces maintain the border with the Greek sector of the island ❍267 **1975** Franco dies; Juan Carlos becomes King of Spain **1977** Czech political reformers form Charter 77 group **1978** Italian Red Brigade terrorists kidnap and murder politician Alberto Moro **1978** Group of Seven (G7) industrialized nations meet to discuss economic policy
1980	**1981** Chinese "Gang of Four" are convicted of treason **1983** Filipino politician Benigno Aquino is assassinated on his return to Manilla **1984** Indian troops storm Sikh protestors in the Golden Temple in Amritsar; Indira Gandhi is assassinated **1984** Chinese government introduces liberal economic reforms	**1980** Iraq invades Iran ❍261 **1980** Rhodesia becomes independent as Zimbabwe; Robert Mugabe is elected prime minister **1980** Libya invades northern Chad **1981** Egyptian president Anwar Sadat is assassinated **1982** Israel returns the Sinai peninsula to Egypt ❍261 **1982** Israel invades Lebanon; PLO evacuates; an international peacekeeping force arrives in Beirut ❍266 **1984** South African president P W Botha grants limited political rights to Asians and "coloureds"	**1980** Following the death of President Tito, Yugoslavia comes under collective leadership **1980** Polish shipyard workers led by Lech Walesa form the Solidarity trade union **1981** Greece joins the EC ❍238 **1981** Nationalist civil guards attempt a coup in Spain **1981–83** Solidarity protests result in martial law being declared in Poland under General Wojciech Jaruzelski **1983** Green Party wins its first parliamentary seats in West German elections **1983** US cruise missiles are located at airbases in Britain **1983** Turkish northern Cyprus declares its independence under President Rauf Denktas
1985	**1986** Marcos flees the Philippines; Corazon Aquino becomes president **1987** Indian troops impose a ceasefire in conflict between the Tamil Tiger guerrillas and Sri Lankan government forces **1988** Azerbaijan and Armenia clash over Nagorno Karabakh ❍262 **1988** Benazir Bhutto becomes prime minister of Pakistan **1988** Soviet Union begins withdrawing troops from Afghanistan **1989** Chinese troops kill pro-democracy protestors in Tiananmen Square in Beijing	**1985** United States and the EC impose economic sanctions against South Africa **1986** US warplanes bomb Libya ❍242 **1987** Chadian forces assisted by the French Foreign Legion expel the Libyans from northern Chad **1987** Syrian troops enter Beirut to keep the peace **1988** Iran–Iraq war ends in stalemate ❍261 **1989** Cuban troops are withdrawn from Angola; a ceasefire is declared in the civil war **1989** F W de Klerk becomes president of South Africa	**1986** Spain and Portugal join the EC ❍238 **1986** An accident at the Chernobyl nuclear reactor in Ukraine releases a radioactive cloud over central and northern Europe **1986** Swedish prime minister Olaf Palme is murdered **1987** Gorbachev announces policies of *glasnost* and *perestroika* in the Soviet Union **1988** United States and Soviet Union sign Intermediate-range Nuclear Forces (INF) Treaty in Moscow **1989** Solidarity wins a majority in Polish elections **1989** Hungary opens its borders; popular protests in East Germany lead to the dismantling of the Berlin Wall; a popular revolution overthrows Ceausescu in Romania; Vaklav Havel forms a government in Czechoslovakia
1990	**1990** Benazir Bhutto is dismissed as prime minister of Pakistan **1991** Rajiv Gandhi is assassinated during an Indian election campaign **1992** Hindu extremists demolish the Ayodhya mosque, igniting widespread violence across India **1992** Afghan Islamic rebels capture Kabul and overthrow the communist government	**1990** Nelson Mandela is released from prison in South Africa **1990** Iraq invades and annexes Kuwait **1991** Coalition of USA, European and Arab nations oust Iraqi troops from Kuwait ❍261 **1992–94** US troops supervise food distribution in Mozambique ❍242 **1994** Presidents of Burundi and Rwanda are killed in air crash; ethnic violence breaks out in Rwanda ❍268 **1994** ANC wins South African elections; Mandela becomes president ❍257 **1994** Palestine National Authority takes control of the Gaza Strip and Jericho	**1990** Lithuania, Latvia, and Estonia declare independence from Soviet Union, leading to invasion by Soviet troops **1990** East and West Germany are reunited ❍264 **1991** Warsaw Pact is dissolved as a military alliance **1991** Croatia and Slovenia declare independence from Yugoslavia; the Yugoslav army invades Croatia **1991** Army officers attempts a coup in Moscow; Gorbachev resigns; the USSR ceases to exist **1992** Bosnia-Herzegovina declares independence from Yugoslavia; Serbs attack Sarajevo **1993** Czech and Slovak republics are established as separate states ❍264 **1993** EC establishes a single market **1994** Chechenia declares independence from Russia
1995	**1995** Russian troops capture the Chechen capital of Grozny **1996** The Afghan Taliban Islamic militia captures Kabul and forms a fundamentalist government **1997** Hong Kong is returned to Chinese sovereignty **1998** The weakness of the Japanese economy causes other Asian economies to decline **1998** Pol Pot dies	**1995** Israeli prime minister Yitzhak Rabin is assassinated by a Jewish extremist **1996** Iraqi aircraft enter a no-fly zone to attack Kurds; the USA launches cruise missiles against Iraq **1997** Zairean rebels rename the country the Democratic Republic of Congo **1997** Israeli troops withdraw from Hebron in Palestine **1998** Civil war breaks out in Sierra Leone **1998** Iraq refuses access to UN inspectors; the US launches cruise missiles against Iraqi military targets ❍242	**1995** Austria, Finland, and Sweden join the EC ❍238 **1995** Fighting in Bosnia leads to NATO airstrikes against the Serbs; Dayton Peace Agreement divides Bosnia ❍267 **1996** Yugoslavian war crimes tribunal opens in The Hague **1998** Ethnic Albanian separatists clash with Serb forces **1998** Protestant and Catholic representatives sign the Good Friday peace agreement in Northern Ireland **1999** "Euro" comes into limited use **1999** NATO aircraft bomb Serbia in support of the Kosovo Albanians; 800,000 people flee Kosovo into neighbouring countries ❍265; Serb forces withdraw from Kosovo; NATO and Russian troops move into Kosovo to perform peacekeeping role

THE AMERICAS	SCIENCE AND TECHNOLOGY	ARTS AND HUMANITIES	

1975 Mexican novelist Carlos Fuentes' *Terra Nostra* is published — **1975**

THE AMERICAS

1976 A coup overthrows Isabel Péron in Argentina

1977 The USA and Panama sign a treaty for the return of the canal zone to Panama

1978 President Sadat of Egypt and Prime Minister Begin of Israel meet at Camp David

1978 Dominica gains independence from Britain

1979 Israel and Egypt sign a peace treaty in Washington

1979 Sandinista guerrillas capture Managua; President Somoza flees Nicaragua

1979 A nuclear accident occurs at Three Mile Island power station in Pennsylvania, USA

1979 Hundreds of US cultists commit mass suicide in the People's Temple at Jonestown, Guyana

1980 Mount St Helens erupts in Washington state

1981 Belize, Antigua, and Barbuda gain independence from Britain

1982 Argentine troops invade the Falkland Islands; British forces invade and recapture the islands

1983 US President Reagan announces the development of a Strategic Defence Initiative (SDI) or "Star Wars" defence system; and support for the Nicaraguan Contras

1983 US marines invade Grenada and overthrow a revolutionary government **◑242**

1983 Martin Luther King Day is inaugurated in the USA

1984 Canadian prime minister Pierre Trudeau resigns

1984 US troops are withdrawn from Lebanon **◑242**

1984 Daniel Ortega is elected President of Nicaragua

1985 Earthquake devastates Mexico City

1986 Space shuttle *Challenger* explodes immediately after launch

1986 Jean Claude "Baby Doc" Duvalier flees Haiti to exile in France

1987 USA secretly sells weapons to Iran, the profits from which are used to arm the anti-Sandinista Contras in Nicaragua, causing a major scandal in the USA

1987 New York Stock Market crashes, triggering a worldwide financial crisis; computerized dealing is blamed for the severity of the collapse

1988 Sandinistas and Contras agree an armistice in Nicaragua

1989–90 US troops invade Panama; capture Noriega **◑242**

1989 Stroessner is overthrown by a coup in Paraguay

1991 Strategic Arms Limitation Talks (START) limits the size of US and Russian nuclear arsenals

1991 A peace agreement ends an 11-year civil war in El Salvador

1992 Canada, the USA and Mexico form the North American Free Trade Association (NAFTA) **◑243**

1992 UN organizes Earth Summit in Rio de Janeiro

1993 Israel and the PLO reach a peace agreement in Washington

1994 Zapatista National Liberation Army leads a revolt in Chiapas state in Mexico **◑269**

1994 US troops invade Haiti **◑242**

1995 US government offices in Oklahoma City are blown up by a terrorist bomb

1995 World Trade Organization succeeds GATT

1995 Louis Farrakhan organizes a "million man march" on Washington

1998 US President Clinton faces charges of misconduct and of authorizing a cover-up of a sex scandal involving Monica Lewinsky

1999 The US Senate declines to impeach President Clinton

SCIENCE AND TECHNOLOGY

1975 *Apollo* and *Soyuz* spacecraft link up while in orbit

1975 A personal computer (PC) in kit form goes on sale in the USA

1976 US *Viking* probes send back pictures from the surface of Mars

1977 Apple II personal computer is introduced

1978 World Health Organization announces that smallpox has been eradicated

1978 First in-vitro fertilization (IVF) baby is born in Britain

1979 US *Voyager* spaceprobes transmit pictures of Jupiter and its moons

1981 US space shuttle makes its first orbital flight

1981 IBM launch a PC using the Microsoft MS-DOS operating system

1981 The disease Acquired Immune Deficiency Syndrome (AIDS) is identified

1982 Compact music discs (CDs) are made available commercially

1983 French doctor Luc Montagnier identifies the human immunodeficiency virus (HIV) as the likely cause of AIDS

1984 A cellphone network is launched in Chicago, USA

1984 British scientist Alec Jeffreys devises a technique for genetic fingerprinting

1985 International Whaling Commission bans commercial whaling

1986 *Voyager II* transmits pictures of Uranus

1987 An international agreement is reached to limit the amounts of chloroflurocarbons (CFCs) released into the atmosphere

1988 US scientists start a project to map the human genome

1989 Convention on the International Trade in Endangered Species (CITES) imposes a worldwide ban on the sale of elephant ivory

1989 *Voyager II* transmits pictures of Neptune

1990 US Hubble space telescope is carried into orbit by the space shuttle

1991 A 5000-year-old body is discovered preserved in ice on the Austrian–Italian border

1992 US COBE satellite detects evidence to support Big Bang theory of the origins of the universe

1994 Fragments of comet Shoemaker-Levy impact on Jupiter

1994 Channel Tunnel opens, creating a rail link between Britain and France

1994 Worldwide web (www) is created

1996 Particles of antimatter are created at CERN (the European Nuclear Research Centre) in Switzerland

1996 The 452 metre-high Petronas Towers is constructed in Kuala Lumpur, Malaysia

1997 A sheep ("Dolly") is cloned in Britain

1997 Deep Blue computer defeats the world chess champion Gary Kasparov

1998 Opposition to genetic modification of food becomes a political issue

ARTS AND HUMANITIES

1975 Mexican novelist Carlos Fuentes' *Terra Nostra* is published

1975 Argentine writer Jorge Louis Borges' *The Book of Sand* is published

1976 US linguist Noam Chomsky's *Reflections on Language* is published

1977 Pompidou Centre, designed by Richard Rogers and Renzo Piano, opens

1978 US novelist John Irving publishes his *The World According to Garp*

1978 French philosopher Jacques Derrida publishes his *Truth in Painting*

1979 Czech novelist Milan Kundera's *The Book of Laughter and Forgetting* is published

1981 Italian author Umberto Eco's novel *The Name of the Rose* is published

1981 US author Toni Morrison's *Tar Baby* is published

1982 Chilean novelist Isabel Allende's *The House of the Spirits* is published

1982 US architect Maya Lin designs the Vietnam Veterans' Memorial in Washington DC

1983 US poet and author Alice Walker's *The Colour Purple* is published

1984 British novelist J G Ballard's *Empire of the Sun* is published

1985 Headquarters of the Hong Kong and Shanghai Bank, designed by British architect Norman Foster, is opened

1986 Japanese novelist Kazuo Ishiguro publishes his *An Artist of the Floating World*

1986 British composer Andrew Lloyd Webber writes his musical *Phantom of the Opera*

1987 US author Tom Woolf's *Bonfire of the Vanities* is published

1988 German composer Karlheinz Stockhausen writes his *Montag aus Licht*

1988 British author Salman Rushdie publishes his *Satanic Verses*

1990 US artist Jeff Koons creates his sculpture *Jeff and Ilona (Made in Heaven)*

1991 British composer Harrison Birtwistle writes his opera *Sir Gawain and the Green Knight*

1992 British artist Damian Hirst creates his sculpture *The Physical Impossibility of Death in the Mind of Someone Living*

1992 US novelist Jung Chang's *Wild Swans* is published

1993 Indian novelist Vikram Seth publishes his *A Suitable Boy*

1996 Restored Globe Theatre opens in London

1997 A branch of the Guggenheim Museum, designed by Canadian architect Frank Gehry, opens in Bilbao, Spain

1998 Film *Titanic* becomes the biggest box-office success to date

1975

1980

1985

1990

1995

EVENTS, PEOPLE AND PLACES

The entries below provide further information on many of the events, people and places referred to elsewhere in the atlas. The numbers at the end of each entry provide cross-references back to the relevant pages, while cross-references to other entries in this section are indicated by SMALL CAPITAL *letters.*

A

ABBAS I (THE GREAT) (1571–1629) Shah of Persia (1588–1629). The outstanding ruler of the Safavid dynasty, he restored Persia as a great power, waging war successfully against the invading Uzbeks and Ottoman Turks and recapturing Ormuz from the Portuguese. Displaying religious tolerance, he encouraged Dutch and English merchants and Christian missionaries. He made Esfahan his capital and turned it into one of the world's most beautiful cities. ❍ *142–43*

ABBASIDS Muslim caliph dynasty (750–1258). They traced their descent from Al-Abbas, the uncle of MUHAMMAD, and came to power by defeating the UMAYYADS. The Abbasids moved the caliphate from Damascus to Baghdad in 862, where it achieved great splendour. From the 10th century Abbasid caliphs ceased to exercise political power, becoming religious figureheads. After the family's downfall in 1258, following the fall of Baghdad to the Mongols, one member was invited by the MAMLUK sultan to Cairo where the dynasty was recognized until the 16th century. ❍ *68–69, 88–89*

ABDUL HAMID II (Ottoman Sultan) Abdul Hamid II (1842–1918) Sultan of Turkey (1876–1909). He suspended the constitution and formed an alliance with Germany after the Treaty of San Stefano (1878) when Russia threatened the continuation of the Ottoman Empire. He was deposed in 1909 after a revolt of the YOUNG TURKS. ❍ *178–79*

ADAMS–ONIS TREATY (1819) Agreement between the USA and Spain. Negotiated by secretary of state John Quincy Adams and Spanish minister Luis de Onis, Spain gave up its land east of the Mississippi River and its claims to the Oregon Territory; the USA assumed debts of $5 million and gave up claims to Texas. ❍ *182–83*

AFGHAN CIVIL WAR In 1978 the government was taken over by a socialist group, who turned to the Soviet Union for aid. Many people rebelled against the pro-communist government and, in late 1979, Soviet troops invaded the country. During the 1980s the Soviet troops fought unsuccessfully to put down the rebel Muslim forces, called the *Mujaheddin*. The Soviet troops withdrew in 1988 and 1989, but the civil war went on and refugees continued to flee from the country. Finally, in 1992, Mujaheddin forces entered Kabul and set up an Islamic government. Conflicts between rival groups within the Mujaheddin continued, and in late 1994 a new militant fundamentalist Islamic movement called the *Taliban* ("students") appeared. By 1996 the *Taliban* controlled large areas of the country. Although rival warlords combined to fight them in the north, the *Taliban* had established a government in Kabul by 1997. ❍ *236–37, 268–69*

AFRICAN NATIONAL CONGRESS (ANC) South African political party. It was formed in 1912 with the aim of securing racial equality and full political rights for non-whites. By the 1950s it was the most important organization opposed to APARTHEID. Banned in 1961, many of its leaders were arrested or fled into exile. A military wing was set up which engaged in economic and industrial sabotage. The ANC was legalized (and ended its armed struggle) in February 1990, when the government committed itself to dismantling apartheid. In April 1994 the party gained two-thirds of the vote in the country's first democratic election and its leader Nelson MANDELA was elected president. ❍ *256–57*

AGRICULTURAL REVOLUTION Series of changes in farming practice in the 18th and early 19th centuries. The main changes comprised crop rotation, new machinery, increased capital investment, scientific breeding, land reclamation and enclosure of common lands. Originating in Britain, these advances led to greatly increased agricultural productivity in Europe. ❍ *210–11*

AKBAR I (THE GREAT) (1542–1605) Emperor of India (1556–1605). Generally regarded as the greatest

ruler of the Mughal Empire, he assumed personal control in 1560 and set out to establish Mughal control of the whole of India, extending his authority as far south as Ahmadnagar. Akbar built a new capital at Fatehpur Sikri and endeavoured to unify his empire by conciliation with Hindus. He also tolerated Christian missionaries. ❍ *144–45*

ALARIC (370–410) King and founder of a group of GOTHS called the VISIGOTHS (395–410). His forces ravaged Thrace, Macedonia and Greece and occupied Epirus in 395–96. He invaded Italy, besieging Rome in 408, and sacking the city in 410 when Emperor Honorius would not grant him a position at court. He planned an invasion of Sicily and Africa, but his fleet was destroyed in a storm. ❍ *56–57*

ALBIGENSES (Cathars) Members of a heretical religious sect that existed in southern France from the 11th to the early 14th centuries and took its name from the French city of Albi. Pope Innocent III ordered a crusade against them in 1209–29, which caused much damage in Languedoc and Provence. ❍ *94–95*

ALEXANDER I (1777–1825) Russian tsar (1801–25). After repulsing Napoleon's attempt to conquer Russia in 1812, he led his troops across Europe and into Paris in 1814. Under the influence of various mystical groups, he helped form the HOLY ALLIANCE with other European powers. He was named constitutional monarch of Poland in 1815 and also annexed Finland, Georgia and Bessarabia to Russia. ❍ *172–73*

ALEXANDER THE GREAT (356–323 BC) King of MACEDONIA (336–323 BC), considered the greatest conqueror of Classical times. Son of PHILIP II of MACEDONIA and tutored by ARISTOTLE, he became king at the age of 20. Destroying rivals, he rapidly consolidated Macedonian power in Greece. In 334 BC he began his destruction of the vast Achaemenid Persian Empire, conquering western Asia Minor in 332 BC. He subdued EGYPT and occupied BABYLON, marching north in 330 BC to MEDIA and then conquering Central Asia in 328 BC. In 327 BC he invaded India but was prevented from advancing beyond the Punjab by the threat of mutiny. He died in Babylon, planning new conquests in Arabia. Although his empire did not outlive him, for he left no heir, he was chiefly responsible for the spread of Greek civilization in the Mediterranean and western Asia. ❍ *42–43*

ALFRED THE GREAT (849–99) King of Wessex (871–99). Warrior and scholar, he saved Wessex from the Danes and laid the foundations of a united English kingdom. After the Danish invasion of 878, he escaped to Athelney in Somerset, returning later to defeat the Danes at Edington and recover the kingdom. In a pact with the Danish leader, Guthrum (who accepted Christian baptism), England was roughly divided in two; the DANELAW occupying the northeast. Although he controlled only Wessex and part of Mercia, Alfred's leadership was widely recognized throughout England after his capture of London in 886. To strengthen Wessex against future attack, he built a fleet of ships, built forts and reorganized the army. ❍ *78–79*

ALGERIAN CIVIL WAR (1954–62) War between the French colonial government, which had ruled Algeria since 1830, and the National Liberation Front (FLN). Guerrilla tactics were employed by the FLN during the war, which resulted in the death of around 250,000 people. Most of the French *colons*, who numbered about one million, opposed Algerian independence and left Algeria in 1961, when the peace talks began. In 1962 an agreement was signed providing for an end to the fighting and the establishment of Algeria as an independent one-party state. ❍ *246–47, 256–57*

ALMOHADS Berber Muslim dynasty (1150–1269) in North Africa and Spain, the followers of a reform movement within ISLAM. It was founded by MUHAMMAD Ibn Tumart, who set out from the Atlas Mountains to purify Islam and oust the ALMORAVIDS from Morocco and eventually Spain. In 1212 Alfonso VIII of Castile routed the Almohads at Las Navas de Tolosa, and in 1269 their capital, Marrakesh, fell to the Marinids. ❍ *88–89*

ALMORAVIDS Berber Muslim dynasty (1054–1143) in Morocco and Spain. They rose to power under Abdullah ibn Yasin, who converted Saharan tribes to ISLAM in a religious revival. ABU BAKR founded Marrakesh as their capital in 1070; his brother, Yusuf Ibn Tashufin, defeated Alfonso VI of Castile in 1086. Almoravid power fragmented after the death of Ali Ibn Yusuf in 1143. Over the subsequent 30 years the Almoravids were ousted by other Muslim rulers and, ultimately, by the ALMOHADS. ❍ *88–89*

AMARU, TUPAC (c. 1742–81) (José Gabriel Condorcanqui) Native American leader in Peru. Between 1780 and 1781 Tupac Amaru led an army of over 10,000 Native Americans in a revolt against Spanish colonial rule. The revolt was ruthlessly quashed and its leader executed. A modern Peruvian guerrilla organization takes its name from this resistance hero. ❍ *190–91*

ANABAPTISTS Radical Protestant sects in the Reformation who shared the belief that infant baptism is not authorized by Scripture, and that it was necessary to be baptized as an adult. The first such baptisms were conducted by the Swiss Brethren sect in Zürich in 1525. The Swiss Brethren were the first to completely separate church from state when they rejected ZWINGLI's Reformed Church. Aided by social upheavals (such as the Peasants War) and the theological arguments of Martin LUTHER and Thomas Münzer, Anabaptism spread rapidly to Germany and the Netherlands. It stressed the community of believers. The communal theocracy established by John of Leiden at Münster was brutally suppressed in 1535. ❍ *154–55*

ANGKOR Ancient Khmer capital and temple complex, northwest Cambodia. The site contains the ruins of several stone temples erected by Khmer rulers, many of which lie within the walled enclosure of Angkor Thom, the capital built 1181–95 by Jayavarman VII. The 12th-century Angkor Wat, the greatest structure in terms of its size and the quality of its carving, lies outside the main Angkor Thom complex. Thai invaders destroyed the Angkor complex in 1431, and it remained virtually neglected until French travellers rediscovered it in 1858. ❍ *64–65*

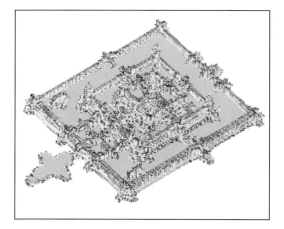

The outer cloister encloses a complex of buildings, which are covered by low-relief carvings of divine dancers, plants and animals. The many towers house images. The walls of the central group are adorned by reliefs depicting Hindu stories and battle scenes.

ANGLO-DUTCH WARS Three 17th-century naval conflicts between HOLLAND and England arising from commercial rivalry. The first war (1652–54) ended inconclusively, but with England holding the advantage. The second war (1665–67) followed England's seizure of New Amsterdam (New York). The Dutch inflicted heavy losses, and destroyed Chatham naval base, England; England modified its trade laws. The third war (1672–74) arose from English support of a French invasion of the Netherlands. The Dutch naval victory forced England to make peace. ❍ *128–29*

ANGLO-SAXONS People of Germanic origin comprising Angles, SAXONS and other tribes who began to invade England from the mid-5th century, when

Roman power was in decline. By 600 they were well established in most of England. They were converted to CHRISTIANITY in the 7th century. Early tribal groups were led by warrior lords whose *thegns* (noblemen) provided military service in exchange for rewards and protection. The tribal groups eventually developed into larger kingdoms, such as Northumbria and Wessex. The term Anglo-Saxon was first used in the late 8th century to distinguish the Saxon settlers in England from the "Old Saxons" of northern Germany, and became synonymous with "English". The Anglo-Saxon period of English history ended with the NORMAN CONQUEST of 1066. ❍ *56–57, 78–79, 92–93*

ANZUS PACT (Australia–New Zealand–United States Treaty Organization) Military alliance organized by the United States in 1951. ANZUS was set up in response to waning British power, the KOREAN WAR, and alarm at increasing Soviet influence in the Pacific. The treaty stated that an attack on any one of the three countries would be regarded as an attack on them all. It was replaced in 1954 by the Southeast Asia Treaty Organization (SEATO). ❍ *202–3, 242–43*

APARTHEID Policy of racial segregation practised by the South African government from 1948 to 1990. Racial inequality and restricted rights for non-whites were institutionalized when the Afrikaner-dominated National Party came to power in 1948. Officially a framework for "separate development" of races, in practice apartheid confirmed white-minority rule. It was based on segregation in all aspects of life including residence, land ownership and education. Non-whites, around 80 per cent of the population, were also given separate political structures and quasi-autonomous homelands or *bantustans*. The system was underpinned by extensive repression, and measures such as pass laws, which severely restricted the movements of non-whites. Increasingly isolated internationally and beset by economic difficulties and domestic unrest, the government pledged to dismantle the system in 1990. The transition to non-racial democracy was completed with the presidential and general elections of April 1994. ❍ *256–57*

ARAB-ISRAELI WARS (1948–49, 1956, 1967, 1973–74) Conflicts between Israel and the Arab states. After Israeli independence on 14 May 1948, troops from Egypt, Iraq, Lebanon, Syria and Jordan invaded the new nation. Initial Arab gains were halted and armistices arranged at Rhodes (January–July 1949). UN security forces upheld the truce until October 1956, when Israeli forces under Moshe Dayan attacked the Sinai Peninsula with support from France and Britain, alarmed at Egypt's nationalization of the Suez Canal. International opinion forced a cease-fire in November. In 1967 guerrilla raids led to Israeli mobilization, and in the ensuing Six Day War, Israel captured Sinai, the Golan Heights on the Syrian border and the Old City of Jerusalem. In the October War of 1973 (after intermittent hostilities) Egypt and Syria invaded on the Jewish holiday of Yom Kippur (6 October). Israel pushed back their advance after severe losses. Fighting lasted 18 days. Subsequent disengagement agreements were supervised by the UNITED NATIONS. In 1979 Israel signed a peace treaty with Egypt, but relations with other Arab states remained hostile. Israeli forces invaded Lebanon in 1982 in an effort to destroy bases of the PALESTINE LIBERATION ORGANIZATION (PLO). They were withdrawn in 1984 after widespread international criticism. After 1988 the PLO renounced terrorism and in the OSLO AGREEMENT of 1993 gained concessions, including limited autonomy in parts of the occupied territories. Israeli relations with some Arab neighbours generally improved after the GULF WAR (1991). ❍ *260–61, 266–67*

ARISTOPHANES (448–380 BC) Greek writer of comedies. Of his more than 40 plays, only 11 survive, the only extant comedies from the period. All follow the same basic plan: caricatures of contemporary Athenians become involved in absurd situations. Graceful choral lyrics frame caustic personal attacks. A conservative, Aristophanes parodied Euripides' innovations in drama, and satirized the philosophical radicalism of Socrates, and the expansionist policies of Athens. The importance of the chorus in early works is reflected in titles, such as *The Frogs* (405 BC), *The Birds* (414) and *The Wasps* (422). Other notable plays include *Lysistrata* (411) and *The Clouds* (423). ❍ *40–41*

ARISTOTLE (384–322 BC) Greek philosopher, founder of the science of logic and one of the greatest figures in Western philosophy, born in MACEDONIA. Aristotle studied (367–347 BC) under PLATO at the Academy in ATHENS. After Plato's death he tutored the young ALEXANDER THE GREAT, before founding the Lyceum in 335 BC. Anti-Macedonian disturbances forced Aristotle to flee in 323 BC to Chalcis on the island of Euboea, where he died. In direct opposition to Plato's idealism, Aristotle's metaphysics is based on the principle that all knowledge proceeds directly from observation of the particular. Aristotle argued that a particular object can only be explained through an understanding of causality. He outlined four causes: the material cause (an object's substance); formal cause (design); efficient cause (maker); and the final cause (function). For Aristotle this final cause was the primary one, and form was inherent in matter. His ethical philosophy stressed the exercise of rationality in political and intellectual life. Aristotle's writings cover nearly every branch of human knowledge, from statecraft to astronomy. His principal works are the *Organon* (six treatises on logic and syllogism); *Politics* (the conduct of the state); *Poetics* (analysis of poetry and tragedy) and *Rhetoric*. After the decline of the Roman Empire, Aristotle was forgotten by the West. But he did have a profound effect on the development of Islamic philosophy, and it was through Arab scholarship that his thought filtered into medieval Christian scholasticism, in particular the work of Thomas Aquinas. ❍ *134–35*

ASANTE Administrative region in West Africa, once the most powerful of the Akan states, reaching its peak around 1820. The forest state of Asante conquered neighbouring states in the savanna from 1701, and then turned towards the coast, eventually creating an empire covering more than 250,000 square kilometres (96,500 square miles). With its capital at Kumasi, Asante was an important trading state, dealing in gold, ivory and slaves with the British and Dutch during the 18th century. 19th-century conflicts with the British led to the sacking of Kumasi in 1874. In 1902 the Asante territories were declared part of the Gold Coast, a British crown colony. ❍ *136–37, 206–7*

ASHIKAGA City in central Japan, 80 kilometres (50 miles) north of Tokyo. An ancient silk-weaving centre, it was the home of the Ashikaga shogunate (hereditary military dictatorship) that dominated Japan during the Muromachi period (1335–1573). Sites include a sacred 12th century temple and an important library of Chinese classics, imported during the 15th century. ❍ *86–87*

ASHOKA (d. c. 231 BC) Indian emperor (c. 272–231 BC). The greatest emperor of the MAURYAN Empire, he at first fought to expand his empire. He was revolted by the bloodshed of war and, renouncing conquest by force, embraced BUDDHISM. He became one of its most fervent supporters and spread its ideas through missionaries to neighbouring countries and through edicts engraved on pillars. His empire encompassed most of India and large areas of Afghanistan. ❍ *46–47*

ASSYRIA Ancient empire of the Middle East. It took its name from the city of Ashur (Assur) on the River Tigris near modern Mosul, Iraq. The Assyrian Empire was established in the 3rd millennium BC and reached its zenith between the 9th and 7th centuries BC, when it extended from the Nile to the Gulf and north into Anatolia. Thereafter it declined and was absorbed by the Persian Empire. Under Ashurbanipal, art (especially bas-relief sculpture) and learning reached their peak. The luxuriance of Ashurbanipal's court at NINEVEH was legendary and, combined with the cost of maintaining his huge armies, fatally weakened the empire. The capture of Nineveh in 612 BC marked the terminal decline of Assyria. ❍ *38–39*

ATAHUALPA (1502–33) Last Inca ruler of Peru. The son of Huayna Capac, upon his father's death he inherited Quito, while his half-brother Huáscar controlled the rest of the Inca kingdom. In 1532 Atahualpa defeated Huáscar, but his period of complete dominance was to be short-lived. In late 1532 Atahualpa was captured by Francisco PIZARRO and subsequently executed. ❍ *120–21*

ATATÜRK (Mustafa Kemal) (1881–1938) Turkish general and statesman, first president (1923–38) of the Turkish republic. As a young soldier he joined the YOUNG TURKS and was chief of staff to Enver Pasha in the successful revolution (1908). He fought against the Italians in Tripoli (1911) and defended Gallipoli in the BALKAN WARS. During the First World War he successfully led resistance to the Allies' Gallipoli Campaign. The defeat of the Ottoman Empire and the capitulation of the sultan in 1919 persuaded Mustafa Kemal to organize the Turkish Nationalist Party and set up a rival government in Ankara. The Treaty of SÈVRES pushed him into attack. His expulsion of the Greeks from Asia Minor (1921–22) forced the sultan to flee Constantinople (Istanbul). The Treaty of LAUSANNE (1923) saw the creation of an independent republic. His dictatorship undertook sweeping reforms, which transformed Turkey into a secular, industrial nation. In 1934 he adopted the title Atatürk (Turkish for father of the Turks). ❍ *178–79, 220–21*

ATHENS (ancient) The ancient city was built around the Acropolis, a fortified citadel, and was the greatest artistic and cultural centre in ancient Greece, gaining importance after the PERSIAN WARS (550–449 BC). The city prospered under Cimon and PERICLES during the 5th century BC and provided a climate in which the great Classical works of philosophy and drama were created. ❍ *40–41*

ATTILA (406–453) King of the Huns (c. 439–53), co-ruler with his elder brother until 445. Attila defeated the Eastern Roman emperor Theodosius II, extorting land and tribute, and invaded GAUL in 451. Although his army suffered heavy losses, he invaded Italy in 452, but disease forced his withdrawal. Attila has a reputation as a fierce warrior, but was fair to his subjects and encouraged learning. On his death the empire fell apart. ❍ *56–57, 76–77*

AUGSBURG, PEACE OF (1555) Agreement reached by the Diet of the HOLY ROMAN EMPIRE in Augsburg ending the conflict between Roman Catholics and Lutherans in Germany. It established the right of each prince to decide on the nature of religions practice in his lands. Dissenters were allowed to sell their lands and move. Free cities and imperial cities were open to both Catholics and Lutherans. The exclusion of other Protestant sects such as CALVINISM proved to be a source of future conflict. ❍ *146–47, 154–55*

AUGUSTUS (63 BC–AD 14) (Gaius Julius Caesar Octavianus) First Roman emperor (27 BC–14 AD), also called Octavian. Nephew and adopted heir of Julius CAESAR, he formed the Second Triumvirate with Mark Antony and Lepidus after Caesar's assassination. They defeated Brutus and Cassius at Philippi in 42 BC and divided the empire between them. Rivalry between Antony and Octavian was resolved by the defeat of Antony at Actium in 31 BC. While preserving the form of the republic, Octavian held supreme power, and was officially called Imperator Caesar Augustus. He introduced peace and prosperity after years of civil war. He built up the power and prestige of Rome, encouraging patriotic literature and rebuilding much of the city in marble. He extended the frontiers and fostered colonization, took general censuses, and tried to make taxation more equitable. He tried to arrange the succession to avoid future conflicts, though had to acknowledge an unloved stepson, TIBERIUS, as his successor. ❍ *54–55*

AURANGZEB (1619–1707) Emperor of India (1658–1707). The last of the great Mughal emperors, Aurangzeb seized the throne from his enfeebled father, SHAH JEHAN. He reigned over an even greater area, and spent most of his reign defending it. Aurangzeb was a devout Muslim, whose intolerance of HINDUISM provoked long wars with the MARATHAS. The empire was already breaking up before his death. ❍ *144–45*

AUSTERLITZ, BATTLE OF (1805) Fought in BOHEMIA on 2 December. The French under NAPOLEON defeated the Austrians and the Russians under Mikhail Kutuzov. The battle, also known as the Battle of the Three Emperors, was one of Napoleon's greatest triumphs. **○** *166–67*

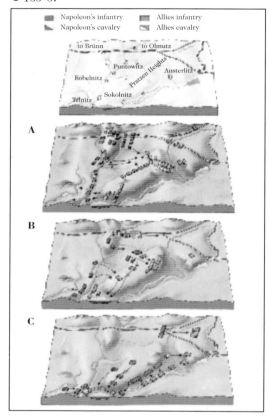

By evacuating Austerlitz and the Pratzen Heights, Napoleon feigned weakness. The Allied forces camped on the heights (A). Their plan was to overwhelm the (deliberately) weak French right flank, before moving north to envelop the French as they headed for Brünn. Under the cover of mist, French forces manoeuvred beneath the Pratzen Heights. The Allies attacked the French right flank. While the Allies were engaged in battle to the south, French forces marched up and occupied the Pratzen Heights (B). Additional support was provided by other French forces that had initially engaged the Allies to the north. Together the French forces dispersed the Allies, driving them onto the frozen lakes near Telnitz, where many drowned as Napoleon ordered his artillery to open fire, breaking the ice (C).

AUSTRIAN SUCCESSION, WAR OF THE (1740–48) Overall name for several related wars. They included the war for the Austrian succession itself, in which France supported Spain's claim to part of the Habsburg domains; the first and second Silesian wars, in which FREDERICK II of Prussia took Silesia from Austria; and the war between France and Britain over colonial possessions, known in North America as King George's War. **○** *156–57, 174–75*

AVIGNON POPES Popes who between 1309 and 1377 resided in Avignon instead of Rome. The papal court was established in Avignon by the French Pope Clement V. In 1348 the city was bought by Clement VI. The GREAT SCHISM occurred shortly after the court returned to Rome. The Papacy held Avignon until 1791, when it was annexed to France by the revolutionary authorities. **○** *106–7*

AWAMI LEAGUE Opposition political party originally formed in Pakistan in 1949; since 1971 the major party of Bangladesh (formerly East Pakistan); renamed Awami Peasants' and Workers' League of Bangladesh in 1975. It was led by Sheikh MUJIBUR RAHMAN from 1953 to his death in a military coup in 1975. From 1966 the party advocated a federation government for Pakistan that would give more autonomy to East Pakistan. In 1970 the party won a majority in the legislature, but

the government cancelled the election and outlawed the party. Civil war broke out in early 1971 and East Pakistan won its independence, becoming Bangladesh. **○** *248–49*

AXIS POWERS Term applied to Germany and Italy after they signed the Rome–Berlin Axis in October 1936. It included Japan after it joined them in the Tripartite Pact (September 1940). Other states that joined the Axis were Hungary and Romania (1940) and Bulgaria (1941). **○** *226–27, 232–33*

AXUM Capital of a powerful kingdom in northern ETHIOPIA 1st–6th centuries AD. It was Christianized in the 4th century and remains a major centre in Ethiopian CHRISTIANITY. A rich trading centre, Axum was known for its ivory. **○** *82–83*

AYYUBIDS Muslim SUNNI dynasty that ruled Egypt and parts of the Middle East from the late 12th century to 1250. The dynasty was founded by SALADIN, who united Syria against the Christian crusaders and displaced the FATIMID dynasty in Egypt. The Ayyubids ruled a decentralized empire prone to internal disputes. The dynasty collapsed when the MAMLUKS seized control of Egypt in 1250. **○** *58–59, 94–95*

AZAÑA, DÍAZ, MANUEL (1880–1940) Spanish politician. He was prime minister 1931–33 and in 1936 until he was elected president of the Second Republic. He was titular head of the Republican government during the SPANISH CIVIL WAR, and resigned in 1939. **○** *230–31*

BABUR (1483–1530) First Mughal emperor of India (1526–30). His name was originally Zahir ud-Din Muhammad; Babur means tiger. He became ruler of Fergana in 1495 and engaged in a long conflict for control of Samarkand, but ultimately lost both territories. Raising an army, he captured Kabul and carved out a new kingdom for himself in Afghanistan. From here he invaded India, gaining Delhi in 1526 and Agra (his future capital) in 1527, and conquering northern India as far as Bengal. **○** *144–45*

BABYLON Ancient city on the River Euphrates in MESOPOTAMIA, capital of the empire of BABYLONIA. It was rebuilt after being destroyed by ASSYRIA c. 689 BC and its new buildings included the Hanging Gardens, one of the SEVEN WONDERS OF THE WORLD. This was the period, under NEBUCHADNEZZAR, of the Babylonian Captivity of the Jews. The city declined after 275 BC, being replaced in importance by Seleucia. **○** *38–39*

BABYLONIA Ancient region and empire of MESOPOTAMIA, based on the city of BABYLON. The Babylonian Empire was first established in the early 18th century BC by HAMMURABI, but declined under the impact of the HITTITES and Kassites in c. 1595 BC. After a long period of weakness and confusion, the empire eventually fell to ASSYRIA in the 8th century BC. Babylon's greatness was restored and in c. 625 BC its independence was won by Nabopolassar, who captured the Assyrian capital of NINEVEH. This new Babylonian (Chaldaean) Empire defeated EGYPT and took the JEWS to captivity in Babylon in 586 BC. In 538 BC it fell to the Persians. **○** *36–37, 38–39*

BALKAN WARS (1912–13) Two wars involving the Balkan states and the Ottoman Empire. In the first, the Balkan League (Serbia, Bulgaria, Greece and Montenegro) conquered most of the European territory of the Ottoman Empire. The second war (mainly between Serbia and Bulgaria) arose out of dissatisfaction with the distribution of these lands. Serbia's victory added to regional tension before the First World War. **○** *216–17*

BASIL I (c. 813–86) Byzantine emperor (r. 867–86) and founder of the Macedonian dynasty. Basil was assisted in his rise to power by Emperor Michael III. After Michael designated him co-emperor, Basil had his former patron murdered and assumed sole power in

Byzantium. Basil's most effective policies concerned the conversion of the Bulgars to Orthodox Christianity; military campaigns against the Paulician religious sect in Asia Minor; and a revision of Roman legal codes. **○** *66–67*

BASIL II (c. 958–1025) Byzantine emperor (976–1025), surnamed Bulgaroctonus ("Bulgar-slayer"). One of Byzantium's ablest rulers, Basil reigned during the heyday of the empire. He is best known for his military victory over the Bulgarian tsar Samuel in 1014, which brought the entire Balkan peninsula under Byzantine control. During Basil's reign, Byzantium's sphere of influence was extended by the conversion of Kievan Russia to ORTHODOX CHRISTIANITY. **○** *66–67*

BENIN Kingdom in western Nigeria that flourished in the 14th–17th centuries. It is remembered chiefly for its bronze sculptures and wood and ivory carvings, considered among the finest African art. Ruled autocratically by a divine sovereign, it was prosperous and peaceful, and was compared favourably with Amsterdam by an early Dutch visitor. Central authority disintegrated in the era of firearms and the slave trade. **○** *80–81, 136–37*

BERLIN AIRLIFT (1948–49) Operation to supply Berlin with food and other necessities after the Soviet Union closed all road and rail links between the city and West Germany. For 15 months British and US aircraft flew more than 270,000 flights, delivering food and supplies. **○** *244–45*

BERLIN, CONFERENCE (1884–85) Meeting of the major European nations, the United States and Turkey in Berlin to discuss the problems of West African colonization and arrange for free trade along the Niger and Congo rivers. The conference affirmed British claims on Nigeria and Belgian claims on the Congo, but other agreements on trade and political neutrality proved untenable in the years to follow. **○** *206–7, 208–9*

BERLIN, CONGRESS OF (1878) Meeting of European powers to revise the Treaty of San Stefano. That treaty had increased Russian power in southeastern Europe to an extent unacceptable to other powers, and the purpose of the Congress, under the presidency of BISMARCK, was to modify its terms. The main territorial adjustment was to reduce the Russian-sponsored Greater Bulgaria. **○** *178–79*

BHUTTO, ZULFIKAR ALI (1928–79) Pakistani statesman, prime minister (1973–77), father of Benazir Bhutto. He founded the Pakistan People's Party in 1967. In the 1970 elections, Bhutto gained a majority in West Pakistan but the AWAMI LEAGUE controlled East Pakistan. Bhutto's refusal to grant autonomy to East Pakistan led to civil war in 1971. Defeat led to the formation of Bangladesh. Bhutto became president of the rump state of Pakistan but was overthrown in a military coup, led by General Zia in 1977. Bhutto was convicted of conspiracy to murder and executed. **○** *248–49*

BILL OF RIGHTS Name given to the first ten amendments to the US Constitution, ratified 1791. Several states had agreed to ratify the Constitution (1787) only after George Washington promised to add such a list of liberties. The main rights confirmed were: freedom of worship, of speech, of the press, and of assembly; the right to bear arms; freedom from unreasonable search and seizure; the right to a speedy trial by jury; and protection from self-incrimination. Powers not granted specifically to the federal government were reserved for the states. **○** *268–69*

BISMARCK, OTTO VON (1815–98) German statesman responsible for 19th-century German unification. Born into a wealthy Prussian family, Bismarck first made an impression as a diehard reactionary during the revolutions of 1848. Keen to strengthen the Prussian army, William I appointed him chancellor of Prussia in 1862. Bismarck dissolved parliament and raised taxes to pay for military improvements. The status of Schleswig-Holstein enabled him to engineer the SEVEN WEEKS WAR (1866) and expel Austria from the GERMAN CONFEDERATION. Bismarck then provoked the Franco-Prussian War (1870–71), in order to bring the south German states into the Prussian-led North German Confederation. Victory saw Bismarck become the first chancellor of the empire in 1871. Through skilful diplomacy and alliance-building he consolidated Germany's position in the heart of Europe. In 1882 he formed the TRIPLE ALLIANCE with Austro-Hungary and Italy. Bismarck's domestic policies were similarly based on the principle of "divide-and-rule". The internal coherence of the empire was disturbed by the *Kulturkampf*, a struggle between the powers of the new state and the traditional strength of German Catholicism. Bismarck relied on a strategic alliance with the liberals to foster a national identity. Bismarck passed an antisocialist law in 1878 to stem the rise of German socialism, but was forced to adopt a paternalist programme of social welfare in 1883–87. The rapid process of industrialization encouraged colonialism and the building of a German overseas empire. The accession of William II in 1888 saw the demise of Bismarck's influence on German and European politics, and in 1890 the "Iron Chancellor" was forced to resign. ◐ *174–75, 176–77, 216–17*

BLACK HOLE OF CALCUTTA Prison in Calcutta, India, where 64 or more British soldiers were placed by the Nawab Siraj-ad-Daula of Bengal in June 1756. The cell was 5.5 x 4.5 metres (18 x 15 feet), and most of the soldiers died of suffocation. ◐ *194–95*

BLITZ Name used by the British to describe the night bombings of British cities by the German Luftwaffe (air force) in 1940–41. It is an abbreviation of *Blitzkrieg* (lightning war), the name used by the German army to describe hard-hitting, surprise attacks on enemy forces. ◐ *232–33*

BOER WARS *See* SOUTH AFRICAN WARS

BOHEMIA Historic region which (with Moravia) now comprises the Czech Republic. Bohemia was first unified in the 10th century when it became part of the HOLY ROMAN EMPIRE, coming under Habsburg control in 1526. It was the centre of occasional religious or nationalistic revolts against Austrian rule, including that of the HUSSITES and the episode that sparked the THIRTY YEARS WAR in 1618. It became part of Czechoslovakia in 1918 and the Czech Republic in 1992. ◐ *70–71, 90–91, 106–7, 146–47, 152–55*

BOLÍVAR, SIMÓN (1783–1830) Latin American revolutionary leader, known as "the Liberator". He was born into a Venezuelan family of Basque origin, and travelled to Napoleonic Europe as a young man. His experiences there influenced his untiring attempts to free South America from Spanish rule. In 1813 he took part in the creation of the first Venezuelan republic and when it was overthrown he continued to lead attempts to oust the Spanish. He achieved no real success until 1819, when his victory at Boyacá led to the liberation of New Granada (later Colombia) in 1821. The liberation of Venezuela (1821), Ecuador (1822), Peru (1824) and Upper Peru (1825) followed, the latter renaming itself Bolivia in his honour. Despite the removal of Spanish hegemony from the continent, his hopes of uniting South America into one confederation were dashed by rivalry between the new states. ◐ *190–91*

BOLSHEVIKS Marxist revolutionaries, led by LENIN, who seized power in the RUSSIAN REVOLUTION of 1917. They narrowly defeated the Mensheviks at the Second Congress of the All-Russian Soviet Democratic Workers' Party in London (1903). The split, on tactics as much as on doctrine, centred on the means of achieving revolution. The Bolsheviks believed it could be obtained only by professional revolutionaries leading the proletariat. The Bolsheviks were able to overthrow the Provisional government of Kerensky through their support in the soviets of Moscow and Petrograd. ◐ *218–19, 222–23*

BONAPARTE, JOSEPH (1768–1844) King of Spain, born in Corsica. He was the eldest brother of NAPOLEON I. He participated in the Italian campaign in 1797 and later served as diplomat for the First Republic of France. Napoleon made him King of Naples in 1806, and he was King of Spain from 1808 to 1813. After Napoleon's defeat at Waterloo, he resided in the USA, and then in Florence. ◐ *166–67*

BOROBUDUR BUDDHIST monument in central Java, built under the Sailendra dynasty c. 800. It comprises a stupa (relic mound), mandalas (ritual diagrams) and the temple mountain, all forms of Indian GUPTA dynasty religious art. Its construction involved the moving of 570,000 cubic metres (2,000,000 cubic feet) of stone from a river bed. ◐ *64–65*

The walls of the Borobudur complex are covered with reliefs relating to Buddhist doctrine and there are more than 500 shrines with seated Buddhas.

BOSNIAN CIVIL WAR (1992–95) War between Serbs, Croats and Bosnian Muslims, whose immediate cause was Bosnia-Herzegovina's declaration of independence from Yugoslavia in 1992. The Serbs, opposed to independence, embarked on war and soon occupied more than two-thirds of the land and were accused of "ethnic cleansing" – the killing or expulsion of other ethnic groups from Serb-occupied areas. The war was later extended when Croat forces seized other parts of the country. The UNITED NATIONS, while attempting a peacekeeping operation, tried to find a way of ending the war by dividing Bosnia-Herzegovina into self-governing provinces, under a unified, multi-ethnic central government. Finally, in 1995 the Dayton Peace Accord ended the war. It affirmed that Bosnia-Herzegovina was a single state, but partitioned it into a Muslim-Croat federation, given 51 per cent of the area, and a Serbian republic, occupying the remaining 49 per cent. ◐ *264–65, 266–67, 268–69*

BOSTON TEA PARTY (1773) Protest by a group of Massachusetts colonists, disguised as Mohawks and led by Samuel Adams, against the Tea Act and, more generally, against "taxation without representation". The Tea Act, passed by the British Parliament in 1773, withdrew duty on tea exported to the colonies. It enabled the ENGLISH EAST INDIA COMPANY to sell tea directly to the colonies without first going to Britain and resulted in colonial merchants being undersold. The protesters boarded three British ships and threw their cargo of tea into Boston harbour. The British retaliated by closing the harbour. ◐ *164–65*

BOURBONS European dynastic family, descendants of the CAPETIANS. The ducal title was created in 1327 and continued until 1527. A cadet branch, the Bourbon-Vendôme line, won the kingdom of Navarre. The Bourbons ruled France from 1589 (when Henry of Navarre became Henry IV) until the French Revolution in 1789. Two members of the family, LOUIS XVIII and CHARLES X, reigned after the restoration of the monarchy. In 1700 the Bourbons became the ruling family of Spain, when PHILIP V (grandson of LOUIS XIV of France), assumed the throne. His descendants mostly continued to rule Spain until the Second Republic was declared in 1931. Juan Carlos I, a Bourbon, was restored to the Spanish throne in 1975. ◐ *122–23, 156–57, 190–91*

BOXER REBELLION (1899–1900) European name for a Chinese revolt aimed at ousting foreigners from China. Forces led by the Society of Righteous and Harmonious Fists (hence the name "Boxers"), with tacit support from the Dowager Empress, attacked Europeans and Chinese Christians and besieged Beijing's foreign legations' enclave for two months in 1900. On 14 August an international expeditionary force captured Beijing, relieving the foreigners besieged there, and suppressed the rising. China agreed to pay an indemnity. ◐ *198–99*

BREST-LITOVSK, TREATY OF (March 1918) Peace treaty between Russia and the CENTRAL POWERS, confirming Russian withdrawal from the First World War. The Ukraine and Georgia became independent and Russian territory was surrendered to Germany and Austria–Hungary. The treaty was declared void when the war ended in November. ◐ *218–19*

BRETTON WOODS CONFERENCE *See* UNITED NATIONS MONETARY AND FINANCIAL CONFERENCE

BREZHNEV, LEONID ILYICH (1906–82) Soviet politician and effective ruler from the mid-1960s until his death. He rose through party ranks to become secretary to the central committee of the Soviet Communist Party (1952) and a member of the presidium (later the Politburo) in 1957. In 1964 he helped plan the downfall of Nikita KHRUSHCHEV and became party general secretary, at first sharing power with Aleksei KOSYGIN. In 1977 he became president of the Soviet Union. He pursued a hard line against reforms at home and in Eastern Europe but also sought to reduce tensions with the West. After the Soviet invasion of Czechoslovakia in 1968, he promulgated the "Brezhnev Doctrine" confirming Soviet domination of satellite states, as seen in the 1979 invasion of Afghanistan in 1979. ◐ *236–37, 242–43*

BRITISH NORTH AMERICA ACT (1867) Act of the British parliament that created the Dominion of Canada. It resulted from a series of conferences and provided a constitution similar to that of Britain. Residual British powers were surrendered in the Canada Act of 1982, when the original act was renamed the Constitution Act. ◐ *188–89*

BUDDHA (Enlightened One) Title adopted by Siddhartha Gautama (c. 563–c. 483 BC), the founder of BUDDHISM. Born at Lumbini, Nepal, Gautama was son of the ruler of the Sakya tribe, and his early years were spent in a life of luxury. At the age of 29, he realized that human life is little more than suffering. He gave up his wealth and comfort, deserted his wife and small son, and took to the road as a wandering ascetic. He travelled south and sought truth in a six-year regime of austerity and self-mortification. After abandoning asceticism as futile, he sought his own middle way towards enlightenment. The moment of truth came c. 528 BC, as he sat beneath a banyan tree in the village of Buddha Gaya, Bihar, India. After this incident, he taught others about his way to truth. The title Buddha applies to those who have achieved perfect enlightenment. Buddhists believe that there have been several buddhas before Gautama, and there will be many to come. The term also serves to describe a variety of Buddha images. ◐ *44–45*

BUDDHISM Religion and philosophy founded in India c. 528 BC by Siddhartha Gautama, the BUDDHA. Buddhism is based on Four Noble Truths: existence is suffering; the cause of suffering is desire; the end of suffering comes with the achievement of *Nirvana*; *Nirvana* is attained through the Eightfold Path: right views, right resolve, right speech, right action, right livelihood, right effort, right mindfulness and right concentration. There are no gods in Buddhism. Alongside the belief in the Four Noble Truths exists *karma*, one of Buddhism's most important concepts: good actions are rewarded and evil ones are punished, either in this life or throughout a long series of lives resulting from *samsara*, the cycle of death and rebirth produced by reincarnation. The achievement of *nirvana* breaks the cycle. Buddhism is a worldwide religion. Its main divisions are Theravada, or Hinayana, in Southeast Asia; Mahayana in north Asia; Lamaism or Tibetan Buddhism in Tibet; and Zen in Japan. ◐ *44–45, 62–63, 248–49*

BUKHARIN, NIKOLAI IVANOVICH (1888–1938) Russian communist political theorist. After the 1917 Revolution he became a leading member of the COMMUNIST INTERNATIONAL (Comintern) and editor of *Pravda*. In 1924 Bukharin became a member of the Politburo. He opposed agricultural collectivization and was executed for treason by STALIN. ◐ *222–23*

BULGARS Ancient Turkic people originating in the region north and east of the Black Sea. In about AD 650

they split into two groups. The western group moved to Bulgaria, where they became assimilated into the Slavic population and adopted CHRISTIANITY. The other group moved to the Volga region and set up a Bulgar state, eventually converting to ISLAM. The Volga Bulgars were conquered by the Kievan Rus in the 10th century. *66–67, 76–77*

BURGUNDY, DUCHY OF Almost independent state that arose after the VALOIS king John the Good of France granted his son, later to become Philip II (1342–1404), the ancient duchy of Burgundy around Dijon in 1363. Under Philip and his successors the duchy acquired extensive territories from France and the HOLY ROMAN EMPIRE – territories that included the wealthy areas that were to become the United Provinces of the Netherlands. The dukes were great patrons of the arts and their court promoted the cult of chivalry. The death in battle of Charles the Bold (1433–77) led to the partition and demise of the state of Burgundy. The original duchy around Dijon went to Louis XI of France while the other territories went to Charles's daughter Mary, and so to her husband, Maximilian von Habsburg. *106–7, 152–53*

BYBLOS Ancient city of the Phoenicians, in Lebanon, 27 kilometres (17 miles) north of Beirut. Byblos was a centre of Phoenician trade with EGYPT from the 2nd millennium BC, and was particularly famous as a source of papyrus. The Greek word for "book" derived from its name. The city was abandoned after its capture by the Christian crusaders in 1103. *38–39*

C

CABOT, JOHN (c. 1450–98) Italian navigator and explorer in English service. Supported by Henry VII, he sailed in search of a western route to India, and reached Newfoundland (1497). His discovery served as the basis for English claims in North America. His account of the Newfoundland fisheries encouraged fishermen from European Atlantic ports to follow his route. *116–17*

CABOT, SEBASTIAN (1476–1557) Italian navigator, explorer and cartographer, son of John CABOT. In 1508 Cabot sailed across the Atlantic in search of a northern passage to China (possibly reaching Hudson Bay) and sailed down the coast of North America. He joined the Spanish navy in 1512 and led an expedition in 1526 to find a route to the Pacific from the Atlantic, reaching the coast of Brazil. In 1547 Cabot returned to England, where he helped to found the Company of Merchant Adventurers for the Discovery of Cathay. As governor of the company, he organized a series of expeditions (1553–56) in search of a northeast passage to China. *116–17, 120–21.*

CAESAR, (GAIUS) JULIUS (100–44 BC) Roman general and statesman. A great military commander and brilliant politician, he defeated formidable rivals to become dictator of Rome. After the death of Sulla, Caesar became military tribune. As *pontifex maximus*, he directed reforms in 63 BC that resulted in the Julian calendar. He formed the First Triumvirate in 60 BC with POMPEY and Crassus, instituted agrarian reforms and created a patrician-plebeian alliance. He conquered GAUL for Rome (58–49 BC) and invaded Britain (54 BC). Refusing Senate demands to disband his army, he provoked civil war with POMPEY. Caesar defeated Pompey at Pharsalus in 48 BC and pursued him to EGYPT, where he made Cleopatra queen. After further victories, he returned to Rome in 45 BC and was received with unprecedented honours, culminating in the title of dictator for life. He introduced popular reforms, but his growing power aroused resentment. He was assassinated in the Senate on 15 March by a conspiracy led by Cassius and Brutus. Caesar bequeathed his wealth and power to his grandnephew, Octavian (later AUGUSTUS) who, together with Mark Antony, avenged his murder. *54–55*

CALVIN, JOHN (1509–64) French theologian of the Reformation. He prepared for a career in the Roman CATHOLIC CHURCH but turned to the study of classics. In c. 1533 he became a Protestant and began work on his *Institutes of the Christian Religion*. In this work he presented the basics of what came to be known as CALVINISM. To avoid persecution, he went to live in Geneva, Switzerland (1536), where he advanced the Reformation. *154–55*

CALVINISM Set of doctrines and attitudes derived from the Protestant theologian John CALVIN. The Reformed and Presbyterian churches were established in his tradition. Rejecting papal authority and relying on the Bible as the source of religious truth, Calvinism stresses the sovereignty of God and predestination. Calvinism usually subordinates state to church, and cultivates austere morality, family piety, business enterprise, education and science. The development of these doctrines, particularly predestination, and the rejection of consubstantiation in its eucharistic teaching, caused a split in PROTESTANTISM between LUTHERANISM and Presbyterianism. The influence of Calvinism spread rapidly. Important Calvinist leaders include John Knox and Jonathan Edwards. *154–55*

CAPETIANS French royal family which provided France with 15 kings. It began with Hugh Capet, Duke of Francia (987), and ended with Charles IV (1328). Hugh Capet was elected king after the death of Louis V, the last of the CAROLINGIANS. Capetians dominated the feudal forces, extending the king's rule over the country. It was succeeded by Philip VI of the House of VALOIS. *92–93*

CAROL II (1893–1953) King of Romania (1930–40), grandnephew of Carol I. In 1925 he renounced the throne. He returned in 1930 and supplanted his son Michael as king, despite Liberal opposition and economic crisis. Carol II supported fascism. He aimed to become dictator, but German pressure forced him to abdicate in favour of his son Michael in 1940, leaving power in the hands of the Roman fascist leader, Ion Antonescu. *230–31*

CAROLINGIANS Second Frankish dynasty of early medieval Europe. Founded in the 7th century by Pippin of Landen, it rose to power under the weak kingship of the MEROVINGIANS. In 732 Charles Martel defeated the Muslims at Poitiers; in 751 his son Pippin III (the Short) deposed the last Merovingian and became king of the Franks. The dynasty reached its peak under Pippin's son CHARLEMAGNE (after whom the dynasty is named), who united the Frankish dominions and much of west and central Europe, and was crowned Holy Roman Emperor by the Pope in 800. His empire was later subdivided and broken up by civil wars. Carolingian rule finally ended in 987. *74–75, 92–93*

CARRANZA, VENUSTIANO (1859–1920) Mexican statesman, president (1914). Carranza supported Francisco Madero's revolution against Porfirio D'az. When Madero was overthrown by Victoriano Huerta, Carranza joined forces with Álvaro Obregón, "Pancho" Villa and Emiliano Zapata to defeat Huerta. Villa and Zapata's refusal to recognize Carranza's authority prolonged the civil war. Carranza supported John Pershing 's expedition against Villa. His attempts to prevent the accession of Obregón led to a revolt. Carranza fled and was murdered. *See also* MEXICAN REVOLUTION. *226–27*

CARTER, JIMMY (JAMES EARL), JR (1924–) 39th US president (1977–81). Carter was a DEMOCRAT senator (1962–66) and governor (1971–74) for the state of Georgia. In 1976 he defeated the incumbent President Gerald Ford. Carter had a number of foreign policy successes, such as the negotiation of the Camp David Agreement (1979). These were overshadowed, however, by the disastrous attempt to free US hostages in Iran (April 1980). Following the Soviet invasion of Afghanistan, Carter backed a US boycott of the 1980 Moscow Olympics. An oil price rise contributed to spiralling inflation, which was dampened only by a large increase in interest rates. In the 1980 presidential election Carter was easily defeated by Ronald Reagan. Since then he has sought to promote human rights and acted as a international peace broker. *242–43*

CARTHAGE Ancient port on a peninsula in the Bay of Tunis, North Africa. Founded in the 9th century BC by Phoenician colonists. It became a great commercial city and imperial power controlling an empire in North Africa, southern Spain and islands of the western Mediterranean. The rise of Rome in the 3rd century resulted in the PUNIC WARS, and in spite of the victories of HANNIBAL, ended with the defeat and destruction of Carthage in the Third Punic War (149–46 BC). It was resettled as a Roman colony, and in the 5th century AD it was the capital of the VANDALS. *38–39, 54–55*

CARTIER, JACQUES (1491–1557) French navigator and explorer who discovered the St Lawrence River in 1535. Cartier was sent to North America in 1534 by FRANCIS I. During this first voyage he discovered the Magdalen Islands and explored the Gulf of St Lawrence. In 1535–36 he sailed up the St Lawrence River to the site of modern Québec and continued on foot to Hochelaga (present-day Montréal). His third voyage was part of an unsuccessful colonization scheme. Although Cartier failed to find the Northwest Passage, his discoveries laid the basis for French settlements in Canada. *116–17*

CASTRO, FIDEL RUZ (1926–) Cuban revolutionary leader and politician, premier since 1959. In 1953 he was sentenced to 15 years' imprisonment after an unsuccessful coup against the Batista regime. Two years later he was granted an amnesty and exiled to Mexico. In January 1959 his guerrilla forces overthrew the regime. He quickly instituted radical reforms, such as collectivizing agriculture and dispossessing foreign companies. In 1961 the USA organized the abortive Bay of Pigs invasion, Castro responded by allying the revolutionary movement more closely with the Soviet Union. In 1962 the Cuban Missile Crisis saw the USA and Soviet Union on the brink of nuclear war. Castro's attempt to export revolution to the rest of Latin America was largely crushed by the capture of his ally "Che" Guevara in 1967. In 1980 he lifted the ban on emigration and 125,000 people left for Florida. While Castro was able to maintain political independence from the Soviet Union, the Cuban economy was heavily dependent on Soviet economic aid. The combined effect of the collapse of European communism and the continuing US trade embargo dramatically worsened the Cuban economy, leading Castro to introduce cautious economic reforms. *244–45, 258–59*

CATHERINE DE MEDICI (1519–89) Queen of France, wife of HENRY II and daughter of Lorenzo de Medici. She exerted considerable political influence after her husband's and first son's deaths in 1559. In 1560 she became regent for her second son, Charles IX, and remained principal adviser until his death (1574). Her initial tolerance of the HUGUENOTS turned to enmity at the beginning of the FRENCH WARS OF RELIGION. Her concern for preserving the power of the monarchy led to a dependence on the CATHOLIC House of Guise, whose growing power she failed to control. Fearing the decline of her own importance at court due to the rise of the Huguenot leader Gaspard de Coligny, she planned the St Bartholomew's Day Massacre (1572). When her third son, Henry III, acceded to the throne in 1574, her effectiveness in policy-making had been compromised. *154–55*

CATHERINE II (THE GREAT) (1729–96) Empress of Russia (1762–96), born in Germany. In 1745 she married the future tsar Peter III, who succeeded to the throne in 1761. With the help of her lover, Grigori Orlov, Catherine overthrew her husband and shortly afterwards he was murdered. Catherine began her reign as an "enlightened despot" with ambitious plans for reform, and in 1762 she secularized the property of the clergy to fund her projects. In 1764 she secured the accession of a former lover to the Polish throne as Stanislaus II. However, her plans for internal reform came to nothing and, after the peasants' revolt of 1773–74, she became increasingly conservative, protecting the interests of the landowners and imposing serfdom on most of those peasants who were not already enslaved. Her centralizing reform of local government in 1775 lasted until the Russian Revolution. In 1785 Catherine extended the powers of the nobility at the expense of the serfs. Catherine's foreign policy, guided by Potemkin, vastly extended

Russian territory chiefly at the expense of the Ottoman Empire. Russia emerged from the first Russo-Turkish War (1768–74) as the dominant power in the Middle East. Crimea was annexed in 1783 and Alaska was colonized. Her dialogue with leading Enlightenment figures such as Voltaire did much to promote her contemporary image in Europe. ❍ *148–49*

CATHOLIC CHURCH Term used in CHRISTIANITY with one of several connotations: (1) It is the Universal church, as distinct from local churches. (2) It is the church holding "orthodox" doctrines, defined by St Vincent of Lérins as doctrines held "everywhere, always, and by all" – in this sense the term is used to distinguish the church from heretical bodies. (3) It is the undivided church as it existed before the schism of East and West in 1054. Following this, the Western church called itself "Catholic", the Eastern church "Orthodox". (4) Since the Reformation, the term has usually been used to denote the Roman Catholic Church. ❍ *154–55, 268–69*

CAVOUR, CAMILLO BENSO, CONTE DI (1810–61) Piedmontese politician, instrumental in uniting Italy under Savoy rule. From 1852 he was prime minister under VICTOR EMMANUEL II. He engineered Italian liberation from Austria with French aid, expelled the French with the help of Giuseppe GARIBALDI, and finally neutralized Garibaldi's influence. This led to the formation of the kingdom of Italy in 1861. ❍ *176–77*

CEAUSESCU, NICOLAE (1918–89) Romanian politician, the country's effective ruler from 1965–89. He became a member of the Politburo in 1955, general secretary of the Romanian Communist Party in 1965 and head of state in 1967. He promoted Romanian nationalism, pursued an independent foreign policy, but instituted repressive domestic policies. He was deposed and executed in the December 1989 revolution. ❍ *264–65*

CELTS Peoples who, after 2000 BC, spread from eastern France and western Germany over much of western Europe, including Britain. They developed a village-based, hierarchical society headed by nobles and Druids. Conquered by the Romans, the Celts were pushed into Ireland, Wales, Cornwall and Brittany by Germanic peoples. Their culture remained vigorous, and Celtic churches were important in the early spread of CHRISTIANITY in northern Europe. ❍ *20–21*

CENTRAL POWERS Alliance of Germany and Austria–Hungary (with Bulgaria and Turkey) during the First World War. The name distinguished them from their opponents (Britain, France and Belgium) in the west, with Russia and others in the east. ❍ *218–19*

CHANDRAGUPTA Founder of the MAURYAN Empire in India (r. c. 321–297 BC) and grandfather of ASHOKA. He seized the throne of Magadha and defeated SELEUCUS, gaining dominion over most of northern India and part of Afghanistan. His reign was characterized by religious tolerance. He established a vast bureaucracy and secret service based at Patna. He abdicated in favour of his son Bindusara and, it is thought, became a Jain monk before dying. ❍ *46–47*

CHARLEMAGNE (742–814) King of the Franks (768–814) and Holy Roman Emperor (800–14). The eldest son of PIPPIN III (THE SHORT), he inherited half the Frankish kingdom (768), annexed the remainder on his brother Carloman's death in 771, and built a large empire. He invaded Italy twice and took the Lombard throne in 773. Between 772 and 804 he undertook a long and brutal conquest of Saxony, annexed the independent duchy of Bavaria in 788 and defeated the Avars of the middle Danube. He undertook campaigns against the Moors in Spain. In 800 he was consecrated as Emperor by Pope Leo III, thus reviving the concept of the Roman Empire, and confirming the separation of the West from the Eastern, Byzantine Empire. Charlemagne encouraged the intellectual awakening of the Carolingian Renaissance, set up a strong central authority and maintained provincial control through court officials. He undertook the reform of the judicial and legal systems, introduced jury courts, established schools and made his capital, Aachen, a centre of learning. His central aim was

Christian reform, both of church and laity, and he promoted missionary work and monastic reforms. He was said to be convinced that God had chosen him to undertake this holy work. ❍ *74–75, 90–91*

Designed to house Charlemagne's throne, his Palace Chapel at Aachen (Aix-la-Chapelle) was the architectural masterpiece of Carolingian Europe.

CHARLES I (1600–49) King of England, Scotland and Ireland (r. 1625–49). Son of James I, he was criticized by Parliament for his reliance on the Duke of Buckingham and for his CATHOLIC marriage to Henrietta Maria. Although he accepted the petition of right, Charles's insistence on the "divine right of kings" provoked further conflict with Parliament, and led him to rule without it for 11 years (1629–40). When attempts to impose Anglican liturgy on Scotland led to the Bishops' War, Charles was forced to recall Parliament to raise revenue. Parliament insisted on imposing conditions, and impeached Charles's chief advisers. Relations steadily worsened, and Charles's attempt to arrest five leading opponents in the Commons precipitated the ENGLISH CIVIL WAR. After the defeat of the Royalists, attempts by the parliamentary and army leaders to reach a compromise with the king failed, and he was tried and executed in January 1649. ❍ *156–57*

CHARLES II (THE BALD) (823–77) King of the West Franks (843–77) and Holy Roman Emperor (875–77). Younger son of Emperor Louis I, he was involved in the ambitious disputes of his elder brothers. The Treaty of VERDUN (843) made him king of the West Franks, in effect the first king of France. Continuing family conflict, rebellion and Viking attacks resulted in territorial losses. Yet, after the death of the Emperor Louis II, he was recognized as Holy Roman Emperor. ❍ *74–75*

CHARLES IV (1316–78) Holy Roman Emperor (1355–78) and King of BOHEMIA (1347–78). Supported by Pope Clement VI, he was a rival of the Wittelsbach Emperor Louis IV, and when Louis died, was elected King of the Germans (Emperor-elect). A skilful diplomat, he blocked or appeased his Wittelsbach and Habsburg rivals and improved relations with the Papacy. In 1356 he introduced a stable system of imperial government. He ruled from Prague, his birthplace, where he founded Charles University (1348) and built the Charles Bridge. Czech culture reached a peak under his patronage. He was succeeded by his son, Wenceslas. ❍ *90–91*

CHARLES IV (1748–1819) King of Spain (1788–1808), son and successor of Charles III. Charles virtually turned over government to his wife Maria Luisa and her lover Manuel de Godoy. Godoy formed an alliance with France, but Spain was nevertheless occupied by French troops in the PENINSULAR WAR. Charles was forced to abdicate in favour of his son FERDINAND VII, who in turn was forced from the throne by NAPOLEON. ❍ *166–67*

CHARLES V (1500–58) Holy Roman Emperor (1519–58) and King of Spain, as Charles I (1516–56).

He ruled the Spanish kingdoms, southern Italy, the Netherlands and the Austrian Habsburg lands by inheritance and, when elected emperor in succession to his grandfather, Maximilian I, headed the largest European empire since CHARLEMAGNE. In addition, the Spanish *conquistadores* made him master of a New World empire. Charles's efforts to unify his possessions were unsuccessful, largely due to the hostility of FRANCIS I of France, the Ottoman Turks in central Europe, and the conflicts arising from the advance of LUTHERANISM in Germany. The struggle with France was centred in Italy: Spanish control was largely confirmed by 1535, but French hostility was never overcome. The Turks were held in check but not defeated, and Charles's attempt to capture Algiers failed in 1541. In Germany, Charles, who saw himself as the defender of the CATHOLIC CHURCH, nevertheless recognized the need for reform, but other commitments prevented him following a consistent policy, and Lutheranism expanded. Charles increasingly delegated power in Germany to his brother FERDINAND I, his successor, and in 1554–56 surrendered his other titles to his son, PHILIP II of Spain. ❍ *146–47, 152–55*

CHARLES VIII (1470–98) King of France (1483–98). He succeeded his father Louis XI and until 1491 was controlled by his sister Anne de Beaujeu and her husband. Obsessed with gaining the kingdom of Naples, he invaded Italy in 1494, beginning the long Italian Wars, and in 1495 he entered Naples. A league of Italian states, the Papacy and Spain forced him to retreat. One positive result was the introduction of Italian RENAISSANCE culture into France. He was succeeded by his cousin Louis XII. ❍ *158–59*

CHARLES X (1622–60) King of Sweden (1654–60). He became king after the abdication of his cousin, Queen Christina. His efforts to dominate the Baltic resulted in a reign of continuous military activity. He established the natural frontiers in Scandinavia, recovering the southern provinces of Sweden from Denmark. He was succeeded by his son, Charles XI. ❍ *150–51*

CHARLES X (1757–1836) King of France (1824–30). Brother of LOUIS XVI and LOUIS XVIII, he fled France at the outbreak of the French Revolution in 1789. He remained in England until the BOURBON restoration in 1814. He opposed the moderate policies of Louis XVIII. After the assassination of his son in 1820, his reactionary forces triumphed. In 1825 he signed a law indemnifying émigrés for land confiscated during the Revolution. In 1830 he issued the July Ordinance, which restricted suffrage and press freedom, and dissolved the newly elected chamber of deputies. The people rebelled and Charles was forced to abdicate. He designated his grandson Henry as successor, but the Duc d'Orléans, LOUIS PHILIPPE, was selected. ❍ *172–73*

CHARLES XII (1682–1718) King of Sweden (1697–1718). His father, Charles XI, trained him in all aspects of administration, and he became king when 14 years old. Charles XII was one of the greatest military leaders in European history. He defeated Denmark, Poland, Saxony and Russia in a series of brilliant campaigns. Leading the battle, he destroyed the army of PETER I (THE GREAT) at Narva in 1700. In 1708 he renewed his assault on Russia, but his army, depleted by the severe winter, was decisively defeated at POLTAVA in 1709. He fled to the Ottomans and persuaded the sultan to attack Russia. The sultan turned against him and, in disguise, he escaped back to Sweden and devoted his energy to the domestic economy. Weakened by war and opposed by numerous enemies, he was killed, while fighting in Norway, and succeeded by his sister, Ulrica Eleanora. ❍ *148–49, 150–51*

CHARLES ALBERT (1798–1849) King of Sardinia-Piedmont (1831–49). He was a liberal reformist who opposed Austria and sought Italian liberation. After defeat by Austria in 1848–49, he abdicated in favour of his son, VICTOR EMMANUEL II. ❍ *176–77*

CHARTISM (1838–48) British working-class movement for political reform. Combining the discontent of industrial workers with the demands of radical artisans, the movement adhered to the People's Charter (1838), which demanded electoral reform including universal male suffrage. As well as local riots and strikes, the Chartists organized mass petitions in 1839, 1842, and 1848. The movement faded away after a major demonstration in 1848. ❍ *172–73*

CHERNOBYL DISASTER Nuclear power station in central Ukraine. On 26 April 1986 an explosion in one of the reactors released 8 tonnes of radioactive material into the atmosphere. Within the first few hours 31 people died. Fallout spread across eastern and northeastern Europe, contaminating much agricultural produce. More than 100,000 people were evacuated from the vicinity of the plant. The reactor was encased in cement and boron. Data on the long-term effects of contamination are inconclusive, though 25,000 local inhabitants have died prematurely. Two of the three remaining reactors were reworking by the end of 1986. In 1991 Ukraine's government pledged to shut down the plant, but energy needs dictated its continued output. In 1994 the West pledged economic aid to ensure the plant's closure. ❍ *236–37, 262–63, 280–81*

CHERNOMYRDIN, VIKTOR Russian prime minister (1992–98). A member of the Central Committee of the Communist Party from 1986 to 1990, Chernomyrdin became prime minister despite the objections of Boris YELTSIN. He broadly supported economic reform, but was critical of the pace of privatization. Yeltsin's illness meant that Chernomyrdin acted as caretaker-president during much of 1996. Yeltsin dismissed him and the entire cabinet in 1998. ❍ *262–63*

CHIANG KAI-SHEK (1887–1975) (Jiang Jieshi) Chinese nationalist leader. After taking part in resistance against the Qing dynasty, he joined the KUOMINTANG, succeeding SUN YAT-SEN as leader in 1925. From 1927 he purged the party of communists, and headed a Nationalist government in Nanjing. The Japanese invasion in 1937 forced a truce between Nationalists and Communists, and during the Second World War, with US support, Chiang led the fight against Japan. Civil war resumed in 1945. Chiang was elected president of China in 1948, but in 1949 the victorious Communists, led by MAO ZEDONG, drove his government into exile in Taiwan. Chiang established a dictatorship and maintained that the KUOMINTANG were the legitimate Chinese government. The United Nations finally accepted the Chinese Communist Party as the official government in 1972. Chiang Kai-shek remained president of Taiwan until his death. ❍ *224–25, 234–35, 244–45*

CHINGGIS KHAN (Ghengis Khan) (c. 1167–1227) Conqueror and founder of the Mongol Empire. He united the Mongol tribes in 1206 and demonstrated his military genius by conquering northern China between 1207 and 1215, annexing Iran and invading Russia. His empire was divided and expanded by his sons and grandsons. ❍ *98–99*

CHRISTIANITY Religion based on belief in JESUS CHRIST as the Son of God. The orthodox Christian faith, summarized in the Apostles' and Nicene Creeds, affirms belief in the Trinity and Christ's incarnation, atoning death on the cross, resurrection and ascension. The moral teachings of Jesus are contained in the New Testament. The history of Christianity has been turbulent and often sectarian. The first major schism took place in 1054, when the eastern and western churches separated. The next occurred in the 16th-century Reformation, with the split of PROTESTANTISM and the Roman CATHOLIC CHURCH. In recent times the ecumenical movement, which aims at the reunion of all Christians, has gained strength. ❍ *44–45, 54–57, 62–63, 90–91, 94–95, 136–37, 154–55*

CHURCHILL, WINSTON LEONARD SPENCER (1874–1965) British statesman. Son of Lord Randolph Churchill, he was a reporter in the SOUTH AFRICAN (BOER) WAR. Elected to parliament in 1900 as a Conservative, he joined the Liberals in 1904 and served in government. In 1924 he resumed allegiance to the Conservatives. Out of office from 1929 to 1939, he spoke out against the rising threat of NAZI Germany. He became First Lord of the Admiralty on the outbreak of the Second World War, and replaced Neville Chamberlain as prime minister in 1940. He was an inspirational war leader. Cultivating close relations with US

President ROOSEVELT, he was the principal architect of the grand alliance of Britain, the United States and the Soviet Union, which eventually defeated the Germans. Rejected by a reform-hungry electorate in 1945, he was prime minister again in 1951–55. ❍ *242–43*

CLAUDIUS I (10 BC–54 AD) (Tiberius Claudius Nero Germanicus) Roman emperor (AD 41–54). The nephew of TIBERIUS, he was the first emperor chosen by the army. He had military successes in Germany, conquered Britain in AD 43, and built both the harbour of Ostia at the mouth of the Tiber and the Claudian aqueduct. Agrippina (his fourth wife) poisoned him and made her son, NERO, emperor. ❍ *54–55*

CLEMENCEAU, GEORGES (1841–1929) French statesman, premier (1906–9, 1917–20). A moderate republican, he served in the Chamber of Deputies from 1876 to 1893, attempted compromise during the revolt of the Paris Commune (1871) and strongly supported Dreyfus. Clemenceau returned to the Senate in 1902. Concerned with the growing power of Germany, his first term as premier saw the strengthening of relations with Britain. After the First World War Clemenceau returned to power and led the French delegation at the Paris Peace Conference. ❍ *220–21*

CLIVE, ROBERT, BARON CLIVE OF PLASSEY (1725–74) British soldier and administrator. He went to India as an official of the English East India Company in 1743, and successfully resisted growing French power. By taking Calcutta and defeating the pro-French Nawab of Bengal at Plassey in 1757 he effectively assured British control of northeastern India. He was governor of Bengal (1757–60, 1765–67). He returned to England in 1773 and was charged with, but acquitted of, embezzling state funds. He committed suicide. ❍ *194–95*

CLOVIS I (465–511) Frankish king (482–511) of the MEROVINGIAN dynasty. He overthrew the Romanized kingdom of Soissons and conquered the Alemmani near Cologne. He and his army converted to CHRISTIANITY in fulfilment of a promise made before the battle. In 507 he defeated the VISIGOTHS under Alaric II near Poitiers. By the time he died he controlled most of GAUL and had firmly established Merovingian power. ❍ *74–75*

CNUT II (c. 994–1035) King of Denmark (1014–28), England (1017–35) and Norway (1028–29). He accompanied his father, SVEIN FORKBEARD, on the Danish invasion of England in 1013. After his father's death in 1014 he was accepted as joint king of Denmark with his brother and later became sole king. He invaded England again in 1015 and divided it with Edmund Ironside, the English king. He was accepted as king after Edmund's death. His rule was a just and peaceful one. He restored the church and codified English law. His reign in Scandinavia was more turbulent. He conquered Norway in 1028, made one son king of Denmark in 1028 and another king of Norway the following year. ❍ *78–79*

COLUMBUS, CHRISTOPHER (1451–1506) Italian explorer incorrectly credited with the European discovery of America. In the belief that the circumference of the Earth is much smaller than it actually is, he thought he could establish a route to China and the East Indies by sailing across the Atlantic After several disappointments, he secured Spanish patronage from FERDINAND V and Isabella. He set out with three ships in 1492 and made landfall in the Bahamas, the first European to reach the Americas since the Vikings, whose achievement was then unknown. Believing he had reached the East, he called the inhabitants "Indians". On a second expedition in 1493 a permanent colony was established in Hispaniola. He made two more voyages (in 1498 and 1502) exploring the Caribbean region without reaching the North American mainland. He never surrendered his belief that he had reached Asia. His discoveries laid the basis for the Spanish empire in the Americas. ❍ *116–17, 120–23*

COMECON See COUNCIL FOR MUTUAL ECONOMIC ASSISTANCE.

COMINTERN See COMMUNIST INTERNATIONAL.

COMMONWEALTH OF INDEPENDENT STATES **(CIS)** Alliance of 12 of the former republics of the Soviet

Union. The CIS was formed in 1991 with Armenia, Azerbaijan, Belarus, Georgia, Kazakstan, Kyrgyzstan, Moldova, Russia, Tajikistan, Turkmenistan, Ukraine and Uzbekistan. The Baltic states (Estonia, Latvia and Lithuania) did not join. All except Ukraine signed a treaty of economic union in 1993, creating a free trade zone. Russia is the dominant power, with an overall responsibility for defence and peacekeeping. It is also the main provider of oil and natural gas. ❍ *262–63*

COMMONWEALTH OF NATIONS Voluntary association of 53 states, consisting of English-speaking countries formerly part of the British Empire. Headed by the British sovereign, it exists largely as a forum for discussion of issues of common concern. A Commonwealth secretariat is located in London. ❍ *246–47*

COMMUNIST INTERNATIONAL (Comintern, Third International) Communist organization founded by Lenin in 1919. He feared the re-emergence of the reformist Second International and wished to secure control of the world socialist movement. The Comintern was made up mainly of Russians, and failed to organize a successful revolution in Europe in the 1920s and 1930s. The Soviet Union abolished the Comintern in 1943 to placate its Second World War allies. ❍ *224–25*

COMPROMISE OF 1850 Set of balanced resolutions by Senator Henry Clay to prevent civil war. The US Congress agreed to admit California as a free state, organize New Mexico and Utah as territories without mention of slavery, provide for a tougher fugitive slave law, abolish the slave trade in Washington, D.C., and assume the Texas national debt. ❍ *184–85*

CONFEDERATE STATES OF AMERICA Southern states that seceded from the Union following the election of Abraham LINCOLN. South Carolina left in December 1860, and was followed closely by Alabama, Florida, Georgia, Louisiana, Mississippi and Texas. In March 1861, Jefferson Davis was elected president and a new constitution protected states' rights and retained slavery. A capital was established at Montgomery, Alabama. On 12 April the US Civil War began, and Arkansas, North Carolina, Tennessee and Virginia joined the Confederacy. The capital was moved to Richmond, Virginia. The Confederacy, despite its cotton trade, received little external support and internal division contributed to its defeat and dissolution in April 1865. ❍ *184–85*

CONFEDERATION OF THE RHINE Alliance of German states proposed by NAPOLEON in January 1806 and agreed to by 16 German princes in July 1806. They renounced their attachment to the HOLY ROMAN EMPIRE, placed themselves under the protection of Napoleon, and pledged 63,000 men to his army. Eventually, the Confederation included Bavaria, Baden, Saxony, Württemberg, Westphalia, Hesse-Darmstadt, and most of the minor states, many of which joined at Napoleon's defeat of the Prussian army on 14 October 1806. A device to control the German princes, the alliance broke apart after Napoleon's defeat in Russia (1812–13). ❍ *176–77*

CONFUCIANISM Philosophy that dominated China until the early 20th century and still has many followers, mainly in Asia. It is based on the Analects, sayings attributed to Confucius. Strictly an ethical system to ensure a smooth-running society, it gradually acquired quasi-religious characteristics. Confucianism views man as potentially the most perfect form of *li*, the ultimate embodiment of good. It stresses the responsibility of sovereign to subject, of family members to one another, and of friend to friend. Politically, it helped to preserve the existing order, upholding the status of the emperors. When the empire was overthrown in 1911, Confucian institutions were ended, but after the Communist Revolution of 1949 many Confucian elements were incorporated into Maoism. ❍ *44–45, 138–39, 196–97*

CONFUCIUS (c. 551–479 BC) Chinese philospher, founder of CONFUCIANISM. Born in Lu, he first worked as a minor official before becoming a teacher. He was an excellent scholar and became an influential teacher of the sons of wealthy families. He is said to have been prime minister of Lu, but he resigned when he realized the post carried no real authority. In his later years he sought a return to the political morality of the early ZHOU dynasty. ❍ *44–45, 48–49*

CONSTANCE, COUNCIL OF (1414–18) Ecumenical council that ended the GREAT SCHISM. It was convoked by the antipope John XXIII. Martin V was elected as a new pope in 1417. The Council also attempted to combat heresy, notably that of Jan HUS. ❍ *106–7*

CONSTANTINE I (THE GREAT) (285–337) Roman emperor (306–37). The first Christian emperor and founder of CONSTANTINOPLE (modern Istanbul). By extending tolerance to Christians throughout the Roman Empire from 313, he may have hoped to achieve political unity. In 324 he won sole control of the empire, and in 325 presided over the first council of the Christian Church at Nicaea. He rebuilt Byzantium as his capital and renamed it CONSTANTINOPLE. ❍ *54–55*

CONSTANTINOPLE Former name of Istanbul, in Turkey. Formally Byzantium, Constantinople was founded as the new capital of the Roman Empire by CONSTANTINE I in AD 330. It was to become one of the great capitals of the world, a centre of power, wealth, commerce and the arts, reaching its zenith under Emperor JUSTINIAN I (527–565). However, in 542 a plague struck the city, killing more than half its population. The city went into a decline from which it did not recover for a further three centuries, and during this period it came under frequent attack from neighbouring kingdoms. In 1204 the city was sacked by the Fourth Crusade and the LATIN EMPIRE was established. The Byzantines retook the city in 1261, but by now the empire was in decline and, in 1453, Constantinople fell to the Ottomans. ❍ *66–69, 94–97*

CONSTANTINOPLE, LATIN EMPIRE OF (1204–61) Empire established after the sacking of CONSTANTINOPLE (now Istanbul) by the leaders of the Fourth Crusade. The empire lay on both sides of the Dardanelles and was divided among the crusaders. Constantinople was placed under the control of Baldwin I. It was constantly under attack from its neighbours, and declined rapidly: Thessalonica fell in 1222 and Asia Minor in 1224. The Latin Empire formally ended in 1261, when Constantinople was recaptured by the Byzantine emperor Michael VIII. ❍ *94–97*

CONSTITUTIONAL ACT 1791 Instituted by British prime minister William Pitt, the act divided French and English Canada, creating a system of government dominated by an appointed executive branch. The elected Legislative Assembly dealt only with local issues. Dissatisfaction with the form of government led to the Rebellion of 1837 and to the Durham Report of 1839. ❍ *188–89*

COOK, JAMES (1728–79) British naval officer and explorer. He demonstrated his remarkable navigational talent charting the approaches to Quebec during the SEVEN YEARS WAR. In 1768–71 he led an expedition to Tahiti to observe an eclipse of the Sun and investigate the strategic and economic potential of the South Pacific. He conducted a survey of the unknown coasts of New Zealand and charted the east coast of Australia, naming it New South Wales and claiming it for Britain. On a second expedition to the south Pacific (1772–75), Cook charted much of the southern hemisphere and circumnavigated Antarctica. On his last voyage (1776–79) he discovered the Sandwich (Hawaiian) Islands, where he was killed in a dispute with the inhabitants. Cook is generally regarded as the greatest European explorer of the Pacific in the 18th century. ❍ *202–3*

COPERNICUS, NICOLAUS (1473–1543) (Mikotay Kopernik) Polish astronomer. He studied in Kraków and Italy, and lectured in Rome. Through his study of planetary motion, Copernicus developed a heliocentric (Sun-centred) theory of the universe in opposition to the accepted geocentric (Earth-centred) theory conceived by Ptolemy nearly 1,500 years before. In the Copernican system (as it is now called) the planet's motions in the sky were explained by their orbit of the Sun. The motion of the sky was simply a result of the Earth turning on its axis. An account of this work, *De*

revolutionibus orbium coelestium, was published in 1543. Most astronomers considered the new system as merely a means of calculating planetary positions, and continued to believe in Aristotle's view of the the world. ❍ *134–35*

CORTÉS, HERNÁN (1485–1547) Spanish *conquistadore* and conqueror of Mexico. After studying law at Salamanca University he became secretary to the governor of Cuba, who sent him on an expedition to Mexico in 1518 with 550 men. Once landed, Cortés declared himself independent of the governor and marched inland toward the Aztec capital, Tenochtitlan (Mexico City), gaining allies among the subject peoples of the Aztec king, MONTEZUMA II. While Cortés was absent conflict broke out. Cortés recaptured the city after a three-month siege in 1521. During the following five years Cortés brought the Aztec lands, Honduras and much of El Salvador and Guatemala under Spanish control. However, his personal power, symbolized by his titles and estates, was gradually eroded by the Crown, and he died in Spain, a private citizen. ❍ *116–17, 120–21*

COSSACKS Bands of Russian adventurers who undertook the conquest of Siberia in the 17th century. Of ethnically mixed origins, they were escaped serfs, renegades and vagabonds who formed independent, semi-military groups on the fringe of society. After the RUSSIAN REVOLUTION of 1917 the Cossacks opposed the BOLSHEVIKS and strongly resisted collectivization. ❍ *148–49, 156–57, 222–23*

COUNCIL FOR MUTUAL ECONOMIC ASSISTANCE (COMECON) International organization (1949–91) aimed at the co-ordination of economic policy among communist states, especially in Eastern Europe. Led by the Soviet Union, its original members were Albania, Bulgaria, Czechoslovakia, East Germany, Hungary, Poland and Romania; it was later joined by Cuba, Mongolia and Vietnam. Co-operation took the form of bilateral trade agreements rather than the establishment of a single market or a uniform price system. ❍ *236–37*

CRÉCY, BATTLE OF (1346) First major battle of the HUNDRED YEARS WAR. The English led by Edward III and his son, the Black Prince, defeated the French forces of Philip VI. The English longbow, as well as superior tactics, accounted for their victory. ❍ *106–7*

CREOLE Person born in the West Indies, Latin America, or southern USA of foreign or mixed descent. Generally a Creole's ancestors were either African slaves or French, Spanish or English settlers. In the USA it also refers to someone of mixed European and African ancestry. A creole language is one that develops as a result of extended contact between two language communities, one of which is usually European (often English, French or Portuguese), and which is adopted as the native language of a community. ❍ *190–91*

CRIMEAN WAR (1853–56) Fought by Britain, France and the Ottoman Turks against Russia. In 1853 Russia occupied Turkish territory and France and Britain, determined to preserve the Ottoman Empire, invaded the Crimea in 1854 to attack Sevastopol. The war was marked on both sides by incompetent leadership and organization. The Charge of the Light Brigade is the best-known example. Sevastopol was eventually captured in 1855. At the Treaty of Paris (1856) Russia surrendered its claims on the Ottoman Empire. ❍ *178–79*

CROESUS (d. c. 546 BC) King of LYDIA in Asia Minor (c. 560–546 BC). Renowned for his wealth, he was overthrown and captured by CYRUS THE GREAT. According to HERODOTUS, Croesus threw himself upon a funeral pyre. ❍ *42–43*

CROMWELL, OLIVER (1599–1658) Ruler of England as Lord Protector (1653–58). Cromwell was a landowner from Huntingdon who entered Parliament in 1628. A committed Puritan, he raised troops for Parliament at the outbreak of the ENGLISH CIVIL WAR, and rose to prominence as the outstanding parliamentary commander, crushing Scots and Irish as well as English Royalists. He tried to reach a compromise with CHARLES I, whose duplicity convinced Cromwell of the

need to execute him. The loyalty of his troops made him powerful in the subsequent quarrels between Parliament and army, and when parliamentary government failed in 1653, army officers drew up the Instrument of Government making Cromwell Lord Protector. As ruler, he pursued a dynamic, anti-Spanish foreign policy while endeavouring to restore social stability at home, suppressing the radical Protestant sects which had emerged during the Civil War. He was offered the crown by Parliament, but refused it. He was succeeded by his son Richard Cromwell. ❍ *156–57*

CULTURAL REVOLUTION (1966–72) The Great Proletarian Cultural Revolution was initiated by MAO Zedong and his wife, Jiang Qing, to purge the Chinese Communist Party of his opponents and to instil correct revolutionary attitudes. Senior party officials were removed from their posts, intellectuals and others suspected of revisionism were victimized and humiliated. A new youth corps, the Red Guards, staged protests, held rallies and violently attacked reactionary ideas. By 1968 China was approaching civil war. The Red Guards were disbanded and the army restored order. ❍ *254–55*

CYPRIOT CIVIL WAR (1964–) Cyprus became an independent country in 1960 with Makarios, the Greek Orthodox Archbishop, as the first president. The constitution of independent Cyprus provided for power sharing between the Greek and Turkish Cypriots. It proved unworkable, however, and fighting broke out between the two communities. In 1964 the UNITED NATIONS sent in a peace-keeping force, but clashes recurred in 1967. In 1974 Greek-led Cypriot forces overthrew Makarios. This led Turkey to invade north Cyprus, occupying about 40 per cent of the island. Many Greek Cypriots fled from the Turkish-occupied area, which, in 1979, was proclaimed to be a self-governing region. In 1983 the Turkish Cypriots declared the north to be an independent state called the Turkish Republic of Northern Cyprus; the only country to recognize it is Turkey. The UN regards Cyprus as a single nation under the Greek Cypriot government. In the south, border clashes continued in the 1990s; the UN reported 900 incidents on the front line in 1996. ❍ *266–67*

CYRIL Greek Christian missionary. With his brother, Methodius, he is one of the two so-called "Apostles to the Slavs" who were sent to convert the Khazars and Moravians to CHRISTIANITY. Cyril is said to have invented the Cyrillic alphabet. ❍ *70–71*

CYRUS THE GREAT (600–529 BC) King of PERSIA, founder of the Achaemenid Persian Empire. He overthrew the Medes, then rulers of Persia, in 549 BC, defeated King CROESUS of LYDIA (c. 546 BC), captured BABYLON (539 BC) and the Greek cities in Asia Minor. Though he failed to conquer EGYPT, his empire stretched from the Mediterranean to India. He delivered the JEWS from their Babylonian captivity, sending them home to PALESTINE. ❍ *42–43*

DACIA Ancient region of Europe (now in Romania). It was colonized by TRAJAN in 101–106. Dacia was later overrun by GOTHS, Huns and Avars. The language is the basis of modern Romanian. ❍ *54–55*

DANELAW Large region of northeast England, occupied by Danes in the late 9th century. Its independent existence was formally confirmed in 886 by ALFRED (THE GREAT) and Guthrum's Pact. Alfred's son, Edward the Elder, and grandson, Athelstan, restored it to English control in the early 10th century. ❍ *78–79*

DAOISM Chinese philosophy and religion considered as being next to CONFUCIANISM in importance. Daoist philosophy is based on the teachings of LAO-TZE as written down, probably in the 3rd century BC, in the *Dao De Jing*. The recurrent theme of this work is the Dao (way or path). To follow the Dao is to follow the path leading to self-realization. Daoist ethics emphasize patience, simplicity and the harmony of nature, achieved through the proper balance of the yin, or female principle, and yang, or male principle. As a religion, Daoism dates from the time of Jang Dao-ling, who organized a group of followers in AD 142. ○ *44–45*

DARIUS I (c. 558–486 BC) King of PERSIA (521–486 BC) of the Achaemenid dynasty. Troubled by revolts, particularly in BABYLON, he restored order by dividing the empire into provinces, allowing some local autonomy and tolerating religious diversity. He also fixed an annual taxation and developed commerce. He was defeated at MARATHON in 490 BC. ○ *42–43*

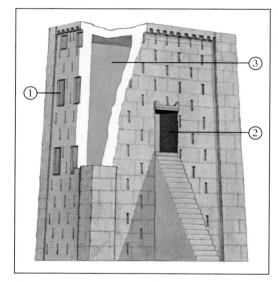

The so-called Fire Temple of Naqsh-i Rustam, near Persepolis, stands in front of a cliff in which the four tombs of Darius I and his successors are carved. It is approximately 11 metres (36 feet) high with blind windows (1) of black limestone and a door (2) leading to an empty room (3).

DARIUS III (380–330 BC) King of PERSIA (336–330 BC). By underestimating the strength of ALEXANDER THE GREAT, he brought about the demise of the Achaemenid Persian empire. Defeated at Issus in 333 BC and Gaugamela in 331 BC, Darius was forced to flee to Ecbatana and then to Bactria, where he was killed by the governor Bessus. ○ *42–43*

DAVID King of ISRAEL (c. 1006–c. 966 BC). His career is related in the Old Testament. He became a hero by defeating Goliath in a duel and was made king of Judah on SAUL's death. He united Judah and Israel and made JERUSALEM his capital. ○ *44–45*

DAYTON PEACE ACCORD See BOSNIAN CIVIL WAR

DECLARATION OF INDEPENDENCE (4 July 1776) Statement of the principles with which the Thirteen Colonies of North America justified the American Revolution and separation from Britain as the United States of America. Its blend of high idealism and practical statements have ensured its place as one of the world's most important political documents. The Declaration was drafted by a committee that included Thomas JEFFERSON, and was based on the theory of natural rights, propounded by John Locke to justify the Glorious Revolution in England. It was approved by the Continental Congress on 4 July. The Declaration states the necessity of government having the consent of the governed, of government's responsibility to its people, and contains the famous paragraph: "We hold these truths to be self-evident, that all men are created equal, that they are endowed by their Creator with certain unalienable Rights, that among these are Life, Liberty, and the Pursuit of Happiness." ○ *164–65*

DE GAULLE, CHARLES ANDRÉ JOSEPH MARIE (1890–1970) French general and statesman, first president (1959–69) of the fifth republic. His experience of the First World War (he was captured in 1916), convinced him of the need to modernize the French army. In 1940 he became undersecretary of war, but fled to London after the German invasion. He organized

French Resistance (Free French) forces, and in June 1944 was proclaimed president of the provisional French government. Following liberation he resigned, disenchanted with the political settlement. In 1958 he emerged from retirement to deal with the war in Algeria. In 1959 a new constitution was signed, creating the French Community. In 1962 De Gaulle was forced to cede Algerian independence. Gaullist foreign policy attempted to reinstate France as a world power. France gained an independent nuclear capability, but alienated the UK and USA by its temporary withdrawal from NATO and by blocking British entry into the EEC. De Gaulle's devaluation of the franc brought relative domestic prosperity. He was re-elected in 1965 but resigned following defeat in a 1969 referendum. ○ *238–39*

DEMOCRATIC PARTY US political party, descendant of the Anti-Federalist Party and Democratic Republican Party. From the election of THOMAS JEFFERSON (1801) until James Buchanan in 1857, the Party was the dominant force in US politics, gathering support from farmers and white-collar workers. The Party was split by the Civil War (1861–65), with support mainly restricted to the South and West. It regained power in 1932 with Franklin D. ROOSEVELT's "New Deal" policy. Democratic presidents were in office 1961–69 (John F. KENNEDY, Lyndon Johnson), a period marked by progressive economic and social policy, such as the passing of civil rights legislation. The 1970s and 1980s were more barren years, during which only Jimmy CARTER (1977–81) held the presidency, and the REPUBLICAN PARTY was seen as the dominant force. Bill Clinton recaptured the centre ground of US politics and was elected in 1992 and re-elected in 1996. ○ *240–41*

DENG XIAOPING (1904–97) Chinese statesman. He took part in the LONG MARCH, served in the Red Army, and became a member of the central committee of the Chinese Communist Party in 1945. After the establishment of the People's Republic in 1949 he held several important posts, becoming general secretary of the Party in 1956. During the CULTURAL REVOLUTION he was denounced for capitalist tendencies and dismissed. He returned to government in 1973, was purged by the Gang of Four in 1976, but reinstated in 1977 after the death of MAO Zedong. Within three years he ousted Hua Guofeng to become the dominant leader of party and government, though without holding the usual titles. He introduced rapid economic modernization, encouraging foreign investment, but without social and political liberalization. Deng officially retired in 1987, but was still essentially in control at the time of the Tiananmen Square massacre (1989). In 1993 Jiang Zemin became president. ○ *254–55*

DIAS, BARTHOLOMEW (c. 1450–1500) Portuguese navigator, the first European to round the Cape of Good Hope. In 1488, under the commission of King John II of Portugal, Dias sailed three ships around the Cape, opening the long-sought route to India. He took part in the expedition of Cabral that discovered Brazil, but was drowned when his ship foundered. ○ *116–17*

DIEM, NGO DINH (1901–63) Vietnamese statesman, prime minister of South Vietnam (1954–63). A nationalist opposed to both the communists and the French, he at first received strong US support, but corruption and setbacks in the war against the communists led to growing discontent. With covert US support, army officers staged a coup in which Diem was murdered. ○ *250–51*

DIET OF WORMS (1521) Conference of the HOLY ROMAN EMPIRE presided over by Emperor CHARLES V. Martin LUTHER was summoned to appear before the Diet to retract his teachings. Luther refused to retract them, and the Edict of Worms (25 May 1521) declared him an outlaw. The Diet was one of the most important confrontations of the early Reformation. ○ *152–55*

DIOCLETIAN (245–313) Roman emperor (284–305). Of low birth, he was made emperor by the army. He reorganized the empire to resist the "barbarians", dividing it into four divisions and sharing power with Maximilian, Constantius I and Galerius. To fund his great armies he carried out tax reforms. He ordered the last great persecution of the Christians in 303. ○ *54–55*

DIRECTORY (1795–99) Government of the First Republic of France, consisting of five directors elected by the Council of Five Hundred and the Council of Ancients. It was established as part of the Thermidorian reaction to the Reign of Terror. Faced with bankruptcy, the Directory instituted monetary reforms but was plagued by corruption and inflation. Successes in the French Revolutionary Wars inspired greater independence among the generals and the coup of 18 Brumaire (November 9) 1799, led to the accession of NAPOLEON I. ○ *166–67*

DRED SCOTT DECISION, 1857 US Supreme Court trial on the issue of Federal jurisdiction over slavery in the territories. In 1834 Dred Scott, a slave of John Emerson, was taken from the slave state of Missouri to the free state of Illinois and then Wisconsin territory, where slavery was prohibited by the MISSOURI COMPROMISE. After Emerson's death, Scott sued for his freedom on the basis that he had lived in a free state. The case went to Federal court because of the diversity of state citizenship, since Mrs Emerson's brother and legal administrator of her property, J.F.A. Sanford, was resident in New York. In the case of Scott v Sanford, Roger B. Taney delivered the verdict that the Missouri Compromise was unconstitutional. Three of the justices also held that slaves were not entitled to the rights of US citizens. ○ *184–85*

DUBČEK, ALEXANDER (1921–92) Czechoslovak statesman, Communist Party secretary (1968–69). Dubček was elected party leader at the start of the Prague Spring, a short period of political and social reorganization. His liberal reforms, which were intended to create "socialism with a human face", led to a Soviet invasion in August 1968. Dubček was forced to resign and was expelled from the party. Following the collapse of Czech communism in 1989, he was publicly rehabilitated and served as speaker of the federal parliament until 1992. ○ *236–37*

DUTCH EAST INDIA COMPANY See EAST INDIA COMPANY

DUTCH WEST INDIA COMPANY Dutch company founded in 1621 and given a monopoly of trade with Dutch colonies in America and Africa, and a monopoly of ownership of the American colonies. It was the dominant power in the early days of the slave trade. It was dissolved in 1794. ○ *130–31*

EARTH SUMMIT (June 1992) United Nations conference on Environment and Development, held in Rio de Janeiro. It was the first serious global acknowledgement of the problems created by the impact of industrial society on the environment. The Rio Declaration laid down principles of environmentally sound development, a "blueprint for action", and imposed limits on the emission of gases responsible for the "greenhouse effect". ○ *280–81*

EAST INDIA COMPANY Name of several organizations set up by European countries in the 17th century to trade east of Africa. The English company was set up in 1600 to compete for the East Indian spice trade, but competition with the Dutch led it to concentrate on India, where it gradually won a monopoly. Although the British government assumed political responsibility in 1773, the company continued to administer the British colony in India until the INDIAN MUTINY of 1857. The Dutch company was founded (1602), with headquarters in Batavia (Jakarta) from 1619, but was dissolved in 1799, and its possessions incorporated into the Dutch Empire. The French East India Company was founded by LOUIS XIV in 1664 and set up colonies on several islands in the Indian Ocean. Established as a trading company in India in the early 18th century, it was defeated by the English company and abolished in 1789. ○ *118–19, 130–31, 196–97*

EGYPT, ANCIENT Civilization that flourished along the River Nile in northwest Africa from around 3500 to

30 BC, when Egypt was annexed to Rome. The dynasties are numbered from 1 to 30, and the kingdoms of Upper and Lower Egypt were united c. 3100 BC by the legendary MENES. Ancient Egyptian history is separated into a number of periods. The **Old Kingdom** (c. 2686–2181 BC) peaked in the 4th Dynasty, when the three PYRAMIDS of Giza were built. After the death of Pepy II in the 6th Dynasty, the central government disintegrated, power devolved to the provinces, and the country was in general chaos. This was the **First Intermediate Period.** Central authority was restored in the 11th Dynasty and the capital was moved to Thebes (now Luxor). The **Middle Kingdom** (c. 2055–1650 BC) saw Egypt develop into a great power. Amenemhet I, founder of the 12th Dynasty (c. 1991 BC), crushed provincial opposition, secured Egypt's borders, and moved to a new city from where he could control affairs. Art, architecture, and literature flourished. At the end of this kingdom, Egypt once again fell into disarray (**Second Intermediate Period**) and control was seized by the Hyksos. The **New Kingdom** (c. 1550– 1069 BC) began when the Hyksos were expelled, and the 18th dynasty was founded by Ahmose I. The New Kingdom (18th, 19th, and 20th Dynasties) brought great wealth. Tombs such as TUTANKHAMUN's in the Valley of the Kings and massive temples were built. However, wars with the HITTITES under Rameses II weakened Egypt and subsequent ineffectual rulers brought about the decline of the New Kingdom. The 21st to 25th dynasties (**Third Intermediate Period**) culminated in Assyrian domination. The Persians ruled from 525 to 404 BC, when the Egyptians revolted, and the last native dynasties appeared. A very brief second Persian period followed, and then Egypt fell to the armies of ALEXANDER THE GREAT in 332 BC, who moved the capital to Alexandria. After Alexander's death, his general, Ptolemy, became ruler of Egypt as PTOLEMY I. The Ptolemies maintained a powerful empire for three centuries, and Alexandria became a great centre of learning. Roman power was on the ascendancy, and when Ptolemy XII asked POMPEY for aid in 58 BC, it marked the end of Egyptian independence. His daughter Cleopatra tried to assert her independence through associations with Julius CAESAR and Mark Antony, but she was defeated at Actium. Her son, Ptolemy XV (whose father was probably Julius Caesar), was the last Ptolemy to rule; he was killed by Octavian (AUGUSTUS), and Egypt became a province of Rome. ❍ *30–31, 36–37, 38–39, 42–43, 48–49*

EIRÍK THE RED (c. 950–1010) Norse explorer of Greenland. Exiled from Norway, he fled in 982 to Greenland. Three years later he brought colonists from Iceland to establish permanent settlements. Returning to Norway, he was converted to Christianity, which he attempted to introduce to the the settlers in Greenland. The two colonies he established there survived until the 14th century. ❍ *78–79*

ELAM Ancient country of Mesopotamia; the capital was Susa. Elamite civilization became dominant c. 2000 BC, with the capture of BABYLON. It continued to flourish until the Muslim conquest in the 7th century. Susa was an important centre under the Achaemenid kings of PERSIA and contained a palace of DARIUS I; archaeological finds include the stele of HAMMURABI, inscribed with his code of law. ❍ *36–37*

ELEANOR OF AQUITAINE (1122–1204) Queen consort of France, and later of England. In 1137 she married Louis VII of France, and accompanied him on the Second Crusade (1147–49). The marriage was annulled in 1152 and she married the future Henry II of England (r. 1154–89). Later she supported their sons, the future RICHARD I and JOHN, in a revolt against Henry, and was imprisoned from 1174 to 1189. ❍ *92–93*

EMANCIPATION PROCLAMATION (1 January 1863) Declaration issued by ABRAHAM LINCOLN abolishing slavery in the CONFEDERATE STATES OF AMERICA. At the start of the Civil War Lincoln refrained from issuing a declaration in order to keep the slaveholding states within the Union. Five days after the first Union victory, at the Battle of Antietam (17 September 1862), Lincoln

issued a warning that if the states that had seceded did not return to the Union by 1 January 1863, he would declare free all Confederate slaves. The South rejected Lincoln's proposal and he issued the proclamation "as a fit and necessary war measure". It was designed to enhance the Union's support from abroad, especially Britain, and reduce the South's fighting force. By the end of the war more than 500,000 slaves had fled to the Union side. Slavery was finally abolished by the 13th Amendment (December 1865). ❍ *184–85*

ENGLISH CIVIL WAR (1642–48) Conflict between King CHARLES I and Parliament. Following years of dispute between the king and state essentially over the power of the Crown, war began when the king raised his standard at Nottingham. Royalist forces were at first successful at Edgehill in 1642 but there were no decisive engagements, and Parliament's position was stronger, as it controlled the southeast and London, the navy, and formed an alliance with Scotland. Parliament's victory at Marston Moor in 1644 was a turning point, and in 1645 Fairfax and CROMWELL won a decisive victory at Naseby with their NEW MODEL ARMY. Charles surrendered in 1646. While negotiating with Parliament, he secretly secured an agreement with the Scots that led to what is usually called the second civil war in 1648. A few local Royalist risings came to nothing, and the Scots, invading England, were swiftly defeated. Charles I was executed in 1649. ❍ *156–59*

ENGLISH EAST INDIA COMPANY See EAST INDIA COMPANY

ENGLISH NAVIGATION ACTS English 17th-century statutes placing restrictions on foreign trade and shipping. The first Navigation Act (1651) declared that English trade should be carried only in English ships, and was the main cause of the first of the ANGLO–DUTCH WARS. Later acts placed restrictions on the trade of the colonies. ❍ *130–31*

ENVER PASHA (1881–1922) Turkish general and political leader. Involved in the revolution of the YOUNG TURKS (1908), he became virtual dictator after a coup in 1913. He was instrumental in bringing Turkey into the First World War as an ally of Germany. He was killed leading an anti-Soviet expedition in Bukhara. ❍ *178–79*

ETHIOPIA, ANCIENT According to tradition the Ethiopian kingdom was founded in c. 1000 BC by SOLOMON's son, Menelik I. Coptic CHRISTIANITY was introduced to the northern kingdom of AXUM in the 4th century. In the 6th century JUDAISM flourished. The expansion of ISLAM in Africa led to the isolation of Axum, and the kingdom fragmented in the 16th century. ❍ *82–83, 136–137*

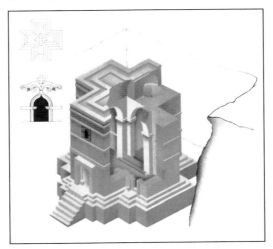

Churches in Ethiopia were sometimes hewn out of solid rock, as was the case with this church at Lalibela, dating from the 13th century.

ETRUSCANS Inhabitants of ancient Etruria (modern Tuscany and Umbria), central Italy. Etruscan civilization flourished in the first millennium BC. HERODOTUS states that they migrated to the region from LYDIA, but others argue that they are indigenous to Italy. Their

sophisticated society was influenced by Greece and organized in city-states. Etruscan civilization reached its peak in the 6th century BC. Their wealth and power was based primarily on their skill at ironworking and their control of the iron trade. They are famed for their naturalistic bronze busts and black *bucchero* pottery. The Etruscan cult of the dead led them to produce elaborate tombs, decorated with vivid two-dimensional frescoes. From the 5th to the 3rd century BC they were gradually overrun by neighbouring peoples, particularly the Romans. Though their alphabet is derived from the Greek, the vocabulary is unknown and cannot be classified. Few written records survive. ❍ *20–21, 54–55*

EUROPEAN COAL AND STEEL COMMUNITY See EUROPEAN COMMUNITY

EUROPEAN COMMUNITY (EC) Economic and political body dedicated to European development. The history of the Community lies in the establishment of the European Coal and Steel Community (ECSC) in 1952, following the Treaty of Paris (1951). The purpose of the ECSC was to integrate the coal and steel industries, primarily of France and West Germany, to create a more unified Europe. The success of the ECSC led to the formation of the European Economic Community (EEC), or Common Market, and the European Atomic Energy Community (EURATOM). Established by the Treaties of Rome (1957/58), the aim was to create a common economic approach to agriculture, employment, trade and social development, and to give Western Europe more influence in world affairs. The original members were France, West Germany, Italy, Belguim, the Netherlands and Luxembourg; the United Kingdom, Ireland and Denmark joined in 1973, Greece in 1981, Spain and Portugal in 1986, and Austria, Finland and Sweden in 1995. The Community's institutional structure comprises the European Commission (responsible for implementing EC legislation), the Council of Ministers (which votes on Commission proposals), the Economic and Social Committee (which advises on draft EC legislation), the European Investment Bank (responsible for all the EC's financial operations), the European Parliament and the European Court of Justice. It was replaced, following the Treaty of Maastricht, by the EUROPEAN UNION. ❍ *238–39, 272–73*

EUROPEAN ECONOMIC COMMUNITY See EUROPEAN COMMUNITY

EUROPEAN FREE TRADE ASSOCIATION (EFTA) Organization seeking to promote free trade among its European members. Established in 1960, it originally comprised Austria, Denmark, Norway, Portugal, Sweden, Switzerland and the UK. By 1995 all but Norway and Switzerland had joined the European Union (EU), while Iceland and Liechtenstein joined EFTA in 1970 and 1991 respectively. ❍ *238–39*

EUROPEAN UNION (EU) Political entity that was established following the ratification of the Maastricht Treaty in 1993. The EU aims to use the existing framework and institutions of the European Community (EC) to implement greater integration of member states, particularly in areas such as foreign and security policies, and internal and judicial policies. Some member states, particularly the UK and Denmark, have resisted moves towards closer integration, particularly in the areas of a single European currency. Arguments persist over loss of national sovereignty and many argue that the union will create a federalist Europe. ❍ *238–39, 272–73*

FATIMIDS SHIITE dynasty who claimed the caliphate on the basis of their descent from Fatima (606–632), the daughter of MUHAMMAD and wife of Ali. The dynasty was founded by Said Ibn Husayn at the close of the 9th century. The Fatimids quickly overthrew the SUNNI rulers in most of northwest Africa. By Ibn Husayn's death in 934 the Fatimid Empire had expanded into southern Europe, and in 969 they captured Egypt and established the Mosque and University of Al-Azhar, which remains one of the most influential educational establishments in contemporary ISLAM. In the late 10th century, Fatimid power began to wane, and by the end of the 11th century, Egypt was all that remained of the empire. ➲ *88–89, 94–95*

FERDINAND I (1503–64) King of BOHEMIA and of Hungary (1526–64), Holy Roman Emperor (1558–64). Ferdinand was continuously at war with the encroaching Ottoman Turks, and also warred against the German Protestants until the Peace of AUGSBURG in 1555, for which he was largely responsible. His brother, CHARLES V, delegated imperial authority to Ferdinand in 1555 and finally abdicated in his favour, although Charles's Spanish and Italian lands, as well as the Spanish Netherlands, went to his son, Philip II. As ruler of the Austrian Habsburg lands, Ferdinand centralized his administration and attempted to revive Catholicism. ➲ *146–47, 152–53*

FERDINAND V (1452–1516) (Ferdinand the Catholic) King of Castile and León 1474–1504, of Aragon (as Ferdinand II) 1479–1516, of Sicily 1468–1516, and of Naples (as Ferdinand III) 1504–16. He became joint king of Castile and León after marrying Isabella I in 1469, and inherited Aragon from his father, John II, in 1479. After he and Isabella conquered the Moorish kingdom of Granada in 1492, they ruled over a united Spain. In 1492 they sponsored the voyage of Christopher COLUMBUS to the New World, expelled the JEWS from Spain, and initiated the Spanish Inquisition. Under Ferdinand, Spain became involved in the Italian wars against France. After Isabella's death in 1504, Ferdinand acted as regent in Castile for their insane daughter, Joanna, and later for her son, Charles I, who succeeded Ferdinand and ruled most of Europe as Holy Roman Emperor CHARLES V. ➲ *106–7, 146–47*

FERDINAND VII (1784–1833) King of Spain (1808–33), son of CHARLES IV. As crown prince he made overtures to NAPOLEON I, and was arrested in 1807 by his father. An uprising in 1808 caused his father's abdication, but the French forced Ferdinand himself off the throne and installed Joseph BONAPARTE. Ferdinand was imprisoned by the French during the PENINSULAR WAR but was restored in 1814. All mainland colonies in the western hemisphere were lost during his reign. He altered the Spanish constitution so his daughter Isabella II could succeed him, which set off the Carlist Wars. ➲ *172–73*

FLANDERS Historic region now divided between Belgium and France. In Belgian Flanders the major language is Flemish. From the 10th century, Flanders grew prosperous on the cloth industry, and the old nobility gradually lost authority to the cities, such as Bruges, Antwerp and Ghent. By 1400 it was part of Burgundy, passing to the Habsburgs in 1482, before becoming part of the Spanish Netherlands. It was frequently fought over by France, Spain, and later Austria, and was the scene of devastating trench warfare in the First World War. ➲ *102–3, 152–53, 218–19*

FORD, GERALD RUDOLPH (1913–) 38th US president (1974–77). Elected in 1948 as a REPUBLICAN to the US House of Representatives, he gained a reputation as an honest and hard-working conservative. He was nominated by President NIXON to replace the disgraced Spiro Agnew as vice president (1973). When Nixon resigned, Ford became president. One of his first acts was to pardon Nixon. His attempts to counter economic recession with cuts in social welfare and taxes were hindered by the DEMOCRAT-dominated Congress. Nominated again in 1976, he narrowly lost the election to Jimmy CARTER. ➲ *242–43*

FORD, HENRY (1863–1947) US industrialist. He developed a petrol-engined car in 1892 and founded Ford Motors in 1903. It achieved huge success with the economical and inexpensive Model T (1908), which from 1913 was produced on an assembly line, lowering production costs enormously. Almost as startling was Ford's introduction of an eight-hour working day and a relatively high basic wage. ➲ *240–41*

FOURTEEN POINTS Programme presented by US president WOODROW WILSON for a just peace settlement of the First World War in January 1918. In general, the programme required greater liberalism in international affairs and supported the principle of national self-determination. It made useful propaganda for the Allies and was the basis on which Germany sued for peace in 1918. Some points found expression in the Treaty of Versailles, others were modified or rejected at the peace conference. The 14th Point laid the basis for the League of Nations. ➲ *220–21*

FRANCIS I (1494–1547) King of France (1515–47), cousin and son-in-law and successor of Louis XII. A leader of the RENAISSANCE, he is best remembered for his patronage of the arts and his palace at Fontainebleau. Persecution of the Waldenses, centralization of monarchical power, and foolish financial policies made Francis unpopular at home. A costly struggle with the Emperor CHARLES V over the imperial crown led to defeat at Pavia (1525). Francis was imprisoned and in the Treaty of Madrid (1526) forced to give up Burgundy and renounce his claims to Italy. Two more wars (1527–29 and 1536–38) ended ingloriously. In 1542 Francis concluded a treaty with SULEIMAN I and attacked Italy for a fourth time. Charles, in alliance with Henry VIII of England, responded by invading France, and Francis lost further territory. He was succeeded by his son, HENRY II. ➲ *146–47*

FRANCIS JOSEPH (1830–1916) (Franz Joseph) Emperor of Austria (1848–1916) and King of Hungary (1867–1916). He succeeded his uncle, the emperor Ferdinand, who abdicated during the revolutions of 1848, and quickly brought the revolutions under control, defeating the Hungarians under KOSSUTH in 1849. In 1851 he abolished the constitution, exercising personal rule until 1867. In the same year he was forced to grant Hungary co-equal status with Austria in the Dual Monarchy of Austria-Hungary. He died in the midst of the First World War, two years before the final collapse of the Habsburg Empire. His son Charles became the last emperor of Austria, abdicating in November 1918. ➲ *174–75*

FRANCO, FRANCISCO (1892–1975) Spanish general and dictator of Spain (1939–75). He joined the 1936 military uprising that led to the SPANISH CIVIL WAR and assumed leadership of the fascist Falange. By 1939, with the aid of Nazi Germany and fascist Italy, he had won the war and become Spain's dictator. He kept Spain neutral in the Second World War, after which he presided over Spain's accelerating economic development, while maintaining rigid control over its politics. In 1947 he declared Spain a monarchy with himself as regent, and in 1969 he designated the future King Juan Carlos as heir to the throne. ➲ *230–31, 238–39*

FRANCO-PRUSSIAN WAR (1870–71) A Prussian victory in the SEVEN WEEKS WAR of 1866 alarmed NAPOLEON III. The Prussian chancellor, BISMARCK, used the prospect of French invasion to frighten the south German states into joining the North German Confederation, dominated by Prussia. The nominal cause of the war was a dispute over the Spanish succession. Prussia was fully prepared for the French declaration of war (14 July 14 1870) and General von Moltke launched a devastating offensive into Alsace. In September the emperor and 100,000 French troops were captured at Sedan. Napoleon III abdicated, and Paris was surrounded and starved into submission. An armistice was agreed in January 1871, and Alsace and Lorraine were ceded to the new German empire under WILLIAM I. Paris refused to surrender its weapons and the Paris Commune was formed. ➲ *170–71, 176–77*

FRANZ FERDINAND (1863–1914) Archduke of Austria. Nephew of the Emperor Franz Joseph, he became heir apparent in 1889 but, having married a woman of "lower rank", had to renounce his children's claim to the throne. On an official visit to Bosnia-Herzegovina in 1914, he and his wife were assassinated by a Serb nationalist, Gavrilo Princip, in Sarajevo (28 June). The incident led directly to the outbreak of the First World War. ➲ *216–17*

FREDERICK II (THE GREAT) (1712–86) King of Prussia (1740–86). Succeeding his father, Frederick William I, he made Prussia a major European force. In the War of the AUSTRIAN SUCCESSION (1740–48), he took the valuable province of Silesia from Austria. During the SEVEN YEARS WAR (1756–63), his brilliant generalship preserved the kingdom from a superior hostile alliance. In 1760 Austro-Russian forces reached Berlin, but Russia's subsequent withdrawal from the war enabled Frederick to emerge triumphant at the peace. He directed Prussia's remarkable recovery from the devastation of war after 1763. Gaining further territory in the first partition of Poland (1772), he renewed the contest against Austria in the War of the Bavarian Succession (1778–79). An artist and an intellectual, he was a friend and patron of Voltaire. He wrote extensively in French, built the palace of Sans Souci, and was a gifted musician. ➲ *156–57*

FREDERICK WILLIAM IV (1795–1861) King of Prussia (1840–61). Son of Frederick William III, he granted a constitution in response to revolutionary demands in 1848, later amending it to eliminate popular influence. He refused the crown of Germany in 1849 because it was offered by the Frankfurt parliament, a democratic assembly. From 1858 the future kaiser William I ruled as regent. ➲ *176–77*

FRENCH WARS OF RELIGION (1562–98) Series of religious conflicts in France. At stake was freedom of worship for HUGUENOTS (Protestants). The conflicts ended after the defeat of the extremist CATHOLIC Holy League by Henry IV. The EDICT OF NANTES (1598) extended toleration to the Huguenots. ➲ *146–47, 154–55*

FRONDES (1648–53) Series of rebellions against oppressive government in France. The Fronde of the Parlement (1648–49) began when Anne of Austria (wife of Louis XIII) tried to reduce the salaries of court officials. It gained some concessions from the regent, LOUIS XIV. The Fronde of the Princes (1650–53) was a rebellion of the aristocratic followers of Condé, and forced the unpopular Cardinal MAZARIN into temporary exile. Condé briefly held Paris, but the rebellion soon collapsed, and promised reforms were withdrawn. The Fronde succeeded in moderating the financial excesses of royal government, but under LOUIS XIV royal absolutism triumphed. ➲ *156–57*

G

GADSDEN PURCHASE Land purchased by the USA from Mexico in 1853. It was a narrow strip, 77,000 sqare kilometres (30,000 square miles) in area, now forming southern Arizona and New Mexico. The deal was negotiated by James Gadsden. ➲ *182–83, 192–93*

GALILEO (1564–1642) (Galileo Galilei) Italian scientist. He studied falling bodies and disproved ARISTOTLE'S view that they fall at different rates according to weight. In 1610, he used one of the first astronomical telescopes to discover sunspots, lunar craters, Jupiter's major satellites and the phases of Venus. In *Sidereus Muncius* (1610) he supported the Copernican view of the Universe, with Earth orbiting the Sun. This was declared a heresy, and in 1633 he was brought before the Inquistion and forced to recant. ➲ *134–35*

GAMA, VASCO DA (c. 1460–1524) Portuguese navigator, diplomat and naval commander. In 1497 he led an expedition around the Cape of Good Hope, following up the voyage of Bartholomew DIAS, who had reached the Cape in 1488. Da Gama's voyage, which took him across the Indian Ocean to Calicut, opened up the sea route to India. In 1502 he led a heavily armed expedition of 20 ships and, employing brutal tactics, secured Portuguese supremacy in the Eastern spice trade. During this expedition he also founded a Portuguese colony in Mozambique, on the east coast of Africa. On his return to Portugal he retired from the navy, but returned to India in 1524 as viceroy, dying there shortly afterwards. ➲ *116–17, 118–19*

GANDHI, INDIRA (1917–84) Indian politician, prime minister (1966–77, 1980–84). The daughter of Jawaharlal NEHRU, she served as president of the INDIAN NATIONAL CONGRESS PARTY between 1959 and 1960.

becoming prime minister in 1966. In 1975, amid growing social disturbance, she was found guilty of breaking electoral rules in her 1971 re-election. She refused to resign, invoked emergency powers and inmprisoned many opponents. When elections took place in 1977, the Congress Party suffered a heavy defeat which split the party. In 1980, leading a faction of the Congress Party, she returned to power. In 1984, after authorizing the use of force against Sikh dissidents in the Golden Temple at Amritsar, she was killed by Sikh bodyguards. ❍ *248–49*

GANDHI, "MAHATMA" (Mohandas Karamchand) (1869–1948) Indian political and spiritual leader, he led the nationalist movement from 1919 to 1947. As a young man he studied law in London, and worked as a lawyer in South Africa between 1893 and 1914. He led equal-rights campaigns in South Africa, and was imprisoned by the authorities for refusing to register as an Indian alien. During his time in South Africa he became convinced of the power of non-violent resistence. He returned to his native India in 1915 and campaigned on behalf of the "untouchables" – the lowest rank in the Hindu caste system. Following the First World War he led a campaign for Indian self-rule and, after the massacre at Amritsar in 1919, he launched a policy of non-violent civil disobedience against the British colonial authorities. Resistance methods included strikes, refusal to pay taxes, and refusal to respect colonial law. In 1924 he became president of the Indian National Congress Party, which office he held until 1935. He was jailed (1930–31) following the famous 388-kilometre (241-mile) protest march against a salt tax (1930). After further frequent imprisonments, he witnessed India gain independence in 1947. As an opponent of partition, he preached unity between Hindus and Muslims. A figure of immense international and moral stature, he was assassinated at a prayer meeting in Delhi by a religious fanatic. ❍ *194–95, 248–49*

GANG OF FOUR Radical faction that tried to seize power in China after the death of MAO ZEDONG. In 1976 both Chairman Mao and Prime Minister Zhou Enlai died, leaving a power vacuum. The Gang of Four, Zhang Chunjao, Wang Hungwen, Yao Wenyuan, and their leader Jiang Qing (Mao's widow), tried to launch a military coup but were arrested for treason by premier Hua Guofeng. They were sentenced to life imprisonment. ❍ *254–55*

GARIBALDI, GIUSEPPE (1807–82) Italian patriot and guerrilla leader who helped to bring about Italian unification. Influenced by MAZZINI, he participated in an unsuccessful republican rising in 1834, and subsequently fled to South America. Returning in 1848, he defended the newly-established Roman Republic against the French, but papal authority was restored with the support of France and Naples. In 1860 he led his 1,000-strong band of "Red Shirts" against the Kingdom of the Two Sicilies, a dramatic episode in the RISORGIMENTO. He handed his conquests over to King VICTOR EMMANUEL II and they were incorporated into the new kingdom of Italy. He retired due to ill health after fighting in the Franco-Prussian War. ❍ *176–77*

GAUL Ancient Roman name for the region roughly equivalent to modern France, Belgium, northern Italy and Germany west of the Rhine. Most of Gaul, which was inhabited by CELTS, was conquered by the Romans under Julius CAESAR in the Gallic Wars (58–51 BC.) From the 3rd century AD, it was under attack by Germanic tribes – the VISIGOTHS, the Franks and the Burgundians – who settled in northern Gaul. ❍ *54–55*

GENERAL AGREEMENT ON TARIFFS AND TRADE (GATT) United Nations agency of international trade, subsumed into the new World Trade Organization in 1995. Founded in 1948, GATT was designed to prevent "tariff wars" (the retaliatory escalation of tariffs) and to work towards the reduction of tariff levels. Most non-communist states were party to GATT. ❍ *272–73*

GENGHIS KHAN *See* CHINGGIS KHAN

GERMAN CONFEDERATION (1815–66) Federation of 39 German principalities set up by the Congress of Vienna to replace the HOLY ROMAN EMPIRE. It had few formal powers, beyond a mutual defence pact, and was dominated by Austria. It collapsed in 1848 but was restored two years later; after the Prussians defeated Austria in 1866 in the SEVEN WEEKS WAR it was dissolved and replaced with the North German Confederation. ❍ *172–73, 176–77*

GHAZVANIDS Muslim Turkish dynasty that ruled a region from eastern Iran to northern India between 977 and 1186. They began as governors of Ghazni in Afghanistan, gaining independence when the Samanid emirs of Bukhara were overthrown. Mahmud Ghazni (r. 998–1030) established Muslim rule for the first time in what is now Pakistan and made devastating raids far into India. Thereafter, the dynasty declined through succession quarrels and attacks by the SELJUK Turks. The empire was much reduced after the Battle of Dandanqan in 1040, Ghazni was sacked by the Ghuri in the 12th century, and the last Ghaznavid retreated to Lahore, which was lost in 1186. ❍ *88–89*

GLASNOST (Russian for "openness") Term adopted by Mikhail GORBACHEV in 1986 to express his more liberal social policy. One result was widespread popular criticism of the Soviet system and the Communist Party, leading to the break-up of the Soviet Union and the fall of Gorbachev. *See also* PERESTROIKA ❍ *236–37, 262–63*

GOLDEN BULL Edict with a golden seal, as issued by medieval Western and Byzantine rulers. Andrew II of Hungary issued a Golden Bull in 1222. This document extended certain rights to the nobility, including tax exemption, freedom to dispose of their property, prohibition of arbitrary imprisonment, and guarantee of annual assembly. The best known Golden Bull is the edict promulgated by Holy Roman Emperor Charles IV in 1356. It defined the procedures for electing the Holy Roman Emperor and provided for election by majority vote of seven princely electors. The procedures remained in effect until the dissolution of the Empire by NAPOLEON in 1806. ❍ *90–91, 146–47*

GOLDEN HORDE Name given to the Mongol state established in southern Russia in the early 13th century. The state derived from the conquests of CHINGGIS KHAN and was extended by his successors, who took over the whole of the Russian state centred on Kiev. It was conquered by TIMUR-LENG in the late 14th century and subsequently split up. ❍ *98–99*

GOLD STANDARD Monetary system in which the gold value of currency is set at a fixed rate and currency is convertible into gold on demand. It was adopted by Britain in 1821, by France, Germany and the USA in the 1870s, and by most of the rest of the world by the 1890s. Internationally it produced nearly fixed exchange rates and was intended to foster monetary stability. The Great Depression forced many countries to depreciate their exchange rates in an attempt to foster trade, and by the mid-1930s all countries had abandoned the gold standard. ❍ *228–29*

GOMULKA, WLADYSLA (1905–82) Polish communist leader. He rose through party ranks to become first secretary and deputy prime minister (1945), but was dismissed during a Stalinist purge in 1948. Reinstated in 1956, he adopted liberal policies that made him popular at home and were accepted by Moscow. As economic problems grew, Gomulka's regime became more oppressive and anti-Semitic. Steep rises in food prices in 1970 led to his resignation. ❍ *236–37*

GORBACHEV, MIKHAIL SERGEYEVICH (1931–) Soviet statesman, president of the Soviet Union (1985–91). Secretary of the Communist Party from 1985, he embarked on a programme of reform based on two principles: PERESTROIKA and GLASNOST. The benefits of radical socio-economic change were slow to take effect, and Gorbachev became extremely unpopular as prices rose. He agreed major arms limitation treaties with the USA and acquiesced in the demolition of the communist regimes in Eastern Europe in 1989 and 1990, effectively ending the Cold War. Having strengthened his presidential powers in 1990, he was forced to resign in 1991 by opponents eager to grant independence to the constituent republics of the Soviet Union. He was awarded the Nobel Peace Prize in 1990. ❍ *236–37, 242–43, 262–63, 264–65*

GOTHIC ARCHITECTURE Architecture of medieval Europe (12th–16th centuries). The style is religious in inspiration and ecclesiastical in nature, and it is characterized by the pointed arch and ribbed vault. Its greatest and most characteristic expression is the cathedral. The introduction of flying buttresses was a technical advance that made the large windows possible. An early prototype is the Abbey Church of St Denis in France (1140–44). Ever higher and lighter structures followed, with increasingly intricate vaulting and tracery. ❍ *102–3*

Cologne Cathedral, begun in 1248 but not completed until the 19th century, is the largest Gothic church in Europe. Its highly decorated twin towers rise to 152 metres (502 feet).

GOTHS Ancient Germanic people, groups of whom settled near the Black Sea in the 2nd–3rd centuries AD. The VISIGOTHS were driven westwards into Roman territory by the Huns in 376, culminating in their sacking Rome under ALARIC in 410. They settled in southwest France, then, after being driven out by the Franks in the early 6th century, in Spain. Some groups united to create the OSTROGOTHS, who conquered Italy under THEODORIC THE GREAT by 493. The Ostrogoths held Italy until conquered by the Byzantines under JUSTINIAN I (535–40). ❍ *56–57*

GRANT, ULYSSES S. (SIMPSON) (1822–85) US Civil War general and 18th US president (1869–77). He served in the MEXICAN WAR (1846–48). In the Civil War he masterminded the Vicksburg Campaign of 1862–63. In 1864 ABRAHAM LINCOLN gave him overall command of the Union forces. He co-ordinated the final campaigns and accepted the surrender of ROBERT E. LEE in 1865. He achieved some foreign policy successes as president, but failed to prevent the growth of domestic corruption. He was re-elected in 1872, but retired at the end of his second term. ❍ *184–85*

GREAT FIRE OF LONDON (2–6 September, 1666) Accidental fire that destroyed most of the City of London, England. The fire started in a baker's shop in Pudding Lane, a site now marked by the Monument, and a strong wind spread it quickly through the closely packed wooden houses. The fire provided an opportunity for rebuilding London on a more spacious plan, but, for the most part, only the famous churches of Christopher Wren (including St Paul's Cathedral) were built. ❍ *132–33*

GREAT LEAP FORWARD Five-year economic plan begun by MAO ZEDONG in China in 1958. It aimed to double industrial production and boost agricultural output in record time. Tens of millions of workers were mobilized to smelt steel in primitive furnaces, but much of the steel proved useless. Collective farms were merged into communes, but progress was dashed by a succession of poor harvests. After four years the government was forced to admit failure. ❍ *254–55*

GREAT NORTHERN WAR (1700–21) Conflict in northern Europe between Sweden and its neighbours. It began with an attack on Sweden by Denmark, Saxony, Poland and Russia. By 1706 CHARLES XII of Sweden had defeated all his opponents, but war was renewed in 1707 when Charles invaded Russia. The Swedes were decisively defeated at POLTAVA (1709). He took refuge with the Ottoman Turks, encouraging their attack on Russia in 1710–11. In the ensuing peace treaties (1719–21), Sweden lost virtually all its northern empire. ❍ *148–49, 150–51, 156–57*

GREAT SCHISM Division within the Roman CATHOLIC CHURCH resulting in the election of rival popes (1378–1417). An Italian line of popes continued in Rome, and a rival "antipope" line in Avignon, France. The schism ended with the COUNCIL OF CONSTANCE (1414–18), which established Martin V as the only pope. ❍ *106–7*

GREAT WALL OF CHINA Defensive frontier and world heritage site, c. 2,400 kilometres (1,500 miles) long, extending from the Huang Hai (Yellow Sea) to the Central Asian desert in northern China. It is an amalgamation of fortifications constructed by various dynasties. Sections of the wall were first built by the Warring States. In 214 BC QIN SHI HUANG DI ordered that they should be joined to form a unified boundary. The present wall was mostly built 600 years ago by the Ming dynasty. It averages 7.6 metres (25 feet) high and up to 9 metres (30 feet) thick. ❍ *48–49, 138–39*

GREAT ZIMBABWE City-state in southern Africa which flourished between the 12th and 15th centuries. Based on the export of gold, which was mined from shafts sunk as deep as 30 metres (100 feet), the kingdom grew wealthy and supported the development of a privileged elite, whose tombs contained precious metal ornaments as well as imported ceramics and textiles. ❍ *82–83*

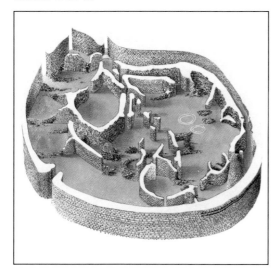

The high walls of the Great Enclosure – built mainly in the 14th and 15th centuries – are among the most important of the sites which can be seen today.

GREEK WAR OF INDEPENDENCE (1821–29) The rebellion of Greece against Ottoman rule that led to the independence of modern Greece. In 1820 Prince Ypsilanti led a premature raid against the Ottomans. He was defeated but other groups staged isolated raids all over Greece, and the Ottomans made brutal reprisals. In 1822–24 there were two civil wars in Greece between the various factions. In 1824 MUHAMMAD ALI of Egypt joined the Ottomans; Athens fell in 1826. Britain, France, and Russia intervened in 1827, the year the

Turkish fleet was crushed at Navarino. By the Treaty of Adrianople (1829), Greek autonomy was established, followed in 1832 by independence. ❍ *172–73*

GUADALUPE-HIDALGO, TREATY OF (1848) Peace settlement ending the MEXICAN WAR. Mexico ceded the present states of Texas, New Mexico, Arizona, California, Nevada, Utah, and parts of Colorado and Wyoming. The USA paid $15 million in compensation. ❍ *182–83, 192–93*

GULF WAR (16 January 1991–28 February 1991) Military action by a US-led coalition of 32 states to expel Iraqi forces from Kuwait. Iraqi forces invaded Kuwait (2 August 1990) and claimed it as an Iraqi province. On 7 August 1990, Operation Desert Shield began a mass deployment of coalition forces to protect Saudi oil reserves. Economic sanctions failed to secure Iraqi withdrawal, and the UN Security Council set a deadline of 15 January 1991 for the peaceful removal of Iraqi forces. Iraqi president Saddam Hussein ignored the ultimatum, and General Norman Schwarzkopf launched Operation Desert Storm. Within a week, extensive coalition air attacks had secured control of the skies and weakened Iraq's military command. Iraqi ground forces were defenceless against the coalition's technologically advanced weaponry. Iraq launched Scud missile attacks on Saudi Arabia and Israel, in the hope of weakening Arab support for the coalition. On 24 February, the ground war was launched. Iraqi troops burned Kuwaiti oil wells as they fled. Kuwait was liberated two days later, and a cease-fire was declared on 28 February. Saddam Hussein remained in power. An estimated 100,000 Iraqi troops were killed, 600,000 wounded, captured or deserted; 300 coalition forces were killed. ❍ *252–53, 260–61, 266–67*

GUPTAS (c. 320– c. 550) Dynasty whose kingdom covered most of northern India. It was founded by CHANDRAGUPTA I. The Gupta dynasty embraced BUDDHISM, and is seen as a golden age. It reached its greatest extent at the end of the 4th century, but declined at the end of the 5th century under concerted attack from the White Huns. ❍ *46–47*

GUSTAV I VASA (1496–1560) King of Sweden (1523–60) and founder of the Vasa dynasty. He led a successful rebellion against the invading Danes in 1520. In 1523 he was elected king and the KALMAR UNION was ended. During his reign, Sweden gained independence, the Protestant church was established, and the Bible was translated into Swedish. ❍ *146–47, 150–51*

GUSTAV II ADOLF (1594–1632) King of Sweden (1611–32). He succeeded his father Charles IX during a constitutional crisis. Gustav's reign was distinguished by constitutional, legal and educational reforms. He ended war with Denmark in 1613 and with Russia in 1617. Hoping to increase Sweden's control of the Baltic and to support PROTESTANTISM, he entered the THIRTY YEARS WAR (1618–48). On 6 November 1632 at Lutzen he defeated the German forces under Wallenstein but was killed in the battle. ❍ *150–51, 158–59*

H

HADRIAN, PUBLIUS AELIUS (AD 76–138) Roman emperor (117–138). Nephew and protégé of Emperor TRAJAN, he adopted a policy of imperial retrenchment, discouraging new conquests, relinquishing territory hard to defend, and ordering the construction of HADRIAN'S WALL in Britain. He made lengthy tours of his empire in 121–25 and 128–33. He erected many fine buildings, notably the vast Hadrian's Villa at Tivoli, and also rebuilt the Pantheon. The erection of a shrine to Jupiter on the site of the Temple in JERUSALEM provoked a Jewish revolt in 132–135, which was ruthlessly suppressed. ❍ *54–55*

HADRIAN'S WALL Defensive fortification in north England, erected 122–36 on the orders of the Roman Emperor HADRIAN. It extended 118.3 kilometres (73.5 miles) and was about 2.3 metres (7.5 feet) thick and 1.8–4.6 metres (6–15 feet) high. Forts were built along

its length at regular intervals . Extensive stretches of the wall survive. ❍ *54–55*

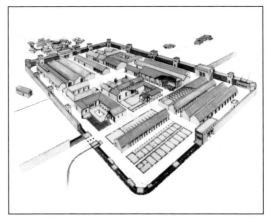

A typical Roman fort, such as those along the length of Hadrian's Wall, held 500 to 1,000 men. From the main gate a main road led to the headquarters building. From there, the other main roads led to the side gateways. Within the fort there were usually barracks, a workshop, granaries, stables, armouries and the commandant's house.

HALLSTATT Small town in west central Austria, believed to be the site of the earliest Iron Age culture in western Europe. Iron was worked there from c. 700 BC. The site contains a large Celtic cemetery and a deep salt mine. Fine bronze and pottery objects have also been discovered. ❍ *20–21*

HAMMURABI King of BABYLONIA (c. 1792–1750 BC) By conquering neighbours such as Sumeria, he extended the empire under the *Code of Hammurabi*, a set of 282 rules covering family life, property and trade. A copy of the Code is in the Louvre, Paris, and provides inform-ation on social and economic conditions in ancient Babylonia. Hammurabi was also an able administrator, improving productivity by commissioning the building of canals and granaries. ❍ *36–37*

HANNIBAL (247–183 BC) Carthaginian general in the second of the PUNIC WARS, son of Hamilcar Barca. One of the greatest generals of ancient times, he fought against the Romans in Spain (221 BC). In 218 BC he invaded northern Italy after crossing the Alps with a force of elephants and 40,000 troops. He won a series of victories in Italy, but was unable to capture Rome. In 203 BC he was recalled to CARTHAGE to confront the in-vasion of Scipio Africanus. Lacking cavalry, he was defeated at Zama in 202 BC. After the war, as chief mag-istrate of CARTHAGE, he alienated the nobility by reduc-ing their power. They sought Roman intervention and Hannibal fled to the SELEUCID kingdom of Antiochus III. He fought under Antiochus against the Romans, was defeated and committed suicide. ❍ *54–55*

HANSEATIC LEAGUE Commercial union of around 160 German, Dutch and Flemish towns established in the 13th century. The League protected its merchants by controlling the trade routes from the Baltic region to the Atlantic. It began to decline in the late 15th century with the opening up of the New World and aggressive trading by the British and Dutch. ❍ *90–91, 106–7, 150–151*

HASTINGS, BATTLE OF (14 October 1066) Fought near Hastings, southeast England, by King Harold II of England against an invading army led by William, Duke of Normandy. The NORMAN victory and death of Harold marked the end of the ANGLO-SAXON monarchy and produced a social revolution. ❍ *92–93*

HAVEL, VACLAV (1936–) Czech playwright, politician and president. Havel was imprisoned several times by the communist regime during the 1970s and 1980s, both for his satirical plays and for his work as a human rights activist. In 1989 he was elected President of Czechoslovakia and tried to preserve a united republic. He resigned in 1992 when break-up became inevitable; he became president of the newly formed Czech Republic in 1993. ❍ *264–65*

HELLENISTIC AGE (323–30 BC) Period of Classical Mediterranean history from ALEXANDER THE GREAT to the reign of AUGUSTUS. Alexander's conquests helped to spread Greek civilization over a wide area east of the Mediterranean. The age was distinguished by remarkable scientific and technological advances, especially in Alexandria, and by more elaborate and naturalistic styles in the visual arts. ○ *42–43*

HENRY II (1519–59) King of France (1547–59). Son and successor of FRANCIS I, he married CATHERINE DE MEDICI but was dominated by his mistress, Diane de Poitiers, and the rival families of Guise and Montmorency. After bankrupting the royal government, the war with Spain ended with the peace of Cateau-Cambrésis in 1559. Henry died after being accidentally wounded in a tournament. ○ *146–47, 152–53*

HERODOTUS (c. 485–c. 425 BC) Greek historian. Regarded as the first true historian, his *Histories* are the first great prose works in European literature. His main theme was the struggle of Greece against the mighty Persian Empire in the PERSIAN WARS, but he also provides an insight into the contemporary Mediterranean world. ○ *40–41, 50–51*

HINDUISM Traditional religion of India, characterized by a philosophy and a way of life rather than by a dogmatic structure. It was not founded by an individual and has been developing gradually since c. 3000 BC, absorbing external influences. There are several schools within Hinduism, but all Hindus recognize the *Vedas* as sacred, believe that all living creatures have souls, follow the doctrine of transmigration of souls and consider *moksha* – liberation from the cycle of suffering and rebirth represented by reincarnation – as the chief aim in life. One of the features of Hindu society is the caste system, but modern Hindu scholars maintain that it is not part of the Hindu religion. ○ *44–45, 62–63, 144–45, 248–49, 268–69*

HIROHITO (1901–89) Emperor of Japan (1926–89). Hirohito was the first crown prince to travel abroad (1921). Although he exercised little political power during his reign, he persuaded the Japanese government to surrender to the Allies in 1945. Under the new constitution of 1946, he lost all power and renounced the traditional claim of the Japanese emperors to be divine. He became mostly an imperial figurehead and was succeeded by his son, Akihito. ○ *200–1, 234–35*

HITLER, ADOLF (1889–1945) German fascist dictator (1933–45), born in. Austria. He served in the German army during the First World War and was decorated for bravery. In 1921 he became leader of the small National Socialist Workers' Party (Nazi Party). While imprisoned for his role in the failed Munich Putsch, he set out his extreme racist and nationalist views in *Mein Kampf*. Economic distress and dissatisfaction with the Weimar government led to electoral gains for NATIONAL SOCIALISM and, by forming an alliance with orthodox Nationalists, Hitler became chancellor in January 1933. He made himself dictator of a one-party state in which all opposition was ruthlessly suppressed by the SS and Gestapo. The racial hatred he incited led to a policy of extermination of JEWS and others in the HOLOCAUST. Hitler pursued an aggressive foreign policy aimed at territorial expansion in eastern Europe. The invasion of Poland finally goaded Britain and France into declaring war on Germany in September 1939. Hitler himself played a large part in determining strategy during the Second World War. In April 1945, with Gemany in ruins, he committed suicide. ○ *230–31, 232–33*

HITTITES People of Asia Minor who controlled a powerful empire in the 14th–13th centuries BC. They founded a kingdom in Anatolia (Turkey) in the 18th century BC; their capital was Hattusas. They expanded east and south in the 15th century BC, and conquered northern Syria before being checked by the Egyptians under Rameses II. Under attack from ASSYRIA, the Hittite Empire disintegrated c. 1200 BC. ○ *36–37*

HOHENZOLLERN German dynasty that ruled Brandenburg, Prussia, and Germany. The family acquired Brandenburg in 1415, and Prussia was added in 1618. Frederick William (the Great Elector) further expanded their territories, and his son, Frederick I,

adopted the title "King in Prussia". Frederick William I built up the famous Prussian army, and FREDERICK II used it to great effect against the Habsburgs. Germany was finally united in 1871 under the Hohenzollern emperor, William I. His grandson William II abdicated at the end of the First World War. ○ *156–57*

HOLLAND Popular name for the Netherlands, but properly referring only to a historic region, now divided into two provinces. A fief of the HOLY ROMAN EMPIRE in the 12th century, Holland was united with the county of Hainaut in 1299. It passed to Burgundy in 1433 and to the Habsburgs in 1482. In the 16th century, Holland led the Netherlands in their long struggle for independence from Spain. ○ *102–3, 128–31, 156–57*

HOLOCAUST Great massacre, in particular the extermination of European JEWS and others by the Nazi regime in Germany (1933–45). The Nazi persecution reached its peak in the "Final Solution", a programme of mass extermination adopted in 1942. Jews, as well as others considered racially inferior by the Nazis, were killed in concentration camps such as Auschwitz, Belsen, Dachau, Majdanek and Treblinka. Total Jewish deaths are estimated at more than 6 million. ○ *232–33*

HOLY ALLIANCE Agreement signed by the crowned heads of Russia, Prussia and Austria in 1815. Its purpose was to re-establish the principle of hereditary rule and to suppress democratic and nationalist movements, which had sprung up in the wake of the French Revolution. The agreement, signed later by every European dynasty except the King of England and the Ottoman Sultan, came to be seen as an instrument of reaction and oppression. ○ *172–73*

HOLY ROMAN EMPIRE European empire centred on Germany (10th–19th centuries), which echoed the empire of ancient Rome. It was founded in 962 when the German king OTTO I (THE GREAT) was crowned in Rome, although some historians date it from the coronation of CHARLEMAGNE in 800. The Emperor, elected by the German princes, claimed to be the temporal sovereign of Christendom, ruling in co-operation with the spiritual sovereign, the Pope. However, the Empire never encompassed all of western Christendom and relations with the Papacy were often stormy. From 1438 the title was virtually hereditary in the Habsburg dynasty. After 1648, the Empire became little more than a loose confederation of hundreds of virtually independent states. It was abolished by NAPOLEON I in 1806. ○ *90–91, 146–7, 152–55*

HOMESTEAD ACT (1862) US federal legislation enacted during the Civil War to encourage westward expansion. The government granted 65 hectares (160 acres) of government land to anyone would live on it and improve it for five years. Along with the Morrill Act for education, and with subsidies for the railroads, the Homestead Act opened the West to widespread settlement. ○ *182–83*

HONECKER, ERICH (1912–94) East German communist leader (1971–89). Imprisoned by the Nazis in 1935 to 1945, he rose rapidly in the East German Communist Party after the Second World War, and succeeded Walter Ulbricht as party leader, pursuing policies approved by Moscow. Unwilling to implement the reforms proposed by the Soviet president, Mikhail GORBACHEV, the ailing Honecker was forced to resign in 1989, shortly before the collapse of European communism. ○ *264–65*

HOPEWELL CULTURE Culture centred in Ohio and Illinois, and reaching its peak in the last centuries BC and the first four centuries AD. The Hopewell people were efficient farmers, built complex earthworks, such as the Great Serpent Mound of Ohio, for ceremonial and business purposes, and traded extensively. ○ *24–25*

HORTHY, MIKLÓS NAGYBÁNAI (1868–1957) Hungarian political leader. He commanded the Austro-Hungarian fleet in the First World War. He took part in the counter-revolution that overthrew Béla Kun, becoming regent and effective head of state (1920–44). His highly conservative regime suppressed political opposition and resisted the return of Charles I. Allied

with the Axis Powers in 1941, he tried but failed to arrange a separate peace with the Allies in 1944. After the war he settled in Portugal. ○ *230–31*

HUDSON'S BAY COMPANY English company chartered in 1670 to promote trade in the Hudson Bay region of North America and to seek a Northwest Passage. The company had a fur trading monopoly and was virtually a sovereign power in the region. Throughout the 18th century it fought with France for control of the bay. In 1763 France ceded control of Canada to England, and the North West Company was formed. Intense rivalry forced the Hudson's Bay Company into a more active role in westward exploration, and in 1771 Samuel Hearne proved the lack of a short Northwest Passage out of the Bay. The companies merged in 1821, with the new company controlling a territory from the Atlantic to the Pacific. After the Confederation of Canada (1867), challenges to its monopoly power increased, and in 1869 it was forced to cede all its territory to Canada for £300,000. As the fur trade declined in the early 20th century, the company diversified, and in 1930 was divided up. ○ *188–89*

HUGUENOTS French Protestants in CATHOLIC France during the Reformation who suffered persecution. In 1559 a national synod of Huguenot congregations met in Paris and adopted a confession of faith and an ecclesiastical structure highly influenced by CALVIN. During the FRENCH WARS OF RELIGION from 1562 to 1598 Huguenots continued to face persecution and thousands died. King Henry IV, a Huguenot, came to the throne in 1589 and, despite adopting the Roman faith in 1593, promulgated the EDICT OF NANTES (1598) which recognized Catholicism as the official religion, but gave Huguenots certain rights. It was revoked by LOUIS XIV in 1685, and thousands of Huguenots fled France. In 1789 their civil rights were restored, and in 1804 the NAPOLEONIC CIVIL CODE (*Code Napoléon*) guaranteed religious equality. ○ *128–29, 154–55*

HUNDRED YEARS WAR Conflict between France and England pursued sporadically between 1337 and 1453. Edward III 's claim to the French crown sparked the war. Early English successes brought territorial gains in the Peace of Brétigny of 1360. The French gradually regained their lost territory and a revival stimulated by Joan of Arc led eventually to the expulsion of the English from all of France except Calais. ○ *106–7*

HUS, JOHN (1369–1415) Bohemian religious reformer. Born at Husinec in southern BOHEMIA, he studied and later taught at Prague, where he was ordained priest. Influenced by the beliefs of the English reformer John WYCLIFFE, he became leader of a reform movement, for which he was excommunicated in 1411. In *De Ecclesia* (1412), Hus outlined his case for reform. In 1415 he was burned at the stake as a heretic. His followers were known as HUSSITES. ○ *106–7,*

HUSSEIN, SADDAM (1937–) Iraqi statesman, president of Iraq (1979–). In 1957 he joined the Ba'ath Socialist Party. In 1959 he was forced into exile after taking part in an attempt to assassinate the Iraqi prime minister. He returned in 1963 and was soon imprisoned. Following his release, he played a prominent role in the 1968 Ba'athist coup, which replaced the civilian government by the Revolutionary Command Council (RCC). In 1979 he became chairman of the RCC and state president. His invasion of Iran marked the beginning of the IRAN–IRAQ WAR (1980–88). Domestically, he ruthlessly suppressed all opposition, including the gassing of Kurdish villagers. His 1990 invasion of Kuwait provoked worldwide condemnation and, in the ensuing GULF WAR (1991), a multinational force expelled Iraqi forces from Kuwait. Further uprisings by Kurds and Iraqi SHIITES were ruthlessly suppressed, and Saddam survived punitive economic sanctions. ○ *260–61*

HUSSITES Followers of the religious reformer Jan HUS in BOHEMIA and Moravia in the 15th century. The execution of Hus in 1415 provoked the Hussite wars against the Emperor Sigismund. Peace was agreed at the Council of Basel in 1431, but it was rejected by the radical wing of the Hussites, the Taborites, who were defeated at the Battle of Lipany in 1434. ○ *106–7*

HU YAOBANG (1915–89) Chinese statesman, general secretary of the Chinese Communist Party (1980–87). The son of peasants, he joined the Communists in 1933 and took part in the Long March. He became closely associated with Deng Xiaoping during the war against Japan (1937–45), during which he served as a political commissar. In 1952 he became head of the Young Communist League, but lost his post after the Cultural Revolution of 1966. He was rehabilitated in 1977 and, after being appointed general secretary, was also made party chairman. In 1987 he was dismissed for alleged mistakes in policy after several weeks of student demonstrations. ➲ *254–55*

I

IBN BATTUTAH (1304–68) Arab traveller and writer. Born in Tangier, Morocco, he began his adventures in 1325 with a pilgrimage to Mecca by way of Egypt and Syria. Travel was to occupy the next 30 years of his life, when he visited parts of Africa, Asia and Europe. He finally returned home to Morocco in 1349. ➲ *62–63*

INDIAN MUTINY (1857–58) Indian rebellion against the British originating among Indian troops (sepoys) in the Bengal army. It is known in India as the First War of Independence. The immediate cause was the introduction of cartridges lubricated with the fat of cows and pigs, a practice offensive to both Hindus and Muslims. A more general cause was resentment at modernization and Westernization. Delhi was captured, and atrocities were perpetrated by both sides. The revolt resulted in the British government taking over control of India from the East India Company in 1858. ➲ *194–95*

INDIAN NATIONAL CONGRESS PARTY Oldest political party in India, whose fortunes have often been intertwined with the Nehru dynasty. It was founded in 1885, but was not prominent until after the First World War, when Mahatma Gandhi transformed it into a mass independence movement, agitating by means of civil disobedience. Jawaharlal Nehru became president of the Congress in1929. In the 1937 provincial elections it gained power in many states. During the Second World War (after the British refused to grant self-government) it remained neutral. At independence in 1947, Nehru became prime minister. Nehru's daughter Indira Gandhi became prime minister in 1966, but in 1969 was challenged by a right-wing faction (lead by Moraji Desai) causing a split in the party. Indira's Congress suffered a landslide defeat at the 1977 elections. They returned to power in 1979, and in 1984 (after Indira's assassination) her son Rajiv Gandhi became leader, securing the party's re-election in that year. Congress were defeated in 1989 and Rajiv was assassinated in 1991. He was succeeded by V. P. Singh, head of the breakaway Janata Dal. The Congress Party, led by Narasimha Rao, heavily lost the 1996 election. ➲ *248–49*

INDOCHINA Peninsula of southeast Asia, including Burma, Thailand, Cambodia, Vietnam, West Malaysia and Laos. The name refers more specifically to the former federation of states of Vietnam, Laos and Cambodia, associated with France within the French Union (1945–54). European penetration of the area began in the 16th century. By the 19th century France controlled Cochin China, Cambodia, Annam and Tonkin, which together formed the union of Indochina in 1887; Laos was added in 1893. By the end of the First World War France had announced plans for a federation within the French Union. Cambodia and Laos accepted the federation, but fighting broke out between French troops and Annamese nationalists, who wanted independence for Annam, Tonkin and Cochin China as Vietnam. The war ended with the French defeat at Dien Bien Phu. French control of Indochina was officially ended by the Geneva Conference of 1954. ➲ *250–51, 254–55*

INKATHA FREEDOM PARTY South African political organization founded in 1975 by Chief Buthelezi. Its initial aim was to work towards a democratic, non-racial political system. In the early 1990s it was involved in violent conflict with the African National Congress (ANC). It ranks third among political parties in terms of representation in the South African National Assembly, and its strongest base is in KwaZulu-Natal. ➲ *256–57*

INTERNATIONAL MONETARY FUND (IMF) Specialized, intergovernmental agency of the United Nations, and administrative body of the international monetary system. Its main function is to provide assistance to member states troubled by balance of payments problems and other financial difficulties. The IMF does not actually lend money to member states; rather, it exchanges the member state's currency with its own Special Drawing Rates (SDR) (a "basket" of other currencies) in the hope that this will alleviate balance of payment difficulties. These loans are usually conditional upon the recipient country agreeing to pursue prescribed policy reforms. The organization is based in Washington, D.C., USA. ➲ *256–57, 258–59, 272–73*

IRAN–IRAQ WAR (1980–88) Contest for supremacy in the Gulf. The war began when Iraq, partly in response to Iranian encouragement of revolt among the Shiites of southern Iraq, invaded Iran, which was disorganized after the Islamic fundamentalist revolution of 1979. Iraq's objective was the Shatt al Arab waterway, but stiff Iranian resistance checked its advance and forced its withdrawal in 1982. The conflict bogged down in stalemate, with sporadic Iranian offensives. US-led intervention in 1987 was seen as tacit support for Iraq. A United Nations cease-fire resolution of 1987 was accepted by Iraq and, after several Iraqi successes, by Iran also. Estimated total casualties were more than one million. ➲ *260–61, 266–67*

IRISH REPUBLICAN ARMY (IRA) Guerrilla organization dedicated to the reunification of Ireland. Formed in 1919, the IRA waged guerrilla warfare against British rule. Some members ("irregulars") rejected the Anglo-Irish settlement of 1921, fighting a civil war until 1923. In 1970 the organization split into an "official" wing (which emphasized political activities), and a "provisional" wing (committed to armed struggle). Thereafter, the provisional IRA became committed to terrorist acts in Northern Ireland and mainland Britain. It declared a ceasefire in 1994, but in 1996 resumed its campaign. A second ceasefire was declared in 1998. ➲ *268–69*

ISLAM (Arabic, submission to God) Monotheistic religion founded by Muhammad in Arabia in the early 7th century. At the heart of Islam stands the *Koran*, considered the divine revelation in Arabic of God to Muhammad. Members of the faith (Muslims) date the beginning of Islam from AD 622, the year of the *Hejira*. Muslims submit to the will of Allah by five basic precepts (pillars). First, the *shahadah*, "there is no God but Allah, and Muhammad is his prophet". Second, *salah*, five daily ritual prayers. At the mosque a Muslim performs ritual ablutions before praying to God in a attitude of submission, kneeling on a prayer mat facing Mecca with head bowed, then rising with hands cupped behind the ears to hear God's message. Third, *zakat* or alms-giving. Fourth, *sawm*, fasting during Ramadan. Fifth, *hajj*, the pilgrimage to Mecca. The rapid growth in Islam during the 8th century can be attributed to the unification of the temporal and spiritual. The community leader (*caliph*) is both religious and social leader. The *Koran* was soon supplemented by the informal, scriptural elaborations of the *Sunna* (Muhammad's sayings and deeds), collated as the *Hadith*. A Muslim must also abide by the *Sharia* or religious law. While Islam stresses the importance of the unity of the *summa* (nation) of Islam, several distinctive branches have developed, such as Sunni, Shiite and Sufism. ➲ *62–63, 68–69, 88–89, 136–37, 142–43, 248–49, 260–61, 268–69*

ISMAIL (1486–1524) Shah of Persia (Iran) (1501–24), founder of the Safavid dynasty. A national and religious hero in Iran, he re-established Persian independence and established Shiite Islam as the state religion. He warred successfully against the Uzbeks in 1510 but was defeated by the Ottoman sultan, Selim I, at the battle of Chaldiran in 1514. ➲ *142–43*

ISRAEL, ANCIENT Name given in the Old Testament to Jacob and to the nation that the Hebrews founded in Canaan. As a geographical name, Israel at first applied to the whole territory of Canaan captured or occupied by the Hebrews after the Exodus from Egypt. This territory was united as a kingdom under David in the early 10th century BC, with its capital at Jerusalem. Following the death of David's son Solomon, the ten northern tribes seceded, and the name Israel thereafter applied to the kingdom they founded in northern Palestine; the remaining two tribes held the southern kingdom of Judah. ➲ *38–39*

IVAN III (THE GREAT) (1440–1505) Grand Duke of Moscow (1462–1505) He laid the foundations of the future empire of Russia. By 1480 Moscow's northern rivals, including Novgorod, were absorbed by conquest or persuasion, domestic rebellion crushed, and the Tatar threat ended permanently. His later years were troubled by conspiracies over the succession. He began to use the title tsar ("caesar") and employed Italian artists in the buildings of the Kremlin. ➲ *148–49*

IVAN IV (THE TERRIBLE) (1530–84) Grand Duke of Moscow (1533–84) and tsar of Russia. Ivan was crowned as tsar in 1547 and married Anastasia, a Romanov. At first, he was an able and progressive ruler, reforming law and government. By annexing the Tatar states of Kazan and Astrakhan, he gained control of the Volga river. He established trade with western European states and began Russian expansion into Siberia. After his wife's death in 1560, he became increasingly unbalanced, killing his own son in a rage. He established a personal dominion, the *oprichnina*, inside Russia. He also created a military force, the *oprichniki*, which he set against the *boyars* (nobility). ➲ *146–47, 148–49*

J

JACOBINS French political radicals belonging to a club that played an important role during the French Revolution. Begun in 1789, the club split in 1791 when the moderates left it. In 1793–94 the club was an instrument of Robespierre and became part of the government's administration. It closed soon after Robespierre's downfall in 1794. ➲ *166–67*

JAGIELLONS Medieval Polish dynasty. It began with the marriage of Grand Duke Jagiellon of Lithuania to Queen Jadwiga of Poland in 1386, uniting Poland and Lithuania. Their combined power enabled them to defeat their common enemy, the Teutonic Knights, at Tannenberg in 1410. Members of the dynasty also reigned in Hungary and Bohemia in the 15th–16th centuries. ➲ *146–47, 152–53*

JAHANGIR (1569–1627) Mughal emperor of India (1605–27). He succeeded his father, Akbar I, and continued the expansion of the Mughal Empire. He granted trading privileges to the Portuguese and the British, and was a patron of poetry and painting. ➲ *144–45*

JAINISM Ancient religion of India originating in the 6th century BC as a reaction against conservative Brahmanism. It was founded by Mahavira (599–527 BC). Jains do not accept Hindu scriptures, rituals, or priesthood, but they do accept the Hindu doctrine of Transmigration of Souls. Jainism lays special stress on *ahimsa* – non-injury to all living creatures. ➲ *44–45, 248–49*

JAMES I (1566–1625) King of England (1603–25) and, as James VI, King of Scotland (1567–1625). Son of Mary, Queen of Scots, James acceded to the Scottish throne as an infant on his mother's abdication. In 1603, following the death of Elizabeth I, he inherited the English throne and thereafter confined his attention to England. He supported the Anglican Church, at the cost of antagonizing the Puritans, and sponsored a translation of the Bible, which became known as the Authorized, or King James Version (1611). His troubled relationship with Parliament weakened his effectiveness as a ruler. ➲ *146–47*

JANISSARIES Elite corps of the Ottoman army, founded in the 14th century. The Janissaries were a highly effective fighting force until the 17th century, when discipline and military prestige declined. They were abolished by Mahmud II in 1826. ➲ *142–43, 178–79*

JAYAWARDENE, JUNUIS RICHARD (1906–96) President of Sri Lanka (1978–88). One of the founders of the United National Party in 1946, he remained actively involved in its leadership. In 1978 Jayawardene introduced a new form of government modelled on that of the French. Elected prime minister in 1977, he became Sri Lanka's first president the next year and was re-elected to a six-year term in 1982. ➲ *248–49*

JEFFERSON, THOMAS (1743–1826) Third US president (1801–09). An accomplished scholar, Jefferson was the primary author of the DECLARATION OF INDEPENDENCE, the first secretary of state (1789–93) under Washington, and vice-president (1797–1801) under John Adams before being elected president in 1800. He was also governor of Virginia (1779–81), US minister to France (1785–89), and the founder of what became the DEMOCRATIC PARTY. He was a slave owner, although in principle opposed to slavery. A notable achievement of his presidency was the LOUISIANA PURCHASE in 1803. ➲ *182–83*

JERUSALEM City in Israel, a sacred site for Christians, JEWS and Muslims. Originally a Jebusite stronghold (2000–1500 BC), the city was captured by King DAVID after 1000 BC. Destroyed by NEBUCHADNEZZAR c. 587 BC, it was rebuilt by Herod the Great c. 35 BC, but was again destroyed by TITUS in AD 70. The Roman colony of Aelia Capitolina was established, and Jews were forbidden within city limits until the 5th century. Christian control was ended by the Persians in AD 614. It was conquered in 1071 by the SELJUKS, whose mistreatment of Christians precipitated the crusades. It was held by the Ottoman Empire Turks from 1244 to 1917, before becoming the capital of the British-mandated territory of PALESTINE. In 1948 it was divided between Jordan (the east) and Israel (the west). In 1967 the Israeli army captured the Old City of East Jerusalem. In 1980 the united city was declared the capital of Israel, although this status is not recognized by the UN. ➲ *36–37, 67–68, 94–95, 98–99, 218–19, 260–61*

JESUITS Members of a Roman CATHOLIC religious order for men, officially known as the Society of Jesus, founded by Ignatius LOYOLA in 1534. They played a significant role in the Counter Reformation. The Jesuits were active missionaries. They antagonized many European rulers because they gave allegiance only to their general in Rome and to the Pope. In 1773 Pope Clement XIV abolished the order, under pressure from the kings of France, Spain and Portugal, but it continued to exist in Russia. The order was re-established in 1814 and remains an influential international religious organization. ➲ *154–55*

JESUS CHRIST (4 BC–AD 29) Hebrew preacher who founded the religion of CHRISTIANITY, hailed and worshipped by his followers as the Son of God. Knowledge of Jesus's life is based mostly on the biblical gospels of St Matthew, St Mark and St Luke. The date of Jesus's birth is now given as around 4 BC but may have been earlier. The birth occurred near the end of the reign of Herod the Great in Bethlehem, Judaea. Jesus grew up in Nazareth, and may have followed his father, Joseph, in becoming a carpenter. In c. AD 26 or 27, Jesus was baptized in the River Jordan by John the Baptist. Thereafter Jesus began his own ministry, preaching to large numbers as he wandered throughout the country. He also taught a special group of 12 of his closest disciples, who were later sent out as his Apostles to bring his teachings to the JEWS. Jesus's basic teaching, summarized in the Sermon on the Mount, was to "love God and love one's neighbour". He also taught that salvation depended on doing God's will rather than adhering to the letter and the contemporary interpretation of the Jewish Law. This angered the hierarchy of the Jewish religion. In c. AD 29 Jesus and his disciples went to JERUSALEM where he was acclaimed by the people as the Messiah. A few days later Jesus was arrested by agents of the Hebrew authorities and summarily tried by the *Sanhedrin*, the Supreme Council of the Jews. He was then handed over to the Roman procurator, Pontius Pilate, on a charge of sedition. Roman soldiers crucified Jesus at Golgotha. According to the gospel accounts, he rose from the dead and appeared to his disciples and to others. Forty days after his resurrection, he is said to have ascended into heaven. ➲ *44–45, 62–63*

JEWS Followers of the religion of JUDAISM, especially those who claim descent from the ancient Hebrews, a Semitic people who settled in PALESTINE toward the end of the 2nd millennium BC. The word Jew arose in medieval times, derived from the Latin word Judaea (Judea), the Romanized name of the region of Palestine. From c. 600 BC, the people of Judah suffered domination by a number of foreign powers, among them the Assyrians, Babylonians, Seleucid rulers, and finally the Romans. After the destruction of JERUSALEM by the Romans in AD 70, they were dispersed throughout the world. The Jews were driven out of England in 1290 and were expelled from Spain in the 15th century. During the Second World War (1939–45), six million Jews were killed in the HOLOCAUST. In 1948, having struggled against British rule in modern Palestine, a group of Jews finally established the modern nation state of Israel there, despite opposition from Arab and other Islamic states. About five million Jews inhabit Israel. Many millions more live in other countries. ➲ *44–45, 230–33, 260–61*

JOHN (1167–1216) King of England (1199–1216). The youngest son of Henry II, he ruled during RICHARD I 's absence on the Third Crusade. Disgraced for intriguing against Richard, John nevertheless succeeded him as king. The loss of vast territories in France in 1204–5 and heavy taxation made him unpopular. In 1215 he was compelled to sign the Magna Carta, and his subsequent disregard of the terms led to the first Barons' War. ➲ *92–93*

JOHN VI (1767–1826) King of Portugal (1816–26). Owing to the insanity of his mother, Queen Maria, he was effectively sovereign from 1792, officially regent from 1799. In 1807 he fled to Brazil to escape the invading French and did not return to claim the throne until 1822, when he accepted the constitutional government proclaimed in 1820. ➲ *190–91*

JOSEPH II (1741–90) Holy Roman Emperor (1765–90). Co-ruler of the Austrian Empire with his mother MARIA THERESA until 1780 and then sole ruler. He introduced sweeping liberal and humanitarian reforms while retaining autocratic powers. Some of his reforms were reversed by his successor, LEOPOLD II. ➲ *174–75*

JUDAISM Monotheistic religion developed by the ancient Hebrews in the Near East during the third millennium BC and practised by modern JEWS. Tradition holds that Judaism was founded by Abraham, who, in c. 20th century BC, was chosen by God to receive favourable treatment in return for obedience and worship. Having entered into this covenant with God, Abraham moved to Canaan, from where centuries later his descendants migrated to EGYPT and became enslaved. God accomplished the Hebrews' escape from Egypt and renewed the covenant with their leader Moses. Through Moses, God gave the Hebrews a set of strict laws. These laws are revealed in the *Torah*, the core of Judaistic scripture. Apart from the *Pentateuch*, the other holy books are the *Talmud* and several commentaries. Local worship takes place in a synagogue, a building where the *Torah* is read in public and preserved in a replica of the Ark of the Covenant. A rabbi undertakes the spiritual leadership and pastoral care of a community. Modern Judaism is split into four large groups: Orthodox, Reform, Conservative and Liberal Judaism. Orthodox Judaism, followed by most of the world's 18 million Jews, asserts the supreme authority of the *Torah* and adheres most closely to traditions, such as the segregation of men and women in the synagogue. Reform Judaism denies the Jews' claim to be God's chosen people, and is more liberal in its interpretation of certain laws and the *Torah*. Conservative Judaism is a compromise between Orthodox and Reform Judaism, adhering to many Orthodox traditions, but seeking to apply modern scholarship in interpreting the *Torah*. Liberal Judaism, also known as Reconstructionism, is a more extreme form of Reform Judaism, seeking to adapt Judaism to the needs of society. ➲ *44–45*

JULIUS II (1443–1513) Pope (1503–13), born Guiliano della Rovere. He tried to recover papal lands and, in 1506, established the Swiss Guard to protect the Pope and Rome. He built up the treasury through the sale of benefices and began the building of St Peter's Basilica. ➲ *146–47*

JULIUS CAESAR *See* CAESAR, (GAIUS) JULIUS

JUSTINIAN I (482–565) Byzantine Emperor (527–565). His troops, commanded by Belisarius, regained much of the old Roman Empire, including Italy, northern Africa and part of Spain. Longer-lasting achievements were the Justinian Code, a revision of the whole body of Roman law, and his buildings in CONSTANTINOPLE, which included Hagia Sophia. Heavy taxation to pay for wars, including defence against SASANIAN PERSIA, drained the strength of the empire. ➲ *66–67*

K

KALMAR, UNION OF (1397–1523) Union of Denmark, Norway and Sweden in 1397. The three crowns were united after Margaret I of Denmark had appointed her grand-nephew Eric VII as her heir. Otherwise the Union had little effect. Sweden withdrew in 1523 although Denmark and Norway remained united until 1814. ➲ *106–7, 150–51*

KANSAS–NEBRASKA ACT (1854) US Congressional measure introduced by Senator Stephen A. Douglas of Illinois which gave states the right to decide for themselves all questions related to slavery. It effectively negated the earlier MISSOURI COMPROMISE of 1820, which declared that all land in the LOUISIANA PURCHASE was to be non-slave, except for the state of Missouri. Written to appease Southern congressmen, the Kansas-Nebraska Act made slavery legally possible in the two vast new territories of Kansas and Nebraska, and revived the bitter slavery controversy ➲ *184–85*

KEMAL, MUSTAFA *See* ATATÜRK, KEMAL

KENNEDY, JOHN F. (FITZGERALD) (1917–63) 35th US president (1961–63). He was elected to Congress as a DEMOCRAT from Massachusetts in 1946, serving in the Senate (1953–60). He gained the presidential nomination in 1960 and narrowly defeated Richard NIXON. He adopted an ambitious and liberal programme, under the title of the "New Frontier", and embraced the cause of civil rights, but his planned legislation was frequently blocked by Congress. In foreign policy he founded the "Alliance for Progress", the aim of which was to improve the image of the USA abroad. Adopting a strong anti-communist line, he was behind the Bay of Pigs disaster (1961) and outfaced KHRUSHCHEV in the ensuing Cuban Missile Crisis, which was followed by a US–Soviet treaty banning nuclear tests. He increased military aid to South Vietnam. He was assassinated in Dallas, Texas, on 22 November 1963. ➲ *242–43*

KHAZARS Turkic people who first appeared in the lower Volga region c. 2nd century AD. Between the 8th and 10th centuries their empire prospered and extended from north of the Black Sea to the River Volga and from west of the Caspian Sea to the River Dnieper. They conquered the Volga BULGARS and fought the Arabs, Russians and Pechenegs. In the 8th century, their ruling class was converted to JUDAISM. Their empire was destroyed in 965 by the army of Sviatoslav, Duke of Kiev. ➲ *76–77*

KHOMEINI, AYATOLLAH RUHOLLAH (1900–89) Iranian religious leader. An Islamic scholar with great influence over his SHIITE students, he published an outspoken attack on Riza Shah Pahlavi in 1941 and remained an active opponent of his son, Muhammad Reza Shah Pahlavi. Exiled in 1964, he returned to Iran in triumph after the fall of the Shah in 1979. His rule was characterized by strict religious orthodoxy, elimination of political opposition, and economic turmoil. In 1988 he issued a *fatwa* (death order) against the author Salman Rushdie. ➲ *260–61*

KHRUSHCHEV, NIKITA SERGEYEVICH (1894–1971) Soviet politician, first secretary of the Communist Party (1953–64) and Soviet prime minister (1958–64). Noted for economic success and ruthless suppression of opposition in the Ukraine, he was elected to the Politburo in 1939. After Stalin died, he made a speech denouncing him and expelled Stalinists from the central committee. Favouring détente with the West, he yielded to the USA in the Cuban Missile Crisis. Economic setbacks and trouble with China led to his replacement by Leonid BREZHNEV and Aleksei KOSYGIN in 1964. ➲ *236–37*

KING, MARTIN LUTHER, JR (1929–68) US Baptist minister and civil rights leader. He led the boycott of segregated public transport in Montgomery, Alabama in 1956. As founder (1960) and president of the Southern Christian Leadership Council, he became a national figure. He opposed the Vietnam War and demanded measures to relieve poverty, organizing a huge march on Washington (1963) where he made his most famous ("I have a dream...") speech. In 1964 he became the youngest person to be awarded the Nobel Peace Prize. He was assassinated in Memphis, Tennessee, where he had gone to support striking workers. ➲ *240–41*

KITCHENER, HORATIO HERBERT, EARL (1850–1916) British soldier and statesman. He took part in the relief of Khartoum (1883–85) and achieved the pacification of the Sudan (1898). After service in the SOUTH AFRICAN WARS (1899–1902) and then in India and Egypt, he was appointed Secretary of State for War in 1914. He was lost at sea when his cruiser sank. ➲ *206–7*

KNIGHTS HOSPITALLERS (Order of the Hospital of St John of JERUSALEM) Military Christian order recognized in 1113 by Pope Paschal II. In the early 11th century a hospital was established in Jerusalem for Christian pilgrims. They adopted a military role in the 12th century to defend Jerusalem. After the fall of Jerusalem in 1187, they moved to Acre, then Cyprus, then Rhodes (1310), from where they were expelled by the Ottoman Turks in 1522. The Pope then gave them Malta, where they remained until driven out by NAPOLEON in 1798. The order still exists as an international, humanitarian charity. ➲ *94–95*

KNIGHTS TEMPLAR Military religious order established in 1118, with headquarters in the supposed Temple of Solomon in JERUSALEM. With the Knights Hospitallers, the Templars protected routes to Jerusalem for Christians during the crusades. The possessions of the Templars in France attracted the envious attention of King Philip IV, who urged Pope Clement V to abolish the order in 1312. ➲ *94–95*

KNOSSOS Ancient palace complex in central Crete, 6.4 kilometres (4 miles) southeast of modern Iráklion. In 1900 Sir Arthur Evans began excavations that revealed that the site had been inhabited before 3000 BC. His main discovery was a palace from the MINOAN CIVILIZATION (built c. 2000 BC and rebuilt c. 1700 BC). Close to the palace were the houses of Cretan nobles. The complex also contains many frescos. Knossos dominated Crete c. 1500 BC, but invaders from MYCENAE occupied the palace c. 1450 BC. ➲ *36–37*

KONGO KINGDOM African state from the 14th century to c. 1700. In the area now included in Zaire and Angola, it was ruled by a king, or *manikongo*. The kingdom began trade with Portugal in 1483. The Portuguese brought CHRISTIANITY, which Manikongo Afonso I tried to spread. However, Afonso was hampered by the greed of the Portuguese, who carried on a brisk slave trade. Under continued depredations by the Portuguese and repeated attacks from interior tribes, the kingdom finally collapsed, and Portugal took control. ➲ *136–37*

KOREAN WAR (1950–53) Conflict between North Korea, supported by China, and South Korea, supported by UN forces dominated by the USA. South Korea was invaded by forces of the North in June 1950. The United Nations Security Council, during a boycott by the Soviet Union, voted to aid South Korea. Major US forces, plus token forces from its allies, landed under the overall command of General DOUGLAS MACARTHUR. The invaders were driven out, but when the UN forces advanced into North Korea, China intervened and drove them back, recapturing Seoul. After more heavy fighting, UN forces slowly advanced until virtual stalemate ensued near the 38th parallel, the border between North and South Korea. Negotiations continued for two years before a truce was agreed in July 1953. Total casualties were estimated at four million. ➲ *244–45, 252–53, 254–55, 272–73*

KOSSUTH, LOUIS (1802–94) Hungarian nationalist leader. Emerging as leader of the Revolution of 1848, he declared Hungarian independence (1849), but Russian intervention led to his defeat. He fled to arouse support for Hungarian independence in Europe and the USA. The compromise of 1867, which created the Dual Monarchy of the Austro-Hungarian empire, put an end to his hopes. ➲ *174–75*

KOSYGIN, ALEXEI NIKOLAI (1904–80) Soviet politician. He was elected to the Communist Party Central Committee in 1939 and the Politburo in 1948. He was removed in 1953 but regained his seat in 1960. After KRUSHCHEV's fall in 1964, he became prime minister, a position he held until his retirement in 1980. ➲ *236–37*

KUBLAI KHAN see QUBILAI KHAN

KUOMINTANG Nationalist Party in China, which was the major political force during and after the creation of a republic in 1911. It was first led by SUN YAT-SEN. It cooperated with the Communist Party until 1927 when Sun's successor, CHIANG KAI-SHEK, turned against the communists, initiating a civil war. Co-operation was renewed in order to repel the Japanese from 1937 to 1945, after which the civil war was resumed. With the communists victorious, Chiang set up a rump state on the island of Taiwan, where the Kuomintang survives. ➲ *224–25*

KURDS Predominantly rural, Islamic population numbering some 18 million, who live in a disputed frontier area of southwest Asia that they call Kurdistan. Traditionally nomadic herdsmen, they are mainly SUNNI Muslims who speak an Iranian dialect. For 3,000 years they have maintained a unique cultural tradition, although internal division and constant invasion have prevented them from uniting into one nation. In recent times, their main conflicts have been with Turkey and Iraq. After the IRAN-IRAQ WAR (1988), Iraq destroyed many Kurdish villages and their inhabitants. The Iraqi response to a Kurdish revolt after the GULF WAR caused 1.5 million Kurds to flee to Iran and Turkey. In 1996 Iraqi troops invaded the region and captured the Kurdish city of Irbil. The USA responded by launching cruise missiles at Iraqi military installations. Currently about 8 million Kurds live in eastern Turkey, 6 million in Iran, 4 million in northern Iraq, 500,000 in Syria and 100,000 in Azerbaijan and Armenia. ➲ *260–61, 268–69*

L

LAJOS II King of Hungary (1456–1516) King of BOHEMIA (1471–1516) and of Hungary (1490–1516). Son of Casimir IV of Poland, he was the first ruler of the JAGIELLON dynasty and a weak king, totally unsuited to his position. His claim to the crown of Hungary was contested by Matthias Corvinus. The aristocracy increased its power during his reign. ➲ *146–47*

LAO-TZE (604–531 BC) Chinese philosopher, credited as the founder of DAOISM. Tradition says that he lived in the 6th century BC and developed Daoism as a mystical reaction to the moral-political concerns of CONFUCIANISM. His teachings were later written down as the *Dao De Jing*, the sacred book of Daoism. ➲ *44–45*

LAUSANNE, TREATY OF (1923) Agreement signed at Lausanne, Switzerland, which abrogated the harsh Treaty of SÈVRES (1920), imposed on the collapsing Ottoman Empire. It ended the war between Greece and Turkey. Turkey was granted full sovereignty over mainland Turkey and renounced claims to Greek islands in the Aegean Sea. Britain obtained Cyprus, and Italy obtained Rhodes and the Dodecanese Islands. ➲ *178–79, 220–21*

LEAGUE OF NATIONS International organization, forerunner of the UNITED NATIONS (UN). Created as part of the Treaty of Versailles (1919) ending the First World War, it was impaired by the refusal of the USA to participate. The threats to world peace from Germany, Italy and Japan caused the League to collapse in 1939. It was dissolved in 1946. ➲ *220–21, 268–69*

LEBANESE CIVIL WAR In the late-1960s Lebanon came under increasing military pressure from Israel to act against Palestinian guerrillas operating in southern Lebanon. In 1975 civil war broke out between MARONITE, SUNNI, SHIITE, and Druse militias. About 50,000 Lebanese died and the economy was devastated. In 1976 Syrian troops imposed a fragile ceasefire. In 1978 Israel invaded southern Lebanon to destroy Palestinian bases. UNITED NATIONS peace-keeping forces were called in to separate the factions. In 1982 Israel launched a full-scale attack on Lebanon. The 1983 deployment of US and European troops in Beirut was met by a terriorist bombing campaign. Multinational forces left in 1984, and Israeli troops withdrew to a buffer zone in southern Lebanon. In 1987 Syrian troops moved into Beirut to quell disturbances. In 1990 an uneasy truce was called and the government began to disarm the militias. Syria maintained troops in West Beirut and the Bekaa Valley. The Syrian-backed Hezbollah and the Israeli-backed South Lebanon Army (SLA) continued to operate in southern Lebanon. ➲ *260–61, 266–67*

LEE, ROBERT E. (EDWARD) (1807–70) Commander of the CONFEDERATE forces in the American Civil War. A professional soldier, he regarded slavery as evil and opposed secession but, as a loyal Virginian, he became military adviser to Jefferson Davis and in 1862 he was appointed commander of the Army of Virginia. He successfully defended Richmond and won the Second Battle of Bull Run. Though checked at Antietam in 1862, he defeated the Union forces at Fredericksburg in 1862 and Chancellorsville in 1863. His invasion of the North ended in decisive defeat at Gettysburg in July 1863, but he proved an equally brilliant general in a defensive campaign. He was finally trapped by GRANT and surrendered at Appomatox Court House in April 1865. ➲ *184–85*

LEIF EIRÍKSSON (c. 970–1020) Norse adventurer and explorer. Son of EIRÍK THE RED, he sailed from Greenland in 1003 to investigate land in the west. Among the places he visited were Helluland (probably Baffin Island), Markland (Labrador) and Vinland. ➲ *78–79*

LENIN, VLADIMIR ILYICH (1870–1924) Russian revolutionary. He evolved a revolutionary doctrine, based principally on Marxism, in which he emphasized the need for a vanguard party to lead the revolution. In 1900 he went abroad, and founded what became the BOLSHEVIKS (1903). After the first part of the RUSSIAN REVOLUTION of 1917, he returned to Russia. He denounced the liberal republican government of Kerensky and demanded armed revolt. After the Bolshevik revolution (November 1917), he became leader of the first Soviet government. He withdrew Russia from the First World War and totally reorganized the government and economy. He founded the third COMMUNIST INTERNATIONAL in 1919. His authority was unquestioned until he was crippled by a stroke in 1922. ➲ *222–23*

LEOPOLD II (1747–92) Grand Duke of Tuscany as Leopold I (1765–90); Holy Roman Emperor (1790–92). The third son of MARIA THERESA, he succeeded his father, FRANCIS I, as ruler in Tuscany and his brother JOSEPH II as Holy Roman Emperor. By reversing many of Joseph's reforms, Leopold pacified much of the Empire. He tried to avoid war with France, but was nevertheless involved in the Declaration of Pillnitz, which aimed to restore Louis XVI and was a primary cause of the French Revolutionary Wars. ➲ *174–75*

LINCOLN, ABRAHAM (1809–65) 16th US president (1861–65). Elected to the Illinois legislature for the Whig Party in 1834. He served in the House of

Representatives (1847–49). He was nominated as Republican candidate for president in 1860. Lincoln's victory made the secession of the Southern, slave-owning states inevitable, and his determination to defend Fort Sumter began the American Civil War. A strong commander-in-chief, he played a leading role in military planning until he found a sufficiently determined general in Ulysses S. GRANT. In September 1862 he issued the EMANCIPATION PROCLAMATION, and in November 1863 delivered his famous Gettysburg Address. He was re-elected in 1864 and saw the war to a successful conclusion. On 14 April 1865, five days after the surrender of Robert E. LEE, he was shot by John Wilkes Booth, a Southern sympathizer. He died the following day. ➲ *184–85, 240–41*

LITTLE ENTENTE (1920–38) Alliance contracted between Romania, Yugoslavia and Czechoslovakia after the First World War to maintain post-war boundaries. Through political and economic unity and the support of France and Poland, the alliance helped to prevent *Anschluss* (the uniting of Germany and Austria) until 1938. ➲ *220–21, 230–31*

LIU SHAO-CH'I (1898–1974) Chinese communist leader. Trained in Moscow (1920–22), he became one of the chief theorists of the Chinese Communist Party, ranking second to MAO ZEDONG. He was author of the influential manual, *How to Be a Good Communist* (1939). Liu was made official head of state in 1959, but was purged in 1968 during the CULTURAL REVOLUTION and died in prison. ➲ *254–55*

LIVINGSTONE, DAVID (1813–73) British explorer of Africa. He went to South Africa as a missionary in 1841 and became famous through his account of his journey, accompanied by Africans, across the continent from Angola to Mozambique (1853–56). He led a major expedition (1858–64) to the Zambezi and Lake Nyasa and set off again in 1866 to find the source of the Nile. He disappeared and was found in 1871 by Henry Morton Stanley on Lake Tanganyika. ➲ *204–5*

LLOYD GEORGE, DAVID (1863–1945) British statesman, prime minister (1916–22). A Welsh Liberal, he sat in the House of Commons from 1890. As chancellor of the exchequer (1908–15), he increased taxation to pay for social measures such as old-age pensions. His "People's Budget" of 1909 provoked a constitutional crisis, which led to a reduction of the powers of the House of Lords. He was an effective minister of munitions (1915). In 1916 he joined with Conservatives to dislodge the prime minister, Asquith, whom he replaced. He won an easy victory for his coalition government in 1918 and was a leading figure at the peace conference at Versailles. He negotiated the treaty creating the Irish Free State (1921), but then fell from power. ➲ *220–21*

LOLLARDS Followers of the 14th-century English religious reformer John WYCLIFFE. They helped to pave the way for the reformation, and challenged many doctrines and practices of the medieval church, including transubstantiation, pilgrimages and clerical celibacy. They rejected the authority of the Papacy and denounced the wealth of the church. The first Lollards appeared at Oxford University, where Wycliffe was a teacher (c. 1377). They went out among the people as "poor preachers", teaching that the Bible was the sole authority in religion. After 1401 many Lollards were burned as heretics, and in 1414 they mounted an unsuccessful uprising in London. ➲ *106–7*

LOMBARD LEAGUE Defensive alliance of the cities of Lombardy in northern Italy (1167). Its purpose was to resist the re-establishment of imperial authority by Emperor Frederick I. Led by Pope Alexander III, the League defeated Federick at Legnano (1176). By the Peace of Constance in 1183 the Italian cities retained independence while paying lip service to Frederick's authority. The League was active again in 1226 against Frederick II. ➲ *90–91*

LOMBARDS Germanic peoples who inhabited the area east of the lower River Elbe until driven west by the Romans in AD 9. In 568 they invaded northern Italy under Alboin and conquered much of the country, adopting CATHOLICISM and Latin customs. The Lombard kingdom reached its peak under Liutprand (d. 744). It went into decline after defeat by the Franks under CHARLEMAGNE in 775. ➲ *56–57*

LONG MARCH (1934–35) Enforced march of the Chinese Red Army during the war against the KUOMINTANG forces. Led by Chu Teh and MAO ZEDONG, 90,000 communist troops, accompanied by around 15,000 civilians, broke through a KUOMINTANG encirclement of their headquarters and marched some 10,000 kilometres (6,000 miles) from Jiangxi province, southeast China, to Shaanxi province in the northwest. Under frequent attack, the Communists suffered 45,000 casualties. This, and other similar marches during 1934–36, prevented the extermination of the Chinese Communist Party by the Kuomintang. ➲ *224–25*

LOUIS I (THE PIOUS) (778–840) Emperor of the Franks (814–840). CHARLEMAGNE 's only surviving son. He struggled to maintain his father's empire, co-operating closely with the Church. The Franks had no law of primogeniture, and Louis' attempts to provide an inheritance for his four sons provoked civil war. ➲ *74–75*

LOUIS XIV (1638–1715) King of France (1643–1715). The first part of his reign was dominated by Cardinal MAZARIN. From 1661, Louis ruled personally as the epitome of absolute monarchy and became known as the "Sun King" for the luxury of his court. He chose men of the junior nobility as ministers, such as the able Colbert, and he reduced the power of the aristocracy in the provinces. After Colbert's death in 1683 decline set in. Louis' wars of aggrandisement in the Low Countries and elsewhere drained the royal treasury. His revocation of the EDICT OF NANTES drove many HUGUENOTS abroad, weakening the economy. In the War of the SPANISH SUCCESSION, the French armies were at last defeated. ➲ *156–57, 174–75*

LOUIS XVI (1754–93) King of France (1774–92). Grandson of Louis XV, he married the Austrian archduchess Marie Antoinette in 1770. Louis' lack of leadership qualities allowed the *parlements* (supreme courts) and aristocracy to defeat the efforts of government ministers, such as Jacques Necker, to carry out vital economic reforms. The massive public debt forced Louis to convoke the Estates-General in order to raise taxation. His indecisiveness on the composition of the Estates-General led the Third (popular) Estate to proclaim itself a National Assembly, signalling the start of the French Revolution. The dismissal of Necker and rumours that Louis intended to forcibly suppress the assembly led to the storming of the Bastille on 14 July 1789. Louis was forced to reinstate Necker, but continued to allow the queen and court to conspire against the revolution. In October 1789 the royal family were forced to return to Paris and were confined to the Tuileries palace. In June 1791 their attempt to flee France failed and Louis was forced to recognize the new constitution. Early French defeats in the war against Austria and Prussia led to the declaration of a republic. Louis was tried for treason by the National Convention and found guilty. He was guillotined on 21 January 1793. ➲ *166–67*

LOUIS XVIII (1755–1824) King of France (1814–24). Brother of Louis XVI, he fled from the French Revolution to England and was recognized by the emigrés as king from 1795. Louis was restored to the throne in 1814, but was forced to flee again during the Hundred Days until NAPOLEON's final defeat at WATERLOO (1815). He agreed to a constitution providing for parliamentary government and a relatively free society, but came increasingly under the influence of extremists. ➲ *166–67, 172–73*

LOUISIANA PURCHASE (1803) Transaction between the USA and France, in which the US bought for 60 million francs ($15 million), 2,144,500 square kilometres (828,000 square miles) of land between the Mississippi River and the Rocky Mountains. In 1801, shortly after THOMAS JEFFERSON's inauguration, a secret treaty between Spain and Napoleonic France was revealed, whereby Spain ceded Louisiana Territory to NAPOLEON. Jefferson was anxious about France gaining control of the vital trade down the Mississippi River, and instructed Robert Livingston to secure New Orleans for the United States. France was faced with a slave revolt in St Domingue in 1802 and the transfer of Louisiana was postponed. Nonetheless, the Spanish governor of Louisiana suspended US traders' rights to store their goods in New Orleans. Jefferson sent James Monroe to assist Livingston's negotiations. Faced with the impending Napoleonic Wars and the possibility of hostilities with the USA, Napoleon resolved to sell all of Louisiana Territory. The treaty, concluded on 30 April, doubled the area of the United States and part or all of 15 states were later formed from the region. ➲ *182–83, 184–85*

LOUIS PHILIPPE (1773–1850) King of France (1830–48). He returned to France from exile in 1814 and acquired a reputation as a liberal. He gained the throne after the July Revolution. Although known as the "Citizen King", he retained much personal power. He abdicated when revolution broke out again and the Second Republic was declared in February 1848. He died in exile in England. ➲ *172–73*

LOYOLA, IGNATIUS (1491–1556) Spanish soldier, churchman and founder of the JESUITS. In 1534, with Francis Xavier and other young men, he made vows of poverty, chastity and obedience. He was ordained to the priesthood in 1537 and moved to Rome where, in 1540, Pope Paul III formally approved his request to found a religious order: the Society of Jesus, or Jesuits. He spent the rest of his life in Rome supervising the rapid growth of the order, which was to become the leading force in the Counter Reformation. ➲ *154–55*

LUTHER, MARTIN (1483–1546) German Christian reformer who was a founder of PROTESTANTISM and leader of the Reformation, which spread from Germany throughout Europe during the 16th century. He was deeply concerned about the problem of salvation, deciding that it could not be attained by good works but was a free gift of God's grace. In 1517 his *Ninety-Five Theses* were posted on the door of the Schlosskirche in Wittenberg. This was a document that included, among other things, statements challenging the sale of indulgences. This led to a quarrel between Luther and church leaders, including the Pope. Luther decided that the Bible was the true source of authority and renounced obedience to Rome. He was excommunicated, but gained followers among churchmen as well as the laity. After the publication of the *Augsburg Confession* in 1530 he gradually retired from the leadership of the Protestant movement. ➲ *152–53, 154–55*

LUTHERANISM Doctrines and Church structure that grew out of the teaching of Martin LUTHER. The principal Lutheran doctrine is that of justification by faith alone (*sola fide*). Luther held that grace cannot be conferred by the Church but is the free gift of God's love. He objected to the CATHOLIC doctrine of transubstantiation. Instead, Luther believed in the real presence of Christ "in, with, and under" the bread and wine (consubstantiation). These and other essentials of Lutheran doctrine were set down in 1530 by Philip Melanchthon in the *Augsburg Confession*, which is still the basic document of the Lutherans. ➲ *154–55*

LYDIA Ancient kingdom of western Asia Minor. Under the Mermnad dynasty (c. 700–547 BC), it was a powerful and prosperous state, the first to issue a coinage, with its capital at Sardis. Its last king was CROESUS, famous for his wealth, who was defeated by the Persians under CYRUS THE GREAT in 547 BC. ➲ *42–43*

M

MACARTHUR, DOUGLAS (1880–1964) US general. A divisional commander in the First World War, he became army chief of staff in 1930 and military adviser to the Philippines in 1935, retiring from the US army in 1937. He was recalled in 1941 and conducted the defence of the Philippines until ordered out to Australia. As supreme Allied commander in the southwest Pacific (1942), he directed the campaigns that led to Japanese defeat. He was appointed commander of UN forces on the outbreak of the KOREAN WAR in 1950. Autocratic and controversial, he was relieved of his command by President Truman in April 1951. ❍ *252–53*

MACEDONIA Ancient country in southeast Europe, roughly corresponding to present-day Macedonia, Greek Macedonia and Bulgarian Macedonia. The Macedonian king Alexander I (d. 420 BC) initiated a process of Hellenization. PHILIP II founded the city of Thessalonica in 348 BC and was acknowledged as king of Greece in 338 BC. His son, ALEXANDER THE GREAT, built a world empire, but this rapidly fragmented after his death in 323 BC. Macedonia was eventually defeated by the Romans in the Macedonian Wars. In 146 BC Thessalonica became capital of the first Roman province. In AD 395 Macedonia became part of the Eastern Roman (Byzantine) Empire. Slavs settled in the 6th century, and from the 9th to the 14th century control of the territory was contested mainly by Bulgaria and the Byzantine Empire. A brief period of Serbian hegemony was followed by Ottoman rule from the 14th to 19th century. In the late 19th century Macedonia was claimed by Greece, Serbia and Bulgaria. In the first of the BALKAN WARS, Bulgaria gained much of historic Macedonia, but it was decisively defeated in the Second BALKAN WAR and the present-day boundaries were established. ❍ *42–43*

MACHU PICCHU Ancient fortified town, 80 kilometres (50 miles) northwest of Cuzco, Peru. The best-preserved of the Inca settlements, it is situated on an Andean mountain saddle 2,057 metres (6,750 feet) above sea-level. A complex of terraces extends over 13 square kilometres (5 square miles), linked by over 3,000 steps. Machu Picchu was discovered in 1911 by the American explorer Hiram Bingham, who dubbed it the "lost city of the Incas". ❍ *110–11*

MAGELLAN, FERDINAND (1480–1521) Portuguese explorer, leader of the first expedition to circumnavigate the globe. He sailed to the East Indies and may have visited the Spice Islands (Moluccas) in 1511. Subsequently he took service with Spain, promising to find a route to the Moluccas via the New World and the Pacific. He set out with five ships and nearly 300 men in 1519. He found the waterway near the southern tip of South America that is now named Magellan's Strait. After severe hardships, the expedition reached the Philippines, where Magellan was killed in a local conflict. Only one ship, the *Victoria*, completed the round-the-world voyage. ❍ *116–17*

MAGYARS People who founded the kingdom of Hungary in the late 9th century. From their homeland in northeast Europe, they moved gradually south over the centuries and occupied the Carpathian basin in 895. Excellent horsemen, they raided the German lands to the west until checked by OTTO I in 955. They adopted CHRISTIANITY and established a powerful state that included much of the northern Balkans, but lost territory to the Ottoman Turks after the battle of MOHÁCS in 1526. The remainder of the kingdom subsequently fell to the Habsburg Empire. ❍ *76–77, 174–75*

MAHATHIR BIN MUHAMMAD (1925–) Malaysian politician and prime minister of Malaysia. Mahathir bin Muhammad practised as a doctor until his election to the House of Representatives in 1964 as a member of the United Malays' National Organization (UMNO), whose leader he became in 1981. From 1974 he held various ministerial posts and acted as deputy prime minister between 1976 and 1981, after which he became prime minister, a position endorsed in subsequent elections. His premiership is noted for its promotion of industrialization, and of Islamic and Malay values. ❍ *250–51*

MALI EMPIRE West African empire covering much of the western Sudan from the 13th to the 16th century. Three great kings (r. c. 1230–c. 1340), Sundiata, Mansa Uli and Mansa Musa, so expanded Mali that it became one of the greatest empires of its time. In 1324 Mansa Musa made a pilgrimage to MECCA , bringing back Egyptian scholars to Timbuktu and Gao. Timbuktu became an important centre of religion, trade and learning, producing many fine Muslim scholars. The Mali Empire gradually declined in importance from the 15th century

as subject peoples successfully rebelled against their Mali overlords. By the mid 16th century it had been overtaken by the SONGHAY Empire. ❍ *80–81, 136–37*

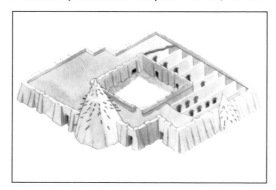

The great mosque at Timbuktu was designed in the 14th century by As-Saheli, one of the Egyptians brought back to Mali by the emperor Mansa Musa after his pilgrimage to Mecca in 1324.

MALPLAQUET, BATTLE OF (1709), engagement between France and the allied English, Austrian and Dutch forces in the War of the SPANISH SUCCESSION. The French attempted to break the Allied siege of the fortress at Mons by concentrating their forces 16 kilometres (10 miles) away, near the village of Malplaquet. The outnumbered French were finally forced into an orderly retreat, but, despite the Allied capture of Mons, the vast Allied losses (22,000 to the French 12,000) ultimately made the battle a French victory, preventing an Allied invasion of Paris. ❍ *158–59*

MAMLUKS Military elite in Egypt and other Arab countries. The Mamluks of Egypt overthrew the AYYUBID dynasty in 1250. They halted the Mongols, defeated the crusaders and crushed the Assassins. Though conquered by the Ottoman Turks in 1517, they continued to control Egypt until suppressed by Muhammad Ali (Mehemet Ali) in 1811. ❍ *88–89, 160–61*

MANDELA, NELSON ROLIHLAHLA (1918–) South African statesman, president (1994–99). Mandela joined the AFRICAN NATIONAL CONGRESS (ANC) in 1944, and for the next 20 years led a campaign of civil disobedience against South Africa's APARTHEID government. Following the Sharpeville Massacre of 1960, Mandela formed Umkhonte We Sizwe (Spear of the Nation), a paramilitary wing of the ANC. The ANC was banned. In 1962 Mandela was acquitted on charges of treason, but in 1964 he was sentenced to life imprisonment for political offences. For the next 27 years in prison, Mandela was a symbol of resistance to apartheid. International sanctions forced F. W. de Klerk to begin the process of dismantling apartheid. In February 1990 Mandela was released to resume his leadership of the newly legalized ANC. In 1993 Mandela and de Klerk shared the Nobel Peace Prize. Mandela gained 66 per cent of the popular vote in South Africa's first democratic general election in 1994. A strong advocate of the need for reconciliation, Mandela made de Klerk deputy president (1994–96) in his government of national unity. ❍ *256–57*

MAORI WARS Two wars (1845–48 and 1860–72) between British settlers and indigenous Maori tribes in New Zealand. They arose when the settlers broke the terms of the Treaty of WAITANGI (1840) that guaranteed the Maoris possession of their lands. As a result of the wars, a Native Land Court was established (1865), a Maori school system formed (1867) and the Maoris allowed to have four elected members in the New Zealand legislature. ❍ *202–3*

MAO ZEDONG (1893–1976) Chinese statesman, founder and chairman (1949–76) of the People's Republic of China. Mao was a founder member of the Chinese Communist Party in 1921. After the nationalist KUOMINTANG, led by CHIANG KAI-SHEK, dissolved the alliance with the Communists in 1927, Mao helped establish rural soviets. In 1931 he was elected chairman of the Soviet Republic of China, based in Jiangxi. The advance of Kuomintang forces forced Mao to lead the Red Army on the LONG MARCH (1934–35). In 1937 the

civil war was suspended as Communists and Kuomintang combined to fight the second Sino–Japanese War. The communists' brand of guerrilla warfare gained hold of much of rural China. Civil war recommenced in 1945, and by 1949 the Kuomintang had been driven out of mainland China. Mao became chairman of the People's Republic while Zhou Enlai acted as prime minister. In 1958 Mao attempted to distinguish Chinese communism from its Soviet counterpart by launching the GREAT LEAP FORWARD. The programme ended in mass starvation and the withdrawal of Soviet aid. Mao's leadership was challenged. The CULTURAL REVOLUTION was an attempt by Mao and his wife, Jiang Qing, to reassert Maoist ideology. The cult of the personality was encouraged, political rivals were dismissed, and Mao became supreme commander of the nation and army (1970). The death of Mao and of Zhou Enlai created a power vacuum. A struggle developed between the GANG OF FOUR, Hua Guofeng, and Deng Xiaoping. ❍ *224–25, 244–45, 254–55*

MARATHAS Hindu warrior people of west central India, who rose to power in the 17th century. They extended their rule throughout western India by defeating the Mughal Empire and successfully resisting British supremacy in India during the 18th century. They were finally defeated in 1818. ❍ *144–45*

MARATHON, BATTLE OF (490 BC) Victory of the Greeks, mainly Athenians, during the PERSIAN WARS. The defeat of a much larger Persian army on the Marathon plain northeast of ATHENS secured Attica from the invasion of CYRUS THE GREAT. ❍ *40–41*

MARCO POLO See POLO, MARCO

MARIA THERESA (1717–80) Archduchess of Austria, ruler of the Austrian Habsburg Empire (1740–80). She succeeded her father, the Emperor Charles VI, but was challenged by neighbouring powers in the War of the AUSTRIAN SUCCESSION (1741–48), losing Silesia to Prussia but securing the imperial title for her husband, Francis I. She initiated a change of alliances, but failed to regain Silesia in the SEVEN YEARS WAR (1756–63). ❍ *174–75*

MARNE, BATTLES OF Two battles on the River Marne, northern France, during the First World War. The first, in September 1914, was a counterattack directed by General Joffre, which checked the German drive on Paris. The second, in July 1918, was another Allied counterstroke, which stopped the last German advance and preceded the final Allied offensive. ❍ *218–19*

MARONITES Members of a Christian community of Syrian origin, which claims its origins both from St Maron, a Syrian hermit, in the late 4th or early 5th centuries, and St John Maro, a patriarch of Antioch (modern Antakiyah, Turkey) from 685–707. The largest group of Maronites (some 400,000 people) live in Lebanon, with other smaller groups in Syria, Cyprus, southern Europe, and North and South America. The Maronites have the status of a uniate Church – that is, an Eastern Church in union with Rome but retaining its own rite and canon law. ❍ *260–61*

MARSHALL PLAN US programme of economic aid to European countries after the Second World War. Promoted by the US secretary of state, General Marshall, its purpose was to repair war damage and promote trade within Europe, while securing political stability. The Soviet Union and East European countries declined to participate. Between 1948 and 1951, 16 countries received a total of $12,000 million under the plan. ❍ *238–39, 242–43, 244–45, 272–73*

MAU MAU Anti-colonial terrorist group of the Kikuyu of Kenya. Following repeated attacks on Europeans, a state of emergency was declared in 1952 and troops brought in. Within four years about 100 Europeans and 2,000 anti-Mau Mau Kikuyu were killed. More than 11,000 Mau Mau died before the state of emergency ended in 1960. Kenya achieved independence in 1963, and the former Mau Mau leader Jomo Kenyatta was elected its first prime minister. ❍ *256–57*

MAURYAN EMPIRE (321–185 BC) Ancient Indian dynasty and state founded by CHANDRAGUPTA (r. c. 321– c. 297 BC). His son Bindusara (r. c. 297–c. 268 BC) conquered the Deccan, and all of northern India was united under ASHOKA, Chandragupta's grandson. After Ashoka's death (c. 231) the empire broke up, the last emperor being assassinated c. 185 BC. ❍ *46–47*

MAXIMILIAN I (1459–1519) Holy Roman Emperor (1493–1519). He was one of the most successful members of the Habsburg dynasty. He gained Burgundy and the Netherlands by marriage, and defended them against France. He was less successful in asserting control over the German princes, and was defeated by the Swiss in 1499, but he strengthened the Habsburg heartland in Austria and organized the marriages that made his grandson, CHARLES V, ruler of a vast empire during the first half of the 16th century. ❍ *152–53*

MAXIMILIAN, FERDINAND JOSEPH (1832–67) Emperor of Mexico (1864–67). An Austrian archduke, brother of the Emperor Francis Joseph, he was offered the throne of Mexico after the French invasion (1862). When the French withdrew in 1867, Maximilian was overthrown by Benito Juárez and executed. ❍ *192–93*

MAYA Outstanding culture of Classic American civilization. Occupying southern Mexico and northern Central America, it was at its height from the 3rd–9th centuries. The Mayans built great temple-cities, with buildings surmounting stepped PYRAMIDS. They were skilful potters and weavers, and productive farmers. They worshipped gods and ancestors, and blood sacrifice was an important element of their religion. Mayan civilization declined after c. 900, and much of what remained was destroyed after the Spanish conquest in the 16th century. ❍ *32–33, 84–85*

MAZARIN, JULES, CARDINAL (1602–61) French statesman, b. Italy. He was the protégé of Cardinal RICHELIEU and Chief Minister under Anne of Austria from 1643. During the civil wars of the FRONDES, he played off one faction against another and though twice forced out of France, emerged in full control. As a former papal diplomat, he was a skilful negotiator of the treaties ending the THIRTY YEARS WAR. ❍ *156–57*

MAZZINI, GUISEPPE (1805–72) Italian patriot and political thinker of the RISORGIMENTO. A member of the *Carbonari* (the Italian republican underground) from 1830, he founded the "Young Italy" movement in 1831, dedicated to the republican unification of Italy. He fought in the revolutionary movement of 1848 and ruled in Rome in 1849, but was then exiled. ❍ *176–77*

MECCA City in western Saudi Arabia and the holiest city of ISLAM. The birthplace of the prophet MUHAMMAD, only Muslims are allowed in the city. Mecca was originally home to an Arab population of merchants. When Muhammad began his ministry here the Meccans rejected him. The flight or *Hejira* of Muhammad from Mecca to MEDINA in 622 marked the beginning of the Muslim era. In 630 Muhammad's followers captured Mecca and made it the centre of the first Islamic empire. The city was controlled by Egypt during the 13th century. The Ottoman Turks held it from 1517 until 1916, finally losing their control after Hussein Ibn Ali secured Arabian independence. Mecca fell in 1924 to the forces of Ibn Saud, who later founded the Saudi Arabian kingdom. ❍ *68–69, 260–61*

MEDIA Ancient country of northwestern Iran. The inhabitants of Media, known as the Medes, frequently clashed with the Assyrians during the 9th century BC, and were conquered by the SCYTHIANS in the 7th century. The Median king Cyaxares drove out the Scythians and also helped the Babylonians destroy the Assyrian Empire; this period (c. 615–c. 585 BC) saw the greatest extent of Median power. CYRUS THE GREAT defeated the last Median king in 550 BC and incorporated MEDIA into his empire. ❍ *38–39*

MEDINA City in Saudi Arabia, north of MECCA. Originally called Yathrib, the city was renamed Medinat an-Nabi ("Prophet's city") after Muhammad fled Mecca and settled here in 622. Medina became his capital. In 661 the UMAYYAD caliphs moved their capital to Damascus, and Medina's importance declined. It came under Turkish rule (1517–1916), after which it briefly formed part of the independent Arab kingdom of the Hejaz. In 1932 it became part of Saudi Arabia. ❍ *68–69, 260–61*

MEIJI RESTORATION Constitutional revolution in Japan in 1867. Opposition to the shogunate built up after Japan's policy of isolation under the Tokugawa regime was ended by US Commodore Perry in 1854. Pressure for modernization resulted in a new imperial government, at first dominated by former SAMURAI, with the young Emperor Meiji as its symbolic leader. ❍ *200–1*

MENELIK II (1844–1913) Emperor of Ethiopia (1889–1913). He became emperor with Italian support, succeeding John IV. He defeated an Italian invasion in 1896, effectively securing Ethiopian independence. Having proved himself an able and aggressive king, he greatly expanded and modernized his empire, establishing a capital at Addis Ababa, increasing imperial power, and constructing a railway. ❍ *204–5*

MENES Egyptian king (c. 3100 BC), regarded as the first king of the First Dynasty. He unified upper and lower EGYPT, establishing the Old Kingdom with its capital at Memphis. ❍ *30–31*

MEROVINGIANS Frankish dynasty (476–750). It was named after Merovech, a leader of the Salian Franks, whose grandson CLOVIS I (r. c. 481–511) ruled over most of France and, converting to CHRISTIANITY, established the common interests of the Frankish rulers and the already Christian population of his new kingdom. The last Merovingian king, Childeric III, was overthrown in 750 by PIPPIN III, who was the founder of the CAROLINGIAN dynasty. ❍ *74–75*

MESOPOTAMIA Ancient region between the rivers Tigris and Euphrates in southwest Asia, roughly corresponding to modern Iraq. It was the setting of one of the earliest human civilizations, resulting from the development of irrigation in the 6th millennium BC and the extreme fertility of the irrigated land. The first cities were established by the Sumerians c. 2500 BC. The first empire builders on a large scale were the people of Akkadia under SARGON I, who conquered the Sumerian cities c. 2300 BC. BABYLONIA gained supremacy in the 18th century BC and was followed by others, notably the Assyrians. Later ruled by foreigners, such as Persians, Greeks and Romans, Mesopotamia gradually lost its distinctive cultural traditions. ❍ *28–29*

METTERNICH, KLEMENS WENZEL LOTHAR, PRINCE VON (1773–1859) Austrian statesman. As foreign minister (1809–48) and chancellor (from 1821), he was the leading European statesman of the post-Napoleonic era. Following Austria's defeat in the Napoleonic Wars (1809), he adopted a conciliatory policy towards France. After Napoleon's retreat from Moscow in 1812, he formed the Quadruple Alliance in 1813, which led to Napoleon's defeat. He was the dominant figure at the CONGRESS OF VIENNA (1814–15). In 1815 he secured international order and thereafter he became increasingly autocratic, pressing for the intervention of the great powers against any revolutionary outbreak. He was driven from power by the Revolution of 1848. ❍ *174–75*

MEXICAN REVOLUTION (1910–40) Extended political revolution that improved the welfare of the Mexican underprivileged. The Mexican Revolution was prompted by the dictatorial, elitist presidency of Porfirio Díaz. In 1910 Díaz, who had agreed not to stand for re-election following the threat of armed revolt led by Francisco Madero, reneged on his agreement and was re-elected. He was forced to resign in 1911 by Madero, who was subsequently elected. Madero intended to make land-ownership more egalitarian, to strengthen labour organizations, and to lessen the influence of the CATHOLIC CHURCH. He was assassinated in 1913, however, by his former general Victoriano Huerta. The repressive regime of Huerta caused massive unrest in the peasant community, who found leaders in Venustiano CARRANZA, Francisco "Pancho" Villa and Emiliano Zapata. In 1914 Huerta resigned and Carranza became president. Although some agrarian, educational and political reforms continued, it was Lázaro Cárdenas (inaugurated in 1934) who finally introduced sweeping measures involving land distribution, support of the labour movement, and improving health and education. ❍ *226–27*

MEXICAN WAR (1846–48) Conflict between Mexico and the USA. It broke out following the US annexation of Texas (1845). The Mexicans were swiftly overwhelmed, and a series of US expeditions effected the conquest of the southwest. The war ended when General Winfield Scott, having landed at Veracruz in March 1847, defeated the army of Santa Anna and entered Mexico City on 8 September. In the Treaty of GUADALUPE-HIDALGO (1848), Mexico ceded sovereignty over California and New Mexico, as well as Texas north of the Rio Grande. ❍ *182–83, 192–93*

MILOSEVIC, SLOBODAN (1941–) Serbian statesman, president (1989–). He became head of the Serbian Communist Party in 1986. As Serbian president, he was confronted with the break-up of the Federation of Yugoslavia. After his re-election in 1992, Milosevic gave moral, financial and technical support to the Serb populations in Croatia and Bosnia-Herzegovina, who fought for a Greater Serbia. Milosevic gradually distanced himself from the brutal activities of the Bosnian Serb leaders Mladic and Karadzic. In November 1995 he signed the Dayton Peace Accord with the Bosnian president Izetbegovic, and the Croatian president Tudjman to end the civil war in former Yugoslavia. In 1999, however, his refusal to grant autonomy to the majority ethnic Albanian population of the province of Kosovo, led to NATO military action against Serbia. ❍ *264–65*

MINOAN CIVILIZATION Ancient Aegean civilization that flourished c. 3000–c. 1100 BC on the Mediterranean island of Crete, named after the legendary King Minos. The Minoan period is divided into three eras: Early (c. 3000–c. 2100 BC), Middle (c. 2100–c.1550 BC), and Late (c. 1550–c. 1100 BC). In terms of artistic achievement, and perhaps power, Minoan civilization reached its height in the Late period. The prosperity of Bronze Age Crete is evident from the works of art and palaces excavated at KNOSSOS, Phaistos and other sites. It was based on trade and seafaring. ❍ *36–37*

MISSOURI COMPROMISE Effort to end the dispute between slave and free states in the USA in 1820–21. Pushed through Congress by Henry Clay, it permitted Missouri to join the Union as a slave state at the same time as Maine was admitted as a free state, preserving an equal balance between slave and free. ❍ *184–85*

MOBUTU SESE SEKO (1930–97) Zairean political leader, born Joseph-Désiré Mobutu. He was defence minister under Patrice Lumumba. With the support of the army, he deposed Lumumba in 1960. In 1965 he led a coup against Joseph Kasavubu, becoming prime minister in 1966, and president in 1967. His calls for Africanization and nationalization of industries were largely publicity stunts. The stark reality of his autocratic rule was a state founded on corruption. Mobutu amassed a huge personal fortune, while Zaireans became increasingly impoverished. With the support of the CIA and the criminal activites of his security forces, Mobutu maintained a dictatorship for over 30 years. In May 1997 he was deposed by a Tutsi-dominated revolt, led by Laurent Kabila, and forced into exile. He died three months later in Morocco. ❍ *256–57*

MOHÁCS, BATTLE OF (1526) Victory of the Ottoman Turks under SULEIMAN I (the Magnificent) over LAJOS II of Hungary. Lajos was killed in the battle, which marked the beginning of Ottoman domination in Hungary. In a later Battle of Mohács (1687), the result was reversed and the Turks withdrew. ❍ *142–43, 146–47, 152–53*

MONTEZUMA Name of two Aztec emperors. Montezuma I (r. 1440–69) expanded the Aztec Empire by conquest. Montezuma II (r. 1502–20) allowed the Spaniards under CORTÉS to enter his capital, Tenochtitlan, unopposed, in 1519, and subsequently became their captive. ❍ *110–11, 120–21*

MOORS Name given to the predominantly Berber people of northwest Africa, from the Latin *Maures*. In Europe the name is applied particularly to the North African Muslims who invaded Spain in 711 and established a distinctive civilization that lasted nearly 800 years. It was at its height under the Córdoba caliphs in the 10th–11th centuries, and the finest surviving example of Hispan-Moorish art is the Alhambra, dating from the 14th century, in Granada. The Christian rulers of northern Spain gradually reconquered the country, and after the ALMOHAD Empire broke up in the 13th century, Granada alone survived, until it fell to the Christians in 1492. **◗** *88–89*

MORMONS Adventist Christian sect, the full name of which is the Church of Jesus Christ of Latter-day Saints. It was established in Manchester, New York, USA, in 1830 by Joseph Smith. Believing that they were to found Zion, or a New Jerusalem, Smith and his followers moved west. They tried to settle in Ohio, Missouri, and Illinois, but were driven out. Joseph Smith was murdered in Illinois in 1844. Brigham Young then rose to leadership and in 1846–47 took the Mormons to Utah. **◗** *182–83*

MUHAMMAD (c. 570–632) Arab prophet and inspirational religious leader who founded ISLAM. He was born in the Arabian city of MECCA. He was orphaned at the age of six and lived first with his grandfather and then with his uncle. At the age of 25, he began working as a trading agent for Khadijah, a wealthy widow of 40, whom he married. For 25 years, she was his closest companion and gave birth to several children. Only one brought him descendants – his daughter Fatima, who became the wife of his cousin Ali. In c. 610, Muhammad had a vision while meditating alone in a cave on Mount Hira, outside Mecca. A voice three times commanded him to "recite", and he felt his body compressed until he could hardly breathe. Then he heard the words of the first of many revelations that came to him in several similar visions over the next two decades. The revelations came from Allah, or God, and Muhammad's followers believe that they were passed to Muhammad through the angel Gabriel. At the core of his new religion was the doctrine that there is no God but Allah and His followers must submit to Him – the word *islam* means "submission". Muhammad gained followers but also many enemies among the Meccans. In 622 he fled to MEDINA. Muslims, followers of Islam, later took this *Hejira* as initiating the first year in their calendar. Thereafter, Muhammad won more followers. He organized rules for the proper worship of Allah and for Islamic society. He also made war against his enemies. He conquered Mecca in 630. Most of the Arab tribes allied with him. In Medina, he married the woman who became his favourite wife, Aishah, the daughter of Abu Bakr, one of his strongest supporters. Muhammad is considered an ideal man, but he never claimed supernatural powers, and is not held to be divine. His tomb is in the Holy Mosque of the Prophet, Medina. **◗** *68–69*

MUHAMMAD ALI (1769–1849) (Mehemet Ali) Albanian soldier who founded an Egyptian dynasty. In 1798 he took part in an Ottoman expeditionary force sent to Egypt to drive out the French. He was unsuccessful, but (after the departure of the French) quickly rose to power. In 1805 he was proclaimed the Ottoman sultan's viceroy. In 1811 he defeated the MAMLUKS, who had ruled Egypt since the 13th century. He conquered the Sudan (1820–30) and put down a rebellion in Greece in 1821 but his fleet was later destroyed by the European powers at the Battle of Navarino in 1827. Muhammad challenged the sultan and began the conquest of Syria in 1831. The European powers again intervened and he was compelled to withdraw. **◗** *172–73, 178–79*

MUJIB-UR-RAHMAN (1920–75) Bangladeshi political leader known as Sheikh Mujib, he worked towards the independence of Pakistan in 1947. In the 1960s he began the East Pakistani separatist movement, which culminated in the civil war of 1971. Mujib became prime minister of the newly formed Bangladesh in 1972, and president in 1975. He was killed in a coup d'etat that overthrew his government. **◗** *248–49*

MUNICH AGREEMENT (September 1938) Pact agreed by Britain, France, Italy and Germany to settle German claims on Czechoslovakia. Hoping to preserve European peace, Britain and France compelled Czechoslovakia, not represented at Munich, to surrender the predominantly German-speaking Sudetenland to Nazi Germany on certain conditions. Hitler ignored the conditions and, six months later, his troops took over the rest of the country, an action that finally ended the Anglo-French policy of appeasement. **◗** *230–31*

MUSSOLINI, BENITO (1883–1945) Italian fascist dictator, prime minister (1922–43). He turned to revolutionary nationalism in the First World War and in 1919 founded the Italian Fascist movement. The fascists' march on Rome in 1922 secured Mussolini's appointment as prime minister. He imposed one-party government with himself as Il Duce (literally "the leader"), or dictator. His movement was a model for HITLER 's NATIONALIST SOCIALIST (Nazi) Party, with whom Mussolini formed an alliance in 1936. Imperial ambitions led to the conquest of Ethiopia (1935–36) and the invasion of Albania (1939). Mussolini delayed entering the Second World War until a German victory seemed probable in 1940. A succession of defeats led to his fall from power. He was briefly restored as head of a puppet government in northern Italy by the Germans, but in April 1945, fleeing Allied forces, he was captured and killed by Italian partisans. **◗** *230–31*

MYCENAE Ancient city in Greece, 11 kilometres (7 miles) north of modern Argos, which gave its name to the Mycenaean civilization. Dating from the 3rd millennium BC, Mycenae was at its cultural peak c. 1580–1120 BC. It was destroyed in the 5th century BC. Later restored, by the 2nd century AD it was in ruins. The ruins of the city were discovered by Heinrich Schliemann in 1874–76. **◗** *36–37*

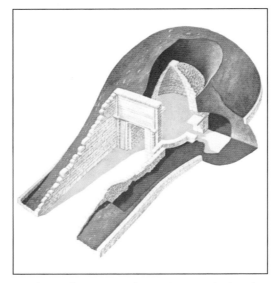

The finest of a number of stone-built tombs found at Mycenae is the so-called Treasury of Atreus, dating from around 1320 BC. A walled passage leads to a monumental door in the mound. Inside, a circular chamber is roofed by a corbelled vault.

MYCENAEAN CIVILIZATION Ancient Bronze Age civilization (c. 1580–1200 BC) centred around MYCENAE in Greece. The Mycenaeans entered Greece from the north, bringing with them advanced techniques particularly in architecture and metallurgy. By 1400 BC, having invaded Crete and incorporated much of MINOAN CIVILIZATION, the Mycenaeans became the dominant power in the Aegean, trading as far as Syria, PALESTINE and EGYPT, and importing luxurious goods for their wealthy and cultured citadel palaces. It is uncertain as to why the Mycenaean civilization collapsed, but one theory is that it was due to invasion by the Dorians, a Greek-speaking people, who settled in northern Greece c. 1200 BC. It is possible that they displaced the culturally more advanced Mycenaean civilization because they had mastered the use of iron while the Mycenaeans used bronze weapons. **◗** *36–37*

MYCENEAN ART Greek art of the Late Bronze Age. The name comes from the fortress-city of MYCENAE and refers to the art and architecture of the late Helladic period (c. 1500–1100 BC). Its greatest achievements came in the fields of architecture, which included both grand fortifications and beehive tombs, and in pottery, precious metalwork and fresco. **◗** *36–39*

NANJING, TREATY OF (Nanking) The Treaty (1842) ended the Opuim War between China and Britain. China was forced to pay an indemnity to Britain, to cede Hong Kong, and to open five Chinese ports to British merchants. **◗** *198–99*

NANTES, EDICT OF (1598) French royal decree establishing toleration for HUGUENOTS (Protestants). It granted freedom of worship and legal equality for Huguenots within limits, and ended the FRENCH WARS OF RELIGION. The Edict was revoked by LOUIS XIV in 1685, causing many Huguenots to emigrate to countries where PROTESTANTISM was tolerated. **◗** *128–29, 154–55*

NAPOLEON I (1769–1821) (Napoleon Bonaparte), Emperor of the French (1804–15), born in Corsica. The greatest military leader of modern times, he became a brigadier in 1793 after driving the British out of Toulon. In 1796 he was given command in Italy, where he defeated the Austrians and Sardinians and in 1798 he launched an invasion of Egypt. French defeats in Europe prompted his return to Paris in 1799, where his coup of 18 Brumaire (9 November) overthrew the DIRECTORY and set up the Consulate, headed by himself. He enacted sweeping administrative and legal reforms with the NAPOLEONIC CIVIL CODE (*Code Napoléon*), while defeating the Austrians at Marengo in 1800 and making peace with the British at Amiens in 1802. In 1804 he crowned himself emperor. His efforts to extend French power provoked the Napoleonic Wars (1803–15). Napoleon's Grand Army shattered his continental opponents but, after Trafalgar (1805), Britain controlled the seas. Napoleon tried to defeat Britain by a commercial blockade, which led indirectly to the PENINSULAR WAR in Portugal and Spain. By 1812 he controlled most of continental Europe. His invasion of Russia in 1812 ended in the destruction of the Grand Army, encouraging a new coalition against France, which captured Paris in March 1814. Napoleon was exiled to Elba, but he returned triumphantly to France in March 1815. The Hundred Days of his renewed reign ended with defeat at WATERLOO in June. Napoleon was exiled to St Helena, where he died. **◗** *166–67, 172–173*

NAPOLEON III (1808–73) (Louis Napoleon) Emperor of the French (1852–70). The nephew of NAPOLEON I, he twice attempted a coup in France in 1836 and 1840. Returning from exile after the February Revolution of 1848, he was elected president of the Second Republic. In 1851 he assumed autocratic powers and established the Second Empire in 1852, taking the title Napoleon III. His attempt to establish a Mexican empire under the Archduke MAXIMILIAN ended in disaster, and in 1870 he was provoked by BISMARCK into declaring war on Prussia. Defeat at Sedan was followed by a republican rising that ended his reign. **◗** *176–77*

NAPOLEONIC CIVIL CODE (*Code Napoléon*) French legal code, operative since 1804. It was intended to end the disunity of French law, and was based on Roman law. It was applied to all countries under NAPOLEON's control. It banned social inequality, permitted freedom of person and contract and upheld the right to own private property. **◗** *166–67*

NASSER, GAMAL ABDEL (1918–70) Egyptian soldier and statesman, prime minister (1954–56) and first president of the republic of Egypt (1956–70). In 1942 he founded the Society of Free Officers, which secretly campaigned against British imperialism and domestic corruption. He led the 1952 army coup against King Farouk, quickly ousted the nominal prime minister General Muhammad Neguib and assumed presidential powers. Nasser's nationalization of the Suez Canal in 1956 prompted an abortive Anglo-French and Israeli invasion. He emerged as champion of the Arab world. Nasser formed the short-lived United Arab Republic (1958–61) with Syria. He briefly resigned after Israel won the Six Day War in 1967, and his regime became increasingly dependent on Soviet aid. The crowning achievement of his brand of Arab socialism was the completion of the Aswan dam in 1970. **◗** *260–61*

NATIONAL SOCIALISM (NAZISM) Doctrine of the National Socialist German Workers' (Nazi) Party, 1921–45. It was biologically racist (believing that the so-called Aryan race was superior to others), anti-Semitic, nationalistic, anti-communist, anti-democratic and anti-intellectual. It placed power before justice and the interests of the state before the individual. These beliefs were stated by the party's leader, ADOLF HITLER, in his book *Mein Kampf* (1925). ❍ *220–21, 230–31, 274–75*

NAZCA Indian civilization in the southernmost coastal valleys of Peru; it flourished 200 BC–AD 600. The Nazca, who developed without outside influence, were a relatively small group of farmers notable for their unique and highly stylized ceramics and textiles. ❍ *34–35*

NAZISM See NATIONAL SOCIALISM

NAZI–SOVIET PACT Treaty of mutual non-aggression between the Soviet Union and Nazi Germany, signed in August 1939. By ensuring Russian neutrality, the pact cleared the way for the German invasion of Poland, which took place a week later. Secret clauses provided for the division of Poland and the Baltic countries between Germany and the Soviet Union. The treaty caused international consternation, especially in Britain and France, which were also trying to negotiate a Soviet pact, and among Communists and Soviet-sympathisers, who had regarded Nazi Germany as their greatest enemy. ❍ *232–33*

NEBUCHADNEZZAR (c. 630–562 BC) Second and greatest king of the Chaldaean (New Babylonian) Empire (605–562 BC) who changed the political map of the ancient Middle East. He subjugated Syria and PALESTINE but was himself defeated by Egyptian forces in 601 BC. He occupied Judah, capturing JERUSALEM in 597 BC and installing the puppet king Zedekiah on the throne of Judah. Following Zedekiah's rebellion, Nebuchadnezzar destroyed the city and Temple of JERUSALEM and deported its population into exile in BABYLON. A brilliant military leader, Nebuchadnezzar continued to follow an expansionist strategy. He was responsible for many buildings in Babylon and, according to legend, built for his Median wife the famous hanging gardens, which became one of the SEVEN WONDERS OF THE WORLD. ❍ *38–39*

NEHRU, JAWAHARLAL (1889–1964) Indian statesman, first prime minister of independent India. Son of a prominent Indian nationalist, and educated in England, he belonged to the more radical wing of the INDIAN NATIONAL CONGRESS and was its president in 1929 and later. Conflicts with the British resulted in frequent spells in prison, but he took a leading part in the negotiations leading to the independence of India and Pakistan and headed the Indian government from 1947 until his death. Internationally he followed a policy of nonalignment, and became a respected leader of the Third World. ❍ *248–49*

NERO (AD 37–68) Roman emperor (AD 54–68). One of the most notorious of rulers, he was responsible for the murders of his half-brother, his mother and his first wife. ROME was burned in AD 64, according to rumour, at Nero's instigation. He blamed it on the Christians and began their persecution. Faced with widespread rebellion, Nero committed suicide. ❍ *44–45*

NEUILLY, TREATY OF (1919) Agreement signed at Neuilly, France between the victorious Entente Powers and defeated Bulgaria. Under the treaty, Bulgaria ceded territory to Yugoslavia and to Greece, thus losing access to the Aegean Sea. ❍ *220–21*

NEW DEAL (1933–39) Programme for social and economic reconstruction in the USA following the Great Depression, launched by President Franklin D. Roosevelt. It was based on massive and unprecedented federal intervention in the economy. Early measures, including extensive public works, were mainly concerned with relief. The New Deal encountered bitter resistance from conservatives and did not avert further recession in 1937–38. Industrial expansion, full employment and agricultural prosperity were achieved less by the New Deal than by the advent of the Second World War. However, the programme laid the basis for

future federal management of the economy and provision of social welfare. ❍ *228–29*

NEW MODEL ARMY Reformed Parliamentary army in the ENGLISH CIVIL WAR. Formed in 1645 largely by Oliver CROMWELL, it was better organized, trained and disciplined than any comparable Royalist force, and it ensured final victory for Parliament. ❍ *156–57*

NEWTON, ISAAC (1642–1727) English scientist. He studied at Cambridge and became professor of mathematics there (1669–1701). His main works were *Philosophiae Naturalis Principia Mathematica* (1687) and *Opticks* (1704). In the former, he outlined his laws of motion and proposed the principle of universal gravitation; in the latter he showed that white light is made up of colours of the spectrum and proposed his particle theory of light. He also created the first system of calculus in the 1660s, but did not publish it until the German mathematician Gottfried Leibniz had published his own system in c. 1684. He built a reflecting telescope in c. 1671. ❍ *134–35*

NICAEA, COUNCILS OF Two important ecumenical councils of the Christian Church held in Nicaea (modern Iznik, Turkey). The first was convoked in AD 325 to resolve the problems caused by the emergence of Arianism. It promulgated a creed, affirming belief in the divinity of Christ. The Second Council of Nicaea, held in 787, was summoned by the patriarch Tarasius to deal with the problem of the worship of icons. ❍ *44–45*

NICHOLAS II (1868–1918) Last tsar of Russia (1894–1917). Torn between the autocracy of his father, Alexander III, and the reformist policies of ministers such as Count Sergei Witte, he lacked the capacity for firm leadership. Defeat in the RUSSO–JAPANESE WAR was followed by the RUSSIAN REVOLUTION OF 1905. Nicholas agreed to constitutional government but, as danger receded, removed most of the powers of the *duma* (parliament). In the First World War he took military command, but defeat in war again provoked revolution in 1917. Nicholas was forced to abdicate, and in July 1918 he and his family were executed by BOLSHEVIKS. ❍ *180–81, 222–23*

NINEVEH Capital of ancient ASSYRIA, on the River Tigris (opposite modern Mosul, Iraq). The site was first occupied in the 6th millennium BC. It became the Assyrian capital under Sennacherib (r. 704–681 BC). The city walls were more than 12 kilometres (7.5 miles) long and contained gardens irrigated by canals. Nineveh was sacked by the Medes in 612 BC, but continued to be occupied until the Middle Ages. ❍ *38–39*

NIXON, RICHARD MILHOUS (1913–94) 37th US president (1969–74), Nixon became the first US president ever to resign from office. Nixon was elected as a REPUBLICAN to the House of Representatives in 1946 and the Senate in 1950. He was vice-president (1953–61) under Eisenhower, but narrowly lost the presidential election to JOHN F. KENNEDY. He was elected president in 1968. In 1969 he began a phased withdrawal of US troops from the VIETNAM WAR, but the USA invaded Cambodia in 1970 and Laos in 1971. In 1972 Nixon became the first US president to visit communist China. He easily won re-election in the same year. Following the blanket bombing of North Vietnam, Nixon withdrew US troops from Vietnam (1973). Congress forced Nixon to stop US bombing in Cambodia. In 1973 members of Nixon's re-election committee were convicted of the burglary of DEMOCRATIC PARTY headquarters in the Watergate building. The scandal sunk his administration. Nixon was forced to submit his secret tape recordings of conversations which revealed that he had attempted to cover up the burglary. On 9 August 1974 Nixon resigned to avoid impeachment. His successor, GERALD FORD, granted him a pardon. ❍ *242–43, 254–55*

NKRUMAH, KWAME (1909–72) Ghanaian statesman, prime minister (1957–60), first president of Ghana (1960–66). Nkrumah was the leading post-colonial proponent of Pan-Africanism. In 1949 he formed the Convention People's Party in the Gold Coast. He was imprisoned in 1950 by the British, but led the Gold Coast to independence in 1957 and became prime

minister. In 1960 Gold Coast became the Republic of Ghana and Nkrumah was made president. Nkrumah formed a loose union with Guinea and Mali and promoted the Charter of African States (1961). Nkrumah gradually assumed absolute power and, following a series of assassination attempts, Ghana became a one-party state in 1964. He undertook extravagant projects and,while on a visit to China in 1966, he was deposed in a military coup. ❍ *256–57*

NORMAN CONQUEST (1066) Invasion of England by WILLIAM I (THE CONQUEROR), Duke of Normandy. William claimed that Edward the Confessor (d. 1066) had recognized him as heir to the throne of England, and he disputed the right of Harold II to be king. William's army defeated and killed Harold at the Battle of HASTINGS, then advanced on London, where William was accepted as king. The ruling class, both lay and ecclesiastical, was gradually replaced by NORMANS, and Norman institutions were imposed. ❍ *92–93*

NORMANS Descendants of Vikings who settled in northwest France in the 9th–10th centuries. They created a powerful state, with a strongly centralized feudal society and warlike aristocracy. In the 11th century, under Robert Guiscard and Robert II, they defeated the Muslims to create an independent kingdom in Sicily. In 1066 Duke William of Normandy conquered England and became WILLIAM I. ❍ *78–79*

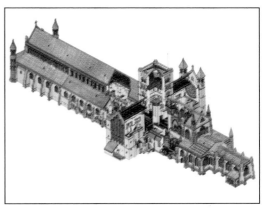

The Abbey Church of St Albans is one of the greatest Norman buildings. The Norman part – the central tower and most of the nave – was built in 1077–93.

NORTH AMERICAN FREE TRADE AGREEMENT (NAFTA) Treaty designed to eliminate trade barriers between Canada, Mexico and the USA. The agreement was signed in 1992, and NAFTA came into effect on 1 January 1994. Some Latin American countries have also applied to join. ❍ *242–43, 272–73*

NORTH ATLANTIC TREATY ORGANIZATION (NATO) Intergovernmental organization, military alliance of the USA, Canada and 17 European countries. The original treaty was signed in Washington in 1949 by Belgium, Britain, Canada, Denmark, France, Iceland, Italy, Luxembourg, Norway, Portugal, Netherlands and the USA. Since then, Greece, Turkey, Spain and Germany have joined. In July 1997 a meeting was held to admit the Czech Republic, Hungary and Poland in 1999. NATO's headquarters is in Brussels. During the Cold War it was the focus of the West's defence against the Soviet Union. Following the end of the Cold War, its air power was used to attack Serbian troops in Bosnia in 1995, and in 1999 in order to halt Serb aggression in Kosovo. ❍ *242–43, 244–45*

NUBIA Ancient state (also called Kush) on the upper Nile in northeast Africa. It was closely associated with EGYPT. At its height, Nubia extended from Egypt to the Sudan. At first ruled by Egypt, it later controlled Egypt in the 8th and 7th centuries BC. It converted to CHRISTIANITY in the 6th century AD and became part of Ethiopia in the 14th century. ❍ *30–31*

NUR AL-DIN (1118–74) (Nureddin) Ruler of Syria. He united Muslim forces in Syria to resist the Christians of the crusades. He recaptured Edessa from the Christians in 1146 and in 1154 took Damascus from the SELJUK Turks. ❍ *94–95*

O

O'HIGGINS, BERNARDO (1778–1842) South American revolutionary leader and ruler of Chile. A member of the revolutionary junta in colonial Chile, he commanded the army against the Spanish. Defeated in 1814, he joined José de SAN MARTÍN in Argentina to defeat the Spanish at Chacabuco (1817). Appointed "supreme director" of Chile, he declared independence in 1818 but resigned in 1823. ➲ *190–91*

OLDUVAI GORGE Site in northern Tanzania where remains of primitive humans have been found. Louis Leakey (1903–72) uncovered four layers of remains dating from c. 2 million years ago to c. 15,000 years ago. In 1964 he announced the discovery of *Homo habilis*, whom he believed to have been a direct ancestor of modern humans. The gorge, which is 40 kilometres (25 miles) long and 100 metres (320 feet) deep, runs through the Serengeti Plain. ➲ *16–17*

OLMEC Early civilization of Mesoamerica, which flourished between the 12th and 4th centuries BC. Its heartland was the south coast of the Gulf of Mexico, but its influence spread more widely. The earliest known site is at San Lorenzo, dated at c. 1500. From the 9th century BC the main Olmec centre was La Venta. Olmec art included high-quality carving of jade and stone, notably giant human heads in basalt. The Olmec heritage can be traced through later civilizations, including the MAYA. ➲ *32–33*

OPIUM WARS Conflict between Britain and China (1840–42), and between an Anglo-French force and China (1856–60). The first war arose following a Chinese ban of the import of opium in 1839. After a British victory, the Treaty of NANJING gave Britain trading rights in certain ports and the grant of Hong Kong. The second war was a successful attempt by the British and French to gain further concessions from China in 1856–60. When China refused to ratify the Treaty of Tientsin (1858), Anglo-French forces occupied Beijing. ➲ *198–99, 208–09*

ORGANIZATION OF AMERICAN STATES (OAS) Organization of 35 member states of the Americas that promotes peaceful settlements to disputes, regional co-operation in the limitation of weapons, and economic and cultural development. An extension of the 1948 Pan American Union held in Colombia, its charter became effective in 1951. The OAS is affiliated to to the UNITED NATIONS (UN). ➲ *242–43, 258–59*

ORGANIZATION OF PETROLEUM EXPORTING COUNTRIES (OPEC) Intergovernmental organization established in 1960 by many of the world's major oil-producing states to safeguard their interests. Its primary purpose is to set production quotas and co-ordinate prices among the 12 members. It was able to control oil prices in the 1970s, but its influence has waned since then, largely because of internal differences and the emergence of major oil-productng coutnries outside OPEC. Its headquarters in in Vienna, Austria. ➲ *260–61, 272–73*

ORTHODOX CHURCH (EASTERN ORTHODOX CHURCH) Family of Christian national churches mostly of eastern Europe, the Middle East and Russia. The churches are independent but acknowledge the primacy of the Patriarch of CONSTANTINOPLE. They developed from the Church of the Byzantine Empire, which separated from that of Rome in 1054. The Russian Orthodox Church is by far the largest in the number of its adherents. ➲ *96–97, 269–69*

OSLO AGREEMENT (1993) Bilateral peace accord between the government of Israel and the PALESTINE LIBERATION ORGANIZATION (PLO). The Agreement was signed in Washington DC by Yasser Arafat (for the PLO) and the Israeli foreign minister, Shimon Peres, in the presence of the Israeli prime minister, Yitzhak Rabin. In return for the PLO recognition of the state of Israel, the Israelis agreed to a military withdrawal from the Gaza Strip, Jericho and areas of the West Bank. They also offered the PLO the opportunity to elect the Palestine National Authority, which took over control of taxation, health, education and social services in these areas. ➲ *260–61*

OSMAN I (1258–1326) Founder of the Ottoman Empire. As ruler of the small Osmanli, or Ottoman, state in northwest Anatolia (Turkey), he declared his independence of the SELJUK sultan around 1290. He expanded his territory in frequent wars against the Byzantine Empire, a process which was continued by his successors. ➲ *96–97*

OSTROGOTHS (East GOTHS) An ancient Germanic people. They were subjected to the Huns in the 4th century and settled in Pannonia (modern Hungary). Under King THEODORIC (r. AD 471–526), the Ostrogoths conquered Italy in 493 and set up a kingdom based at Ravenna. They were defeated by Byzantium (535–40) and expelled from Italy. They were soon absorbed by other peoples. ➲ *56–57*

OTTO I (THE GREAT) (912–73) King of the Germans (936–73) and Holy Roman Emperor (962–73). He succeeded his father, Henry I, in Germany and defeated rebellious princes and their ally, Louis IV of France. Royal power was further augmented by Otto's close control of the Church. In 955 he crushed the MAGYARS at Lechfeld. He invaded Italy to aid Adelaide, the widowed queen of Lombardy, and married her, declaring himself King of Lombardy. In 962 he was crowned as Roman Emperor (the "Holy", meaning "Christian", was added later). ➲ *90–91*

P

PACIFIC, WAR OF THE, 1879–83 Pacific, war of the (1879–83) Conflict between Chile and the allied nations of Peru and Bolivia. The threatened take-over of Chilean holdings in the nitrate-rich province of Antofagasta induced Chile to send troops into the area, precipitating the war. Peru and Bolivia were thoroughly vanquished. Under the Treaty of Ancón (1883), Peru lost the provinces of Tarapacá and, ultimately, Tacna and Arica. Bolivia became a landlocked nation with the loss of its only coastal province. ➲ *192–93*

PALESTINE Territory in the Middle East, on the eastern shore of the Mediterranean Sea; considered a Holy Land by JEWS, Christians and Muslims. Palestine has been settled continuously since 4000 BC. The Jews settled in Palestine c. 1250 BC but were subjects of the PHILISTINES until 1020 BC, when SAUL, DAVID and SOLOMON established Hebrew kingdoms. The region was then under Assyrian and, later, Persian control before coming under Roman rule in 63 BC. In succeeding centuries Palestine became a focus of Christian pilgrimage. It was conquered by the Muslim Arabs in 640. In 1099 Palestine fell to the crusaders, but in 1291 they in turn were routed by the MAMLUKS. The area was part of the Ottoman Empire from 1516 to 1918, when British forces defeated the Turks at Megiddo. Jewish immigration was encouraged by the Balfour Declaration. After the First World War the British held a LEAGUE OF NATIONS mandate over the land west of the River Jordan (now once again called Palestine). Tension between Jews and the Arab majority led to an uprising in 1936. The Second World War and Nazi persecution brought many Jews to Palestine, and in 1947 Britain, unable to satisfy both Jewish and Arab aspirations, consigned the problem to the UNITED NATIONS. The UN proposed a plan for separate Jewish and Arab states. This was rejected by the Arabs, and in 1948 (after the first of several Arab-Israeli Wars) most of ancient Palestine became part of the new state of Israel; the Gaza Strip was controlled by Egypt and the West Bank of the River Jordan by Jordan. These two areas were subsequently occupied by Israel in 1967. From the 1960s, the PALESTINE LIBERATION ORGANIZATION (PLO) led Palestinian opposition to Israeli rule, which included acts of terrorism and an uprising in the occupied territories. In 1993 Israel signed the OSLO AGREEMENT with the PLO, and in 1994 the Palestine National Authority took over nominal administration of the Gaza Strip and the West Bank. ➲ *36–37, 38–39, 44–45, 94–95, 178–79, 220–21, 260–61*

PALESTINE LIBERATION ORGANIZATION (PLO) Organization of Palestinian parties and groups, widely recognized as the representative of the Palestinian people. It was founded in 1964 with the aim of dissolving the state of Israel and establishing a Palestinian state to enable Palestinian refugees to return to their ancestral land. Many of its component guerrilla groups were involved in political violence against Israel and, in the 1970s, in acts of international terrorism to further their cause. Dominated by the al-Fatah group led by Yasir Arafat, in 1974 the PLO was recognized as a government in exile by the Arab League and the UNITED NATIONS. In the early 1990s PLO representatives conducted secret negotiations with the government of Israel, culminating in the OSLO AGREEMENT signed in 1993. ➲ *260–61*

PALMYRA (TADMUR, CITY OF PALMS) Ancient oasis city, in the Syrian Desert. By the 1st century BC it had become a city-state by virtue of its control of the trade routes between MESOPOTAMIA and the Mediterranean. In c. AD 30 the city-state became a Roman dependency under local rule. By the 2nd century AD Palmyra's influence had spread to Armenia. In 267 Zenobia became queen, the empire expanded and she severed the state's links with Rome. In AD 273, the Roman emperor Aurelian laid waste to the city, which was afterwards largely forgotten. ➲ *44–45*

PARIS, TREATIES OF Name given to several international agreements made in Paris. The most notable include: the treaty of 1763, which ended the SEVEN YEARS WAR; the treaty of 1783, in which Britain recognized the independence of the USA; the treaty of 1814, which settled the affairs of France after the first abdication of Napoleon; the treaty of 1815, after Napoleon's final defeat; the treaty of 1856 ending the CRIMEAN WAR; the treaty of 1898 ending the Spanish-American War and giving the Philippines to the USA; and the main international settlement (1919) after the First World War, more often called the Treaty of Versailles. Certain territorial adjustments after the Second World War were signed in Paris in 1946–47, as was the truce in the Vietnam War (1973), in which the USA agreed to withdraw its forces. ➲ *164–65, 188–89, 196–97, 220–21*

PARTHIA Region in ancient PERSIA. It was the seat of the Parthian Empire, founded after a successful revolt against the SELEUCIDS (238 BC). Under the Arsacid dynasty, the Parthian Empire extended, at its peak, from Armenia to Afghanistan. In AD 224 the Parthians were defeated by the rising power of the SASANIANS and the empire rapidly crumbled. ➲ *46–47*

PEARL HARBOR US naval base in Hawaii. On 7 December 1941, the base (headquarters of the US Pacific fleet) was attacked by aircraft from a Japanese naval task force, which had approached unobserved. About 2,400 people were killed and 300 aircraft and 18 ships were destroyed or severely damaged. The attack provoked US entry into the Second World War. ➲ *234–235*

PEASANTS' REVOLT (1381) Rebellion in England. The immediate provocation was the poll tax of 1380. Fundamental causes were resentment at feudal restrictions and statutory control of wages, which were held down artificially, despite the shortage of labour caused by the Black Death. Roused by a rebel priest, John Ball, and led by Wat Tyler, the men of Kent marched into London, where they were pacified by Richard II. Promises to grant their demands were broken after they dispersed. ➲ *106–7*

PEDRO I (1798–1834) Emperor of Brazil (1822–31). Son of the future JOHN VI of Portugal, he fled with the rest of the royal family to Brazil on Napoleon's invasion of Portugal in 1807. When his father reclaimed the Portuguese crown in 1821, he became Prince Regent of Brazil and declared it an independent monarchy (1822). His reign was marked by opposition from right and left, military failure against Argentina (1825–28) and revolt in Rio de Janeiro (1831). He abdicated in 1831 and returned to Portugal, where he secured the succession of his daughter, Maria II, to the Portuguese throne. ➲ *190–91*

PELOPONNESIAN WAR (431–404 BC) Conflict in ancient Greece between ATHENS and SPARTA. The underlying cause was Sparta's fear of Athenian hegemony, while Athenian hostility towards Corinth, Sparta's chief ally, provoked the Spartan declaration of war. Having a stronger army, Sparta regularly invaded Attica, while Athens, under PERICLES, avoided land battles and relied on its navy. The early stages were inconclusive. The Peace of Nicias (420 BC) proved temporary. Neither side kept to the agreement, and in 415 BC Athens launched a disastrous attack on Syracuse, which encouraged Sparta to renew the war. With Persian help, Sparta built up a navy which, under Lysander, defeated Athens in 404 BC. Besieged and blockaded, Athens surrendered. ❍ *40–41*

PENINSULAR WAR (1808–14) Campaign of the Napoleonic Wars in Portugal and Spain. A British force commanded by the future Duke of Wellington supported Portuguese and Spanish rebels against the French. Initally on the defensive, Wellington's forces gradually drove the French out of the Iberian peninsula and, after the victory of Vitoria (1813), invaded southern France. Napoleon's abdication (1814) brought the campaign to an end. ❍ *166–67*

PERESTROIKA (Rus. reconstruction) Adopted by Soviet prime minister, Mikhail GORBACHEV in 1986, perestroika was linked with GLASNOST. The restructuring included reform of government and the bureaucracy, decentralization and abolition of the Communist Party monopoly. Liberalization of the economic system included the introduction of limited private enterprise and freer movement of prices. ❍ *236–37, 262–63*

PERICLES (490–429 BC) Athenian statesman. He dominated ATHENS from c. 460 BC to his death, overseeing its golden age. He is associated with achievements in art and literature, including the building of the Parthenon, while strengthening the Athenian empire and government. Believing war with SPARTA to be inevitable, he initiated the PELOPONNESIAN WAR (431–404 BC) but died of plague at the outset. ❍ *40–41*

PERÓN, JUAN DOMINGO (1895–1974) President of Argentina (1946–55 and 1973–74). He was an army officer who took part in the coup of 1943 and became the leading figure in the military junta (1943–46). He cultivated the trade unions and earned support from the poor by social reforms, greatly assisted by his wife, "Evita" Perón. Though briefly ousted in 1945, he won the presidential election in 1946 and was re-elected in 1952. Perón's populist programme was basically nationalistic and totalitarian. Changing economic circumstances and the death of his wife reduced Perón's popularity after 1951, and he was overthrown in 1955. He retired to Spain, but returned to regain the presidency in 1973. His rule, marked by violence, was cut short by his death. ❍ *258–59*

PERSEPOLIS City of ancient PERSIA (Iran), c. 60 kilometres (37 miles) northeast of Shiraz. It was the capital (539–330 BC) of the Achaemenid Empire, and was renowned for its splendour. It was destroyed by the forces of ALEXANDER THE GREAT in 330 BC, during their pursuit of the Persian king Darius. ❍ *42–43*

PERSIA, ANCIENT Former name of Iran, in southwest Asia. The earliest empire in the region was that of MEDIA (c. 700–550 BC). It was overthrown by the Persian king, CYRUS THE GREAT, who established the much larger Achaemenid dynasty (550–330 BC), destroyed by ALEXANDER THE GREAT. Alexander's successors, comprising the SELEUCIDS, were replaced by people from PARTHIA in the 3rd century BC. The Persian SASANIAN dynasty was established by Ardashir I in AD 224. Weakened by defeat by the Byzantines under Heraclius, it was overrun by the Arabs in the 7th century. ❍ *40–43*

PERSIAN WARS (499–479 BC) Conflict between the ancient Greeks and Persians. In 499 BC the Ionian cities of Asia Minor rebelled against Persian rule. ATHENS sent a fleet to aid them. Having crushed the rebellion, the Persian emperor, DARIUS I, invaded Greece but was defeated at MARATHON in 490 BC. In 480 BC his successor, XERXES, burned Athens but withdrew after defeats at SALAMIS and PLATAEA (479 BC). Under Athenian leadership, the Greeks fought on, regaining territory in Thrace and Anatolia, until the outbreak of the PELOPONNESIAN WAR in 431 BC. ❍ *40–41*

PÉTAIN, HENRI PHILIPPE (1856–1951) French general and political leader. In the First World War his defence of Verdun (1916) made him a national hero. He was appointed commander-in-chief in 1917 and held high positions between the wars. With the defeat of France in 1940, he was recalled as prime minister. He signed the surrender and became head of the collaborationist VICHY regime. He was charged with treason after the liberation of France in 1945 and died in prison. ❍ *232–33*

PETER I (THE GREAT) (1672–1725) Russian tsar (1682–1725), regarded as the founder of modern Russia. After ruling jointly with his half-brother Ivan (1682–89) he gained sole control in 1689. He employed foreign experts to modernize the army, transport, and technology, visiting European countries to study developments. He compelled the aristocracy and the Church to serve the interests of the state, eliminating ancient tradition in favour of modernization. In the GREAT NORTHERN WAR, Russia replaced Sweden as the dominant power in northern Europe and gained lands on the Baltic, where Peter built his new capital, St Petersburg. In the east, he warred against Turks and Persians and initiated the exploration of Siberia. ❍ *148–49, 150–51*

PETERLOO MASSACRE (1819) Violent suppression of a political protest in Manchester, England. A large crowd demonstrating for reform of parliament was dispersed by soldiers. Eleven people were killed and 500 injured. ❍ *172–73*

PHILIP II (382–336 BC) King of MACEDONIA (359–336 BC). He conquered neighbouring tribes and gradually extended his rule over the Greek states, defeating the Athenians at Chaeronea in 338 BC and gaining reluctant acknowledgment as king of Greece. He was preparing to attack the Persian Empire when he was assassinated, leaving the task to his son, ALEXANDER THE GREAT. ❍ *42–43*

PHILIP II (1165–1223) (Philip Augustus) King of France (1180–1223). Greatest of French medieval kings, he increased the royal domain by marriage, by exploiting his feudal rights, and by war. His main rival was Henry II of England. Philip supported the rebellions of Henry's sons, fought a long war against RICHARD I, and during the reign of JOHN, occupied Normandy and Anjou. English efforts to regain them were defeated at Bouvines in 1214. Philip persecuted JEWS and Christian heretics, joined the Third Crusade but swiftly withdrew, and opened the crusade against the ALBIGENSES in southern France. ❍ *92–93*

PHILIP II (1527–98) King of Spain (1556–98), King of Portugal (1580–98). From his father, the Emperor CHARLES V, he inherited Milan, Naples and Sicily, the Netherlands, as well as Spain with its huge new empire in the New World. War with France ended at Cateau-Cambrésis (1559), but the revolt of the Netherlands began in 1566. A defender of Roman Catholicism, Philip launched the unsuccessful Armada of 1588 to crush the English who, as fellow PROTESTANTS, aided the Dutch. In the Mediterranean, the Ottoman Turks presented a continuing threat, in spite of their defeat at Lepanto in 1571. ❍ *146–47, 152–53*

PHILIP V (1683–1746) King of Spain (1700–46). Because he was a grandson and a possible successor of Louis XIV of France, his accession to the Spanish throne provoked the WAR OF THE SPANISH SUCCESSION. By the Treaty of UTRECHT (1713), he kept the Spanish throne at the price of exclusion from the succession in France and the loss of Spanish territories in Italy and the Netherlands. ❍ *174–75*

PHILISTINE Member of a non-Semitic people who lived on the south coast of modern Israel, known as Philistia, from c. 1200 BC. They clashed frequently with the Hebrews. ❍ *38–39*

PHOENICIA Greek name for an ancient region bordering the eastern Mediterranean coast. The Phoenicians were related to the Canaanites. Famous as merchants and sailors, they never formed a single political unit, and Phoenicia was dominated by EGYPT before c. 1200 BC and by successive Near Eastern empires from the 9th century BC. The Phoenician city-states such as TYRE, Sidon and BYBLOS reached the peak of their prosperity in the intervening period, but the Phoenicians dominated trade in the Mediterranean throughout the Bronze Age. Expert navigators, they traded for tin in Britain and sailed as far as West Africa. The Phoenicians founded colonies in Spain and North Africa, notably CARTHAGE. ❍ *38–39*

PHRYGIA Historic region of west central Anatolia. A prosperous kingdom was established by the Phrygians, immigrants from southeast Europe, early in the 1st millennium BC, with its capital at Gordion. Midas was a legendary Phrygian king. In the 6th century BC Phrygia was taken over by LYDIA, then by PERSIA and later empires. ❍ *38–39*

PIPPIN III (THE SHORT) (714–768) First CAROLINGIAN king of the Franks (750–768). He and his brother Carloman inherited the office of "mayor of the palace" (*de facto* ruler) in 741. They suppressed a series of revolts by 747, when Carloman entered a monastery. In 750 Pippin deposed the last MEROVINGIAN king, Childeric III, and was annointed King of the Franks. He promised to change Carolingian policy by supporting the Papacy against the LOMBARDS, and defeated the Lombards in 754 and 756. He ceded the conquered territories (the future Papal States) to the Papacy in what was known as the Donation of Pippin. Pippin was the father of CHARLEMAGNE. ❍ *74–75*

PIUS IX (1792–1878) Pope (1846–78), born Giovanni Maria Mastai-Ferrett. He was driven from Rome in 1848–50, but restored by NAPOLEON III in 1850. The Papal States were seized by the Italian nationalists in 1860, and Rome itself incorporated into the kingdom of Italy, which in 1870 Pius refused to accept. He defended German Catholics from persecution by BISMARCK. In 1869 Pius convened the First Vatican Council, which proclaimed the principle of Papal Infallibility. ❍ *176–77*

PIZARRO, FRANCISCO (1471–1541) Spanish *conquistadore* of the Inca empire of Peru. He served under CORTÉS and led expeditions to South America (1522–28). Having gained royal support, he led 180 men to Peru in 1531. They captured and later murdered the Inca ATAHUALPA, and took Cuzco, the capital. Pizarro acted as governor of the conquered territory, founding Lima in 1535. ❍ *116–17, 120–21*

PLANTAGENET English royal dynasty (1154–1485). The name encompasses the Angevins (1154–1399) and the houses of Lancaster and York. They are descended from Geoffrey of Anjou and Matilda, daughter of Henry I. The name was adopted by Richard, duke of York and father of Edward IV, during the Wars of the Roses. ❍ *106–7*

PLATAEA, BATTLE OF (479 BC) Decisive battle of the PERSIAN WARS. The Greeks under the Spartan Pausanias and the Athenian Aristides won a total victory. The Persian army was almost destroyed and its commander, Mardonius, killed. The battle ended the ambitions of XERXES to conquer Greece. ❍ *40–41*

PLATO (427–347 BC) Ancient Greek philosopher who formulated an ethical and metaphysical system based upon philosophical idealism. From c. 407 BC he was a disciple of Socrates, from whom he may have derived many of his ideas about ethics. Following the trial and execution of Socrates in 399 BC, Plato withdrew to Megara, after which he is believed to have travelled extensively in EGYPT, Italy and Sicily. He visited Syracuse in Sicily three times, in about 388, 367 and 361–360 BC, during the reigns of the tyrants Dionysius I and II. Plato sought to educate Dionysius II as a philosopher-king and set up an ideal political system under him, but the venture failed. Meanwhile, in ATHENS, Plato set up his famous Academy (c. 387 BC). In the Academy he taught several young people, including ARISTOTLE. In addition to being a philosopher of great influence, Plato wrote in the form of dialogues, in which Socrates genially interrogates another person, demolishing their arguments. All of Plato's 36 works survive. His most famous dialogues include *Gorgias* (on rhetoric as an art of flattery), *Phaedo* (on death and the immortality of the soul) and the *Symposium* (a discussion on the nature of love). Plato's greatest work was the *Republic*, an extended dialogue on justice, in which he outlined his view of the ideal state. ❍ *40–41*

PLATT AMENDMENT (1901) US legislation effectively making Cuba a US protectorate after the Spanish-American War. It was included in the Cuban constitution of 1901, and provided the legal basis for the US occupation of Cuba (1906–9). It was repealed in 1934, although the USA retained the right to maintain its naval base on Guantánamo Bay. ❍ *192–93*

POL POT (1928–98) Cambodian ruler. He became leader of the communist Khmer Rouge, which overthrew the US-backed government of Lon Nol in 1975. He instigated a reign of terror in Cambodia (renamed Kampuchea). The intellectual elite were massacred, and the people in the cities driven into the countryside. Estimates of the dead range from one to four million. Pol Pot's regime was overthrown by a Vietnamese invasion in 1979. He continued to lead the Khmer Rouge until after the Vietnamese withdrawal (1989), but became a shadowy figure in the 1990s until his death. ❍ *212–13*

POLO, MARCO (1254–1324) Venetian traveller in Asia. In 1274 he accompanied his father and uncle on a trading mission to the court of QUBILAI KHAN, the Mongol emperor of China. According to his account, he remained in the Far East more than 20 years, becoming the confidant of Qubilai Khan and travelling throughout China and beyond. His account, *The Description of the World*, became the chief source of European knowledge of China for centuries. ❍ *86–87*

POLTAVA, BATTLE OF (1709) Victory of the Russians over the Swedes in the GREAT NORTHERN WAR, fought near Poltava in the Ukraine. All but 1,500 of the Swedish army were either killed or captured. The battle marked the end of CHARLES XII's conquests and also of Swedish supremacy in northern Europe. Russia emerged as the dominant power in the region. ❍ *148–49, 150–51*

POMPEY (106–48 BC) (Gnaeus Pompeius Magnus) Roman general. He fought for Sulla in 83 BC and campaigned in Sicily, Africa and Spain. He was named consul with Crassus in 70 BC and fought a notable campaign against Mithridates VI of Pontus in 66 BC. In 59 BC he formed the first triumvirate with Crassus and his great rival, Julius CAESAR. After the death of Crassus, Pompey joined Caesar's enemies, and civil war broke out in 49 BC. Driven out of Rome by Caesar's advance, Pompey was defeated at Pharsalus in 48 BC and fled to Egypt, where he was murdered. ❍ *54–55*

PRAGMATIC SANCTION Term of Byzantine origin meaning a public decree, used later by European monarchs for state documents such as those defining their powers or settling the royal succession. Unless qualified, it usually refers to the pragmatic sanction of 1713 by which Charles VI of Austria settled the succession to his throne upon his daughter MARIA THERESA. ❍ *174–75*

PREMYSLIDS Bohemian dynasty established in the 10th century by Premysl. The conversion of BOHEMIA to CHRISTIANITY was completed at this time by German priests. Under Bratislav I (1034–55) a stable period ensued. Problems of succession and tribute exacted by the HOLY ROMAN EMPIRE contributed to the decline of the Premyslids, and the dynasty ended with the death of Wenceslaus in 1306. ❍ *70–71*

PRIMO DE RIVERA, MIGUEL (1870–1930) Spanish dictator (1923–30). He staged a coup in 1923 with the support of King Alfonso XIII. He dissolved parliament and established a military dictatorship modelled on the government of MUSSOLINI. He restored order, stimulated economic improvement and helped to end the revolt of Abd-el-Krim in Morocco (1926). He was forced from power shortly before his death. ❍ *230–31*

PROTESTANTISM Branch of CHRISTIANITY formed in protest against the practices and doctrines of the old Roman CATHOLIC CHURCH. Protestants sought a vernacular Bible to replace the Latin Vulgate, and to express individual elements of nationalism. The movement is considered to have started when Martin LUTHER'S *Ninety-Five Theses* were nailed to a Wittenberg church door. His predecessors included John WYCLIFFE and Jan HUS. Later supporters included Huldreich ZWINGLI and John CALVIN, whose interpretation of the Bible and concept of predestination had great influence. The Protestants held the Eucharist to be a symbolic celebration, as opposed to the Roman Catholic dogma of transubstantiation, and claimed that because Christ is the sole medium between God and man, his function cannot be displaced by the priests of the Church. ❍ *154–55, 268–69*

PTOLEMY I (367–283 BC) (Ptolemy Soter) King of ancient EGYPT, first ruler of the Ptolemaic dynasty. A leading Macedonian general of ALEXANDER THE GREAT, he was granted Egypt in the division of Alexander's empire upon the latter's death in 323 BC. After many struggles for control, he assumed the title of King in 305 BC. He made his capital at Alexandria, where the the famous library was created. He abdicated in 284 BC in favour of his son, Ptolemy II. The dynasty continued until 30 BC. ❍ *42–43*

PUEBLO Generic name for the several Native American peoples inhabiting the Mesa and Rio Grande regions of Arizona and New Mexico. They belong to several language families, including Keresan, Tewa, Hopi and Zuni. The multi-storeyed buildings of the Zuni gave rise to the legendary "Seven Cities of C'bola" eagerly sought by the Spaniards. ❍ *108–9*

PUNIC WARS (264–146 BC) Series of wars between ROME and CARTHAGE. In the First Punic War (264–241 BC), Carthage was forced to surrender Sicily and other territory. In the Second Punic War (218–201 BC), the Carthaginians under Hannibal invaded Italy and won a series of victories. They were eventually forced to withdraw, whereupon the Romans invaded North Africa and defeated HANNIBAL. The Third Punic War (149–146 BC) ended in the destruction of CARTHAGE. ❍ *54–55*

PYRAMIDS Monuments on a square base with sloping sides rising to a point. They are associated particularly with ancient EGYPT, where some of the largest have survived almost intact. They served as burial chambers for pharaohs. The earliest Egyptian pyramid was a step pyramid (ziggurat), built c. 2700 BC for Djoser. Pyramid building in Egypt reached its peak during the 4th Dynasty, the time of the Great Pyramid at Giza. The largest pyramid in the world, it was built for Pharaoh Khufu (c. 2589–2566 BC) and stands 146 metres (480 feet) high with sides 231 metres (758 feet) long at the base. The largest New World ziggurat pyramid was built in TEOTIHUACAN, Mexico, in the 1st century AD; it was 66 metres (216 feet) high with a surface area of 50,600 square metres (547,200 square feet). ❍ *30–33*

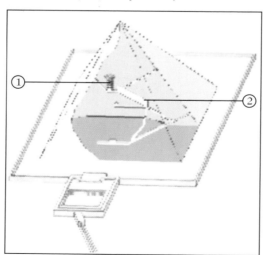

The Great Pyramid at Giza is built of limestone blocks and was originally faced with fine white limestone. The final resting place of Pharaoh Khufu's body, called the King's Chamber (1), was constructed of granite and approached via the Grand Chamber (2).

PYRENEES, PEACE OF THE (1659) Treaty between France and Spain after the THIRTY YEARS WAR. France gained territory in Artois and Flanders, and Philip IV of Spain reluctantly agreed to his daughter's marriage to LOUIS XIV. She was to renounce her claim to the Spanish throne in exchange for a subsidy. Because Spain could not pay the subsidy, the renunciation became void, giving Louis a claim on the Spanish Netherlands and resulting in the War of Devolution. ❍ *152–53*

Q

QIN Imperial dynasty of China (221 BC–206 BC). Originating in northwest China, the Qin emerged after the collapse of the ZHOU dynasty. Its founder was QIN SHI HUANG DI. The first centralized imperial administration was established, with the country divided into provinces, each under a governor. Uniformity was encouraged in every sphere, including law, language, coinage, weights and measures. The GREAT WALL OF CHINA took permanent shape during this period. ❍ *48–49*

QIN SHI HUANG DI (259–210 BC) Emperor of China (221–210 BC). The first emperor of the QIN dynasty, he reformed the bureaucracy and consolidated the GREAT WALL OF CHINA. Excavations of his tomb on Mount Li (near Xi-an) during the 1970s revealed, among other treasures, an "army" of around 7,500 life-size terracotta guardians. ❍ *48–49*

QUBILAI KHAN (1215–94) Mongol emperor of China (1260–94). Grandson of CHINGGIS KHAN, he completed the conquest of China in 1279, establishing the YUAN dynasty, which ruled until 1368. He conquered the southern Song dynasty and extended operations into Southeast Asia, although his attempt to invade Japan was thwarted by storms. He conducted correspondence with European rulers and apparently employed Marco POLO. ❍ *98–99*

QUEBEC ACT (1774) Act of British parliament creating a government for Quebec. It set up a council to assist the governor and recognized the Roman CATHOLIC CHURCH and the French legal and landholding systems in the former French colony. Quebec's boundary was extended south to the Ohio River, a cause of resentment among the thirteen North American colonies. ❍ *164–65, 188–89*

R

RAMESES III Egyptian king of the 20th dynasty (r. 1184–1153 BC). He defended EGYPT from attacks by Libya and the Sea Peoples. Later in his reign, however, Egypt withdrew into political and cultural isolation. ❍ *36–37*

REAGAN, RONALD WILSON (1911–) 40th US president (1981–89). A well-known film actor, he joined the REPUBLICAN PARTY (1962) and won a landslide victory to become governor of California (1966–74). Nominated as presidential candidate at his third attempt in 1980, he defeated Carter. Surviving an assassination attempt (1981), he introduced large tax cuts and reduced public spending, except on defence. By the end of his second term, budget and trade deficits had reached record heights. Fiercely anti-communist, he adopted strong measures against opponents abroad, invading Grenada (1983) and undermining the Sandinista regime in Nicaragua. While pursuing his Strategic Defence Initiative (SDI, or "Star Wars"), he reached an arms control agreement with Soviet leader GORBACHEV (1987) and welcomed improved relations with the Soviet Union. ❍ *242–43*

RED GUARDS Chinese youth movement active between 1966 and 1969, in the CULTURAL REVOLUTION They were named after the groups of armed workers who took part in the BOLSHEVIK revolution in Russia of 1917. The Chinese Red Guards attacked revisionists,

Westerners and alleged bourgeois influences. Originally encouraged by MAO, they caused severe social disorder and were suppressed after 1968. ➲ *254–55*

REFORM ACTS A series of British acts of Parliament extending the right to vote. The Great Reform Bill (1832) redistributed seats in the House of Commons to include large cities that were previously unrepresented. It also gave the vote to adult males occupying premises worth at least £10 a year. The second Reform Act (1867) extended the franchise to include better-off members of the working class. The acts of 1884 and 1885 gave the vote to most adult males. Women over 30 gained the vote in 1918, and the Representation of the People Act (1928) introduced universal adult suffrage. ➲ *172–73*

RENAISSANCE Period of European history lasting roughly from the mid-15th century to the end of the 16th century. The word was used by late 15th-century Italian scholars to describe the revival of interest in classical learning. It was helped by the fall of CONSTANTINOPLE to the Ottoman Turks in 1453, which resulted in the transport of Classical texts to Italy. In Germany, the invention of a printing press with moveable type assisted the diffusion of the new scholarship. In religion, the spirit of questioning led to the Reformation. In politics, the Renaissance saw the rise of assertive sovereign states – Spain, Portugal, France and England – and the expansion of Europe beyond its own shores, with the building of trading empires in Africa, the East Indies and America. The growth of a wealthy urban merchant class led to a tremendous flowering of the arts. ➲ *134–35*

REPUBLICAN PARTY (USA) US political party. It was organized in 1854 as an amalgamation of Whigs and Free-Soilers, with workers and professional people who had formerly been known as Independent Democrats, Know-Nothings, Barnburners or Abolitionists. Its first successful presidential candidate was ABRAHAM LINCOLN (elected 1860). During the the 20th century, the Republicans were generally the minority party to the DEMOCRATIC PARTY in Congress, especially in the House of Representatives. From the 1970s, however, the party shifted to the right and gained substantial support. ➲ *240–41*

RICHARD I (1157–99) King of England (1189–99), known as Richard the Lion-Heart, or *Coeur de Lion*. He was involved in rebellions against his father, Henry II, before succeeding him. A leader of the Third Crusade (1189–92), he won several victories but failed to retake JERUSALEM. While returning from the crusade he was captured by the Duke of Austria and handed over to the Emperor Henry VI (1192); he was held prisoner until 1194. During this period, his brother JOHN conspired against him in England, while in France PHILIP II invaded Richard's territories. The revolt in England was contained, and from 1194 until his death, Richard endeavoured to restore the Angevin Empire in France. He was killed in battle. ➲ *92–93*

RICHELIEU, ARMAND JEAN DU PLESSIS, DUC DE (1585–1642) French cardinal and statesman. A protégé of Marie de Médicis and a cardinal from 1622, he became chief of the royal council in 1624. He suppressed the military and political power of the HUGUENOTS, but tolerated Protestant religious practices. He alienated many powerful Catholics by his policy of placing the interests of the state above all else. He survived several aristocratic plots against him. In the THIRTY YEARS WAR, he formed alliances with German Protestant princes and GUSTAV ADOLF of Sweden against the Habsburgs. His more scholarly interests resulted in the foundation of the Académie Française (1635). ➲ *156–57*

RIEL, LOUIS (1844–85) French-Canadian revolutionary, leader of the *Métis* (people of mixed French and native descent) in the Red River Rebellion in Manitoba (1869–70). When it collapsed he fled to the USA. In 1884 he led resistance to Canada's western policies in Saskatchewan and set up a rebel government in 1885. He was captured and subsequently executed for treason. ➲ *188–89*

RISORGIMENTO (Italian for "resurgence") Nationalist movement resulting in the unfication of Italy between 1859 and 1870. With the restoration of Austrian and Bourbon rule in 1815, revolutionary groups formed, notably the Young Italy movement of MAZZINI, whose aim was a single, democratic republic. Mazzini's influence was at its peak in the revolutions of 1848. In Sardinia–Piedmont (the only independent Italian state), the aim of the chief minister, CAVOUR, was a parliamentary monarchy under the royal house of Savoy. Securing the support of France under NAPOLEON III in a war against Austria, he acquired much of Austrian-dominated north Italy in 1859. In 1860 GARIBALDI conquered Sicily and Naples. Although Garibaldi belonged to the republican tradition of Mazzini, he co-operated with Cavour, and the kingdom of Italy was proclaimed in 1861 under VICTOR EMMANUEL II of Savoy. Other regions were acquired later. Rome, the future capital, was seized when the French garrison withdrew in 1870. ➲ *176–77*

ROBESPIERRE, MAXIMILIEN FRANÇOIS MARIE ISIDORE DE (1758–94) French revolutionary leader. Elected to the National Assembly in 1789, he advocated democracy and liberal reform. In 1791 he became leader of the JACOBINS, and gained credit when his opposition to war with Austria was justified by French defeats. With the king and the Girondins discredited, Robespierre led the republican revolution of 1792 and was elected to the National Convention. His election to the Committee of Public Safety (June 1793) heralded the Reign of Terror, in which hundreds died on the guillotine. Ruthless methods seemed less urgent after French victories in war, and Robespierre was arrested in the coup of 9th Thermidor (27 July) 1794 and executed. ➲ *166–67*

ROMANOVS Russian imperial dynasty (1613–1917). Michael, the first Romanov tsar, was elected in 1613. His descendants, especially PETER I (THE GREAT) and CATHERINE II (THE GREAT), a Romanov only by marriage, transformed Russia into the largest empire in the world. The last Romanov emperor, NICHOLAS II, abdicated in 1917 and was later murdered with his family by the BOLSHEVIKS. ➲ *148–49, 222–23*

ROME, ANCIENT Capital of the Roman republic. According to tradition, Rome was founded in 753 BC by Romulus and Remus. By 509 BC the Latin-speaking Romans had thrown off the rule of ETRUSCAN kings and established an independent republic dominated by an aristocratic elite. Its history was one of continual expansion, and by 340 BC Rome controlled Italy south of the River Po. By the 3rd century BC the plebeian class had largely gained political equality. The PUNIC WARS gave it dominance of the Mediterranean in the 2nd century BC, when major eastward expansion began with the conquest of the Greek lands around the Aegean. The republican constitution was strained by social division and military dictatorship. Spartacus's slave revolt was crushed by POMPEY, who emerged as Sulla 's successor. Pompey and Julius CAESAR formed the First Triumverate in 60 BC. Caesar emerged as leader and greatly extended Rome's territory and influence. His assassination led to the formation of the Roman Empire under AUGUSTUS in 27 BC. ➲ *54–55*

ROOSEVELT, FRANKLIN D. (1882–1945) 32nd US president (1933–45). He served in the New York Senate as a DEMOCRAT, as assistant secretary of the navy under WOODROW WILSON (1913–20) and was vice-presidential candidate in 1920. In 1921 he lost the use of his legs as a result of polio. He was governor of New York (1928–32) and won the Democratic candidacy for president. He was elected in 1932. In response to the Great Depression, he embarked upon his NEW DEAL, designed to restore the economy through direct government intervention. He was re-elected in 1936 and won an unprecedented third term in 1940 and a fourth in 1944. When the Second World War broke out in Europe, he gave as much support to Britain as a neutral government could until the Japanese attack on PEARL HARBOR ended US neutrality. He died in office and was succeeded by Harry S. Truman. ➲ *228–29, 240–41*

RUSSIAN REVOLUTION OF 1905 Series of violent strikes and protests against tsarist rule in Russia. It was provoked mainly by defeat in the RUSSO-JAPANESE WAR (1904–05). It began on Bloody Sunday (22 January), when a peaceful demonstration in St Petersburg was fired on by troops. Strikes and peasant risings spread, culminating in a general strike in October, which forced the tsar to institute a democratically elected *duma* (parliament). By the time the *duma* met in 1906, the government had regained control. Severe repression followed. ➲ *180–81*

RUSSO-JAPANESE WAR (1904–5) Conflict arising from the rivalry of Russia and Japan for control of Manchuria and Korea. The war opened with a Japanese attack on Port Arthur (Lushum). Russian forces suffered a series of defeats on land and at sea, culminating in the Battle of Mukden (February–March 1905) and the annihilation of the Baltic fleet at Tsushima (May). Russia was forced to surrender Korea, the Liaotung Peninsula and south Sakhalin to Japan. ➲ *180–81, 200–1, 216–17*

S

ST GERMAIN, TREATY OF (1919) Part of the peace settlement after the First World War. It established the new republic of Austria and other independent states from the old Austro-Hungarian Empire. The treaty also forbade the union of Germany and Austria – a condition that was broken by the Anschluss of 1938 – and stipulated that the Austro-Hungarian navy should be disbanded and the army limited to 30,000 volunteers. ➲ *174–75, 220–21*

SALADIN (1138–93) (Salah ad-din) Muslim general and founder of the Egyptian AYYUBID dynasty. From 1152, he was a soldier and administrator in Egypt. Appointed grand vizier in 1169, he overthrew the FATIMIDS in 1171 and made himself sultan. After conquering most of Syria, he gathered widespread support for a *jihad* to drive the Christians from PALESTINE in 1187. He reconquered JERUSALEM, provoking the Third Crusade (1188). Saladin's rule restored Egypt as a major power and introduced a period of stability and growth. ➲ *94–95*

SALAMIS, BATTLE OF Decisive naval victory of the Greeks in the PERSIAN WARS (480 BC). By skilful tactics, the Athenian commander, THEMISTOCLES, lured the larger Persian fleet into battle off the island of Salamis and destroyed 200 Persian ships. By depriving the army of XERXES of its seaborne supplies, the defeat halted the Persian invasion of Greece. ➲ *40–41*

SAMURAI Member of the elite warrior class of feudal Japan. Beginning as military retainers in the 10th century, the samurai came to form an aristocratic ruling class. They conformed to a strict code of conduct, known as *bushido*, "the way of the warrior". Under the Tokugawa shogunate (1603–1867), they were divided into hereditary subclasses and increasingly became bureaucrats and scholars. ➲ *86–87, 140–41*

SAN MARTÍN, JOSÉ DE (1778–1850) South American revolutionary. He led revolutionary forces in Argentina, Peru and Chile, gaining a reputation as a bold commander and imaginative strategist. After defeating the Spaniards in Argentina, he took them by surprise in Chile during 1817–18 by crossing the Andes, and won Peru in 1821 after an unexpected naval attack. He surrendered his effective rule of Peru to Simón BOLÍVAR in 1822 and retired to Europe. ➲ *190–91*

SARGON I (c. 2334–c. 2279 BC) King of Akkad (c. 2316–c. 2279 BC). One of the first of the great Mesopotamian conquerors, he was a usurper who founded his capital at Agade (Akkad), from which his kingdom took its name. He conquered Sumeria and upper MESOPOTAMIA and extracted tribute from lands as far west as the Mediterranean. ➲ *28–29*

SARGON II (d. 705 BC) King of ASSYRIA (721–705 BC). He conquered Samaria in 721 BC, subdued the Medes and, according to tradition, dispersed those Israelites who became the "lost tribes" of ISRAEL. He established an imperial administration and defeated his enemies before being killed in battle against the Cimmerians. ➲ *38–39*

SASANIANS Royal dynasty of PERSIA (AD 224–651). The dynasty was founded by Ardashir I (r. 224–244) who defeated the Parthians and established his capital at Ctesiphon. The Sasanians revived the native Persian traditions of the Achaemenids, confirming ZOROASTRIANISM as the state religion. Ardashir was succeeded by Shapur I (r. 244–72) who defeated the Roman emperor Valerian. There were about 30 Sasanian rulers, the most important after Shapur I being Shapur II (309–79), Khoshru I (531–79) and Khoshru II (590–628), whose conquest of Syria, PALESTINE and Egypt marked the height of the dynasty's power. The Sasanians were finally overthrown by the Arabs. ➲ *44–45, 68–69*

SAUL (active late 11th century BC) First king of the Hebrew state of ancient ISRAEL (c. 1020– c.1000 BC). He was the son of Kish, a member of the tribe of Benjamin. Saul was anointed by the prophet Samuel and acclaimed as king by all Israel. For much of his reign he waged war against Israel's threatening neighbours, notably the PHILISTINES, the Ammonites, and the Amalekites. He and his sons eventually died in battle against the Philistines on Mount Gilboa. The story of Saul is contained in the First Book of Samuel, the ninth book of the Old Testament. ➲ *38–39*

SAXONS Ancient Germanic people. They appear to have originated in northern Germany and perhaps southern Denmark. By the 5th century they were settled in northwest Germany, northern GAUL and southern Britain. In Germany they were eventually subdued by CHARLEMAGNE. In Britain, together with other Germanic tribes, known collectively as ANGLO-SAXONS, they evolved into the English. ➲ *74–75*

SCYTHIANS Nomadic people who inhabited the steppes north of the Black Sea in the 1st millennium BC. In the 7th century BC their territory extended into MESOPOTAMIA, the Balkans and Greece. Powerful warriors, their elaborate tombs contain evidence of great wealth. Pressure from the Sarmatians confined them to the Crimea (c. 300 BC) and their culture eventually disappeared. ➲ *42–43, 50–51*

SELEUCIDS Hellenistic dynasty founded by SELEUCUS, a former general of ALEXANDER THE GREAT, in 306–281 BC. Centred on Syria, it included most of the Asian provinces of Alexander's empire, extending from the eastern Mediterranean to India. War with the Ptolemies of EGYPT and, later, the Romans, steadily reduced its territory. In 63 BC its depleted territory became the Roman province of Syria. ➲ *42–43*

SELEUCUS Name of two kings of Syria. Seleucus I (c. 355–281 BC) was a trusted general of ALEXANDER THE GREAT and founder of the SELEUCID dynasty. By 281 BC he had secured control of BABYLONIA, Syria and all of Asia Minor, founding a western capital at Antioch to balance the eastern capital of Seleucia, in BABYLON. He appeared to be on the brink of restoring the whole of Alexander's empire under his rule when he was murdered. Seleucus II (r. 247–226 BC) spent his reign fighting Ptolemy III of EGYPT and Antiochus Hierax, his brother and rival, losing territory to both. ➲ *42–43*

SELIM I (THE GRIM) (1470–1520) Ottoman sultan (1512–20). The son of Bayezid II, who was dethroned in his favour, Selim began his reign by killing all his brothers and their sons, as well as four of his own sons, leaving only SULEIMAN as his heir. He then launched a vigorous campaign against ISMAIL, the founder of the Persian Safavid dynasty, defeating him in 1514 at the Battle of Chaldiran. In 1516–17 Selim defeated the Egyptian Mamluk armies, bringing Egypt, Syria and Palestine under Ottoman rule and doubling the size of his empire. Recognized as the leader of the Islamic world, he was presented with the keys to the holy city of MECCA. ➲ *142–43*

SELJUKS Nomadic tribesmen from Central Asia who adopted ISLAM in the 7th century and founded the Baghdad sultanate in 1055. Their empire included Syria, Mesopotamia and Persia. Under Alp Arslan, they defeated the Byzantines at Manzikert in 1071, which led to their occupation of Anatolia. They revived SUNNI administration and religious institutions, checking the spread of SHIITE Islam and laying the organizational basis for the future Ottoman administration. In the early 12th century the Seljuk Empire began to disintegrate, and by 1156 the Seljuk era was over. ➲ *88–89*

SEPHARDIM Descendants of the JEWS of medieval Spain and Portugal and others who follow their customs. Iberian Jews followed the Babylonian rather than the Palestinian Jewish tradition and developed their own language, Ladino. After the expulsion of the Jews from Spain in 1492, many settled in parts of the Middle East and North Africa under the Ottoman Empire. Continuing persecution led many of them to form colonies in Amsterdam and other cities of northwest Europe. ➲ *142–43*

SEVEN WEEKS WAR (Austro-Prussian War) (1866) Conflict between Prussia and Austria. Otto von BISMARCK engineered the war to further Prussia's supremacy in Germany and reduce Austrian influence. Defeat at Sadowa forced Austria out of the GERMAN CONFEDERATION. ➲ *176–77*

SEVEN WONDERS OF THE WORLD Group of fabled sights that evolved from various ancient Greek lists. They were, in chronological order: the PYRAMIDS of EGYPT; the Hanging Gardens of BABYLON; the statue of Zeus by Phidias at Olympia; the temple of Artemis at Ephesus; the mausoleum at Halicarnassus; the Colossus of Rhodes; and the Pharos at Alexandria. ➲ *30–31*

SEVEN YEARS WAR (1756–63) Major European conflict that established Britain as the foremost maritime and colonial power and ensured the survival of Prussia as a major power in central Europe. The war was a continuation of the rivalries involved in the War of the AUSTRIAN SUCCESSION (1740–48). Britain and Prussia were allied, with Prussia undertaking nearly all the fighting in Europe, against Austria, Russia, France and Sweden. FREDERICK II (THE GREAT) of Prussia fought a defensive war against superior forces. Only his brilliant generalship and the withdrawal of Russia from the war in 1762 saved Prussia from being overrun. Overseas, Britain and France fought in North America, India and West Africa, with the British gaining major victories. At the end of the war, the Treaty of PARIS (1763) confirmed British supremacy in North America and India, while the Treaty of Hubertusberg left Prussia in control of Silesia. ➲ *122–23, 124–25, 128–29, 164–65, 188–89*

SÈVRES, TREATY OF (1920) Peace treaty between Turkey and its European opponents in the First World War that imposed harsh terms on the Ottoman sultan. It was not accepted by the Turkish nationalists led by ATATÜRK, who fought a war for Turkish independence (1921–22). The treaty, never ratified, was replaced by the Treaty of LAUSANNE (1923). ➲ *178–79, 220–21*

SHAH JAHAN (1592–1666) Mughal emperor of India (1627–58). Third son of JAHANGIR, he secured his succession by killing most of his male relatives. His campaigns expanded the Mughal dominions and accumulated great treasure. Though relatively tolerant of HINDUISM, he made ISLAM the state religion. Shah Jahan was responsible for the Taj Mahal, and the vast ornamental chambers of the Red Fort at Delhi, which he made the capital. ➲ *144–45*

SHEVARDNADZE, EDUARD AMBROSIEVICH (1928–) Georgian political leader. Head of the Georgian Communist Party, he was made Soviet foreign minister in 1985 by Mikhail GORBACHEV. He resigned in 1990, warning that Gorbachev was becoming authoritarian. In 1992 he became head of state in Georgia. With Russian help, he overcame supporters of the deposed leader, Zviad Gamsakhurdia. He escaped an assassination attempt in 1995, and later that year won a sweeping victory in elections. ➲ *262–63*

SHIITE Second-largest branch of ISLAM. Shiites believe that the true successor of Muhammad was Ali, whose claim to be caliph was not recognized by SUNNI Muslims. It rejects the Sunna (the collection of teachings outside the Koran) and relies instead on the pronouncements of a succession of holy men called imams. The Safavid dynasty in Iran were the first to adopt Shiism as a state religion. One of the principal causes of the Iranian revolution was Shah Pahlavi's attempt to reduce clerical influence on government. Ayatollah KHOMEINI's Shiite theocracy stresses the role of Islamic activism in liberation struggles. After the 1991 GULF WAR the Iraqi Shiites opposed the oppressive régime of SADDAM HUSSEIN, but were defeated in the marshlands of southern Iraq. The Ismailis of the Indian subcontinent are the other major Shiite grouping. ➲ *142–43, 260–61*

SHOGUN Title of the military ruler of Japan, first conferred upon Yoritomo in 1192. The Minamoto (1192–1333), Ashikaga (1338–1568) and Tokugawa (1603–1868) shogunates in effect ruled feudal Japan, although an emperor retained ceremonial and religious duties. The shogunate ended with the MEIJI RESTORATION in 1868. ➲ *86–87, 140–41*

SIKHISM Indian religion founded in the 16th century by Nanak, the first Sikh guru. Combining Hindu and Muslim teachings, it is a monotheistic religion whose adherents believe that their one God is the immortal creator of the universe. All human beings are equal, and Sikhs oppose any caste system. The path to God is through prayer and meditation, but nearness to God is only achievable through divine grace. Sikhs believe in reincarnation, and are taught to seek spiritual guidance from their guru or leader. Begun in Punjab as a pacifist religion, Sikhism (under Nanak's successors) became an activist military brotherhood and a political force. All Sikh men came to adopt the surname Singh ("lion"). Since Indian independence, Sikh extremists have periodically agitated for an independent Sikh state (Khalistan). In 1984 the leader of a Sikh fundamentalist revival, Sant Jarnail Singh Bhindranwale (1947–84), was killed by government forces at the Golden Temple of Amritsar, and in retaliation Indira GANDHI was assassinated by Sikh bodyguards. More than 1,000 Sikhs died in the ensuing riots. ➲ *248–49, 268–69*

SILK ROAD Ancient trade route linking China with Europe, the major artery of all Asian land exploration before AD 1500. A variety of goods were transported along the route, but silk was the most important. For 3,000 years the manufacture of silk was a secret closely guarded by the Chinese. Silk fetched extravagant prices in Greece and Rome, and the trade became a major source of income for the Chinese ruling dynasties. By 100 BC 12 caravan trains were making the annual journey along the Silk Road, and the tax on the trade provided a third of all the Han dynasty's revenue. The silk trade began to decline in the 6th century, when the silk-worm's eggs were smuggled to CONSTANTINOPLE and the secret was exposed. During the 13th century, European merchants (including Marco POLO in 1271) travelled along the Silk Road when it was controlled by the Mongols. ➲ *52–51, 116–17*

SINHALESE People who make up the largest ethnic group of Sri Lanka. They speak an Indo-European language and practise Theravada BUDDHISM. ➲ *248–49*

SINO-JAPANESE WARS Two wars between China and Japan marking the beginning and the end of Japanese imperial expansion on the Asian mainland. The first (1894–95) arose from rivalry for control of Korea. In 1894 Japanese influence helped to provoke a rebellion in Korea. Both states intervened, and Japanese forces swiftly defeated the Chinese. China was forced to accept Korean independence and ceded territory including Taiwan and the Liaotung peninsula. The latter was returned after European pressure. The second war (1937–45) developed from Japan's seizure of Manchuria (1931), where it set up the puppet state of Manchukuo. Further incidents of Japanese aggression led to war, in which the Japanese swiftly conquered eastern China, driving the government out of Beijing. US and British aid was despatched to China from 1938 and the conflict merged into the Second World War, ending with the final defeat of Japan in 1945. ➲ *200–1, 224–25, 234–35*

SMITH, ADAM (1723–90) Scottish philosopher, regarded as the founder of modern economics. His book *The Wealth of Nations* (1776) was enormously influential in the development of Western, capitalist society. Smith formulated the doctrine of *laissez-faire*: that governments should not interfere in economic affairs and that free trade increases wealth. ➲ *128–29*

SOLOMON (d. 922 BC) King of ISRAEL (c. 966–926 BC), son of DAVID and Bathsheba. His kingdom prospered thanks partly to economic relations with the Egyptians and Phoenicians, enabling Solomon to build the JERUSALEM Temple. His reputation for wisdom reflected his interest in literature, although the works attributed to him, including the *Song of Solomon*, were probably written by others. ➲ *38–39*

SOMME, BATTLE OF THE Major First World War engagement along the River Somme, northern France. It was launched by the British commander Douglas Haig on 1 July 1916. On the first day the British suffered over 60,000 casualties in a futile attempt to break through the German lines. A desperate trench war of attrition continued until the offensive was abandoned on 19 November 1916. Total casualties were over one million, and the British had advanced only 16 kilometres (10 miles). A second battle around St Quentin (March–April 1918) is sometimes referred to as the Second Battle of the Somme. A German offensive designed to secure a victory before the arrival of US troops, it was halted by Anglo-French forces. ➲ *218–19*

SONGHAY EMPIRE West African trading empire which reached its peak between 1464 and 1492 under Sonni Ali Ber (d. 1492). The empire was centred on the middle reaches of the Niger River and controlled the major West African trade routes. It began to disintegrate because of factional in-fighting and in 1591 Timbuktu fell to Moroccan invaders. ➲ *80–81, 136–37*

SOUTH AFRICAN WARS Two wars between the Afrikaners (Boers) and the British in South Africa. The first (1880–81) arose from the British annexation of the Transvaal in 1877. Under Paul Kruger, the Transvaal regained autonomy, but further disputes, arising largely from the discovery of gold and diamonds, provoked the second, greater conflict (1899–1902), known to Afrikaners as the Second War of Freedom and to the British as the Boer War. It was a civil war between whites; black Africans played little part on either side. In 1900 the British gained the upper hand, defeating the Boer armies and capturing Bloemfontein and Pretoria. Boer commandos fought a determined guerrilla campaign but were forced to accept British rule in the peace treaty signed at Vereeniging (1902).
➲ *188–89, 206–7, 216–17*

SOPHOCLES (496–406 BC) Greek playwright. Of his 100 plays, only seven tragedies and part of a Satyr play remain. These include *Ajax*, *Antigone* (c. 442–441 BC), *Electra* (409 BC), *Oedipus Rex* (c. 429 BC) and *Oedipus at Colonus* (produced posthumously). His works introduced a third speaking actor and increased the members of the chorus from 12 to 15. ➲ *40–41*

SPANISH-AMERICAN WAR (1898) Conflict fought in the Caribbean and the Pacific, between Spain and the USA. The immediate cause was the explosion of the US battleship *Maine* at Havana. Fighting lasted ten weeks (April–July). The Spanish fleets in the Philippines and Cuba were destroyed. Spanish troops in Cuba surrendered after defeat at San Juan Hill, where Theodore Roosevelt led the Rough Riders. The USA also seized Guam and Wake Island, annexed Hawaii and the Philippines, and, at the Treaty of PARIS, forced Spain to cede Puerto Rico. ➲ *192–93, 226–27*

SPANISH CIVIL WAR (1936–39) Conflict that developed from a military rising against the republican government in Spain. The revolt began in Spanish Morocco, led by General FRANCO. It was supported by conservatives and reactionaries of many kinds, collectively known as the Nationalists and including the fascist Falange. The leftist Popular Front government was supported by republicans, socialists and a variety of ill-co-ordinated leftist groups, collectively known as Loyalists or Republicans. The Nationalists swiftly gained control of most of rural western Spain but not the main industrial regions. The war, fought with great savagery, became a serious international issue, representing the first major clash between the forces of the extreme right and the extreme left in Europe. Franco received extensive military support, especially aircraft, from the fascist dictators MUSSOLINI and HITLER. The Soviet Union provided more limited aid for the Republicans. Liberal and socialist sympathizers from countries such as Britain and France fought as volunteers for the Republicans, but their governments remained neutral. The Nationalists extended their control in 1937, while the Republicans were increasingly weakened by internal quarrels. In 1938, despite some Republican gains, the Nationalists reached the Mediterranean, splitting the Republican forces. The fall of Madrid, after a long siege, in March 1939 brought the war to an end, with Franco supreme. More than one million Spaniards had been killed. ➲ *230–31*

SPANISH SUCCESSION, WAR OF THE (1701–14) Last of the series of wars fought by European coalitions to contain the expansion of France under LOUIS XIV. It was precipitated by the death of the Spanish king, Charles II, without an heir. He willed his kingdom to the French PHILIP V of Anjou, Louis' grandson. England and the Netherlands supported the Austrian claimant to the Spanish throne, the Archduke (later Emperor) Charles. The ensuing war marked the emergence of Britain as a maritime and colonial power. The Spanish succession was settled by a compromise in the Peace of UTRECHT, with Philip attaining the Spanish throne on condition that he renounced any claim to France, and Britain and Austria receiving substantial territorial gains. Exhaustion of the participants, especially France, helped ensure peace in Europe until the outbreak of the War of the AUSTRIAN SUCCESSION in 1740. ➲ *174–75*

SPARTA City-state of ancient Greece. Founded after c. 1100 BC, Sparta conquered Laconia (southeast Peloponnese) by the 8th century BC and headed the Peloponnesian League against PERSIA in 480 BC. In the PELOPONNESIAN WAR (431–404 BC) it defeated its great rival, ATHENS, but was defeated by Thebes in 371 BC and failed to withstand the invasion of PHILIP II of MACEDONIA. In the 3rd century BC Sparta struggled against the Achaean League, subsequently joining it but coming under Roman dominance after 146 BC. The ancient city was destroyed by ALARIC and the GOTHS in AD 395. Sparta was famous for its remarkable social and military organization. ➲ *40–41*

STALIN, JOSEPH (1879–1953) (Joseph Vissarionovich Dzhugashvili) Leader of the Soviet Union (1924–53). He supported LENIN and the BOLSHEVIKS from 1903, adopting the name Stalin ("man of steel") while editing *Pravda*, the party newspaper. Exiled to Siberia between 1913 and 1917, he returned to join in the Russian Revolution in 1917 and became secretary of the central committee of the party in 1922. On Lenin's death in 1924, he achieved supreme power through his control of the party organization. He outmanoeuvred rivals such as TROTSKY and BUKHARIN and drove them from power. From 1929 he was virtually dictator. He enforced collectivization of agriculture and intensive industrialization, brutally suppressing all opposition, and, in the 1930s, exterminated all possible opponents in a series of purges of political and military leaders. During the Second World War Stalin controlled the armed forces and negotiated skilfully with the chief Allied leaders, Churchill and ROOSEVELT. After the war he reimposed severe repression and forced puppet communist governments on the states of Eastern Europe.➲ *222–23, 236–37, 242–43, 262–63, 274–75*

STALINGRAD, BATTLE OF (1942–43) Decisive conflict marking the failure of the German invasion of the Soviet Union during the Second World War. The city (now Volgograd) withstood a German siege from August 1942 to February 1943, when the German 6th Army under Friedrich von Paulus, surrounded with no prospect of relief, surrendered to the Russian General Zhukov. Total casualties at Stalingrad exceeded 1.5 million. ➲ *232–33*

STOLYPIN, PETER ARKADIEVICH (1862–1911) Russian statesman. Interior minister and prime minister (1906–11), after the RUSSIAN REVOLUTION OF 1905. He introduced agricultural reforms favouring the peasants, but was a conservative whose regime crushed political dissent and enforced changes to reduce the electorate, ensuring a more compliant *duma*. Negotiating between mutually hostile interests, he eventually antagonized the *duma*, fellow ministers and the emperor. In 1911 he was assassinated. ➲ *180–81*

STONEHENGE Circular group of prehistoric standing stones within a circular earthwork on Salisbury Plain, in southern England. The largest and most precisely constructed megalith in Europe, Stonehenge dates from the early 3rd millennium BC, although the main stones were erected c. 2000–1500 BC. The large standing bluestones were brought from southwest Wales c. 2100 BC. The significance of the structure is unknown. ➲ *20–21*

STUARTS Scottish royal house, which inherited the Scottish crown in 1371 and the English crown in 1603. The Stuarts, or Stewarts, descended from Alan, whose descendants held the hereditary office of steward in the royal household. Walter (d. 1326), the sixth steward, married a daughter of King Robert I, and their son, Robert II, became the first Stuart king (1371). The crown descended in the direct male line until the death of James V (1542), who was succeeded by his infant daughter, Mary, Queen of Scots. Her son, James VI, succeeded Elizabeth I of England in 1603 as JAMES I. His son, CHARLES I was executed in 1649 following the ENGLISH CIVIL WAR, but the dynasty was restored with the Restoration of Charles II in 1660. His brother, James II, lost the throne in the Glorious Revolution (1688) and was replaced by the joint monarchy of William III and Mary II, James' daughter. On the death of James's second daughter, who died in 1714 without an heir, the House of Hanover succeeded. The male descendants of James II made several unsuccessful attempts to regain the throne, culminating in the Jacobite rebellion of 1745, during which Prince Charles briefly won Scotland. ➲ *156–57*

SUCRE, ANTONIO JOSÉ DE (1795–1830) South American revolutionary leader and first president of Bolivia (1826–28). He joined the fight for independence in 1811 and played a key role in the liberation of Ecuador, Peru and Bolivia, winning the final, decisive battle at Ayacucho (1824). With BOLÍVAR 's support he was elected first constitutional president of what became Bolivia, but local opposition forced his resignation. He was assassinated while working to preserve the unity of Colombia. ➲ *190–91*

SUHARTO, RADEN (1921–) Indonesian military and political figure. Suharto seized power from President Sukarno, averting an alleged communist coup, in 1966. He was formally elected president in 1968 and subsequently re-elected at regular intervals. Under military rule, Indonesia experienced rapid economic development, but Suharto's autocratic rule was denounced by civil-rights groups. He was forced to resign from office in 1998. ➲ *250–51*

SULEIMAN I (THE MAGNIFICENT) (1494–1566) Ottoman sultan (1520–66). He succeeded his father, SELIM I. Suleiman captured Rhodes from the KNIGHTS HOSPITALLERS and launched a series of campaigns against the Austrian Habsburgs, defeating the Hungarians at the Battle of MOHÁCS in 1526, and subsequently controlling most of Hungary. His troops besieged Vienna in 1529. Suleiman's admiral, Barbarossa, created a navy that dominated the Mediterranean and ensured Ottoman control of much of the North African coastal region. In the east, Suleiman won victories against the Safavids of Persia and conquered Mesopotamia. ➲ *142–43*

SUMERIA World's first civilization, dating from before 3000 BC, in southern MESOPOTAMIA. The Sumerians are credited with inventing cuneiform writing, the first basic socio-political institutions, and a money-based economy. Major cities were UR, Kish and Lagash. During the 3rd millennium Sumeria developed into an imperial power. In c. 2340 BC, the Semitic peoples of Akkadia conquered Mesopotamia, and by c.1950 BC the ancient civilization had disintegrated. ➲ *28–29*

SUNNI Traditionalist orthodox branch of ISLAM, the followers of which are called Ahl as-Sunnah (People of the Path). It is followed by 90 per cent of Muslims. Sunnis accept the Hadith, the body of orthodox teachings based on MUHAMMAD's spoken words outside the Koran. The Sunni differ from the SHIITE sect, in that they accept the first four caliphs (religious leaders) as the true successors of Muhammad. ➲ *142–43, 260–61*

SUN YAT-SEN (1866–1925) Chinese nationalist leader, first president of the Chinese Republic (1912). In exile (1895–1911), he adopted his "Three Principles of the People": nationalism, democracy and prosperity. After the revolution of 1911 he became provisional president, but soon resigned in favour of the militarily powerful Yüan Shikai. As Yüan's leadership became increasingly autocratic, Sun lent his support to the opposition KUOMINTANG (Nationalist Party). ➲ *198–99, 224–25*

SVEIN FORKBEARD (d. 1014) King of Denmark (987–1014), son of Harald Bluetooth. His attack on England in 994 forced King Ethelred II to pay tribute and, after his invasion in 1013, Ethelred fled to Normandy and Svein became king. Svein died before his coronation and his son CNUT II eventually succeeded him. ➲ *78–79*

T

TAIPING REBELLION (1851–64) Revolt in China against the Manchu Qing dynasty, led by a Hakka fanatic, Hong Xiuquan. The fighting laid waste to some of China's most fertile land and resulted in more than 20 million casualties. The Qing never fully recovered their ability to govern all of China. ➲ *198–99*

TAMIL TIGERS Militant Tamil group in Sri Lanka that seeks independence from the Sinhalese majority. Located mainly in the north and east of the island, the three million Tamils are Hindus, unlike the Buddhist Sinhalese. In the 1980s the Tamil Tigers embarked on a campaign of civil disobedience and terrorism. Autonomy for the Tamils was agreed by India and Sri Lanka in 1986, but no date was fixed. The Indian army was sent to the island in 1987, but withdrew in 1990, having failed to stop the violence. ➲ *248–49*

TARTARS See TATARS

TASMAN, ABEL JANSZOON (1603–59) Dutch maritime explorer who made many discoveries in the Pacific. On his voyage of 1642–43 he discovered Tasmania. He reached New Zealand, but was attacked by Maoris in Golden Bay. He landed on Tonga and Fiji and sailed along the northern coast of New Ireland. Although he circumnavigated Australia, he never sighted the mainland coast. ➲ *202–3*

TATARS (Tartars) Turkic-speaking people of Central Asia. In medieval Europe the name Tatar was given to many different Asiatic invaders. True Tatars originated in eastern Siberia. They were converted to ISLAM in the 14th century, and became divided into two groups: one in southern Siberia, who came under Russian rule, the other in the Crimea, which was part of the Ottoman Empire until annexed by Russia in 1783. ➲ *98–99, 138–39, 148–49, 236–37*

TEOTIHUACAN Ancient city of Mexico, about 48 kilometres (30miles) north of Mexico City. It flourished between c. 100 BC and c. AD 700 and contained huge and impressive buildings, notably the PYRAMID of the Sun. At its greatest, around AD 600, the city may have housed as many as 200,000 people and was the centre of a considerable empire. ➲ *32–33*

TET OFFENSIVE Campaign in the VIETNAM WAR. North Vietnamese and Vietcong troops launched attacks on numerous towns and cities of South Vietnam early in 1968. Although of little strategic effect, the offensive discredited current US military reports that victory over North Vietnam was imminent. ➲ *250–51*

TEUTONIC KNIGHTS German military and religious order, founded in 1190. Its members, of aristocratic class, took monastic vows of poverty and chastity. During the 13th century the knights waged war on non-Christian peoples, particularly those of Prussia, whom they defeated, annexing their land. They were defeated in 1242 by Alexander Nevski, Grand Dule of Novgorod, and, in 1410, by the Poles and Lithuanians at Tannenberg. ➲ *94–95, 150–51*

THEMISTOCLES (528–460) Athenian statesman. He created the Athenian navy and secured the crucial victory over the Persian fleet at SALAMIS in 480 BC. He was subsequently accused of conspiring with the Persians and left ATHENS for Argos. A strong opponent of SPARTA, his associations with the former Spartan ruler, Pausanias, led to his condemnation as a traitor, but he escaped to PERSIA, where he was given lands by Artaxerxes I. ➲ *40–41*

THEODERIC THE GREAT (THE AMAL) (454–526) King of the OSTROGOTHS and ruler of Italy. He drove King Odoacer from Italy in 488 and attempted to recreate the Western Roman Empire with himself as emperor. His capital was Ravenna. Religious differences and political rivalries frustrated his empire-building, and after his death his kingdom was destroyed and absorbed into the Byzantine Empire by JUSTINIAN I. ➲ *56–57*

THIRTY YEARS WAR (1618–48) Conflict fought mainly in Germany, arising out of religious differences and developing into a struggle for power in Europe. It began with a Protestant revolt in BOHEMIA against the Habsburg emperor, Ferdinand II. Both sides sought allies and the war spread to much of Europe. The Habsburg generals Tilly and Wallenstein registered early victories and drove the Protestant champion, Christian IV of Denmark, out of the war (1629). A greater champion appeared in GUSTAV II ADOLF of Sweden, who waged a series of victorious campaigns before being killed in 1632. In 1635 France, fearing Habsburg dominance, declared war on Habsburg Spain. Negotiations for peace were not successful until the Peace of WESTPHALIA was concluded in 1648. War between France and Spain continued until the Peace of the PYRENEES (1659), and other associated conflicts continued for several years. The chief loser in the war, apart from the German peasants, was the emperor, Ferdinand III, who lost control of Germany. Sweden was established as the dominant state in northern Europe, while France replaced Spain as the greatest European power. ➲ *158–59*

THUTMOSE Name of four kings of the 18th dynasty in ancient EGYPT. Thutmose I (r. c. 1504–c. 1492 BC) extended his kingdom southward into NUBIA, and campaigned successfully in the Near East. He was succeeded by his son, Thutmose II (r. c. 1492–c. 1479 BC), who married his half-sister, Hatshepsut. She ruled as regent for his son, Thutmose III (r. c. 1479–c. 1425 BC). Thutmose III expanded the kingdom to its greatest extent, defeating the Mitanni kingdom on the River Euphrates and pushing the southern frontier beyond the fourth cataract of the Nile. His grandson, Thutmose IV (r. c. 1425–c. 1416 BC), continued an expansive policy but also sought to strengthen the empire by peaceful means, marrying a Mitanni princess. ➲ *36–37*

TIBERIUS (42 BC–AD 37) (Tiberius Julius Caesar Augustus) Roman emperor (AD 14–37). He was the stepson of AUGUSTUS, who adopted him as his heir in AD 4. Initially, his administration was just and moderate, but he became increasingly fearful of conspiracy and had many people executed for alleged treason. He left Rome and spent his last years in seclusion on Capri. ➲ *54–55*

TIBET (Xizang) Autonomous region in southwest China. Tibet flourished as an independent kingdom in the 7th century, and in the 8th century Padmasambhava developed the principles of Mahayana BUDDHISM and founded Lamaism. In 1206 CHINGGIS KHAN conquered the region, and it remained under nominal Mongol rule until 1720, when the Chinese Qing dynasty claimed sovereignty. At the close of the 19th century the Tibetan areas of Ladakh and Sikkim were incorporated into British India, and in 1906 Britain recognized Chinese sovereignty over Tibet. In 1912 the fall of the Qing dynasty prompted the Tibetans to reassert their independence. China, however, maintained its right to govern, and in 1950 the new communist regime sent its forces to invade. In 1951 Tibet was declared an autonomous region of China, nominally governed by the Dalai Lama. The Chinese government began a series of repressive measures principally targeting the Buddhist monasteries. In March 1959 a full-scale revolt against Chinese rule was suppressed by the Chinese People's Liberation Army. The Dalai Lama fled to northern India and established a government-in-exile at Dharamsala. The CULTURAL REVOLUTION banned religious practice and 4,000 monasteries were destroyed. Many thousands of Tibetans were forced into exile by the brutality of the communist regime. Despite the restoration of some of the desecrated monasteries and the reinstatement of Tibetan as an official language, human rights violations continued. Pro-independence rallies in 1987–89 were violently suppressed by the Chinese army. ➲ *254–55*

TIMURID DYNASTY (1370–1506) Turkicized Mongol family which ruled Transoxania, approximating to modern Uzbekistan. It was founded by TIMUR-LENG who became sole ruler in 1369 and made his capital at Samarqand. His campaigns extended his empire, which was largely to die with him, into Iraq, Syria, Caucasia and India. In the following century, the Timurid kingdom was twice divided and came to an end when the Uzbeks expelled BABUR, who went on to found the Mughal Empire in India.➲ *98–99*

TIMUR-LENG [Tamerlane] (1336–1405) Mongol conqueror, b. Uzbekistan. He claimed descent from CHINGGIS KHAN. By 1369 Timur-leng had conquered present-day Turkestan and established Samarqand as his capital. He extended his conquests to the region of the GOLDEN HORDE between the Caspian and Black seas. In 1398 he invaded northwest India and defeated the Delhi Sultanate, sacking Delhi. He then turned toward the MAMLUK empire, capturing Syria and Damascus. In 1402 he captured the Ottoman sultan Beyazid I at Angora. His death at the head of a 200,000-strong invasion force of China enabled the reopening of the SILK ROAD. His vast empire was divided among the TIMURID DYNASTY. ➲ *98–99*

TITO, JOSIP BROZ (1892–1980) Yugoslav statesman. As a Croatian soldier in the Austro-Hungarian army, he was captured by the Russians in 1915 but released by the BOLSHEVIKS in 1917. He helped to organize the Yugoslav Communist Party and adopted the name Tito in 1934. He led the successful campaign of the Partisans against the Germans in Yugoslavia during the Second World War. In 1945 he established a communist government, holding the office of prime minister from 1945 to 1953 and thereafter president, although he was virtually a dictator. Soviet efforts to control Yugoslavia led to a split between the two countries in 1948. At home, Tito sought to balance the deep ethnic and religious divisions in Yugoslavia and to develop an economic model of communist "self-management". Abroad, he became an influential leader of the Non-Aligned Movement. As later events confirmed, his greatest achievement was to hold the Yugoslavian federation together. ➲ *236–37, 264–65*

TOLTECS Ancient Native American civilization, whose capital was Tollan (Tula), Mexico. The Toltecs were the dominant people in the region from AD 900 to 1200. Their architecture is characterized by PYRAMID building. Although theirs was considered a polytheistic culture, images of the god Quetzalcoatl ("The Feathered Serpent") predominate. ➲ *84–85*

TORDESILLAS, TREATY OF Agreement of 1494 to divide the newly discovered Atlantic lands into Portuguese and Spanish spheres. The dividing line ran from pole to pole, through a point 100 leagues, c. 500 kilometres (300 miles), west of the Cape Verde Islands, with the area to the west belonging to Spain. ➲ *116–17, 122–23*

TRAJAN (AD 53–117) Roman emperor (AD 98–117), born in Seville, Spain. He distinguished himself as a general and administrator and was made junior co-emperor by Nerva in AD 97. With army support, he became emperor on Nerva's death. He conducted major compaigns in DACIA in 101–102 and 105–106, and against the people of PARTHIA in 113–117, enlarging the Roman Empire to its greatest extent. ➲ *54–55*

TRANS-SIBERIAN RAILWAY Russian railway from Moscow to Vladivostok. The world's longest railway, the major part, east from Chelyabinsk, was built between 1891 and 1904, giving Russia access to the Pacific via a link with the Chinese Eastern Railway in Manchuria. After the RUSSO-JAPANESE WAR of 1904–5, however, the

Russians, fearing a Japanese takeover of Manchuria, embarked upon the Amur Railway, which was routed to the north of Manchuria and was completed in 1916. The total length of the Trans-Siberian Railway is around 9,000 kilometres (5,750miles). ❍ *180–81*

TRENT, COUNCIL OF (1545–63) Nineteenth ecumenical council of the Roman CATHOLIC CHURCH, which provided the main impetus of the Counter Reformation in Europe. It met at Trent, in northern Italy, in three sessions under three popes (Paul III, Julius III and Pius IV). It clarified Catholic doctrine and refused concessions to the Protestants, while also instituting reform of many of the abuses that had provoked the Reformation. ❍ *154–55*

TRIANON, TREATY OF (1920) Agreement between victorious Entente Powers and Hungary, signed in the Grand Trianon Palace, Versailles, France. Under the terms of the treaty Hungary lost large areas of its territory and half of its population to surrounding countries. Limitations were placed on the size of its army (35,000 men) and reparations were demanded in the form of livestock and coal. Financial reparations were also to be made. ❍ *220–21*

TRIPLE ALLIANCE Name given to several alliances involving three states. They include the anti-French alliance of Britain, the Netherlands and Sweden of 1668, and the alliance of Britain, France and the Netherlands of 1717, directed against Spanish ambitions in Italy. The most recent was the Triple Alliance of 1882, when Italy joined the Dual Alliance of Austria-Hungary and Germany. In South America, Argentina, Brazil and Uruguay formed a triple alliance in the war against Paraguay (1865–70). ❍ *216–17*

TRIPLE ENTENTE Name given to the alliance of Britain, France and Russia before the First World War. It developed from the Franco-Russian Alliance (1894) formed to counterbalance the threat posed by the Triple Alliance of Germany, Austria and Italy. In 1904 Britain became allied with France in the Entente Cordiale, and the Anglo-Russian Convention of 1907 completed the Triple Entente. ❍ *216–17, 218–19*

TROTSKY, LEON (1879–1940) Russian revolutionary leader and theoretician, born Lev Davidovich Bronstein. A Marxist revolutionary from 1897, he headed the workers' soviet in St Petersburg in the Russian Revolution of 1905. Arrested, he escaped abroad and embarked on the work that made him, with LENIN, the leading architect of the Russian Revolution of 1917. Trotsky returned to Russia after the March revolution (1917), and joined the BOLSHEVIKS. As chairman of the Petrograd Soviet, he set up the Military Revolutionary Committee to seize power, ostensibly for the Soviet, actually for the Bolsheviks. After the Bolshevik success, he negotiated the Treaty of BREST-LITOVSK, withdrawing Russia from the First World War. As commissar of war (1918–25), he created the Red Army, which won the civil war and made the Bolshevik revolution safe. However, he criticized the growth of bureaucracy in the party, the lack of democracy, and failure to expand industrialization. He disapproved of Lenin's dictatorial tendencies in power. He fiercely objected to STALIN's adoption of a policy of "socialism in one country", rather than the world revolution in which Trotsky believed. He was driven from power, from the party and eventually from the country. In exile he continued to write prolifically on many subjects. His ideas, though rejected in the Soviet Union, were extremely influential internationally, especially in Third World countries. In 1936 he settled in Mexico, where he was assassinated by a Stalinist agent. ❍ *222–23*

TRUMAN, HARRY S. (1884–1972) 33rd US president (1945–53). From a farming and small-business background in Missouri, he entered politics in the 1920s and won election to the Senate in 1934. He was vice-presidential candidate with Franklin D. ROOSEVELT in 1944. He became president on Roosevelt's death, and was faced with many difficulties abroad. He approved the use of the atomic bomb to force Japanese surrender (1945), ending the Second World War, and adopted a robust policy towards the Soviet Union during the Cold War that followed. He approved the MARSHALL PLAN

(1947) and the creation of the NORTH ATLANTIC TREATY ORGANIZATION (NATO) (1949). Lacking Roosevelt's charisma, Truman was expected to lose the election of 1948, but won narrowly. In the KOREAN WAR he was forced to dismiss the US commander, General MACARTHUR. He declined to run for a second full term in 1952. ❍ *242–43, 244–45*

TRUMAN DOCTRINE Principle of US foreign policy under President TRUMAN. It promised US support for any democratic country threatened by foreign domination. In practice, application of the Truman Doctrine was limited. The USA did not act against communist takeovers in Eastern Europe, although it did resist the invasion of South Korea by North Korean forces in the KOREAN WAR. ❍ *244–45*

TUTANKHAMUN (active c. 1350 BC) Egyptian pharaoh of the New Kingdom's 18th Dynasty (1550–1307 BC). The revolutionary changes that had been introduced by his predecessor, Akhnaten, were reversed during his reign. The capital was re-established at Thebes (Luxor) and worship of Amun reinstated. Tutankhamun's fame is due to the discovery of his tomb by Howard Carter in 1922. The only royal tomb of ancient EGYPT not completely stripped by robbers, it contained magnificent treasures. ❍ *36–37*

TYRE Historic city on the coast of modern Lebanon. Built on an island, it was a major commercial port of ancient PHOENICIA. It supplied both craftsmen and raw materials, especially cedarwood, for the building of the temple in JERUSALEM in the 10th century BC, and established colonies, including CARTHAGE, around the eastern Mediterranean. Tyre was never successfully besieged until ALEXANDER THE GREAT built a causeway linking the island to the mainland in 332 BC. Ruled by successive empires, including the Romans, it was captured by the Arabs in AD 638 and destroyed by the MAMLUKS in 1291. It never regained its former eminence. ❍ *38–39*

UMAYYADS Dynasty of Arabian Muslim caliphs (661–750). From their capital at Damascus, the Umayyads ruled what was basically an Arab empire, which stretched from Spain to India. They made little effort to convert conquered peoples to ISLAM, but there was great cultural exchange, and Arabic became established as the language of Islam. They were overthrown by the ABBASIDS. ❍ *68–69*

UN CONFERENCE ON ENVIRONMENT AND DEVELOPMENT (1992) *See* EARTH SUMMIT

UNION, ACTS OF Series of acts uniting England with Wales (1536) and Scotland (1707), and Britain with Ireland (1800). In addition, the 1840 Act of Union united French-speaking Lower Canada and English-speaking Upper Canada. The Welsh acts incorporated Wales within the kingdom of England, provided Welsh parliamentary representation and made English the official language. The Scottish act united the kingdoms of England and Scotland forming Great Britain. Scotland retained its own legal system and Presbyterian Church. Under the Irish act, the Irish legislature was abolished and Ireland was given 32 peers and 100 seats in the British parliament. The established churches of the two countries were united, and free trade was introduced. The Canadian act led to the establishment of a parliament for the province. ❍ *188–89*

UNITED NATIONS International organization set up to enable countries to work together for peace and mutual development. It was established by a charter signed in San Francisco in June 1945 by 50 countries. In 1998 the UN had 185 members, essentially all the world's sovereign states except for North and South Korea and Switzerland. ❍ *248–49, 244–45, 254–55, 258–59, 260–61, 266–67, 268–69, 278–79*

UNITED NATIONS MONETARY AND FINANCIAL CONFERENCE (Bretton Woods Conference) It met at Bretton Woods, New Hampshire, in July 1944. It was

summoned on the initiative of US President Franklin ROOSEVELT to establish a system of international monetary co-operation and prevent severe financial crises such as that of 1929, which had precipitated the Great Depression. Representatives of 44 countries agreed to establish the INTERNATIONAL MONETARY FUND (IMF), to provide cash reserves for member states faced by deficits in their balance of payments, and the International Bank for Reconstruction and Development, or WORLD BANK, to provide credit to states requiring financial investment in major economic projects. ❍ *272–73*

UR (Ur of the Chaldees) Ancient city of MESOPOTAMIA that flourished in the 3rd millennium BC and, during the Akkadian period initiated by SARGON I, witnessed the integration of Semitic and Sumerian cultures. King Ur-Nammu (r. 2113–2096 BC) began building the great ziggurat. In c. 2004 BC much of the city was detroyed by the invading Elamites. In the 6th century BC NEBUCHADNEZZAR briefly restored Ur as a centre of Mesopotamian civilization, but by the 5th century BC it had fallen into terminal decline. ❍ *28–29*

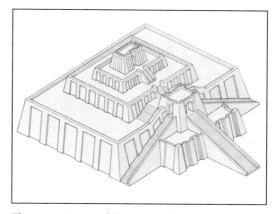

The great ziggurat of the moon god at Ur was probably begun in the 3rd millennium BC. Additions were made in the 6th century BC.

UTRECHT, PEACE OF (1713–14) Series of treaties that ended the War of the SPANISH SUCCESSION. It confirmed the BOURBON King PHILIP V on the Spanish throne on condition that he renounced any claim to the throne of France. Austria received the Spanish Netherlands and extensive Italian territories; Britain gained Gibraltar, Minorca and an area of eastern Canada. ❍ *156–57, 174–75*

UZBEKS Turkic-speaking people, originally of Persian culture, who form 71 per cent of the population of the Republic of Uzbekistan. They took their name from Uzbeg Khan (d. 1340), a chief of the GOLDEN HORDE. By the end of the 16th century, the Uzbeks had extended their rule to parts of Persia, Afghanistan and Chinese Turkestan. Their empire was never united and in the 19th century its various states were absorbed by Russia. ❍ *98–99*

VALOIS DYNASTY Royal house of France (1328–1589). Valois kings survived the HUNDRED YEARS WAR (1337–1453) and challenges by Burgundian and Armegnac rivals and consolidated royal strength over feudal lords. They established the Crown's sole right to tax and to wage war and extended parliaments throughout France. Louis XI (r. 1461–83) is considered the founder of French royal absolutism. The direct Valois line ended with the death of CHARLES VIII in 1498, when the dynasty was continued by Louis XII (Valois-Orléans) and, after his death in 1515, by FRANCIS I and the Valois-Angouleme line. With the death in 1589 of Henry III, the house of BOURBON, descending from a younger son of Louis IX, ascended the throne in the person of Henry IV. ❍ *106–107, 152–53*

VANDALS Germanic tribe who attacked the Roman Empire in the 5th century AD. They looted Roman

GAUL and invaded Spain in 409. Defeated by the GOTHS, they moved farther south and invaded North Africa in 429, establishing a kingdom from which they controlled the western Mediterranean. They sacked ROME in 455. The Vandal kingdom was destroyed by the Byzantine general Belisarius in 533–534. ➲ *56–57*

VASCO DA GAMA *See* GAMA, VASCO DA

VAUBAN, SEBASTIEN LE PRESTRE DE (1633–1707) French military engineer who revolutionized the art of siege warfare and defensive fortifications, working on 160 fortresses during his lifetime. He joined the army in 1653 and took part in all LOUIS XIV's wars. Among his achievements were the siege of Luxembourg in 1684 and the subsequent rebuilding of the defences of the city. During peacetime he inspected the French frontier fortifications and introduced radical new designs. In 1703 Vauban was appointed as a marshall of France in recognition of his service to the king. ➲ *158–59*

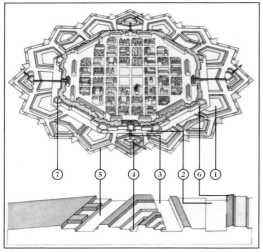

Among the fortifications designed by Vauban was Neuf Brisach fortress in Alsace. Its defences included detached bastions (1), four moats (2, 3, 4 and 5) and an inner rampart (6) commanded by angled "bastion towers" (7).

VERDUN, BATTLE OF Campaign of the First World War, February–December 1916. A German offensive made initial advances, but was checked by the French under General Pétain. After renewed German assaults, the Allied offensive on the Somme drew off German troops and the French regained the lost territory. Total casualties were estimated at one million. ➲ *218–19*

VERDUN, TREATY OF (843) Treaty dividing the CAROLINGIAN empire between the three surviving sons of LOUIS I (THE PIOUS). CHARLES II (THE BALD) received western France, Louis (the German) received Germany east of the Rhine, and Lothair I the central region from the Low Countries to Italy, as well as the imperial title. ➲ *74–75, 146–47*

VICHY GOVERNMENT During the Second World War, regime of southern France after the defeat by Germany in June 1940. Its capital was the town of Vichy, and it held authority over French overseas possessions as well as the unoccupied part of France. After German forces occupied Vichy France in November 1942, it became little more than a puppet government. ➲ *232–33*

VICTOR EMMANUEL II (1820–78) King of Italy (1861–78). He succeeded his father, CHARLES ALBERT, as king of Piedmont–Sardinia in 1849. Guided by his minister, CAVOUR, he strengthened his kingdom, formed a French alliance, and defeated Austria in 1859–61. In 1861 he assumed the title of King of Italy, making Rome his new capital in 1870. ➲ *176–77*

VIENNA, CONGRESS OF (1814–15) European conference that settled international affairs after the Napoleonic Wars. It attempted to restore the Europe of pre-1789, and it disappointed the hopes of nationalists and liberals. Among steps to prevent future European wars, it established the German Confederation. ➲ *166–67, 172–73, 174–75, 176–77*

VIETCONG Nickname for the Vietnamese communist guerrillas who fought against the US-supported regime in South Vietnam during the VIETNAM WAR. After earlier, isolated revolts against the government of Ngo Dinh Diem, the movement was unified as the National Liberation Front (NLF) in 1960, modelled on the VIETMINH. ➲ *250–51*

VIETMINH Vietnamese organization that fought for independence from the French (1946–54). It resisted the Japanese occupation of French INDOCHINA during the Second World War. After the war, when the French refused to recognize it as a provisional government, it began operations against the colonial forces. The French were forced to withdraw after their defeat at Dien Bien Phu (1954). ➲ *250–51, 254–55*

VIETNAM WAR (1959–75) Conflict between US-backed South Vietnam and insurgents known as the VIETCONG, who had the support of communist North Vietnam. It arose following the partition of Vietnam in 1954 and was fuelled by the USA's fear of the spread of communism. The USA received token support from its allies in the Pacific region, and North Vietnam was supplied by China and the Soviet Union. As fighting intensified, US troops were committed in increasing numbers: by 1968 there were more than 500,000. In spite of US technological superiority and command of the air, military stalemate ensued. The instability of the unrepresentative South Vietnamese government, heavy casualties and daily television coverage made the war highly unpopular in the USA. A peace agreement, negotiated by Henry Kissinger and Le Duc Tho, was signed in Paris in 1973; it provided for the withdrawal of US troops. In 1975 South Vietnam was overrun by North Vietnamese forces, and the country was united under communist rule. ➲ *242–43, 244–45*

VILLA, "PANCHO" (Francisco) (1877–1923) Mexican revolutionary leader. He began as an outlaw and joined the forces of Francisco Madero in 1909. He sided with VENUSTIANO CARRANZA for some time, but later supported EMILIANO ZAPATA. Angered by US recognition of Carranza's government, Villa murdered US citizens in northern Mexico and New Mexico. In 1920 he was granted a pardon in return for agreeing to retire from politics. He was assassinated in 1923. ➲ *226–27*

VISIGOTHS (West GOTHS) An ancient Germanic people. In 376 the Visigoths were driven across the Danube by the Huns, into the Roman Empire. At the end of the 4th century, they began to revolt against the Romans, defeating the emperor Valens at Adrianople in 378. After settling until 395 in the Balkans, they moved into Greece and Italy where, under the leadership of ALARIC, they sacked ROME in 410. Under Alaric's successor, Ataulphus, they settled in GAUL and Spain, where they went on to establish kingdoms. ➲ *56–57*

WAITANGI, TREATY OF (1840) Pact between Britain and several New Zealand Maori tribes. The agreement protected and provided rights for Maoris, guaranteeing them possession of certain tracts of land, while permitting Britain formally to annex the islands and purchase other land areas. ➲ *202–3*

WANG MANG (33 BC–AD 23) Emperor of China. A usurper, he overthrew the Han dynasty and proclaimed the Xin (New) dynasty in AD 6. Opposition from landowners and officials forced him to withdraw the reforms and his one-emperor dynasty, which divides the Early Han from the Later Han, ended with his assassination. ➲ *48–49*

WAR OF 1812 (1812–15) Conflict between the USA and Britain. The main source of friction was British maritime policy during the Napoleonic Wars, in which US merchant ships were intercepted. Difficulties on the border with Canada also contributed. The US invasion of Canada failed, although Britain suffered defeats on Lake Erie and on the western frontier. The end of the Napoleonic Wars (1814) freed more British forces.

They imposed a naval blockade and captured Washington, burning the White House. A US naval victory on Lake Champlain ended the British threat to New York, and New Orleans was saved by the victory of Andrew Jackson, after peace had been agreed, unknown to the combatants, in Europe. ➲ *182–83, 188–89*

WARSAW PACT (1955) Agreement creating the Warsaw Treaty Organization, a defensive alliance of the Soviet Union and its communist allies in Eastern Europe. It was founded after the admission of West Germany to the NORTH ATLANTIC TREATY ORGANIZATION (NATO), the equivalent organization of Western Europe. Its headquarters were in Moscow and it was effectively controlled by the Soviet Union. Attempts to withdraw by Hungary in 1956 and Czechoslovakia in 1968 were forcibly denied. It was officially dissolved in 1991 after the collapse of the Soviet Union. ➲ *236–37, 244–45*

WATERLOO, BATTLE OF (1815) Final engagement of the Napoleonic Wars, fought about 20 kilometres (12 miles) from Brussels, Belgium. Allied forces were commanded by the Duke of Wellington, against NAPOLEON's slightly larger French forces. Fighting was even until the Prussians arrived to overwhelm the French flank, whereupon Wellington broke through the centre. The battle ended Napoleon's Hundred Days and resulted in his second and final abdication. ➲ *166–67*

WATT, JAMES (1736–1819) Scottish engineer. In 1765 Watt invented the condensing steam engine. In 1782 he invented the double-acting engine, in which steam pressure acted alternately on each side of a piston. With Matthew Boulton, Watt coined the term "horse-power". The unit of power is called the Watt in his honour. ➲ *134–35, 170–71*

WESTPHALIA, PEACE OF (1648) Series of treaties among the states involved in the THIRTY YEARS WAR. Peace negotiations began in 1642, and meetings were held in cities of Westphalia, leading to the final settlement. In Germany, the peace established the virtual autonomy of the German states, which diminished the authority of the Holy Roman Emperor. In Europe, the peace established the ascendancy of France, the power of Sweden in northern Europe, and the decline of Spain. ➲ *146–47, 150–51, 152–53, 154–55, 156–57*

WILLIAM I (THE CONQUEROR) (1027–87) King of England (1066–87) and Duke of Normandy. He succeeded to the dukedom in 1035. Supported initially by Henry I of France, he consolidated his position in Normandy against hostile neighbours and acquired Maine by conquest in 1063. On the death of Edward the Confessor, he claimed the throne, having allegedly gained the agreement of King Harold, in 1064. More importantly, he defeated and killed Harold at the Battle of HASTINGS (1066) and subsequently enforced his rule over the whole kingdom. He rewarded his followers by grants of land, eventually replacing almost the entire feudal ruling class, and intimidated potential rebels by rapid construction of castles. He invaded Scotland (1072), extracting an oath of loyalty from Malcolm III Canmore, and Wales (1081), although he spent much of the reign in France. He ordered the famous survey known as the *Domesday Book* (1086). ➲ *92–93*

WILLIAM, PRINCE OF ORANGE (1533–84) Governor of Holland under PHILIP II of Spain from 1559 and leader of the 1572 revolt against Spanish rule in the Netherlands. He became a Protestant and led the northern provinces, which formed the United Provinces of the Netherlands in 1579. He was assassinated by a Spanish agent in 1584. ➲ *156–57*

WILLIAM II (1626–50) Prince of Orange, stadtholder of several Dutch provinces (1647–50). He married Mary, daughter of CHARLES I of England in 1650. His dynastic ambitions were scotched by early death, and he was succeeded by his son, the future William III of England. ➲ *156–57*

WILSON, (THOMAS) WOODROW (1856–1924) 28th US president (1913–21). As governor of New Jersey (1910–12), he gained a reputation as a progressive, liberal Democrat. As president, his "New Freedom" reforms included the establishment of the Federal

Reserve System in 1913. Several amendments to the Constitution were introduced, including prohibition and the extension of the franchise to women. When the MEXICAN REVOLUTION brought instability to the southern border of the States, Wilson ordered General Pershing's intervention. Despite Wilson's efforts to maintain US neutrality at the start of the First World War, the failure of diplomacy and continuing attacks on US shipping led Wilson to declare war on Germany in April 1917. His FOURTEEN POINTS represented US war aims and became the basis of the peace negotiations at Versailles (1919). He was forced to accept some compromises in the final settlement, which included the formation of the LEAGUE OF NATIONS. The Republican-dominated Senate refused to ratify the Versailles Treaty, and Wilson suffered a stroke in 1919 that permanently reduced his effectiveness. ➲ *220–21*

WOLFE, JAMES (1727–59) British general. He commanded the force that captured Quebec by scaling the cliffs above the St Lawrence River and defeating the French in 1759 on the Plains of Abraham. This feat, the importance of the victory, (resulting in Britain's acquisition of Canada), and Wolfe's death in action made him an almost legendary hero. ➲ *188–89*

WORLD BANK Popular name for the International Bank for Reconstruction and Development (IBRD). It is an intergovernmental organization established in 1944, and has been a specialized agency of the UNITED NATIONS since 1945, based in Washington D.C. Its role is to make long-term loans to member governments to aid their economic development. The major part of the Bank's resources are derived from the world's capital markets. ➲ *272–73*

WORLD HEALTH ORGANIZATION (WHO) Intergovernmental, specialized agency of the UNITED NATIONS. Founded in 1948, it collects and distributes medical and scientific information and promotes the establishment of international standards for drugs and vaccines. WHO has made major contributions to the prevention of diseases such as malaria, polio, leprosy and tuberculosis, and the eradication of smallpox. ➲ *276–77*

WYCLIFFE, JOHN (1330–84) English religious reformer. Under the patronage of John of Gaunt, he attacked corrupt practices in the Church and the authority of the Pope, condemning in particular the Church's landed wealth. His criticism became increasingly radical, questioning the authority of the Pope and insisting on the primacy of scripture, but he escaped condemnation until after his death. His ideas were continued by the LOLLARDS in England and influenced Jan HUS in BOHEMIA. ➲ *106–7*

X

XERXES I (519–465 BC) King of PERSIA (486–465 BC). Succeeding his father, DARIUS I, he regained EGYPT and crushed a rebellion in BABYLON before launching his invasion of Greece in 480 BC. After his fleet was destroyed at the Battle of SALAMIS he retired, and the defeat of the Persian army in Greece at the Battle of PLATAEA in 481 BC ended his plans for conquest. He was later assassinated by one of his own men. ➲ *40–41*

XIONGNU Ancient Central Asian nomadic people. Their repeated invasions of northern Chinese kingdoms from the 5th century BC led to the building of defenses, later to become the GREAT WALL OF CHINA. In the 3rd century BC they formed a tribal league and became a serious threat to China. Their territory included much of Central Asia, from the Pamirs to western Manchuria, from where they periodically raided China. During the QIN period (221–206 BC) the Great Wall was completed, but Xiongnu invasions continued. In the 2nd century BC the Han emperor began an agressive policy against the raiders, which led eventually to the expansion of the Chinese Empire into southern Manchuria and Choson. The Xiongnu split into eastern and western hordes in 51 BC and, although their raids continued into the 4th century AD, there is no record of them after the 5th century. ➲ *40–41, 50–51, 52–53*

Y

YELTSIN, BORIS NIKOLAYEVICH (1931–) First democratically elected president of the Russian Federation (1991–). He was Communist Party leader in Ekaterinburg (Sverdlovsk) before joining GORBACHEV'S government in 1985, and becoming party chief in Moscow. His criticism of the slow pace of *perestroika* led to demotion in 1987, but Yeltsin's popularity saw him elected as president of the Russian Republic in 1990. His prompt denunciation of the attempted coup against Gorbachev in August 1991 established his political supremacy. Elected president of the Russian Federation, Yeltsin presided over the dissolution of the Soviet Union and the end of Communist Party rule. His popularity was damaged by economic disintegration, rising crime, and internal conflicts (notably in Chechenia). Despite failing health he was re-elected in 1996. He dismissed the prime minister and the entire cabinet in 1998, and again in 1999. ➲ *262–63*

YOUNG TURKS Group of Turks who wished to remodel the Ottoman Empire as a modern European state with a liberal constitution. Their movement began in the 1880s with unrest in the army and universities. In 1908 a rising, led by ENVER PASHA and his chief of staff Mustafa Kemal (ATATÜRK), deposed Sultan ABDUL HAMID II and replaced him with his brother, Muhammad V. After a coup d'etat in 1913 Enver Pasha became a virtual dictator. Under Atatürk, the Young Turks merged into the Turkish Nationalist Party. ➲ *178–79*

YPRES, BATTLES OF Several battles of the First World War fought around the Belgian town of Ypres. The first (October–November 1914) stopped the German "race to the sea" to capture the Channel ports, but resulted in the near destruction of the British Expeditionary Force. The second (April–May 1915), the first battle in which poison gas was used, resulted in even greater casualties without victory to either side. The third (summer 1917) was a mainly British offensive which culminated in the Passchendaele campaign, the costliest campaign in British military history. ➲ *218–19*

YUAN (1246–1368) Mongol dynasty in China. Continuing the conquests of CHINGGIS KHAN, QUBILAI KHAN established his rule over China, eliminating the last Song claimant in 1279. Qubilai Khan returned the capital to Beijing and promoted construction and commerce. Chinese literature took new forms during the Mongol period. Native Chinese were excluded from government, and foreign visitors, including merchants such as Marco POLO, were encouraged. Among the Chinese, resentment of alien rule was aggravated by economic problems and the less competent successors of Qubilai were increasingly challenged by rebellion, culminating in the victory of the Ming. ➲ *98–99*

Z

ZAPATA, EMILIANO (1880–1919) Mexican revolutionary leader. Of peasant origin, he became leader of the growing peasant movement in 1910. His demands for radical agrarian reform, such as the return of haciendas (great estates) to native Mexican communal ownership, led to the MEXICAN REVOLUTION. In pursuit of "Land and Liberty" he opposed, successively, Porfirio D'az, Francisco Madero, Victoriano Huerta, and (with Francisco "Pancho" VILLA) Venustiano CARRANZA. His guerrilla campaign ended with his murder. ➲ *226–27*

ZAPOTECS Central American Indian group in part of the Mexican state of Oaxaca. The Zapotec built great pre-Conquest urban centres at Mitla and Monte Alban (which in around 200 BC had a population of up to 16,000), and fought to preserve their independence from the rival Mixtecs and Aztecs until the arrival of the Spanish. ➲ *32–33, 84–85*

ZHENG HE (1371–1435) Chinese admiral, explorer and diplomat. He came from the Muslim community that had established itself in the trading communitiies of the Chinese coastal ports and was popularly known as the "Three-Jewelled Eunuch". Between 1405 and 1433 Zheng He led seven naval expeditions across the China Sea and the Indian Ocean to gather treasures and unusual tributes for the Imperial court. His voyages reached as far west as the Gulf, visiting ports in southeastern Asia, India, eastern Africa and Egypt, and they prepared the way for Chinese colonization of Southeast Asia. ➲ *116–17, 138–39*

ZHOU Chinese dynasty (1050–221 BC). After the nomadic Zhou overthrew the Shang dynasty, Chinese civilization spread to most parts of modern China, although the dynasty never established effective control over the regions. The Late Zhou, from 772 BC, was a cultural golden age, marked by the writings of CONFUCIUS and LAO-TZE, and a period of rising prosperity. As the provincial states grew in power, the Zhou dynasty disintegrated. ➲ *48–49*

ZIA-UL-HAQ, MUHAMMAD (1924–88) President of Pakistan. Born in Punjab and trained in the Indian military, in 1966 Zia was appointed to the Command and Staff College in Quetta, Pakistan. In 1976 Prime Minister BHUTTO made him chief of staff of the army He ousted Bhutto in a coup in 1977, establishing Pakistan's fourth military dictatorship since 1947. Zia became president of his country in 1978. During his administration better relations between Pakistan and India were fostered. He died in a plane crash. ➲ *248–49*

ZOLLVEREIN German customs union formed in 1834 by 18 German states under Prussian leadership. By reducing tariffs and improving transport, it promoted economic prosperity. Nearly all other German states had joined the Zollverein by 1867, despite Austrian opposition. It represented the first major step towards the creation of the German Empire in 1871. ➲ *176–77*

ZOROASTER (c. 628–c. 551 BC, but possibly 10th century BC). Ancient Persian religious reformer and founder of ZOROASTRIANISM. At the age of 30, he is said to have seen the divine being Ahura Mazdah in the first of many visions. Unable to convert the chieftains of his native region, Zoroaster went to eastern PERSIA where, in Chorasmia, he converted the royal family. By the time of his death (tradition says that he was murdered while at prayer) his new religion had spread to a large part of PERSIA. ➲ *44–45*

ZOROASTRIANISM Religion founded by ZOROASTER in the 6th century BC, although it may have been as early as the 10th century BC. It was the state religion of PERSIA from the middle of the 3rd century AD until the mid-7th century. Viewing the world as divided between the spirits of good and evil, Zoroastrians worship Ahura Mazdah as the supreme deity, who is forever in conflict with Ahriman, the spirit of evil. They also consider fire sacred. The rise of ISLAM in the 7th century led to the decline and near disappearance of Zoroastrianism in Persia. Today the Parsi comprise most of the adherents of Zoroastrianism, which has its main centre in Bombay, India. ➲ *44–45*

ZWINGLI, HULDREICH (1484–1531) Swiss Protestant theologian and reformer. He was ordained as a Roman CATHOLIC priest in 1506, but his studies of the New Testament in Erasmus's editions led him to become a reformer. By 1522 he was preaching reformed doctrine in Zürich, a centre for the Reformation. More radical than LUTHER, he saw communion as mainly symbolic and commemorative. He died while serving as a military chaplain with the Zürich army during a battle against the Catholic cantons at Kappel. ➲ *154–55*

ZYUGANOV, GENNADY (1944–) Russian politician. He moved up the Soviet Communist Party hierarchy in the 1970s and 80s and in 1993 he became chairman of the executive committee of the reconstituted Russian Communist Party and was elected to the state *duma* (the lower house of the Russian parliament). In the 1995 parliamentary elections, the Communist Party gained the largest number of votes and, with the support of other groups, controlled around 30 per cent of the seats in the *duma*. Zyuganov mounted a strong challenge in the 1996 presidential elections, but was defeated by a coalition of Boris YELTSIN and Aleksander Lebed. ➲ *262–63*

INDEX

THE SPELLING OF PLACE NAMES

While every effort has been made to standardize the place names in this atlas, the fact that they can differ so much over time – as well as with language – means that variations inevitably exist. (These variations are given in the index.) In applying the basic guidelines outlined below, a commonsense approach has been adopted that allows for deviations where they serve a purpose.

The conventional Anglicized spelling, without accents, is used for large and familiar places (e.g. Munich rather that München, Mecca rather than Makkah). For smaller places in countries that use the Roman alphabet, the local form is given (e.g. Kraków). However, in keeping with current academic practice in the United States, accents are omitted from the Spanish forms of American-Indian place names dating from before the 16th-century conquest.

Where a name has changed due to political creed or ownership, this is often reflected in the maps. Thus St Petersburg is sometimes shown as Petrograd or Leningrad in maps of 20th-century Russia, and Strasbourg is spelt Strassburg when it was under German rather than French control.

If a country was once known by a name that differs from the one it holds at present, this is used where appropriate. Thus Thailand appears as Siam on many of the maps dating from before 1938 when it adopted its present name.

For Chinese names the increasingly familiar Pinyin form is used throughout (e.g. Beijing rather than Peking). However, where appropriate, the former spelling adopted under the Wade-Giles system also appears (e.g. Guangzhou is also labelled Canton on maps relating to European colonial activity in China). For the sake of clarity, diacritics are generally omitted from names derived from other non-Roman scripts by transliteration – notably Arabic and Japanese.

THE INDEX

The index includes the names of people and events as well as place names. To avoid unhelpful references to maps, place names are indexed only when the place is associated with a particular event or is marked by a symbol included in the key.

Alternative place names are given wherever appropriate, either in brackets or after the words "see also". References to maps are indicated by italics (e.g. 119/3 refers to map 3 on page 119), as are references to pictures.

D

I

J

K

BIBLIOGRAPHY

The books listed below are recommended by the contributors to this atlas as sources of further information on the topics covered by the maps and text.

GENERAL WORLD HISTORY

Bentley, J.H. and Zeigler, H.F. *Traditions and Encounters: A Global Perspective on the Past* McGraw-Hill 1999

Bulliet, R. et al. *The Earth and its Peoples: A Global History since 1500* Houghton Mifflin 1997

Clark, R.P. *The Global Imperative: An Interpretive History of the Spread of Humankind* Westview Press, Oxford/Boulder 1997

Crosby, A. W. *Ecological Imperialism: The Biological Expansion of Europe* Cambridge University Press, Cambridge 1993

Curtin, P. *Cross-Cultural Trade in World History* Cambridge University Press, Cambridge 1984

Frank, A.G. *ReOrient: Global Economy in the Asian Age* University of California Press, Berkeley, California 1998

Goody, J. *The East in the West* Cambridge University Press, Cambridge 1996

Heiser, C.B. *From Seed to Civilization: The Story of Food* Harvard University Press, Cambridge MA 1990

Huff, T. *The Rise of Early Modern Science: Islam, China and the West* Cambridge University Press, Cambridge 1993

Hugill, P.J. *World Trade since 1431: Geography, Technology and Capitalism* John Hopkins University Press, Baltimore/London 1993

Jones, E.L. *Growth Recurring: Economic Change in World History* Clarendon Press, Oxford 1988

Landes, D. *The Wealth and Poverty of Nations* Little, Brown and Co/W. W. Norton London/New York 1998

Livi-Bacci, M. *Concise History of World Population: An Introduction to Population Processes* Blackwell, Oxford/Cambridge, MA, 1992

McNeill, W. *The Human Condition: An Ecological and Historical View* Princeton University Press, Princeton, NJ/Guildford 1980

Mokyr, J. *The Lever of Riches: Technological Creativity and Economic Progress* Oxford University Press, Oxford/New York, 1992

O'Brien, P.K. (ed.) *Industrialisation: Critical Perspectives on the World Economy* Routledge, London 1998

Ponting, C. *A Green History of the World: The Environment and the Collapse of Great Civilizations* Penguin, London 1991

Roberts J.M. *Penguin History of the World*, Penguin, Harmondsworth 1995

Roberts, J.M. *Shorter Illustrated History of the World* Helicon, Oxford 1996

1 THE ANCIENT WORLD

World
Johanson, D. and Edgar, B. *From Lucy to Language* Weidenfeld and Nicolson, London 1996

Sherratt, A. *The Cambridge Encyclopedia of Archaeology* Cambridge University Press, Cambridge 1980

Smart, N. *The World's Religions* Cambridge University Press, Cambridge 1998

Asia and Australasia
Adams, R.M. *Heartland of Cities* Chicago University Press, Chicago 1981

Allchin, B. and Allchin, F.R. *The Rise of Civilization in India and Pakistan* Cambridge University Press, Cambridge 1982

Allchin, F.R. *The Archaeology of Early Historic South Asia: The Emergence of Cities and States* Cambridge University Press, Cambridge 1995

Aubet Semmler, M.E. *The Phoenicians and the West Politics, Colonies and Trade* Cambridge University Press, Cambridge 1993

Barnes, G.L. *China, Korea and Japan* Thames and Hudson, London 1993

Chang, K.C. *Shang Civilization* Yale University Press, New Haven/London 1980

Dani, A.H. and Masson, V.M. (eds) *History of Civilizations of Central Asia Vol. 1 The Dawn of Civilization; Earliest Times to 700 BC* UNESCO, Paris 1992

Ebrey, P.B. *The Cambridge Illustrated History of China* Cambridge University Press, Cambridge 1996

Fa-hsien (trans. Giles, H.A.) *The Travels of Fa-Hsien* Cambridge University Press, Cambridge 1923

Flood, J. *The Archaeology of Dreamtime* Collins, Sydney 1989

Harmatta, J., Puri, B.N. and Etemadi, G.F. (eds) *History of Civilizations of Central Asia Vol. 2 The Development of Sedentary and Nomadic Civilizations* UNESCO, Paris 1994

Jansen, M. and Urban, G. *Mohenjo Daro* Vol. 1 Brill, Leiden 1985

Kohl, P.L. *Central Asia. Palaeolithic Beginnings to the Iron Age* Editions Recherche sur les Civilisations, Paris 1984

Kozintsev et al. (eds) *New Archaeological Discoveries in Asiatic Russia and Central Asia* Russian Academy of Sciences, St Petersburg 1994

Maisels, C.K. *The Emergence of Civilization* Routledge, London 1993

Moorey, P. *Biblical Lands* Elsevier-Phaidon, London 1975

Nelson, S.M. *The Archaeology of Korea* Cambridge University Press, Cambridge 1993

Phillips, E.D. *The Royal Hordes: Nomad Peoples of the Steppes* Thames and Hudson, London 1965

Possehl, G.L. (ed.) *Ancient Cities of the Indus* Vikas, New Delhi 1979

Possehl, G.L. (ed.) *Harappan Civilization: A Recent Perspective* American Institute of Indian Studies, New Delhi/IBH, Oxford 1993

Postgate, J.N. *Early Mesopotamia* Routledge, London 1995

Potts, T. *Mesopotamia and the East* Oxford University Committee for Archaeology, Oxford 1994

Ratnagar, S. *Encounters: The Westerly Trade of the Harappa Civilization* Oxford University Press, Delhi/Oxford 1981

Rawson, J. *Ancient China* British Museum Publications, London 1980

Ray, H.P. *The Winds of Change: Buddhism and the Maritime Links of Early South Asia* Cambridge University Press, Cambridge 1994

Rice, T.T. *Ancient Arts of Central Asia* Thames and Hudson, London 1965

Stein, G. and Rothman, M.S. (eds) *Chiefdoms and Early States in the Near East* Madison Prehistory Press, Madison 1994

Waldron, A. *The Great Wall of China* Cambridge University Press, Cambridge 1990

Xinru, L. *Ancient India and Ancient China* Oxford University Press, Oxford 1988

Zvelebil, M. "The Rise of Nomads in Central Asia", in Sherratt, A. (ed.) *The Cambridge Encyclopedia of Archaeology* Cambridge University Press, Cambridge 1982

Africa
Ki-Zerbo, J. (ed.) *General History of Africa* Vol. 1 UNESCO/Heinemann, Edinburgh 1981

Lhote, H. *The Search for the Tassili Frescoes: The Story of the Pre-historic Rock Paintings of the Sahara* Hutchinson, London 1973

Phillipson, D. *African Archaeology* Cambridge University Press, Cambridge 1993

Quirke, S. and Spencer, J. *The British Museum Book of Ancient Egypt* British Museum Press, London 1992

Shaw, T. et al. (ed.) *The Archaeology of Africa: Food, Metal and Towns* Routledge, London 1993

Shinnie, P. *The African Iron Age* Clarendon Press, Oxford 1971

Tattersall, I. "Out of Africa Again ... and Again?", in *Scientific American* Vol. 276 No. 4, pp. 46-53 April 1997

Vinnicombe, P. *People of the Eland* University of Natal Press, Pietermaritzburg 1976

Europe
Boardman, J. *The Greeks Overseas* Thames and Hudson, London 1980

Boardman, J., Griffin, J. and Murray, O. (eds) *The Oxford History of the Classical Greece* Oxford University Press, Oxford 1986

Champion, T. et al. *Prehistoric Europe* Academic Press, London 1984

Collis, J. *The European Iron Age* Batsford, London 1984

Culican, W. *The Medes and Persians* Thames and Hudson, London 1965

Cunliffe, B. (ed.) *The Oxford Illustrated Prehistory of Europe* Oxford University Press, Oxford 1994

Cunliffe, B. *Rome and her Empire* McGraw-Hill Book Company, Maidenhead 1994

Garnsey, P. and Saller, R. *The Roman Empire, Economy, Society and Culture* Duckworth, London 1987

Harding, D. *Prehistoric Europe* Elsevier-Phaidon, Oxford 1978

Heather, P.J. *The Goths* Blackwell, Oxford 1996

Hood, S. *The Minoans* Thames and Hudson, London 1971

Isager, S. and Skydsgaard, J.E. *Ancient Greek Agriculture* Routledge, London/New York 1992

James, S. *Exploring the World of the Celts* Thames and Hudson, London 1993

Lane Fox, R. *The Search for Alexander* Allen Lane, London 1980

Taylor, W. *The Mycenaeans* Thames and Hudson, London 1983

Thompson, E.A. *Romans and Barbarians: The Decline of the Western Empire* Wisconsin, 1982

Todd, M. *The Early Germans* Blackwell, Oxford 1992

Walbank, F.W. *The Hellenistic World* Fontana, London 1992

Whittle, A. *Europe in the Neolithic* Cambridge University Press, Cambridge 1996

The Americas
Bruhns, K.O. *Ancient South America* Cambridge University Press, Cambridge 1994

Burger, R.L. *Chavin and the Origins of Andean Civilization* Thames and Hudson, London 1995

Culbert, T.P. *The Lost Civilization: The Story of the Classic Maya* Harper and Row, New York/London 1974

Fagan, B. *Ancient North America* Thames and Hudson, London 1995

Fiedel, S.J. *Prehistory of The Americas* Cambridge University Press, Cambridge 1992

Freidel, D., Schele, L. and Parker, J. *Maya Cosmos: Three Thousand Years on the Shaman's Path* William Morrow, New York 1993

Hammond, N. *Ancient Maya Civilization* Cambridge University Press, Cambridge 1982

Harrison, P. and Turner, B. *Pre-Hispanic Maya Agriculture* University of New Mexico, Albuquerque 1978

Isbell, W.H. and McEwan, G.F. (eds) *Huari Administrative Structure: Prehistoric Monumental Architecture and State Government* Dumbarton Oaks Research Library and Collection, Washington DC 1991

Schele, L. and Freidel, D. *A Forest of Kings* William Morrow, New York 1990

Schele, L. and Miller, M. *The Blood of Kings* Thames and Hudson, London 1992

2 THE MEDIEVAL WORLD

World

Harvey, P. *An Introduction to Buddhism. Teachings, History and Practices* Cambridge University Press, Cambridge 1990

McNeill, W.H. *Plagues and Peoples* Penguin, Harmondsworth 1979

Asia

Boyle, J.A. (ed.) *The Cambridge History of Iran* Vol. 5 *The Saljuq and Mongol Periods* Cambridge University Press, Cambridge 1968

Ebrey, P.B. *The Cambridge Illustrated History of China* Cambridge University Press, Cambridge 1996

Goeper, R. and Whitfield, R. *Treasures from Korea: Art through 5000 Years* British Museum Publications, London 1984

Golden, P.B. *An Introduction to the History of the Turkic Peoples* Otto Harrassowitz, Wiesbaden 1992

Higham, C. *The Archaeology of Mainland Southeast Asia* Cambridge University Press, Cambridge 1989

Holt, P.M., Lambton, A.K.S. and Lewis, B. *The Cambridge History of Islam* Vol. 1A *The Central Islamic Lands from pre-Islamic Times to the First World War* Cambridge University Press, Cambridge 1977

Hourani, A. *A History of the Arab Peoples* Faber and Faber, London 1991

Kennedy, H. *The Prophet and the Age of the Caliphates*, Longman, Harlow 1986

Morris, I. *The World of the Shining Prince: Court Life in Ancient Japan* Penguin, Harmondsworth 1969

Nahm, A.C. *Korea: Tradition and Transformation. A History of the Korean People* Hollym International, Elizabeth 1988

Tampoe, M. *Maritime Trade Between China and the West* BAR International series 555, Oxford 1989

Thapar R.A. *History of India* Vol. 1 Penguin, London 1990

Vryonis, S.J.R. *The Decline of Medieval Hellenism in Asia Minor and the Process of Islamization from the Eleventh through to the Fifteenth Century* University of California Press, Berkeley/London 1986

Africa

Connah, G. *African Civilizations* Cambridge University Press, Cambridge 1987

Insoll, T.A. "The archaeology of Islam in sub-Saharan Africa: A review", in *Journal of World Prehistory* Vol. 10 No. 4, pp. 439–504 December 1996

Levtzion, N. "The early states of the Western Sudan to 1500", in Ajayi, J.F.A. and Crowder, M. (eds) *History of West Africa* Vol. 1 Longman, Harlow 1985

Levtzion, N. *Ancient Ghana and Mali* Methuen, London 1973

Lovejoy, P.E. "The internal trade of West Africa before 1800", in Ajayi, J.F.A. and Crowder, M. (eds) *History of West Africa* Vol. 1 Longman, Harlow 1985

Mabogunje, A.L. and Richards, P. "The land and peoples of West Africa", in Ajayi, J.F.A. and Crowder, M. (eds) *History of West Africa* Vol. 1 Longman, Harlow 1985

Ryder, A.F.C. "From the Volta to Cameroon", in Niane, D.T. (ed.) *General History of Africa* Vol. 4 *Africa from the Twelfth to the Sixteenth Century* Heinemann Educational, London 1983

Europe

Hay, D. *Europe in the Fourteenth and Fifteenth Centuries* Longman, London/New York 1989

Amitai-Preiss, R. *Mongols and Mamluks: The Mamluk–Ilkhanid War, 1260–1281* Cambridge University Press, Cambridge/New York 1995

Arnold, B. *Medieval Germany 500–1300* Macmillan, Basingstoke 1997

Clanchy, M.T. *England and its Rulers 1066–1272* Blackwell, Oxford 1998

Crawford, B. *Scandinavian Scotland* Leicester University Press, Leicester 1993

Ennen, E. *The Medieval Town* North Holland Publishing Company, Amsterdam 1979

Fletcher, R.A. *The Conversion of Europe* Harper Collins, London 1998

Franklin, S. and Shepard, J. *The Emergence of Rus 750–1200* Longman, London 1996

Gojda, M. *The Ancient Slavs: Settlement and Society* Edinburgh University Press, Edinburgh 1991

Graham-Campbell, J. and Kidd, D. *The Vikings* British Museum Publications, London 1980

Hallam, E.M. *Capetian France 987–1328* Longman, London 1983

Holmes, G. *Europe: Hierarchy and Revolt (1320–1450)* Harvester Press, Hassocks 1975

James, E. *The Franks* Blackwell, Oxford 1988

Kennedy, H. *Muslim Spain and Portugal: A Political History of al-Andalus* Longman, London 1996

Khazanov, A.M. *Nomads and the Outside World* University of Wisconsin Press, Madison/ London 1994

Lopez, R.S. and Raymond, I.W. *Medieval Trade in the Mediterranean World* Columbia University Press, New York 1990

Mayer, H.E., *The Crusades* Oxford University Press, Oxford 1988

Nelson, J.L. *Charles the Bald* Longman, London 1992

Nicholas, D. *Medieval Flanders* Longman, London/New York 1992

Pryor, J.H. *Geography, Technology and War: Studies in the Maritime History of the Mediterranean 649–1571* Cambridge University Press, Cambridge 1988

Reilly, B.F. *The Medieval Spains* Cambridge University Press, Cambridge 1993

Roesdahl, E. *The Vikings* Penguin, London 1998

Sewter, E.R.A. (ed.) *The Alexiad of Anna Comnena* Penguin, Harmondsworth 1969

Shaw, S.J. *History of the Ottoman Empire and Modern Turkey* Vol. 1 *Empire of the Gazis* Cambridge University Press, Cambridge 1976

Spufford, P. *Money and its Use in Medieval Europe* Cambridge University Press, Cambridge 1989

Tabacco, G. *The Struggle for Power in Medieval Italy* Cambridge University Press, Cambridge 1989

Thompson, E.A. *The Huns* Blackwell, Oxford 1996

Waley, D. *The Italian City Republics* Longman, London 1988

Wood, I.N. *The Merovingian Kingdoms 450–751* Longman, London 1994

Ziegler, P. *The Black Death* Stroud, Sutton 1997

The Americas

Berdan, F.F., Blanton, R.E., Boone, E.H., Hodge, MG., Smith, M.E., Umberger, E. *Aztec Imperial Strategies* Dumbarton Oaks Research Library and Collection, Washington DC 1996

Bruhns, K.O. *Ancient South America* Cambridge University Press, Cambridge 1994

Carrasco, P. *Estructura politico-territorial del Imperio tenochca, La Triple Alianza de Tenochtitlan, Tetzcoco y Tlacopan* Fondo de Cultura Economica, Mexico 1996

Coe, M.D. *The Maya* Thames and Hudson, London 1993

Collier, G.A., Rosaldo, R.I. and Wirth, J.D. *The Inca and Aztec States, 1400–1800, Anthropology and History* Academic Press, NewYork/London 1982

Conrad, G.W. and Demarest, A.A. *Religion and Empire: The Dynamics of Aztec and Inca Expansionism* Cambridge University Press, Cambridge 1984

Fagan, B. *Ancient North America* Thames and Hudson, London 1995

Fiedel, S.J. *Prehistory of the Americas* Cambridge University Press, Cambridge 1992

Isbell, W.H. and McEwan, G.F. (eds) *Huari Administrative Structure: Prehistoric Monumental Architecture and State Government* Dumbarton Oaks Research Library and Collection, Washington DC 1991

Kolata, A. *The Tiwanaku: Portrait of an Andean Civilization* Blackwell, Oxford 1993

Morris, C. and Thompson, D.E. *Huanuco Pampa: An Inca City and its Hinterland* Thames and Hudson, London 1985

Moseley, M.E. and Cordy-Collins, A. (eds) *The Northern Dynasties Kingship and Statecraft in Chimor* Dumbarton Oaks Research Library and Collection, Washington DC 1990

Moseley, M.E. *The Incas and their Ancestors: The Archaeology of Peru* Thames and Hudson, London 1992

Scarre, C. ed. *Timelines of the Ancient World* Dorling Kindersley, London 1993

Weaver, M.P. The Aztecs, Maya and Their Predecessors Academic Press, New York 1993

3 THE EARLY MODERN WORLD

Asia and Africa

Alam, M. *Crisis of Empire in Mughal North India* Oxford University Press, Delhi 1986

Chaudhuri, K.N. *The Trading World of Asia and the English East India Company, 1650–1750* Cambridge University Press, Cambridge 1978

Fairbank, J.K. (ed.) *Chinese Thought and Institutions* University of Chicago Press, Chicago 1957

Fisher, S.N.L. and Ochsenwald, W. *The Middle East: A History* McGraw-Hill, New York/London

Gascoigne, B. *The Great Moghuls* Jonathan Cape, London 1971

Holt, P.M., Lambton, A.K.S. and Lewis, B. *The Cambridge History of Islam* Vol. 1A *The Central Islamic Lands from pre-Islamic Times to the First World War* Cambridge University Press, Cambridge 1977

Middleton, J. (ed.) *Encyclopedia of Africa South of the Sahara* C. Scribner's Sons, New York/Simon & Schuster and Prentice Hall International, London 1997

Morgan, D.O. *Medieval Persia 1040–1797* Longman, London 1988

Nakane, C. and Oishi, S. (eds) *Tokugawa Japan: The Social and Economic Antecedents of Modern Japan* University of Tokyo Press, Tokyo 1990

Oliver, R. and Fage, J.D. (eds.) *The Cambridge History of Africa* Cambridge University Press, Cambridge 1985

Rawski, T.G. and Li, L.M. (eds) *Chinese History in Economic Perspective* University of California Press, Berkeley/Oxford 1992

Richards, J.F. *The Mughal Empire* Cambridge University Press, Cambridge 1993

Spence, J.D. and Wills, J.E. (eds) *From Ming to Ch'ing: Conquest, Religion, and Continuity in Seventeenth-Century China* Yale University Press, New Haven 1979

Subrahmanyam, S. *The Portuguese Empire in Asia, 1500–1700* Longman, London 1993

Europe

Aston, T. (ed.) *Crisis in Europe, 1560–1660* Routledge & Kegan Paul, London 1974

Berenger, J.A. *History of the Habsburg Empire, 1243–1700* Longman, London 1994

Black, J. A *Military Revolution? Military Change in European Society 1550–1800* Macmillan Education, Basingstoke 1991

Bonney, R. *The European Dynastic States 1494–1660* Oxford University Press, Oxford 1991

Boxer, C.R. *The Portuguese Seaborne Empire* Penguin, London 1973

Brandi, K. *The Emperor Charles V: The Growth and Destiny of a Man and of a World-Empire* Harvester, Brighton 1980

Braudel, F. *Civilisation and Capitalism, 15th–18th Centuries* Vol. 3 *The Perspective of the World* Collins, London 1984

Cameron, E. *The European Reformation* Clarendon Press, Oxford 1991

Clay, C.G.A. *Economic Expansion and Social Change: England 1500–1700* Cambridge University Press 1984

Crummey, R.O. *The Formation of Muscovy, 1304–1613* Longman, London 1987

Davidson, N.S. *The Counter Reformation* Blackwell, Oxford 1987

Davies, N. *God's Playground: A History of Poland* Vol. 1 *The Origins to 1795* Clarendon, Oxford 1982

Delumeau, J. *Catholicism from Luther to Voltaire: A New View of the Counter Reformation* Burns & Oates, London 1977

Dukes, P. *The Making of Russian Absolutism 1613–1801* Longman, London 1990

Elliott, J. *The Old World and the New, 1492–1650* Cambridge University Press, Cambridge 1992

Evans, R.J.W. *The Making of the Habsburg Monarchy 1550–1700* Clarendon, Oxford 1979

Fichtner, P.S. *The Habsburg Empire: From Dynasticism to Multinationalism* Krieger, Malabar 1997

Greengrass, M. (ed.) *Conquest and Coalescence: The Shaping of the State in Early Modern Europe* Edward Arnold, London 1991

Hall, A.R. *The Revolution in Science, 1500–1750* Longman, Harlow 1983

Hohenberg, P.M. and Hollen Lees, L. *The Making of Urban Europe 1000–1994* Harvard University Press, Cambridge, Mass./London 1995

Lantzeff, G.V. and Pierce, R.A. *Eastward to Empire: Exploration and Conquest of the Russian Open Frontier, to 1750* McGill-Queen's University Press, Montreal/London 1973

Mackenney, R. *Sixteenth-century Europe: Expansion and Conflict* Macmillan, Basingstoke 1993

Pagden, A. *Lords of all the World. Ideologies of Empire in Spain, Britain and France, 1500–1800* Cambridge University Press, Cambridge 1995

Parker, G. and Smith, L.M. (eds) *The General Crisis of the Seventeenth Century* Routledge, London 1997

Parker, G. *The Military Revolution; Military Innovation and the Rise of the West 1500–1800* Cambridge University Press, Cambridge 1996

Parry, J.H. *The Spanish Seaborne Empire* Penguin, London 1973

Pohl, H. (ed.) *The European Discovery of the World and its Economic Effects on Pre-industrial Society, 1500–1800* Steiner, Stuttgart 1990

Rabb, T.D. *The Struggle for Stability in Early Modern Europe* Oxford University Press, New York 1975

Rich, E.E. and Wilson, C.H. (eds) *The Cambridge Economic History of Europe* Vol. 5 *The Economic Organisation of Early Modern Europe* Cambridge University Press, Cambridge 1977

Roberts, M. *Sweden as a Great Power, 1611–97* Edward Arnold, London 1968

Roberts, M. *The Early Vasas: A History of Sweden, 1523–1611* Cambridge University Press, Cambridge 1968

Shapin, S. and Schaffer, S. *Leviathan and the Air-Pump: Hobbes, Boyle, and the Experimental Life* Princeton University Press, Princeton/Guildford 1985

Shaw, S.J. and Shaw, E.K. *History of the Ottoman Empire and Modern Turkey* Vol. 1 *Empire of the Gazis; The Rise and Decline of the Ottoman Empire* Cambridge University Press, Cambridge 1977

Stewart, L. *The Rise of Public Science: Rhetoric, Technology and Natural Philosophy in Newtonian Britain* Cambridge University Press, Cambridge 1992

Tallett, F. *War and Society in Early Modern Europe 1495–1715* Routledge, London/New York 1992

Tilly, C. and Blockmans, W.P. (eds) *Cities and the Rise of States in Europe, AD 1000 to 1800* Westview Press, Boulder/Oxford 1994

Tracy, J.D. (ed.) *The Political Economy of Merchant Empires. State Power and World Trade, 1350–1750* Cambridge University Press, Cambridge 1991

Tracy, J.D. (ed.) *The Rise of Merchant Empires. Long Distance Trade in the Early Modern World, 1350–1750* Cambridge University Press, Cambridge 1990

Vries, J. de and Woude, A. van der *The First Modern Economy. Success, Failure and Perseverence of the Dutch Economy, 1500–1815* Cambridge University Press, Cambridge 1997

Vries, J. de *European Urbanisation, 1500–1800* Methuen, London 1984

The Americas

Bethell, L. (ed.) *Colonial Brazil* Cambridge University Press, Cambridge 1987

Bethell, L. (ed.) *Colonial Spanish America* Cambridge University Press, Cambridge 1987

Blackburn, R. *The Making of New World Slavery: From the Baroque to the Modern 1492–1800* Verso, London 1997

Curtin, P.D. *The Atlantic Slave Trade: A Census* University of Wisconsin Press, Madison/London 1969

Davies, K.G. *The North Atlantic World in the Seventeenth Century* University of Minnesota Press, Minneapolis/Oxford University Press, Oxford 1974

Davis, R. *The Rise of the Atlantic Economies* Weidenfeld and Nicolson, London 1973

Morrison, S.E. *The European Discovery of America: The Southern Voyages, AD 1492–1616* Oxford University Press, New York 1974

Rawley, J.A. *The Transatlantic Slave Trade. A History* Norton, London/New York 1981

Savelle, M. *Empires to Nations: Expansion in America, 1713–1824* University of Minnesota Press, Minneapolis/Oxford University Press, London 1974

Williamson, E. *The Penguin History of Latin America* Penguin, London 1992

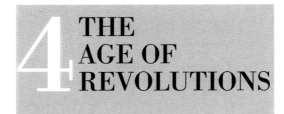

4 THE AGE OF REVOLUTIONS

World

Chamberlain, M.E. *The New Imperialism* Historical Association, London 1984

Cipolla, C.M. *The Economic History of World Population* Penguin, Harmondsworth 1970

Fieldhouse, D.K. *Economics and Empire 1830–1914* Macmillan, Basingstoke 1984

Foreman-Peck, J. *A History of the World Economy International Economic Relations since 1850* Wheatsheaf, Brighton 1983

Livi-Bacci, M. *A Concise History of World Population* Blackwell, Oxford 1997

World Bank *World Development Report* Oxford University Press for the World Bank, Washington 1980

Asia, Africa and Australasia

Bayly, C.A. *Indian Society and the Making of the British Empire* Cambridge University Press, Cambridge 1988

Beasley, W.G. *The Rise of Modern Japan* Weidenfeld and Nicolson, London 1995

Bose, S. and Jalal, A. *Modern South Asia: History, Culture, Political Economy* Routledge, London 1998

Cameron, M.E *The Reform Movement in China, 1898–1912* Stanford University Press, Stanford 1974

Clark, C.M.H. *A History of Australia* Melbourne University Press, Parkville/Cambridge University Press, Cambridge 1987

Fay, P.W. *The Opium War, 1840–1842* The University of North Carolina Press, Chapel Hill 1975

Fromkin, D.A. *Peace to End all Peace: Creating the Modern Middle East 1914–1922* Penguin, London 1991

Hall, D.G.E. *A History of South-East Asia* Macmillan Education, London 1987

Jansen, M.B. *The Emergence of Meiji Japan* Cambridge University Press, Cambridge 1995

Mazrui, A.A. *The African Condition: A Political Diagnosis* Heinemann Educational, London 1980

Metcalf, T. *Ideologies of the Raj* Cambridge University Press, Cambridge 1994

Ogot, B.A. (ed.) *General History of Africa* Heinemann, Oxford 1992

Oliver, R.A. *The African Experience* Weidenfeld and Nicolson, London 1991

Rice, G.W. (ed.) *The Oxford History of New Zealand* Oxford University Press, Auckland/Oxford 1993

Scalapina, R.A. and Yu, G.T. *Modern China and its Revolutionary Process: Recurrent Challenges to the Traditional Order 1850–1920* University of California Press, Berkeley 1985

Steinberg, D.J. et al. *In Search of Southeast Asia: A Modern History* University of Hawaii Press, Honolulu 1987

Tarling, N. (ed.) *The Cambridge History of Southeast Asia* Vol. 2 *The Nineteenth and Twentieth Centuries* Cambridge University Press, Cambridge 1992

Yapp, M.E. *The Making of the Modern Near East 1792–1923* Longman, London 1988

Europe

Beales, D. *The Risorgimento and the Unification of Italy* Longman, London 1981

Berg, M. *The Age of Manufactures 1700–1820* Blackwell, London 1985

Broers, M. *Europe under Napoleon, 1799–1815* Arnold, London/New York 1996

Carr, W. *The Origins of the Wars of German Unification* Longman, London 1991

Chandler, D. *The Campaigns of Napoleon* Weidenfeld, London 1993

Clerel de Tocqueville, A. de, *The Ancient Regime and the French Revolution* Collins, London 1966

Duggan, C. *A Concise History of Italy* Cambridge University Press, Cambridge 1994

Fichtner, P.S. *The Habsburg Empire: From Dynasticism to Multinationalism* Krieger, Malabar 1997

Fromkin, D.A. *Peace to End all Peace: Creating the Modern Middle East 1914–1922* Penguin, London 1991

Fulbrook, M. *A Concise History of Germany* Cambridge University Press, Cambridge 1990

Furet, F. *La Revolution* Hachette, Paris 1988

Gooch, J. *The Unification of Italy* Methuen, London 1986

Hudson, P. *The Industrial Revolution* Edward Arnold, London 1992

Jelavich B. *A History of the Balkans* Vol. 1 *Eighteenth and Nineteenth Centuries* Cambridge University Press, Cambridge 1983

Jelavich, C. and Jelavich, B. *The Balkans* Prentice-Hall, Englewood Cliffs 1965

Matthias, P. *The First Industrial Nation* Methuen, London 1983

Milward, A.S. and Saul, S.B. *The Economic Development of Continental Europe, 1780–1870* George Allen & Unwin, London 1979

Mosse, W.E. *An Economic History of Russia 1856–1914* I.B. Tauris, London 1996

Palmer, A. *The Decline and Fall of the Ottoman Empire* John Murray, London 1992

Pollard, S. *Peaceful Conquest: the Industrializaton of Europe, 1760–1970* Oxford University Press, Oxford 1981

Porter, A.N. *European Imperialism 1860–1914* Macmillan, Basingstoke 1994

Seton-Watson, H. *The Russian Empire 1801–1917* Clarendon Press, Oxford 1967

Shaw, S.J. and Shaw, E.K. *History of the Ottoman Empire and Modern Turkey* Vol. 2 *Reform, Revolution and Republic: The Rise of Modern Turkey* Cambridge University Press 1977

Soboul, A. *Dictionnaire Historique de la Revolution Française* Presses Universitaires de France, Paris 1989

Stavrianos, L.S. *The Balkans since 1453* Holt, Rinehart & Winston, New York 1961

Sylla, R. and Toniolo, G. *Patterns of European Industrialization: The Nineteenth Century*, Routledge, London 1991

Tunzelmann, G.N. von *Technology and Industrial Progress: The Foundations of Economic Growth* Edward Elgar, Aldershot 1995

Waller, B. *Bismarck* Blackwell, Oxford 1985

The Americas

Bailyn, B. *The Ideological Origins of the American Revolution* Cambridge University Press, Cambridge 1992

Bell, D. and Tepperman, L. *The Roots of Disunity: A Look at Canadian Political Culture* McLelland and Stewart, Toronto, 1979

Bethell, L. (ed.) *Spanish America after Independence, c.1820–1870* Cambridge University Press, Cambridge/New York 1987

Bethell, L. (ed.) *The Independence of Latin America* Cambridge University Press, Cambridge/New York 1987

Billington, R.A. and Ridge, M. *Westward Expansion: A History of the American Frontier* Macmillan, New York 1982

Bodnar, J. *The Transplanted: A History of Immigrants in Urban America* Indiana University Press, Bloomington 1985

Bushnell, D. and Macaulay, N. *The Emergence of Latin America in the Nineteenth Century* Oxford University Press, New York/London 1988

Conway, S. *The War of American Independence 1775–1783* Edward Arnold, London 1995

Cook, R., Ricker, J. and Saywell, J. *Canada: A Modern Study* Clarke, Irwin & Co., Toronto 1977

Donald, D.H *Lincoln* Cape, London 1995

Freehling, W.W. *The Road to Disunion* Oxford University Press, Oxford 1990

Hafen, L.R. et al. *Western America* Prentice-Hall, New York 1970

Halperin Donghi, T. *The Aftermath of Revolution in Latin America* Harper and Row, New York 1973

Lynch, J. *The Spanish American Revolutions 1808–1826* W.W. Norton, New York 1986

Marr, W.L. and Paterson, D.G. *Canada: An Economic History* Clarke, Irwin, Gage, Toronto 1981

Martin, A. *Railroads Triumphant* Oxford University Press, Oxford 1991

McNaught, K. *The Penguin History of Canada* Penguin, London 1988

McPherson, J.M. *Battle Cry of Freedom: The Civil War Era* Oxford University Press, Oxford 1988

Porter, G. *The Rise of Big Business 1860–1920* Harlan Davidson, Arlington Heights 1992

Utley, R.M. *The Indian Frontier of the American West, 1846–1890* University of New Mexico Press, Albuquerque 1984

Ward, H.M. *The American Revolution: Nationhood Achieved 1763–1788* St Martin's Press, New York 1995

5 THE TWENTIETH CENTURY

World

Benton, B. *Soldiers of Peace: Fifty Years of U.N. Peace-keeping* Facts on File, New York 1996

Brown, L. *Vital Signs: The Environmental Trends that are Shaping our Future* W W Norton, New York/Earthscan, London 1999

Calvocoressi, P., Wint, G., Pritchard, J. *Total War: The Causes and Courses of the Second World War* Penguin, London 1995

Foot, M.R.D. (ed.) *The Oxford Companion to the Second World War* Oxford University Press, Oxford 1995

Gilbert, M. *The First World War: A Complete History* Harper Collins, London 1995

Hauchler, I. and Kennedy, P.M. *Global Trends: The World Almanac of Development and Peace* Continuum Publishing, New York 1994

Joll, J. *The Origins of the First World War* Longman, London/New York 1992

Kennedy, P. *The Rise and Fall of the Great Powers* Fontana, London 1989

Kenwood, A.G. and Lougheed, A.L. *Growth of the International Economy, 1820–1990* Routledge, London 1992

Kindleberger, C.P. *The World in Depression* Allen Lane, London 1973

Luard, E. *The United Nations: How it Works and What it Does* Macmillan, London 1994

Martel, G. *The Origins of the First World War* Longman, London/New York 1996

Overy, R. *Why the Allies Won* Jonathan Cape, London 1995

Robbins, K. *The First World War* Oxford University Press, Oxford/New York 1984

United Nations The Blue Helmets: A Review of United Nations Peace-keeping United Nations Department of Public Information, New York 1996

Weinberg, G.L. *A World at Arms. A Global History of World War II* Cambridge University Press, Cambridge 1994

Weiss, T.G. *The United Nations and Changing World Politics* Westview Press, Boulder/Oxford 1997

Dockrill, M.L. and Douglas Goold, J. *Peace Without Promise: Britain and the Peace Conferences 1919–23* Batsford Academic and Educational, London 1981

Asia, Africa and Australasia

Beinart, W. *Twentieth-Century South Africa* Oxford University Press, Oxford 1994

Bowring, R. and Kornicki, P. (eds) *The Cambridge Encyclopedia of Japan* Cambridge University Press, Cambridge 1993

Brown, J.M. *Modern India. The Origins of an Asian Democracy* Oxford University Press, Oxford 1985

Chen, Y. *Making Revolution: The Communist Movement in Eastern and Central China, 1937–1945* University of California Press, Berkeley 1986

Clark, C.M.H. *A History of Australia* Melbourne University Press, Parkville/Cambridge University Press, Cambridge 1987

Fei, H. *Peasant Life in China: A Field Study of Country Life in the Yangtze Valley* Routledge and Kegan Paul, London 1980

Freund, B. *The Making of Contemporary Africa* Longman, London 1998

Iriye, A. *The Origins of the Second World War in Asia and the Pacific* Longman, London 1987

Myers, R.H. (ed.) *Two Societies in Opposition: The Republic of China and the People's Republic of China* Hoover Institution Press, Stanford 1991

Ovendale, R. *The Middle East since 1914* Longman, London/New York 1992

Pannell, C.W. and Ma, L.J.C. *China, the Geography of Development and Modernization* Edward Arnold, London 1983

Rawski, T.G. *Economic Growth in pre-War China* University of California Press, Berkeley 1989

Rice, G.W. (ed.) *The Oxford History of New Zealand* Oxford University Press, Auckland/Oxford 1992

Sargent, J. *Perspectives on Japan: Towards the Twenty-First Century* Curzon Press, London 1999

Sarkar, S. *Modern India 1885–1947* Macmillan, Basingstoke 1989

Spence, J.D. *The Search for Modern China* W.W. Norton, New York 1990

Steinberg, D.J. et al. *In Search of Southeast Asia: A Modern History* University of Hawaii Press, Honolulu 1987

Tarling, N. (ed.) *The Cambridge History of Southeast Asia* Vol. 2 *The Nineteenth and Twentieth Centuries* Cambridge University Press, Cambridge 1992

Witherick, M. and Carr, M. *The Changing Face of Japan: A Geographical Perspective* Hodder and Stoughton, Sevenoaks 1993

Wolpert, S. *A New History of India* Oxford University Press, Oxford 1997

Yapp, M.E. *The Near East since the First World War. A History to 1995* Longman, Harlow 1996

Europe

Chamberlain, M.E. *Decolonization. The Fall of the European Empires* Blackwell, Oxford, 1985

Crockatt, R. *The Fifty Year War* Routledge, London 1995

Dunbabin, J.P.D. *The Cold War* Longman, London 1994

Feinstein, C.H., Temin, P. and Toniolo, G. *The European Economy between the Wars* Oxford University Press, Oxford 1997

Figes, O. *A People's Tragedy. The Russian Revolution 1891–1924* Pimlico, London 1997

Galeotti, M. *Gorbachev and his Revolution* Macmillan, Basingstoke 1997

Grimal, H. *Decolonization: The British, French, Dutch and Belgian Empires* Routledge, London 1978

Holland, R.F. *European Decolonization 1918-1981. An Introductory Survey* Macmillan, Basingstoke 1985

Hosking, G. *A History of the Soviet Union* Fontana, London 1992

Keep, J. *Last of the Empires. A History of the Soviet Union 1945–1991* Oxford University Press, Oxford 1995

Kitchen, M. *Europe between the Wars. A Political History* Longman, London/New York 1988

Lee, S.J. *The European Dictatorships 1918–1945* Routledge, London/New York 1987

McCauley, M. *Gorbachev* Longman, London 1998

Nove, A. *An Economic History of the USSR* Penguin, London 1992

Pryce-Jones, D. *The War that Never Was: The Fall of the Soviet Empire 1985–1991* Weidenfeld and Nicolson, London 1995

Sakwa, R. *Russian Politics and Society* Routledge, London 1996

Schulze, M.-S. (ed.) *Western Europe. Economic and Social Change since 1945* Longman, London 1999

Service, R. *A History of the Twentieth Century Russia* Allen Lane, London 1997

Steele, J. *Eternal Russia* Faber, London 1994

Stewart, J.M. (ed.) *The Soviet Environment: Problems, Policies and Politics* Cambridge University Press, Cambridge 1992

Swain, G. and Swain, N. *Eastern Europe Since 1945* Macmillan, Basingstoke 1993

Ward, C. *Stalin's Russia* Edward Arnold, London 1993

White, S. *After Gorbachev* Cambridge University Press, Cambridge 1993

White, S., Pravda, A. and Gitelman, Z. *Developments in Russian Politics* Macmillan, Basingstoke 1997

Williams, B. *The Russian Revolution* Blackwell, Oxford, 1995

Woolf, S.J. (ed.) *Fascism in Europe* Methuen, London, 1981

Young, J.W. *The Longman Companion to the Cold War and Detente* Longman, London 1993

The Americas

Bethell, L. (ed.) *Brazil: Empire and Republic 1922–30* Cambridge University Press Cambridge/New York 1989

Bethell, L. (ed.) *Cambridge History of Latin America* Vol. 6 Parts 1 and 2, Cambridge University Press, Cambridge/New York 1994

Bethell, L. (ed.) *Latin America: Economy and Society 1870–1930* Cambridge University Press, Cambridge/New York 1989

Calvert, P. and Calvert, S. *Latin America in the Twentieth Century* Macmillan, Basingstoke, 1993

LaFeber, W. *Inevitable Revolutions: The United States in Central America* W.W. Norton, New York/London 1993

LaFeber, W. *The American Age, United States Foreign Policy at Home and Abroad Since 1750* W.W. Norton, London 1994

Lukacs, J. *Outgrowing Democracy: A History of the United States in the Twentieth Century* University of America Press, Lanham, MD 1986

Williamson, E. *The Penguin History of Latin America* Part 3 Penguin, London 1992

Wright, R.O. *A Twentieth-century History of United States Population* Scarecrow Press, London/Lanham, MD 1996

ACKNOWLEDGEMENTS

MAP ACKNOWLEGEMENTS

The map of trench warfare on page 218 is based on a map in the *Atlas of the First World War* by Martin Gilbert (Weidenfeld and Nicolson, 1970), by permission of Routledge. The maps of European urbanization on pages 132–33 are based on statistics supplied in *European Urbanization 1500–1800* by J. de Vries (Methuen, 1984), by permission of Routledge.

Among the atlases consulted by authors and editors in preparing the maps in this atlas are the following:

Ajayi, J.F.A and Crowder, M. (eds) *Historical Atlas of Africa* Longman, Harlow 1985

Baines, J. and Malek, J. *Atlas of Ancient Egypt* Phaidon, Oxford 1980

Banks, A. *A World Atlas of Military History* Seeley, London 1978

Blunden, C. and Elvin, M. *Cultural Atlas of China* Phaidon, Oxford 1983

Coe, M., Snow, D. and Benson, E. *Atlas of Ancient America* Facts on File, New York 1986

Collcutt, M., Jansen, M. and Isao, K. *Cultural Atlas of Japan* Phaidon, Oxford 1988

Cornell, T. and Matthews, J. *Atlas of the Roman World* Phaidon, Oxford 1982

Darby, H.C. and Fullard, H. (eds) *The New Cambridge Modern History Atlas* Cambridge University Press, Cambridge 1978

Castello-Cortes, I. (ed.) *The Economist Atlas* Economist Books, London 1991

Engel, J. (ed.) *Großer Historischer Weltatlas* Vols 1–4 Bayerischer Schulbuch-Verlag, Munich 1953–96

Fage, J.D. *An Atlas of African History* Edward Arnold, London 1978

Flon, C. (ed.) *The World Atlas of Archaeology* Mitchell Beazley, London 1985/Portland House, New York 1988

Gardner, J.L. (ed.) *Reader's Digest Atlas of the Bible: An Illustrated Guide to the Holy Land* Reader's Digest Association, Pleasantville 1981

Gilbert, M. *Atlas of the First World War* Routledge, London 1970

Gilbert, M. *The Dent Atlas of Russian History* Dent, London 1993

Hall, D.G.E. *Atlas of South-East Asia* Djambatan, Amsterdam 1964

Johnson, G. *Cultural Atlas of India* Facts on File, New York 1996

Kinder, H. and Hilgemann, W. *The Penguin Atlas of World History* Vols 1–2 Penguin, Harmondsworth 1995

Levi, P. *Atlas of the Greek World* Phaidon, Oxford 1970

Manley, B. *The Penguin Historical Atlas of Ancient Egypt* Penguin, London 1996

Moore, R.I. (ed.) *Philip's Atlas of World History* Philip's, London 1994

Morkot, R. *The Penguin Historical Atlas of Ancient Greece* Penguin, London 1996

Muir, R. *Muir's Historical Atlas: Ancient, Medieval and Modern* Philip's, London 1963

Parker, G.I. (ed.) *The Times Atlas of World History* Times Books, London 1993

Pluvier, J.M. *Historical Atlas of South-East Asia* E.J. Brill, New York 1995

Pritchard, J. *Times Atlas of the Bible* Times Books, London 1996

Riley Smith, J. (ed.) *The Atlas of the Crusades* Times Books, London 1991

Roaf, M. *Cultural Atlas of Mesopotamia and The Ancient Near East* Facts on File, New York 1990

Scarre, C. (ed.) *Past Worlds – The Times Atlas of Archaeology* Times Books, London 1995

Schwartzberg, J.E. *A Historical Atlas of South Asia* Oxford University Press, New York 1993

Segal, A. *An Atlas of International Migration* Hans Zell, London 1993

Stier, H.-E. et al. (eds) *Westermann großer Atlas zur Weltgeschichte*, Westermann, Braunschweig 1976

Thomas, A. et al. *The Third World Atlas* Open University Press, Buckingham 1994

Unwin, T. *Atlas of World Development* Wiley, Chichester 1994

PHOTOGRAPH ACKNOWLEDGEMENTS

AKG London 60, 179, /Erich Lessing 40; **Bridgeman Art Library** 30, 49, 143, /Artephot, Private Collection 165, /Bibliotheque Nationale, Paris 95, /Bode-Museum, Berlin 36, /British Museum, London 20, 26, 111, 114, /Chester Beatty Library and Gallery of Oriental Art, Dublin 69, /Christie's Images 112, /Christie's, London 58, 174, /Gavin Graham Gallery, London 156, /Giraudon, Civico Museo Correr, Venice 131, /Guildhall Art Gallery, Corporation of London 132, /Guildhall Library, Corporation of London 160, /Heini Schneebeli 82, /Heini Schneebeli, National Commission for Museums and Monuments, Ife, Nigeria 81, /Johnny Van Haeften Gallery, London 129, /Kunsthistorisches Museum, Vienna 114, /Lauros-Giraudon, Bibliotheque Nationale, Paris 161, /Lauros-Giraudon, Galerie Nationale, Palermo, Sicily 104, /Lauros-Giraudon, Louvre, Paris 56, /National Museum of India, New Delhi (detail) 14, /National Museum of Iceland, Reykjavik 78, /Nationalmuseet, Copenhagen 137, /Novosti 99, /Peter Willi/Louvre, Paris 54, /Private Collection 135, 140, 162, 165, 170, 177, 211, 254, /Roger-Viollet/Museo E Gallerie Nazionale Di Capodimonte, Naples 43, /Roudnice Lobkowica Collection, Nelahozeves Castle, Czech Republic 70, /Victoria & Albert Museum, London 59, 61, 73, 113, 163, 198, /Wallace Collection, London 156; **Peter Carey** 197; **Corbis** 241; **E.T.Archive** 15, 79, 86, 96, 113, 150, 201, 204, /Amano Museum, Lima 35, /Arteaga Collection Peru 110, /British Museum, London 31, /Canning House Library 191, /Imperial War Museum, London 243, /Mjolnir 233, /Museo Amano, Lima 35; **Robert Harding** 13, 15, 62, 244, 270, /Gavin Hellier 61, /M.J. Howell 215; Michael Holford /Musee Guimet, Paris 145; **Hulton Getty Picture Collection** /Hulton Getty 225, 230, /Hulton-Deutsch Collection 222; **Peter Newark's American Pictures** 84, 121, 182, 188, 226, 259; **Panos Pictures** /Peter Barker, 274, /Caroline Penn 279, /Paul Smith 273, /Chris Stowers 214, /Liba Taylor 276; Popperfoto /Mike Segar/Reuters 267; **Rex Features** 213, 221, 236, 249, 257, 263, 282, /Sipa-Press, Paris 212, /Markus Zeffler 214; **Werner Forman Archive** 41, /Anthropology Museum, Veracruz University, Jalapa 32, /Beijing Museum 59, /Dallas Museum of Art, USA 12.